ORGANIZING A STATISTICAL PROBLEM: THE FOUR-STEP PROCESS

STATE: What is the practical question, in the context of the real-world setting?
PLAN: What specific statistical operations does this problem call for?
SOLVE: Make the graphs and carry out the calculations needed for this problem.
CONCLUDE: Give your practical conclusion in the setting of the real-world problem.

CONFIDENCE INTERVALS: THE FOUR-STEP PROCESS

STATE: What is the practical question that requires estimating a parameter?

PLAN: Identify the parameter and choose a level of confidence.

SOLVE: Carry out the work in two phases:

1. **Check the conditions** for the interval you plan to use.
2. Calculate the **confidence interval** or use technology to obtain it.

CONCLUDE: Return to the practical question to describe your results in this setting.

TESTS OF SIGNIFICANCE: THE FOUR-STEP PROCESS

STATE: What is the practical question that requires a statistical test?

PLAN: Identify the parameter, state null and alternative hypotheses, and choose the type of test that fits your situation.

SOLVE: Carry out the test in three phases:

1. **Check the conditions** for the test you plan to use.
2. Calculate the **test statistic.**
3. Find the ***P*-value** using a table of Normal probabilities or technology.

CONCLUDE: Return to the practical question to describe your results in this setting.

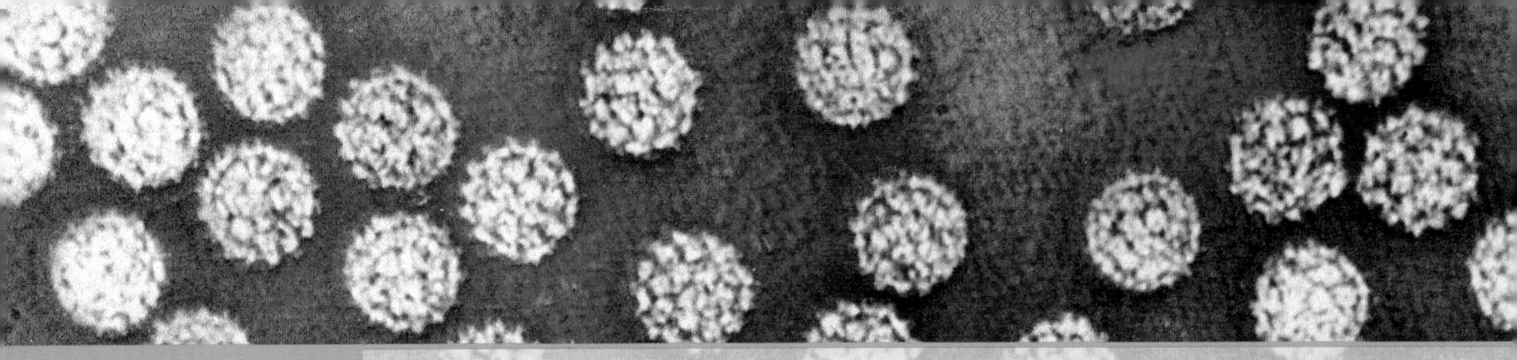

THIRD EDITION

The Practice of Statistics in the Life Sciences

BRIGITTE BALDI
University of California, Irvine

DAVID S. MOORE
Purdue University

W. H. Freeman and Company
A Macmillan Higher Education Company

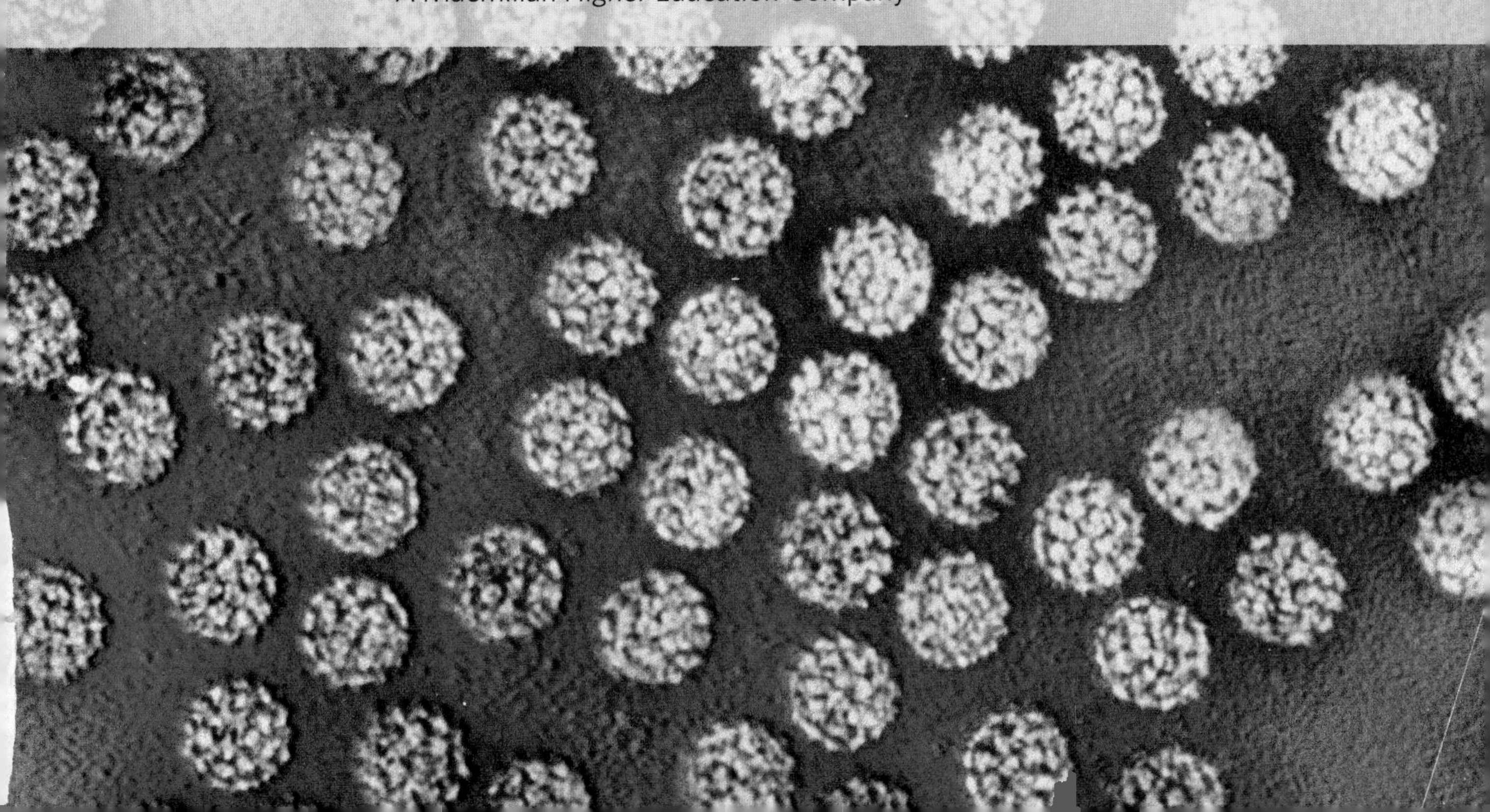

Senior Publisher: Ruth Baruth
Acquisitions Editor: Karen Carson
Marketing Manager: Steve Thomas
Developmental Editor: Katrina Wilhelm
Senior Media Editor: Laura Judge
Media Editor: Catriona Kaplan
Associate Editor: Jorge Amaral
Assistant Media Editor: Liam Ferguson
Photo Editor: Bianca Moscatelli
Photo Researcher: Ramón Rivera Moret
Cover and Text Designer: Blake Logan
Project Editor: Elizabeth Geller
Production Coordinator: Ellen Cash
Composition: MPS Limited
Printing and Binding: RR Donnelley
Cover Photo: Kwangshin Kim/Science Source

Library of Congress Control Number: 2013953572

Student Edition Hardcover (packaged with EESEE/CrunchIt! access card):
ISBN-13: 978-1-4641-7536-7
ISBN-10: 1-4641-7536-5

Student Edition Looseleaf (packaged with EESEE/CrunchIt! access card):
ISBN-13: 978-1-4641-7534-3
ISBN-10: 1-4641-7534-9

Instructor Complimentary Copy:
ISBN-13: 978-1-4641-3318-3
ISBN-10: 1-4641-3318-2

Printed in China

Second printing

W. H. Freeman and Company
41 Madison Avenue
New York, NY 10010
Houndmills, Basingstoke RG21 6XS, England
www.whfreeman.com

BRIEF CONTENTS

*Starred material is optional.

CONTENTS

*Starred material is optional.

TO THE INSTRUCTOR: About this Book

The Practice of Statistics in the Life Sciences, third edition (*PSLS 3e*), is an introduction to statistics for college and university students interested in the quantitative analysis of life science problems. Statistics has penetrated the life sciences pervasively with a specific set of application challenges, such as observational studies with confounding variables or experiments with limited sample sizes. Consequently, students can clearly benefit from a teaching of statistics that is explicitly applied to their major. All examples and exercises in *PSLS* are drawn from diverse areas of biology, such as physiology, brain and behavior, epidemiology, health and medicine, nutrition, ecology, and microbiology. Instructors can choose to either cover a wide range of topics or select examples and exercises related to a particular field.

PSLS focuses on the applications of statistics rather than the mathematical foundation. The book is adapted from David Moore's best-selling introductory statistics textbook *The Basic Practice of Statistics* (*BPS*). *BPS* was the pioneer in presenting a modern approach to statistics in a genuinely elementary text. Like *BPS*, *PSLS* emphasizes a balanced content, working with real data, and statistical ideas. It does not require any specific mathematical skills beyond being able to read and use simple equations and can be used in conjunction with almost any level of technology for calculating and graphing.

In the following we describe in further detail for instructors the guiding principles and features of *PSLS 3e*.

GAISE guiding principles

Student audiences and access to technology have changed substantially over the years, and educational guidelines in statistics have evolved accordingly. The American Statistical Association offers a set of recommendations for introductory statistics courses at the college level described in the Guidelines for Assessment and Instruction in Statistics Education (GAISE).[1] The guiding principles of *PSLS 3e* closely follow the six GAISE recommendations for the teaching of introductory statistics.

1. **Emphasize statistical literacy and develop statistical thinking.** Students should understand the basic ideas of statistics, including the need for data, the importance of data production, the omnipresence of variability, and the quantification and explanation of variability. To this end, *PSLS* begins with data analysis (Chapters 1 to 6), then moves to data production (Chapters 7 and 8), and then to probability (Chapters 9 to 13) and inference (Chapters 14 to 28). In studying data analysis, students learn useful skills immediately and get over some of their fear of statistics. Data analysis is a necessary preliminary step to inference in practice, because inference requires suitable data. Designed data production is the surest foundation for inference, and the deliberate use of chance in random sampling and randomized comparative experiments motivates the study of probability in a course that emphasizes data-oriented statistics. *PSLS* gives a full presentation of basic probability and inference (20 of the

28 chapters) but places it in the context of statistics as a whole. Furthering this approach, each chapter contains a summary section titled "This Chapter in Context" that highlights how the concepts from the chapter relate to concepts introduced in earlier chapters and how they will figure in following chapters.

Students should also understand the general statistical approach used to solve scientific problems. A discussion box in Chapter 15 describes this approach in the context of the Nobel Prize-winning discovery of the bacterial origin of most peptic ulcers. The detailed historical and scientific account helps students see how the concepts they learn throughout the book come together to form a coherent science of data. In addition, many of the examples and exercises in *PSLS* are presented in the context of a "four-step process" (State, Plan, Solve, Conclude) intended to teach students how to work on realistic statistical problems. Figure 1 provides an overview. The process emphasizes a major theme in *PSLS*: Statistical problems originate in a real-world setting ("State") and require conclusions in the language of that setting ("Conclude"). Translating the problem into the formal language of statistics ("Plan") is a key to success. The graphs and computations needed ("Solve") are essential but are not the whole story. An icon in the margin helps students see the four-step process as a thread throughout the text. The four-step process appears whenever it fits the statistical content. Its repetitive use should foster the ability to address statistical problems independently.

2. **Use real data.** The study of statistics is supposed to help students work with data in their varied academic disciplines and later employment. This is particularly important for students in the life sciences, because they are asked to collect and analyze data in their laboratory courses and elective undergraduate research. *PSLS* prepares students by providing real (not merely realistic) data from many areas of the life sciences, with sources cited at the back of the book. Data are more than mere numbers—they are numbers with a context that should play a role in making sense of the numbers and in stating conclusions. Examples and exercises in *PSLS* give enough background to allow students to consider the meaning of their calculations.

   PSLS 3e provides about 50 examples and exercises per chapter, with both small data sets for in-class use and large data sets (with several variables and a fairly large number of subjects) more suitable for lab work or assignments. Some data sets recur throughout the book, providing an opportunity for comprehensive analysis spanning a range of statistical topics. The wealth of exercises allows instructors to emphasize some statistical topics or biological themes to tailor the content to their specific learning objectives. In addition, two discussion boxes address in greater depth some important issues when dealing with real data: The discussion box in Chapter 1 exposes students to some of the challenges of data entry and validation, while the discussion box in Chapter 2 explains how to recognize different types of outliers and how to deal with them legitimately.

ORGANIZING A STATISTICAL PROBLEM: THE FOUR-STEP PROCESS

STATE: What is the practical question, in the context of the real-world setting?

PLAN: What specific statistical operations does this problem call for?

SOLVE: Make the graphs and carry out the calculations needed for this problem.

CONCLUDE: Give your practical conclusion in the setting of the real-world problem.

CONFIDENCE INTERVALS: THE FOUR-STEP PROCESS

STATE: What is the practical question that requires estimating a parameter?

PLAN: Identify the parameter and choose a level of confidence.

SOLVE: Carry out the work in two phases:

1. **Check the conditions** for the interval you plan to use.
2. Calculate the **confidence interval** or use technology to obtain it.

CONCLUDE: Return to the practical question to describe your results in this setting.

TESTS OF SIGNIFICANCE: THE FOUR-STEP PROCESS

STATE: What is the practical question that requires a statistical test?

PLAN: Identify the parameter, state null and alternative hypotheses, and choose the type of test that fits your situation.

SOLVE: Carry out the test in three phases:

1. **Check the conditions** for the test you plan to use.
2. Calculate the **test statistic.**
3. Find the ***P*-value** using a table of Normal probabilities or technology.

CONCLUDE: Return to the practical question to describe your results in this setting.

FIGURE 1 Overview of the "four-step process" used in *PSLS 3e*.

3. **Stress conceptual understanding rather than mere knowledge of procedures.** A first course in statistics introduces many skills, from making a histogram to calculating a correlation to choosing and carrying out a significance test. In practice (even if not always in the course), calculations and graphs are automated. Moreover, anyone who makes serious use of statistics will need some specific procedures not taught in his or her college statistics course. *PSLS* therefore aims to make clear the larger patterns and big ideas of statistics in the context of learning specific skills and working with specific data. Many of the big ideas are summarized in graphical outlines in Statistics in Summary figures

within the review chapters. Review chapters also offer a comprehensive summary of the important concepts and skills that students should have mastered, along with an opportunity to select the appropriate statistical analysis without the obvious prompting from a chapter title.

Throughout the text, numerous cautionary statements are included to warn students about common confusions and misinterpretations. A handy "Caution" icon in the margin calls attention to these warnings. In addition, two discussion boxes address how a meaningful interpretation must rely on a comprehensive analysis of the data available: The discussion box in Chapter 10 discusses the interpretation of conditional probabilities in the context of diagnostic and screening tests, while the discussion box in Chapter 20 helps students assess and interpret health risks beyond the P-value of a significance test.

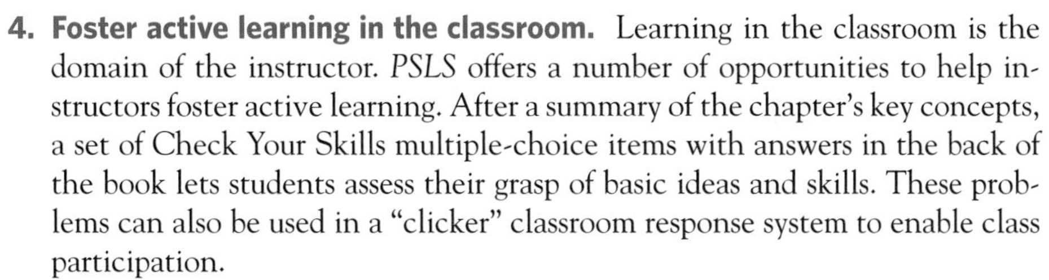

4. **Foster active learning in the classroom.** Learning in the classroom is the domain of the instructor. *PSLS* offers a number of opportunities to help instructors foster active learning. After a summary of the chapter's key concepts, a set of Check Your Skills multiple-choice items with answers in the back of the book lets students assess their grasp of basic ideas and skills. These problems can also be used in a "clicker" classroom response system to enable class participation.

   PSLS also provides many examples and exercises, based on small data sets or summary statistics, that can be solved during class with a simple calculator and a table or with a graphing calculator. For courses that include a computer lab component, the large data set exercises at the end of most chapters offer an opportunity for hands-on analysis with statistical software. There is no short answer given to students for these specific exercises, so that instructors can elect to assign them for a grade.

   Graphical representations of new concepts can also help students learn through experience. *PSLS* and *BPS* offer on their companion websites a set of interactive applets created to our specifications. These applets are designed primarily to help in learning statistics rather than in doing statistics. We suggest using selected applets for classroom demonstrations even if you do not ask students to work with them. The *Correlation and Regression*, *Confidence Interval*, and *P-value* applets, for example, convey core ideas more clearly than any amount of chalk and talk.

5. **Use technology for developing conceptual understanding and analyzing data.** Automating calculations increases students' ability to complete problems, reduces their frustration, and helps them concentrate on ideas and problem recognition rather than mechanics. *All students should have at least a "two-variable statistics" calculator* with functions for correlation and the least-squares regression line as well as for the mean and standard deviation. Because students have calculators, the text doesn't discuss out-of-date "computing formulas" for the sample standard deviation or the least-squares regression line.

   Many instructors will take advantage of more elaborate technology. And many students will find themselves using statistical software on the job. *PSLS* does not assume or require use of software, except in the last few chapters where the work is otherwise too tedious. *PSLS* does accommodate technology use, however, and shows students that they are gaining knowledge that will enable them to read and use output from almost any source. There are regular "Using Technology" sections throughout the text. These sections display and comment on output from different technologies, representing graphing calculators (the Texas Instruments TI-83 or TI-84), spreadsheets (Microsoft Excel), and statistical software (CrunchIt!, JMP, Minitab, R, SPSS). The output always concerns one of the main teaching examples, so that students can compare text and output.

A quite different use of technology appears in the interactive applets available on the companion website. Some examples and exercises in the text use applets, and they are marked with a dedicated icon in the margin.

6. **Use assessments to improve and evaluate student learning.** *PSLS* is structured to help students learn through practice and self-assessment. Within chapters, exercises progress from straightforward applications to comprehensive review problems, with short answers to odd-numbered exercises revealed at the back of the book. To facilitate the learning process, content is broken into digestible bites of material followed by a few Apply Your Knowledge exercises for a quick check of basic mastery. After a summary of the chapter's key concepts, a set of Check Your Skills multiple-choice items with answers in the back of the book lets students assess their grasp of basic ideas and skills (or they can be used in class in a "clicker" response system). End-of-chapter exercises integrate all aspects of the chapter. For chapters dealing with quantitative data, a set of exercises using large data sets is provided to serve as the basis for comprehensive assignments or for use within a computer lab teaching environment. Exercises in the three review chapters enlarge the statistical context beyond that of the immediate lesson. (Many instructors will find that the review chapters appear at the right points for pre-exam review.)

   PSLS 3e also comes with a new instructor test bank designed to offer more comprehensive testing options that span multiple chapters. The objective is to reinforce the idea that statistics is the science of data and that data analysis is a comprehensive approach.

What's new in the third edition?

This third edition of *PSLS* brings many new examples and exercises throughout the book, as well as opportunities to polish the exposition in ways intended to help students learn. Here are, specifically, some of the major changes:

- **New exercises and examples.** *PSLS 3e* has over 300 new or updated exercises and examples, representing nearly one-quarter of the exercises and examples in the second edition. Why such a large turnover? One reason is practical: Instructors can get bored teaching the same old material, and homework assignments need to be varied over time to avoid "recycled paper" issues. The other reason is pedagogical: Statistics *is* an exciting and relevant scientific field, and students should see this for themselves through interesting and current problems. All surveys cited in *PSLS 3e* provide the most recent data available at the time of publication, and all new problems based on research are derived from articles published in the last few years.

PSLS also strives to provide examples of statistical applications in various areas of the life sciences to accommodate different student audiences and a wide range of interests. New problems with a human focus include test performances of individuals with Highly Superior Autobiographical Memory, the relationship between body mass index and testosterone levels among adolescent males, the comparative efficiency of aerobic and resistance training for reducing body fat, and the recently approved over-the-counter OraQuick HIV test. Interesting new animal studies include the righting behavior of aphids in free fall, the paw preference of tree shrews in grasping tasks, and the effect of access to junk food on the body weights of lab rats. New plant and ecological studies include the monitoring of worldwide cases of herbicide-resistant weeds, an ecological approach to control algae bloom, and the genetics of heat resistance in rice.

- **This Chapter in Context.** Students often struggle to understand how concepts covered in different chapters are related and complementary. We created a new section in each chapter to help integrate learning across chapters. Following the Chapter Summary, a new This Chapter in Context section reminds students of the elements of previous chapters that are directly relevant to the current chapter. It also highlights how the new material will come into play in following chapters, providing a road map to guide students' learning.

  For example, this section in Chapter 24, on ANOVA, draws attention to the similarities with the one-sample and two-sample t tests from Chapters 17 and 18, as they all address inference on the parameter μ for a quantitative variable. Because of that, they share a commonality of approaches for assessing the conditions for inference, drawing on descriptive techniques from Chapter 1. Which procedure to use and what conclusions can be made depend on the study design, something discussed in Chapters 7 and 8. A reference is also made to follow-up analyses and inference for more complex designs, described in Chapter 26.

- **New discussion box.** The first two editions of *PSLS* contain a number of discussion boxes that address more leisurely and in greater depth some important conceptual issues in statistics. The third edition adds a new discussion on the challenges of data entry. Statistical consultants often lament that many problems they encounter when helping scientists stem from poor data entry choices (as well as really poor choices of design . . .). This new discussion box addresses the need for keeping detailed records of the data collection process and of all the computations performed, explains ways to organize data for easier software analysis, and describes simple methods that should be used to check for errors and missing values.

- **Tables versus technology.** We asked the *PSLS 3e* reviewers about their use of technology versus printed tables for probability and statistical computations—and found essentially an even split. Some instructors use printed tables in class because technology is not an option for exams, but others use printed tables as a pedagogical preference. Some instructors—in increasing numbers—completely forgo printed tables. Furthering changes initiated in the second edition, *PSLS 3e* uses a modular organization within the relevant chapters that offers instructors the flexibility to teach using either approach.

Why did you do that?

There is no single best way to organize the presentation of statistics to beginners. That said, our choices reflect thinking about both content and pedagogy. Here are comments on several "frequently asked questions" about the order and selection of material in *PSLS 3e*.

Why does the distinction between population and sample not appear in Part I? The concepts of populations and samples are briefly introduced in Chapter 1, but the distinction between them is not emphasized until much later in the book. This is a sign that there is more to statistics than inference. In fact, statistical inference is appropriate only in rather special circumstances. The chapters in Part I present tools and tactics for describing data—any data. Many data sets in these chapters do not lend themselves to inference, because they represent an entire population. John Tukey of Bell Labs and Princeton, the philosopher of modern data analysis, insisted that the population-sample distinction be avoided when it is not relevant, and we agree with him.

Why not begin with data production? It is certainly reasonable to do so—the natural flow of a planned study is from design to data analysis to inference. But in their future employment most students will use statistics mainly in settings other than planned research studies. We place the design of data production (Chapters 7 and 8) after data analysis to emphasize that data-analytic techniques apply to any data. One of the primary purposes of statistical designs for producing data is to make inference possible, so the discussion in Chapters 7 and 8 opens Part II and motivates the study of probability.

Why not delay correlation and regression until late in the course, as is traditional? *PSLS 3e* begins by offering experience working with data and gives a conceptual structure for this nonmathematical but essential part of statistics. Students profit from more experience with data and from seeing the conceptual structure worked out in relations among variables as well as in describing single-variable data. Correlation and least-squares regression are very important descriptive tools, and they are often used in settings where there is no population-sample distinction, such as studies based on state records or average species data. Perhaps most important, the *PSLS* approach asks students to think about what kind of relationship lies behind the data (confounding, lurking variables, association doesn't imply causation, and so on), without overwhelming them with the demands of formal inference methods. Inference in the correlation and regression setting is a bit complex, demands software and a close examination of residuals, and often comes right at the end of the course. Delaying all mention of correlation and regression to that point often impedes the mastering of basic uses and properties of these methods. We consider Chapters 3 and 4 (correlation and regression) essential and Chapter 23 (regression inference and residual plots) optional when time constraints limit the amount of material that can be taught. For similar reasons, two-way tables are introduced first in the context of exploratory data analysis before moving on to inference with the chi-square test in Part III.

What about probability? Chapters 9, 11, and 13 present in a simple format the ideas of probability and sampling distributions that are needed to understand inference. These chapters go from the idea of probability as long-term regularity through concrete ways of assigning probabilities to the idea of the sampling distribution of a statistic. The central limit theorem appears in the context of discussing the sampling distribution of a sample mean. What is left with the *optional* Chapters 10 and 12 is mostly "general probability rules" (including conditional probability) and the binomial and Poisson distributions.

We suggest that you omit these optional chapters unless they represent important concepts for your particular audience. Experienced teachers recognize that students find probability difficult, and research has shown that this is true even for professionals. If a course is intended for med or premed students, for instance, the concept of conditional probability is very relevant because it is a key part of diagnosis that both doctors and patients have difficulty interpreting.[2] However, attempting to present a substantial introduction to probability in a data-oriented statistics course for students who are not mathematically trained is a very difficult challenge. Instructors should keep in mind that formal probability does not help students master the ideas of inference as much as we teachers often imagine, and it depletes reserves of mental energy that might better be applied to essential statistical ideas.

Why use the *z* procedures for a population mean to introduce the reasoning of inference? This is a pedagogical issue, not a question of statistics in practice. Some time in the golden future, we will start with resampling methods, as permutation tests make the reasoning of tests clearer than any traditional approach. For now, the main choices are z for a mean and z for a proportion.

The z procedures for means are pedagogically more accessible to students. We can say up front that we are going to explore the reasoning of inference in an overly simple setting. Remember, exactly Normal populations and true simple random samples are as unrealistic as known σ, especially in the life sciences. All the issues of practice—robustness against lack of Normality and application when the data aren't an SRS, as well as the need to estimate σ—are put off until, with the reasoning in hand, we discuss the practically useful t procedures. This separation of initial reasoning from messier practice works well.

On the contrary, starting with inference for p introduces many side issues: no exact Normal sampling distribution, but a Normal approximation to a discrete distribution; use of $\hat{p}$ in both the numerator and the denominator of the test statistic to estimate both the parameter p and $\hat{p}$'s own standard deviation; loss of the direct link between test and confidence interval. In addition, we now know that the traditional z confidence interval for p is often grossly inaccurate, as explained in the following section. Lastly, major polling organizations like Gallup and Pew now report substantially different margins of error (likely accounting for data weighing), making it increasingly challenging to show students real-life applications of the z method for a proportion.

Why does the presentation of inference for proportions go beyond the traditional methods? Recent computational and theoretical work has demonstrated convincingly that the standard confidence intervals for proportions can be trusted only for very large sample sizes. It is hard to abandon old friends, but the graphs in section 2 of the paper by Brown, Cai, and DasGupta in the May 2001 issue of *Statistical Science* are both distressing and persuasive.[3] The standard intervals often have a true confidence level much less than what was requested, and requiring larger samples encounters a maze of "lucky" and "unlucky" sample sizes until very large samples are reached. Fortunately, there is a simple cure: Just add two successes and two failures to your data. (Therefore, no additional software tool is required for this procedure.) We present these "plus four intervals" in Chapters 19 and 20, along with guidelines for use.

Why didn't you cover Topic X? Introductory texts ought not to be encyclopedic. Including each reader's favorite special topic results in a text that is formidable in size and intimidating to students. The topics covered in *PSLS* were chosen because they are the most commonly used in the life sciences and they are suitable vehicles for learning broader statistical ideas. Three chapters available on the companion website cover more advanced inference procedures. Students who have completed the core of *PSLS* will have little difficulty moving on to more elaborate methods.

ACKNOWLEDGMENTS

We are grateful to colleagues from two-year and four-year colleges and universities who commented on successive drafts of the manuscript. Special thanks are due to John Samons of Florida State College at Jacksonville, who read the manuscript and checked its accuracy. We also wish to thank those who reviewed the third edition:

Flavia Cristina Drumond Andrade
University of Illinois at Urbana–Champaign
Christopher E. Barat
Stevenson University
Gregory C. Booton
The Ohio State University
Jason Brinkley
Eastern Carolina University
Margaret Bryan
University of Missouri–Columbia
Stephen Dinda
Youngstown State University
Chris Drake
University of California–Davis
Sally Freels
University of Illinois at Chicago
Leslie Gardner
University of Indianapolis
Lisa Gardner
Drake University
Bud Gerstman
San Jose State University
Diane Gilbert-Diamond
Geisel School of Medicine at Dartmouth
Olga G. Goloubeva
University of Maryland
Steven Green
University of Miami
Patricia Humphrey
Georgia Southern University
Ravindra P. Joshi
Old Dominion University
Karry Kazial
SUNY Fredonia
Robert Kushler
Oakland University
Heather Ames Lewis
Nazareth College
Lia Liu
University of Illinois at Chicago
Maria McDermott
University of Rochester
Monica Mispireta
Idaho State University
Sumona Mondal
Clarkson University
David Poock
Davenport University
Tiantian Qin
Purdue University
Rachel Rader
Ohio Northern University
Daniel Rand
Winona State University
Bonnie J. Ripley
Grossmont College
Guogen Shan
University of Nevada–Las Vegas
Laura Shiels
University of Hawai'i at Mānoa
Jenny Shook
Penn State University
Thomas H. Short
John Carroll University
Albert F. Smith
Cleveland State University
Manfred Stommel
Michigan State University
Ming T. Tan
Georgetown University
Stephen D. Van Hooser
Brandeis University
Michael H. Veatch
Gordon College
Karen M. Villarreal
Loyola University New Orleans
Andria Villines
Bellevue College
Barbara Ward
Belmont University
Ken Yasukawa
Beloit College

We are also particularly grateful to Ruth Baruth, Karen Carson, Katrina Wilhelm, Liam Ferguson, Pamela Bruton, Elizabeth Geller, Blake Logan, Ellen Cash, Bianca Moscatelli, and the other editorial and design professionals who have helped make this textbook a reality.

Finally, our thanks go to the many students whose compliments and complaints have helped improved our teaching over the years.

Brigitte Baldi and David S. Moore

LaunchPad

W. H. Freeman's new online homework system, **LaunchPad,** offers our quality content curated and organized for easy assignability in a simple but powerful interface.

We've taken what we've learned from thousands of instructors and hundreds of thousands of students to create a new generation of W. H. Freeman/Macmillan technology.

Curated Units. Combining a curated collection of videos, homework sets, tutorials, applets, and e-Book content, LaunchPad's interactive units give you a building block to use as is or as a starting point for your own learning units. Thousands of exercises from the text can be assigned as online homework, including many algorithmic exercises. An entire unit's worth of work can be assigned in seconds, drastically reducing the amount of time it takes for you to have your course up and running.

Easily customizable. You can customize the LaunchPad Units by adding quizzes and other activities from our vast wealth of resources. You can also add a discussion board, a dropbox, and RSS feed, with a few clicks. LaunchPad allows you to customize your students' experience as much or as little as you like.

Useful analytics. The gradebook quickly and easily allows you to look up performance metrics for classes, individual students, and individual assignments.

Intuitive interface and design. The student experience is simplified. Students' navigation options and expectations are clearly laid out at all times, ensuring they can never get lost in the system.

Assets integrated into LaunchPad include:

Interactive e-Book. Every LaunchPad e-Book comes with powerful study tools for students, video and multimedia content, and easy customization for instructors. Students can search, highlight, and bookmark, making it easier to study and access key content. And teachers can ensure that their classes get just the book they want to deliver: customize and rearrange chapters, add and share notes and discussions, and link to quizzes, activities, and other resources.

LEARNINGCurve

LearningCurve provides students and instructors with powerful adaptive quizzing, a game-like format, direct links to the e-Book, and instant feedback. The quizzing system features questions tailored specifically to the text and adapts to students responses, providing material at different difficulty levels and topics based on student performance.

FREEMAN SolutionMaster

SolutionMaster offers an easy-to-use web-based version of the instructor's solutions, allowing instructors to generate a solution file for any set of homework exercises.

New! Stepped Tutorials are centered on algorithmically generated quizzing with step-by-step feedback to help students work their way toward the correct solution. These new exercise tutorials (two to three per chapter) are easily assignable and assessable.

Statistical Video Series consists of StatClips, StatClips Examples, and Statistically Speaking "Snapshots." View animated lecture videos, whiteboard lessons, and documentary-style footage that illustrate key statistical concepts and help students visualize statistics in real-world scenarios.

New! Video Technology Manuals available for TI-83/84 calculators, Minitab, Excel, JMP, SPSS, R, Rcmdr, and CrunchIT!® provide brief instructions for using specific statistical software.

Updated! StatTutor Tutorials offer multimedia tutorials that explore important concepts and procedures in a presentation that combines video, audio, and interactive features. The newly revised format includes built-in, assignable assessments and a bright new interface.

Updated! Statistical Applets give students hands-on opportunities to familiarize themselves with important statistical concepts and procedures, in an interactive setting that allows them to manipulate variables and see the results graphically. Icons in the textbook indicate when an applet is available for the material being covered.

CrunchIT!® is a web-based statistical program that allows users to perform all the statistical operations and graphing needed for an introductory statistics course and more. It saves users time by automatically loading data from *PSLS3e*, and it provides the flexibility to edit and import additional data.

Stats @ Work Simulations put students in the role of the statistical consultant, helping them better understand statistics interactively within the context of real-life scenarios.

EESEE Case Studies (Electronic Encyclopedia of Statistical Examples and Exercises), developed by The Ohio State University Statistics Department, teach students to apply their statistical skills by exploring actual case studies using real data.

Data files are available in ASCII, Excel, TI, Minitab, SPSS (an IBM Company),[1] and JMP formats.

Student Solutions Manual provides solutions to the odd-numbered exercises in the text. Available electronically within LaunchPad, as well as in print form.

Interactive Table Reader allows students to use statistical tables interactively to seek the information they need.

[1]SPSS was acquired by IBM in October 2009.

Instructor's Guide with Solutions includes teaching suggestions, chapter comments, and detailed solutions to all exercises. Available electronically within LaunchPad, and in print form.

Test Bank offers hundreds of multiple-choice questions. Also available on CD-ROM (for Windows and Mac), where questions can be downloaded, edited, and resequenced to suit each instructor's needs.

Lecture PowerPoint Slides offer a detailed lecture presentation of statistical concepts covered in each chapter of *PSLS*.

Additional Resources Available with PSLS3e

Companion Website `www.whfreeman.com/psls3e` This open-access website includes statistical applets, data files, and self-quizzes. The website also offers three optional companion chapters covering ANOVA, nonparametric tests, and multiple and logistic regression.

Instructor access to the Companion Website requires user registration as an instructor and features all of the open-access student web materials, plus:

- Instructor version of **EESEE** with solutions to the exercises in the student version.
- **PowerPoint Slides** containing all textbook figures and tables.
- **Lecture PowerPoint Slides**

Special Software Packages Student versions of JMP and Minitab are available for packaging with the text. Contact your W. H. Freeman representative for information or visit `www.whfreeman.com`.

Course Management Systems W. H. Freeman and Company provides courses for Blackboard, Angel, Desire2Learn, Canvas, Moodle, and Sakai course management systems. These are completely integrated solutions that you can easily customize and adapt to meet your teaching goals and course objectives. Visit `macmillanhighered.com/Catalog/other/Coursepack` for more information.

i>clicker

i-Clicker is a two-way radio-frequency classroom response solution developed by educators for educators. Each step of i-clicker's development has been informed by teaching and learning. To learn more about packaging i-clicker with this textbook, please contact your local sales rep or visit `www.iclicker.com`.

TO THE STUDENT: Statistical Thinking

Statistics is about data. Data are numbers, but they are not "just numbers." **Data are numbers with a context.** The number 10.5, for example, carries no information by itself. But a newborn weighing 10.5 pounds is an indication of a healthy size at birth. The context engages our background knowledge and allows us to make judgments; it makes the number informative.

Statistics is the science of data. To gain insight from data, we make graphs and do calculations. But graphs and calculations are guided by ways of thinking that amount to educated common sense. Let's begin our study of statistics with an informal look at some principles of statistical thinking.

DATA BEAT ANECDOTES

An anecdote is a striking story that sticks in our minds exactly because it is striking. Anecdotes humanize an issue, but they can be misleading.

Does living near power lines cause leukemia in children? The National Cancer Institute spent 5 years and $5 million gathering data on this question. The researchers compared 638 children who had leukemia with 620 who did not. They went into the homes and measured the magnetic fields in the children's bedrooms, in other rooms, and at the front door. They recorded facts about power lines near the family home and also near the mother's residence when she was pregnant. Result: no connection between leukemia and exposure to magnetic fields of the kind produced by power lines. The editorial that accompanied the study report in the *New England Journal of Medicine* thundered, "It is time to stop wasting our research resources" on the question.[1]

Now compare the effectiveness of a television news report of a 5-year, $5 million investigation against a televised interview with an articulate mother whose child has leukemia and who happens to live near a power line. In the public mind, the anecdote wins every time. A statistically literate person knows better. **Data are more reliable than anecdotes, because they systematically describe an overall picture rather than focus on a few incidents.**

WHERE THE DATA COME FROM IS IMPORTANT

The advice columnist Ann Landers once asked her readers, "If you had it to do over again, would you have children?" A few weeks later, her column was headlined "70% OF PARENTS SAY KIDS NOT WORTH IT." Indeed, 70% of the nearly 10,000 parents who wrote in said they would not have children if they could make the choice again. *Do you believe that 70% of all parents regret having children?*

You shouldn't. The people who took the trouble to write Ann Landers are not representative of all parents. Their letters showed that many of them were angry at their children. All we know from these data is that there are some unhappy parents out there. A statistically designed poll, unlike Ann Landers's appeal, targets specific people chosen in a way that gives all parents the same chance to be asked. Such a poll showed that 91% of parents *would* have children again. Where data come from matters a lot. If you are careless about how you get your data, you may announce 70% "no" when the truth is about 90% "yes."

Here's another question: *Should episiotomy be a routine part of childbirth?* Episiotomy is the surgical cut of the skin and muscles between the vagina and the rectum sometimes performed during childbirth. Until recently, it was one of the most common surgical procedures in women in the United States, performed routinely to speed up childbirth and in the hope that it would help prevent tearing of the mother's tissue and possible later incontinence. Episiotomy rates vary widely by hospital and by physician, based principally on personal beliefs about the procedure's benefits.

However, recent clinical trials and epidemiological studies have shown no benefit of episiotomy unless the baby's health requires accelerated delivery or a large natural tearing seems likely. In fact, these studies indicate that episiotomy is associated with longer healing times and increased rates of complications, including infection, extensive tearing, pain, and incontinence.[2]

To get convincing evidence on the benefits and risks of episiotomy, we need unbiased data. Proper clinical trials and epidemiological studies rely on randomness of patient or treatment selection to avoid bias. The careful studies of the risks and benefits of routine episiotomy could be trusted because they had sound data collection designs. As a result, rates of episiotomy in the United States have dropped substantially.

The most important information about any statistical study is how the data were produced. Only statistically designed opinion polls and surveys can be trusted. Only experiments can give convincing evidence that an alleged cause really does account for an observed effect.

BEWARE THE LURKING VARIABLE

Energy drinks contain high levels of caffeine, which we know can temporarily boost alertness but can also cause sleep problems later on. A survey of U.S. service members in a combat environment in Afghanistan showed that those consuming energy drinks three or more times a day slept less, had more sleep disruptions, and were more likely to fall asleep on duty than those consuming fewer or no energy drinks.[3] What shall we make of this finding? Among servicemen in combat deployment, does the frequent consumption of energy drinks impair sleep or do individuals with impaired sleep consume more energy drinks to stay alert on the job? The answer may well be both, in a self-reinforcing loop. It is also possible that other factors, such as high stress, both interfere with sleep and create a desire for heightened alertness from energy drinks. A statistician knows that a causal link cannot be established here, because the data from this survey are simply observations.

Should women take hormones such as estrogen after menopause, when natural production of these hormones ends? In 1992, several major medical organizations said "yes." In particular, women who took hormones seemed to reduce their risk of a heart attack by 35% to 50%. But in 2002, the National Institutes of Health declared these findings wrong. Use of hormones after menopause immediately plummeted. Both recommendations were based on extensive studies. What happened?

The evidence in favor of hormone replacement came from a number of *observational studies* that compared women who were taking hormones with others who were not. But women who choose to take hormones are very different from women who do not: They are richer and better educated and see doctors more often. These women do many things to maintain their health. It isn't surprising that they have fewer heart attacks.

We can't conclude that hormone replacement reduces heart attacks just because we see a relationship in the data. Education and affluence are *lurking variables*, background factors that help explain the relationship between hormone replacement and good health.

Almost all relationships between two variables are influenced by other variables lurking in the background. To understand the relationship between two variables, you must often look at other variables. Careful statistical studies try to think of and measure possible lurking variables in order to correct for their influence. As the hormone saga illustrates, this is not always done. News reports often ignore possible lurking variables that might ruin a good headline. The habit of asking "What might lie behind this relationship?" is part of thinking statistically.

Another way to address the effect of lurking variables is by designing careful *experiments*. Several experiments randomly assigned volunteer women either to hormone replacement or to dummy pills that looked and tasted the same as the hormone pills. By 2002, these experiments agreed that hormone replacement does *not* reduce the risk of heart attacks, at least for older women. Faced with this better evidence, medical authorities changed their recommendations.[4]

ALWAYS LOOK AT THE DATA

Yogi Berra said it: "You can observe a lot by just watching." That's a motto for learning from data. **A few carefully chosen graphs are often more instructive than great piles of numbers.**

Let's look at some data. Figure 1 displays a histogram of the body lengths of 56 perch (*Perca fluviatilis*) caught in a Finnish lake.[5] Each bar in the graph represents how many perch had a body length between two values on the horizontal axis. For example, the tallest bar indicates that 17 perch had a body length between 20 and 25 centimeters (cm).

We see great variability in perch body lengths, from about 5 to 50 cm, but we also notice two clear clusters: One group of smaller fish and one group of larger fish. This suggests that the data include two different groups of perch, possibly males

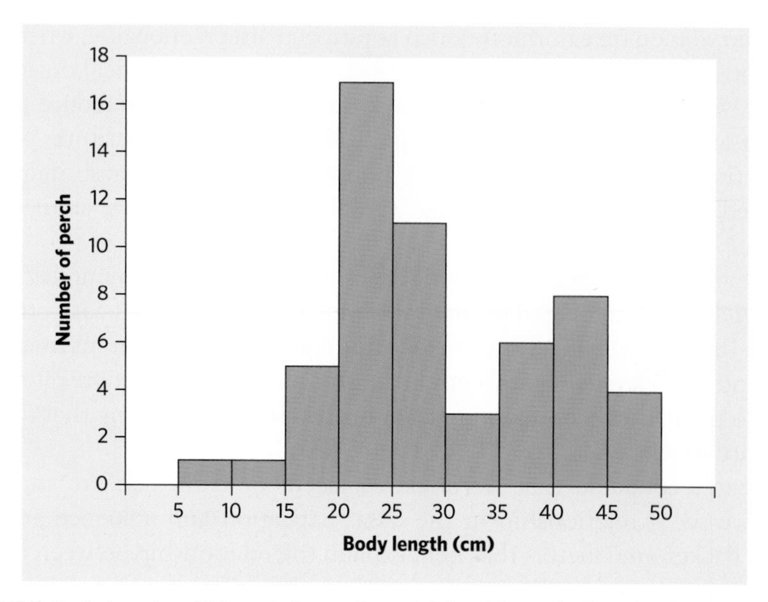

FIGURE 1 Body lengths of 56 perch from a Finnish lake. Always look at the data: Two separate clusters are clearly visible.

and females or perch of different ages. In fact, observations in the wild indicate that perch can become much larger if they survive to a very old age (up to 22 years).

Any attempt to summarize the data in Figure 1 without a close inspection first would miss the two distinct clusters. Failure to examine any data carefully can lead to misleading or absurd results. As humorist Des McHale put it, "The average human has one breast and one testicle."

VARIATION IS EVERYWHERE

Many students in biology lab courses are surprised to obtain somewhat different results when they repeat an experiment. Yet, variability is ubiquitous.

Figure 2 plots the ventilation rates of seven goldfish placed in tanks of varying temperature.[6] Each goldfish is represented by one color. While there is a clear overall pattern in this figure, there is also a lot of variability, with ventilation rates ranging overall from 6 to 119 opercular movements per minute. Some of that variability can be attributed to lack of precision in measurements, some to fish physiology, and some to slight differences in circumstances at the time of each measurement (such as movements around the fish tanks that might stress the fish). Before the experiment, the fish were kept in tanks set to 22 degrees Celsius. At that temperature, the fish ventilation rates range from about 60 to 100 opercular movements per minute. Yet when the same seven fish are transferred to containers with warmer water, their ventilation rates increase overall. And when the fish are transferred to containers at lower temperatures, the ventilation rates decrease quite dramatically. These data show that ventilation rate in goldfish varies both from fish to fish and as a function of water temperature.

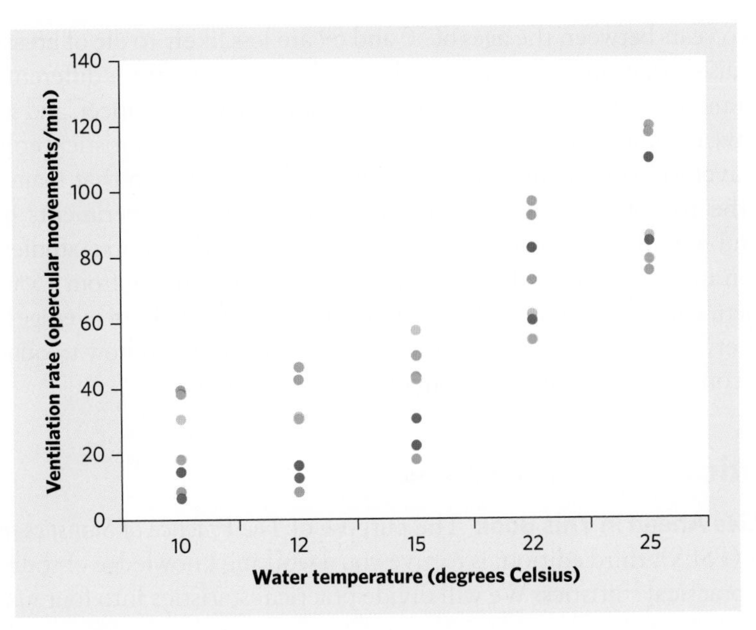

FIGURE 2 Variation is everywhere: the ventilation rates of seven goldfish placed in tanks with varying water temperature.

Students are not the only ones with a tendency to underestimate variability in real data. Here is Arthur Nielsen, head of the country's largest market research firm, describing his experience:

> *Too many business people assign equal validity to all numbers printed on paper. They accept numbers as representing Truth and find it difficult to work with the concept of probability. They do not see a number as a kind of shorthand for a range that describes our actual knowledge of the underlying condition.*[7]

Variation is everywhere. Individuals vary; repeated measurements on the same individual vary; almost everything varies over time. One reason we need to know some statistics is that statistics helps us deal with variation.

CONCLUSIONS ARE NOT CERTAIN

Most women who reach middle age have regular mammograms to detect breast cancer. *Do mammograms reduce the risk of dying of breast cancer?* Doctors rely on clinical trials that compare different ways of screening for breast cancer. The conclusion of the U. S. Preventive Services Task Force after examining all the data available is that biennial mammograms reduce the risk of death in women aged 50 to 69 years by about 16.5% compared with no screening.[8]

Because variation is everywhere, conclusions are uncertain. Statistics gives us a language for talking about uncertainty that is used and understood by statistically literate people everywhere. On average, women who have mammograms every two years between the ages of 50 and 69 are less likely to die of breast cancer. But because variation is everywhere, the results are different for different women. Some women who have biennial mammograms die of breast cancer, and some who never have mammograms live long, breast-cancer-free lives. Statistical conclusions are "on average" statements only. In addition, the conclusion that mammograms reduce the risk of death from breast cancer comes from experiments involving some, and not all, women aged 50 to 69 years. In fact, the various studies found a reduction in breast cancer deaths versus no screening ranging from 15% to 23%. Well then, can we be certain that mammograms reduce risk on average? No. We can be very confident, but we can't be certain. We will learn how to quantify our level of confidence in statistical conclusions.

Statistical Thinking and You

What Lies Ahead in This Book The purpose of *The Practice of Statistics in the Life Sciences* (*PSLS*), third edition, is to give you a working knowledge of the ideas and tools of practical statistics. We will divide practical statistics into four main areas:

1. **Data analysis** concerns methods and strategies for exploring, organizing, and describing data using graphs and numerical summaries. Only organized data can illuminate reality. Only thoughtful exploration of data can defeat the lurking variable. Part I of *PSLS* (Chapters 1 to 6) discusses data analysis.
2. **Data production** provides methods for producing data that can give clear answers to specific questions. Where the data come from really is important. Basic concepts about how to select samples and design experiments are the most influential ideas in statistics. These concepts are the subject of Chapters 7 and 8.
3. **Probability** is the language that describes variation and uncertainty. Because we are concerned with practice rather than theory, we need only a limited knowledge of probability. Chapters 9, 11, and 13 introduce us to the most important probability rules and models. Chapters 10 and 12 offer more probability for those who want it.
4. **Statistical inference** moves beyond the data in hand to draw conclusions about some wider universe, taking into account that variation is everywhere and that conclusions are uncertain. Chapters 14 and 15 discuss the reasoning of statistical inference. These chapters are the key to the rest of the book. Chapters 17 to 21 present inference as used in practice in the most common settings. Chapters 22 to 24, and the Optional Companion Chapters 26 to 28 on the companion website, concern more advanced or specialized kinds of inference.

Statistics, however, is an integrated science of data, and you will see that concepts from one area often relate directly to concepts from other areas. To help you see this, each chapter contains a summary section titled This Chapter in Context. These sections highlight how the concepts from the current chapter relate to concepts introduced in earlier chapters and how they will figure in following chapters.

Because data are numbers with a context, doing statistics means more than manipulating numbers. You must **state** a problem in its real-world context, **plan** your specific statistical work in detail, **solve** the problem by making the necessary graphs and calculations, and **conclude** in context by explaining what your findings say about the real-world setting. We'll make regular use of this four-step process to encourage good habits that go beyond graphs and calculations to ask, "What do the data tell me?"

Statistics does involve lots of calculating and graphing. The text presents the techniques you need, but you should use a calculator or software to automate calculations and graphs as much as possible. Because the big ideas of statistics don't depend on any particular level of access to computing, *PSLS* does not require software. Even if you make little use of technology, you should look at the Using Technology sections throughout the book. You will see at once that you can read and use the output from almost any technology used for statistical calculations. The ideas really are more important than the details of how to do the calculations.

Unless you have constant access to software or a graphing calculator, *you will need a calculator with some built-in statistical functions*. Specifically, your calculator should find means and standard deviations and calculate correlations and regression lines. Look for a calculator that claims to do "two-variable statistics" or mentions "regression." You will get the most, though, out of a graphing calculator or statistical software.

Because graphing and calculating are automated in statistical practice, the most important assets you can gain from the study of statistics are an understanding of the big ideas and the beginnings of good judgment in working with data. *PSLS* aims to explain the most important ideas of statistics, not just teach methods. Some examples of big ideas that you will see are "always plot your data," "randomized comparative experiments," and "statistical significance." Some particularly important ideas (such as how to treat outliers and the scientific approach) are given more extensive treatment with real-life examples in discussions placed throughout the book.

You learn statistics by doing statistical problems. As you read, you will see several levels of exercises, arranged to help you learn. Short Apply Your Knowledge problem sets appear after each major idea. These are straightforward exercises that help you solidify the main points as you read. Be sure you can do these exercises before going on. The end-of-chapter exercises begin with multiple-choice Check Your Skills exercises (with all answers in the back of the book). Use them to check your grasp of the basics. The regular Chapter Exercises help you combine all the ideas of a chapter. At the end of chapters that deal with the analysis of quantitative data, you will find additional exercises marked with the "Large data set" icon; these provide an opportunity to apply your statistical skills to the kind of large and complex data sets encountered very often in the life sciences. Finally, the three Part Review chapters look back over major blocks of learning, with many review exercises. At each step, you are given less advance knowledge of exactly what

statistical ideas and skills the problems will require, so each type of exercise requires more understanding.

The Part Review chapters (and the individual optional chapters on the companion website) include point-by-point lists of specific things you should be able to do. Go through that list, and be sure you can say "I can do that" to each item. Then try some of the review exercises. Every odd-numbered exercise in the book (except for exercises marked "Large data set") has a short answer available at the back of the book (or on the companion website for optional chapters). You can use this to check your work.

The key to learning is persistence. The main ideas of statistics, like the main ideas of any important subject, took a long time to discover and take some time to master. The gain will be worth the pain.

ABOUT THE AUTHORS

Brigitte Baldi is a graduate of France's Ecole Normale Supérieure in Paris. In her academic studies, she combined a love of math and quantitative analysis with wide interests in the life sciences. She majored in math and biology and obtained a master's in molecular biology and biochemistry and a master's in cognitive sciences. She earned her PhD in neuroscience from the Université Paris VI, studying multisensory integration in the brain; and as a postdoctoral fellow at the California Institute of Technology, she used computer simulations to study patterns of brain reorganization after lesion. She then worked as a management consultant advising corporations before returning to academia to teach statistics.

Dr. Baldi is currently a lecturer in the Department of Statistics at the University of California, Irvine. She is actively involved in statistical education. She was a local and later national adviser in the development of the *Statistically Speaking* video series. She developed UCI's first online statistics courses and is always interested in ways to integrate new technologies in the classroom to enhance participation and learning. She is currently serving as an elected member to the Executive Committee At Large of the section on Statistical Education of the American Statistical Association.

David S. Moore is Shanti S. Gupta Distinguished Professor of Statistics, Emeritus, at Purdue University and was 1998 president of the American Statistical Association. He received his AB from Princeton and his PhD from Cornell, both in mathematics. He has written many research papers in statistical theory and served on the editorial boards of several major journals. Professor Moore is an elected fellow of the American Statistical Association and of the Institute of Mathematical Statistics and an elected member of the International Statistical Institute. He has served as program director for statistics and probability at the National Science Foundation.

In recent years, Professor Moore has devoted his attention to the teaching of statistics. He was the content developer for the Annenberg/Corporation for Public Broadcasting college-level telecourse *Against All Odds: Inside Statistics* and for the series of video modules *Statistics: Decisions through Data,* intended to aid the teaching of statistics in schools. He is the author of influential articles on statistics education and of several leading texts. Professor Moore has served as president of the International Association for Statistical Education and has received the Mathematical Association of America's national award for distinguished college or university teaching of mathematics.

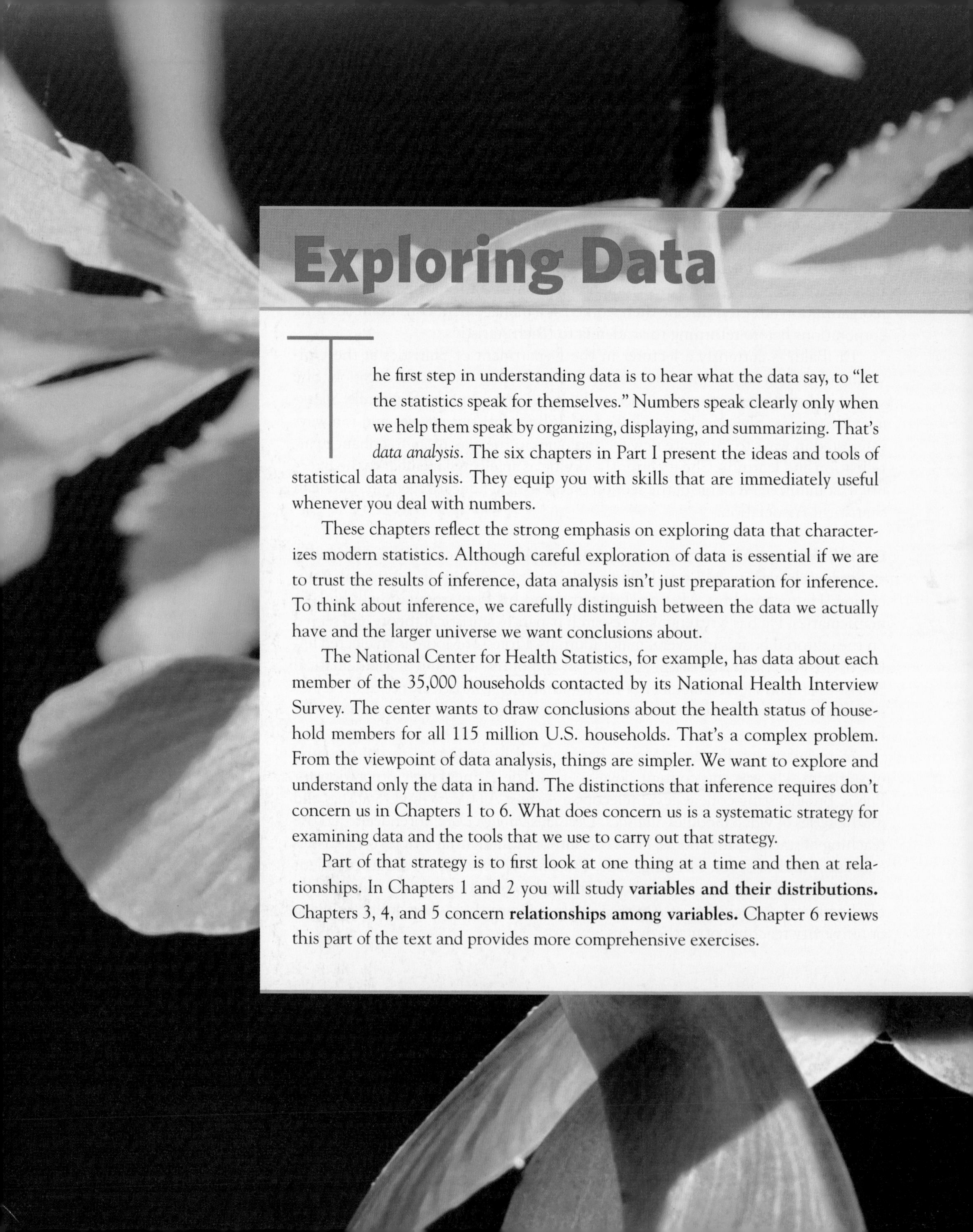

Exploring Data

The first step in understanding data is to hear what the data say, to "let the statistics speak for themselves." Numbers speak clearly only when we help them speak by organizing, displaying, and summarizing. That's *data analysis*. The six chapters in Part I present the ideas and tools of statistical data analysis. They equip you with skills that are immediately useful whenever you deal with numbers.

These chapters reflect the strong emphasis on exploring data that characterizes modern statistics. Although careful exploration of data is essential if we are to trust the results of inference, data analysis isn't just preparation for inference. To think about inference, we carefully distinguish between the data we actually have and the larger universe we want conclusions about.

The National Center for Health Statistics, for example, has data about each member of the 35,000 households contacted by its National Health Interview Survey. The center wants to draw conclusions about the health status of household members for all 115 million U.S. households. That's a complex problem. From the viewpoint of data analysis, things are simpler. We want to explore and understand only the data in hand. The distinctions that inference requires don't concern us in Chapters 1 to 6. What does concern us is a systematic strategy for examining data and the tools that we use to carry out that strategy.

Part of that strategy is to first look at one thing at a time and then at relationships. In Chapters 1 and 2 you will study **variables and their distributions.** Chapters 3, 4, and 5 concern **relationships among variables.** Chapter 6 reviews this part of the text and provides more comprehensive exercises.

PART I

VARIABLES AND DISTRIBUTIONS

RELATIONSHIPS

OGphoto/Getty Images

Oscar Domínguez/age fotostock/SuperStock

CHAPTER 1 Picturing Distributions with Graphs

Statistics is the science of data. The volume of data available to us is overwhelming. In 2003, the Human Genome Project uncovered, after 13 years of international cooperation and billions of dollars, the complete sequence of the 3 billion DNA bases of the human genome. A decade later, individuals can have their own genome sequenced within weeks for just thousands of dollars. In the United States, the Census Bureau collects extensive data on the nation from numerous surveys, such as the yearly National Health Interview Survey collecting health and socioeconomic information on each member of 35,000 households. Gallup, a private polling firm, surveys roughly 1000 U.S. adults per day for its Gallup-Healthways Well-Being Index covering a whole range of physical and mental health conditions. Cell phones have become so prominent that they are used to monitor the ongoing malaria epidemics in Kenya. And satellites provide detailed daily records of climatic conditions at the planetary level. Overall, data collection has increased tremendously in the new millennium with no sign of slowing down. The first step in dealing with such a flood of data is to organize our thinking about data.

IN THIS CHAPTER WE COVER...

Individuals and variables

Any set of data contains information about some group of *individuals*. The information is organized in *variables*. The techniques of descriptive statistics covered in Part I of this book apply equally to data sets obtained from a given **population** (the entire group of individuals about which we want information) or from only those individuals in a smaller **sample.** The distinction between population and sample is an important one for inference, and we will address it in more detail in Chapter 7.

population

sample

> **INDIVIDUALS AND VARIABLES**
>
> **Individuals** are the objects described by a set of data. Individuals may be people, but they may also be animals, plants, or things.
>
> A **variable** is any characteristic of an individual. A variable can take different values for different individuals.

A botanist's plant database, for example, includes data about various aspects of the plants examined. The plants are the individuals described by the database. For each individual (each plant), the data contain the values of variables such as stem length, number of flower petals, and flower color. An example of what such a database might look like is shown in Figure 1.1, with each row dedicated to one individual and each column to a variable. **Spreadsheet** programs that have rows and columns readily available are commonly used to store data and do simple calculations.

spreadsheet

Some variables, like a plant's flower color, simply place individuals into categories. Others, like stem length and number of flower petals, take numerical values with which we can do arithmetic. It makes sense to give an average stem length for plants of a given environment, but it does not make sense to give an "average" flower color. We can summarize our findings on plant color by obtaining the counts of yellow, pink, and white flowers in the database, or by obtaining the proportion that each color type represents. *It is important, though, not to confuse these numerical summaries (average, count, or proportion) with the variables themselves.*

Microsoft Excel

	A	B	C	D
1	Specimen ID	Stem length (cm)	Number of flower petals	Flower color
2	1	3.1	28	yellow
3	2	2.6	5	white
4	3	8.2	8	red
5	4	5.9	12	pink
6	5	14.7	5	yellow

FIGURE 1.1 Example of a botanical database displayed in a spreadsheet. Each column includes data about a different variable. Each row represents data for one plant specimen.

CATEGORICAL AND QUANTITATIVE VARIABLES

A **categorical variable** places an individual into one of several groups or categories.

A **quantitative variable** takes numerical values for which arithmetic operations such as adding and averaging make sense. The values of a quantitative variable are usually recorded in a **unit of measurement** such as seconds or kilograms.

Further distinction is sometimes made beyond simply categorical or quantitative. Some quantitative variables, like stem length, are **continuous** variables that can take any real numerical value over an interval. **Discrete** variables, on the other hand, are quantitative variables that can take only a limited, finite number of values, like the number of petals in a flower.[1] Categorical variables can also be broken down into nominal and ordinal. **Nominal** variables are purely qualitative and unordered, like flower color, whereas **ordinal** data can be ranked, like star ratings or the Likert scales commonly used in psychology (for example, "do you strongly disagree, disagree, agree, or strongly agree with the following statement?"). Although ordinal data can be ranked, they are not true quantitative variables, because the intervals between consecutive ranks are often not identical.

continuous

discrete

nominal

ordinal

EXAMPLE 1.1 Paw preference in tree shrews

Oscar Domínguez/age fotostock/SuperStock

Tree shrews, *Tupaia belangeri*, are small omnivorous mammals phylogenetically related to primates. A research team examined paw preference during grasping tasks among 36 tree shrews born and raised in captivity. Here are some of the data they reported:[2]

Subject	Sex	Age	PI	Bias
Abel	m	6	−1.00	L
Anna	f	5	−0.95	L
Aragorn	m	6	−1.00	L
Barbossa	m	1	−0.25	A
Bea	f	4	1.00	R
Beatrice	f	1	0.56	R
Berta	f	1	0.13	A
...				

In addition to the animal's name (an identifier), four variables are displayed. The first one, sex, is a categorical variable because each animal can be only male (m) or female (f). *Sex would still be a categorical variable if it had been recorded with a "0" for male and a "1" for female, because these would be labels rather than true numerical values*. The second variable, age, is a quantitative variable measured in years. Age values displayed in the table are rounded to whole numbers of years, but animals age continuously: this means that age is continuous and not discrete. The third variable, PI, stands for "pawedness index," a unitless continuous variable taking values between −1 and 1, depending on the relative use of the right and left paws. The last variable, bias, is categorical and represents the animal's general paw preference: right paw (R), left paw (L), or ambidextrous (A). ■

Gesundheit! Researchers who study treatments against the common cold must find ways to assess their effectiveness. Effectiveness may be either preventing a cold or reducing its symptoms. Researchers can record simply whether or not each person exposed to the common-cold virus developed cold symptoms. They can also count the number of cold symptoms detectable, count the number of tissues used in a day, or record the duration of symptoms. Smart studies often look at a problem from more than just one angle.

Few scientific publications display their raw data the way the study in Example 1.1 does. Instead, information about a study's individuals and variables must often be retrieved from the abstract, the methods section, or a figure legend, typically presented in summary form. When you explore data from someone else's work—or when planning your own statistical study—ask the following questions:

1. **Who** or what are the **individuals** studied? **How many** individuals appear in the data set?
2. **What** do we record for each individual? How many **variables** do the data contain? What are the **exact definitions** of these variables? In what **units of measurement,** if any, is each variable recorded? Lengths, for example, might be recorded in inches, in yards, or in meters.
3. **Why? What purpose** do the data have? Do we hope to answer some specific questions? Do we want to draw conclusions about individuals other than the ones we actually have data for? Are the variables suitable for the intended purpose?

EXAMPLE 1.2 Tumor-specific antibodies for cancer treatment

Could cancer cells be selectively targeted by using antibodies recognizing a tumor-specific protein marker? Researchers grafted human cancerous cells onto healthy adult mice. Some of the mice were then injected with tumor-specific antibodies (treatment group) while the other mice were not (control group). A figure in the publication displays the following findings: "the number of mice exhibiting lymph node metastases in each group" (there were 10 mice in the treatment group and 10 mice in the control group) and "the number of secondary lymph nodes detected in each mouse."[3]

First we ask who or what are the individuals studied. Here the individuals are the mice injected with human cancerous cells. There were 20 mice in total, divided into two groups of 10 each. Then we ask what was recorded for each individual. The figure describes two findings. While both findings make reference to a "number of . . . ," the first number represents the summary finding for each group, whereas the second number describes individual mice. Therefore, the first variable is the presence or absence of lymph node metastases, a categorical variable. The second variable is the number of secondary lymph nodes, a quantitative discrete variable. Lastly we ask what purpose the data serve. The first variable answers the question of whether the treatment can prevent the formation of metastases (yes or no), while the second variable answers the question of how effective the treatment is (how many nodes are found). ■

The publication referenced in Example 1.2 presented a lot more findings. This is not particularly unusual. *Be aware that some scientific publications combine the findings from a series of related but distinct studies*. In such cases, you need to identify each study before you can determine their specific individuals and variables.

APPLY YOUR KNOWLEDGE

1.1 A medical study. Data from a medical study contain values of many variables for each of the people who were the subjects of the study. Which of the following variables are categorical and which are quantitative?

(a) Gender (female or male)

(b) Age (years)

(c) Weight group (underweight, normal weight, overweight, obese)

(d) Smoker (yes or no)

(e) Systolic blood pressure (millimeters of mercury)

(f) Level of calcium in the blood (micrograms per milliliter)

1.2 **Cereal content.** Here is a small part of an EESEE data set (available on the companion website), "Nutrition and Breakfast Cereals," that describes the nutritional content per serving of 77 brands of breakfast cereal:[4]

Brand name	Manufacturer	Cold/hot	Calories (Cal)	Sugar (grams)	Fiber (grams)
All Bran	K	C	70	5	9
All Bran with Extra Fiber	K	C	50	0	14
Almond Delight	R	C	110	8	1
Apple Cinnamon Cheerios	G	C	110	10	2
Apple Jacks	K	C	110	14	1
⋮					

(a) What are the individuals in this data set?

(b) For each individual, what variables are given? Which of these variables are categorical and which are quantitative?

Categorical variables: pie charts and bar graphs

Statistical tools and ideas help us examine data in order to describe their main features. This examination is called **exploratory data analysis.** Like an explorer crossing unknown lands, we want first to simply describe what we see. Here are two principles that help us organize our exploration of a set of data.

exploratory data analysis

EXPLORING DATA

1. Begin by examining each variable by itself. Then move on to study the relationships among the variables.
2. Begin with a graph or graphs. Then add numerical summaries of specific aspects of the data.

We will also follow these principles in organizing our learning. Chapters 1 and 2 present methods for describing a single variable. We study relationships among several variables in Chapters 3 to 5. In each case, we begin with graphical displays, then add numerical summaries for more complete description.

The proper choice of graph depends on the nature of the variable. To examine a single variable, we usually want to display its *distribution*.

DISTRIBUTION OF A VARIABLE

The **distribution** of a variable tells us what values it takes and how often it takes these values.

The values of a categorical variable are labels for the categories. The **distribution of a categorical variable** lists the categories and gives either the count or the percent of individuals that fall in each category.

EXAMPLE 1.3 Leading causes of death

What are the leading causes of death in America? The National Center for Health Statistics reports such information on a yearly basis. Here is a breakdown of the top 10 leading causes of death for the year 2011 in the United States:[5]

Top 10 causes of death	Count of deaths	Percent of top 10 causes
Heart disease	596,339	32
Cancer	575,313	31
Chronic respiratory disease	143,382	8
Cerebrovascular disease	128,931	7
Accidental death	122,777	7
Alzheimer's disease	84,691	5
Diabetes	73,282	4
Flu and pneumonia	53,667	3
Kidney disorder	45,731	2
Suicide	38,285	2
Total	1,862,398	100

frequency
relative frequency

The data table provides both the count of deaths for each category and the percent that each category represents in the set. Counts are also sometimes referred to as **frequencies,** and percents as **relative frequencies.**

roundoff error

It's a good idea to check data for consistency. The counts should add to 1,862,398, the total number of deaths from the top 10 causes in the year 2011. They do. The percents should add to 100% or, because each percent is rounded to the nearest integer, very nearly 100%. **Roundoff errors** don't point to mistakes in our work, just to the effect of rounding off results. ■

pie chart

Columns of numbers take time to read and interpret. The **pie chart** in Figure 1.2 shows the distribution of the top 10 causes of death more vividly. For example, the "cancer" slice makes up 31% of the pie because 31% of all those who died from a top 10 cause in the United States in 2011 died from cancer. Pie charts are awkward to make by hand, but software will do the job for you. *A pie chart must include all the categories that make up a whole. Use a pie chart only when you want to emphasize each category's relation to the whole.* Here the pie chart displays all

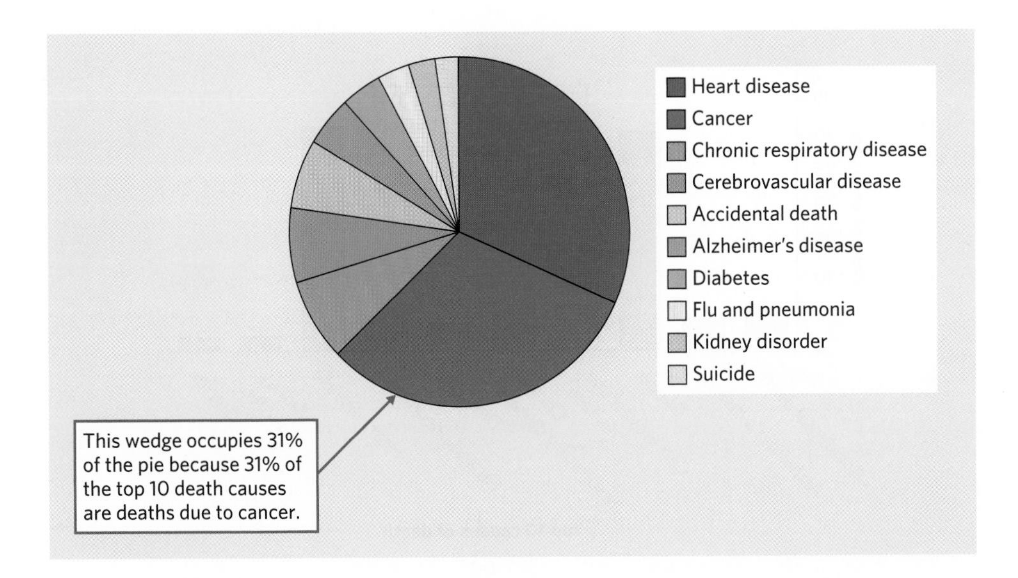

FIGURE 1.2 You can use either a pie chart or a bar graph to display the distribution of a categorical variable. Here is a pie chart of the top 10 causes of death in the United States for 2011.

10 categories making up the top 10 causes of death. The graph shows more clearly than the table of raw numbers the predominance of heart disease and cancer as leading causes of death.

bar graph

We could also make a **bar graph** that represents each cause of death by the height of a bar. Pie charts must include all the categories that make up a whole, but bar graphs are more flexible: They can compare any set of quantities that are measured in the same units. Bar graphs are particularly clear in pointing to the order and the relative importance of the different categories. Figure 1.3 shows two possible bar graphs of the data in Example 1.3. The bar graph shown in Figure 1.3(a) is sorted alphabetically: Accidental death is the first bar in the graph because this category starts with "A." A more interesting and more intuitive way of representing the same data is shown in Figure 1.3(b). This bar graph is sorted by magnitude instead: The tallest bar appears first, followed by the second-tallest bar, and so on. It makes very clear that heart disease and cancer are by far the most important causes of death in the modern-day United States.

EXAMPLE 1.4 Who uses marijuana?

The 2009 National Household Survey on Drug Use and Health reported the percent of current users of cannabis (marijuana and hashish) in each of four age groups:[6]

Age group (years)	Percent who are current users
12–17	10.0
18–25	21.2
26–34	9.6
35+	3.4

It's clear that young adults between 18 and 25 years of age show the most interest in consuming cannabis. ■

Emilio Ereza/Alamy

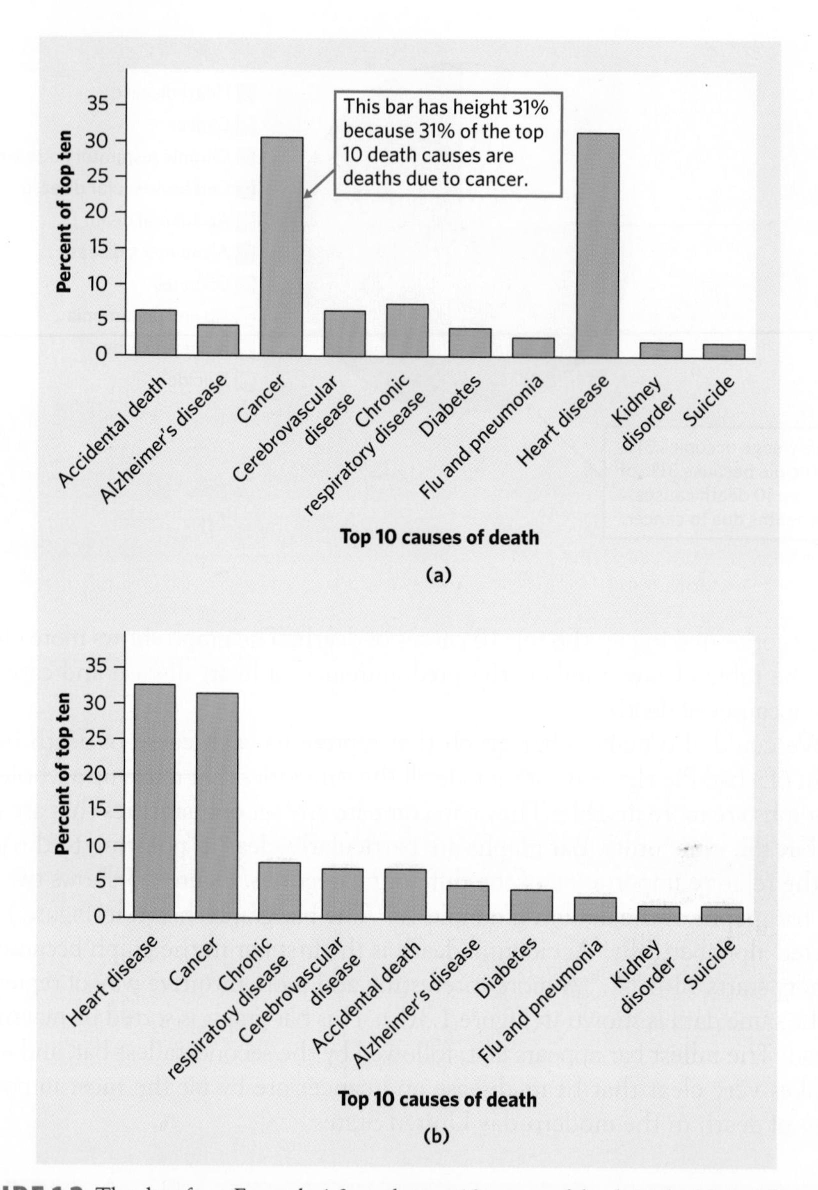

FIGURE 1.3 The data from Example 1.3 on the top 10 causes of death in the United States for 2011 are displayed in a bar graph format. (a) The bars in the graph are sorted alphabetically. (b) The bars are now sorted according to their height (by order of importance).

Statistics in Your World **Convenient graphs** The movie *An Inconvenient Truth* won the Oscar for Best Documentary in 2006, and it is the third-highest-grossing documentary in U.S. history. The movie features scientific evidence of global warming, supported by a detailed presentation of about 30 graphs and data tables.

Figure 1.4 is a bar graph of the data in Example 1.4. We see at a glance that, among adults, current cannabis use declines with age.

We can't make a pie chart to display the data in Example 1.4, because each percent in the table refers to a different age group, not to parts of a single whole. It is important to understand this distinction in order to interpret the data correctly. Notice that the 35+ age group represents a much larger share of the American population than the other age groups listed, and 3.4% of this large age group could constitute a large number of cannabis users. So we can conclude from the data

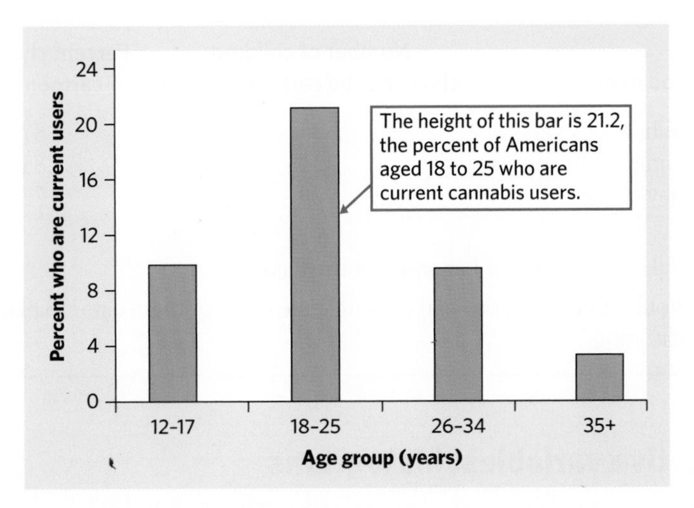

FIGURE 1.4 Bar graph showing the percent of Americans in each of four age groups who are current cannabis users, for Example 1.4.

that young adults aged 18 to 25 have the highest rate of cannabis usage, but not necessarily that they are the most common cannabis users.

Example 1.4 illustrates the fact that bar graphs are more versatile than pie charts in displaying categorical data. One important consequence of this versatility is that bar graphs should be examined very carefully to avoid misinterpreting the findings they display.

APPLY YOUR KNOWLEDGE

1.3 Children's food choices. Do popular cartoon characters on food packages influence children's food choices? A study asked 40 young children (ages four to six) to taste graham crackers presented in a package either with or without a popular cartoon character, and to indicate whether the two foods tasted the same or one tasted better. Unknown to the children, the crackers were the same both times. Here are the findings:[7]

Taste preference	Number of children	Percent
Without character	3	7.5
Taste the same	15	37.5
With character	22	55.0

Present these data in a well-labeled bar graph. Would it also be correct to make a pie chart? What do the data suggest about the influence of cartoon characters on graham cracker preference in young children?

1.4 More on children's food choices. The study in Exercise 1.3 also asked these 40 children to taste gummy fruit snacks and baby carrots presented in packages either with or without a popular cartoon character. For each food type, the children selected which of the two options they would prefer to eat for a snack. (Note that this is a different question from the one asked in the previous exercise.) The number and percent of children choosing the version with a cartoon on the package are displayed in the table on the next page:

Food item	Number of children choosing the cartoon version	Percent choosing the cartoon version
Graham crackers	35	87.5
Gummy fruit snacks	34	85.0
Baby carrots	29	72.5

(a) Make a well-labeled bar graph of the data.

(b) Would it be correct to make a single pie chart of these data? Explain your reasoning.

Quantitative variables: histograms

Quantitative variables often take many values. The distribution of a variable tells us what values the variable takes and how often it takes these values. When there are more than just a few data points, a graph of the distribution is often easier to interpret if nearby values are grouped together. The most common graph of the distribution of one quantitative variable is a **histogram.**

histogram

Stephen Frink Collection/Alamy

EXAMPLE 1.5 Making a histogram: sharks

The great white shark, *Carcharodon carcharias*, is a large ocean predator at the top of the food chain. Here are the lengths in feet of 44 great whites:[8]

18.7	12.3	18.6	16.4	15.7	18.3	14.6	15.8	14.9	17.6	12.1
16.4	16.7	17.8	16.2	12.6	17.8	13.8	12.2	15.2	14.7	12.4
13.2	15.8	14.3	16.6	9.4	18.2	13.2	13.6	15.3	16.1	13.5
19.1	16.2	22.8	16.8	13.6	13.2	15.7	19.7	18.7	13.2	16.8

The *individuals* in this data set are individual sharks, and the *variable* is body length in feet. To make a histogram of the distribution of this variable, proceed as follows:

Step 1. Choose the classes. Divide the range of the data into classes of equal width. The data range from 9.4 to 22.8 feet, so we decide to use these classes:

$$9.0 < \text{individuals with body length} \leq 11.0$$

$$11.0 < \text{individuals with body length} \leq 13.0$$

$$\vdots$$

$$21.0 < \text{individuals with body length} \leq 23.0$$

In this example, we chose to exclude the lower bound and include the upper bound in making the histogram classes. Choosing, instead, to include the lower bound and exclude the upper bound would also have been a valid option. What matters is to specify the classes precisely so that each individual falls into exactly one class. You can explain the nature of the class boundaries in the legend accompanying your histogram.

Based on the chosen class definition, the smallest shark, measuring 9.4 feet, falls into the first class; a shark measuring exactly 11.0 feet would still fall in that first class, but a shark measuring 11.1 feet would fall into the second class.

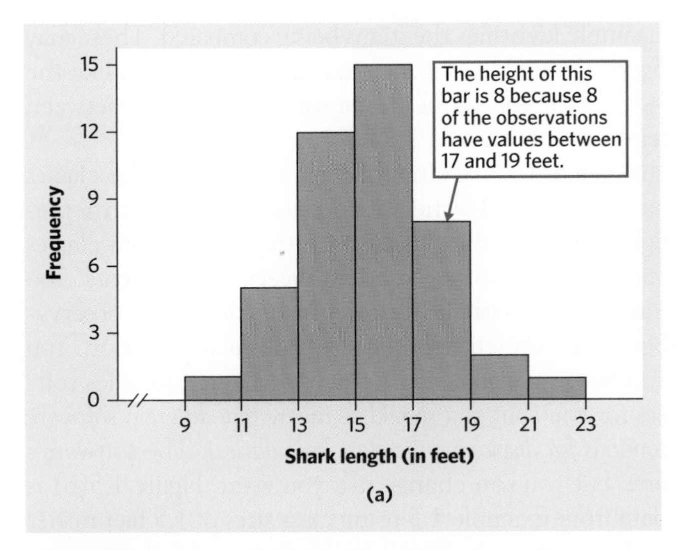

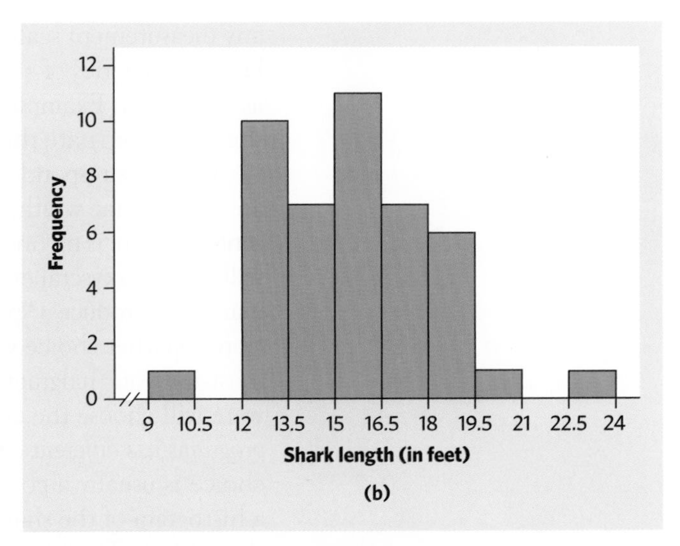

FIGURE 1.5 Histograms of the body length in feet for 44 great white sharks, for Example 1.5. (a) This histogram has 7 classes. (b) This histogram has 10 classes and shows more detail.

Step 2. Count the individuals in each class. Here are the counts:

Class	Count
9.1 to 11.0	1
11.1 to 13.0	5
13.1 to 15.0	12
15.1 to 17.0	15
17.1 to 19.0	8
19.1 to 21.0	2
21.1 to 23.0	1

Check that the counts add to 44, the number of individuals in the data (the 44 great white sharks).

Step 3. Draw the histogram. Mark the scale for the variable whose distribution you are displaying on the horizontal axis. That's the shark body length in feet. The scale runs from 9 to 23 feet because that is the span of the classes we chose. The vertical axis contains the scale of counts. Each bar represents a class. The base of the bar covers the class, and the bar height is the class count. The horizontal axis represents a continuum of values (here body lengths), and therefore histograms do not leave any horizontal space between the bars unless a class is empty (in which case the bar has height zero). Figure 1.5(a) is our histogram. ■

Although histograms resemble bar graphs in some aspects, their details and uses are very different. A histogram displays the distribution of one quantitative variable. The horizontal axis of a histogram is marked in the units of measurement for the variable. A histogram should be drawn with no extra space between consecutive classes, to indicate that all values of the variable are covered. In contrast, a bar graph compares different items. The horizontal axis of a bar graph need not have any measurement scale but simply identifies the items being compared. These may be the categories of a categorical variable, but they may also be separate, like the age groups in Example 1.4. A bar graph should leave some blank space between the bars to separate the items being compared.

Our eyes respond to the *area* of the bars in a histogram.[9] Because the classes are all the same width, area is determined by height and all classes are fairly represented. There is no one "right" choice of the classes in a histogram. Too few classes will give a "skyscraper" graph, with all values in a few classes with tall bars. Too many will produce a "pancake" graph, with most classes having one or no observations. Neither choice will give a good picture of the shape of the distribution. You must use your judgment in choosing classes to display the shape. Statistics software will choose the classes for you (but *you should be aware that different software programs use different conventions for displaying the class boundaries*). The software's choice is usually a good one, but you can change it if you want. Figure 1.5(b) is a histogram of the shark data from Example 1.5 using class sizes of 1.5 feet rather than 2 feet. The histogram function in the *One-Variable Statistical Calculator* applet on the companion website allows you to change the number of classes by dragging with the mouse, so that it is easy to see how the choice of classes affects the histogram.

APPLY YOUR KNOWLEDGE

1.5 Healing of skin wounds. Biologists studying the healing of skin wounds measured the rate at which new cells closed a razor cut made in the skin of an anesthetized newt. Here are the sorted data from 18 newts, measured in micrometers (millionths of a meter) per hour:[10]

11	12	14	18	22	22	23	23	26
27	28	29	30	33	34	35	35	40

Make a histogram of the healing rates using classes of width 5 micrometers/hour starting at 10 micrometers/hour. (Make this histogram by hand even if you have software, to be sure you understand the process. You may want to compare your histogram with your software's choice.)

1.6 Choosing classes in a histogram. The data set menu that accompanies the *One-Variable Statistical Calculator* applet includes a data set called "healing of skin wounds." Choose these data, then click on the "Histogram" tab to see a histogram.

(a) How many classes does the applet choose to use? (You can click on the graph outside the bars to get a count of classes.)

(b) Click on the graph and drag to the left. What is the smallest number of classes you can get? What are the lower and upper bounds of each class? (Click on each bar to find out.) Make a rough sketch of this histogram.

(c) Click and drag to the right. What is the largest number of classes you can get? How many observations does the class with the highest count have?

(d) You see that the choice of classes changes the appearance of a histogram. Drag back and forth until you get the histogram you think best displays the distribution. How many classes did you use?

Interpreting histograms

Making a statistical graph is not an end in itself. The purpose of the graph is to help us understand the data. After you make a graph, always ask, "What do I see?" Once you have displayed a distribution, you can see its important features as follows.

> **EXAMINING A HISTOGRAM**
>
> In any graph of data, look for the **overall pattern** and for striking **deviations** from that pattern.
>
> You can describe the overall pattern of a histogram by its **shape, center,** and **spread.**
>
> An important kind of deviation is an **outlier,** an individual value that falls outside the overall pattern.

We will learn how to describe center and spread numerically in Chapter 2. For now, we can describe the center of a distribution by its *midpoint,* the value with roughly half the observations taking smaller values and half taking larger values. We can describe the spread of a distribution by giving the *smallest and largest values*.

EXAMPLE 1.6 Describing a distribution: sharks

Look again at the histogram in Figure 1.5 (a).

SHAPE: The distribution is **unimodal.** That is, it has a **single peak,** which represents sharks with a body length between 15 and 17 feet. The distribution is also *symmetric.* In mathematics, the two sides of symmetric patterns are exact mirror images. Real data are almost never exactly symmetric. We are content to describe the histogram in Figure 1.5(a) as roughly symmetric.

unimodal
single-peaked

CENTER: The counts in Example 1.5 show that 18 of the 44 sharks have a body length of 15 feet or less, but the count increases to 33 out of 44 for lengths of 17 feet or less. So the midpoint of the distribution is about 15 to 17 feet.

SPREAD: The spread is from 9.4 to 22.8 feet, but only 1 shark is shorter than 11 feet and only 1 is more than 21 feet long.

OUTLIERS: In Figure 1.5(a), the observations greater than 21 feet or less than 11 feet are part of the continuous range of body lengths and do not stand apart from the overall distribution. This histogram, with only 7 classes, hides some of the detail in the distribution. Look at Figure 1.5(b), a histogram of the same data with 10 classes instead. It reveals that the shortest and longest sharks, at 9.4 and 22.8 feet respectively, do stand apart from the other observations.

Once you have spotted possible outliers, look for an explanation. Some outliers are due to mistakes, such as typing 19.4 as 9.4. Other outliers point to the special nature of some observations. The smallest shark might be a juvenile exhibiting some adult features, for instance. An outlier could also simply be an unusual but perfectly legitimate observation. For instance, the largest shark could be just that: an unusually large shark. Human height, for instance, has a somewhat homogeneous distribution within a gender and ethnicity, but some individuals, such as the famous basketball player Shaquille O'Neal, clearly stand out from the norm. ■

Comparing Figures 1.5(a) and (b) reminds us that *the choice of classes in a histogram can influence the appearance of a distribution*. Both histograms portray a symmetric distribution with one peak, but only Figure 1.5(b) shows the mild outliers.

When you describe a distribution, concentrate on the main features. Look for major peaks, not for minor ups and downs in the bars of the histogram. Look for clear outliers, not just for the smallest and largest observations. Look for rough *symmetry* or clear *skewness*.

SYMMETRIC AND SKEWED DISTRIBUTIONS

A distribution is **symmetric** if the right and left sides of the histogram are approximately mirror images of each other.

A distribution is **skewed to the right,** or positively skewed, if the right side of the histogram (containing the half of the observations with larger values) extends much farther out than the left side. It is **skewed to the left,** or negatively skewed, if the left side of the histogram extends much farther out than the right side.

Here are more examples of describing the overall pattern of a histogram.

EXAMPLE 1.7 Guinea pig survival times

Figure 1.6 displays the survival times in days (reported below) of 72 guinea pigs after they were injected with infectious bacteria in a medical experiment.[11]

43	45	53	56	56	57	58	66	67	73	74	79
80	80	81	81	81	82	83	83	84	88	89	91
91	92	92	97	99	99	100	100	101	102	102	102
103	104	107	108	109	113	114	118	121	123	126	128
137	138	139	144	145	147	156	162	174	178	179	184
191	198	211	214	243	249	329	380	403	511	522	598

The distribution is *single-peaked* and obviously *skewed to the right*. Most guinea pigs have a short survival time, between 50 and 150 days, and they make up the peak of the distribution (a single peak can span several classes, just like the top of a mountain can spread across country borders). However, some animals survive longer, so that the graph extends to the right of its peak much farther than it extends to the left. Survival times, whether of machines under stress or cancer patients after treatment, usually have distributions that are skewed to the right. Here, the center (half above, half below) lies between 100 and 149 days (the third class). The survival times range from 43 to 598 days. However, almost all infected guinea pigs die within 250 days. A few guinea pigs are outliers and survive for much longer, maybe because they had some immunity to the bacteria used for the experiment.

Notice that the vertical scale in Figure 1.6 is not the *count* of guinea pigs but the *percent* of guinea pigs from the experiment falling in each histogram class. A histogram of percents rather than counts is convenient when we want to compare several distributions. It would, for instance, allow us to compare the survival times obtained in this experiment with the results of another experiment or with the survival times of a control group of guinea pigs not injected with the infectious bacteria. We will see in Chapter 8 that using a control group is very helpful in interpreting the findings of an experiment. ■

The vital few

Skewed distributions can show us where to concentrate our efforts. Ten percent of the cars on the road account for half of all carbon dioxide emissions. A histogram of CO_2 emissions would show many cars with small or moderate values and a few with very high values. Cleaning up or replacing these cars would reduce pollution at a cost much lower than that of programs aimed at all cars. Statisticians who work at improving quality in industry make a principle of this: distinguish "the vital few" from "the trivial many."

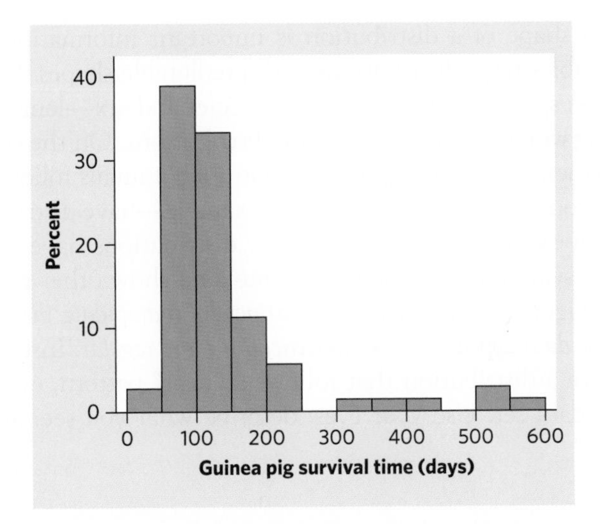

FIGURE 1.6 Histogram of guinea pig survival times (in days) after inoculation with a pathogen, for Example 1.7. Classes include the lower-bound value but exclude the upper-bound value.

EXAMPLE 1.8 Lyme disease

Lyme disease is a bacterial infection spread through the bite of an infected blacklegged tick. Left untreated, it can cause lifelong complications. Figure 1.7 displays data about the age in years of 241,931 individuals diagnosed with Lyme disease in the United States between 1992 and 2006.[12] We can't call this irregular distribution either symmetric or skewed. The big feature of the overall pattern is the presence of two main peaks: a **bimodal** distribution corresponding to two **clusters** of individuals—children and adults.

bimodal

clusters

Clusters suggest that several types of individuals are mixed in the data set. The first cluster has a clear peak around 5 to 10 years while the peak of the second cluster is at 45 to 50 years. Because ticks are found outdoors in woody or grassy areas, these two clusters might reflect the outdoor activities of children and the adults accompanying them.

Giving a single center and spread for this distribution would be misleading, because the data highlight two age groups. It would be better to describe the two groups separately. ■

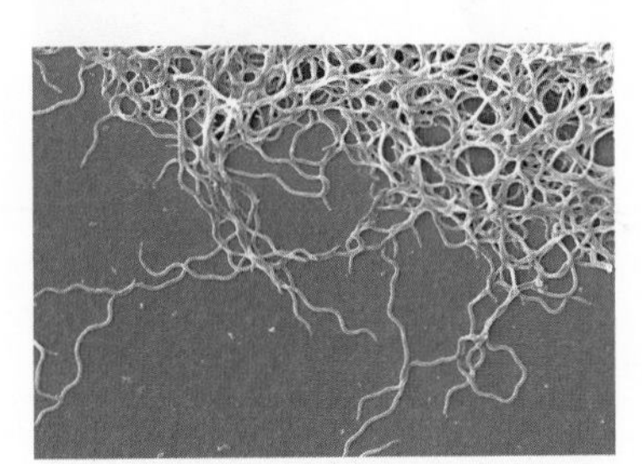

CDC/Claudia Molins/Janice Haney Carr

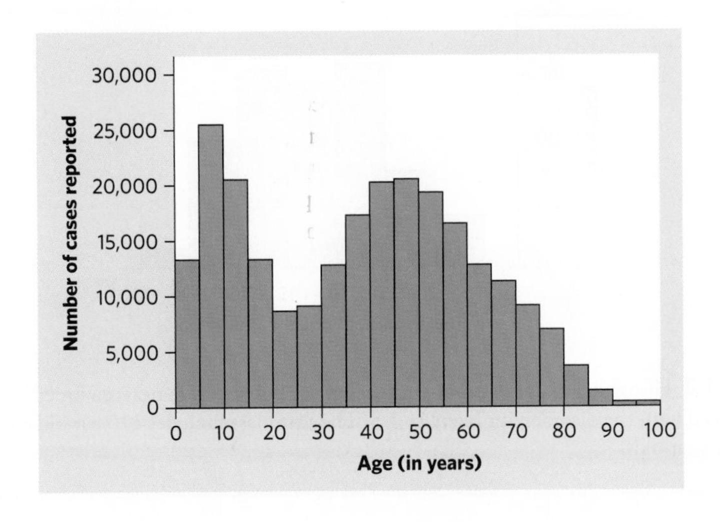

FIGURE 1.7 Histogram of patient age in years for 241,931 cases of Lyme disease reported in the United States between 1992 and 2006, for Example 1.8. Notice the two separate peaks around 10 years and 45 years.

The overall shape of a distribution is important information about a variable. Some variables have distributions with predictable shapes. Many biological measurements on specimens from the same species and sex—lengths of bird bills, heights of young women—have symmetric distributions. On the other hand, survival times—patients after an organ transplant, lab animals following an experimental inoculation such as in the guinea pig example—have distributions that are typically strongly skewed to the right. Many distributions have irregular shapes that are neither symmetric nor skewed. Some data show other patterns, such as the clusters in Figure 1.7. *Do not try to artificially manipulate the classes of a histogram so that the data appear more symmetrical or more regular.* Instead, accept that not all data have a distribution that follows a "neat" pattern, even if you could obtain a larger data set. Use your eyes, describe what you see, and then try to explain it.

APPLY YOUR KNOWLEDGE

1.7 Healing of skin wounds, continued. In Exercise 1.5, you made a histogram of the healing rates of skin wounds in newts. The shape of the distribution is a bit irregular. Is it closer to symmetric or skewed? Would you say that it is unimodal or bimodal?

1.8 Age of onset for anorexia nervosa. Anorexia nervosa is an eating disorder characterized, in part, by the fear of becoming fat and by voluntary starvation; it often leads to death from suicide or organ failure. A Canadian study surveyed 691 adolescent girls diagnosed with anorexia nervosa and their families.[13] Figure 1.8 is a histogram of the distribution of age at onset of anorexia for the 691 girls in the study (note that, if the onset occurred on the 12th birthday, for example, the onset is reported as 12.01 for the purpose of the histogram, to be coherent with customary age descriptions). Describe the shape of this distribution. Within which class does the midpoint of the distribution lie?

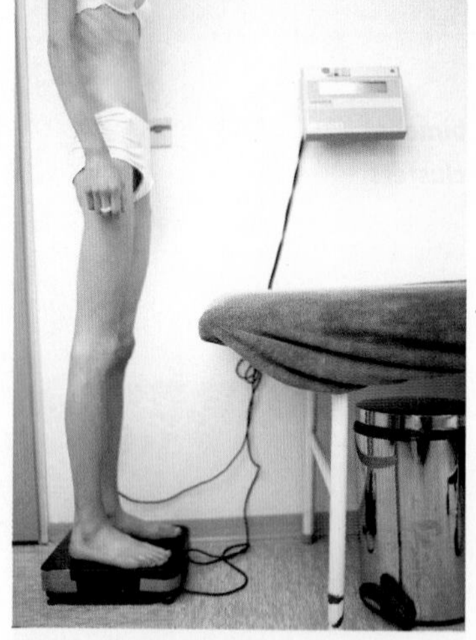

Mette/laif/Redux

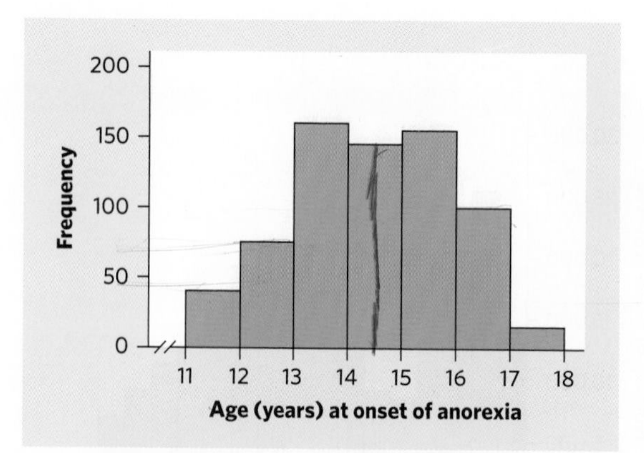

FIGURE 1.8 Histogram of age at onset (in years of age) of anorexia nervosa for 691 Canadian girls diagnosed with the disorder, for Exercise 1.8. The first class includes 11-year-old girls but excludes 12-year-olds.

Quantitative variables: stemplots and dotplots

Histograms are not the only graphical display of quantitative distributions. For small data sets, *stemplots* and *dotplots* are quicker to make and easier to interpret. They also have the advantage of displaying the **raw data;** that is, they show each one of the values in the data set.

raw data

STEMPLOT

To make a **stemplot:**

1. Separate each observation into a **stem,** consisting of all but the final (rightmost) digit, and a **leaf,** the final digit. Stems may have as many digits as needed, but each leaf shows only a single digit.
2. Write the stems in a vertical column with the smallest at the top, and draw a vertical line at the right of this column.
3. Write each leaf in the row to the right of its stem, in increasing order out from the stem.

EXAMPLE 1.9 Making a stemplot: skin wounds

Exercise 1.5 presents the rate at which new cells closed a razor cut made in the skin of 18 anesthetized newts (in micrometers per hour). The data are all two digits long. To make a stemplot of these sorted data, take the tens part of the value as the stem and the final digit (ones part) as the leaf. The slowest healing rate, 11 micrometers per hour, has stem 1 and leaf 1. There are three more values in the 10s (12, 14, 18). Place the final digit of these values next to the first leaf so that the first stem (1) has a total of four leaves (1, 2, 4, and 8) all on the same row in the stemplot. Next, there are eight values in the 20s, and these all go on the second row, with stem 2. The first two values in the 20s range are 22 and 22. Both must be listed, as 2 and 2, on the leaves side. Here is the complete stemplot for the healing rates data:

```
1 | 1 2 4 8
2 | 2 2 3 3 6 7 8 9
3 | 0 3 4 5 5
4 | 0
```

You can also **split stems** to double the number of stems when all the leaves would otherwise fall on just a few stems. Each stem then appears twice. Leaves 0 to 4 go on the upper stem, and leaves 5 to 9 go on the lower stem. If you split the stems in the stemplot above, the new stemplot becomes

splitting stems

```
1 | 1 2 4
1 | 8
2 | 2 2 3 3
2 | 6 7 8 9
3 | 0 3 4
3 | 5 5
4 | 0
4 |
```

■

On the other hand, data with many significant digits (for example, 21,422) need to be **rounded** or truncated (for example, to the nearest hundred, 21,400; then drop the last two zeros) before they can be used in a stemplot. Rounding and splitting stems are matters for judgment, like choosing the classes in a histogram.

rounding

In fact, a stemplot looks very much like a histogram turned on end: the stems serve the same function as the classes of a histogram. One distinction is that you can choose the classes in a histogram at will, but the classes (the stems) of a stemplot are given to you by the actual numerical values. Compare the stemplots in the healing rates example above with the histograms of the same data that you built in Exercises 1.5 and 1.6. The *One-Variable Statistical Calculator* applet on the companion website allows you to decide whether to split stems, so that it is easy to see the effect.

Histograms are more flexible than stemplots because you can choose the starting lower bound and the size of the classes. But a stemplot, unlike a histogram, preserves the actual value of each observation. *Stemplots do not work well for large data sets, where each stem must hold a large number of leaves.*

DOTPLOT

To make a **dotplot:**

1. Sort the data set and plot each observation according to its numerical value along a labeled scaled axis.
2. Identical observations are typically stacked.

EXAMPLE 1.10 Making a dotplot: skin wounds

Exercise 1.5 presents the rate (in micrometers per hour) at which new cells closed a razor cut made in the skin of 18 anesthetized newts. To make a dotplot of these sorted data, create a one-dimensional graph with healing rate on the horizontal axis (in micrometers per hour). This is the only axis on this graph. Figure 1.9 shows the dotplot of this data set. Data points with the same numerical value are shown as stacked dots. For example, two newts had a healing rate of 35 micrometers per hour.

Like a histogram or a stemplot, the dotplot shows the shape, center, and spread of the distribution. However, because the dotplot is one-dimensional, shape and center are indicated by the density of dots rather than the height of the histogram bar or the width of the stemplot. Notice how the dots in Figure 1.9 are fairly evenly spread, with a slightly higher density in the center of the graph. ■

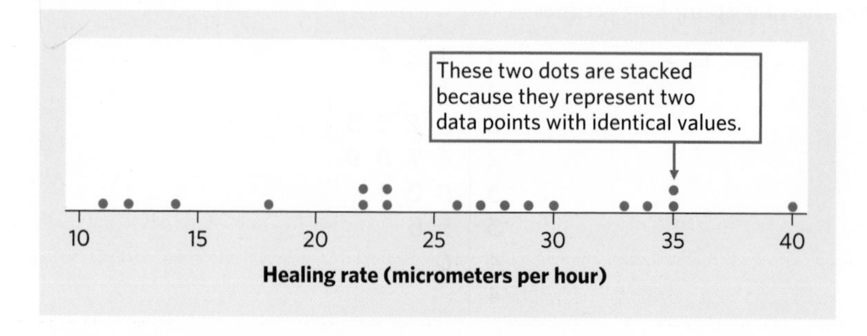

FIGURE 1.9 Dotplot of healing rate (in micrometers per hour) of skin wounds in 18 anesthetized newts, for Example 1.10.

Like the stemplot, the dotplot has the advantage of retaining the information about individual observations. However, it is not well suited for very large data sets, which benefit from the summarizing offered by a histogram.

Florence Nightingale

Florence Nightingale (1820–1910) dedicated her life to promoting and improving the field of nursing. She recorded data and displayed the information in graphs that policymakers could understand. During the Crimean war, she showed that more soldiers died because of poor hospital conditions than from battle wounds. She also documented how casualties drastically dropped after sanitation was improved. Nightingale was the first woman elected to the Royal Statistical Society, honoring her outstanding contributions.

APPLY YOUR KNOWLEDGE

1.9 Glucose levels. People with diabetes must monitor and control their blood glucose level. The goal is to maintain "fasting plasma glucose" between about 90 and 130 milligrams per deciliter (mg/dl). Here are the fasting plasma glucose levels for 18 diabetics enrolled in a diabetes control class, five months after the end of the class:[14]

141	158	112	153	134	95	96	78	148
172	200	271	103	172	359	145	147	255

(a) Make a stemplot of these data and describe the main features of the distribution. (You will want to round and also split stems.) Are there any outliers? How well is the group as a whole achieving the goal of controlling glucose level?

(b) Because the stemplot preserves the actual values of the observations, it is easy to find the midpoint (9th and 10th of the 18 observations in order) and the spread. What are they?

(c) Make a dotplot of these data. Compare with the stemplot and explain the differences and similarities.

1.10 More on glucose levels. The study described in the previous exercise also measured the fasting plasma glucose levels of 16 diabetics who were given individual, rather than group, instruction on diabetes control. Here are the data:

128	195	188	158	227	198	163	164
159	128	283	226	223	221	220	160

Make a **back-to-back stemplot** to compare the two samples. That is, use one set of stems with two sets of leaves, one to the right and one to the left of the stems. (Draw a line on either side of the stems to separate stems and leaves.) Order both sets of leaves from smallest at the stem to largest away from the stem. How do the distribution shapes and success in achieving the glucose control goal differ?

back-to-back stemplot

Time plots

Many variables are measured at intervals over time. We might, for example, measure the height of a growing child or water precipitation at the end of each month. In these examples, our main interest is in changes over time. To display change over time, make a *time plot*.

TIME PLOT

A **time plot** of a variable plots each observation against the time at which it was measured. Always put time on the horizontal scale of your plot and the variable you are measuring on the vertical scale. Connecting the data points by lines helps emphasize any change over time.

EXAMPLE 1.11 Water salinity in the Shark River

The Shark River runs through the southwestern portion of Everglades National Park in Florida and flows into the Gulf of Mexico. The U.S. Geological Survey closely monitors this important natural habitat. Figure 1.10 is a time plot of water salinity at the Gunboat Island station on the Shark River over a seven-day period in the fall of 2009. Salinity was recorded every 15 minutes from 12 A.M. (midnight) of September 30 to 12 A.M. (midnight) of October 7.[15] ■

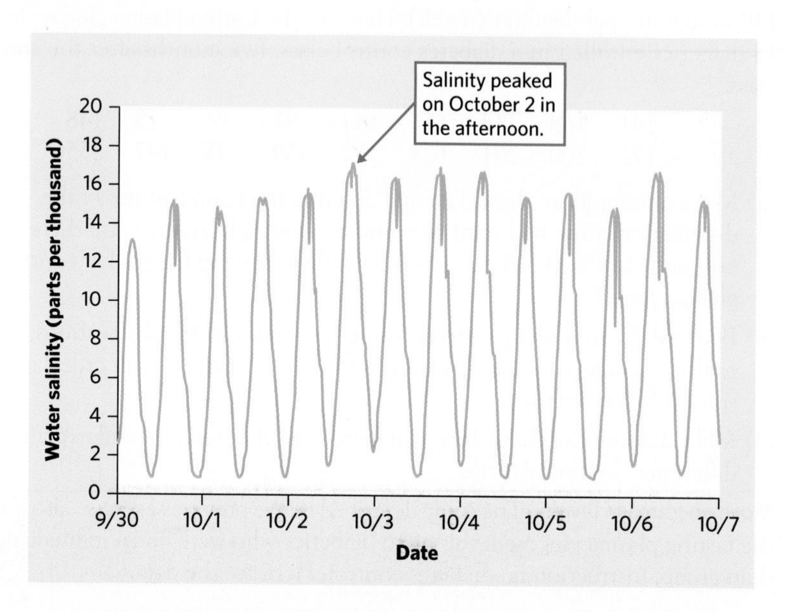

FIGURE 1.10 Time plot of Shark River water salinity in Everglades National Park over a seven-day period in 2009. The daily cycles reflect the influence of the tides from the Gulf of Mexico, into which the Shark River discharges.

cycles

When you examine a time plot, look once again for an overall pattern and for strong deviations from the pattern. Figure 1.10 shows strong **cycles,** regular up-and-down movements in water salinity. The cycles show the effects of the ocean tides on the salinity of the Shark River water. Water salinity in the river is highest twice a day, every day, coinciding with the high tides. This cyclical pattern is very clear and consistent, despite some variations in salinity from one day to the next.

© Biosphoto/Eric Lefranc

EXAMPLE 1.12 Global warming

Global warming is an important planet-wide ecological issue that has been hotly debated. Table 1.1 lists the annual global temperature anomaly (in degrees Celsius) from 1880 to 2012 based on data from recording stations around the world.[16] Individual annual temperature anomalies are computed locally by comparing the local annual sea surface temperature average with the local temperature reference (the 1951–1980 average). Both graphs in Figure 1.11 describe these data.

The histogram in Figure 1.11(a) shows the distribution of annual global temperature anomalies. The histogram is unimodal (single-peaked) and skewed to the right, with center around −0.04 to 0.08 degrees Celsius. We might think that the data show just chance year-to-year fluctuation in global temperature anomalies, with a small proportion of larger anomalies.

Figure 1.11(b) is a time plot of the same data. For example, the first point lies above 1880 on the "Year" scale at height −0.25, the global temperature anomaly for 1880.

TABLE 1.1 Annual global temperature anomaly (degrees Celsius)

Year	Anomaly	Year	Anomaly	Year	Anomaly	Year	Anomaly	Year	Anomaly	Year	Anomaly
1880	−0.25	1903	−0.31	1926	−0.01	1949	−0.06	1972	0	1995	0.38
1881	−0.19	1904	−0.34	1927	−0.13	1950	−0.15	1973	0.14	1996	0.29
1882	−0.22	1905	−0.24	1928	−0.11	1951	−0.04	1974	−0.08	1997	0.39
1883	−0.23	1906	−0.2	1929	−0.24	1952	0.03	1975	−0.05	1998	0.56
1884	−0.3	1907	−0.38	1930	−0.06	1953	0.11	1976	−0.16	1999	0.32
1885	−0.3	1908	−0.34	1931	−0.01	1954	−0.1	1977	0.13	2000	0.33
1886	−0.25	1909	−0.35	1932	−0.06	1955	−0.1	1978	0.01	2001	0.48
1887	−0.35	1910	−0.33	1933	−0.17	1956	−0.17	1979	0.08	2002	0.56
1888	−0.26	1911	−0.33	1934	−0.05	1957	0.07	1980	0.18	2003	0.54
1889	−0.15	1912	−0.34	1935	−0.1	1958	0.08	1981	0.26	2004	0.48
1890	−0.36	1913	−0.32	1936	−0.04	1959	0.06	1982	0.05	2005	0.62
1891	−0.28	1914	−0.15	1937	0.08	1960	−0.01	1983	0.26	2006	0.54
1892	−0.32	1915	−0.09	1938	0.11	1961	0.08	1984	0.09	2007	0.56
1893	−0.31	1916	−0.3	1939	0.03	1962	0.04	1985	0.05	2008	0.44
1894	−0.33	1917	−0.4	1940	0.05	1963	0.08	1986	0.12	2009	0.59
1895	−0.27	1918	−0.32	1941	0.11	1964	−0.21	1987	0.26	2010	0.66
1896	−0.16	1919	−0.2	1942	0.04	1965	−0.11	1988	0.31	2011	0.54
1897	−0.12	1920	−0.18	1943	0.1	1966	−0.03	1989	0.19	2012	0.56
1898	−0.24	1921	−0.13	1944	0.2	1967	0	1990	0.38		
1899	−0.17	1922	−0.24	1945	0.07	1968	−0.04	1991	0.35		
1900	−0.1	1923	−0.2	1946	−0.04	1969	0.08	1992	0.12		
1901	−0.15	1924	−0.21	1947	0.01	1970	0.03	1993	0.14		
1902	−0.27	1925	−0.16	1948	−0.04	1971	−0.1	1994	0.23		

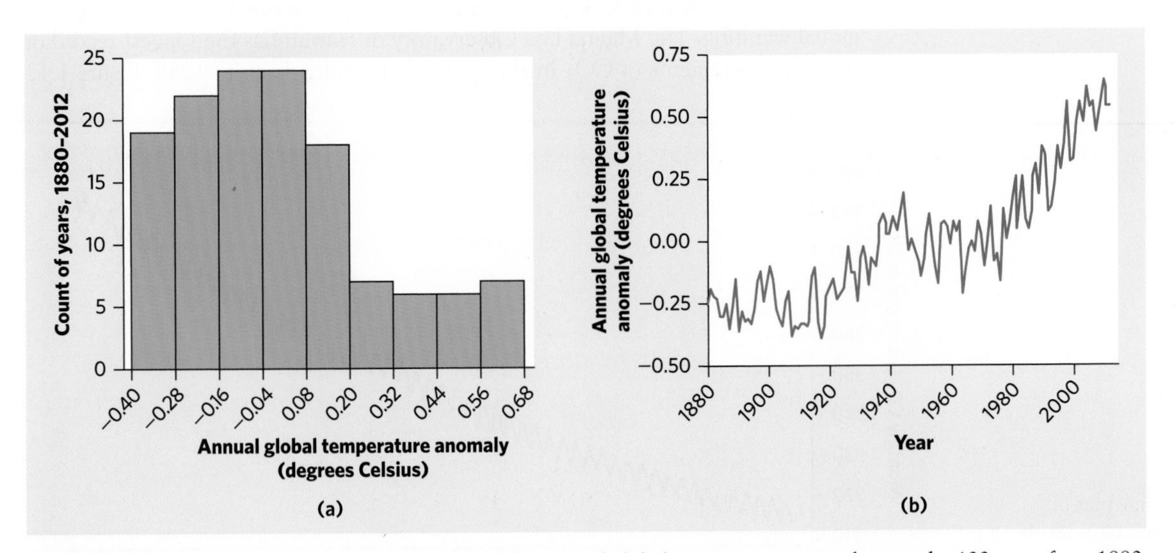

FIGURE 1.11 Histogram (a) and time plot (b) of the annual global temperature anomaly over the 133 years from 1880 to 2012, for Example 1.12. Data from Table 1.1. Anomalies are relative to the 1951–1980 reference period.

trend

The time plot tells a more interesting story than the histogram. There is a great deal of year-to-year variation, but there is also a clear increasing **trend** over time. That is, there is a long-term rise in annual global temperature anomalies. A trend in a time plot is a long-term upward or downward movement over time. The trend in the temperature data reflects a climate change: Global temperatures have been rising fairly steadily for over a century, despite substantial year-to-year variations. ■

time series

Histograms and time plots give different kinds of information about a variable. The time plot in Figure 1.11(b) presents **time series data** that show the change in annual global temperature anomalies over time. A histogram displays the distribution of data, such as the temperature anomalies, regardless of time.

APPLY YOUR KNOWLEDGE

1.11 Routine episiotomy. An episiotomy is a surgical incision through a woman's perineum made to assist childbirth. Episiotomies have been performed routinely (that is, even when not medically necessary) in the United States and all over the world under the assumption that they may prevent later damage caused by natural tearing of the perineum. Studies have challenged this assumption, however, by showing more damage with routine episiotomy than without. Here are episiotomy rates in Virginia in the years 1999 to 2008:[17]

Year	1999	2000	2001	2002	2003
Percent of childbirths	35.6	31.9	29.0	25.9	23.8

Year	2004	2005	2006	2007	2008
Percent of childbirths	22.6	19.0	17.2	15.5	14.0

Make a time plot of the data from 1999 till 2008. Describe the trend that your plot shows. What does the trend emphasize?

1.12 Atmospheric CO_2 levels. Elevated atmospheric CO_2 levels have been linked to global warming. The Mauna Loa Observatory in Hawaii has the longest record of direct measurements of CO_2 in the atmosphere, going back to 1958. Figure 1.12

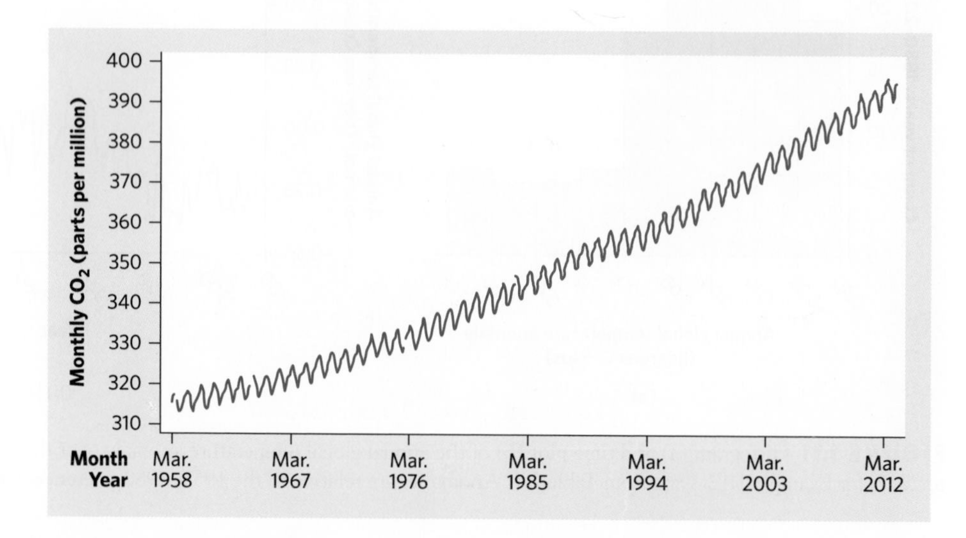

FIGURE 1.12 Time plot of monthly atmospheric CO_2 levels recorded at Mauna Loa Observatory, from March 1958 to January 2013, for Exercise 1.12.

shows the trends in monthly atmospheric CO_2 levels over Hawaii between March 1958 and January 2013.[18] Describe the patterns in the data. What do the data suggest about controlling our carbon footprint?

DISCUSSION (Mis)adventures in data entry

Data are at the center of statistics, but how do you record data so that they can be analyzed easily? And what are some issues you should be concerned about? This discussion provides a few pointers. Failure to pay attention to them can lead to disastrous results, from having to redo your whole analysis, to needing to collect new data or, worse yet, publishing incorrect findings and having to rescind them later.

Keeping detailed records

Scientific inquiries must be documented and you will be required to keep clear, detailed records of your findings. This is a necessary safeguard against dishonest conduct, but it is also a great help in data entry and proofing. To avoid accusations of data falsification or disputes over who discovered something first, bound notebooks with page numbers and written dates are good practice. You might also keep photographs of your findings or actual specimens. The discussion in Chapter 2 on dealing with outliers highlights how the first step after identifying an outlier is to check your data records and stored evidence to see if the outlier could have been a typographical or other simple data entry error.

Along with the actual data, you should record anything relevant to how the data were obtained, measured, or computed. For example, are values of weight recorded in grams, kilograms, ounces, or pounds, and are they self-reported (common in telephone surveys) or actually measured with a scale? And if you computed body mass index from values of weight and height, exactly how did you perform that computation and did it involve rounding numerical values? This step will help you communicate your findings after your analysis is done, and it will be very useful if you ever find puzzling values during the course of your analysis.

Organizing data for use with statistical software

Whether you write your scientific findings on paper or record them electronically, at some point you will want to use statistical software for graphing and statistical analysis. Spreadsheet programs like Microsoft Excel or LibreOffice Calc are commonly used because they are widely available and because they allow data, notes, and even pictures to be saved into one convenient space. You can think of this approach as the electronic equivalent of your paper-based notebook. Proper statistical analysis, however, often requires that your data be organized in very specific ways.

Statistical software packages differ quite a bit on what data format they require, sometimes even requiring different formats for different analyses. Let's just say that nobody in the software business seems to have given serious consideration to the painful experiences of their users . . . . Statisticians, however, generally agree that the best way to record your data is to use one row for each individual and one column for each variable. The table in Example 1.1 follows this format: each tree shrew is given its own row, and every variable recorded about these tree shrews is given its own column (including a unique identifier, here the animal's given name).

Why is the "one individual, one row" approach optimal for data analysis? Let's consider the alternatives. (1) The researchers in Example 1.1 probably had information about each animal's sex and age before they obtained the data on paw usage, so it might have been tempting to use a different spreadsheet for the animals' personal information and for the experimental findings. However, this would make statistical analysis a lot more difficult if they wanted to, say, compare the male and female shrews. (2) Inversely, the researchers could have recorded their findings on paw usage separately for the male and the female shrews, creating two different data tables. The challenge, then, would be performing an analysis using all shrews (regardless of sex).

When entering categorical values, be sure to be consistent with your choice of naming convention. Statistical software will consider "m" and "male" two different things, even if they both mean to you that the shrew is male. Some software packages won't even accept text and require that you use a numerical code, such as 0 for male and 1 for female. In such a case it would be better to use "Female" rather than "Sex" as the variable name and to record it as 0 for no and 1 for yes, which would be easier to interpret.

Lastly, there will be times when you have to deal with missing data. Maybe a tree shrew died before you collected all the data or it refused to perform one of the food tasks in your study—or perhaps you forgot to write that particular piece of data in your notebook . . . . These things happen. But how do you handle it in your electronic data records? The answer, unfortunately, depends on the particular statistical software you use. Some mark missing data with an asterisk, others with a dot, the letters "NA," or a blank space. This can create serious complications if you need to transfer your data analysis from one software application to another. Statisticians sometimes recommend using a special code for missing data, such as 0 or −99 (using different codes for different causes for the missing data). This is a great strategy if the person recording the data is also the person performing the data analysis; but it can be disastrous if the person running the data analysis is not aware of it.

Checking for obvious errors, inconsistencies, and missing values

Data entry does not end with the creation of an electronic file of data values. You need to make sure your data are correct. Even data entry professionals are not 100% accurate. Every now and then you can expect to face problems like spelling differences (as in the "m" and "male" mentioned above for categorical data), typos (notice how the values 9 and 0 are right next to each other on the keyboard's top row, or 9 and 6 on the numerical keypad), or a skipped row (this mistake is almost too easy to make). The real issue is catching them.

All your data values should be realistic or at least plausible biologically. In an online class survey asking for height in inches, a student wrote "6.0" as the answer—this is definitely *not* realistic. If you do not catch this obvious error, all your analyses will be affected. In Chapters 2, 3, and 4, we discuss at length how outliers impact various statistical computations.

The best way to catch a mistake like an impossible height value is to plot your data. Stemplots, dotplots, and histograms are excellent ways to *see* unusual or implausible values. Numerical summaries and boxplots (described in Chapter 2) work too. Time plots for data collected over a period of time are also useful for revealing events that influenced the data collection process (for example, data values becoming systematically larger after switching to a new measuring instrument or when taken by a different person). If your data are categorical, obtain a summary frequency table: you would see, for example, four values for Sex if you carelessly recorded it as either m or male, and f or female.

Missing data can create substantial problems, especially if they go unnoticed. The publicly available Pima Indians diabetes data set, for example, contains biological data for 768 female members of the Pima Indian tribe. The data set is listed as having no missing data values. Yet, careful graphical exploration of the data reveals large numbers of zeros that are not biologically plausible (for example, for values of blood pressure or skinfold thickness). Perhaps even more troubling, it appears that many published analyses using this data set failed to identify the problem and instead treated such zeros as actual biological values.[19]

Don't make the same mistake. Note that "I am not the only one to miss data issues" is not an excuse to be careless. In fact, if you do not plot your data and carefully check them before running further statistical analyses, you could waste a lot of time and energy. Be smart. Plot your data!

CHAPTER 1 SUMMARY

- A data set contains information on a number of **individuals.** Individuals may be people, animals, or things. For each individual, the data give values for one or more **variables.** A variable describes some characteristic of an individual, such as a person's height, sex, or age.
- Some variables are **categorical** and others are **quantitative.** A categorical variable places each individual into a category, like male or female. A quantitative variable has numerical values that measure some characteristic of each individual, like height in centimeters or age in years.
- **Exploratory data analysis** uses graphs and numerical summaries to describe the variables in a data set and the relations among them.
- After you understand the background of your data (individuals, variables, units of measurement), the first thing to do almost always is **plot your data.**
- The **distribution** of a variable describes what values the variable takes and how often it takes these values. **Pie charts** and **bar graphs** display the distribution of a categorical variable. Bar graphs can also compare any set of

quantities measured in the same units. **Histograms, stemplots,** and **dotplots** display the distribution of a quantitative variable.

- When examining any graph, look for an **overall pattern** and for notable **deviations** from the pattern.
- **Shape, center, and spread** describe the overall pattern of the distribution of a quantitative variable. Some distributions have simple shapes, such as **symmetric** or **skewed.** Not all distributions have a simple overall shape. Describing the shape of a distribution when there are few observations can be particularly challenging.
- **Outliers** are observations that lie outside the overall pattern of a distribution. Always look for outliers and try to explain them.
- When observations on a variable are taken over time, make a **time plot** that graphs time horizontally and the values of the variable vertically. A time plot can reveal **trends, cycles,** or other changes over time.

THIS CHAPTER IN CONTEXT

Practical statistics is the science of extracting meaning out of validly collected data. We will see in Chapters 7 and 8 that the way we collect data drastically impacts what we may conclude. Some studies examine data from an entire population of interest, as in Example 1.3 on the leading causes of death in the United States. Other studies select a sample from a given population, as did the study in Example 1.1 about the paw usage preferences of 36 tree shews. The ultimate objective, though, was to draw conclusions about the wider population of all tree shrews—a process called statistical inference that we will describe starting in Chapter 14 of this textbook.

Regardless of the objective, understanding data starts with exploratory data analysis: the use of graphs and numerical summaries to reveal patterns. This may be done purely for a descriptive purpose—as highlighted in Part I of this book. Or it may be done to check whether the data we have are suitable for a specific inference procedure—something we will cover in Parts III and IV of this book. In addition, the discussion box on page 27 illustrates how graphs and numerical summaries are useful tools for finding inconsistencies in the data (typos and other errors) before any real analysis can begin.

In this chapter we showed how categorical data can be summarized graphically using pie charts and bar graphs and how the distribution of a quantitative variable can be displayed with histograms, dotplots, and stemplots—or, for time series, with time plots. These are the simplest and most common types of graphs for inspecting one variable at a time. In Chapters 3 through 5 we will also discuss how graphs can be used to examine the relationship between two quantitative variables or two categorical variables. There are more complex graphs used in the life sciences, such as graphs mapping data to a geographical location to picture patterns of epidemic spread, animal migration, or climate change. Their study is beyond the scope of this introductory textbook.

CHECK YOUR SKILLS

1.13 A study records the sex and weight (in kilograms) of 30 recently born bear cubs in Alaska. Which of the following statements is true?

(a) Sex and weight are both categorical variables.
(b) Sex and weight are both quantitative variables.
(c) Sex is a categorical variable and weight is a quantitative variable.

The Statistical Abstract of the United States, *prepared by the Census Bureau, provides the number of single-organ transplants for the year 2010, by organ. The following two exercises are based on this table:*

Heart	2,333
Lung	1,770
Liver	6,291
Kidney	16,898
Pancreas	350
Intestine	151

1.14 The data on single-organ transplants can be displayed in

(a) a pie chart but not a bar graph.
(b) a bar graph but not a pie chart.
(c) either a pie chart or a bar graph.

1.15 Kidney transplants represented what percent of single-organ transplants in 2010?

(a) Nearly 61%.
(b) One-sixth (nearly 17%).
(c) This percent cannot be calculated from the information provided in the table.

Figure 1.13 is a histogram of the takeoff angles of 51 videotaped jumps of adult hedgehog fleas, Archaeopsyllus erinacei.[20] *The following three exercises are based on this histogram.*

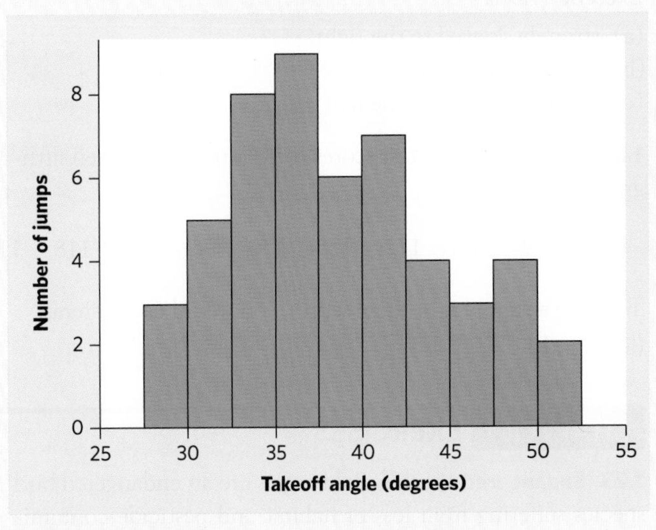

FIGURE 1.13 Histogram of the takeoff angles (measured in degrees) of 51 flea jumps, for Exercises 1.16–1.18.

1.16 The number of flea jumps with an angle covered by the rightmost bar in the histogram is

(a) 1. (b) 2. (c) 10.

1.17 The rightmost bar in the histogram covers angles ranging from about

(a) 25 to 55 degrees. (b) 50 to 55 degrees.
(c) 50 to 52.5 degrees.

1.18 The shape of the distribution of takeoff angles in Figure 1.13 is

(a) skewed to the right. (b) roughly symmetric.
(c) skewed to the left.

1.19 The 2010 U.S. census reported the percent of individuals under the age of 18 in each of the 50 states and the District of Columbia. The data are shown in the stemplot below, with stems representing whole percents and leaves tenths of a percent.

Stem	Leaves
16	8
17	
18	
19	
20	7 7 9
21	3 3 7 8
22	0 3 3 3 6 6 9 9
23	2 4 4 5 5 6 6 6 7 7 7 8 9 9
24	0 2 4 4 4 6 7 8 8 9
25	0 1 2 5 5 5 7
26	4
27	3 4
28	
29	
30	
31	5

There are two outliers: the District of Columbia has the lowest percent of individuals under 18, and Utah has the highest. What is the percent for Utah?

(a) 16.8% (b) 31.5% (c) 315%

1.20 Ignoring the outliers, the shape of the distribution in Exercise 1.19 is

(a) strongly skewed to the right.
(b) roughly symmetric.
(c) strongly skewed to the left.

1.21 Here are the IQ test scores of 10 randomly chosen fifth-grade students:

145 139 126 122 125 130 96 110 118 118

To make a stemplot of these scores, you would use as stems

(a) 0 and 1.
(b) 09, 10, 11, 12, 13, and 14.
(c) 96, 110, 118, 122, 125, 126, 130, 139, and 145.

1.22 Multiple myeloma is a cancer of the bone marrow currently without effective cure. It affects primarily older individuals: It is rarely diagnosed in individuals under 40 years old, and its incidence rate (the number of diagnosed malignant cases per 100,000 individuals in the population) is highest among individuals 70 years of age and older.[21] The distribution of incidence rate of multiple myeloma by age at diagnosis is

(a) skewed to the left.
(b) roughly symmetric.
(c) skewed to the right.

CHAPTER 1 EXERCISES

1.23 Endangered species. Bald eagles are an endangered bird species suffering from loss of habitat and pesticide contamination of rivers. A field biologist studying the reproduction of bald eagles records the following variables. Which of these variables are categorical, and which are quantitative?

(a) Number of eggs laid
(b) Incubation period (in days)
(c) Parental care (mostly mother, mostly father, both parents)
(d) Nest size (in centimeters)
(e) Presence of pesticides in samples of local waters (yes/no)

1.24 Eating habits. You are preparing to study the eating habits of elementary school children. Describe two categorical variables and two quantitative variables that you might record for each child. Give the units of measurement for the quantitative variables.

1.25 Mercury in lakes. Mercury is a metal highly toxic to the nervous system. Below is a small part of an EESEE data set ("Mercury in Bass") from a study that assessed the water quality of 53 representative lakes in Florida:[22]

Lake name	pH	Chlorophyll (mg/l)	Avg. mercury in fish (parts per million)	Number of fish sampled	Age of data
Alligator	6.1	0.7	1.23	5	year old
Annie	5.1	3.2	1.33	7	recent
Apopka	9.1	128.3	0.04	6	recent
Blue Cypress	6.9	3.5	0.44	12	recent
Brick	4.6	1.8	1.20	12	year old
Bryant	7.3	44.1	0.27	14	year old
⋮					

(a) What individuals does this data set describe?
(b) In addition to the lake's name, how many variables does the data set contain? Which of these variables are categorical and which are quantitative?

1.26 Deaths among young people. Here are the number of deaths among persons aged 15 to 24 years in the United States in 2010 due to the leading causes of death for this age group: accidents, 12,015; homicide, 4651; suicide, 4559; cancer, 1594; heart disease, 984; congenital defects, 401.[23]

(a) Make two bar graphs of these data, one with bars ordered alphabetically (by death type) and the other with bars in order from tallest to shortest. Comparisons are easier if you order the bars by height.
(b) What additional information do you need to make a pie chart?

1.27 Manatee deaths. Manatees are an endangered species of herbivorous, aquatic mammals found primarily in the rivers and estuaries of Florida. As part of its conservation efforts, the Florida Fish and Wildlife Commission records the cause of death for every recovered manatee carcass. Here is a breakdown of the dead manatee counts in Florida for 2012, by cause of death:[24]

Cause of death	Manatees recovered
Watercraft collisions	81
Perinatal	68
Natural	65
Cold stress	28
Flood gate /canal lock	11
Other human	9
Undetermined	
Total	392

(a) Most mortalities recorded as "Undetermined" correspond to manatee carcasses too badly decomposed to make any determination as to the cause of death. How many manatee carcasses had an "undetermined" cause of death?
(b) What is the percent of total manatee deaths caused by collisions with watercraft?
(c) Make a bar graph sorted by manatee count. What does it show?
(d) Could you display these data in a pie chart? Explain why or why not.

1.28 The overweight problem. The 2011 National Health Interview Survey by the National Center for Health Statistics (NCHS) provides weight categorization for adults 18 years old and over based on their body mass index:

Weight category	Percent of adults
Underweight	1.6
Healthy weight	36.3
Overweight	34.2
Obese	27.9

(a) Make a bar graph of these percents. What do you conclude about the weight problem in America?
(b) Could you make a pie chart for these data? (Explain why or why not.) If so, what would it emphasize?

1.29 More on the overweight problem. The same NCHS report (see previous exercise) breaks down the sampled individuals by age group. Here are the percents of obese individuals in the 2011 survey for each age group:

Age group	Percent who are obese
18 to 44	26.2
45 to 64	32.2
65 to 74	31.6
75 and over	19.5

(a) Make a bar graph of these percents. What do you conclude about the weight problem in America?
(b) Could you make a pie chart for these data? (Explain why or why not.) If so, what would it emphasize?

1.30 Do you have Highly Superior Autobiographical Memory? Some individuals have the ability to recall accurately vast amounts of autobiographical information without mnemonic tricks or extra practice. This ability is called HSAM, for Highly Superior Autobiographical Memory. A research team administered a quiz on past public events to 115 individuals claiming to have HSAM. The histogram in Figure 1.14(a) shows the distribution of scores on this quiz, expressed in percent of correct answers.[25]

(a) Describe the distribution of quiz scores for individuals claiming to have HSAM.
(b) What does this histogram suggest about the HSAM claims?

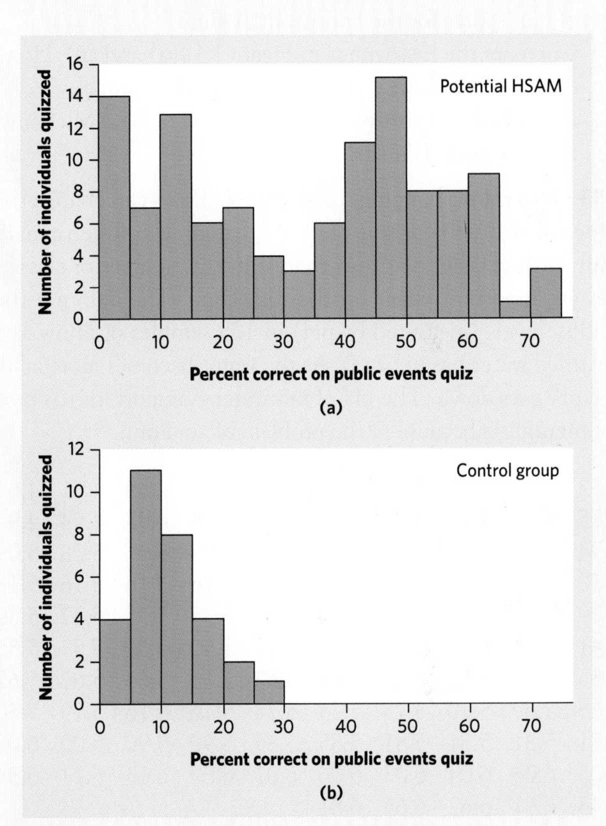

FIGURE 1.14 Histograms of scores on a quiz of past public events (a) for 115 individuals claiming to have Highly Superior Autobiographical Memory (HSAM) and (b) for 30 control individuals.

1.31 What graph? Of the following data sets, which would you display with a bar graph and which would you display with a histogram? Explain why.

(a) A record of the gender of selected individuals (labeled as male = 0, female = 1)
(b) A record of the age of selected parents (in years) at birth of first child
(c) A record of the height of selected individuals (in inches)
(d) A record of the blood type of selected individuals (A, B, AB, or O)

1.32 Highly Superior Autobiographical Memory, continued. The researchers in Exercise 1.30 also administered the quiz of past public events to 30 control individuals who did not claim to have any unusual memory abilities. Their scores are displayed in Figure 1.14(b), using the same horizontal axis scale as that of the histogram of Figure 1.14(a) for easier comparison.

(a) Describe the shape, center, and spread of the distribution of quiz scores for the control individuals.

(b) Compare the histograms in Figure 1.14(a) and (b). How does the distribution of scores among the control individuals support your interpretation of HSAM claims from Exercise 1.30(b)?

1.33 Acid rain. Changing the choice of classes can change the appearance of a histogram. Here is an example in which a small shift in the classes, with no change in the number of classes, has an important effect on the histogram. The data are the acidity levels (measured by pH) in 105 samples of rainwater. Distilled water has pH 7.00. As the water becomes more acid, the pH goes down. The pH of rainwater is important to environmentalists because of the problem of acid rain.[26]

4.33	4.38	4.48	4.48	4.50	4.55	4.59	4.59	4.61	4.61
4.75	4.76	4.78	4.82	4.82	4.83	4.86	4.93	4.94	4.94
4.94	4.96	4.97	5.00	5.01	5.02	5.05	5.06	5.08	5.09
5.10	5.12	5.13	5.15	5.15	5.15	5.16	5.16	5.16	5.18
5.19	5.23	5.24	5.29	5.32	5.33	5.35	5.37	5.37	5.39
5.41	5.43	5.44	5.46	5.46	5.47	5.50	5.51	5.53	5.55
5.55	5.56	5.61	5.62	5.64	5.65	5.65	5.66	5.67	5.67
5.68	5.69	5.70	5.75	5.75	5.75	5.76	5.76	5.79	5.80
5.81	5.81	5.81	5.81	5.85	5.85	5.90	5.90	6.00	6.03
6.03	6.04	6.04	6.05	6.06	6.07	6.09	6.13	6.21	6.34
6.43	6.61	6.62	6.65	6.81					

(a) Make a histogram of pH with 14 classes, using class boundaries 4.2, 4.4, …, 7.0. Describe this histogram. How many peaks does it show? The presence of more than one peak suggests that the data contain groups that have different distributions.

(b) Make a second histogram, also with 14 classes, using class boundaries 4.14, 4.34, …, 6.94. The classes are those from (a) moved 0.06 to the left. Describe this new histogram. How many peaks does it show?

1.34 Food oils and health. Fatty acids, despite their unpleasant name, are necessary for human health. Two types of essential fatty acids, called omega-3 and omega-6, are not produced by our bodies and so must be obtained from our food. Food oils, widely used in food processing and cooking, are major sources of these compounds. There is some evidence that a healthy diet should have more omega-3 than omega-6. Table 1.2 gives the ratio of omega-3 to omega-6 in some common food oils.[27] Values greater than 1 show that an oil has more omega-3 than omega-6.

(a) Make a histogram of these data, using classes bounded by the whole numbers from 0 to 6.

(b) What is the shape of the distribution? How many of the 30 food oils have more omega-3 than omega-6? What does this distribution suggest about the possible health effects of modern food oils?

(c) Table 1.2 contains entries for several fish oils (cod, herring, menhaden, salmon, sardine). How do these values support the idea that eating fish is healthy?

1.35 Carbon dioxide emissions. Burning fuels in power plants or motor vehicles emits CO_2, which contributes to global warming. Table 1.3 displays CO_2 emissions per person for countries with populations of at least 20 million.[28]

TABLE 1.2 Omega-3 fatty acids as a fraction of omega-6 fatty acids in food oils

Oil	Ratio	Oil	Ratio	Oil	Ratio
Perilla	5.33	Margarine	0.05	Herring	2.67
Walnut	0.20	Olive	0.08	Soybean, hydrogenated	0.07
Wheat germ	0.13	Shea nut	0.06	Rice bran	0.05
Mustard	0.38	Sunflower (oleic)	0.05	Butter	0.64
Sardine	2.16	Sunflower (linoleic)	0.00	Sunflower	0.03
Salmon	2.50	Flaxseed	3.56	Corn	0.01
Mayonnaise	0.06	Canola	0.46	Sesame	0.01
Cod liver	2.00	Soybean	0.13	Cottonseed	0.00
Shortening (household)	0.11	Grape seed	0.00	Palm	0.02
Shortening (industrial)	0.06	Menhaden	1.96	Cocoa butter	0.04

TABLE 1.3 Carbon dioxide emissions (metric tons per person)

Country	CO_2	Country	CO_2	Country	CO_2	Country	CO_2	Country	CO_2
Algeria	2.6	Ethiopia	0.1	Kenya	0.3	Peru	1.0	Thailand	3.3
Argentina	3.6	France	6.2	Korea, North	3.3	Philippines	0.9	Turkey	3.0
Australia	18.4	Germany	9.9	Korea, South	9.3	Poland	7.8	Ukraine	6.3
Bangladesh	0.3	Ghana	0.3	Malaysia	5.5	Romania	4.2	United Kingdom	8.8
Brazil	1.8	India	1.1	Mexico	3.7	Russia	10.8	United States	19.6
Canada	17.0	Indonesia	1.6	Morocco	1.4	Saudi Arabia	13.8	Uzbekistan	4.2
China	3.9	Iran	6.0	Myanmar	0.2	South Africa	7.0	Venezuela	5.4
Colombia	1.3	Iraq	2.9	Nepal	0.1	Spain	7.9	Vietnam	1.0
Congo	0.2	Italy	7.8	Nigeria	0.4	Sudan	0.3		
Egypt	2.0	Japan	9.5	Pakistan	0.8	Tanzania	0.1		

(a) Why do you think we choose to measure emissions per person rather than total CO_2 emissions for each country?

(b) Make a dotplot or a stemplot to display the data of Table 1.3. Describe the shape, center, and spread of the distribution. Which countries are outliers?

1.36 Fur seals on Saint George Island. Every year hundreds of thousands of northern fur seals return to their haul-out territory in the Pribilof Islands in Alaska to breed, give birth, and teach their pups to swim, hunt, and survive in the Bering Sea. U.S. commercial fur sealing operations continued until 1983, but despite a reduction in harvest, the population of fur seals has continued to decline. Here are data on the number of fur seal pups born on Saint. George Island (in thousands) from 1975 to 2006:[29]

Year	Pups born (thousands)	Year	Pups born (thousands)
1975	53.70	1991	24.28
1976	56.16	1992	25.16
1977	43.41	1993	23.70
1978	47.25	1994	22.24
1979	47.47	1995	24.82
1980	39.34	1996	27.39
1981	38.15	1997	24.74
1982	39.29	1998	22.09
1983	31.44	1999	21.13
1984	33.44	2000	20.18
1985	28.87	2001	18.89
1986	32.36	2002	17.59
1987	33.12	2003	17.24
1988	24.82	2004	16.88
1989	33.11	2005	16.97
1990	23.40	2006	17.07

Make a dotplot or a histogram to display the distribution of pups born per year. Describe the shape, center, and spread of the distribution. Are there any outliers?

1.37 Nanomedicine. Researchers examined a new treatment for advanced ovarian cancer in a mouse model. They created a nanoparticle-based delivery system for a suicide gene therapy to be delivered directly to the tumor cells. The grafted tumors were injected either with the new treatment or with only some buffer solution to serve as a comparison. The data below give the fold increase in tumor size after two weeks in 20 mice. A 1 represents no change; a 2 represents a doubling in volume of the tumor.[30]

Buffer solution									
9.1	8.1	7.8	7.0	6.8	5.4	5.4	4.1	3.8	3.3
Nanoparticle-delivered gene therapy									
4.1	3.5	2.1	2.1	1.8	1.8	1.4	1.2	1.1	1.1

(a) Make a back-to-back stemplot to compare the two samples. That is, use one set of stems with two sets of leaves, one to the right and one to the left of the stems. (Draw a line on either side of the stems to separate stems and leaves.) Order both sets of leaves from smallest at the stem to largest away from the stem.

(b) Make two dotplots, one for each group, using the same scale on the horizontal axis for both. How do your dotplots compare with the stemplots you created in (a)?

(c) Report the approximate midpoints of both groups. What are the most important differences between the two groups? What can you conclude?

1.38 Fur seals on Saint George Island, continued. Make a time plot of the number of fur seals born per year from Exercise 1.36. What does the time plot show that your plot in Exercise 1.36 does not show? When you have data collected over time, a time plot is often needed to understand what is happening.

1.39 Herbicide resistance in weeds. Farmers use herbicides to limit the growth of weeds among their crops. Eventually, this technique places evolutionary pressure on weed species, resulting in herbicide resistance. The International Survey of Herbicide Resistant Weeds keeps track of documented cases of herbicide resistance in 80 countries worldwide. Each case is examined to identify the weed species and particular mutation allowing herbicide resistance. Table 1.4 gives the number of unique cases of herbicide resistance (corresponding to a specific weed species and a specific mutation) documented worldwide every year since 1950.[31]

(a) Make a time plot of the data. Does the time plot illustrate only year-to-year variation or are there other patterns apparent? Specifically, is there a trend over any period of years? What about cyclical fluctuation? Explain in words the change in the worldwide number of unique cases of herbicide resistance in weeds over this 63-year period.

(b) Do you think it would be appropriate to display these data in a histogram without first plotting them in a time plot? Explain why or why not.

1.40 Birth methods. During birth, obstetricians sometimes use forceps or vacuum extractors to assist with vaginal deliveries. The Centers for Disease Control and Prevention track the percent of vaginal deliveries using these methods. Here are the data for the United States between 1990 and 2010:[32]

Year	1990	1994	1998	2002	2006	2010
Forceps	6.61	4.87	3.23	2.00	1.29	0.98
Vacuum	5.05	7.21	7.66	5.93	5.45	4.40

Make two time plots in the same graph to compare the percents of vaginal deliveries using either forceps or vacuum extraction. What can you conclude from your graph?

1.41 Orange prices. Figure 1.15 is a time plot of the average price of fresh oranges each month from January 2000 to August 2009.[33] The prices are "index numbers" given as percents of the average price during 1982 to 1984.

(a) The most notable pattern in this time plot is yearly cycles. At what season of the year are orange prices highest? Lowest? (To read the graph, note that the tick mark for each year is at the beginning of the year.) The cycles are explained by the time of the orange harvest in the United States.

(b) Is there a longer-term trend visible in addition to the cycles? If so, describe it.

1.42 To split or not to split. APPLET The data sets in the *One-Variable Statistical Calculator* applet on the companion website include the "guinea pig survival times" data

TABLE 1.4 Number of unique cases of herbicide resistance in weeds worldwide, by year

Year	Number of unique cases	Year	Number of unique cases	Year	Number of unique cases	Year	Number of unique cases	Year	Number of unique cases
1950	0	1963	2	1976	12	1989	115	2002	289
1951	0	1964	3	1977	14	1990	129	2003	293
1952	0	1965	3	1978	18	1991	139	2004	311
1953	0	1966	3	1979	27	1992	151	2005	331
1954	0	1967	3	1980	41	1993	170	2006	335
1955	0	1968	3	1981	49	1994	184	2007	344
1956	0	1969	3	1982	60	1995	191	2008	350
1957	2	1970	4	1983	65	1996	211	2009	365
1958	2	1971	4	1984	72	1997	225	2010	377
1959	2	1972	5	1985	76	1998	239	2011	391
1960	2	1973	7	1986	80	1999	256	2012	397
1961	2	1974	7	1987	92	2000	269		
1962	2	1975	9	1988	99	2001	278		

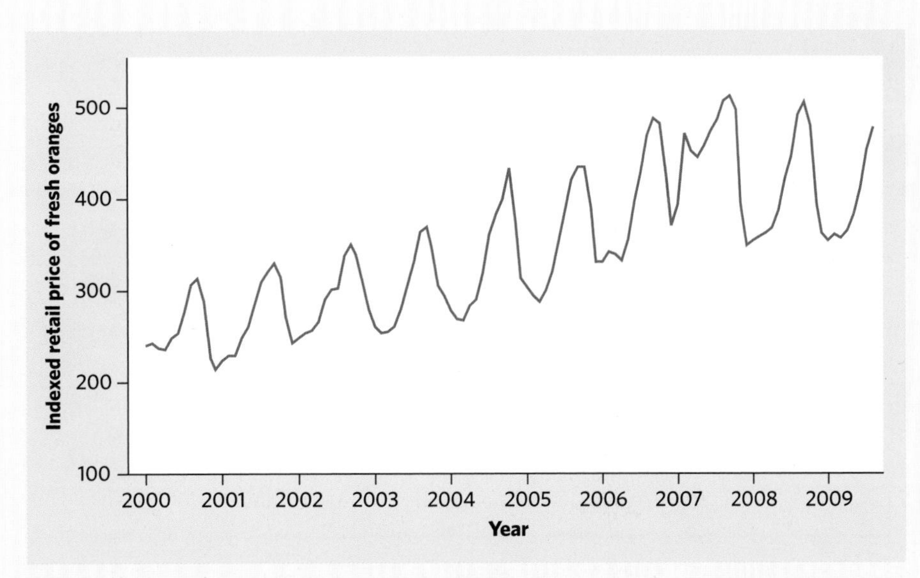

FIGURE 1.15 Time plot of the indexed monthly retail price of fresh oranges from January 2000 to August 2009, for Exercise 1.41. Data are indexed to the 1982–1984 average price (100).

from Example 1.7. Use the applet to make stemplots with and without split stems. Which stemplot do you prefer? Explain your choice.

1.43 Everglades. Everglades National Park is the largest subtropical wilderness in the United States and has been designated a Wetland of International Importance. This important ecosystem is closely monitored by the U.S. Geological Survey. The *Large.Everglades* data file contains data on five variables describing the properties of the Shark River at the Gunboat Island station (on the coast of the Gulf of Mexico) over a seven-day period ranging from November 25 to December 2, 2009. The data are recordings taken every 15 minutes.[34]

LARGE DATA SET

(a) Make a histogram for each of the five variables and describe each distribution.
(b) Make a time plot for each of the five variables and describe any pattern in the data.
(c) Summarize your findings and describe some of the key features of the river at this station by the ocean.

1.44 Lung capacity of children. The forced expiratory volume (FEV, measured in liters) is a primary indicator of lung function and corresponds to the volume of air that can forcibly be blown out in the first second after full inspiration. The *Large.FEV* data file contains the FEV values of a large sample of children, along with some categorical descriptors of each individual.[35]

LARGE DATA SET

(a) Make a histogram of FEV and describe the distribution.
(b) Make separate histograms of FEV for girls ($sex = 0$) and for boys ($sex = 1$), using the same horizontal scale for both. Describe the differences between the two groups.
(c) Now create a new variable based on age: *preschool* for ages 3 to 5, *elementary* for ages 6 to 10, *middle* for ages 11 to 13, and *highschool* for ages 14 and above. Then make separate histograms using the same horizontal scale for each of the four age groups. What do you notice?
(d) Lastly, create eight separate histograms with the same horizontal scale for the eight combinations of sex and age group. Does this last analysis change your previous conclusions? If so, explain how.

1.45 Estimating body fat in men. The data file *Large.Bodyfat* contains data on the percent body fat, age, weight, height, and 10 body circumference measurements for 252 adult men.[36]

LARGE DATA SET

(a) Make a histogram of the men's body weights (in pounds). Describe the distribution and any deviations from the main pattern.
(b) Make a histogram of the men's heights (in inches). Describe the distribution and any deviations from the main pattern.

CHAPTER 2 Describing Distributions with Numbers

Examining nature closely reveals surprising variability. The needles in a given pine tree species are not all the same size, for instance. So it is the distribution of needle lengths that characterizes the pine species. Here are the lengths in centimeters (cm) of 15 needles taken at random from different parts of several Aleppo pine trees located in Southern California:[1]

7.2 7.6 8.5 8.5 8.7 9.0 9.0 9.3 9.4 9.4 10.2 10.9 11.3 12.1 12.8

Here is a dotplot of these data:

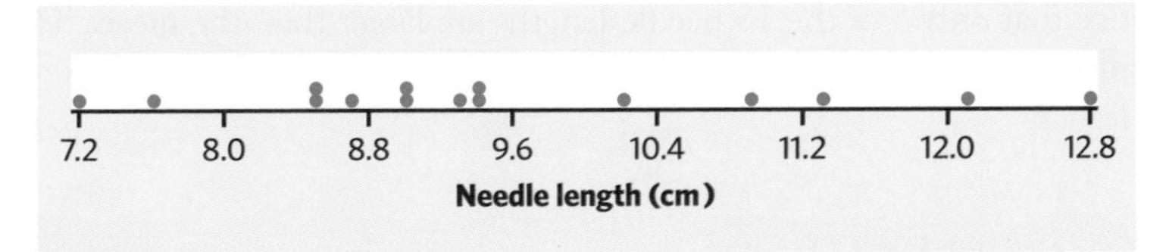

The distribution is single-peaked and slightly skewed to the right without outliers. Our goal in this chapter is to describe with numbers the center and spread of this and other distributions of *quantitative variables*. Categorical variables are simply described by the proportion (or percent) of the data falling into each category, as we have already seen in Chapter 1.

IN THIS CHAPTER WE COVER...

Measuring center: the mean

The most common measure of center is the ordinary arithmetic average, or *mean*.

THE MEAN $\bar{x}$

To find the **mean** of a set of observations, add their values and divide by the number of observations. If the n observations are $x_1, x_2, \ldots, x_n$, their mean is

$$\bar{x} = \frac{x_1 + x_2 + \cdots + x_n}{n}$$

or, in more compact notation,

$$\bar{x} = \frac{1}{n}\sum x_i$$

The $\sum$ (capital Greek sigma) in the formula for the mean is short for "add them all up." The subscripts on the observations x_i are just a way of keeping the n observations distinct. They do not necessarily indicate order or any other special facts about the data. The bar over the x indicates the mean of all the x-values. Pronounce the mean $\bar{x}$ as "x-bar." This notation is very common. When writers who are discussing data use $\bar{x}$ or $\bar{y}$, they are talking about a mean.

Craig Tuttle/CORBIS

EXAMPLE 2.1 Pine needle lengths

The mean needle length of our 15 Aleppo pine needles is

$$\begin{aligned}\bar{x} &= \frac{x_1 + x_2 + \cdots + x_n}{n} \\ &= \frac{7.2 + 7.6 + \cdots + 12.8}{15} \\ &= \frac{143.9}{15} = 9.59 \text{ cm}\end{aligned}$$

In practice, you can key the data into your calculator or software and ask for $\bar{x}$. You don't have to actually add and divide. But you should know that this is what the calculator or software is doing.

Notice that only 5 of the 15 needle lengths are larger than the mean. This unbalanced allocation of data around the mean mirrors the slight right-skew in the distribution. ■

APPLY YOUR KNOWLEDGE

2.1 Spider silk. Spider silk is the strongest known material, natural or man-made, on a weight basis. A study examined the mechanical properties of spider silk using 21 female golden orb weavers, *Nephila clavipes*. Here are data on silk yield stress, which represents the amount of force per unit area needed to reach permanent deformation of the silk strand. The data are expressed in megapascals (MPa).[2]

164.0	478.7	251.3	351.7	173.0	448.9	300.6
362.0	272.4	740.2	329.0	327.2	270.5	332.1
288.8	176.1	282.2	236.1	358.2	270.5	290.7

(a) Make either a stemplot or a dotplot of these data. Describe the shape, center, and spread of the distribution.

(b) Find the mean yield stress. How many of the 21 spiders had a yield stress less than the mean?

Measuring center: the median

In Chapter 1, we used the midpoint of a distribution as an informal measure of center. The *median* is the formal version of the midpoint, with a specific rule for calculation.

THE MEDIAN *M*

The **median M** is the midpoint of a distribution, the number such that half the observations are smaller and the other half are larger. To find the median of a distribution:

1. Arrange all observations in order of size, from smallest to largest.
2. If the number of observations n is odd, the median M is the center observation in the ordered list. Find the location of the median by counting $(n + 1)/2$ observations up from the smallest observation in the list.
3. If the number of observations n is even, the median M is the mean of the two center observations in the ordered list. The location of the median is again $(n + 1)/2$, counting from the smallest observation in the list.

Note that the formula $(n + 1)/2$ does *not* give the median, just the location of the median in the ordered list. Medians require little arithmetic, so they are easy to find by hand for small sets of data. Arranging even a moderate number of observations in order is very tedious, however, so that finding the median by hand for larger sets of data is unpleasant. Even simple calculators have an $\overline{x}$ button, but you will need to use software or a graphing calculator to automate finding the median.

EXAMPLE 2.2 Finding the median: odd *n*

What is the median length for our 15 Aleppo pine needles? Here are the data again, arranged in order:

7.2 7.6 8.5 8.5 8.7 9.0 9.0 **9.3** 9.4 9.4
10.2 10.9 11.3 12.1 12.8

The number of observations $n = 15$ is odd. The bold **9.3** is the center observation in the ordered list, with 7 observations to its left and 7 to its right. This is the median, M = 9.3 cm.

Because $n = 15$, our rule for the location of the median gives

$$\text{location of M} = \frac{n+1}{2} = \frac{16}{2} = 8$$

That is, the median is the 8th observation in the ordered list. Use this rule to locate the center in the ordered list or even on the dotplot. ■

EXAMPLE 2.3 Finding the median: even *n*

We also have the lengths (in cm) of 18 needles from trees of the noticeably different Torrey pine species.[3] What is the median length for these 18 pine needles? The ordered data are now:

21.2	21.6	21.7	23.1	23.7	24.2	24.2	25.5	**26.6**	**26.8**
28.9	29.0	29.7	29.7	30.2	32.5	33.7	33.7		

There is no center observation, but there is a center pair. These are the bold **26.6** and **26.8**, with 8 observations before them in the ordered list and 8 after them. The median is midway between these two observations:

$$M = \frac{26.6 + 26.8}{2} = 26.7 \text{ cm}$$

With $n = 18$, the rule for locating the median in the list gives

$$\text{location of M} = \frac{n+1}{2} = \frac{19}{2} = 9.5$$

The location 9.5 means "halfway between the 9th and 10th observations in the ordered list." ■

Comparing the mean and the median

In Examples 2.1 and 2.2, we studied the lengths of 15 Aleppo pine needles. We found that the mean and median were fairly similar, at 9.59 cm and 9.30 cm, respectively. The distribution of the 15 pine needles was only very slightly skewed and did not have any outliers. But what would happen to the relationship between the mean and the median of a data set with a marked skew or extreme outliers?

EXAMPLE 2.4 Guinea pig survival times

Figure 1.6 (page 19) shows the survival times of 72 guinea pigs after inoculation with infectious bacteria. The distribution is noticeably skewed to the right and has some potential high outliers. The mean survival time is 141.9 days, whereas the median survival time is only 102.5 days. The mean is pulled toward the right tail of this right-skewed distribution.

If the longest survival time were increased tenfold to 5980 days rather than 598 days, the mean would increase to more than 216 days but the median would not change at all. The artificially created outlier would count as just one observation above the center, no matter how far above the center it might lie. In contrast, the mean uses the actual value of each observation and is, therefore, very sensitive to any extreme values. ■

Example 2.4 illustrates an important fact about the mean as a measure of center: It is sensitive to the influence of a few extreme observations, such as in a skewed distribution or when there are outliers. Because the mean cannot resist the influence of extreme observations, we say that it is not a **resistant measure** of center. In contrast, the median is influenced only by the total number of data points and the numerical value of the point or points located at the center of the distribution. The *Mean and Median* applet is an excellent way to compare the resistance of M and $\overline{x}$.

resistant measure

COMPARING THE MEAN AND THE MEDIAN

The mean and median of a symmetric distribution are close together. If the distribution is exactly symmetric, the mean and median are exactly the same. In a skewed distribution, the mean is usually farther out in the long tail than is the median.[4]

Many biological variables have distributions that are skewed to the right. Survival times, such as in the guinea pig inoculation experiment, are typically right-skewed. Most individuals die within a certain period of time, but a few will survive much longer. When dealing with strongly skewed distributions, it is somewhat customary to report the median ("midpoint") rather than the mean ("arithmetic average"). However, a health organization or a government agency may need to include all survival times, and thus calculate the mean, to estimate the cost of medical care for a given disease and to plan medical staffing appropriately. Relying only on the median would result in underestimating the medical and financial needs. The mean and median measure center in different ways, and both are useful. *Don't confuse the "average" value of a variable (the mean) with its "typical" value, which we might describe by the median.*

APPLY YOUR KNOWLEDGE

2.2 Deep-sea sediments. Phytopigments are a marker of the amount of organic matter that settles in sediments. Phytopigment concentrations in deep-sea sediments collected worldwide showed a very strong right-skew.[5] Of these two summary statistics, 0.015 and 0.009 grams per square meter of bottom surface, which one is the mean and which one is the median? Explain your reasoning.

2.3 Spider silk, continued. In Exercise 2.1 you calculated the mean silk yield stress for 21 female golden orb weaver spiders. Find the median for the yield stress values. Compare the mean and median for these data. What general fact does your comparison illustrate? (Refer to the plot you made in Exercise 2.1.)

2.4 Glucose levels. Exercise 1.9 (page 23) provides the fasting plasma glucose levels for 18 diabetics enrolled in a diabetes control class, five months after the end of the class. The data set has one high outlier. Calculate the mean and median glucose level with and without the outlier. How does the outlier affect each measure of center? What general fact about the mean and median does your result illustrate?

Jonathan Heger/istockphoto

Measuring spread: the quartiles

The mean and median provide two different measures of the center of a distribution. But a measure of center alone can be misleading. The mean and median survival times of guinea pigs after inoculation in Example 2.4 are 141.9 and 102.5 days, respectively. The mean is higher because the distribution of survival times is skewed to the right. But the median and mean don't tell the whole story. Ten percent of the 72 guinea pigs survived less than 2 months while a handful survived for over a year. We are interested in the *spread* or *variability* of survival times as well as their center. *The simplest useful numerical description of a distribution requires both a measure of center and a measure of spread.*

One way to measure spread is to give the smallest and largest observations. For example, the guinea pig survival times range from 43 to 598 days. These single observations show the full spread of the data, but they may be outliers. We can improve our description of spread by also looking at the spread of the middle half of the data. The *quartiles* mark out the middle half. Count up the ordered list of observations, starting from the smallest. The *first quartile* lies one-quarter of the way up the list. The *third quartile* lies three-quarters of the way up the list. In other words, the first quartile is larger than 25% of the observations, and the third quartile is larger than 75% of the observations. The second quartile is the median, which is larger than 50% of the observations. That is the idea of quartiles. We need a rule to make the idea exact. The rule for calculating the quartiles uses the rule for the median.

THE QUARTILES Q_1 AND Q_3

To calculate the **quartiles:**

1. Arrange the observations in increasing order and locate the median M in the ordered list of observations.
2. The **first quartile Q_1** is the median of the observations whose position in the ordered list is to the left of the location of the overall median.
3. The **third quartile Q_3** is the median of the observations whose position in the ordered list is to the right of the location of the overall median.

Here are examples that show how the rules for the quartiles work for both odd and even numbers of observations.

EXAMPLE 2.5 Finding the quartiles: odd *n*

Our sample of 15 Aleppo pine needles (Example 2.2), arranged in increasing order, is:

7.2 7.6 8.5 8.5 8.7 9.0 9.0 **9.3** 9.4 9.4
10.2 10.9 11.3 12.1 12.8

We have an odd number of observations, so the median is the middle one, the bold **9.3** in the list. The first quartile is the median of the 7 observations to the left of the median. This is the 4th of these 7 observations, so $Q_1 = 8.5$ cm. If you want, you can use the recipe for the location of the median with $n = 7$:

$$\text{location of } Q_1 = \frac{n+1}{2} = \frac{7+1}{2} = 4$$

The third quartile is the median of the 7 observations to the right of the median, $Q_3 = 10.9$ cm.

The quartiles are resistant to outliers. For example, Q_3 would still be 10.9 if the largest needle length were 50 cm rather than 12.8 cm. ■

EXAMPLE 2.6 Finding the quartiles: even *n*

Here are the 18 Torrey pine needles (Example 2.3), arranged in increasing order:

21.2	21.6	21.7	23.1	23.7	24.2	24.2	25.5	26.6	*	26.8
28.9	29.0	29.7	29.7	30.2	32.5	33.7	33.7			

We have an even number of observations, so the median lies midway between the middle pair, the 9th and 10th in the list. Its value is $M = 26.7$ cm and its location is marked by a star. The first quartile is the median of the first 9 observations, because these are the observations to the left of the location of the median. Check that $Q_1 = 23.7$ cm and $Q_3 = 29.7$ cm. ■

Be careful when, as in these examples, several observations take the same numerical value. Write down all the observations and apply the rules just as if they all had distinct values.

The five-number summary and boxplots

The smallest and largest observations tell us little about the distribution as a whole, but they give information about the tails of the distribution that is missing if we know only Q_1, M, and Q_3. To get a quick summary of both center and spread, combine all five numbers.

THE FIVE-NUMBER SUMMARY

The **five-number summary** of a distribution consists of the smallest observation, the first quartile, the median, the third quartile, and the largest observation, written in order from smallest to largest. In symbols, the five-number summary is

$$\text{Minimum} \quad Q_1 \quad M \quad Q_3 \quad \text{Maximum}$$

These five numbers offer a reasonably complete description of center and spread. The five-number summaries from Examples 2.5 and 2.6 are

Aleppo pine	7.2	8.5	9.3	10.9	12.8
Torrey pine	21.2	23.7	26.7	29.7	33.7

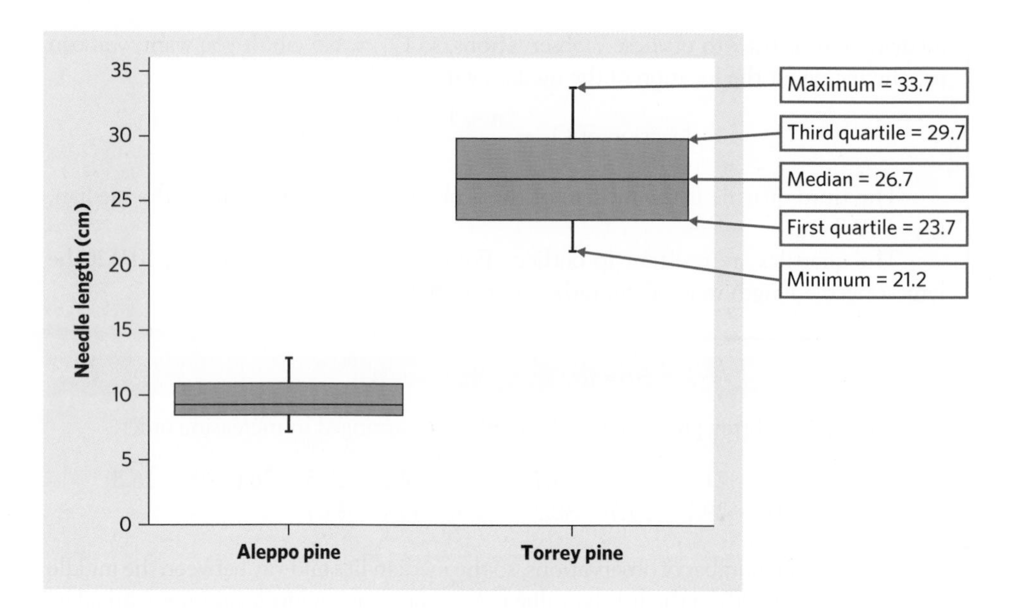

FIGURE 2.1 Boxplots comparing the lengths of 15 Aleppo pine needles and 18 Torrey pine needles.

The five-number summary of a distribution leads to a new graph, the *boxplot*. Figure 2.1 shows boxplots comparing needle lengths for the Aleppo pine and the Torrey pine species.

BOXPLOT

A **boxplot** is a graph of the five-number summary.

- A central box spans the quartiles Q_1 and Q_3.
- A line in the box marks the median M.
- Lines extend from the box out to the smallest and largest observations.

Because boxplots show less detail than histograms or dotplots, they are best used for side-by-side comparison of more than one distribution, as in Figure 2.1. Be sure to include a numerical scale in the graph. When you look at a boxplot, first locate the median, which marks the center of the distribution. Then look at the spread. The box shows the spread of the middle half of the data, and the extremes (the smallest and largest observations) show the spread of the entire data set. We see from Figure 2.1 that needles are all longer in the Torrey pine set than in the Aleppo pine set. The median, both quartiles, and the minimum and maximum are all larger in the Torrey pine. Torrey pine needle lengths are also more variable, as shown by the spread of the box and the spread between the extremes.

Finally, the data for the Torrey pine are symmetrical but the data for the Aleppo pine are mildly right-skewed. In a symmetric distribution, the first and third quartiles are equally distant from the median. In contrast, in most distributions that are skewed to the right, the third quartile will be farther above the median than the first quartile is below it. The extremes behave the same way, but remember that they are just single observations and may say little about the distribution as a whole.

APPLY YOUR KNOWLEDGE

2.5 **Spider silk, continued.** In Exercise 2.1 you plotted the silk yield stress for 21 female golden orb weaver spiders. Give the five-number summary of the distribution of yield stresses (in MPa). (The plot helps because it arranges the data in order, but make sure to use the exact numerical values provided.) Does the five-number summary reflect what you see in your plot? Remember that only a graph gives a clear picture of the shape of a distribution.

2.6 **Glucose levels, continued.** People with diabetes must monitor and control their blood glucose level. The goal is to maintain a "fasting plasma glucose" between about 90 and 130 milligrams per deciliter (mg/dl). Exercise 1.9 (page 23) provides the fasting plasma glucose levels for 18 diabetics enrolled in a diabetes control class, five months after the end of the class. Here are the data for 16 diabetics who received individual instruction instead:

128	195	188	158	227	198	163	164
159	128	283	226	223	221	220	160

CDC/Amanda Mills

(a) Calculate the five-number summary for each of the two data sets.

(b) Make side-by-side boxplots comparing the two groups (as in Figure 2.1). What can you say from the graph about the differences between the two diabetes control training methods?

Spotting suspected outliers*

Figure 2.2 shows the dotplot of the sizes in cubic centimeters (cm^3) of 11 acorns from oak trees found on the Pacific coast.[6] The five-number summary for this distribution is

0.4   1.6   4.1   6.0   17.1

How shall we describe the spread of this distribution? The largest observation is clearly an extreme that doesn't represent the majority of the data. The distance between the quartiles (the range of the center half of the data) is a more resistant measure of spread. This distance is called the *interquartile range*.

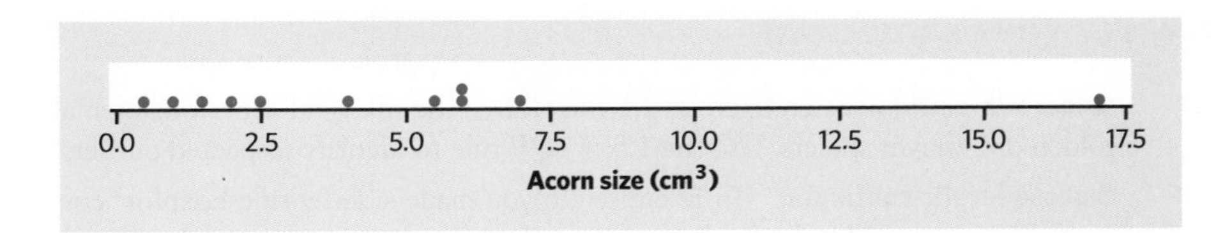

FIGURE 2.2 Dotplot of the sizes of 11 acorns.

*This short section is optional.

THE INTERQUARTILE RANGE *IQR*

The **interquartile range IQR** is the distance between the first and third quartiles,

$$IQR = Q_3 - Q_1$$

For our data on acorn sizes, $IQR = 6.0 - 1.6 = 4.4 \text{ cm}^3$. However, *no single numerical measure of spread, such as IQR, is very useful for describing skewed distributions*. The two sides of a skewed distribution have different spreads, so one number can't summarize them. The interquartile range is mainly used as the basis for a rule of thumb for identifying suspected outliers.

THE 1.5 × *IQR* RULE FOR OUTLIERS

Call an observation a suspected outlier if it falls more than $1.5 \times IQR$ above the third quartile or below the first quartile.

EXAMPLE 2.7 Spotting suspected outliers

For the acorn sizes data, with five-number summary 0.4, 1.6, 4.1, 6.0, 17.1,

$$1.5 \times IQR = 1.5 \times 4.4 = 6.6$$

Any values not falling between

$$Q_1 - (1.5 \times IQR) = 1.6 - 6.6 = -5.0 \quad \text{and}$$
$$Q_3 + (1.5 \times IQR) = 6.0 + 6.6 = 12.6$$

are flagged as suspected outliers. Look again at the dotplot of the data in Figure 2.2: the only suspected outlier is the largest acorn, 17.1 cm^3. The $1.5 \times IQR$ rule suggests that this high value is indeed an outlier. ■

The $1.5 \times IQR$ rule is not a replacement for looking at the data. It is most useful when large volumes of data are scanned automatically. Some software programs create **modified boxplots,** which display any value outside of the $1.5 \times IQR$ interval around either quartile by a star (rather than including the value within the long tail going to the minimum or maximum).

modified boxplot

APPLY YOUR KNOWLEDGE

2.7 Spider silk, continued. In Exercise 2.1 you plotted the silk yield stress for 21 female golden orb weaver spiders. Use the $1.5 \times IQR$ rule to identify suspected outliers.

2.8 Glucose levels, continued. In Exercise 2.6 you made side-by-side boxplots comparing the group and individual training methods for diabetes control. Look at the raw data to see if there are unusually high or low values in either data set. Use the $1.5 \times IQR$ rule to identify any suspected outliers.

Measuring spread: the standard deviation

The five-number summary is not the most common numerical description of a distribution. That distinction belongs to the combination of the mean to measure center and the *standard deviation* to measure spread. The standard deviation and its close relative, the *variance*, measure spread by looking at how far the observations are from their mean.

THE STANDARD DEVIATION s

The **variance s^2** of a sample set of observations is an average of the squares of the deviations of the observations from their mean. In symbols, the variance of n sample observations $x_1, x_2, \ldots, x_n$ is

$$s^2 = \frac{(x_1 - \overline{x})^2 + (x_2 - \overline{x})^2 + \cdots + (x_n - \overline{x})^2}{n - 1}$$

or, more compactly,

$$s^2 = \frac{1}{n-1} \sum (x_i - \overline{x})^2$$

The **standard deviation s** is the square root of the variance s^2:

$$s = \sqrt{\frac{1}{n-1} \sum (x_i - \overline{x})^2}$$

In practice, use software or your calculator to obtain the standard deviation from keyed-in data. Doing an example step-by-step will help you understand how the variance and standard deviation work, however.

EXAMPLE 2.8 Calculating the standard deviation

A person's metabolic rate is the rate at which the body consumes energy. Metabolic rate is important in studies of weight gain, dieting, and exercise. Here are the metabolic rates of 7 men who took part in a study of dieting. The units are kilocalories (Cal) for a 24-hour period. These are the same calories used to describe the energy content of foods.

1792 1666 1362 1614 1460 1867 1439

The researchers reported $\overline{x}$ and s for these men. First find the mean:

$$\overline{x} = \frac{1792 + 1666 + 1362 + 1614 + 1460 + 1867 + 1439}{7}$$

$$= \frac{11{,}200}{7} = 1600 \text{ Cal}$$

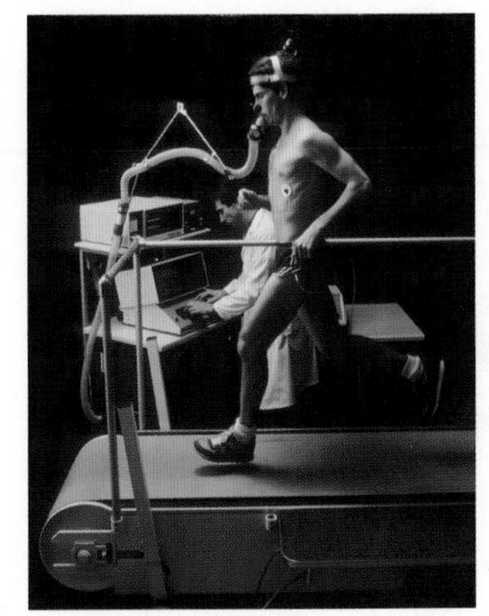
Tom Tracy Photography/Alamy

Figure 2.3 displays the data as points above the number line, with their mean marked by an asterisk (*). The arrows mark two of the deviations from the mean. These deviations show how spread out the data are about their mean. They are the starting point for calculating the variance and the standard deviation.

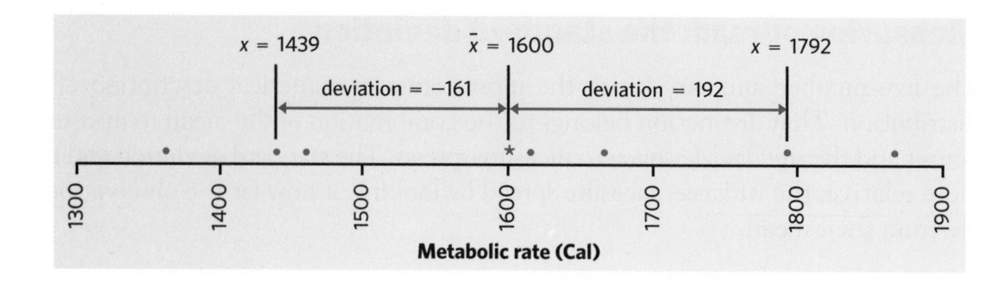

FIGURE 2.3 Metabolic rates for 7 men, with their mean (*) and the deviations of two observations from the mean indicated.

Observations x_i	Deviations $x_i - \overline{x}$			Squared deviations $(x_i - \overline{x})^2$		
1792	1792 − 1600	=	192	192^2	=	36,864
1666	1666 − 1600	=	66	66^2	=	4,356
1362	1362 − 1600	=	−238	$(-238)^2$	=	56,644
1614	1614 − 1600	=	14	14^2	=	196
1460	1460 − 1600	=	−140	$(-140)^2$	=	19,600
1867	1867 − 1600	=	267	267^2	=	71,289
1439	1439 − 1600	=	−161	$(-161)^2$	=	25,921
	sum	=	0	sum	=	214,870

The variance is the sum of the squared deviations divided by 1 less than the number of observations:

$$s^2 = \frac{214{,}870}{6} = 35{,}811.67 \text{ Cal}^2$$

The standard deviation is the square root of the variance:

$$s = \sqrt{35{,}811.67} = 189.24 \text{ Cal}$$ ■

Notice that the "average" in the variance s^2 divides the sum by 1 less than the number of observations, that is, $n - 1$ rather than n. The reason is that the deviations $x_i - \overline{x}$ always sum to exactly 0, so that knowing $n-1$ of them determines the last one. Only $n - 1$ of the squared deviations can vary freely, and we average by dividing the total by $n - 1$. The number $n - 1$ is called the **degrees of freedom** of the variance or standard deviation. Many calculators offer a choice between dividing by n and dividing by $n - 1$, so be sure to use $n - 1$.

degrees of freedom

More important than the details of hand calculation are the properties that determine the usefulness of the standard deviation:

- s measures *spread about the mean*[7] and should be used only when the mean is chosen as the measure of center.
- s is *always zero or greater than zero*. $s = 0$ only when there is no spread. This happens only when all observations have the same value. Otherwise, $s > 0$. As the observations become more spread out about their mean, s gets larger.
- s has the *same units of measurement as the original observations*. For example, if you measure metabolic rates in kilocalories, both the mean $\overline{x}$ and the

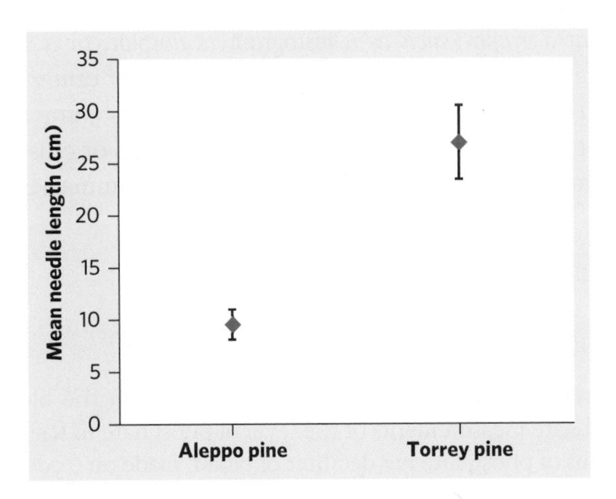

FIGURE 2.4 The same data as in Figure 2.1 now displayed as the mean plus and minus one standard deviation (error bars).

standard deviation s are also in kilocalories. This is one reason to prefer s to the variance s^2, which is in squared kilocalories (Cal^2).

- Like the mean $\bar{x}$, s is *not resistant*. A few outliers can make s very large.

The use of squared deviations renders s even more sensitive than $\bar{x}$ to a few extreme observations. For example, the standard deviation of the 11 Pacific acorn sizes displayed in Figure 2.2 is 4.66 cm^3. If we omit the high outlier, the standard deviation drops to 2.40 cm^3.

You may rightly feel that the importance of the standard deviation is not yet clear. We will see in later chapters that the standard deviation is the natural measure of spread for an important class of symmetric distributions, the Normal distributions.

error bars

Many scientific publications provide summary statistics for two or more groups in the form of one graph displaying each mean with **error bars** extending on either side to show the standard deviation in each group. Figure 2.4 shows such a graph for the needle lengths of two species of pine trees from Examples 2.2 and 2.3. *Always check the legend of such graphs, because error bars may represent some other measure also related to the spread of the data.*

Choosing measures of center and spread

We now have a choice between two descriptions of the center and spread of a distribution: either the five-number summary or $\bar{x}$ and s. Because $\bar{x}$ and s are sensitive to extreme observations, they can be misleading when a distribution is strongly skewed or has outliers. In fact, because the two sides of a skewed distribution have different spreads, no single number such as s describes the spread well. The five-number summary, with its two quartiles and two extremes, does a better job.

CHOOSING A SUMMARY

The five-number summary is usually better than the mean and standard deviation for describing a skewed distribution or a distribution with strong outliers. Use $\bar{x}$ and s only for reasonably symmetric distributions that are free of outliers.

Remember that a graph (such as a histogram, a dotplot, or a stemplot) gives the best overall picture of a distribution. Numerical measures of center and spread report specific facts about a distribution, but they do not describe its entire shape. Numerical summaries do not disclose the presence of multiple peaks or clusters, for example. Exercise 2.10 below shows how misleading numerical summaries can be. **Always plot your data.**

APPLY YOUR KNOWLEDGE

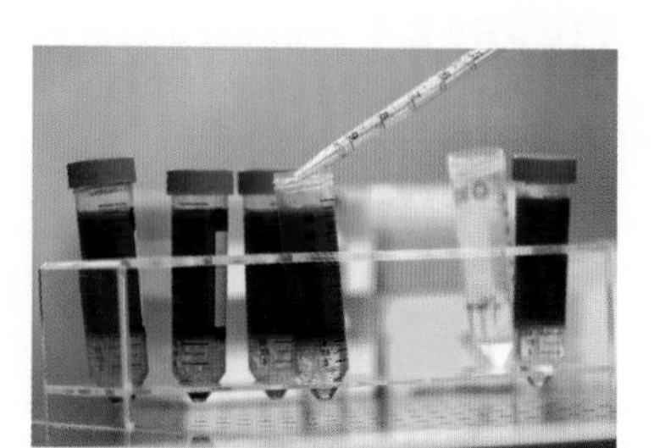

Claudio Gallone/AgeFotostock

2.9 Blood phosphate. The level of various substances in the blood influences our health. Here are measurements of the level of phosphate in the blood of a patient, in milligrams of phosphate per deciliter of blood, made on 6 consecutive visits to a clinic:

5.6 5.2 4.6 4.9 5.7 6.4

A graph of only 6 observations gives little information, so we proceed to compute the mean and standard deviation.

(a) Find the mean from its definition. That is, find the sum of the 6 observations and divide by 6.

(b) Find the standard deviation from its definition. That is, find the deviation of each observation from the mean, square the deviations, then obtain the variance and the standard deviation. Example 2.8 shows the method.

(c) Now enter the data into your calculator and use the mean and standard deviation buttons to obtain $\bar{x}$ and s. Do the results agree with your hand calculations?

2.10 $\bar{x}$ and s are not enough. The mean $\bar{x}$ and standard deviation s measure center and spread but are not a complete description of a distribution. Data sets with different shapes can have the same mean and standard deviation. To demonstrate this fact, use your calculator to find $\bar{x}$ and s for these two small data sets. Then make a stemplot of each and comment on the shape of each distribution.

Data A	9.14	8.14	8.74	8.77	9.26	8.10	6.13	3.10	9.13	7.26	4.74
Data B	6.58	5.76	7.71	8.84	8.47	7.04	5.25	5.56	7.91	6.89	12.50

2.11 Choose a summary. The shape of a distribution is a rough guide to whether the mean and standard deviation are a helpful summary of center and spread. For which of these distributions would $\bar{x}$ and s be useful? In each case, give a reason for your decision.

(a) Length of great white sharks, Figure 1.5 (page 15)

(b) Age of onset for anorexia nervosa, Figure 1.8 (page 20)

(c) Takeoff angles of flea jumps, Figure 1.13 (page 31)

DISCUSSION Dealing with outliers: recognition and treatment

When collecting data, we sometimes come across wild observations that clearly fall outside the overall pattern of data distribution. In this chapter, we have presented a way to identify suspected outliers and discussed how outliers can affect numerical summaries. But what are these wild observations and how do we deal with them? Entire books have been written on the topic. Here we describe three major types of wild observations and offer some general guidance on how to handle each type.

Human error in recording information

Perhaps the most famous example of an outlier caused by a data-recording error lies in the story of Popeye the Sailor. Created in 1929, Popeye is a friendly cartoon character who attains immediate superhuman strength whenever he eats iron-rich spinach. Indeed, an 1870 scientific publication reported that spinach had by far the highest iron content of any other green leafy vegetable, about 10 times more than lettuce or cabbage. This claim remained unquestioned until a 1937 study showed that spinach's iron content was similar to that of other leafy greens. It turns out that the iron content of spinach in the 1870 publication had a misplaced decimal point! The spinach value had been correctly identified as a wild observation, but for over half a century nobody had questioned its nature.

How many was that? Good causes often breed bad statistics. An advocacy group claims, without much evidence, that 150,000 Americans suffer from the eating disorder anorexia nervosa. Soon someone misunderstands and says that 150,000 people *die* from anorexia nervosa each year. This wild number gets repeated in countless books and articles. It really is a wild number: only about 55,000 women aged 15 to 44 (the main group affected) die of *all causes* each year.

Errors in data recording are not that uncommon. Typos are an obvious concern. But the data themselves are not always clean. Surveys of individuals are particularly prone to errors. People may forget, lie, or simply misunderstand a question. In an online survey, undergraduate students enrolled in a biostatistics course were asked to record their heights in inches. Of the 149 numerical values submitted, two wild observations appeared as 5.3 and 6. These are obvious errors. Maybe the students had meant 5 feet 3 inches and 6 feet, respectively, or maybe the 5.3 value was a typo for 53 inches.

What should you do with wild observations that you have clearly identified as being errors in data entry? The obvious answer is that these values do not belong and should not be included with the whole data set. You might be able to correct the mistakes by checking your original records (notes, data tables, photos). Good scientific practice always includes keeping clear and extensive records of data and how they were obtained.

Human error in experimentation or data collection

Sloppy experimentation methods can lead to unexpected results. If you forgot to add bacteria to one of your Petri dishes and find that nothing grew in it, that's just a silly mistake. But some experimental blunders lead to more interesting results.

Fleming, for instance, had less-than-ideal lab techniques. A few of his Petri dishes ended up being contaminated with a mold. Instead of simply throwing these away, he noticed a halo around the mold where no bacteria grew. He went on to cultivate the mold and discover its antibiotic properties, revolutionizing medicine and later earning a Nobel Prize. When Pasteur was studying chicken cholera, his assistant left some bacterial cultures out while he went on vacation. The dried-out cultures failed to kill inoculated chickens, as other cultures usually did. The assistant's first impulse was to discard the data, but Pasteur decided to take a closer look and explore the reason for this unexpected result. This work led him to understand the workings of the immune system and develop the vaccine.

Not all technical errors lead to fame or a Nobel Prize. Some mistakes are indeed not worth a second look. But sometimes the wild observations can be even more interesting than the original study. Again, you should always keep detailed notes of everything you do when gathering data, as this might help you identify mistakes and understand how they arose. Either way, these kinds of wild observations do not belong with the rest of your data (it would be like comparing apples and oranges). If you suspect an experimental error, the wild observation should be either ignored or studied separately.

Unexplainable but apparently legitimate wild observations

Most studies in the life sciences are conducted by collecting data about a small sample taken from the whole population of interest. Because of this, it can be difficult to determine whether a suspected outlier in a sample truly is a wild observation or just the consequence of studying only a small subset of the population. When you find a suspected outlier in a sample and have ruled out human error, you are faced with the challenging task of deciding what to do with it.

This is an important step because, for many statistical procedures, outliers are influential and can distort conclusions. Running the analysis with and then again without the outlier can help you determine whether the outlier affects your conclusions substantially or not. We will see in future chapters that some statistical methods are robust against mild outliers. That is, the method will be valid despite the presence of a mild outlier. Extreme outliers, however, are always a cause of concern. Many complex statistical approaches have been devised to deal with outliers. Most are well beyond the scope of this introductory textbook. Nonparametric tests, described in optional Chapter 27, are just a few examples. For simple studies, however, deciding what to do with a suspected outlier typically boils down to deciding whether to include the wild observation with the rest of the data or not.

How extreme is too extreme? In 1949, Mr. Hadlum sought divorce on the basis that his wife had given birth 349 days after his departure to war. He postulated that it was evidence of adultery because the average gestation time is 280 days. The judges ruled that such a long gestation time was unlikely but not impossible and denied his appeal. For comparison, while the average men's height in the United States is 1.76 m (5′9″), the *Guinness Book of World Records* lists the tallest man in recorded history as a 2.72 m tall American (8′11″).

To some extent, deciding whether to include or exclude wild observations in your analysis depends on the purpose of your study. Are you more interested in describing a population in its entire extent or only in its most typical, general pattern? For example, before Pasteur invented the pasteurization method to prevent the spoilage of liquids, wine and beer producers often had some batches that developed a troublesome sour taste. If you were studying the performance of various wine production methods, you would want to include bad batches in your analysis because they too reflect the performance of the method. On the other hand, if you were studying the chemical composition of wines and beers to better understand their health effects, you would not include sour batches, because these would never be sold for consumption. Obviously, you should never discard wild observations simply because they lead to conclusions you do not like.

Lastly, and perhaps most importantly, whether you choose to keep or to discard them, all outliers should be disclosed. Outliers can be very influential or have a tremendous scientific relevance, and you should not act as if they did not exist. You should also explain why you chose to keep or discard them. Honesty and full disclosure are the foundation of a good scientific methodology.

Using technology

Although a "two-variables statistics" calculator will do the basic calculations we need, more elaborate tools are helpful. Graphing calculators and computer software will do calculations and make graphs as you command, freeing you to concentrate on choosing the right methods and interpreting your results. Figure 2.5 displays output describing the 15 Aleppo pine needle lengths of Example 2.2. Can you find $\bar{x}$, s, and the five-number summary in each output? The big message of this section is: *Once you know what to look for, you can read output from any technological tool.*

The displays in Figure 2.5 come from the TI-83 graphing calculator, three statistical programs, and the Microsoft Excel spreadsheet program. The statistical programs are CrunchIt!, Minitab, and SPSS. Statistical programs allow you to choose what descriptive measures you want. Excel gives some things we don't need. Just ignore the extras. Excel's "Descriptive Statistics" menu item doesn't give the quartiles. We used the spreadsheet's separate quartile function to get Q_1 and Q_3.

EXAMPLE 2.9 What are the quartiles?

In Example 2.5, we saw that the quartiles of the Aleppo pine needle lengths are $Q_1 = 8.5$ and $Q_3 = 10.9$. Look at the output displays in Figure 2.5. The TI-83, Minitab, and CrunchIt! all agree with our work. Excel says that $Q_1 = 8.6$ and $Q_3 = 10.55$. What happened? *There are several rules for finding the quartiles. Some software packages use rules that give results different from ours for some sets of data.* SPSS provides the results from both methods. Our rule is simplest for hand computation. Results from the various rules are always close to each other, so *to describe data you should just use the answer your technology gives you.* ■

Organizing a statistical problem

Most of our examples and exercises have aimed at helping you learn basic tools (graphs and calculations) for describing and comparing distributions. You have also learned principles that guide the use of these tools, such as "always start with

TI-83

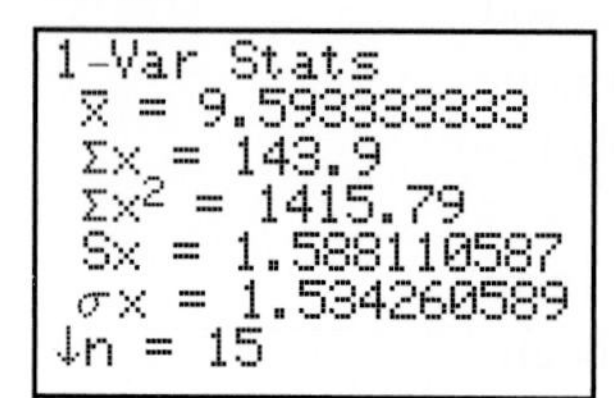
1-Var Stats
$\bar{x}$ = 9.593333333
Σx = 143.9
Σx^2 = 1415.79
Sx = 1.588110587
σx = 1.534260589
↓n = 15

1-Var Stats
↑n = 15
minX = 7.2
Q_1 = 8.5
Med = 9.3
Q_3 = 10.9
maxX = 12.8

SPSS

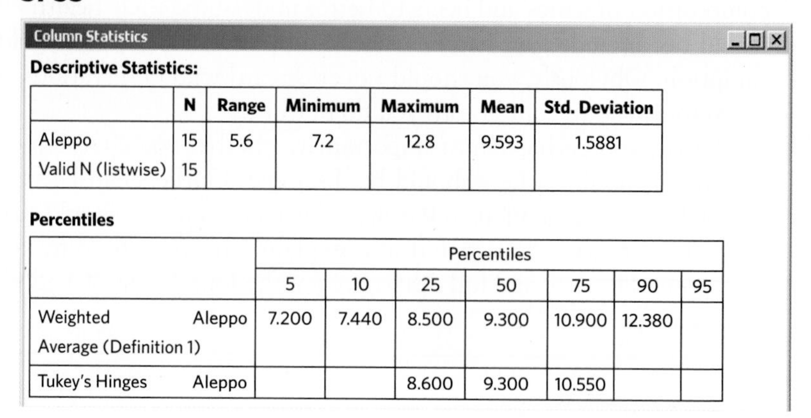
Column Statistics

Descriptive Statistics:

	N	Range	Minimum	Maximum	Mean	Std. Deviation
Aleppo	15	5.6	7.2	12.8	9.593	1.5881
Valid N (listwise)	15					

Percentiles

		Percentiles						
		5	10	25	50	75	90	95
Weighted Average (Definition 1)	Aleppo	7.200	7.440	8.500	9.300	10.900	12.380	
Tukey's Hinges	Aleppo			8.600	9.300	10.550		

Minitab

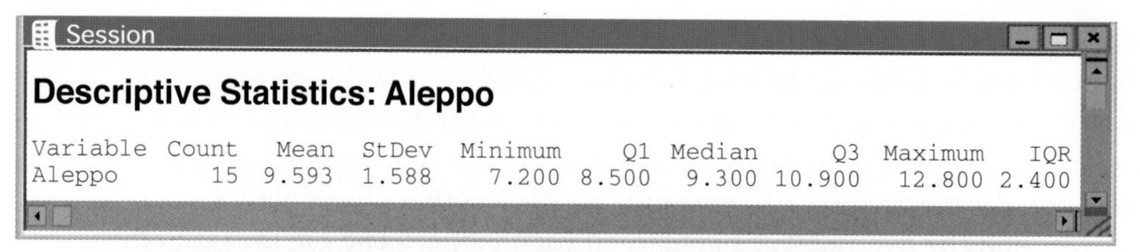
Session

Descriptive Statistics: Aleppo

Variable	Count	Mean	StDev	Minimum	Q1	Median	Q3	Maximum	IQR
Aleppo	15	9.593	1.588	7.200	8.500	9.300	10.900	12.800	2.400

Microsoft Excel

	A	B	C	D	
1	*Aleppo pine*				
2					
3	Mean	9.593333			
4	Standard Error	0.410048			
5	Median	9.3			
6	Mode	8.5			
7	Standard Deviation	1.588111			
8	Sample Variance	2.522095			
9	Kurtosis	-0.15687			
10	Skewness	0.632432			
11	Range	5.6			
12	Minimum	7.2			
13	Maximum	12.8			
14	Sum	143.9			
15	Count	15			
16	Q1	8.6	= QUARTILE(A2:A16,1)		
17	Q3	10.55	= QUARTILE(A2:A16,3)		

CrunchIt!

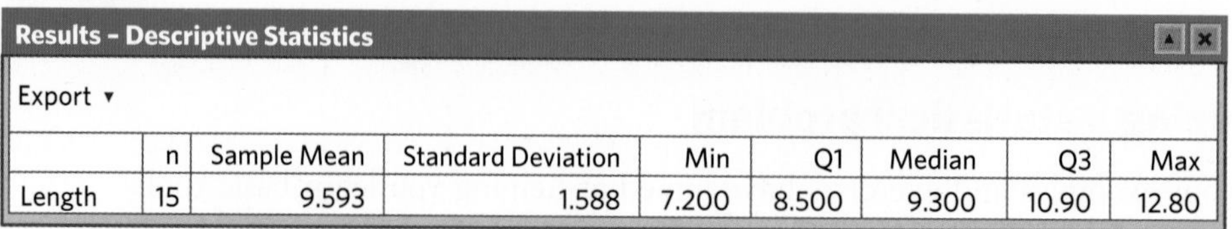
Results – Descriptive Statistics

Export ▾

	n	Sample Mean	Standard Deviation	Min	Q1	Median	Q3	Max
Length	15	9.593	1.588	7.200	8.500	9.300	10.90	12.80

FIGURE 2.5 Output from a graphing calculator, a spreadsheet program, and three software packages describing the Aleppo pine needle length data.

a graph" and "look for the overall pattern and striking deviations from the pattern." The data you work with are not just numbers. They describe specific settings, such as water salinity in the Everglades or needle length for two species of pine trees. Because data come from a specific setting, the final step in examining data is a conclusion for that setting. Water depth in the Everglades has a yearly cycle that reflects Florida's wet and dry seasons. Needle lengths are longer in Torrey pine trees than in Aleppo pine trees.

As you learn more statistical tools and principles, you will face more complex statistical problems. Although no framework accommodates all the varied issues arising in applying statistics to real settings, the following four-step thought process gives useful guidance. In particular, the first and last steps emphasize that statistical problems are tied to specific real-world settings and therefore involve more than doing calculations and making graphs.

ORGANIZING A STATISTICAL PROBLEM: THE FOUR-STEP PROCESS

STATE: What is the practical question, in the context of the real-world setting?

PLAN: What specific statistical operations does this problem call for?

SOLVE: Make the graphs and carry out the calculations needed for this problem.

CONCLUDE: Give your practical conclusion in the setting of the real-world problem.

To help you master the basics, many of our exercises will continue to tell you what to do—make a histogram, find the five-number summary, and so on. Real statistical problems don't come with detailed instructions. You will meet some exercises that are more realistic, especially in the later chapters of the book. Use the four-step process as a guide to solving and reporting these problems. They are marked with the four-step icon, as the following example illustrates.

EXAMPLE 2.10 Highly Superior Autobiographical Memory

STATE: Some individuals have the ability to recall accurately vast amounts of autobiographical information without mnemonic tricks or extra practice. This ability is called HSAM, for Highly Superior Autobiographical Memory. A study recruited 11 adults with confirmed HSAM and 15 control individuals of similar age without HSAM. All study participants were given a battery of cognitive and behavioral tests in the hope of finding out what might explain this extraordinary ability.[8]

To begin with, autobiographical memory was assessed by asking each participant to recall in detail 5 important personal events. These events were selected by the researchers, and the participants did not know ahead of time what events they would have to recall. Answer accuracy was verified from documents and interviews, and each correct detail was scored as one point. Individual total scores on this verifiable autobiographical memory task are displayed below.

HSAM	22	23	26	26	33	34	38	39	39	46	47				
Control	5	5	5	6	7	7	10	10	11	12	13	16	18	22	23

Do individuals with HSAM and without HSAM display distinct distributions of verifiable autobiographical memory scores? How do the mean scores compare?

PLAN: Use graphs and numerical descriptions to describe and compare these two distributions of verifiable autobiographical memory scores.

SOLVE: Dotplots work best for data sets of these sizes. Figure 2.6 displays stacked dotplots to facilitate comparison. The control group has a somewhat right-skewed distribution, so we might choose to compare the five-number summaries. But because the researchers plan to use $\overline{x}$ and s for further analysis, we instead calculate these measures:

Group	Mean score	Standard deviation
HSAM	33.9	8.8
Control	11.3	6.0

CONCLUDE: The two groups differ so much in verifiable autobiographical memory scores that there is little overlap among them. Overall, individuals with HSAM have higher scores than control individuals without HSAM. The mean scores are 33.9 for the HSAM group compared with only 11.3 for the control group, despite similar variation in individual scores (standard deviations 8.8 and 6.0, respectively). This first test confirms that individuals with HSAM have better recall of autobiographical events than ordinary individuals do. The researchers can now proceed to compare cognitive and behavioral traits in both groups. ■

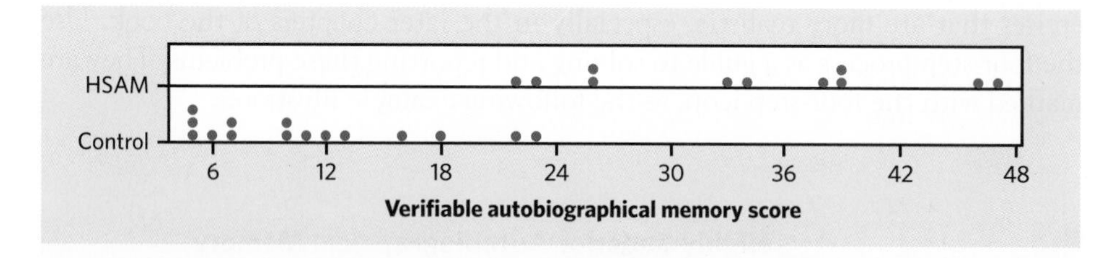

FIGURE 2.6 Dotplots comparing the distributions of scores on a test of verifiable autobiographical memory, for Example 2.10.

APPLY YOUR KNOWLEDGE

2.12 **Highly Superior Autobiographical Memory, continued.** The study in Example 2.10 assessed common obsessional symptoms in the HSAM and the control individuals, as previous findings had suggested a possible obsessional component to HSAM. The study participants completed the Leyton Obsessional Inventory test, scored out of 30 points. The results for the 11 HSAM individuals and for 14 of the controls are displayed below (1 individual in the control group left before completing this test).

HSAM	2	4	5	8	8	9	9	10	11	11	12			
Control	1	2	2	2	2	4	4	4	5	7	8	8	12	12

How do individuals with and without HSAM compare in terms of obsessional score? Follow the four-step process in reporting your work.

CHAPTER 2 SUMMARY

- A numerical summary of a distribution should report at least its **center** and its **spread** or **variability.**
- The **mean** $\bar{x}$ and the **median** M describe the center of a distribution in different ways. The mean is the arithmetic average of the observations, and the median is the midpoint of the values.
- When you use the median to indicate the center of the distribution, describe its spread by giving the **quartiles.** The **first quartile** Q_1 has one-fourth of the observations below it, and the **third quartile** Q_3 has three-fourths of the observations below it.
- The **five-number summary** consisting of the median, the quartiles, and the smallest and largest individual observations provides a quick overall description of a distribution. The median describes the center, and the quartiles and extremes show the spread.
- **Boxplots** based on the five-number summary are useful for comparing several distributions. The box spans the quartiles and shows the spread of the central half of the distribution. The median is marked within the box. Lines extend from the box to the extremes and show the full spread of the data.
- The **variance** s^2 and especially its square root, the **standard deviation** s, are common measures of spread about the mean as center of a sample data set. The standard deviation s is zero when there is no spread and gets larger as the spread increases.
- A **resistant measure** of any aspect of a distribution is relatively unaffected by changes in the numerical value of a small proportion of the total number of observations, no matter how large these changes are. The median and quartiles are resistant, but the mean and the standard deviation are not.
- The mean and standard deviation are good descriptions for symmetric distributions without outliers. They are most useful for the Normal distributions, which we will describe in Chapter 11. The five-number summary is a better description for skewed distributions.
- Numerical summaries do not fully describe the shape of a distribution. Always plot your data.
- A statistical problem has a real-world setting. You can organize many problems using the four steps **State, Plan, Solve,** and **Conclude.**

THIS CHAPTER IN CONTEXT

In this chapter we have continued our study of exploratory data analysis. Numerical summaries can be useful for describing a single distribution as well as for comparing the distributions of several groups of observations.

Two important features of a distribution are its center and its spread. The mean and standard deviation are excellent numerical summaries for distributions that are approximately symmetric without outliers. If a distribution is not symmetric, has outliers, or both, the five-number summary provides a better, more comprehensive description. But if a distribution is complex, with clusters or multiple peaks, for instance, reducing the information of the distribution to a few numbers would be misleading. This is one reason why data should always be graphed before computing and communicating numerical summaries.

We will see later, starting with Chapter 14, how numerical summaries from random samples are used for estimating population parameters or for comparing several population distributions. Because of this, it will be important to distinguish the notation for the mean $\bar{x}$ and standard deviation s of a sample from the notation for the mean and standard deviation of a population (μ and σ, respectively).

CHECK YOUR SKILLS

2.13 Cesium-137 is a waste product of nuclear reactors. A study examined the cesium-137 tissue concentration of a random sample of 15 Pacific bluefin tuna, *Thunnus orientalis*, captured off the coast of California four months after the Fukushima (Japan) nuclear reactor meltdown of 2011. Here are the findings, in becquerels per kilogram of dry tissue:[9]

4.8 7 6.2 7.3 6.0 7.3 3.7 9.4 8.4 6.0 5.0 5.1 6.0 7.3 4.7

The mean of these data is

(a) 5.70. (b) 6.00. (c) 6.28.

2.14 The median of the data in Exercise 2.13 is

(a) 5.70. (b) 6.00. (c) 6.28.

2.15 (Optional) The interquartile range, *IQR*, of the data in Exercise 2.13 is

(a) 1.3. (b) 2.0. (c) 2.3.

2.16 If a distribution is clearly skewed to the right,

(a) the mean is less than the median.
(b) the mean and median are equal.
(c) the mean is greater than the median.

2.17 What percent of the observations in a distribution lie between the first quartile and the third quartile?

(a) 25% (b) 50% (c) 75%

2.18 To make a boxplot of a distribution, you must know

(a) all the individual observations.
(b) the mean and the standard deviation.
(c) the five-number summary.

2.19 What are all the values that a standard deviation s can possibly take?

(a) $0 \le s$
(b) $0 \le s \le 1$
(c) $-1 \le s \le 1$

2.20 The standard deviation of the 15 cesium-137 values in Exercise 2.13 (use your calculator) is about

(a) 1.47. (b) 1.52. (c) 2.32.

2.21 The correct units for the standard deviation of the 15 cesium-137 values in Exercise 2.13 are

(a) no units—it's just a number.
(b) becquerels per kilogram of dry tissue.
(c) becquerels squared per kilogram squared of dry tissue.

2.22 Which of the following is least affected if an extreme high outlier is added to your data?

(a) the median
(b) the mean
(c) the standard deviation

CHAPTER 2 EXERCISES

2.23 Florida lakes. Alkalinity of lake waters is related to carbon content, which is mostly derived from ground and algal sources, and plays an important role in aquatic ecosystems. Water analysis of 53 Florida lakes[10] resulted in a mean alkalinity of 37.53 milliequivalents per liter (meq/l) and a median alkalinity of 19.60 meq/l. What aspect, or aspects, of the data distribution might explain the difference between these two measures of center?

2.24 Food oils and health. Table 1.2 (page 34) gives the ratio of omega-3 to omega-6 fatty acids in common food oils. Exercise 1.34 asked you to plot the data. The distribution is strongly right-skewed with a high outlier. Do you expect the mean to be greater than the median, about equal to the median, or less than the median? Why? Calculate $\bar{x}$ and M and verify your expectation.

2.25 Lyme disease. Figure 1.7 (page 19) shows the age in years of 241,931 patients diagnosed with Lyme disease in the United States between 1992 and 2006. Give a brief description of the important features of the distribution. Explain why no numerical summary would appropriately describe this distribution.

2.26 Making resistance visible. In the *Mean and Median* applet, place three observations on the line by clicking below it: two close together near the center of the line, and one somewhat to the right of these two.

APPLET

(a) Pull the single rightmost observation out to the right. (Place the cursor on the point, hold down a mouse button, and drag the point.) How does the mean behave? How does the median behave? Explain briefly why each measure acts as it does.

(b) Now drag the single rightmost point to the left as far as you can. What happens to the mean? What happens to the median as you drag this point past the other two (watch carefully)?

2.27 Anorexia nervosa. Figure 1.8 (page 20) is a histogram of the distribution of age at onset of anorexia for 691 Canadian girls diagnosed with the disorder. If you round the age to whole numbers of years, the first bar of the histogram (the first class) would include all girls diagnosed during their 11th year. With a little care, you can find the median and the quartiles from the histogram. What are these numbers? How did you find them?

2.28 Weight of newborns. Here is the distribution of the weight at birth for all babies born in the United States in 2010:[11]

Weight (grams)	Count	Weight (grams)	Count
Less than 500	5,980	3,000 to 3,499	1,566,755
500 to 999	22,015	3,500 to 3,999	1,055,004
1,000 to 1,499	29,846	4,000 to 4,499	262,997
1,500 to 1,999	63,427	4,500 to 4,999	36,706
2,000 to 2,499	204,295	5,000 to 5,499	4,216
2,500 to 2,999	744,181		

(a) For comparison with other years and with other countries, we prefer a histogram of the *percents*, rather than the counts, in each weight class. Explain why.

(b) How many babies were there? Make a histogram of the distribution, using percents on the vertical scale.

(c) What are the positions of the median and quartiles in the ordered list of all birth weights? In which weight classes do the median and quartiles fall?

2.29 Metabolic rate. In Example 2.8 you examined the metabolic rates of 7 men. Here are the metabolic rates for 12 women from the same study:

995 1425 1396 1418 1502 1256 1189
913 1124 1052 1347 1204

(a) The most common methods for formal comparison of two groups use $\bar{x}$ and s to summarize the data. What kinds of distributions are best summarized by $\bar{x}$ and s?

(b) Make a summary graph comparing the metabolic rates of the 7 men and 12 women, as in Figure 2.4. What can you conclude about these two groups from your graph?

2.30 Behavior of the median. Place five observations on the line in the *Mean and Median* applet by clicking below it.

APPLET

(a) Add one additional observation *without changing the median*. Where is your new point?

(b) Use the applet to convince yourself that when you add yet another observation (there are now seven in all), the median does not change no matter where you put the seventh point. Explain why this must be true.

2.31 Nanomedicine. In Exercise 1.37 (page 35) you graphed the distribution of ovarian tumor increases under two experimental conditions: a new nanoparticle-based delivery system for a suicide gene therapy or an inactive buffer solution.

(a) Make a boxplot comparing tumor increase under the two conditions and compute the mean and standard deviation for each condition.

(b) Write a short description of the experimental results based on your work in (a).

2.32 Guinea pig survival times. Example 1.7 (page 18) provides the numerical values for the guinea pig survival times after inoculation plotted in Figure 1.6. As often with survival times, this distribution is strongly skewed to the right.

(a) Which numerical summary would you choose for these data? Explain why. Does it show the expected right-skew?

(b) Calculate your chosen summary (Example 2.4 already provides the mean and median for this data set). How does it reflect the skewness of the distribution?

2.33 A standard deviation contest. This is a standard deviation contest. You must choose four numbers from the whole numbers 0 to 10, with repeats allowed.

(a) Choose four numbers that have the smallest possible standard deviation.

(b) Choose four numbers that have the largest possible standard deviation.

(c) Is more than one choice possible in either (a) or (b)? Explain.

2.34 You create the data. Create a set of five positive numbers (repeats allowed) that have median 10 and mean 7. What thought process did you use to create your numbers?

2.35 You create the data. Give an example of a small set of data for which the mean is larger than the third quartile.

2.36 Does breast-feeding weaken bones? 4 STEP Breast-feeding mothers secrete calcium into their milk. Some of the calcium may come from their bones, so mothers may lose bone mineral content. Researchers compared 47 breast-feeding women with 22 women of similar age who were neither pregnant nor lactating. They measured the percent change in the mineral content of the women's spines over three months. A negative value indicates a loss in mineral content. Here are the data:[12]

Breast-feeding women							
−4.7	−0.3	−2.0	−6.5	−2.7	1.7	−4.4	−5.7
−8.3	0.4	−4.9	−6.5	−4.3	−2.2	−5.9	−3.6
−3.1	−2.5	−6.2	−2.1	−1.8	−0.8	0.3	−3.3
−7.0	−2.1	−5.3	−4.7	−1.0	2.2	−5.1	−2.5
−5.2	−1.0	−4.9	−6.8	−5.6	−5.2	−5.3	−2.3
−4.0	−2.2	−6.8	0.2	−3.8	−3.0	−7.8	

Other women							
2.4	2.2	−1.6	1.1	−0.2	−0.1	0.7	−0.4
2.9	0.0	−0.4	−0.1	−0.1	1.0	1.7	
1.2	−0.6	0.9	−2.2	−1.5	−0.4	0.3	

Do the data show distinctly greater bone mineral loss among the breast-feeding women? Be sure to consider in your interpretation that a bone mineral loss is reflected by a negative value here. Follow the four-step process illustrated by Example 2.10.

2.37 Compressing soil. 4 STEP Farmers know that driving heavy equipment on wet soil compresses the soil and is detrimental to future crops. Table 2.1 gives data on the "penetrability" of the same soil at three levels of compression.[13] Penetrability is a measure of how much resistance plant roots will meet when they try to grow through the soil. How does increasing compression affect penetrability? Follow the four-step process in your work.

2.38 Daily activity and obesity. 4 STEP People gain weight when they take in more energy from food than they expend. Table 2.2 compares volunteer subjects who were lean with others who were mildly obese. None of the subjects followed an exercise program. The subjects wore sensors that

TABLE 2.1 Penetrability of soil at three compression levels

Compressed	2.86	2.68	2.92	2.82	2.76	2.81	2.78	3.08	2.94	2.86
	3.08	2.82	2.78	2.98	3.00	2.78	2.96	2.90	3.18	3.16
Intermediate	3.13	3.38	3.10	3.40	3.38	3.14	3.18	3.26	2.96	3.02
	3.54	3.36	3.18	3.12	3.86	2.92	3.46	3.44	3.62	4.26
Loose	3.99	4.20	3.94	4.16	4.29	4.19	4.13	4.41	3.98	4.41
	4.11	4.30	3.96	4.03	4.89	4.12	4.00	4.34	4.27	4.91

TABLE 2.2 Time (minutes per day) active and lying down by lean and obese subjects

Lean Subjects			Obese Subjects		
Subject	Stand/walk	Lie	Subject	Stand/walk	Lie
1	511.100	555.500	11	260.244	521.044
2	607.925	450.650	12	464.756	514.931
3	319.212	537.362	13	367.138	563.300
4	584.644	489.269	14	413.667	532.208
5	578.869	514.081	15	347.375	504.931
6	543.388	506.500	16	416.531	448.856
7	677.188	467.700	17	358.650	460.550
8	555.656	567.006	18	267.344	509.981
9	374.831	531.431	19	410.631	448.706
10	504.700	396.962	20	426.356	412.919

recorded every move for 10 days. The table shows the average minutes per day spent in activity (standing and walking) and in lying down.[14] Compare the distributions of time spent actively for lean and obese subjects and also the distributions of time spent lying down. How does the behavior of lean and mildly obese people differ? Follow the four-step process in your work.

2.39 Logging in the rain forest. "Conservationists have despaired over destruction of tropical rainforest by logging, clearing, and burning." These words begin a report on a statistical study of the effects of logging in Borneo.[15] Researchers compared forest plots that had never been logged (Group 1) with similar plots nearby that had been logged 1 year earlier (Group 2) and 8 years earlier (Group 3). All plots were 0.1 hectare in area. Here are the counts of trees for plots in each group:

Group 1	27	22	29	21	19	33	16	20	24	27	28	19
Group 2	12	12	15	9	20	18	17	14	14	2	17	19
Group 3	18	4	22	15	18	19	22	12	12			

To what extent has logging affected the count of trees? Follow the four-step process in reporting your work.

Exercises 2.40 to 2.43 make use of the optional material on the $1.5 \times IQR$ rule for suspected outliers.

2.40 Carbon dioxide emissions. Table 1.3 (page 35) gives carbon dioxide (CO_2) emissions per person for countries with populations of at least 20 million. A dotplot (or a stemplot) shows that the distribution is strongly skewed to the right. The United States and several other countries appear to be high outliers.

(a) Give the five-number summary. Explain why this summary suggests that the distribution is right-skewed.

(b) Which countries are suspected outliers according to the $1.5 \times IQR$ rule? Make a dotplot or a stemplot of the data (or look at your graph from Exercise 1.35, page 34). Do you agree with the rule's suggestions about which countries are and are not outliers?

2.41 Aggression and social status in macaque monkeys. The boxplots in Figure 2.7 summarize the number of aggressive behaviors directed at female macaque monkeys of differing social status living in experimentally arranged groups.[16] These are modified boxplots that indicate suspected outliers by a dot, using the $1.5 \times IQR$ rule. Based on these plots, describe how "aggressions received" by female macaque monkeys vary depending on their dominance rank within the group.

2.42 Golden orb weavers. Exercise 2.1 described a study of female golden orb weaver spiders. The study also reported the body mass (in grams) for each of the 21 spiders. Here are the data:

0.04	0.11	0.16	0.07	0.13	0.1	0.17
0.25	0.36	0.33	0.29	0.14	0.32	0.57
0.31	0.79	0.49	0.64	0.6	0.99	0.81

(a) Give the five-number summary, mean, and standard deviation of this distribution. How do the mean and the median compare?

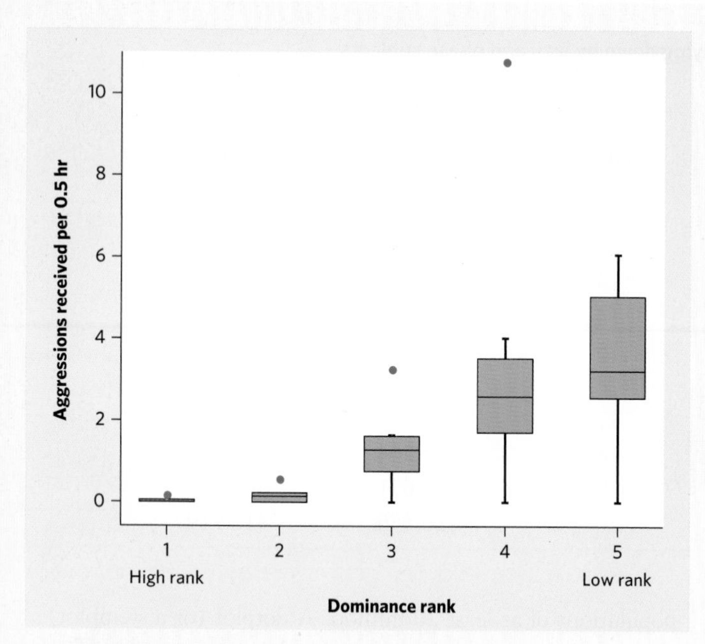

FIGURE 2.7 Boxplots of the distributions of number of aggressions received per half hour, for female macaque monkeys of differing social status, for Exercise 2.41.

(b) Does the 1.5 × IQR rule identify any suspected outliers?

(c) What do the results in (a) and (b) suggest about the distribution of body mass in female golden orb weavers? Make a dotplot or a stemplot to confirm your interpretation.

2.43 Young Americans. The stemplot in Exercise 1.19 (page 31) displays the distribution of the percents of individuals under the age of 18 in each of the 50 states and the District of Columbia. Stemplots help you find the five-number summary because they arrange the observations in increasing order.

(a) Give the five-number summary of this distribution.

(b) Does the 1.5 × IQR rule identify the District of Columbia and Utah as suspected outliers? Does it also flag any other states?

(c) Find the mean percent of young Americans in the 50 states and DC, and compare it with the median. Explain your result.

2.44 Lung capacity of children. Continue your work from Exercise 1.44 (page 37) with the data file *Large.FEV*.

LARGE DATA SET

(a) Obtain the mean, standard deviation, and five-number summary for each of the eight combinations of sex and age group.

(b) Make a side-by-side boxplot comparing the eight combinations of sex and age group.

(c) Write a short paragraph describing the differences between girls and boys by age group, referencing your work in (a) and (b).

2.45 Elderly health. A study examined the medical records of elderly patients to determine whether there are differences between men and women in their calcium or inorganic phosphorus blood levels (both in millimoles per liter, mmol/l). The data file *Large.Calcium* contains the data.[17]

LARGE DATA SET

(a) Make side-by-side boxplots comparing the calcium levels for men ($sex = 1$) and for women ($sex = 2$). Do the same for the inorganic phosphorus levels. What do you conclude?

(b) Another purpose of the study was to determine whether different laboratories yield substantially different results. Use boxplots to compare the calcium levels obtained by the six different labs. Do you see notable differences? Do the same for the inorganic phosphorus levels. Note that the labs had each received blood samples from a different set of patients. How does that impact your interpretation of the boxplots?

2.46 Estimating body fat in men. Continue your work from Exercise 1.45 (page 37) with the data file *Large.Bodyfat*.

LARGE DATA SET

(a) Make a boxplot for the men's body weights (in pounds). Look for suspected outliers using the 1.5 × IQR rule. Based on your histogram in Exercise 1.45(a), which are actual outliers? There is one extreme outlier. Is this nonetheless a plausible value (that is, do individuals like that exist)? Check all the values given for that man before you answer.

(b) Make a boxplot for the men's heights (in inches). Look for individual values that qualify as suspected outliers using the 1.5 × IQR rule. Based on your histogram in Exercise 1.47(b), which are actual outliers? There is one extreme outlier. Is this nonetheless a plausible value (that is, do individuals like that exist)? This was actually a typo made in recording the man's height, and this data point should be excluded when performing further analysis involving height.

Ragnar Th. Sigurdsson/age fotostock/SuperStock

CHAPTER 3 Scatterplots and Correlation

A medical study finds that lung capacity decreases with the number of cigarettes smoked per day. The Department of Motor Vehicles warns that alcohol consumption reduces reflex time and that the effect is larger as more alcohol is consumed. These and many other statistical studies look at the *relationship between two variables*. Statistical relationships are overall tendencies, not ironclad rules. They allow individual exceptions. Although smokers on the average die younger than nonsmokers, some people live to 90 while smoking three packs a day.

To understand a statistical relationship between two variables, we measure both variables on the same individuals. Often, we must examine other variables as well. To conclude that smoking undermines lung capacity, for example, the researchers had to eliminate the effect of other variables such as each person's size and exercise habits. In this chapter we begin our study of relationships between variables. One of our main themes is that the relationship between two variables can be strongly influenced by other variables that are lurking in the background.

IN THIS CHAPTER WE COVER...

- Explanatory and response variables
- Displaying relationships: scatterplots
- Interpreting scatterplots
- Adding categorical variables to scatterplots
- Measuring linear association: correlation
- Facts about correlation

Explanatory and response variables

We think that blood alcohol content helps explain variations in reflex time and that smoking influences lung capacity and, eventually, life expectancy. In each of these relationships, the two variables play different roles: One explains or influences the other.

RESPONSE VARIABLE, EXPLANATORY VARIABLE

A **response variable** measures an outcome of a study. An **explanatory variable** may explain or influence changes in a response variable.

independent variable
dependent variable

You will often find explanatory variables called **independent variables,** and response variables called **dependent variables.** The idea behind this language is that the response variable depends on the explanatory variable. Because "independent" and "dependent" have other meanings in statistics that are unrelated to the explanatory-response distinction, we prefer to avoid those words in this textbook.

It is easiest to identify explanatory and response variables when we actually set values of one variable in order to see how it affects another variable.

EXAMPLE 3.1 Beer and blood alcohol

How does drinking beer affect the level of alcohol in our blood? The legal limit for driving is now 0.08% in all states. Student volunteers at Ohio State University drank different numbers of cans of beer. Thirty minutes later, a police officer measured their blood alcohol content. Number of beers consumed is the explanatory variable, and percent of alcohol in the blood is the response variable. ■

When we don't set the values of either variable but just observe both variables, there may or may not be explanatory and response variables. Whether there are depends on how we plan to use the data.

EXAMPLE 3.2 Height

The National Center for Health Statistics (NCHS) surveys the American population and collects, for each individual in the survey, information about body height, age, sex, and a long list of attributes. The purpose of the NCHS survey is to document the characteristics of the American population. There is no explanatory or response variable in this context.

A pediatrician looks at the same data with an eye to using age and sex, along with other variables such as ethnicity, to discuss a child's growth. Now age and sex are explanatory variables, and height is the response variable. ■

In many studies, the goal is to show that changes in one or more explanatory variables actually *cause* changes in a response variable. But many explanatory-response relationships do not involve direct causation. The age and sex of a child can help predict future height, but they certainly do not cause a particular height.

Most statistical studies examine data on more than one variable. Fortunately, statistical analysis of several-variable data builds on the tools we used to examine individual variables. The principles that guide our work also remain the same:

- Plot your data. Look for overall patterns and deviations from those patterns.
- Based on what your plot shows, choose numerical summaries for some aspects of the data.

APPLY YOUR KNOWLEDGE

3.1 Explanatory and response variables? In each of the following situations, is it more reasonable to simply explore the relationship between the two variables or to view one of the variables as an explanatory variable and the other as a response variable? In the latter case, which is the explanatory variable and which is the response variable?

(a) The typical amount of calories a person consumes per day and that person's percent of body fat

(b) The weight in kilograms and height in centimeters of a person

(c) Inches of rain in the growing season and the yield of corn in bushels per acre

(d) A person's leg length and arm length, in centimeters

3.2 Coral reefs. How sensitive to changes in water temperature are coral reefs? To find out, scientists examined data on summer sea surface temperatures and coral growth per year at locations in the Red Sea.[1] What are the explanatory and response variables? Are they categorical or quantitative?

3.3 Beer and blood alcohol. Example 3.1 describes a study in which college students drank different amounts of beer. The response variable was their blood alcohol content (BAC). BAC for the same amount of beer might depend on other facts about the students. Name two other variables that could influence BAC.

Displaying relationships: scatterplots

The most useful graph for displaying the relationship between two quantitative variables is a *scatterplot*.

EXAMPLE 3.3 An endangered species: the manatee

Manatees are large, herbivorous, aquatic mammals found primarily in the rivers and estuaries of Florida. This endangered species suffers from cohabitation with human populations, and many manatees die each year from collisions with powerboats. Following our four-step process (page 57), let's look at the influence of the number of powerboats registered on manatee deaths from collisions with powerboats. We examine the relationship between the number of manatee deaths from powerboat collisions and the number of powerboats registered in any given year between 1977 and 2012, as displayed in Table 3.1.[2]

STATE: The number of powerboats registered in Florida varies from year to year. Does it help explain the differences from year to year in the number of manatee deaths from collisions with powerboats?

PLAN: Examine the relationship between powerboats registered and manatee deaths from collision. Choose the explanatory and response variables (if any). Make a scatterplot to display the relationship between the variables. Interpret the plot to understand the relationship.

SOLVE (FIRST STEPS): We suspect that "powerboats registered" will help explain "manatee deaths from collisions." So "powerboats registered" is the explanatory variable, and "manatee deaths from collisions" is the response variable. Time (year the data were gathered) is not a variable of interest here. We want to see how the variable manatee deaths changes when the variable powerboat registrations changes, so we put powerboat

Getty Images/Discovery Channel Images

TABLE 3.1 Powerboat registrations (in thousands) and manatee deaths from powerboat collisions in Florida

Year	Powerboats	Deaths	Year	Powerboats	Deaths	Year	Powerboats	Deaths	Year	Powerboats	Deaths
1977	447	13	1986	614	33	1995	713	42	2004	983	69
1978	460	21	1987	645	39	1996	732	60	2005	1010	79
1979	481	24	1988	675	43	1997	755	54	2006	1024	92
1980	498	16	1989	711	50	1998	809	66	2007	1027	73
1981	513	24	1990	719	47	1999	830	82	2008	1010	90
1982	512	20	1991	681	55	2000	880	78	2009	982	97
1983	526	15	1992	679	38	2001	944	81	2010	942	83
1984	559	34	1993	678	35	2002	962	95	2011	922	87
1985	585	33	1994	696	49	2003	978	73	2012	902	81

registrations (the explanatory variable, expressed in thousands) on the horizontal axis. Figure 3.1 is the scatterplot. Each point represents a single year. In 1997, for example, there were 755,000 powerboats registered and 54 manatee deaths due to powerboat collisions. Find 755 on the x (horizontal) axis and 54 on the y (vertical) axis. The year 1997 appears as the point (755, 54) above 755 and to the right of 54, as shown on Figure 3.1. ■

SCATTERPLOT

A **scatterplot** shows the relationship between two quantitative variables measured on the same individuals. The values of one variable appear on the horizontal axis, and the values of the other variable appear on the vertical axis. Each individual in the data appears as the point in the plot fixed by the values of both variables for that individual.

Always plot the explanatory variable, if there is one, on the horizontal axis (the x axis) of a scatterplot. As a reminder, we usually call the explanatory variable x and the response variable y. If there is no explanatory-response distinction, either variable can go on the horizontal axis.

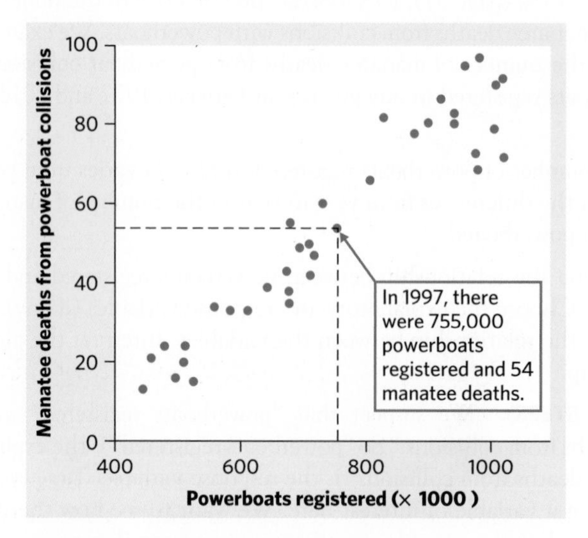

FIGURE 3.1 Scatterplot of the number of manatee deaths due to powerboat collisions in Florida each year against the number of powerboats registered (in thousands) that same year, for Examples 3.3 and 3.4. The dotted lines intersect at the point (755, 54), the data for year 1997.

APPLY YOUR KNOWLEDGE

3.4 Chickadee alarm calls. The black-capped chickadee (*Poecile atricapilla*) is a small songbird commonly found in the northern United States and Canada. Chickadees often live in cooperative flocks, using a complex language to communicate about food sources and predator threats. In an experiment, researchers recorded chickadee vocalizations in an aviary when presented with predators of various sizes. The following data represent the average number of D notes per chickadee warning call for each type of predator, along with the predator wingspan (in centimeters):[3]

Predator	Predator wingspan (cm)	Number of D notes per call	Predator	Predator wingspan (cm)	Number of D notes per call
Pygmy owl	31.2	3.96	Gyrfalcon	115.1	2.25
Saw-whet owl	38.8	4.09	Peregrine falcon	120.0	2.80
Kestrel	57.6	2.76	Red-tailed hawk	120.0	2.56
Merlin	60.6	3.04	Great horned owl	120.4	2.46
Cooper's hawk	80.6	3.18	Great gray owl	132.2	2.06
Short-eared owl	89.2	2.28	Rough-legged hawk	138.0	1.36
Prairie falcon	109.9	2.20			

Plot the number of D notes per call (response) against the predator wingspan (explanatory).

Do big skulls house smart brains? Nineteenth-century scientists thought that the volume of a human skull might be related to the intelligence of the skull's owner. Without imaging technology, it was difficult to measure a skull's volume accurately. Paul Broca, a professor of surgery, showed that filling an empty skull with small lead shot, then pouring out the shot and weighing it, gave quite accurate measurements of the skull's volume.

Interpreting scatterplots

To interpret a scatterplot, apply the strategies of data analysis learned in Chapters 1 and 2.

> **EXAMINING A SCATTERPLOT**
>
> In any graph of data, look for the **overall pattern** and for striking **deviations** from that pattern.
>
> You can describe the overall pattern of a scatterplot by the **direction, form,** and **strength** of the relationship.
>
> An important kind of deviation is an **outlier,** an individual value that falls outside the overall pattern of the relationship.

EXAMPLE 3.4 An endangered species: the manatee

SOLVE (INTERPRET THE PLOT): Figure 3.1 shows a clear *direction:* The overall pattern moves up, from lower left to upper right. That is, years in which powerboat registrations were higher tend to have higher counts of manatee deaths from collisions. We call this a positive association between the two variables. The *form* of the relationship is **linear.**

linear relationship

That is, the overall pattern follows a straight line from lower left to upper right. The *strength* of a relationship in a scatterplot is determined by how closely the points follow a clear form. The overall relationship in Figure 3.1 is strong.

CONCLUDE: The number of powerboat registrations explains much of the variation among manatee deaths from collisions. Years that had fewer powerboats registered also tended to have fewer accidental manatee deaths. To preserve this endangered species, restricting the number of powerboats registered might be helpful. However, the scatterplot in Figure 3.1 does not take into consideration other factors, such as speed limits, fines, or driver education, that might also be influential. Our conclusions are limited to the available data and what they say about the relationship between powerboat registrations and manatee accidents. ■

POSITIVE ASSOCIATION, NEGATIVE ASSOCIATION

Two variables are **positively associated** when above-average values of one tend to accompany above-average values of the other, and below-average values also tend to occur together.

Two variables are **negatively associated** when above-average values of one tend to accompany below-average values of the other, and vice versa.

Robert Daly/Getty Images

EXAMPLE 3.5 Does fidgeting keep you slim?

Obesity is a growing problem around the world. Surprisingly, some people don't gain weight even when they overeat. Perhaps fidgeting and other "nonexercise activity" (NEA) explains why—some people may spontaneously increase nonexercise activity when fed more. Researchers deliberately overfed 16 healthy young adult volunteers for 8 weeks. They measured fat gain (in kilograms, kg) and, as an explanatory variable, change in energy use (in kilocalories, or Calories, Cal) from activity other than deliberate exercise—fidgeting, daily living, and the like. Here are the data:[4]

NEA change (Cal)	−94	−57	−29	135	143	151	245	355
Fat gain (kg)	4.2	3.0	3.7	2.7	3.2	3.6	2.4	1.3
NEA change (Cal)	392	473	486	535	571	580	620	690
Fat gain (kg)	3.8	1.7	1.6	2.2	1.0	0.4	2.3	1.1

Do people with larger increases in NEA tend to gain less fat? To see the pattern in the data, we make a scatterplot with NEA on the horizontal axis (the explanatory variable), as displayed in Figure 3.2.

The plot shows a negative association. That is, people with larger increases in NEA tend to gain less fat. The form of the association between NEA and fat gain is linear. The association is moderately strong because the points make up a linear pattern but deviate quite a lot from the line. ■

Of course, not all relationships have a simple form and a clear direction that we can describe as positive association or negative association. Exercise 3.6 gives an example that does not have a single direction.

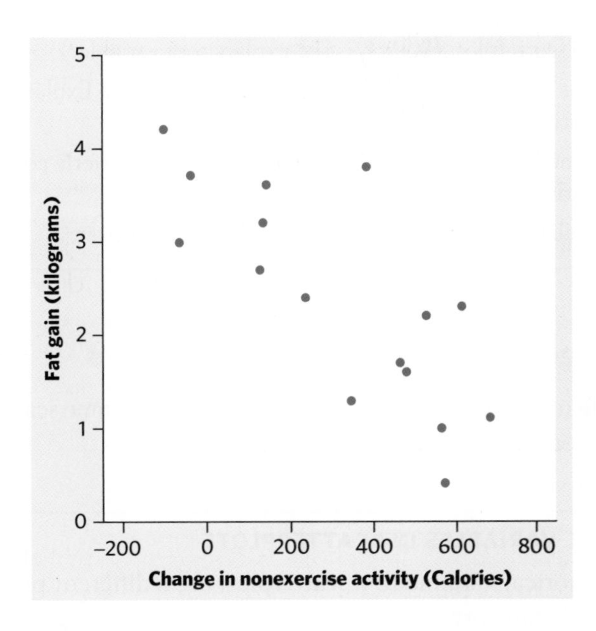

FIGURE 3.2 Scatterplot of fat gain after 8 weeks of overeating plotted against increase in nonexercise activity over the same period, for Example 3.5.

APPLY YOUR KNOWLEDGE

3.5 **Chickadee alarm calls, continued.** Describe the form, direction, and strength of the relationship between the number of D notes per call and predator wingspan, as displayed in your plot from Exercise 3.4. What does the relationship suggest about chickadee warning calls?

3.6 **Enzyme activity and temperature.** Enzymatic activity is known to be affected by temperature. A study examined the activity rate (in micromoles per second, μmol/s) of the digestive enzyme acid phosphatase in vitro at varying temperatures (measured in kelvins, K). The findings are displayed below.[5] (For reference, the range of 253 to 298 K corresponds to roughly 25 to 80 degrees Celsius, or 77 to 176 degrees Fahrenheit.)

Temperature (K)	Rate (μmol/s)	Temperature (K)	Rate (μmol/s)	Temperature (K)	Rate (μmol/s)
298	0.04	318	0.34	338	1.05
298	0.05	318	0.34	338	1.10
298	0.05	318	0.35	338	1.06
303	0.08	323	0.47	343	0.96
303	0.08	323	0.48	343	0.98
303	0.08	323	0.48	343	0.99
308	0.11	328	0.78	348	0.72
308	0.13	328	0.78	348	0.73
308	0.11	328	0.79	348	0.75
313	0.18	333	0.97	353	0.54
313	0.20	333	0.98	353	0.58
313	0.19	333	1.03	353	0.62

(a) Make a scatterplot. (Which is the explanatory variable?)

(b) Describe the form of the relationship. It is not linear. Explain why the form of the relationship makes sense.

(c) It does not make sense to describe the variables as strictly positively associated or negatively associated. Why?

(d) Is the relationship strong or weak? Explain your answer.

Adding categorical variables to scatterplots

It is also possible to plot two or more relationships on the same scatterplot by using a different color or a different symbol for each.

CATEGORICAL VARIABLES IN SCATTERPLOTS

To add a categorical variable to a scatterplot, use a different plot color or symbol for each category.

Doing so allows us to examine visually the effect of a third variable on the relationship between x and y. First, describe the relationship between x and y separately for each category (that is, for each color or symbol). Then compare and contrast these relationships.*

Herman Eisenbeiss/Photo Researchers

EXAMPLE 3.6 The cost of reproduction in male fruit flies

Longevity in male fruit flies is positively associated with adult size. But do other factors, such as sexual activity, also matter? The cost of reproduction is well documented for the females of the species. A study looks at the association between longevity and adult size in male fruit flies kept under one of two conditions. One group is kept with sexually active females over their life span. The other group is cared for in the same way but kept with females that are not sexually active.[6]

Figure 3.3 shows the scatterplot of longevity (response) versus adult thorax length (indicative of body size, explanatory) for both conditions. The conditions are coded so that individual fruit flies from the sexually active group are represented by triangles, while the sexually inactive fruit flies from the second group are represented by circles. This coding introduces a third variable into the scatterplot. "Condition" is a categorical variable that has two values, identified by the two different plotting symbols.

The scatterplot is quite clear. Both groups show a positive linear relationship between thorax length and longevity, as expected. However, the sexually active fruit flies have, on the whole, a lower longevity. That is, fruit flies of a given size tend to die sooner in the sexually active group than in the inactive group. In fact, only one male in the sexually active group outlived a male in the inactive group with similar thorax length. ■

---

*Optional Chapter 28 on multiple regression provides more detailed coverage of this topic.

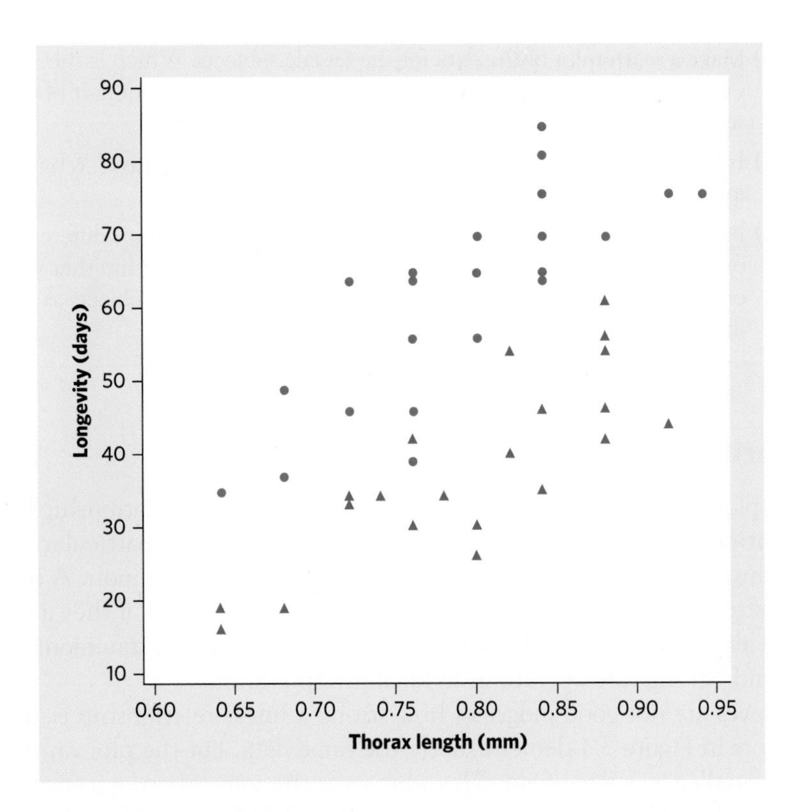

FIGURE 3.3 Cost of reproduction: scatterplot of thorax length (in millimeters) and longevity (in days) for sexually active (triangles) or inactive (circles) male fruit flies, for Example 3.6.

APPLY YOUR KNOWLEDGE

3.7 **Do heavier people burn more energy?** Metabolic rate, the rate at which the body consumes energy, is important in studies of weight gain, dieting, and exercise. Here are data on the lean body mass and resting metabolic rate for 12 women and 7 men who are subjects in a study of dieting. Lean body mass, given in kilograms, is a person's weight leaving out all fat. Metabolic rate is measured in kilocalories (Cal) burned per 24 hours (the same calories used to describe the energy content of foods). Researchers believe that lean body mass has an important influence on metabolic rate.

Subject	Sex	Mass (kg)	Rate (Cal)	Subject	Sex	Mass (kg)	Rate (Cal)
1	M	62.0	1792	11	F	40.3	1189
2	M	62.9	1666	12	F	33.1	913
3	F	36.1	995	13	M	51.9	1460
4	F	54.6	1425	14	F	42.4	1124
5	F	48.5	1396	15	F	34.5	1052
6	F	42.0	1418	16	F	51.1	1347
7	M	47.4	1362	17	F	41.2	1204
8	F	50.6	1502	18	M	51.9	1867
9	F	42.0	1256	19	M	46.9	1439
10	M	48.7	1614				

(a) Make a scatterplot of the data for the female subjects. Which is the explanatory variable? Explain why the subject number is not part of the scatterplot.

(b) Is the association between these variables positive or negative? What is the form of the relationship? How strong is the relationship?

(c) Now add the data for the male subjects to your graph, using a different color or a different plotting symbol. Does the pattern of relationship that you observed for women hold for men also? How do the male subjects as a group differ from the female subjects as a group?

Body mass index and body fat
Obesity and, more specifically, body fat have been linked to cardiovascular disease, strokes, and type 2 diabetes. Methods that directly measure body fat content, like underwater weighing, dual-energy X-ray absorptiometry, or computerized tomography, are costly and require professional intervention. On the other hand, the body mass index (BMI) is calculated only from a person's height and weight. It has been shown to be strongly correlated with the results of these more complex methods.

Measuring linear association: correlation

A scatterplot displays the direction, form, and strength of the relationship between two quantitative variables. Linear (straight-line) relations are particularly important because a straight line is a simple pattern that is quite common. A linear relation is strong if the points lie close to a straight line, and weak if they are widely scattered about a line. Like all qualitative judgments, however, statements such as "weak" and "strong" are open to individual interpretation.

Our eyes are not good judges of how strong a linear relationship is. The two scatterplots in Figure 3.4 depict exactly the same data, but the plot on the right is drawn smaller in a large field. This plot gives the impression of a stronger linear relationship. Create your scatterplots so that they focus on the relationships, avoiding extra blank space. Similarly, you should draw your scatterplots so that both axes are given the same emphasis, resulting in plots that are roughly square

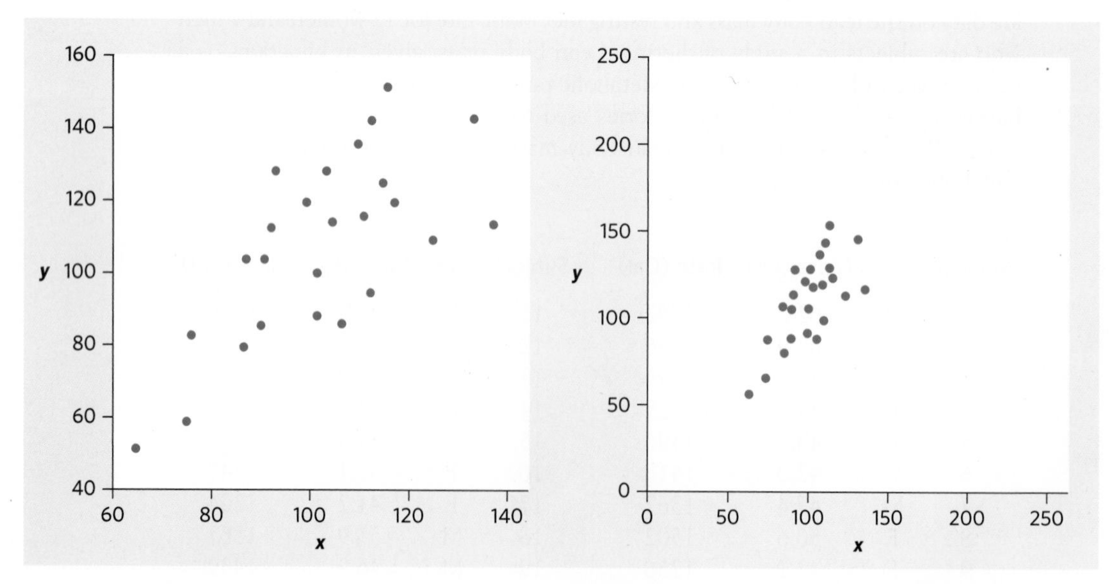

FIGURE 3.4 Two scatterplots of the same data. The straight-line pattern in the plot on the right appears stronger because of the surrounding space.

rather than rectangular in shape. *When creating or studying a scatterplot, keep in mind that our eyes can be fooled by changing the plotting scales or the amount of white space around the cloud of points.*[7]

We need to follow our strategy for data analysis by using a numerical measure to supplement the graph. *Correlation* is the measure we use.

CORRELATION

The **correlation** measures the direction and strength of the linear relationship between two quantitative variables. Correlation is usually written as r.

Suppose that we have data on variables x and y for n individuals. The values for the first individual are x_1 and y_1, the values for the second individual are x_2 and y_2, and so on. The means and standard deviations of the two variables are $\overline{x}$ and s_x for the x-values, and $\overline{y}$ and s_y for the y-values. The correlation r between x and y is

$$r = \frac{1}{n-1}\sum\left(\frac{x_i - \overline{x}}{s_x}\right)\left(\frac{y_i - \overline{y}}{s_y}\right)$$

As always, the summation sign $\sum$ means "add these terms for all the individuals." Computing r by hand is a lengthy process. The formula for r helps you to see what correlation is, and the next exercise asks you to calculate a correlation step-by-step from the definition to solidify its meaning. In practice, you should use software or a calculator that finds r from keyed-in values of two variables x and y. Be aware that some software programs refer to r as **Pearson's correlation** coefficient, or Pearson's r.

Pearson's correlation

The formula for r begins by standardizing the observations. Suppose, for example, that x is height in centimeters and y is weight in kilograms and that we have height and weight measurements for n people. Then $\overline{x}$ and s_x are the mean and standard deviation of the n heights, both in centimeters. The value

$$\frac{x_i - \overline{x}}{s_x}$$

is the standardized height of the ith person. The standardized height says how many standard deviations above or below the mean a person's height lies. Standardized values have no units—in this example, they are no longer measured in centimeters. Standardize the weights also. Then, for each person, multiply the standardized height and the standardized weight. The correlation r is, roughly speaking, an average of these products for the n people.

APPLY YOUR KNOWLEDGE

3.8 **Coral reefs, continued.** Exercise 3.2 discusses a study in which scientists examined data on mean summer sea surface temperature (in degrees Celsius, °C) and mean coral growth (in millimeters per year, mm/y) over a several-year period at locations in the Red Sea. Here are the data:[8]

Amar and Isabelle Guillen-Guillen Photography/Alamy

Temperature (°C)	29.68	29.87	30.01	30.25	30.47	30.65	30.90
Growth (mm/y)	2.63	2.57	2.67	2.60	2.47	2.39	2.25

(a) Make a scatterplot. Which is the explanatory variable? The plot shows a negative linear pattern.

(b) As a pedagogical exercise, find the correlation r step-by-step. You may wish to round off to two decimal places in each step. First, find the mean and standard deviation of each variable. Second, find the seven standardized values for each variable. Finally, use the formula for r. Explain how your value for r matches your graph in (a).

(c) Enter these data into your calculator or software and use the correlation function to find r. Check that you get the same result as in (b), up to roundoff error.

Facts about correlation

The formula for correlation helps us see that r is positive when there is a positive association between the variables. Height and weight, for example, have a positive association. People who are above average in height tend to also be above average in weight. For these people, both the standardized height and the standardized weight tend to be positive. People who are below average in height tend to also have below-average weight. For them, both standardized height and standardized weight tend to be negative. Overall, the products in the formula for r are mostly positive and so r is positive. In the same way, we can see that r is negative when the association between x and y is negative. More detailed study of the formula gives more detailed properties of r. Here is what you need to know in order to interpret correlation.

1. *Correlation makes no distinction between explanatory and response variables*. It makes no difference which variable you call x and which you call y in calculating the correlation.
2. Because r uses the standardized values of the observations, *r does not change when we change the units of measurement of x, y, or both*. Measuring height in inches rather than centimeters and weight in pounds rather than kilograms does not change the correlation between height and weight. The correlation r itself has no unit of measurement; it is just a number.
3. *Positive r indicates positive association between the variables, and negative r indicates negative association*.
4. *The correlation r is always a number between -1 and 1, inclusively*. Values of r near 0 indicate a very weak linear relationship. The strength of the linear relationship increases as r moves away from 0 toward either -1 or 1. Values of r close to -1 or 1 indicate that the points in a scatterplot lie close to a straight line. The extreme values $r = -1$ and $r = 1$ occur only in the case of a perfect linear relationship, when the points lie exactly along a straight line.

EXAMPLE 3.7 From scatterplot to correlation

The scatterplots in Figure 3.5 illustrate how values of r closer to 1 or -1 correspond to stronger linear relationships. To make the meaning of r clearer, the standard deviations of both variables in these plots are equal, and the horizontal and vertical scales are the same. In general, it is not so easy to guess the value of r from the appearance of a scatterplot. Remember that changing the plotting scales in a scatterplot may mislead our eyes, but it does not change the correlation.

The real data we have examined also illustrate how correlation measures the strength and direction of linear relationships. Figure 3.1 shows a very strong positive linear relationship between manatee deaths and powerboat registrations in Florida. The correlation is $r = 0.953$. Figure 3.2 shows a weaker but still fairly strong negative linear relationship between fat gain and nonexercise activity when overeating. The correlation is $r = -0.779$. Figure 3.3 shows a reasonably strong positive association between thorax length and longevity in the group of sexually active fruit flies (displayed as triangles). The correlation is $r = 0.806$. Notice how the strength of the association depends on the absolute value of r (its numerical value irrespective of its sign). ■

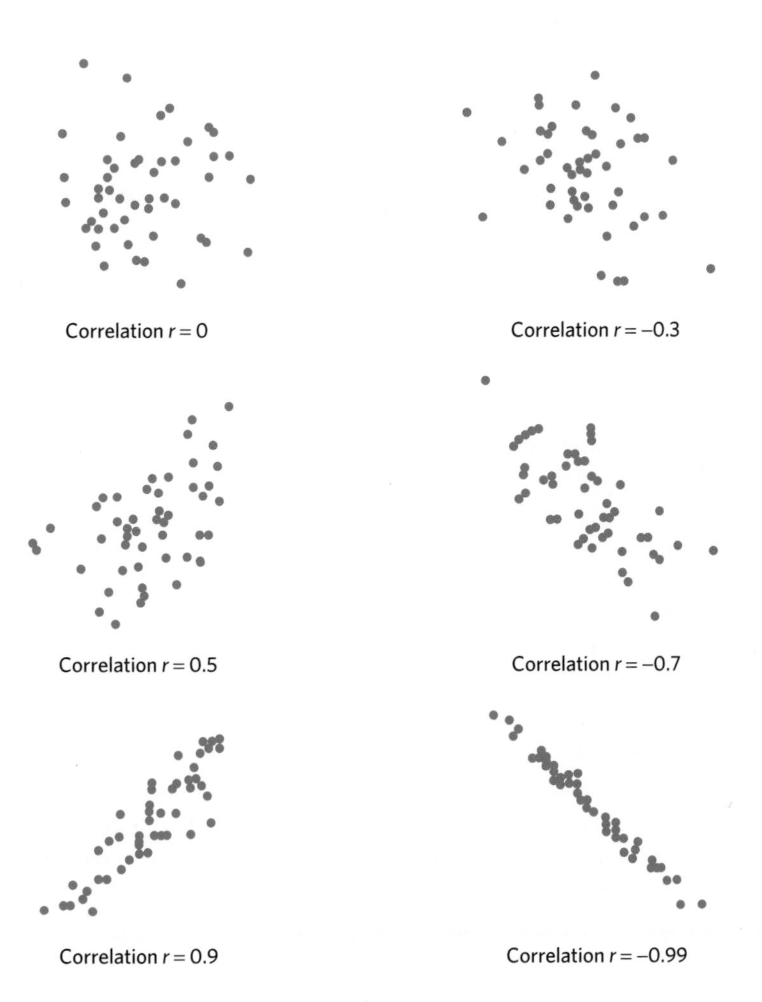

FIGURE 3.5 How correlation measures the strength of a linear relationship, for Example 3.7. Patterns closer to a straight line have correlations closer to 1 or -1.

Describing the relationship between two variables is a more complex task than describing the distribution of one variable. Here are some more facts about correlation, cautions to keep in mind when you use r.

1. *Correlation requires that both variables be quantitative, so that it makes sense to do the arithmetic indicated by the formula for* r. We cannot calculate a correlation between the amount of fat in the diets of a group of people and their ethnicity, because ethnicity is a categorical variable.

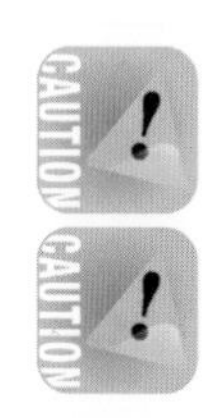

2. Correlation measures the strength of only the linear relationship between two variables. *Correlation does not describe curved relationships between variables, no matter how strong they are.* Exercise 3.11 illustrates this important fact.
3. *Like the mean and standard deviation, the correlation is not resistant:* r *is strongly affected by a few outlying observations.* Use r with caution when outliers appear in the scatterplot. To explore how extreme observations can influence r, use the *Correlation and Regression* applet.

4. *Correlation calculated from averaged data is typically much stronger than correlation calculated from the raw individual data points* because averaged values mask some of the individual-to-individual variations that make up scatter in the scatterplot.

5. *Correlation is not a complete summary of two-variable data,* even after you have established that the relationship between the variables is linear. You should give the means and standard deviations of both x and y along with the correlation. Of course, these numerical summaries do not point out outliers or **clusters** in the scatterplot. Numerical summaries complement plots of data, but they don't replace them.

clusters

Lastly, we may come across data that have been collected over **time** instead of over a number of individuals. How do we handle these data?

Most often in biology, time is not used as an explanatory variable, even though time might be recorded during data collection. For instance, in Table 3.1, manatee deaths and powerboats registered are listed for each year between 1977 and 2012. It would be interesting to study the trends over time of each variable separately—say, to find out how fast the number of powerboats registered in Florida has been increasing in the past few decades. However, that would not help us understand how powerboats affect manatee deaths. That is why we chose to build a scatterplot of the relationship between the number of powerboats registered and the number of manatee deaths from powerboat collisions in any given year. Both approaches provide very different kinds of information.

Some research fields, such as ecology and evolution, do include time in their correlation analyses as a means to understand patterns of change. Although time in itself does not explain or cause the changes, it can help reveal the phenomenon of interest, such as global warming, habitat loss, or extinction rates. *Always examine carefully the question you are asking and match the proper analytic method to the question.*

APPLY YOUR KNOWLEDGE

3.9 Changing the units. In Exercise 3.8 you calculated the correlation coefficient for the relationship between mean sea temperature in degrees Celsius and mean coral growth measured in millimeters per year. If we measured growth in centimeters per year (1 centimeter equals 10 millimeters), would r change? Explain your answer.

3.10 Changing the correlation. Exercise 3.4 gives data on the number of D notes per chickadee warning call for predators with various wingspans.

(a) Use your calculator to find the correlation between the number of D notes per call and wingspan.

(b) The researchers also recorded chickadee calls in response to the introduction of a bobwhite, a harmless quail. The bobwhite has a wingspan of 48.1 cm, and the chickadees responded with an average of 1.74 D notes per call. Make a scatterplot of the data with this additional point. Find the new correlation when the data point for the bobwhite is included.

(c) Measuring the chickadee response to the introduction of a bobwhite was meant as a contrast to the response to various predators. It is, therefore, not surprising that the bobwhite data point falls outside the pattern of responses to predators. Explain in terms of what correlation measures why adding the bobwhite data point makes the correlation weaker (closer to 0).

3.11 Correlation measures linear associations only. In Exercise 3.6 you made a scatterplot of acid phosphatase activity rate at different temperatures and described the strength of the relationship. If you calculate the correlation based on the data provided in that exercise, you find $r = 0.81$ (check for yourself). Explain why the numerical value of the correlation is only moderately strong when the scatterplot indicates that the effect of temperature on enzymatic activity is clear and very reliable.

3.12 Antarctic ice cores document our past. As part of a larger effort to document trends of global warming and its causes, researchers took ice core samples obtained in the Antarctic. They were able to assess from them the year of each snow deposit as well as the concentrations of carbon dioxide and of methane in the air in each particular year. One data set provided information spanning 1978 to 1995.[9]

Ragnar Th. Sigurdsson/age footstock/SuperStock

(a) To study graphically the evolution of carbon dioxide concentrations over time, what variables would you use for the horizontal and vertical axes? What point would you be trying to make?

(b) To study graphically how concentrations of carbon dioxide and methane are related in any given year, what variables would you use for the horizontal and vertical axes? What point would you be trying to make?

CHAPTER 3 SUMMARY

- To study relationships between variables, we must measure the variables on the same group of individuals.
- If we think that a variable x may explain or even cause changes in another variable y, we call x an **explanatory variable** (also called an independent variable) and y a **response variable** (also called a dependent variable).

- A **scatterplot** displays the relationship between two quantitative variables measured on the same individuals. Mark values of one variable on the horizontal axis (x axis) and values of the other variable on the vertical axis (y axis). Plot each individual's data as a point on the graph. Always plot the explanatory variable, if there is one, on the x axis of a scatterplot.
- Plot points with different colors or symbols to see the effect of a categorical variable in a scatterplot.
- In examining a scatterplot, look for an overall pattern showing the **form, direction,** and **strength** of the relationship, and then for **outliers** or other deviations from this pattern.
- **Form: Linear relationships,** where the points show a straight-line pattern, are an important form of relationship between two variables. **Curved** relationships and **clusters** are other forms to watch for.
- **Direction:** If the relationship has a clear direction, we speak of either **positive association** (high values of the two variables tend to occur together) or **negative association** (high values of one variable tend to occur with low values of the other variable).
- **Strength:** The **strength** of a relationship is determined by how close the points in the scatterplot lie to a simple form such as a line.
- The **correlation r** measures the strength and direction of the linear association between two quantitative variables x and y. Although you can calculate a correlation for any scatterplot, r measures only straight-line relationships.
- Correlation indicates the direction of a linear relationship by its sign: $r > 0$ for a positive association and $r < 0$ for a negative association. Correlation always satisfies $-1 \leq r \leq 1$ and indicates the strength of a relationship by how close it is to -1 or 1. Perfect correlation, $r = \pm 1$, occurs only when the points on a scatterplot lie exactly on a straight line.
- Correlation ignores the distinction between explanatory and response variables. The value of r is not affected by changes in the unit of measurement of either variable. Correlation is not resistant, so outliers can greatly change the value of r.

THIS CHAPTER IN CONTEXT

In this and the following two chapters we continue our study of exploratory data analysis but for the purpose of examining relationships *between* variables. Here we show how scatterplots are used to visualize patterns in the relationship between two quantitative variables. When this pattern is roughly linear, we can compute the quantity r, the correlation coefficient, to quantify the strength and direction of the observed relationship. Because this value loses its meaning when the form of a relationship is not linear, it is important to always plot and carefully examine your data first.

In the next chapter we will see how to obtain the equation of a straight line that best represents an observed linear relationship, the least-square regression line, and how this equation can be used to make predictions. We will also discuss the importance of remembering that an observed association does not imply a causal relationship.

In Chapter 23, we will expand correlation and regression into the area of statistical inference, asking whether the patterns we observe in our data hold for individuals that we have not observed. That is, we will ask whether the observed relationship could be explained by chance alone (like seeing a pattern in the clouds) or if it represents an actual association in an entire population of individuals.

CHECK YOUR SKILLS

3.13 You have data for many individuals on their walking speed and their heart rate after a 10-minute walk. When you make a scatterplot, the explanatory variable on the x axis

(a) is walking speed.
(b) is heart rate.
(c) doesn't matter.

3.14 You have data for many individuals on their walking speed and their heart rate after a 10-minute walk. You expect to see

(a) a positive association.
(b) very little association.
(c) a negative association.

3.15 Figure 3.6 is a scatterplot of the age (in months) at which a child begins to talk and the child's later score on a test of mental ability (Gesell Adaptive Score), for each of 21 children (2 sets of 2 children happen to have the same values).[10] One child clearly falls outside the overall pattern of the scatterplot and thus might be an outlier. Which child is it? The child with

(a) age 42 months; score 57.
(b) age 26 months; score 71.
(c) age 17 months; score 121.

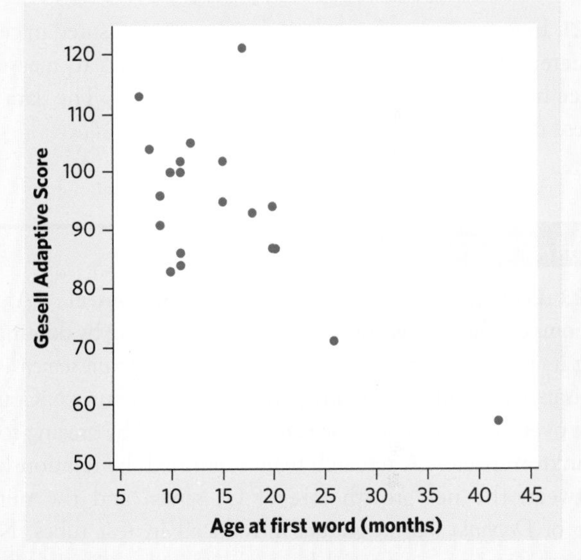

FIGURE 3.6 Scatterplot of Gesell Adaptive Score versus the age at first word for 21 children, for Exercise 3.15.

3.16 To describe the relationship in Figure 3.6, you would say that

(a) it is a positive curved association.
(b) it is a negative linear association.
(c) it shows no association.

3.17 What are all the values that a correlation r can possibly take?

(a) $r \geq 0$
(b) $0 \leq r \leq 1$
(c) $-1 \leq r \leq 1$

3.18 If mothers were always 2 years younger than the fathers of their children, the correlation between the ages of mother and father would be

(a) 1.
(b) 0.5.
(c) Can't tell without seeing the data.

3.19 For a class project, you measure the weight in grams and the tail length in millimeters of a group of mice. The correlation is $r = 0.7$. The scatterplot shows one outlier of the relationship, falling clearly below the overall pattern of points. It turns out that this mouse was sick, and you decide to exclude it from the data set. With the outlier removed, you can expect the value of r to

(a) stay the same. (b) increase. (c) decrease.

3.20 Because elderly people may have difficulty standing to have their height measured, a study looked at the relationship between overall height and height to the knee. Here are data (in centimeters) for five elderly men:

Knee height x (cm)	57.7	47.4	43.5	44.8	55.2
Overall height y (cm)	192.1	153.3	146.4	162.7	169.1

Use your calculator: the correlation between knee height and overall height is about

(a) $r = 0.88$.
(b) $r = 0.09$.
(c) $r = 0.77$.

3.21 In the exercise above, both heights are measured in centimeters. Just for the fun of it, someone decides to measure knee height in millimeters and height in meters. The data in these units are

Knee height x (mm)	577	474	435	448	552
Overall height y (m)	1.921	1.533	1.464	1.627	1.691

The correlation for the data using these units would

(a) be very close to the value calculated in the previous exercise.
(b) be exactly the same as in the previous exercise.
(c) be exactly 10 times smaller (1 cm = 10 mm; 1 cm = 1/100 m).

3.22 In Exercise 3.20 you calculated a correlation for knee height (x, in cm) and height (y, in cm). If, instead, you used height (in cm) as the x variable and knee height (in cm) as the y variable, the new correlation would

(a) have the inverse value (1 over).
(b) have the opposite value (a different sign).
(c) remain the same.

CHAPTER 3 EXERCISES

3.23 Ecological approach to algal bloom control. Algal blooms can have negative effects on an ecosystem by dominating its phytoplankton communities. *Gonyostomum semen* is a nuisance alga infesting many parts of northern Europe. Could the overall biomass of *G. semen* be controlled by grazing zooplankton species? A research team examined the relationship between the net growth rate of *G. semen* and the number of *Daphnia magna* grazers introduced in test tubes. Net growth rate was computed by comparing the initial and final abundance of *G. semen* in the experiment, with a negative value indicative of a decrease in abundance. Here are the findings:[11]

Number of grazers	1	2	3	4	5	6
Net growth rate	−1.9	−2.5	−2.2	−3.9	−4.1	−4.3

(a) Make a scatterplot of number of grazers and net growth rate. Do you think that *D. magna* is an effective grazer of the *G. semen* alga?
(b) Find the correlation r. How does it support your interpretation?

3.24 The amygdala and memory. The amygdala is a brain structure involved in the processing and memory of emotional reactions. In a research project, 10 subjects were shown emotional video clips. They then had their brains scanned with positron emission tomography (PET), and their memory of the clips was assessed quantitatively. Here are the relative amygdala activity and the memory score for each subject:[12]

Relative activity	−0.417	−0.258	−0.234	−0.249	−0.156
Memory score	31	29	29	30	33
Relative activity	−0.031	0.120	0.240	0.342	0.654
Memory score	32	31	35	34	33

(a) Make a scatterplot that shows the relationship between relative amygdala activity and memory score.
(b) Describe the form of the relationship. Is there a linear pattern? How strong is the relationship? Compute the correlation coefficient.
(c) Is the direction of the association positive or negative? Explain in simple language what this suggests about the involvement of the amygdala in the memory of emotional events.

3.25 More on algal bloom control. The study described in Exercise 3.23 also examined the grazing effect of another species, *Daphnia pulex*, on the abundance of the nuisance alga *G. semen* in the lab. Here are the findings:

Number of grazers	0	1	2	3	4	5	6
Net growth rate	−0.7	−0.4	−0.6	−0.3	−0.5	−1.0	−0.2

(a) Make a scatterplot of number of grazers and net growth rate. Do you think that *D. pulex* is an effective grazer of the *G. semen* alga?

(b) Find the correlation r. How does it support your interpretation?

(c) The correlation between number of *D. magna* and net growth rate of *G. semen* in Exercise 3.23 was $r = -0.93$. Contrast the two findings and state your conclusions about the potential effectiveness of *D. magna* and *D. pulex* as ecological methods to control *G. semen* algal blooms.

3.26 Global warming and beer quality. Hops are the flower clusters of a perennial herbaceous plant, *Humulus lupulus*, used in brewing to facilitate fermentation and add a bitter flavor to beer. Hops with a higher α-acid content provide more bitterness, and α-acid content is used as a criterion for hops quality. A study examined how climate change is affecting α-acid content in hops. Temperature records and hop production in the Czech Republic over the 1954–2006 period were examined. Table 3.2 shows the annual average daily temperature from April to August (in degrees Celsius) and the annual percent of α-acids in hops during that time period. The data are sorted by temperature, not by year.[13]

(a) The researchers examined the pattern of change over time in annual percent of α-acids. What type of graph do you think they used? Explain your reasoning. They also plotted the pattern of change over time in annual average daily temperature.

(b) Make a graph showing the relationship between average daily temperature and percent of α-acids in the same year. Which variable would you assign to be the explanatory variable, if any? Describe the relationship as revealed by your graph.

(c) If appropriate, calculate a numerical summary of the strength of the relationship between average daily temperature and percent of α-acids. What are your conclusions about the impact of higher average daily temperature on hops quality?

3.27 Correlation is not resistant. Go to the *Correlation and Regression* applet. Click on the scatterplot to create a group of 10 points in the lower-left corner of the scatterplot with a strong straight-line pattern (correlation about 0.9).

APPLET

(a) Add 1 point at the upper right that is in line with the first 10. How does the correlation change?

(b) Drag this last point down until it is opposite the group of 10 points. How weak can you make the correlation? Can you make the correlation negative? You see that a single outlier can greatly strengthen or weaken a

TABLE 3.2 Annual average daily temperature and annual percent of α-acids in hops

Temperature (°C)	Percent α-acids	Temperature (°C)	Percent α-acids	Temperature (°C)	Percent α-acids	Temperature (°C)	Percent α-acids
13.34	4.39	14.53	6.09	15.21	6.21	16.06	3.49
13.59	4.39	14.53	5.59	15.22	4.10	16.13	3.19
13.66	5.63	14.62	4.82	15.33	4.10	16.14	6.00
13.83	6.30	14.69	5.12	15.36	4.69	16.22	3.50
13.95	5.49	14.71	5.23	15.40	5.60	16.34	2.80
14.01	4.10	14.76	4.39	15.44	3.51	16.42	3.61
14.13	4.10	14.82	5.81	15.53	5.91	16.71	2.00
14.19	5.29	14.90	5.50	15.56	3.99	16.74	3.50
14.24	5.10	14.91	5.84	15.61	3.59	16.84	3.00
14.24	4.41	14.92	4.36	15.65	4.18	17.17	3.61
14.28	5.01	15.02	4.91	15.80	3.80	17.51	2.97
14.37	5.80	15.06	4.99	15.91	3.61		
14.45	4.75	15.09	3.70	15.98	3.30		
14.47	5.48	15.19	6.21	16.02	3.80		

correlation. Always plot your data to check for outlying points.

3.28 Grading oysters. Oysters are categorized for retail as small, medium, or large based on their volume. The grading process is slow and expensive when done by hand. A computer program estimates oyster volume based on the pixel area of two-dimensional (2D) images of the oysters. Table 3.3 shows the actual volume (in cm^3) and 2D reconstruction (in thousands of pixels) for 30 oysters.[14]

(a) We want to know whether the 2D imaging program is a good predictor of oyster volume. Therefore, 2D reconstruction is our explanatory variable here. Make a scatterplot of 2D reconstruction and actual volume.
(b) Describe the overall pattern of the relationship: its form, direction, and strength. If appropriate, give the correlation coefficient. Would you say that the 2D system is an accurate way to assess oyster volume?

3.29 More on grading oysters. The previous exercise describes an automated grading system for oysters. Engineers were given the task of improving on the 2D reconstruction program. They designed a new program that estimates oyster volume using three-dimensional (3D) digital image processing. The results are displayed in Table 3.3.

(a) Make a scatterplot of 3D volume reconstruction (in millions of volume pixels) and actual volume (in cm^3), using 3D reconstruction as the explanatory variable. Describe the overall pattern of the relationship and, if appropriate, give the correlation coefficient.
(b) Compare your analysis for this 3D system with that for the 2D system from the previous exercise. Is the 3D reconstruction program an improvement over the 2D version? Explain your reasoning.

3.30 Attracting beetles. To detect the presence of harmful insects in farm fields, we can put up boards covered with a sticky material and examine the insects trapped on the boards. Which colors attract insects best? Experimenters placed boards of several colors at random locations in a field of oats (four colors and six boards of each color). Here are the counts of cereal leaf beetles trapped by each board:

Color	Insects trapped					
Blue	16	11	20	21	14	7
Green	37	32	20	29	37	32
White	21	12	14	17	13	20
Yellow	45	59	48	46	38	47

(a) Make a plot of the number of beetles trapped on each board against the board's color, arranging the colors to show blue first, then white, then green, and last yellow (space the four colors equally on the horizontal axis).
(b) Does it make sense to speak of a positive or negative association between beetles trapped and board color? Why?
(c) Based on your graph, which of the four colors seems to best attract cereal leaf beetles?

TABLE 3.3 Actual volume and computer reconstruction of 30 oysters

Actual (cm^3)	2D (thousand pixels)	3D (million voxels)	Actual (cm^3)	2D (thousand pixels)	3D (million voxels)	Actual (cm^3)	2D (thousand pixels)	3D (million voxels)
13.04	47.907	5.136699	10.53	31.216	3.942783	10.95	37.156	4.707532
11.71	41.458	4.795151	10.84	41.852	4.052638	7.97	29.070	3.019077
17.42	60.891	6.453115	13.12	44.608	5.334558	7.34	24.590	2.768160
7.23	29.949	2.895239	8.48	35.343	3.527926	13.21	48.082	4.945743
10.03	41.616	3.672746	14.24	47.481	5.679636	7.83	32.118	3.138463
15.59	48.070	5.728880	11.11	40.976	4.013992	11.38	45.112	4.410797
9.94	34.717	3.987582	15.35	65.361	5.565995	11.22	37.020	4.558251
7.53	27.230	2.678423	15.44	50.910	6.303198	9.25	39.333	3.449867
12.73	52.712	5.481545	5.67	22.895	1.928109	13.75	51.351	5.609681
12.66	41.500	5.016762	8.26	34.804	3.450164	14.37	53.281	5.292105

3.31 Attracting beetles, continued.

(a) Using the data from the previous exercise, now make another plot of beetles trapped against board color, but this time arranging the colors in the order used in the table (alphabetical). Notice how different the two graphs look, although the same information is displayed. Explain why.

(b) What do you think a graph of the same data would look like if you arranged the colors in the reverse order of the previous exercise? Is one graph more acceptable than the others?

(c) Some statistical software packages do not allow the use of text for data and force you to use numerical codes instead (for example, blue = 1, green = 2, and so on). Point out the potential danger when creating graphs like the ones you made for this beetle problem.

3.32 Energy cost of running. If you run on a computerized treadmill at the gym and check the calories you have just burned, chances are that the machine's answer is based on the following 1963 study. Researchers asked athletes to run on a treadmill at various speeds and inclines and assessed the athletes' energy expenditure (computed indirectly via oxygen consumption and individual body measurements). Figure 3.7 shows 25 measurements of energy expenditure measured under various conditions of speed (in km/h) and treadmill incline (in percent). Each incline is coded by a different color on the graph.[15]

(a) Describe the relationship between energy expenditure and speed (form, direction, and strength) separately for each incline.

(b) How does the treadmill incline affect the relationship between energy expenditure and speed?

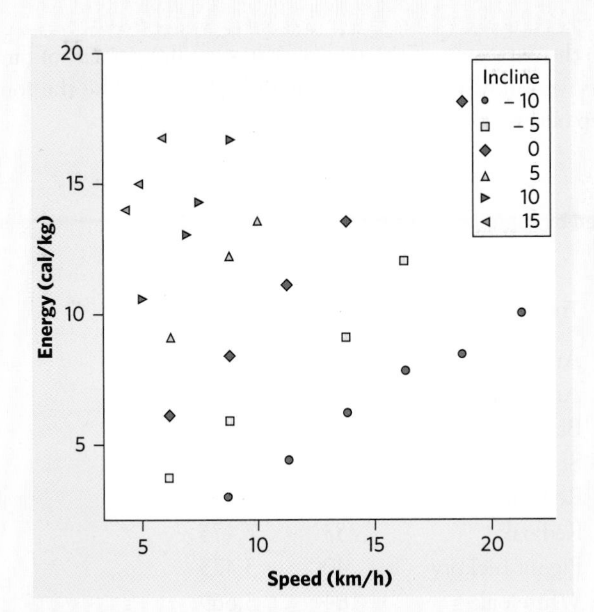

FIGURE 3.7 Scatterplot of energy expenditure against running speed, using different colors and symbols to represent various running inclines, for Exercise 3.32.

3.33 Neural mechanism of nociception. The roundworm *Caenorhabditis elegans* is a widely studied animal model, in part because of its small number of neurons and easily manipulated genome. Nociception is the neural perception of an actually or potentially harmful stimulus. In *C. elegans*, it evokes a self-preserving withdrawal behavior. However, repeated stimulation can result in reduced withdrawal response, or habituation. Researchers compared the withdrawal response to disturbing light stimuli in wild-type *C. elegans* and a mutant *C. elegans* line that exhibits a slower response of PVD sensory neurons. Figure 3.8 shows the scatterplot of the percent of animals tested that exhibited a withdrawal reaction to a noxious stimulus consisting of varying numbers of consecutive light pulses. Failure to react indicates habituation. Circles represent wild-type *C. elegans*, and squares represent the mutant line.[16]

(a) What does the scatterplot show about the pattern of withdrawal responses in wild-type *C. elegans* for

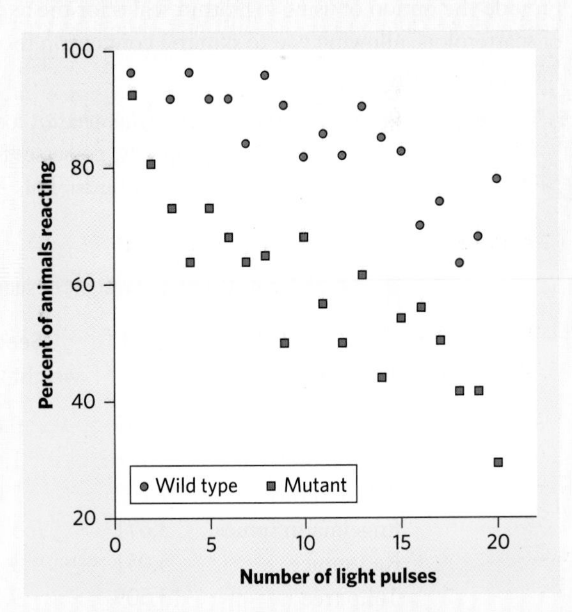

FIGURE 3.8 Scatterplot of the percent of animals reacting to a noxious luminous stimulus consisting of varying numbers of consecutive light pulses, for Exercise 3.33. Circles represent wild-type *C. elegans*, and squares represent mutant *C. elegans*.

increasing numbers of light pulses? How does your answer fit in the context of habituation?

(b) What does the scatterplot show about the difference between wild-type and mutant C. *elegans* lines? Explain why these results suggest an involvement of PVD sensory neurons in nociception and habituation.

3.34 Transforming data: counting seeds. Table 3.4 gives data on the mean number of seeds produced in a year by several common tree species and the mean weight (in milligrams) of the seeds produced. (Some species appear twice because their seeds were counted in two locations.) We might expect that trees with heavy seeds produce fewer of them, but what is the form of the relationship?[17]

(a) Make a scatterplot showing how the weight of tree seeds helps explain how many seeds the tree produces. Describe the form, direction, and strength of the relationship.

(b) When dealing with sizes and counts, the logarithms of the original data are often a better choice of variable. Use your calculator or software to obtain the logarithms of both the seed weights and the seed counts in Table 3.4. Make a new scatterplot using these new variables. Now what are the form, direction, and strength of the relationship? Most software programs provide the option of using logarithm scales for the axes of scatterplots, allowing you to skip the conversion to logarithms of the original data.

3.35 The toucan's beak. 4 STEP The toco toucan, *Ramphastos toco*, the largest member of the toucan family, possesses the largest beak relative to body size of all birds. This exaggerated feature has received various interpretations, such as being a refined adaptation for feeding. However, the large surface area may also be an important mechanism for radiating heat (and hence cooling the bird) as outdoor temperature increases. Here are data for beak heat loss, as a percent of total body heat loss, at various temperatures in degrees Celsius:[18]

Temperature (°C)	15	16	17	18	19	20	21	22
Percent heat loss from beak	32	34	35	33	37	46	55	51

Temperature (°C)	23	24	25	26	27	28	29	30
Percent heat loss from beak	43	52	45	53	58	60	62	62

Investigate the relationship between outdoor temperature and beak heat loss as a percent of total body heat loss.

3.36 Merlins breeding. 4 STEP Often the percent of an animal species in the wild that survive to breed again is lower following a successful breeding season. This is part of nature's self-regulation to keep population size stable. A study of merlins (small falcons) in northern Sweden observed the number of breeding pairs in an isolated area and the percent of males (banded for identification) who returned the next breeding season. Here are data for nine years:[19]

Breeding pairs	28	29	29	29	30	32	33	38	38
Percent return	82	83	70	61	69	58	43	50	47

Do the data support the theory that a smaller percent of birds survive following a successful breeding season? Use the four-step outline (page 57) in your answer.

TABLE 3.4 Count and weight of seeds produced by common tree species

Tree species	Seed count	Seed weight (mg)	Tree species	Seed count	Seed weight (mg)
Paper birch	27,239	0.6	American beech	463	247
Yellow birch	12,158	1.6	American beech	1,892	247
White spruce	7,202	2.0	Black oak	93	1,851
Engelmann spruce	3,671	3.3	Scarlet oak	525	1,930
Red spruce	5,051	3.4	Red oak	411	2,475
Tulip tree	13,509	9.1	Red oak	253	2,475
Ponderosa pine	2,667	37.7	Pignut hickory	40	3,423
White fir	5,196	40.0	White oak	184	3,669
Sugar maple	1,751	48.0	Chestnut oak	107	4,535
Sugar pine	1,159	216.0			

3.37 Bushmeat. Bushmeat, the meat of wild animals, is widely traded in Africa, but its consumption threatens the survival of some animals in the wild. Bushmeat is often not the first choice of consumers—they eat bushmeat when other sources of protein are in short supply. Researchers looked at declines in 41 species of mammals in nature reserves in Ghana and at catches of fish (the primary source of animal protein) in the same region. The data appear in Table 3.5.[20] Fish supply is measured in kilograms per person. The other variable is the percent change in the total "biomass" (weight in tons) for the 41 animal species in six nature reserves. Most of the yearly percent changes in wildlife mass are negative because most years saw fewer wild animals in West Africa. Follow the four-step outline (page 57) to examine whether the data support the idea that more animals are killed for bushmeat when the fish supply is low.

3.38 Does social rejection hurt? We often describe our emotional reaction to social rejection as "pain." A clever study asked whether social rejection causes activity in areas of the brain that are known to be activated by physical pain. If it does, we really do experience social and physical pain in similar ways. Subjects were first included and then deliberately excluded from a social activity while changes in brain activity were measured. After each activity, the subjects filled out questionnaires that assessed how excluded they felt. Here are data for 13 subjects.[21]

Subject	Social distress	Change in brain activity	Subject	Social distress	Change in brain activity
1	1.26	−0.055	8	2.18	0.025
2	1.85	−0.040	9	2.58	0.027
3	1.10	−0.026	10	2.75	0.033
4	2.50	−0.017	11	2.75	0.064
5	2.17	−0.017	12	3.33	0.077
6	2.67	0.017	13	3.65	0.124
7	2.01	0.021			

The explanatory variable is "social distress" measured by each subject's questionnaire score after exclusion relative to the score after inclusion. (So values greater than 1 show the degree of distress caused by exclusion.) The response variable is change in activity in a region of the brain that is activated by physical pain. Discuss what the data show. Follow the four-step process (page 57) in your discussion.

3.39 Match the correlation. You are going to use the *Correlation and Regression* applet to make scatterplots with 10 points that have correlation close to 0.7. The lesson is that many patterns can have the same correlation. Always plot your data before you trust a correlation.

(a) Stop after adding the first 2 points. What is the value of the correlation? Why does it have this value?

TABLE 3.5 Fish supply and wildlife decline in West Africa

Year	Fish supply (kg per person)	Biomass change (percent)	Year	Fish supply (kg per person)	Biomass change (percent)
1971	34.7	2.9	1985	21.3	−5.5
1972	39.3	3.1	1986	24.3	−0.7
1973	32.4	−1.2	1987	27.4	−5.1
1974	31.8	−1.1	1988	24.5	−7.1
1975	32.8	−3.3	1989	25.2	−4.2
1976	38.4	3.7	1990	25.9	0.9
1977	33.2	1.9	1991	23.0	−6.1
1978	29.7	−0.3	1992	27.1	−4.1
1979	25.0	−5.9	1993	23.4	−4.8
1980	21.8	−7.9	1994	18.9	−11.3
1981	20.8	−5.5	1995	19.6	−9.3
1982	19.7	−7.2	1996	25.3	−10.7
1983	20.8	−4.1	1997	22.0	−1.8
1984	21.1	−8.6	1998	21.0	−7.4

(b) Make a lower-left to upper-right pattern of 10 points with correlation about $r = 0.7$. (You can drag points up or down to adjust r after you have 10 points.) Make a rough sketch of your scatterplot.
(c) Make another scatterplot with 9 points in a vertical stack at the left of the plot. Add 1 point far to the right and move it until the correlation is close to 0.7. Make a rough sketch of your scatterplot.
(d) Make yet another scatterplot with 10 points in a curved pattern that starts at the lower left, rises to the right, then falls again at the far right. Adjust the points up or down until you have a quite smooth curve with correlation close to 0.7. Make a rough sketch of this scatterplot also.

3.40 Lung capacity of children. Continue your work from Exercise 1.44 (page 37) with the data file *Large.FEV*.

LARGE DATA SET

(a) How is lung capacity affected by children's growth? Plot the relationship between FEV and age for all the individuals in the data set, using FEV as the response variable. Describe the relationship. Do the same for the relationship between FEV and height.
(b) Display the relationships in (a) separately for boys and for girls (on separate graphs or on the same graph using different symbols). How do boys and girls differ in lung capacity as they age and grow?
(c) Let's examine the impact of smoking on lung capacity for children in middle and high school (ages 11 and up). Display the relationships in (a) separately for smokers and nonsmokers for children aged 11 and up. How does smoking affect the change of lung capacity as children grow and age?

3.41 Everglades. Continue your work from Exercise 1.43 (page 37) with the data file *Large.Everglades*.

LARGE DATA SET

(a) Make a scatterplot matrix displaying the relationship for each combination of 5 variables. There should be 10 different graphs (20 if each variable is used on the x and then on the y axis for each combination). Which relationships, if any, are linear?
(b) The Shark River ends in the Gulf of Mexico. When the tide comes in, ocean water flows into the river; when the tide goes out, water flows from the river into the ocean. The velocity and discharge variables take positive values when the tide is out, and negative values when the tide is in. Describe again the scatterplot matrix you created in (a), this time taking into account the tides' impact (that is, describe the tide in and tide out separately).

3.42 Estimating body fat in men. Continue your work from Exercises 1.45 and 2.46 (pages 37 and 64) with the data file *Large.Bodyfat*.

LARGE DATA SET

(a) A man's weight depends on a number of factors, including his height. Plot and describe the relationship between height (explanatory) and weight (response). In Exercise 2.47 you identified two extreme outliers. Point them out on your scatterplot.
(b) The study made two accurate measurements of the percent body fat (percent fat1 and percent fat2) based on two computation methods using underwater weighing. Make a scatterplot of the two variables (neither is explanatory here) and describe the relationship. Find the correlation coefficient r. Do the two methods give similar results?

CHAPTER 4 Regression

Linear (straight-line) relationships between two quantitative variables are easy to understand and quite common. In Chapter 3 we found linear relationships in settings as varied as manatee deaths, weight gain, and fruit fly longevity. Correlation measures the direction and strength of these relationships. When a scatterplot shows a linear relationship, we may want to summarize the overall pattern by drawing a line on the scatterplot.

IN THIS CHAPTER WE COVER...

- The least-squares regression line
- Using technology
- Facts about least-squares regression
- Outliers and influential observations
- Cautions about correlation and regression
- Association does not imply causation

The least-squares regression line

regression line

A **regression line** is a straight line that summarizes the relationship between two variables, but only in a specific setting: when one of the variables is thought to help explain or predict the other. That is, regression describes a relationship between an explanatory variable (x) and a response variable (y). We often use a regression line to predict the value of y for a given value of x.

In Example 3.5 (page 70) we examined the relationship between fat gain (in kilograms, kg) and change in nonexercise activity (NEA in kilocalories, or Calories, Cal) in young adult volunteers who had overeaten for 8 weeks. We found that those with larger NEA increases gained less fat. To add to this conclusion, we must study regression lines in more detail.

Figure 4.1 repeats Figure 3.2, with the addition of a regression line for predicting fat gain from NEA change. We can start using this regression line without

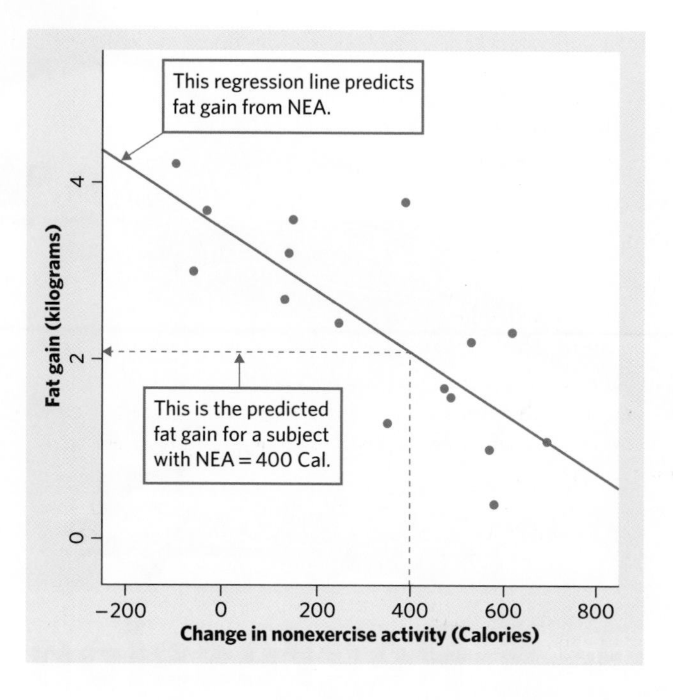

FIGURE 4.1 Fat gain after 8 weeks of overeating, plotted against increase in nonexercise activity over the same period.

knowing about its properties. Suppose that an individual's NEA increases by 400 Cal when she overeats. Go "up and over" on the graph in Figure 4.1. From 400 Cal on the x axis, go up to the regression line and then over to the y axis. The graph shows that the predicted gain in fat is a bit more than 2 kg.

Many calculators and software programs will give you the equation of a regression line from keyed-in data. Understanding and using the line are more important than the details of where the equation comes from.

Regression toward the mean To "regress" means to go backward. Why are statistical methods for predicting a response from an explanatory variable called "regression"? Sir Francis Galton (1822–1911), who was the first to apply regression to biological and psychological data, looked at examples such as the heights of children versus the heights of their parents. He found that the taller-than-average parents tended to have children who were also taller than average but not as tall as their parents. Galton called this fact "regression toward the mean," and the name came to be applied to the statistical method.

REVIEW OF STRAIGHT LINES

Suppose that y is a response variable (plotted on the vertical axis) and x is an explanatory variable (plotted on the horizontal axis). A straight line relating y to x has an equation of the form

$$y = a + bx$$

In this equation, b is the **slope,** the amount by which y changes when x increases by one unit. The number a is the **intercept,** the value of y when $x = 0$.

EXAMPLE 4.1 Using a regression line

Any straight line describing the NEA data has the form

$$\text{fat gain} = a + (b \times \text{NEA change})$$

The line in Figure 4.1 is the regression line with the equation

$$\text{fat gain} = 3.505 - (0.00344 \times \text{NEA change})$$

Be sure you understand the role of the two numbers in this equation:

- The slope, $b = -0.00344$, tells us that fat gained goes down by 0.00344 kg for each added Calorie of NEA. The slope of a regression line is the expected *rate of change* in the response variable as the explanatory variable changes.
- The intercept, $a = 3.505$ kg, is the estimated fat gain if NEA does not change when a person overeats.

The slope of a regression line is an important numerical description of the relationship between the two variables. Although we need the value of the intercept to draw the line, this value is statistically meaningful only when, as in this example, the explanatory variable can actually take values close to zero.

The equation of the regression line makes it easy to predict fat gain. If a person's NEA increases by 400 Cal when she overeats, substitute $x = 400$ in the equation. The predicted fat gain is

$$\text{fat gain} = 3.505 - (0.00344 \times 400) = 2.13 \text{ kg}$$

plotting a line

To **plot the line** on the scatterplot, use the equation to find the predicted y for two values of x, one near each end of the range of x in the data. Plot each predicted y above its x and draw the line through the two points. ■

The slope $b = -0.00344$ in Example 4.1 is small. This does *not* mean that change in NEA has little effect on fat gain. The size of the slope depends on the units in which we measure the two variables. In this example, the slope is the change in fat gain in kilograms when NEA increases by 1 Cal. There are 1000 grams in a kilogram. If we measured fat gain in grams, the slope would be 1000 times larger, $b = 3.44$. *You can't say how important a relationship is by looking at the size of the slope of the regression line*.

logarithm transformation

Logarithm transformations are quite frequent in biology to study problems ranging from population growth to animal physiology. A regression line can be computed using the transformed data, and prediction within the range of the data can be performed as with simple linear relationships. The difference is that we will need to transform the predicted y back into its original (nonlogarithmic) unit.

EXAMPLE 4.2 Using a regression line: transformed data

Humans have large brains compared with other mammals, but they are also a relatively large mammal species. How does an animal's brain relate to its body size? Figure 4.2(a) shows the scatterplot of brain weight in grams against body weight in kilograms for 96 species of mammals.[1] Figure 4.2(b) shows the scatterplot for the same data after a logarithm transformation. The line is the least-squares regression line for predicting y from x.

The regression line for the untransformed data in Figure 4.2(a) is clearly unsatisfactory because most mammals are so small compared to elephants. The correlation between brain weight and body weight is $r = 0.86$, but this is misleading. If we remove the elephant data point, the correlation for the other 95 species falls to $r = 0.50$. Humans, dolphins, and hippos are also clear outliers of the relationship.

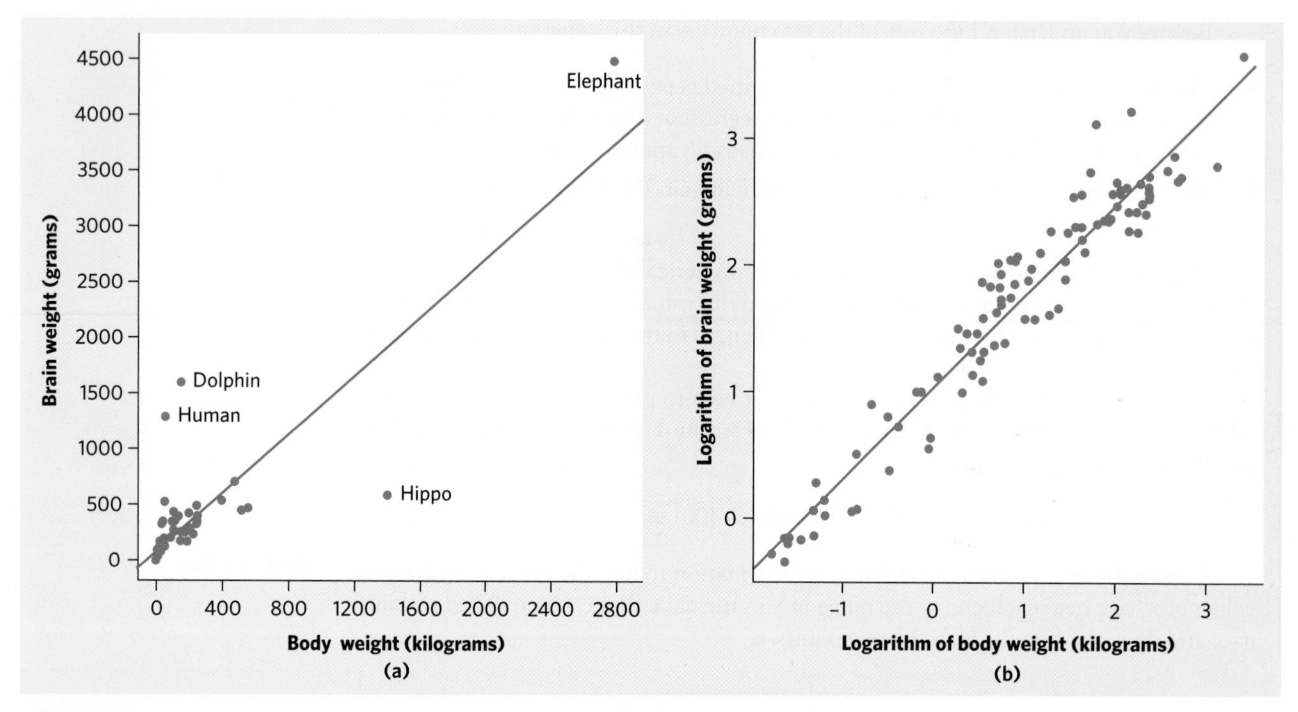

FIGURE 4.2 (a) Scatterplot of body weight and brain weight for 96 species of mammals in the original numerical values of kilograms and grams, for Example 4.2. (b) Scatterplot of the data in (a) using the logarithm of brain weight and the logarithm of body weight, for Example 4.2.

The scatterplot using transformed data in Figure 4.2(b) has no obvious outliers and is clearly linear, with correlation $r = 0.96$. The least-squares regression for predicting the logarithm of brain weight from the logarithm of body weight has the equation

$$\log(\text{brain weight}) = 1.01 + [0.72 \times \log(\text{body weight})]$$

For a mammal species weighing, on average, 100 kg, the logarithm is $\log(100) = 2$, and the predicted logarithm of brain weight is

$$\log(\text{brain weight}) = 1.01 + (0.72 \times 2) = 2.45$$

To undo the logarithm transformation, remember that for common logarithms with base 10, $y = 10^{\log(y)}$. Thus, the predicted brain weight, in grams, for the 100 kg species is

$$\text{brain weight} = 10^{2.45} = 282 \text{ g}$$

Verify on Figure 4.2 that this prediction makes sense on both scatterplots, although the scatterplot of transformed data is much easier to use. ■

How do we draw a regression line on a scatterplot that we can all agree on and that does not depend on an approximate guess? Because we use the line to predict y from x, the prediction errors we make are errors in y, the vertical direction in the scatterplot. A *good regression line makes the vertical distances of the points from the line as small as possible*. The most common way to make the collection of vertical distances "as small as possible" is the *least-squares* method.

LEAST-SQUARES REGRESSION LINE

The **least-squares regression line** of y on x is the line that makes the sum of the squares of the vertical distances of the data points from the line as small as possible.

One reason for the popularity of the least-squares regression line is that the problem of finding the line has a simple answer. We can give the equation for the least-squares line in terms of the means and standard deviations of the two variables and the correlation between them.

EQUATION OF THE LEAST-SQUARES REGRESSION LINE

We have data on an explanatory variable x and a response variable y for n individuals. From the data, calculate the means $\bar{x}$ and $\bar{y}$ and the standard deviations s_x and s_y of the two variables and their correlation r. The least-squares regression line is the line

$$\hat{y} = a + bx$$

with **slope**

$$b = r\frac{s_y}{s_x}$$

and **intercept**

$$a = \bar{y} - b\bar{x}$$

We write $\hat{y}$ (read "y hat") in the equation of the regression line to emphasize that the line gives a *predicted* response $\hat{y}$ for any x. Because of the scatter of points about the line, the predicted response will usually not be exactly the same as the actually *observed* response y.

EXAMPLE 4.3 Finding the regression line

Example 4.1 gives us the equation for the least-squares regression line between NEA change and fat gain. We can obtain this equation fairly quickly. From the raw data given in Example 3.5 (page 70), we find that the means and standard deviations for the two variables are

	NEA change	Fat gain
Mean	324.75	2.388
Standard deviation	257.66	1.139

The correlation between them is $r = -0.779$. This is all we need to find the slope and intercept:

$$\text{slope} = r\frac{s_y}{s_x} = -0.779\frac{1.139}{257.66} = -0.00344$$

$$\text{intercept} = \bar{y} - b\bar{x} = 2.388 - (-0.00344)(324.75) = 3.505$$ ■

You should make sure to keep at least 3 significant digits, especially for the slope. *Rounding the slope too much can lead to substantial inaccuracies when trying to predict y using the equation for the least-squares regression line.*

If all you have available is summary data such as means, standard deviations, and the correlation, you can use these formulas to find the equation for the least-squares regression line. In practice, however, you usually don't need these calculations. Software or your calculator will give the slope b and intercept a of the least-squares line from the values of the variables x and y. You can then concentrate on understanding and using the regression line.

APPLY YOUR KNOWLEDGE

4.1 Don't drink and drive. The Department of Motor Vehicles warns of the effect of drinking alcohol on one's blood alcohol content (BAC, in percent of volume) and behavior. At a BAC of 0.08 or higher, driving skills and response time are impaired, and the risk of accidental death is greatly increased. One "drink" is defined as 12 ounces (oz) of beer or 4 oz of table wine or 1.25 oz of 80-proof liquor. For a 160-pound man, the blood alcohol content as a function of the number of drinks consumed in one hour can be expressed by the regression line

$$\text{BAC} = 0.023 \times \text{ number of drinks}$$

for predicting BAC from the number of drinks.[2]

(a) Say in words what the slope of this line tells you. Why do you think the intercept is zero?

(b) Find the predicted BAC score for a 160-pound male who drank four 12 oz beers over the last hour.

(c) Draw a graph of the regression line for a range of number of drinks between 0 and 10. (Be sure to show the scales for the x and y axes.)

4.2 Don't drink and drive, continued. The impact of a single drink on a person's BAC depends on sex and weight. For a woman weighing 120 pounds, the BAC increases by 0.038 for each drink taken in one hour.[3]

(a) What is the equation of the regression line for predicting BAC from number of drinks for a 120-pound woman?

(b) The BAC limit in the United States is 0.08 for drivers over 21 years old. What BAC would we expect for a 120-pound woman having 2 drinks in one hour? Would her BAC be over the legal limit, assuming she is over 21 years old? Would it be safe for her to drive?

OGphoto/Getty Images

4.3 Predicting seed weight. Exercise 3.34 examined the relationship between mean seed weight (in milligrams, mg) and mean number of seeds produced in a year by several common tree species. After converting the data from Table 3.4 into a logarithm scale, a scatterplot of the converted data shows a linear relationship that can be described by the least-square regression equation

$$\hat{y} = 4.238 - 0.567x$$

where $\hat{y}$ is the log of seed count and x is the log of seed weight (both in base 10).

(a) Using this equation, predict the log of seed count $\hat{y}$ for a tree species with a mean seed weight of 1000 mg ($x = 3$).

(b) Following Example 4.2, undo the logarithm transformation to find the predicted mean annual seed count for a tree species with a mean seed weight of 1000 mg.

Using technology

Least-squares regression is one of the most common statistical procedures. Any technology you use for statistical calculations will give you the least-squares line and related information. Figure 4.3 displays the regression output for the data of Examples 4.1 and 4.2 from a graphing calculator, three statistical programs, and

TI-83

```
LinReg
 y=a+bx
 a=3.505122916
 b=-.003441487
 r²=.6061492049
 r=-.7785558457
```

R

```
lm (formula = fat ~ nea, data = eg04_01)

Coefficients :
               Estimate    Std. Error t value  Pr (>|t|)
 (Intercept) 3.5051229    0.3036164  11.545    1.53e-08 ***
 nea        -0.0034415    0.0007414  -4.642    0.000381 ***

Residual standard error : 0.7399 on 14 degrees of freedom
Multiple R-squared: 0.6061, Adjusted R-squared: 0.578
F-statistic: 21.55 on 1 and 14 DF, p-value: 0.0003810
```

Minitab

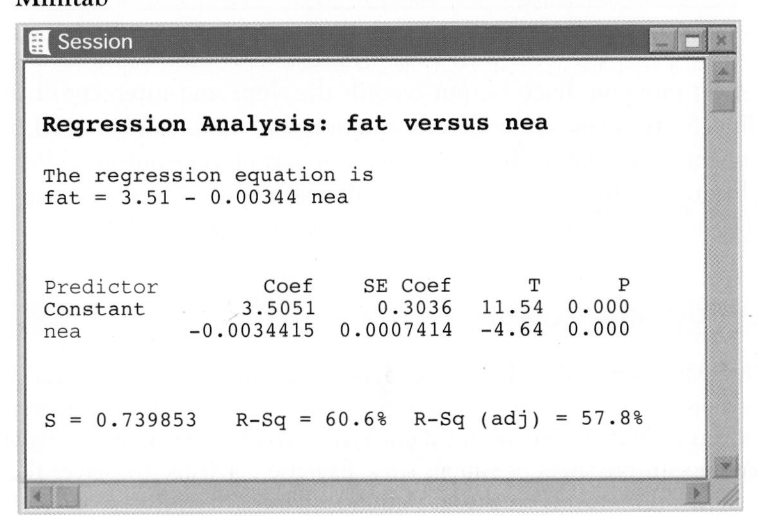

Session

Regression Analysis: fat versus nea

```
The regression equation is
fat = 3.51 - 0.00344 nea


Predictor        Coef    SE Coef      T      P
Constant       3.5051     0.3036  11.54  0.000
nea        -0.0034415  0.0007414  -4.64  0.000


S = 0.739853   R-Sq = 60.6%   R-Sq (adj) = 57.8%
```

FIGURE 4.3 Least-squares regression for the nonexercise activity data: output from a graphing calculator, three statistical programs, and a spreadsheet program. (*Continued on next page*)

Microsoft Excel

	A	B	C	D	E
1	SUMMARY OUTPUT				
2					
3	*Regression statistics*				
4	Multiple R	0.778555846			
5	R Square	0.606149205			
6	Adjusted R Square	0.578017005			
7	Standard Error	0.739852874			
8	Observations	16			
9					
10		*Coefficients*	*Standard Error*	*t Stat*	*P-value*
11	Intercept	3.505122916	0.303616403	11.54458	1.53E-08
12	nea	-0.003441487	0.00074141	-4.64182	0.000381
13					

SPSS

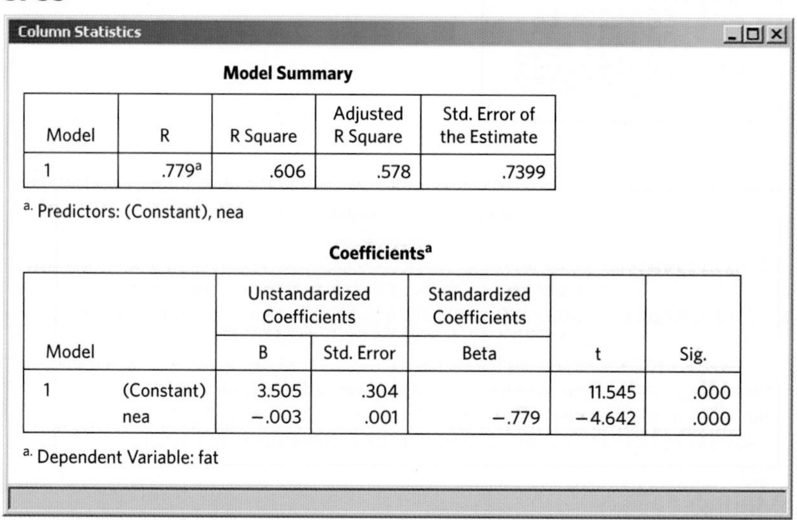

Column Statistics

Model Summary

Model	R	R Square	Adjusted R Square	Std. Error of the Estimate
1	.779[a]	.606	.578	.7399

[a] Predictors: (Constant), nea

Coefficients[a]

Model		Unstandardized Coefficients		Standardized Coefficients		
		B	Std. Error	Beta	t	Sig.
1	(Constant)	3.505	.304		11.545	.000
	nea	−.003	.001	−.779	−4.642	.000

[a] Dependent Variable: fat

FIGURE 4.3 *(Continued)*

a spreadsheet program. Each output records the slope and intercept of the least-squares line. Software also provides information that we do not yet need, although we will use much of it later. (In fact, we left out part of some outputs.) Be sure that you can locate the slope and intercept on all five outputs. *Once you understand the statistical ideas, you can read and work with almost any software output.*

APPLY YOUR KNOWLEDGE

4.4 Chickadee alarm calls. The black-capped chickadee (*Poecile atricapilla*) is a small songbird that uses complex language to communicate about food sources and predator threats. Researchers recorded chickadee vocalizations in an aviary when presented with predators of various sizes. Exercise 3.4 (page 69) gives the average number of D notes per chickadee warning call for each type of predator, along with the predator wingspan (in centimeters) for that experiment. As you found out in Exercise 3.4, the relationship between these two variables is linear.

(a) Find the least-squares regression line for predicting y from x. Make a scatterplot and draw your line on the plot.

(b) Explain in words what the slope of the regression line tells us.

4.5 **Coral reefs.** Exercise 3.8 discusses a study in which scientists examined data on mean summer sea surface temperatures (in degrees Celsius) and mean coral growth (in millimeters per year) over a several-year period at locations in the Red Sea. Here are the data:[4]

Temperature (°C)	29.68	29.87	30.01	30.25	30.47	30.65	30.90
Growth (mm/y)	2.63	2.57	2.67	2.60	2.47	2.39	2.25

(a) Use your calculator to find the mean and standard deviation of both sea surface temperature x and growth y and the correlation r between x and y. Use these basic measures to find the equation of the least-squares line for predicting y from x.

(b) Enter the data into your software or calculator and use the regression function to find the least-squares line. The result should agree with your work in (a) up to roundoff error.

Facts about least-squares regression

One reason for the popularity of least-squares regression lines is that they have many convenient special properties. Here are some facts about least-squares regression lines. Facts 2, 3, and 4 are special properties of least-squares regression. They are not true for other methods of fitting a line to data.

Fact 1. The distinction between explanatory and response variables is essential in regression. Least-squares regression makes the distances of the data points from the line small only in the y direction. If we reverse the roles of the two variables, we get a different least-squares regression line.

Fact 2. There is a close connection between correlation and the slope of the least-squares line. The slope is

$$b = r\frac{s_y}{s_x}$$

You see that **the slope and the correlation always have the same sign.** For example, if a scatterplot shows a positive association, then both b and r are positive. The formula for the slope b says more: along the regression line, **a change of one standard deviation in x corresponds to a change of r standard deviations in y.** When the variables are perfectly correlated ($r = 1$ or $r = -1$), the change in the predicted response $\hat{y}$ is the same (in standard deviation units) as the change in x. Otherwise, because $-1 \leq r \leq 1$, the change in $\hat{y}$ (in standard deviation units) is less than the change in x. As the correlation grows less strong, the prediction $\hat{y}$ moves less in response to changes in x.

Fact 3. **The least-squares regression line always passes through the point $(\bar{x}, \bar{y})$** on the graph of y against x. Remember that the intercept is $a = \bar{y} - b\bar{x}$. So, when $x = \bar{x}$, $y = (\bar{y} - b\bar{x}) + b\bar{x} = \bar{y}$.

Fact 4. The correlation r describes the strength of a straight-line relationship. In the regression setting, this description takes a specific form: **The square of the correlation, r^2, is the fraction of the variation in the values of y that is explained by the least-squares regression of y on x.**

The idea is that when there is a linear relationship, some of the variation in y is accounted for by the fact that as x changes it pulls y along with it. More exactly, the predicted response $\hat{y}$ moves along the regression line as we move x from one end of its scale to the other. This is the variation in y that is explained by regression. Unless all the data points lie exactly on a line, there is additional variation in the actual response y that appears as the scatter of points above and below the line. Although we will skip the algebra, it's possible to write

$$r^2 = \frac{\text{variation in } \hat{y} \text{ as } x \text{ pulls it along the line}}{\text{total variation in observed values of } y}$$

Look again at Figure 4.1, the scatterplot of the NEA data. The variation in y corresponds to the spread of fat gains from 0.4 to 4.2 kg. Some of this variation is explained by the fact that x (change in NEA) varies from a loss of 94 Cal to a gain of 690 Cal. As x moves from -94 to 690, it pulls y along the line. But the straight-line tie of y to x doesn't explain *all* the variation in y. The remaining variation appears as the scatter of points above and below the line.

EXAMPLE 4.4 Using r^2

For the NEA data, $r = -0.7786$ and $r^2 = 0.6062$. About 61% of the variation in fat gained is accounted for by the linear relationship with change in NEA. The other 39% is individual variation among subjects that is not explained by the linear relationship.

Figure 3.1 (page 68) shows a stronger linear relationship in which the points are more tightly concentrated along a line. Here, $r = 0.9525$ and $r^2 = 0.9073$. Nearly 91% of the variation in manatee deaths from collisions with powerboats in a given year is explained by the number of powerboats registered that year. Only about 9% is variation among years with the same number of powerboat registrations. ■

When you report a regression, give r^2 as a measure of how successful the regression was in explaining the response. All the outputs in Figure 4.3 include r^2, either in decimal form or as a percent. Perfect correlation ($r = -1$ or $r = 1$) means that the points lie exactly on a line. Then $r^2 = 1$, and all the variation in one variable is accounted for by the linear relationship with the other variable. If $r = -0.7$ or $r = 0.7$, then $r^2 = 0.49$, and about half the variation is accounted for by the linear relationship. In the r^2 scale, correlation ± 0.7 is about halfway between 0 and ± 1.

Many students are puzzled by the use of both r and r^2 in linear regression. Almost anything can be computed, such as $5r - 3$, $r^3 + 10$, $1/2r$, and so on, but few of these values will be practically meaningful. The values of r and r^2 are commonly cited because both are informative and easy to interpret: r is a measure of the strength and direction of a linear relationship, whereas r^2 represents the fraction of the variation in y that can be explained by the linear regression model. They are related yet different pieces of information.

APPLY YOUR KNOWLEDGE

4.6 **Tara gum.** Food additives, such as the plant-based extract tara gum, are used in industrial food preparation as thickening agents to stabilize emulsions and create an impression of thickness and creaminess without the extra calories. The effect of tara gum on the viscosity (in millipascal-seconds, mPa.s) of fruit yogurt drinks was assessed for various concentrations of tara gum (in grams per 100 grams of drink, g/100 g). Here are the findings:[5]

Concentration (g/100 g)	0.00	0.03	0.06	0.09	0.13	0.16	0.19	0.22
Viscosity (mPa.s)	11.7	20.9	25.7	31.4	42.4	53.8	60.8	81.5
Concentration (g/100 g)	0.25	0.28	0.31	0.34	0.38	0.41	0.44	0.47
Viscosity (mPa.s)	105.4	126.5	150.3	181.2	210.8	216.0	303.5	316.2

Monte dei Paschi di Siena A.p.A.

(a) Make a scatterplot of the data. Find the least-squares regression line for predicting drink viscosity from tara gum concentration, and add this line to your plot. Why should we *not* use the regression line for prediction in this setting?

(b) Even regression lines that make no practical sense obey Facts 2, 3, and 4. Use the equation of the regression line you found in (a) to show that when x is the mean tara gum concentration, the predicted viscosity $\hat{y}$ is the mean of the observed viscosities. Always plot your data before using regression.

4.7 **How useful is regression?** Figure 3.3 for Example 3.6 (page 72) displays the longevity and thorax length of two groups of male fruit flies, one sexually active and one sexually inactive. The correlation is $r = 0.806$ for the group of sexually active fruit flies. A biologist grows tomato plants under various conditions and records the plants' potassium and phosphorus levels.[6] The relationship between potassium and phosphorus levels is appropriate for linear regression, and the correlation is $r = -0.151$. Explain in simple language why knowing only these correlations enables you to say that prediction of longevity from thorax length for sexually active male fruit flies will be much more accurate than prediction of phosphorus level from potassium level in tomato plants.

Outliers and influential observations

One of the first principles of data analysis is to look for an overall pattern and also for striking deviations from the pattern. A regression line describes the overall pattern of a linear relationship between an explanatory variable and a response variable. We see deviations from this pattern by looking at the scatter of the data points about the regression line. The vertical distances from the points to the least-squares regression line are as small as possible, in the sense that they have the smallest possible sum of squares. Because they represent "left-over" variation in the response after fitting the regression line, these distances are called **residuals.** We will discuss residuals in greater detail in Chapter 23, which covers inference for linear regression.

residuals

When examining scatterplots, look for points with large residuals—outliers of the relationships—and any other unusual observations.

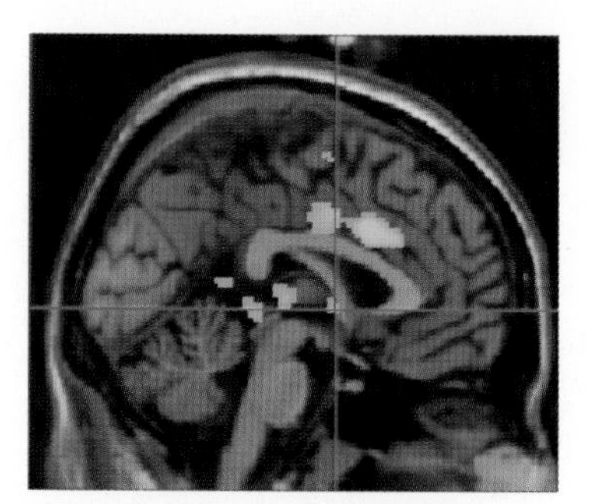
Philippe Psalia/Science Source/Photo Researchers

EXAMPLE 4.5 I feel your pain

"Empathy" means being able to understand what others feel. To see how the brain expresses empathy, researchers recruited 16 couples in their midtwenties who were married or had been dating for at least two years. They zapped the man's hand with an electrode while the woman watched and while they measured the activity in several parts of the woman's brain that would respond to her own pain. Brain activity was recorded as a fraction of the activity observed when the woman herself was zapped with the electrode. The women also completed a psychological test that measures empathy. Will women who score higher in empathy respond more strongly when their partner has a painful experience? Here are data for one brain region:[7]

Subject	1	2	3	4	5	6	7	8
Empathy score	38	53	41	55	56	61	62	48
Brain activity	−0.120	0.392	0.005	0.369	0.016	0.415	0.107	0.506
Subject	**9**	**10**	**11**	**12**	**13**	**14**	**15**	**16**
Empathy score	43	47	56	65	19	61	32	105
Brain activity	0.153	0.745	0.255	0.574	0.210	0.722	0.358	0.779

Figure 4.4 is a scatterplot of these data, with empathy score as the explanatory variable x and brain activity as the response variable y. The plot shows a positive association. That is, women who score higher in empathy do indeed react more strongly to their partners' pain. The overall pattern is moderately linear, correlation $r = 0.515$.

The line on the plot is the least-squares regression line of brain activity on empathy score. Its equation is

$$\hat{y} = -0.0578 + 0.00761x$$ ■

Figure 4.4 shows one unusual observation. Subject 16 is an outlier in the x direction, with empathy score 40 points higher than any other subject. Because of its extreme position on the empathy scale, this point has a strong influence on the correlation. Dropping Subject 16 reduces the correlation from $r = 0.515$ to $r = 0.331$. You can see that this point extends the linear pattern in Figure 4.4 and so increases the correlation. We say that Subject 16 is *influential* for calculating the correlation.

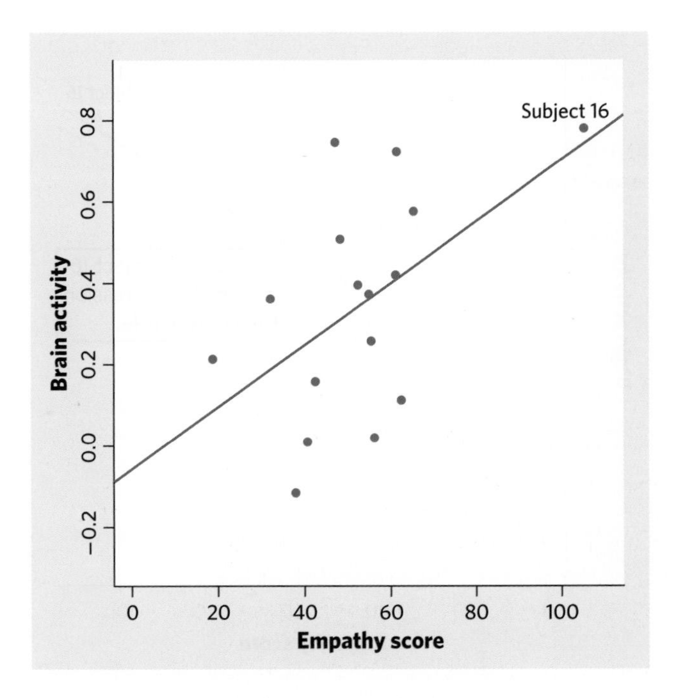

FIGURE 4.4 Scatterplot of activity in a region of the brain that responds to pain versus score on a test of empathy, for Example 4.5. Brain activity is measured as the subject watches her partner experience pain. The line is the least-squares regression line.

INFLUENTIAL OBSERVATIONS

An observation is **influential** for a statistical calculation if removing it would markedly change the result of the calculation.

The result of a statistical calculation may be of little practical use if it depends strongly on one or more influential observations.

Points that are outliers in either the x or y direction of a scatterplot are often influential for the correlation. Points that are outliers in the x direction are often influential for the least-squares regression line.

EXAMPLE 4.6 An influential observation?

In Example 4.5, Subject 16 is influential for the correlation because removing it greatly reduces r. Is this observation also influential for the least-squares line? Figure 4.5 shows that it is not. The regression line calculated without Subject 16 (dashed) differs little from the line that uses all the observations (solid). The reason that this outlier in the x direction has little influence on the regression line is that it lies close to the dashed regression line calculated from the other observations. ■

To see why points that are outliers in the x direction are often influential for regression, let's try an experiment. Suppose we pull Subject 16's point in the scatterplot straight down. This point is now an outlier of the relationship in addition to being an outlier in the x direction. What happens to the regression line? Figure 4.6 shows the result. The dashed line is the regression line with Subject 16's point in

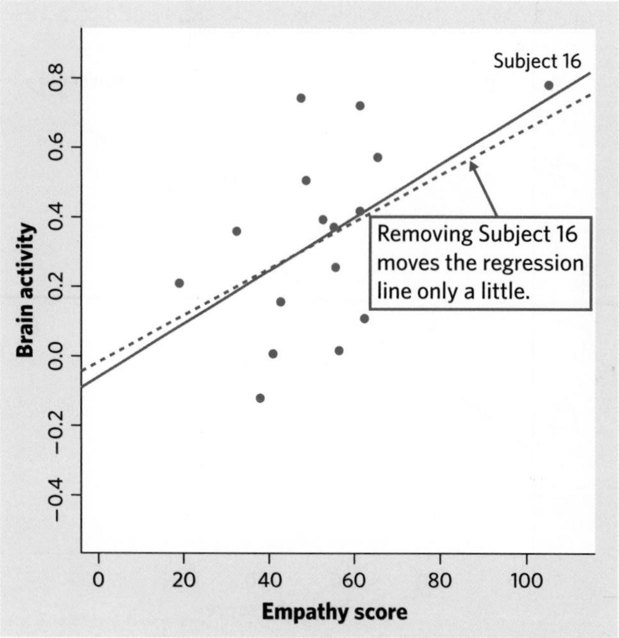

FIGURE 4.5 Subject 16 is an outlier in the x direction. The outlier is not influential for least-squares regression, because removing it moves the regression line only a little.

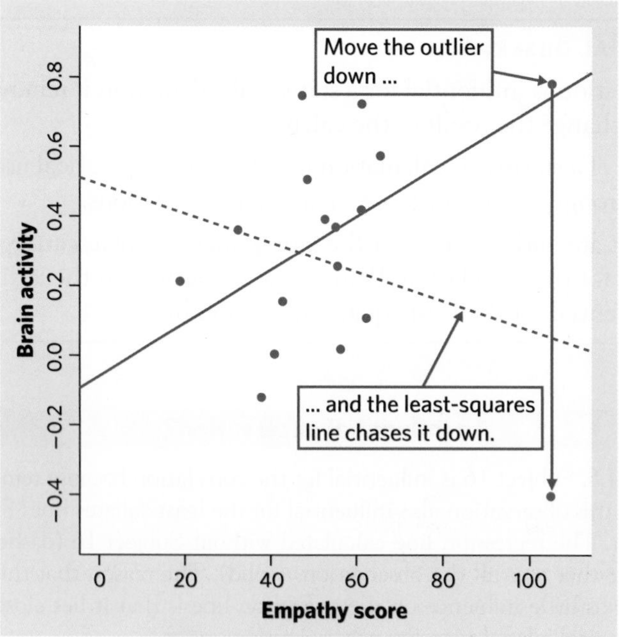

FIGURE 4.6 An outlier in the x direction pulls the least-squares line to itself because there are no other observations with similar values of x to hold the line in place. When the outlier moves down, the original regression line (solid) chases it down to the dashed line.

its new, lower position. Because there are no other points with similar x-values, the line chases the outlier. An outlier in x pulls the least-squares line toward itself. If the outlier does not lie close to the line calculated from the other observations, it will be influential. You can use the *Correlation and Regression* applet to animate Figure 4.6 and perform your own experiment.

We did not need the distinction between outliers and influential observations in Chapter 2. When studying one variable by itself, an outlier that lies far above the other values pulls up the mean $\overline{x}$ toward it and is thus always influential. In the regression setting, however, not all outliers are influential.

APPLY YOUR KNOWLEDGE

4.8 Chickadee alarm calls, continued. In Exercise 4.4 you obtained the regression line for the relationship between the average number of D notes per chickadee warning call and a predator's wingspan (in centimeters). The researchers also recorded chickadee calls in response to the introduction of a bobwhite, a harmless quail. The bobwhite has a wingspan of 48.1 cm, and the chickadees responded with an average of 1.74 D notes per call.

Frank Leung/iStockphoto

(a) Make a scatterplot of the data with this additional point. Find the equation for the new regression line when the bobwhite data point is included and draw it on your graph.

(b) Is the bobwhite data point an outlier of the relationship? Is it an outlier in the distribution of the x variable? Is it an outlier in the distribution of the y variable? Is it influential? Explain why the bobwhite data point is interesting as a comparison but should not be included to describe the relationship between D notes per warning call and predator wingspan.

4.9 Bird colonies. One of nature's patterns connects the percent of adult birds in a colony that return from the previous year and the number of new adults that join the colony. Here are data for 13 colonies of sparrowhawks:[8]

Percent returning	New adults	Percent returning	New adults	Percent returning	New adults
74	5	62	15	46	18
66	6	52	16	60	19
81	8	45	17	46	20
52	11	62	18	38	20
73	12				

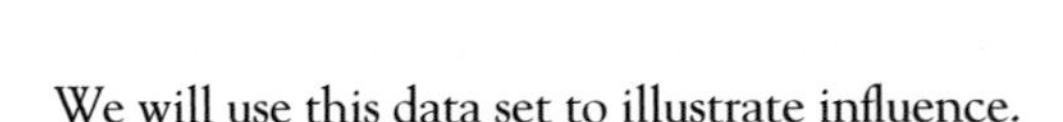

We will use this data set to illustrate influence.

(a) Make a scatterplot of the data suitable for predicting new adults from percent of returning adults. Then add two new points. Point A: 10% return, 15 new adults. Point B: 60% return, 28 new adults. In which direction is each new point an outlier?

(b) Add three least-squares regression lines to your plot: for the original 13 colonies, for the original colonies plus Point A, and for the original colonies plus Point B. Which new point is more influential for the regression line? Explain in simple language why each new point moves the line in the way your graph shows.

Cautions about correlation and regression

Correlation and regression are powerful tools for describing the relationship between two variables. When you use these tools, you must be aware of their limitations. You already know:

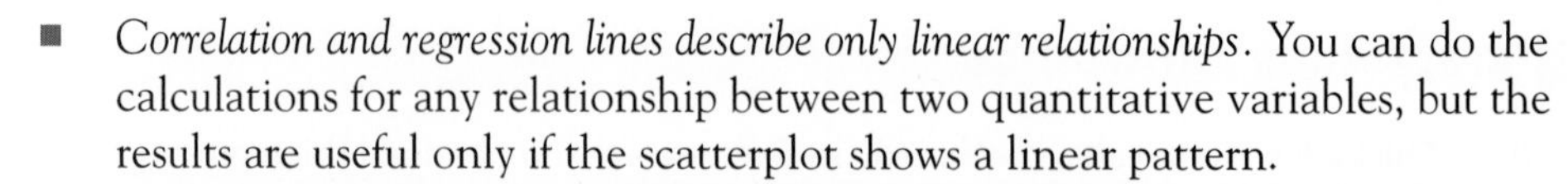

- *Correlation and regression lines describe only linear relationships.* You can do the calculations for any relationship between two quantitative variables, but the results are useful only if the scatterplot shows a linear pattern.

- *Correlation and least-squares regression lines are not resistant.* Always plot your data and look for observations that may be influential.

Here are two more things to keep in mind when you use correlation and regression.

Beware extrapolation. Suppose that you have data on a child's growth between 3 and 8 years of age. You find a strong linear relationship between age x and height y. If you fit a regression line to these data and use it to predict height at age 25 years, you will predict that the child will be 8 feet tall. Growth slows down and then stops at maturity, so extending the straight line to adult ages is foolish. *Few relationships are linear for all values of x. Don't make predictions for values of x outside the range that actually appears in your data.*

> **EXTRAPOLATION**
>
> **Extrapolation** is the use of a regression line for prediction well outside the range of values of the explanatory variable x that you used to obtain the line. Such predictions are often not accurate.

Beware the lurking variable. Another caution is even more important: *The relationship between two variables can often be understood only by taking other variables into account. Lurking variables* can make a correlation or regression misleading.

> **LURKING VARIABLE**
>
> A **lurking variable** is a variable that is not among the explanatory or response variables in a study and yet may influence the interpretation of relationships among those variables.

You should always think about possible lurking variables before you draw conclusions based on correlation or regression.

Inversely, it makes no sense to study the relationship between any two variables just to see if something might come up. For instance, the worldwide AIDS epidemic and Internet use have both exploded over the past two decades, and thus you could expect to find a positive association between both variables. However, this association is purely coincidental and meaningless, or spurious. Remember that statistics is the analysis of data in context.

EXAMPLE 4.7 Nature, nurture, and lurking variables

The Kalamazoo (Michigan) Symphony once advertised a "Mozart for Minors" program with this statement: "Question: Which students scored 51 points higher in verbal skills and 39 points higher in math? Answer: Students who had experience in music."[9]

The advertisement was worded in a way suggesting that children's brains are hard-wired to respond to music developmentally. However, we could probably just as well answer "Students who played soccer." Why? Children with prosperous and well-educated parents are more likely than poorer children to have experience with music and also to play soccer. They are also likely to attend good schools, get good health care, and be encouraged to study hard. These advantages lead to high test scores. Family background is a lurking variable that explains, at least in part, why test scores are related to experience with music. ■

Spurious correlation A series of research papers described a puzzling, unexplainable positive correlation between the prevalence of doctors and infant mortality rate in developed countries. Notice, however, that both doctor prevalence and infant mortality rate are ratios of counts divided by the population size. In fact, a later paper reported a significant negative association between the number of physicians and infant mortality, after adjusting for population size. The first series of papers had failed to take into account population size as a lurking variable.

Attempts, such as this one by the Kalamazoo Symphony, to tease out the relative influence of large networks of genes and complex environmental factors typically come up against numerous lurking variables that must be carefully evaluated before any conclusion can be formed.

Lastly, an association can sometimes be markedly different for different groups. The failure to recognize such differences can lead to an incorrect interpretation of the association between two variables.

EXAMPLE 4.8 Men's physiology, women's physiology

High blood pressure and arterial stiffness are risk factors for cardiovascular disease, the leading cause of death worldwide. Understanding the physiology of cardiovascular health is important for developing preventive and curative approaches to the disease.

The sympathetic nervous system is known to affect vasoconstriction. A research team investigated the relationship between sympathetic nerve activity (measured in number of bursts per 100 heart beats) and an indicator of arterial stiffness (augmented aortic pressure, in mm Hg) in 44 healthy young adults.[10] The results, displayed in Figure 4.7(a), show no clear relationship between sympathetic nerve activity and the arterial stiffness indicator ($r = -0.17$).

Men and women have a somewhat different cardiovascular physiology, so the researchers decided to examine their findings separately for men and for women. The results are displayed in Figures 4.7(b) and (c), respectively. To their surprise, a moderate linear relationship was noticeable in both cases, but the relationship was positive for men ($r = 0.53$) and negative for women ($r = -0.58$).

In this case, combining data from men and women created the *appearance* of no relationship between sympathetic nerve activity and the arterial stiffness indicator. Gender is a lurking variable that drastically affects the relationship between the two quantitative variables studied. ■

APPLY YOUR KNOWLEDGE

4.10 The endangered manatee. Table 3.1 gives data on powerboats registered in Florida and manatees killed by collisions with powerboats from 1977 to 2012. Figure 3.1 (page 68) shows a strong positive linear relationship.

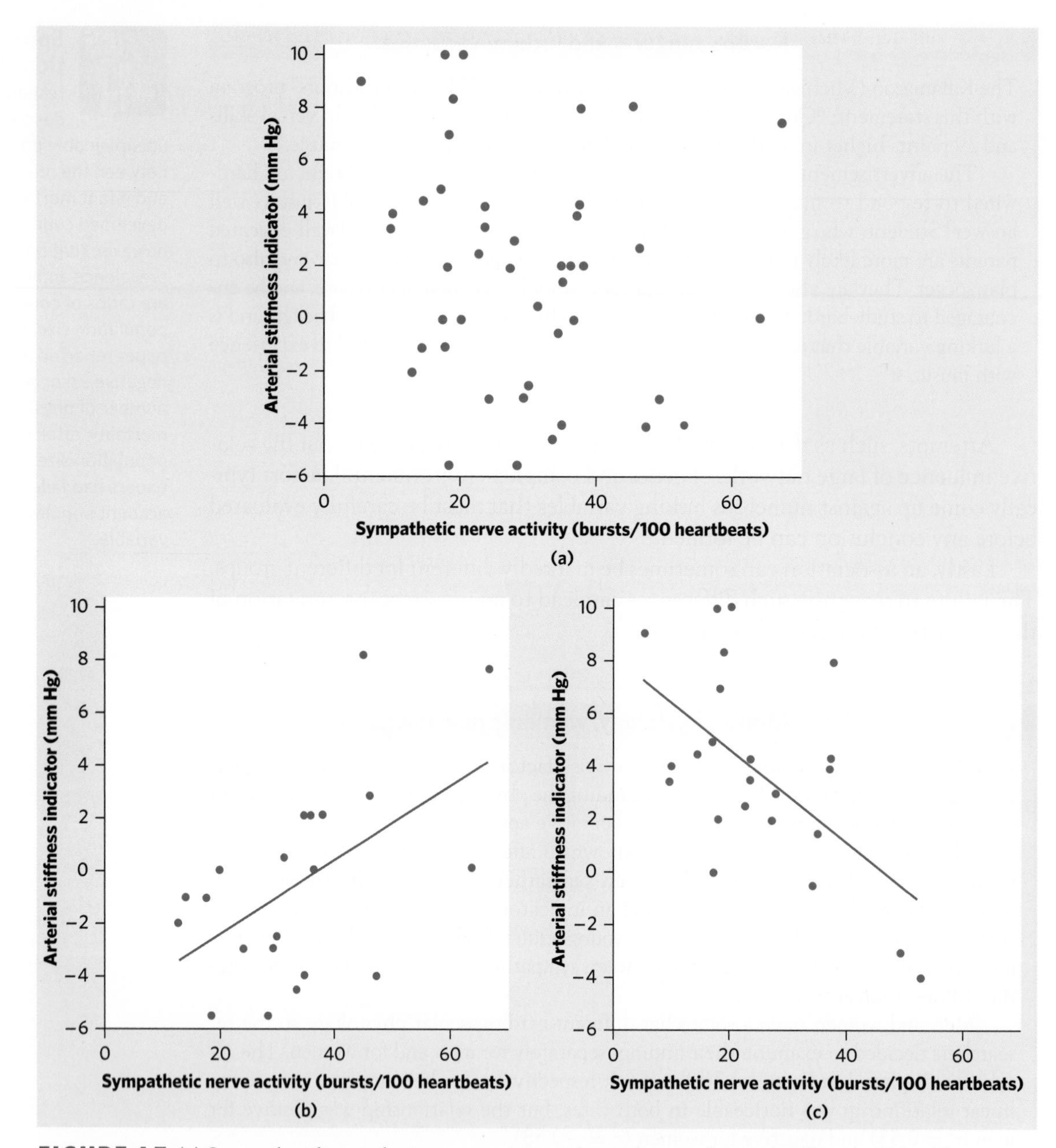

FIGURE 4.7 (a) Scatterplot of sympathetic nerve activity and an indicator of arterial stiffness in 44 healthy young adults, for Example 4.8. Notice the lack of any obvious association. (b)–(c) Scatterplot of the data in (a) after separating the data for men (b) and for women (c). Notice the opposite trends for men (positive linear association) and for women (negative linear association).

(a) Find the equation of the least-squares line for predicting manatees killed from thousands of boats registered. Because the linear pattern is so strong, we expect predictions from this line to be quite accurate—but only if conditions in Florida remain similar to those of the past 36 years.

(b) Manatees are an endangered species. Imagine that Florida limited powerboat registration in a given year to 800,000. Predict the number of manatees killed in such a year. (Pay attention to the fact that registrations in this data set are recorded in thousands.) Explain why we can trust this prediction.

(c) Now imagine that Florida took a drastic measure and limited the number of powerboat registrations to 200,000 per year. Explain why you shouldn't use the regression equation you found in (a) to predict the number of manatee deaths in this condition. Make the calculation anyway and comment on how this prediction indeed makes no sense at all.

4.11 **Blame the pharmacists?** Across American cities, the number of pharmacists and the number of deaths in a given year are positively associated. Should we limit the number of pharmacists allowed to work in our cities? What might influence the number of deaths in American cities? What might influence the number of pharmacists in American cities? What can you conclude from the positive association between pharmacists and deaths?

Association does not imply causation

Thinking about lurking variables leads to the most important caution about correlation and regression. When we study the relationship between two variables, we are often interested in whether changes in the explanatory variable *cause* changes in the response variable. A *strong association between two variables is not enough to draw conclusions about cause and effect*. Sometimes an observed association really does reflect cause and effect. A household that heats with natural gas uses more gas in colder months because cold weather requires burning more gas to stay warm. In other cases, an association is explained by lurking variables, and the conclusion that x causes y is either wrong or not proved.

EXAMPLE 4.9 Chocolate and Nobel Prizes

Even reputable publications like the *New England Journal of Medicine* can display a certain sense of humor ... Right around Nobel Prize season, the journal published a satirical editorial describing an analysis of the correlation between chocolate consumption x (in kilograms per year per capita) and the number of Nobel laureates y (per 10 million population) among nations that have ever received a Nobel Prize. A scatterplot in the editorial shows a positive linear trend with $r = 0.791$.[11]

The basic meaning of causation is that by changing x we can bring about a change in y. Could a country consume more chocolate to boost its number of Nobel laureates? No. As the editorial points out, chocolate consumption is computed at the population level and there is no indication of how much chocolate Nobel laureates actually consume. This makes it difficult to justify a plausible cause-and-effect mechanism between per capita chocolate consumption and per capita number of Nobel laureates. ■

Correlations such as that in Example 4.9 are sometimes called "nonsense correlations." The correlation is real. What is nonsense is the conclusion that changing one of the variables causes changes in the other. A lurking variable—such as possibly a country's wealth in Example 4.9—that influences both x and y can create a high correlation even though there is no direct connection between x and y.

ASSOCIATION DOES NOT IMPLY CAUSATION

An association between an explanatory variable x and a response variable y, even if it is very strong, is not by itself good evidence that changes in x actually cause changes in y.

EXAMPLE 4.10 Obesity in mothers and daughters

Obese parents tend to have obese children. But is obesity determined genetically? The results of a study of Mexican American girls aged 9 to 12 years are typical. The investigators measured body mass index (BMI), a measure of weight relative to height, for both the girls and their mothers. People with high BMI are overweight or obese. The correlation between the BMI of daughters and the BMI of their mothers was $r = 0.506$.[12]

Body type is in part determined by heredity. Daughters inherit half their genes from their mothers. There is therefore the potential for a direct causal link between the BMI of mothers and daughters. But perhaps mothers who are overweight also set an example of little exercise, poor eating habits, and lots of television. Their daughters may pick up these habits, so the influence of heredity is mixed up with influences from the girls' environment. Both likely contribute to the mother-daughter correlation. ■

The lesson of Example 4.10 is more subtle than just "association does not imply causation." *Even when direct causation is present, it may not be the whole explanation for a correlation.* You must still worry about lurking variables. Careful statistical studies try to anticipate lurking variables and measure them. The mother-daughter study did measure TV viewing, exercise, and diet. Elaborate statistical analysis can remove the effects of these variables to come closer to the direct effect of mother's BMI on daughter's BMI. This remains a second-best approach to causation. The best way to get good evidence that x causes y is to do an **experiment** in which we change x and keep lurking variables under control. We will discuss experiments in Chapter 8.

experiment

Cereals to lower your cholesterol? Many cereal packages and health foods claim that they may lower your cholesterol. Cheerios went a step further and sponsored a randomized, double-blind experiment, concluding from the study that "Cheerios is the only leading cold cereal clinically proven to help lower cholesterol in a low fat diet."

When experiments cannot be done, explaining an observed association can be difficult and controversial. Many of the sharpest disputes in which statistics plays a role involve questions of causation that cannot be settled by experiment. Does using cell phones cause brain tumors? Are our extreme levels of carbon dioxide emissions causing global warming? Is human activity triggering a new, planet-wide mass extinction? All these questions have become public issues. All concern associations among variables. And all have this in common: They try to pinpoint cause and effect in a setting involving complex relations among many interacting variables.

EXAMPLE 4.11 Does smoking cause lung cancer?

Despite the difficulties, it is sometimes possible to build a strong case for causation in the absence of experiments. The evidence that smoking causes lung cancer is about as strong as nonexperimental evidence can be.

Doctors had long observed that most lung cancer patients were smokers. Comparison of smokers and "similar" nonsmokers showed a very strong association between smoking and death from lung cancer. Could the association be explained by lurking variables? Might there be, for example, a genetic factor that predisposes people both to nicotine addiction and to lung cancer? Smoking and lung cancer would then be positively associated even if smoking had no direct effect on the lungs. How were these objections overcome? ■

James Leynse/CORBIS

Let's answer this question in general terms: What are the criteria for establishing causation when we cannot do an experiment?

- *The association is strong.* The association between smoking and lung cancer is very strong.
- *The association is consistent.* Many studies of different kinds of people in many countries link smoking to lung cancer. That reduces the chance that a lurking variable specific to one group or one study explains the association.
- *Higher doses are associated with stronger responses.* People who smoke more cigarettes per day or who smoke over a longer period get lung cancer more often. People who stop smoking reduce their risk.
- *The alleged cause precedes the effect in time.* Lung cancer develops after years of smoking. The number of men dying of lung cancer rose as smoking became more common, with a lag of about 30 years. Lung cancer kills more men than any other form of cancer. Lung cancer was rare among women until women began to smoke. Lung cancer in women rose along with smoking, again with a lag of about 30 years, and has now passed breast cancer as the leading cause of cancer death among women.
- *The alleged cause is plausible.* Experiments with animals show that tars from cigarette smoke do cause cancer.

Cigarette warning labels The history of cigarette warning labels illustrates how challenging establishing causation from evidence of association can be. Since the first Surgeon General's report on smoking and health there have been four successive warnings on packages sold in the United States:
1965—"Caution: Cigarette Smoking May Be Hazardous to Your Health."
1967—"Warning: Cigarette Smoking Is Dangerous to Health and May Cause Death from Cancer and Other Diseases."
1969—"Warning: The Surgeon General Has Determined That Cigarette Smoking Is Dangerous to Your Health."
1984—"SURGEON GENERAL'S WARNING: Smoking Causes Lung Cancer, Heart Disease, Emphysema, and May Complicate Pregnancy." The last is still in effect, along with three other warnings.

Medical authorities do not hesitate to say that smoking causes lung cancer. The U.S. Surgeon General has long stated that cigarette smoking is "the largest avoidable cause of death and disability in the United States."[13] The evidence for causation is overwhelming—but it is not as strong as the evidence provided by well-designed experiments.

APPLY YOUR KNOWLEDGE

4.12 **The health and wealth of nations.** There is a positive relationship between the wealth of nations (measured in dollars of income per capita adjusted for purchasing power) and the life expectancy at birth (measured in years) in these nations.[14] In part, this association reflects causation—wealth enables health through better nutrition and increased access to medical infrastructures. Suggest other explanations for the association between the wealth and health of nations. (Ask yourself how people's health can influence the economy.)

4.13 **Predicting the risk of breast cancer.** A large study of the detailed lifestyle of California Seventh-Day Adventists found that a woman's level of education is one of the better predictors of her risk of getting breast cancer later in life, with higher levels of education associated with higher risks of breast cancer.[15]

(a) Do you think that we should ask women to drop out of school in order to reduce the incidence of breast cancer? Is education a plausible candidate to be a direct cause of breast cancer? Explain.

(b) Another variable associated with the risk of breast cancer is the age at which a woman has her first child, with later ages associated with higher cancer rates. Is the age at first child a plausible cause of breast cancer? Explain. How does knowing about the association between age at first child and breast cancer help us understand the association between education and breast cancer?

CHAPTER 4 SUMMARY

- A **regression line** is a straight line that describes how a response variable y changes as an explanatory variable x changes. You can use a regression line to **predict** the value of y for any value of x by substituting this x into the equation of the line.
- The **slope** b of a regression line $\hat{y} = a + bx$ is the rate at which the predicted response $\hat{y}$ changes along the line as the explanatory variable x changes. Specifically, b is the change in $\hat{y}$ when x increases by 1.
- The **intercept** a of a regression line $\hat{y} = a + bx$ is the predicted response $\hat{y}$ when the explanatory variable $x = 0$. This prediction is of no statistical interest unless x can actually take values near 0.
- The most common method of fitting a line to a scatterplot is least squares. The **least-squares regression line** is the straight line $\hat{y} = a + bx$ that minimizes the sum of the squares of the vertical distances of the observed points from the line.
- The least-squares regression line of y on x is the line with slope rs_y/s_x and intercept $a = \bar{y} - b\bar{x}$. This line always passes through the point $(\bar{x}, \bar{y})$.
- **Correlation and regression** are closely connected. The correlation r is the slope of the least-squares regression line when we measure both x and y in standardized units. The **square of the correlation,** r^2, is the fraction of the variation in one variable that is explained by least-squares regression on the other variable.
- Correlation and regression must be **interpreted with caution. Plot the data** to be sure the relationship is roughly linear and to detect outliers and influential observations.
- Look for **influential observations,** individual points that substantially change the correlation or the regression line. Outliers in the x direction are often influential for the regression line.
- Avoid **extrapolation,** the use of a regression line for prediction for values of the explanatory variable well outside the range of the data from which the line was calculated.

- **Lurking variables** may explain the relationship between the explanatory and response variables. Correlation and regression can be misleading if you ignore important lurking variables.
- Most of all, be careful not to conclude that there is a cause-and-effect relationship between two variables just because they are strongly associated. **High correlation does not imply causation.** The best evidence that an association is due to causation comes from an **experiment** in which the explanatory variable is directly changed and other influences on the response are controlled.

THIS CHAPTER IN CONTEXT

In this chapter we use the least-squares regression line to describe the straight-line relationship between two quantitative variables when such a pattern is seen in a scatterplot (first introduced in Chapter 3). The equation of the least-squares regression line can be used to make predictions, as long as you are careful not to do so outside the range of values of the explanatory variable that you used to obtain the line. This approach rests on the assumption that the relationship is valid in some broader sense (an issue we will explore more carefully in Chapter 23 when reviewing simple linear regression in the context of statistical inference).

It is important to remember that association, as indicated by a strong correlation, does not imply that there is an underlying cause-and-effect relation between the response and explanatory variables. There may, in fact, be a lurking variable that influences the interpretation of any association between the response and explanatory variables. This is a concept we will revisit in Chapters 7 and 8 when we discuss the limitations arising from data collected in an observational, as opposed to an experimental, setting.

Lastly, in optional Chapter 28 we will expand the concept of simple linear regression to include multiple explanatory variables used to describe and predict one response variable (multiple linear regression). We will also see how a categorical variable can be coded with zeros and ones to allow it into the regression model (in multiple linear regression and in logistic regression).

CHECK YOUR SKILLS

4.14 Figure 4.8 is a scatterplot of tree frog mating-call frequency (in notes per second) against outside temperature (in degrees Celsius, °C) for 16 American bird-voiced tree frogs.[16] The line is the least-squares regression line for predicting call frequency from temperature. If another frog of this species initiates a mating call when the outside temperature is 24°C, you predict the call frequency (in notes/s) to be close to

(a) 5. (b) 7. (c) 9.

4.15 The slope of the line in Figure 4.8 is closest to (in notes/s/°C)

(a) –0.5. (b) 0. (c) 0.5.

4.16 In the frog mating-call data of Figure 4.8, temperature explains about 75% of the variability in call frequency. The correlation between call frequency and temperature is close to

(a) 0.5. (b) 0.75. (c) 0.85.

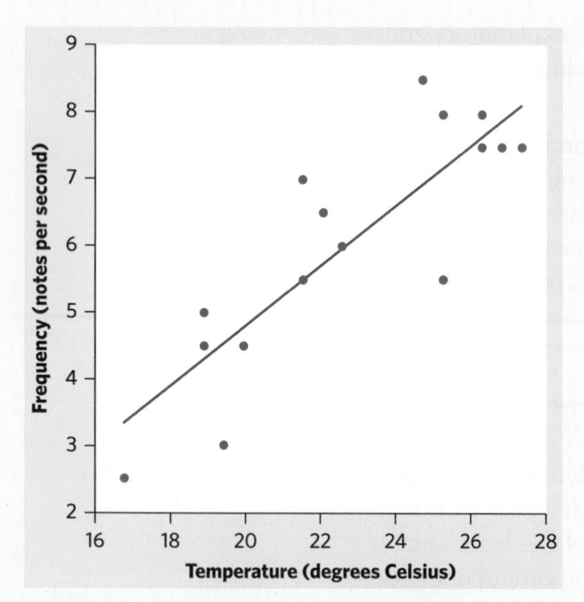

FIGURE 4.8 Outside temperature and frequency of mating calls for 16 tree frogs, for Exercise 4.14.

4.17 Given that the relationship between temperature and call frequency in Figure 4.8 is clearly linear and that temperature explains about 75% of the variability in call frequency, you conclude that

(a) temperature may influence mating-call frequency.
(b) temperature causes 75% of the variation in mating-call frequency.
(c) temperature and mating-call frequency are unrelated.

4.18 Marco's mom has measured his height in centimeters (cm) every few months between ages 4.5 and 7.5. The recorded values lie close to the line whose equation is $y = 80.46 + 6.25x$. Using this equation to find out whether Marco will be taller than his mom by age 14 (she is 168 cm tall) is an example of

(a) an influential point.
(b) prediction within range.
(c) extrapolation.

4.19 For a class project, you measure the weight in grams (g) and the tail length in millimeters (mm) of a group of mice. The equation of the least-squares line for predicting tail length from weight is

$$\text{predicted tail length} = 20 + 3 \times \text{weight}$$

How much (on average) does tail length increase for each additional gram of weight?

(a) 3 mm (b) 20 mm (c) 23 mm

4.20 According to the regression line in Exercise 4.19, the predicted tail length for a mouse weighing 18 g is

(a) 74 mm. (b) 54 mm. (c) 34 mm.

4.21 By looking at the equation of the least-squares regression line in Exercise 4.19, you can see that the correlation between weight and tail length is

(a) greater than zero.
(b) less than zero.
(c) Can't tell without seeing the data.

4.22 If you had measured the tail length in Exercise 4.19 in centimeters instead of millimeters, what would be the slope of the regression line? (There are 10 millimeters in a centimeter.)

(a) $3/10 = 0.3$ (b) 3 (c) $(3)(10) = 30$

4.23 Because elderly people may have difficulty standing to have their heights measured, a study looked at predicting overall height from height to the knee. Here are data (in centimeters, cm) for five elderly men:

Knee height x (cm)	57.7	47.4	43.5	44.8	55.2
Overall height y (cm)	192.1	153.3	146.4	162.7	169.1

Use your calculator or software: What is the equation of the least-squares regression line for predicting height from knee height?

(a) $\hat{y} = 2.4 + 44.1x$
(b) $\hat{y} = 44.1 + 2.4x$
(c) $\hat{y} = -2.5 + 0.32x$

CHAPTER 4 EXERCISES

4.24 Penguins diving. A study of king penguins looked for a relationship between how deep the penguins dive to seek food and how long they stay underwater.[17] For all but the shallowest dives, there is a linear relationship that is different for different penguins. The study report gives a scatterplot for one penguin titled "The relation of dive duration (DD) to depth (D)." Duration DD is measured in minutes, and depth D is in meters. The report then says, "The regression equation for this bird is: DD = 2.69 + 0.0138D."

(a) What is the slope of the regression line? Explain in specific language what this slope says about this penguin's dives.
(b) The dives varied from 40 to 300 meters in depth. According to the regression line, how long does a typical dive to a depth of 200 meters last?
(c) Plot the regression line from $x = 40$ to $x = 300$.

4.25 More on tara gum. The study described in Exercise 4.6 also examined the impact of tara gum on the perceived creaminess of fruit yogurt drinks. Participants were asked to rate creaminess on a scale of 0 to 100 (0 meaning not at all) for fruit yogurt drinks containing various concentrations of tara gum ranging from 0 to 0.47 g per 100 g of drink. The observed relationship between perceived creaminess and tara gum concentration was clearly linear and very strong, with $r^2 = 0.95$. Here is the equation of the least-squares regression line:

$$\text{creaminess} = 29.38 + 101.98 \text{ concentration}$$

(a) What is the slope of the regression line? Is the relationship between creaminess and tara gum concentration positive or negative?
(b) What is the predicted perceived creaminess for a tara gum concentration of 0.2 g/100 g? Use r^2 to assess the reliability of this prediction.
(c) Using this model, what would be the predicted perceived creaminess for a tara gum concentration of 1 g/100 g? Explain why this is *not* a legitimate prediction.

4.26 Grading oysters. Exercise 3.28 (page 84) describes an automated computerized grading system for oysters based on 2D visual reconstruction. Table 3.3 gives the actual volume (in cm^3) of 30 oysters and the pixel area reconstruction (in thousands of pixels) produced by the computer program. The relationship is clearly linear.

(a) If you haven't done so already, make a scatterplot of the relationship between actual volume (response) and 2D volume reconstruction (explanatory). Find the equation for the corresponding least-squares regression line.
(b) Use the regression equation to predict the actual volume of an oyster with a 2D reconstruction of 35 thousand pixels.
(c) How accurate is this prediction? What percent of the variations in actual oyster volume can be explained by the regression model?

4.27 More on grading oysters. The previous exercise describes an automated grading system for oysters. Engineers were given the task of improving on the 2D reconstruction software. They designed a new program that estimates oyster volume using 3D image processing. Table 3.3 (page 84) gives the actual volume (in cm^3) of 30 oysters and the 3D volume reconstruction (in millions of volume pixels).

(a) If you haven't done so already, make a scatterplot of the relationship between actual volume (response) and 3D volume reconstruction (explanatory). Find the equation for the corresponding least-squares regression line.
(b) Use the regression equation to predict the actual volume of an oyster with a 3D reconstruction of 4.5 million volume pixels.
(c) How accurate is this prediction? What percent of the variations in actual oyster volume can be explained by the 3D regression model? How does this compare with the 2D system from the previous exercise?

4.28 Keeping water clean. Keeping water supplies clean requires regular measurement of levels of pollutants. The measurements are indirect—a typical analysis involves forming a dye by a chemical reaction with the dissolved pollutant, then passing light through the solution and measuring its "absorbance." To calibrate such measurements, the laboratory measures known standard solutions and uses regression to relate absorbance and pollutant concentration. This is usually done every day. Here is one series of data on the absorbance for different levels of nitrates. Nitrates are measured in milligrams per liter of water.[18]

Nitrates	50	50	100	200	400	800	1200	1600	2000	2000
Absorbance	7.0	7.5	12.8	24.0	47.0	93.0	138.0	183.0	230.0	226.0

(a) Chemical theory says that these data should lie on a straight line. If the correlation is not at least 0.997, something went wrong, and the calibration procedure is repeated. Plot the data and find the correlation. Must the calibration be done again?
(b) The calibration process sets nitrate level and measures absorbance. Once established, the linear relationship will be used to estimate the nitrate level in water from a measurement of absorbance. What is the equation of the line used for estimation? What is the estimated nitrate level in a water specimen with absorbance 40?
(c) Do you expect estimates of nitrate level from absorbance to be quite accurate? Why?

4.29 Counting carnivores. Ecologists look at data to learn about nature's patterns. One pattern they have found relates

TABLE 4.1 Size and abundance of carnivores

Carnivore species	Body mass (kg)	Abundance	Carnivore species	Body mass (kg)	Abundance
Least weasel	0.14	1656.49	Eurasian lynx	20.0	0.46
Ermine	0.16	406.66	Wild dog	25.0	1.61
Small Indian mongoose	0.55	514.84	Dhole	25.0	0.81
Pine marten	1.3	31.84	Snow leopard	40.0	1.89
Kit fox	2.02	15.96	Wolf	46.0	0.62
Channel Island fox	2.16	145.94	Leopard	46.5	6.17
Arctic fox	3.19	21.63	Cheetah	50.0	2.29
Red fox	4.6	32.21	Puma	51.9	0.94
Bobcat	10.0	9.75	Spotted hyena	58.6	0.68
Canadian lynx	11.2	4.79	Lion	142.0	3.40
European badger	13.0	7.35	Tiger	181.0	0.33
Coyote	13.0	11.65	Polar bear	310.0	0.60
Ethiopian wolf	14.5	2.70			

the size of a carnivore (body mass in kilograms) to how many of those carnivores there are in an area. One way to measure "how many" is to count carnivores per 10,000 kilograms of their prey in the area. Table 4.1 gives average data for 25 carnivore species.[19]

(a) Plot and describe the relationship between carnivore abundance (response) and body mass (explanatory). Would it be appropriate to conduct a linear regression analysis for the data as is? Explain your answer.

(b) When a relationship is clearly not linear, transforming the data can sometimes help. For each species, compute the logarithm of body mass and the logarithm of abundance. Plot and describe the relationship between the log of abundance (response) and the log of body mass (explanatory). Is linear regression appropriate for the transformed data?

(c) The least-squares regression for the transformed data gives

$$\log(\text{abundance}) = 1.95 - [1.05 \times \log(\text{body mass})]$$

Predict the abundance of a carnivore species whose individuals weigh about 1 kg. Make sure to transform the units both ways (remember that $y = 10^{\log(y)}$).

4.30 The intriguing cerebellum. The cerebellum is a highly convoluted brain structure sitting underneath the cerebral hemispheres. This intriguing structure is thought to facilitate the acquisition and use of sensory data by the rest of the brain, particularly the motor areas. Studies suggest that the cerebellum may scale up with the size of animals' bodies and brains, whereas other parts of the brain are clearly represented differentially across species. Here are data on 15 mammal species showing the weights in grams for their bodies and cerebellums, along with the logarithm-transformed values:[20]

Species	Body (g)	Cerebellum (g)	Log (body)	Log (cerebellum)
Mouse	58	0.09	1.76	−1.05
Bat	30	0.09	1.48	−1.05
Flying fox	130	0.30	2.11	−0.52
Pigeon	500	0.40	2.70	−0.40
Guinea pig	485	0.90	2.69	−0.05
Squirrel	350	1.50	2.54	0.18
Chinchilla	500	1.70	2.70	0.23
Rabbit	1,800	1.90	3.26	0.28
Hare	3,000	2.30	3.48	0.36
Cat	3,500	5.30	3.54	0.72
Dog	3,500	6.00	3.54	0.78
Macaque	6,000	7.80	3.78	0.89
Sheep	25,000	21.50	4.40	1.33
Bovine	300,000	35.70	5.48	1.55
Human	60,000	142.00	4.78	2.15

(a) Make a scatterplot of cerebellum weight (y) against body weight (x) and another scatterplot using the transformed values. Describe and compare both scatterplots. What did you learn about the relationship between cerebellum weight and body weight from these graphs?

(b) Find the equation for the regression line using the log-transformed values and add it to the scatterplot. In the logarithm scales, what proportion of the variation in cerebellum weight can be explained by the variation in body weight?

(c) Predict the cerebellum weight in grams of a species weighing 100,000 g (100 kg, or about 220 pounds). Make sure to transform the grams into log units first and then back into grams (remember that $y = 10^{\log(y)}$).

4.31 Predicting tree height. Measuring tree height is not an easy task. How well does trunk diameter predict tree height? A survey of 958 live trees in an old-growth forest in Canada provides us with the following information: The mean tree height is 15.6 meters (m), with standard deviation 13.4 m; the mean diameter, measured at "breast height" (1.3 m aboveground), is 23.4 centimeters (cm), and the standard deviation is 23.5 cm. The correlation between the height and diameter is very high: $r = 0.96$.[21]

(a) What are the slope and intercept of the regression line to predict tree height from trunk diameter? Draw a graph of this regression line.

(b) The tree diameters ranged from 1 to 101 cm. Predict the height of a tree in this area that is 50 cm in diameter at breast height. Use r^2 to assess the reliability of this prediction.

4.32 A toucan's beak. Exercise 3.35 (page 86) gives data on toucan beak heat loss, as a percent of total body heat loss from all sources, at various temperatures. The data show that beak heat loss is higher at higher temperatures and that the relationship is roughly linear. Figure 4.9 shows Minitab regression output for these data.

(a) What is the equation of the least-squares regression line for predicting beak heat loss, as a percent of total body heat loss, from temperature? Use the equation to predict beak heat loss, as a percent of total body heat loss from all sources, at a temperature of 25 degrees Celsius.

(b) What percent of the variation in beak heat loss is explained by the straight-line relationship with temperature?

(c) Use the information in Figure 4.9 to find the correlation r between beak heat loss and temperature. How do you know whether the sign of r is + or −?

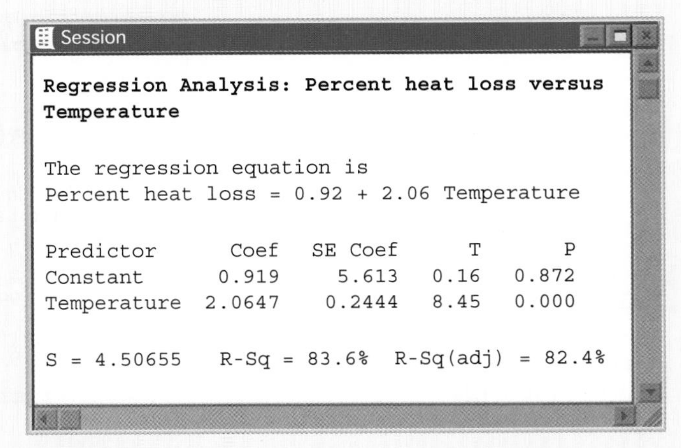

Session

Regression Analysis: Percent heat loss versus Temperature

The regression equation is
Percent heat loss = 0.92 + 2.06 Temperature

Predictor	Coef	SE Coef	T	P
Constant	0.919	5.613	0.16	0.872
Temperature	2.0647	0.2444	8.45	0.000

S = 4.50655   R-Sq = 83.6%   R-Sq(adj) = 82.4%

FIGURE 4.9 Minitab regression output for a study of the effect of temperature on toucan beak heat loss, for Exercise 4.32.

4.33 Does social rejection hurt? Exercise 3.38 (page 87) gives data from a study that shows that social exclusion causes "real pain." That is, activity in an area of the brain that responds to physical pain goes up as distress from social exclusion goes up. A scatterplot shows a moderately strong linear relationship. Figure 4.10 shows regression output from software for these data.

Results - Simple Linear Regression

Export ▾

Fitted Equation: Brain = −0.1261 + 0.06078 * Distress

	Estimate	Std. Error	t value	Pr(>|t|)	CI
(Intercept)	−0.1261	0.02465	−5.116	0.0003357	(−0.1803, −0.07184)
Distress	0.06078	0.009979	6.091	<0.0001	(0.03882, 0.08275)

estimated sigma: 0.02509

FIGURE 4.10 CrunchIt! regression output for a study of the effects of social rejection on brain activity, for Exercise 4.33.

Results - Simple Linear Regression

Export ▾

Fitted Equation: Pct = 157.7 − 2.993 * Pairs

	Estimate	Std. Error	t value	Pr(>|t|)	CI
(Intercept)	157.7	27.68	5.696	0.0007385	(92.22, 223.1)
Pairs	−2.993	0.8655	−3.459	0.01057	(−5.040, −0.9469)

estimated sigma: 9.463

FIGURE 4.11 CrunchIt! regression output for a study of how breeding success affects survival in birds, for Exercise 4.34.

(a) What is the equation of the least-squares regression line for predicting brain activity from social distress score? Use the equation to predict brain activity for a social distress score of 2.0.

(b) What percent of the variation in brain activity among these subjects is explained by the straight-line relationship with social distress score?

4.34 Merlins breeding. Exercise 3.36 (page 86) gives data on the number of breeding pairs of merlins in an isolated area in each of nine years and the percent of males who returned the next year. The data show that the percent returning is lower after successful breeding seasons and that the relationship is roughly linear. Figure 4.11 shows software regression output for these data.

(a) What is the equation of the least-squares regression line for predicting the percent of males that return from the number of breeding pairs? Use the equation to predict the percent of returning males after a season with 30 breeding pairs.

(b) What percent of the year-to-year variation in percent of returning males is explained by the straight-line relationship with number of breeding pairs the previous year?

4.35 Always plot your data! Table 4.2 presents four sets of data prepared by the statistician Frank Anscombe to illustrate the dangers of calculating without first plotting the data.[22]

(a) Without making scatterplots, find the correlation and the least-squares regression line for all four data sets. What do you notice? Use the regression line to predict y for $x = 10$.

(b) Make a scatterplot for each of the data sets and add the regression line to each plot.

(c) In which of the four cases would you be willing to use the regression line to describe the dependence of y on x? Explain your answer in each case.

4.36 Toxicology of aspartame. The National Toxicology Program evaluates the toxicology of substances suspected of

TABLE 4.2 Four data sets for exploring correlation and regression

Data Set A	x	10	8	13	9	11	14	6	4	12	7	5
	y	8.04	6.95	7.58	8.81	8.33	9.96	7.24	4.26	10.84	4.82	5.68
Data Set B	x	10	8	13	9	11	14	6	4	12	7	5
	y	9.14	8.14	8.74	8.77	9.26	8.10	6.13	3.10	9.13	7.26	4.74
Data Set C	x	10	8	13	9	11	14	6	4	12	7	5
	y	7.46	6.77	12.74	7.11	7.81	8.84	6.08	5.39	8.15	6.42	5.73
Data Set D	x	8	8	8	8	8	8	8	8	8	8	19
	y	6.58	5.76	7.71	8.84	8.47	7.04	5.25	5.56	7.91	6.89	12.50

carcinogenicity (causing cancer). One such study evaluated the impact of aspartame added to the diet on the survival rates of mice. For each aspartame concentration studied (in parts per million, ppm), 30 mice were fed and monitored for 40 weeks, and the group's survival rate was recorded as the percent of mice still alive at the end of the 40 weeks. Here are the results of this study:[23]

Concentration (ppm)	0	3125	6250	12,500	25,000	50,000
Survival rate (%)	67	73	57	70	73	60

(a) Make a scatterplot of mice survival rate (response) against aspartame concentration (explanatory) and add the regression line. Describe the direction, form, and strength of the relationship. What percent of the variation in mice survival rate can be explained by variations in aspartame concentrations?

(b) Make another scatterplot, this time ignoring the last data point (an aspartame concentration of 50,000 ppm). Add the regression line. What happened to the regression line? What does this say about the data point corresponding to the aspartame concentration of 50,000 ppm?

(c) Make another scatterplot, this time with all the data points except the one for an aspartame concentration of 6250 ppm. Add the regression line. What happened to the regression line? What does this say about the data point you just removed?

4.37 Drilling into the past. Drilling down beneath a lake in Alaska yields chemical evidence of past changes in climate. Biological silicon, left by the skeletons of single-celled creatures called diatoms, is a measure of the abundance of life in the lake. A rather complex variable based on the ratio of certain isotopes relative to ocean water gives an indirect measure of moisture, mostly from snow. As we drill down, we look further into the past. Here are data from 2300 to 12,000 years ago:[24]

Isotope (%)	Silicon (mg/g)	Isotope (%)	Silicon (mg/g)	Isotope (%)	Silicon (mg/g)
−19.90	97	−20.71	154	−21.63	224
−19.84	106	−20.80	265	−21.63	237
−19.46	118	−20.86	267	−21.19	188
−20.20	141	−21.28	296	−19.37	337

(a) Make a scatterplot of silicon (response) against isotope (explanatory). Ignoring the outlier, describe the direction, form, and strength of the relationship. The researchers say that this and relationships among other variables they measured are evidence for cyclic changes in climate that are linked to changes in the sun's activity.

(b) The researchers single out one point: "The open circle in the plot is an outlier that was excluded in the correlation analysis." Circle this outlier on your graph. What is the correlation with and without this point? The point strongly influences the correlation. Explain why the outlier moves r in the direction revealed by your calculations.

4.38 Managing diabetes. People with diabetes must manage their blood sugar levels carefully. They measure their fasting plasma glucose (FPG; in milligrams per milliliter of plasma) several times a day with a glucose meter. Another measurement, made at regular medical checkups, is called HbA. This is roughly the percent of red blood cells that have a glucose molecule attached. It measures average exposure to glucose over a period of several months. Table 4.3 gives data on both HbA and FPG for 18 diabetic people five months after they had completed a diabetes education class.[25]

TABLE 4.3 Two measures of glucose level in diabetics

Subject	HbA (%)	FPG (mg/ml)	Subject	HbA (%)	FPG (mg/ml)	Subject	HbA (%)	FPG (mg/ml)
1	6.1	141	7	7.5	96	13	10.6	103
2	6.3	158	8	7.7	78	14	10.7	172
3	6.4	112	9	7.9	148	15	10.7	359
4	6.8	153	10	8.7	172	16	11.2	145
5	7.0	134	11	9.4	200	17	13.7	147
6	7.1	95	12	10.4	271	18	19.3	255

(a) Make a scatterplot with HbA as the explanatory variable. There is a positive linear relationship, but it is surprisingly weak.

(b) Subject 15 is an outlier in the y direction. Subject 18 is an outlier in the x direction. Find the correlation for all 18 subjects, for all except Subject 15, and for all except Subject 18. Are either or both of these subjects influential for the correlation? Explain in simple language why r changes in opposite directions when we remove each of these points.

4.39 Drilling into the past, continued. Is the outlier in Exercise 4.37 also strongly influential for the regression line? Calculate and draw on your graph two regression lines, then discuss what you see. Explain why adding the outlier moves the regression line in the direction shown on your graph.

4.40 Managing diabetes, continued. Add three regression lines for predicting FPG from HbA to your scatterplot from Exercise 4.38: for all 18 subjects, for all except Subject 15, and for all except Subject 18. Is either Subject 15 or Subject 18 strongly influential for the least-squares line? Explain in simple language what features of the scatterplot explain the degree of influence.

4.41 Climate change and ecosystems. Many studies have documented a clear pattern of climate change toward warmer temperatures globally. What has been the subject of controversies, however, is the role of human activity and the impact of climate change on ecosystems. A survey of the North Sea over a 25-year period (1977 to 2001) shows that, as winter sea temperatures have warmed up, fish populations have moved farther north. Here are the data for mean winter sea bottom temperature (in degrees Celsius) and mean northern latitude (in degrees) of anglerfish populations each year over the 25-year study period:

Temperature (°C)	Latitude (°)	Temperature (°C)	Latitude (°)	Temperature (°C)	Latitude (°)
6.26	57.20	6.52	57.72	7.09	58.07
6.26	57.96	6.68	57.83	7.13	58.49
6.27	57.65	6.76	57.87	7.15	58.28
6.31	57.59	6.78	57.48	7.29	58.49
6.34	58.01	6.89	58.13	7.34	58.01
6.32	59.06	6.90	58.52	7.57	58.57
6.37	56.85	6.93	58.48	7.65	58.90
6.39	56.87	6.98	57.89		
6.42	57.43	7.02	58.71		

(a) Make a scatterplot of the relationship between latitude of anglerfish populations (y) and temperature (x). Describe the form and strength of that relationship.

(b) Give the equation for the regression line to predict latitude from temperature and the value of the correlation. Add the regression line to your scatterplot.

(c) The scatterplot shows one outlier of the relationship. Take that point out of the data set and calculate the new regression line and correlation. Add the new regression line to your original scatterplot, making sure to use a different color or line thickness.

(d) How do the regression line and correlation omitting the outlier compare with the original ones in (b)? Use your findings to discuss how influential the outlier is.

4.42 Influence in regression. The *Correlation and Regression* applet allows you to animate Figure 4.6. Click to create a group of 10 points in the lower-left corner of the scatterplot with a strong straight-line pattern (correlation about 0.9). Click the "Show least-squares line" box to display the regression line.

(a) Add one point at the upper right that is far from the other 10 points but exactly on the regression line. Why does this outlier have no effect on the line even though it changes the correlation?

(b) Now use the mouse to drag this last point straight down. You see that one end of the least-squares line chases this single point, while the other end remains near the middle of the original group of 10. What makes the last point so influential?

4.43 Another reason not to smoke? A stop-smoking booklet says, "Children of mothers who smoked during pregnancy scored nine points lower on intelligence tests at ages three and four than children of nonsmokers." Suggest some lurking variables that may help explain the association between smoking during pregnancy and children's later test scores. The association by itself is not good evidence that mothers' smoking *causes* lower scores.

4.44 How's your self-esteem? People who do well tend to feel good about themselves. Perhaps helping people feel good about themselves will help them do better in school and life. Raising self-esteem became for a time a goal in many schools. California even created a state commission to advance the cause, hoping that it would help reduce crime, drug use, and teen pregnancy (the commission was ended in 1995). Can you think of explanations for the association between high self-esteem and good school performance other than "Self-esteem causes better work in school"?

4.45 Are big hospitals bad for you? A study shows that there is a positive correlation between the size of a hospital (measured by its number of beds x) and the median number of days y that patients remain in the hospital. Does this mean that you can shorten a hospital stay by choosing a small hospital? Why?

4.46 Beavers and beetles. Ecologists sometimes find rather strange relationships in our environment. For example, do beavers benefit beetles? Researchers laid out 23 circular plots, each 4 meters in diameter, in an area where beavers were cutting down cottonwood trees. In each plot, they counted the number of stumps from trees cut by beavers and the number of clusters of beetle larvae. Ecologists think that the new sprouts from stumps are more tender than other cottonwood growth, so that beetles prefer them. If so, more stumps should produce more beetle larvae. Here are the data:[26]

Stumps	2	2	1	3	3	4	3	1	2	5	1	3
Beetle larvae	10	30	12	24	36	40	43	11	27	56	18	40

Stumps	2	1	2	2	1	1	4	1	2	1	4
Beetle larvae	25	8	21	14	16	6	54	9	13	14	50

Analyze these data to see if they support the "beavers benefit beetles" idea. Follow the four-step process (page 57) in reporting your work.

4.47 Insect reproduction. Reproduction is very taxing on an individual's physiology. What is the impact of better nutrition on reproduction? Insects are an easy model with which to answer this question because they produce large numbers of eggs and reach maturity very fast. Female fruit flies were kept in closed glass vials and fed a diet containing varying amounts (in milligrams per vial) of yeast, a good source of proteins for fruit flies. The average number of eggs produced per female was then computed for each vial. Here are the data:[27]

Yeast (mg)	0	0	0	0	0	1	1	1	1	1
Eggs	9.85	6.25	10.05	5.2	7.15	21.35	25.95	21.1	14.94	20.6

Yeast (mg)	3	3	3	3	3	7	7	7	7	7
Eggs	36.75	58.05	67.05	42.85	49.2	63.8	86.45	82.6	73.35	85.5

Analyze these data to see if they support the notion that better nutrition (here, more yeast) improves the reproductive capacity of fruit flies. Follow the four-step process (page 57) in reporting your work.

4.48 Is regression useful? In Exercise 3.39 (page 87) you used the *Correlation and Regression* applet to create three scatterplots having correlation about $r = 0.7$ between the horizontal variable x and the vertical variable y. Create three similar scatterplots again and click the "Show least-squares line" box to display the regression lines. Correlation $r = 0.7$ is considered reasonably strong in many areas of work. Because there is a reasonably strong correlation, we might use a regression line to predict y from x. In which of your three scatterplots does it make sense to use a straight line for prediction?

4.49 Lung capacity of children. Continue your work from Exercises 1.44 and 3.40 (pages 37 and 88) with the data file *Large.FEV*. Based on the graphs produced in Exercise 3.40, we now study adolescents (middle school and high school: ages 11 and older) separately from the younger children (preschool and elementary school: ages 10 and younger).

(a) Among the younger children, boys and girls are similar enough to be studied together. Plot the relationship between FEV and height for these younger children and obtain the least-squares regression equation. What percent of the variation in FEV is explained by this model? Predict the FEV of a child who is 5 feet tall (60 inches).

(b) The graphs you created in Exercise 3.40 suggest that, among adolescents, boys and girls should be studied separately. Plot the relationship between FEV and height for adolescent girls and obtain the least-squares regression equation. What percent of the variation in FEV is explained by this model? Predict the FEV of an adolescent girl who is 68 inches tall.

(c) Plot the relationship between FEV and height for adolescent boys and obtain the least-squares regression equation. What percent of the variation in FEV is explained by this model? Predict the FEV of an adolescent boy who is 68 inches tall.

4.50 Children's growth. You can use the data file *Large.FEV* from the previous exercise to study growth patterns in children.

(a) Plot the relationship between age in years (explanatory) and height in inches (response) in boys and girls in two separate graphs. Compare both graphs. How do boys and girls differ? In what way are their growth patterns similar?

(b) As in the previous exercise, the nonlinear pattern in growth over the whole age range suggests that we should study younger children (ages 10 and younger) separately from the adolescents (ages 11 and older). Plot the relationship between age and height for the younger children and obtain the least-squares regression equation. What percent of the variation in height is explained by this model?

(c) Plot the relationship between age and height for adolescent girls and obtain the least-squares regression equation. What percent of the variation in height is explained by this model? Do the same for adolescent boys. Describe the main differences between the growth patterns of adolescent boys and girls.

4.51 Estimating body fat in men. Continue your work from Exercise 3.42 (page 88) with the data file *Large.Bodyfat*. Accurate measurements of the percent body fat require underwater weighing. This is time-consuming and expensive. We'd like to know if there are simpler body measurements that can provide a good approximation of a person's percent body fat.

LARGE DATA SET

(a) Make a series of scatterplots with percent fat1 on the y axis (response) and each of the 10 body circumference variables, in turn, on the x axis. If appropriate, calculate r^2 in each case. Which of these 10 body circumference variables seems like the best candidate to use as a predictor for percent body fat?

(b) Find the equation for the least-squares regression predicting percent fat1 from abdomen circumference. What percent of the variation in percent fat1 is explained by this model? What percent body fat would you predict for an adult man with an abdomen circumference of 100 cm?

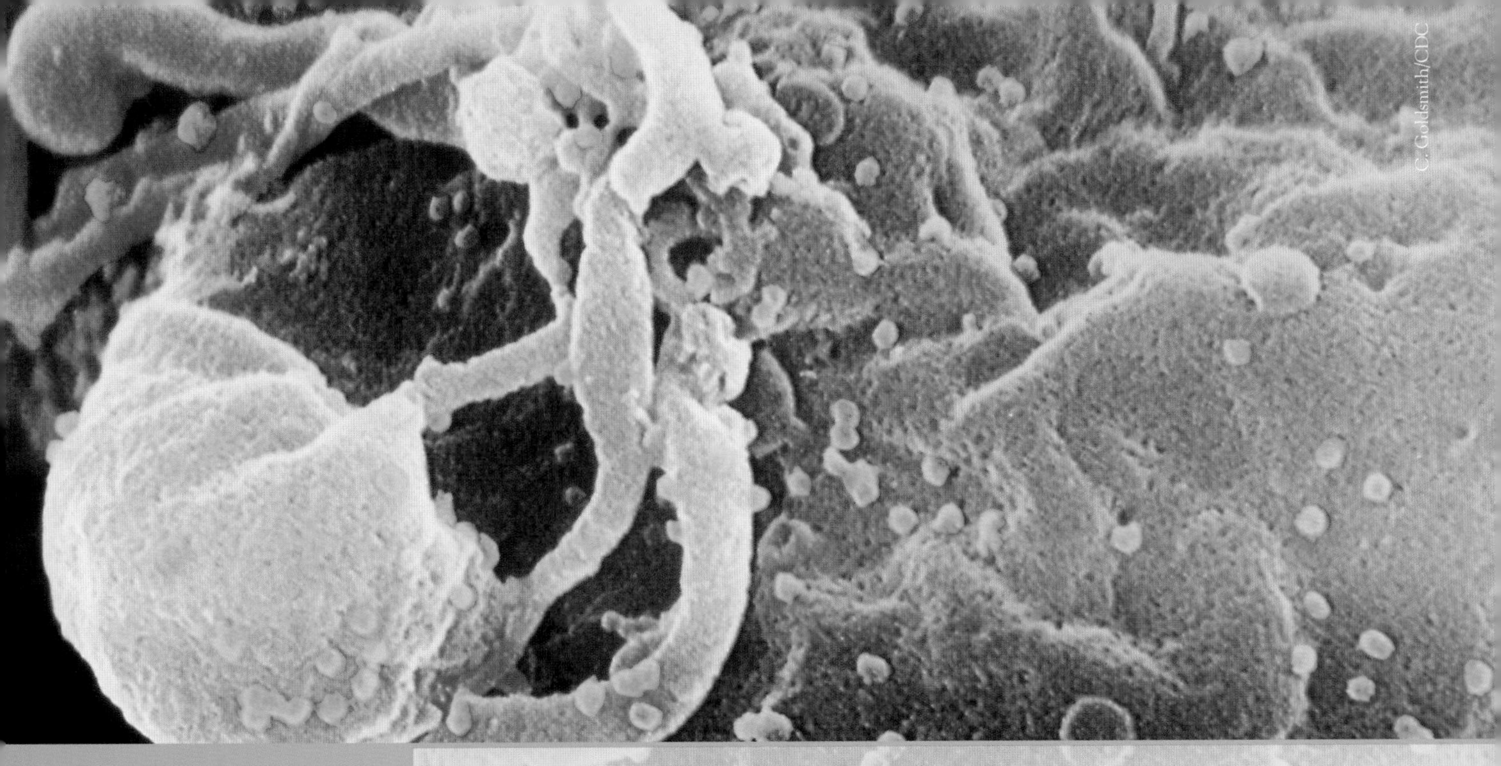

CHAPTER 5 Two-Way Tables*

We have concentrated on relationships in which at least the response variable is quantitative. Now we will describe relationships between two categorical variables. Some variables—such as sex, species, and color—are categorical by nature. Other categorical variables are created by grouping values of a quantitative variable into classes—like age groups, for example. Published data often appear in grouped form to save space. To analyze categorical data, we use the *counts* or *percents* of individuals that fall into various categories.

IN THIS CHAPTER WE COVER...

- Marginal distributions
- Conditional distributions
- Simpson's paradox

EXAMPLE 5.1 Foot health

Foot pain is a very common musculoskeletal complaint in the United States, especially among older adults. Foot pain is more frequent among women than among men, possibly because men and women wear different types of shoes over their lifetimes.

A research group surveyed individuals from a large established cohort of residents from Framingham, Massachusetts, a group now made up of mainly older individuals (65 years of age, on average). The participants were asked to select from a list which type of footwear they had worn most regularly in the past. The answers were then categorized as shoes providing good, average, or poor structural foot support. Table 5.1 presents the results for the men and the women in the study.[1]

*This material is important in statistics, but it is needed later in this book only for the section on conditional probabilities and the chapter on the chi-square test. You may omit it if you do not plan to cover these topics, or delay reading it until you reach them.

TABLE 5.1 Study participants by shoe type and gender

	Gender		
Shoe type	Men	Women	Total
Good support	94	137	231
Average support	1348	581	1929
Poor support	30	1182	1212
Total	1472	1900	3372

two-way table

row and column variables

Both variables, shoe type and gender, are categorical. Table 5.1 is a **two-way table** because it describes the relationship between these two categorical variables. Shoe type is the **row variable** because each row in the table describes one type of shoe classified in the study by the structural support it provides. Gender is the **column variable** because each column describes either men or women. The entries in the table are the counts of study participants in each support-by-gender class. ■

Marginal distributions

How can we best grasp the information contained in Table 5.1? First, *look at the distribution of each variable separately*. The distribution of a categorical variable says how often each outcome occurred. The "Total" column at the right of the table contains the totals for each of the rows. These row totals give the distribution of shoe types (the row variable) in the study: 231 participants had worn mainly shoes providing good support; 1929, average support; and 1212, poor support. In the same way, the "Total" row at the bottom of the table gives the gender distribution: the study interviewed 1472 men and 1900 women.

marginal distribution

If the row and column totals are missing, the first thing to do in studying a two-way table is to calculate them. The distributions of shoe type alone and gender alone are called **marginal distributions** because they appear at the right and bottom margins of the two-way table.

Percents are often more informative than counts. We can display the marginal distribution of shoe types in terms of percents by dividing each row total by the table total and converting to a percent.

EXAMPLE 5.2 Calculating a marginal distribution

The percent of these study participants who wore mostly shoes providing good support is

$$\frac{\text{good support total}}{\text{table total}} = \frac{231}{3372} = 0.0685 = 6.9\%$$

Do two more such calculations to obtain the marginal distribution of shoe types in percents. Here is the complete distribution:

Shoe type	Percent
Good support	$\frac{231}{3372} = 6.9\%$
Average support	$\frac{1929}{3372} = 57.2\%$
Poor support	$\frac{1212}{3372} = 35.9\%$

It seems that only a small percent of individuals in the study tended to wear mainly shoes providing good support. The total sums to 100% because everyone in the study selected one type of shoe as the one they tended to wear most often. ■

Sometimes percents add up to slightly less or more than 100% because we round the computed values. This is called **roundoff error.**

roundoff error

Each marginal distribution from a two-way table is a distribution for a single categorical variable. As we saw in Chapter 1, we can use a bar graph or a pie chart to display such a distribution. Figure 5.1 is a bar graph of the distribution of shoe types for all the study participants.

In working with two-way tables, you must calculate lots of percents. Here's a tip to help decide what fraction gives the percent you want. Ask, "What group represents the total that I want a percent of?" The count for that group is the denominator of the fraction that leads to the percent. In Example 5.2, we want a percent of "study participants," so the count of study participants (the table total) is the denominator.

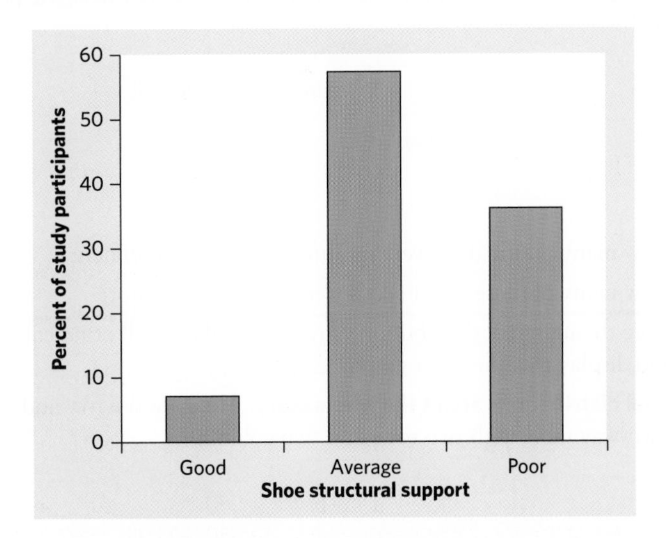

FIGURE 5.1 A bar graph of the distribution of shoes worn by the type of structural support they offer. This is one of the marginal distributions for Table 5.1.

APPLY YOUR KNOWLEDGE

5.1 **HIV vaccine boost.** Creating an effective HIV vaccine has been very challenging. A new approach combines a DNA vaccine with a delivery system using electrical impulses (electroporation) to boost the immune response. A preliminary study of the method's efficacy was conducted on adult volunteers. Each volunteer received one of five possible treatment options and a follow-up clinical examination to assess his or her immune response to the vaccine. Here is a two-way table that shows the results of the clinical exam after two immunization sessions:[2]

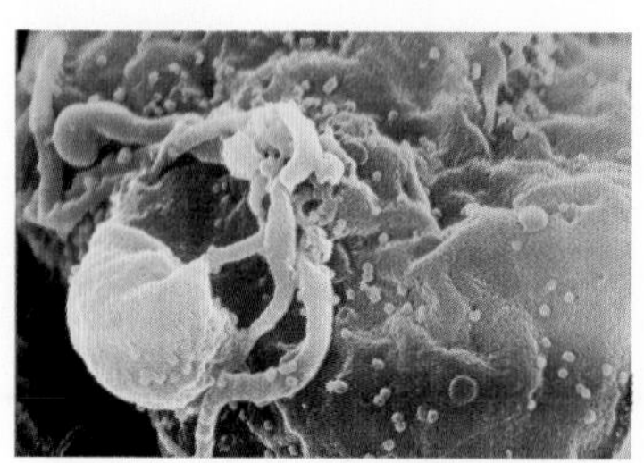

C. Goldsmith/CDC

	Immune response	No immune response
0.2 mg vaccine via electroporation	3	5
1 mg vaccine via electroporation	7	1
4 mg vaccine via electroporation	6	2
4 mg vaccine intramuscularly	0	8
Saline solution via electroporation	0	8

(a) How many people do these data describe?

(b) How many of these people received the vaccine intramuscularly?

(c) Give the marginal distribution of immune response, both as counts and as percents.

5.2 **Altruism in prairie dogs.** Prairie dogs are social rodents with sophisticated calls capable of identifying various kinds of predators to alert their kin. The warning calls are reminiscent of the barks of dogs, hence the species' name. Raising the alarm when a predator is spotted is the group's only defense mechanism, but it places the caller at a higher risk of predation. A study examined whether proximity to the predator (a stuffed badger controlled remotely by an experimenter) influences the likelihood of a prairie dog's raising the alarm. The table below shows the count of prairie dogs raising the alarm or not when they are either near the predator or far from it:[3]

	Alarm call	No call
Near	29	80
Far	55	78

(a) How many prairie dogs were monitored in this experiment?

(b) How many of these prairie dogs were near the predator?

(c) Give the marginal distribution of reaction (alarm call or no call) as percents and display them in a bar graph.

5.3 **Marginal distributions aren't the whole story.** Here are the row and column totals for a two-way table with two rows and two columns:

a	b	50
c	d	50
60	40	100

Find *two different* sets of counts a, b, c, and d for the body of the table that give these same totals. This shows that the relationship between two variables cannot be obtained from the two individual distributions of the variables.

Conditional distributions

Table 5.1 contains much more information than the two marginal distributions of shoe type alone and gender alone. *Marginal distributions tell us nothing about the relationship between two variables*. To describe a relationship between two categorical variables, we must calculate some well-chosen percents from the counts given in the body of the table. We use percents because counts are often hard to compare. For example, there are 1900 women in the study but only 1472 men, so comparing counts for men and women would obviously be misleading.

Let's say that we want to compare the types of shoes worn by men and women. To do this, compare percents for women alone with percents for men alone. First, to study the type of shoes that men say they tend to wear most often, we look only at the "Men" column in Table 5.2. To find the percent *of men* who wore mostly shoes providing good support, divide the count of such men by the total number of men (the column total):

$$\frac{\text{men wearing shoes with good support}}{\text{men's column total}} = \frac{94}{1472} = 0.064 = 6.4\%$$

Doing this for all three entries in the "Men" column gives the *conditional distribution* of shoe types among men. We use the term "conditional" because this distribution describes only study participants who satisfy the condition that they are male.

TABLE 5.2 Shoe type by gender: men's group

	Gender		
Shoe type	Men	Women	Total
Good support	94	137	231
Average support	1348	581	1929
Poor support	30	1182	1212
Total	1472	1900	3372

MARGINAL AND CONDITIONAL DISTRIBUTIONS

The **marginal distribution** of one of the categorical variables in a two-way table of counts is the distribution of values of that variable among all individuals described by the table.

A **conditional distribution** of a variable is the distribution of values of that variable among only individuals who have a given value of the other variable. There is a separate conditional distribution for each value of the other variable.

EXAMPLE 5.3 Comparing men and women: conditional distribution of shoe type, given a particular gender

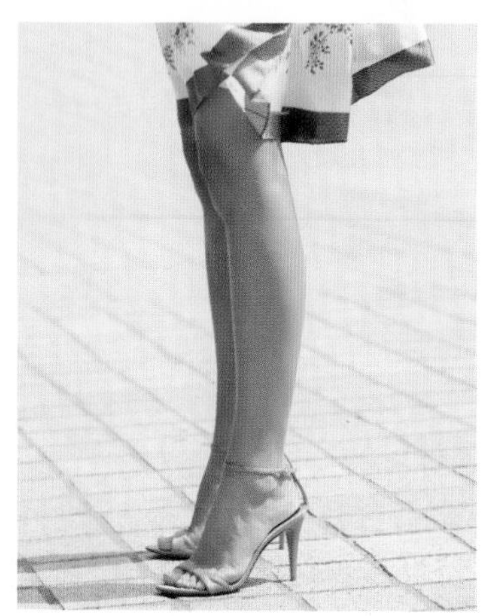
Fabrice Lerouge/ONOKY/Getty Images

STATE: Foot pain is a common ailment in older individuals, especially among women. Poor footwear is thought to be a contributing factor of foot pain. We want to know how older men and women differ in the type of shoes they wore most regularly in the past.

PLAN: Make a two-way table of shoe type by gender. Find the conditional distributions of shoe type for men alone and for women alone. Compare these two distributions.

SOLVE: Comparing conditional distributions reveals the nature of the association between shoe type and gender. We use the data displayed in Table 5.1. Look first at just the "Men" column to find the conditional distribution for men, then at just the "Women" column to find the conditional distribution for women. Here are the calculations and the two conditional distributions:

Shoe type	Men	Women
Good support	$\frac{94}{1472} = 6.4\%$	$\frac{137}{1900} = 7.2\%$
Average support	$\frac{1348}{1472} = 91.6\%$	$\frac{581}{1900} = 30.6\%$
Poor support	$\frac{30}{1472} = 2.0\%$	$\frac{1182}{1900} = 62.2\%$

Each set of percents within a gender condition adds to 100% because everyone within the gender group selected one of the three shoe types. The bar graph in Figure 5.2(a) compares the percents of shoe types (offering good, average, and poor structural support) among men and among women in the study. Figure 5.2(b) displays the same percents in a "stacked" bar graph, which helps visualize the fact that the conditional distribution within each gender adds up to 100% (a format closely related to that of the pie chart).

CONCLUDE: Men overwhelmingly wore shoes providing average foot support. In contrast, women overwhelmingly wore shoes with either average or poor foot support. In particular, women were much more likely than men to have worn shoes providing poor support (62.2% compared with only 2.0%). This might explain why foot pain is more common among women than among men, although we should be careful not to assign causality to an observed association. ■

Software will do these calculations for you once you understand how to enter and label the data for your particular program. Most programs allow you to choose which conditional distributions you want to compare. The outputs from Minitab and R in Figure 5.3 compare the conditional distributions of shoe type, given gender. These percents agree with the results in Example 5.3. Minitab also provides the marginal distribution of shoe type for all study participants, matching the values we computed in Example 5.2.

Remember that there are two sets of conditional distributions for any two-way table. Example 5.3 looked at the conditional distributions of shoe type for each gender. We could also examine the conditional distributions of gender for the three shoe types. Starting with the shoe type providing good support, we find the conditional distribution of gender by looking only at the first row of the two-way

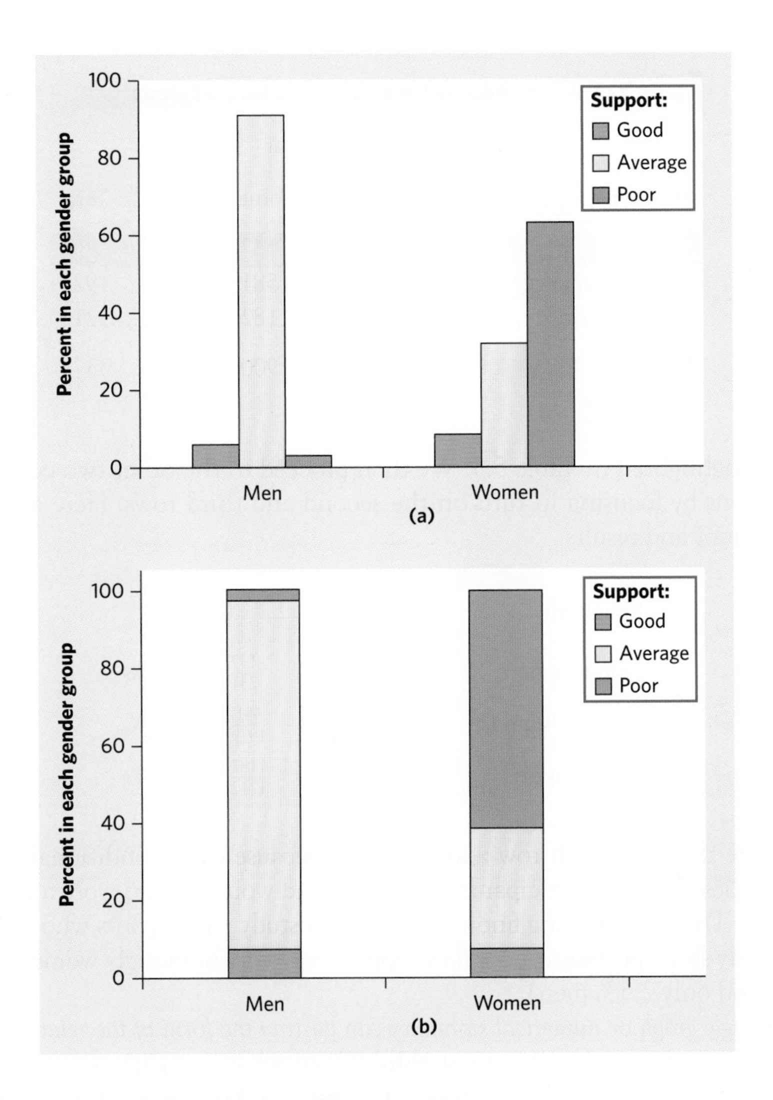

FIGURE 5.2 (a) Bar graph comparing the percents of shoe types (offering good, average, and poor structural support) among men and among women in the study. Men and women wore all three types of shoes, but the percent of shoes offering poor structural support was higher among the women. (b) The same percents are displayed in a "stacked" bar graph, which helps visualize the fact that the conditional distribution within each gender adds up to 100%.

Minitab

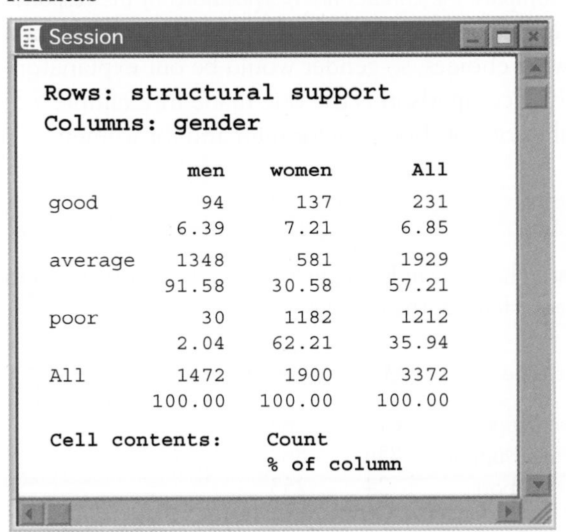

Session

Rows: structural support
Columns: gender

	men	women	All
good	94	137	231
	6.39	7.21	6.85
average	1348	581	1929
	91.58	30.58	57.21
poor	30	1182	1212
	2.04	62.21	35.94
All	1472	1900	3372
	100.00	100.00	100.00

Cell contents: Count
% of column

R

```
> .Table      #   Counts
            men    women
good         94      137
average    1348      581
poor         30     1182
> colPercents (.Table) #
Column Percentages
            men    women
good        6.4      7.2
average    91.6     30.6
poor        2.0     62.2
Total     100.0    100.0
Count    1472.0   1900.0
```

FIGURE 5.3 Minitab and R outputs of the two-way table of shoe support types for men and women with each entry as a percent of its column total. The percents in each column give the conditional distribution of shoe support type for one gender. The "All" column in the Minitab output also gives the marginal distribution of shoe support type for the whole study (both genders combined).

TABLE 5.3 Gender by shoe type: good support group

	Gender		
Shoe type	**Men**	**Women**	**Total**
Good support	94	137	231
Average support	1348	581	1929
Poor support	30	1182	1212
Total	1472	1900	3372

table, as highlighted in Table 5.3. We then proceed to the other two conditional distributions by focusing in turn on the second and third rows. Here are all the computations and results:

Shoe type	Men	Women
Good support	$\frac{94}{231} = 40.7\%$	$\frac{137}{231} = 59.3\%$
Average support	$\frac{1348}{1929} = 69.9\%$	$\frac{581}{1929} = 30.1\%$
Poor support	$\frac{30}{1212} = 2.5\%$	$\frac{1182}{1212} = 97.5\%$

Notice that now each row adds to 100% because each conditional distribution includes all study participants (both men and women) who wore a particular shoe type. The most striking finding here is that study participants who tended to wear mostly shoes providing poor foot support are overwhelmingly women (97.5% women and only 2.5% men).

No single graph or numerical summary can portray the form of the relationship between categorical variables. You must think about what comparisons you want to display graphically and which percents to compute. Here is a hint: *If there is an explanatory-response relationship, compare the conditional distributions of the response variable for the separate values of the explanatory variable.* Here we know that men and women make different footwear choices, so gender would be our explanatory variable. Thus, the most interesting comparison is the one made in Example 5.3: comparing the conditional distributions of shoe type for men and for women.

APPLY YOUR KNOWLEDGE

5.4 Foot health: current footwear. The study in Example 5.1 also asked participants to describe their current footwear. Here are the findings:

Shoe type	Men	Women
Good support	612	785
Average support	836	862
Poor support	24	253

How do older men and women from the Framingham cohort differ in the type of shoes they currently wear most regularly? Follow the four-step process, as illustrated in Example 5.3.

Tyler Mallory/Alamy

5.5 Altruism in prairie dogs, continued. The two-way table in Exercise 5.2 describes a study of warning-call behavior in prairie dogs exposed to different predation risks as defined by the level of proximity to a predator.

(a) Predator proximity is the explanatory variable in this study. Find the conditional distribution of call behavior, given predator proximity.

(b) Based on your calculations, describe the differences in call behavior when a predator is near or far. Use both a graph and a written explanation.

5.6 HIV vaccine boost, continued. The two-way table in Exercise 5.1 describes a study of a clinical trial testing a DNA vaccine for HIV delivered via electroporation to boost the immune response. The researchers hoped that the vaccine delivered via electroporation would be more effective (result in a higher percent of immune responses) than either the vaccine delivered intramuscularly or an inactive saline solution delivered via electroporation. Do the data support this presumption? Follow the four-step process, as illustrated in Example 5.3.

Simpson's paradox

As is the case with quantitative variables, the effects of lurking variables can change or even reverse relationships between two categorical variables. Here is an example that demonstrates the surprises that can await the unsuspecting user of data.

EXAMPLE 5.4 Do medical helicopters save lives?

Accident victims are sometimes taken by helicopter from the accident scene to a hospital. Helicopters save time. Do they also save lives? Let's compare the percents of accident victims who die with helicopter evacuation and with the usual transport to a hospital by road. Here are hypothetical (yet realistic) data that illustrate a practical difficulty:[4]

	Helicopter	Road
Victim died	64	260
Victim survived	136	840
Total	200	1100

We see that 32% (64 out of 200) of helicopter patients died, compared with only 24% (260 out of 1100) of the others. That seems discouraging.

The explanation is that the helicopter is sent mostly to serious accidents, so that the victims transported by helicopter are more often seriously injured. They are more likely to die with or without helicopter evacuation. Here are the same data broken down by the seriousness of the accident:

Serious Accidents

	Helicopter	Road
Died	48	60
Survived	52	40
Total	100	100

Less Serious Accidents

	Helicopter	Road
Died	16	200
Survived	84	800
Total	100	1000

Inspect these tables to convince yourself that they describe the same 1300 accident victims as the original two-way table. For example, 200 (100 + 100) were moved by helicopter, and 64 (48 + 16) of these died.

Among victims of serious accidents, the helicopter saves 52% (52 out of 100), compared with 40% for road transport. If we look only at less serious accidents, 84% of those transported by helicopter survive, versus 80% of those transported by road. Both groups of victims have a higher survival rate when evacuated by helicopter. ■

At first, it seems paradoxical that the helicopter does better for both groups of victims but worse when all victims are lumped together. This is called *Simpson's paradox*. Examining the data closely provides the explanation for this paradox. Half the helicopter transport patients are from serious accidents, compared with only 100 of the 1100 road transport patients. So helicopters carry patients who are more likely to die. The seriousness of the accident is a *lurking variable* that, until we uncovered it, made the relationship between survival and mode of transport to a hospital hard to interpret.

SIMPSON'S PARADOX

An association or comparison that holds for all of several groups can reverse direction when the data are combined to form a single group. This reversal is called **Simpson's paradox.**

Do left-handers die early? Yes, said a study of 1000 deaths in California. Left-handed people died at an average age of 66 years; right-handers, at 75 years of age. Should left-handed people fear an early death? No—the lurking variable has struck again, resulting in Simpson's paradox. Older people grew up in an era when many natural left-handers were forced to use their right hands. So left-handers are more common among the young. When we look at deaths, the left-handers who die are younger on the average because left-handers in general are younger. Mystery solved.

The lurking variable in Simpson's paradox is categorical. That is, it breaks the individuals into groups, as when accident victims are classified as injured in either a "serious accident" or a "less serious accident." Simpson's paradox is just an extreme form of the fact that observed associations can be misleading when there are lurking variables.

Many medical studies examine categorical variables such as "dead/still alive" or "condition improved/not improved." Because pooling data from heterogeneous groups can lead to Simpson's paradox, data gathered in different studies are not pooled together even if the studies looked at the same set of variables. Instead, a more complex method called *meta-analysis*, which compares the conclusions reached by each study, has been developed and is often found in the biomedical literature. This method is beyond the scope of our book.

APPLY YOUR KNOWLEDGE

5.7 **Kidney stones.** A study compared the success rates of two different procedures for removing kidney stones: open surgery and percutaneous nephrolithotomy (PCNL), a minimally invasive technique. Here are the number of procedures that were successful or not at getting rid of patients' kidney stones, by type of procedure. A separate table is given for patients with small kidney stones and for patients with large stones:[5]

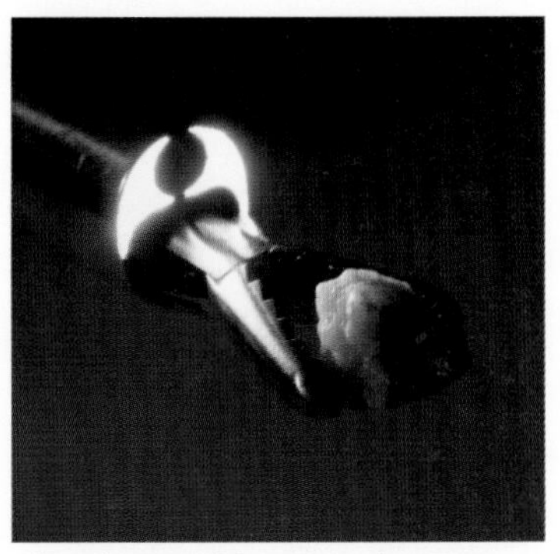

Alexander Tsiaras/Photo Researchers, Inc.

Small Stones

	Open surgery	PCNL
Success	81	234
Failure	6	36

Large Stones

	Open surgery	PCNL
Success	192	55
Failure	71	25

(a) Find the percent of kidney stones, combining the data for small and large stones, that were successfully removed for each of the two medical procedures. Which procedure had the higher overall success rate?

(b) What percent of all small kidney stones were successfully removed? What percent of all large kidney stones were successfully removed? Which type of kidney stone appears to be easier to treat?

(c) Now find the percent of successful procedures of each type for small kidney stones only. Do the same for large kidney stones. PCNL performed worse for *both* small and large kidney stones, yet it did better overall. That sounds impossible. Explain carefully, referring to the data, how this paradox can happen.

5.8 Smoking and staying alive. In the mid-1970s, a medical study contacted randomly chosen people in a district in England. Here are data on the 1314 women contacted who were either current smokers or who had never smoked. The tables classify these women by their smoking status and age at the time of the survey and whether they were still alive 20 years later.[6]

Ages 18 to 44

	Smoker	Not
Dead	19	13
Alive	269	327

Ages 45 to 64

	Smoker	Not
Dead	78	52
Alive	167	147

Ages 65+

	Smoker	Not
Dead	42	165
Alive	7	28

(a) From these data make a two-way table of smoking (yes or no) by dead or alive. What percent of the smokers stayed alive for 20 years? What percent of the nonsmokers survived? It seems surprising that a higher percent of smokers stayed alive.

(b) The age of the women at the time of the study is a lurking variable. Show that within each of the three age groups in the data, a higher percent of nonsmokers remained alive 20 years later. Explain why this is an example of Simpson's paradox.

(c) The study authors give this explanation: "Few of the older women (over 65 at the original survey) were smokers, but many of them had died by the time of follow-up." Compare the percents of smokers in the three age groups to verify the explanation.

CHAPTER 5 SUMMARY

- A **two-way table** of counts organizes data about two categorical variables. Levels of the **row variable** label the rows that run across the table, and levels of the **column variable** label the columns that run down the table. Two-way tables are often used to summarize large amounts of information by grouping outcomes into categories.
- The **row totals** and **column totals** in a two-way table give the **marginal distributions** of the two individual variables. It is clearer to present these distributions as percents of the table total. Marginal distributions tell us nothing about the relationship between the variables.
- There are two sets of **conditional distributions** for a two-way table: the distributions of the row variable for each level of the column variable, and the distributions of the column variable for each level of the row variable. Comparing one set of conditional distributions is one way to describe the association between the row and the column variables.
- To find the **conditional distribution** of the row variable for one specific level of the column variable, look only at that one column in the table. Find each entry in the column as a percent of the column total.
- **Bar graphs** are a flexible means of presenting categorical data. There is no single best way to describe an association between two categorical variables.
- A comparison between two variables that holds for each individual value of a third variable can be changed or even reversed when the data for all values of the third variable are combined. This is **Simpson's paradox.** Simpson's paradox is an example of the effect of lurking variables on an observed association.

THIS CHAPTER IN CONTEXT

In Chapters 3 and 4 we considered relationships between two quantitative variables. In this chapter we use two-way tables to describe relationships between two *categorical* variables. To explore relationships between two categorical variables, we examine their conditional distributions. The conditional distribution of one of the categorical variables is the distribution of that variable among only individuals who have a given value of the other variable.

We will revisit conditional distributions in Chapter 10 in the context of probability. We will see that the probability of an event can be computed given some known information, resulting in the concept of conditional probability. In fact, two-way tables are a useful tool to represent probabilities and help compute conditional probabilities.

When exploring relationships between two categorical variables, changes in the pattern of the conditional distribution of one variable as the value of the other varies provide information about the nature of the relationship. No change in this pattern suggests that there is no relationship. In Chapters 20 and 22, we will use statistical inference to test whether the observed relationship can be explained by chance alone or whether we can infer that the relationship is valid in some broader sense.

As in Chapters 3 and 4, we must be careful not to assume that an observed association implies a cause-and-effect relation. A lurking variable may influence both variables studied and create the appearance of an association or change the nature of the association. Simpson's paradox is an example of how such a lurking variable can mislead us. We will revisit the concept of lurking variables in Chapters 7 and 8 in the context of data collection.

CHECK YOUR SKILLS

A study was designed to assess the effect of echinacea extracts on the rates of infection of individuals voluntarily exposed to the common-cold rhinovirus. Some subjects took an echinacea extract daily starting 7 days before viral exposure until the end of the study. Another group of subjects received the extract from the time of viral exposure onward only. A last group received just an inactive solution ("placebo group"). Below is a two-way table of the diagnosed clinical colds among infected subjects in each treatment group.[7] *Exercises 5.9 to 5.17 are based on this table.*

	Outcome	
Treatment	Cold	No cold
7 days before viral exposure and onward	73	59
From time of viral exposure onward	88	43
Placebo (no echinacea)	58	30

5.9 How many infected subjects have been diagnosed with a cold?
(a) 219 (b) 351 (c) need more information

5.10 How many individuals are described by this table?
(a) 219 (b) 351 (c) need more information

5.11 The percent of infected subjects diagnosed with a cold is
(a) about 22%. (b) about 38%. (c) about 62%.

5.12 Your percent from the previous exercise is part of
(a) the marginal distribution of outcome (cold or no cold diagnosed).
(b) the marginal distribution of treatment.
(c) the conditional distribution of outcome, given treatment.

5.13 What percent of the infected subjects in the placebo group have been diagnosed with a cold?
(a) about 58% (b) about 62% (c) about 66%

5.14 Your percent from the previous exercise is part of
(a) the marginal distribution of outcome.
(b) the conditional distribution of outcome, given treatment.
(c) the conditional distribution of treatment, given outcome.

5.15 What percent of those diagnosed with a cold are in the placebo group?
(a) about 26% (b) about 58% (c) about 66%

5.16 Your percent from the previous exercise is part of
(a) the marginal distribution of outcome.
(b) the conditional distribution of outcome, given treatment.
(c) the conditional distribution of treatment, given outcome.

5.17 A bar graph showing the conditional distribution of treatment, given that the infected subjects have been diagnosed with a cold, has
(a) 2 bars. (b) 3 bars. (c) 6 bars.

5.18 To help consumers make informed decisions about health care, the government releases data about patient outcomes in hospitals. A large regional hospital and a small private hospital both serve your community. The regional hospital receives a lot of patients in critical condition because of its state-of-the-art emergency room and diverse set of medical specialties. The private hospital specializes in scheduled surgeries and admits few patients in critical condition. The counts of patients who survived surgery or did not are provided for both hospitals, for all surgeries performed in the previous year. The regional hospital had the higher surgery survival rate (the percent of patients still alive six weeks after surgery) for both patients admitted in critical condition and patients coming in for scheduled surgery. Yet, the private hospital had the higher overall survival rate when considering both types of surgery patients together. This finding is

(a) not possible: If the regional hospital had had higher survival rates for each type of patient separately, then it must also have had a higher overall survival rate when both types of patient are combined.
(b) an example of Simpson's paradox: The regional hospital performed better with each type of patient but it did worse overall because it took in a lot more patients in critical condition, who came in with a lower chance of survival due to their condition.
(c) due to comparing two conditional distributions that should not be compared.

CHAPTER 5 EXERCISES

Sex ratios in geese. *Most species have a similar proportion of males and females born, maximizing the chances of genetic mix from both sexes. However, some disparities have been found in the conditions under which a male or a female is born. A field biologist surveyed the eggs of wild lesser snow geese, from egg production to hatching. The two-way table below presents data from 29 clutches, each containing 4 eggs numbered 1 through 4 to reflect the order in which they were laid. When an egg hatched, the sex of the gosling was assessed. A few eggs in the 29 clutches did not hatch, explaining why the column totals are not equal to 29.*[8] *Exercises 5.19 to 5.22 are based on these data.*

	Order in Which Egg Is Laid				
Sex	1	2	3	4	Total
Male	17	16	7	5	45
Female	10	9	17	14	50
Total	27	25	24	19	95

5.19 Marginal distributions. Give (in percents) the two marginal distributions, for sex and for egg order.

5.20 Percents. What percent of male goslings are first eggs? What percent of first eggs are male?

5.21 Conditional distribution. Give (in percents) the conditional distribution of egg order among males. Should your percents add to 100% (up to roundoff error)?

5.22 Sex and egg order. One way to see the relationship between sex and egg order is to look at which goslings were first eggs.

(a) There are 17 male first eggs and only 10 female first eggs. Explain why these counts by themselves don't describe the relationship between sex and egg order.
(b) Find the percent of goslings of each sex among the first eggs. Then find the percent of goslings of each sex among the second eggs, among the third eggs, and among the fourth eggs. What do these percents say about the relationship?

5.23 Botox for excessive sweating. Hyperhidrosis is a stressful medical condition characterized by chronic excessive sweating. Primary hyperhidrosis is inherited and typically starts during adolescence. Botox was approved for treatment of hyperhidrosis in adults in 2004. A clinical trial examined the effectiveness of Botox for excessive armpit sweating in teenagers aged 12 to 17. Participants received either a Botox injection or a placebo injection (a simple saline solution) and were examined four weeks later to see if their sweating had been reduced by 50% or more. Here are the study's findings:[9]

	At Least 50% Reduction		
Treatment	Yes	No	Total
Botox	84	20	104
Placebo	44	64	108
Total	128	84	212

(a) Find the marginal distribution of outcome (sweating reduced or not). What does this distribution tell us?
(b) How does the outcome of the teenagers in the study differ depending on what treatment they received? Use conditional distributions as a basis for your answer. What can you conclude about the effectiveness of Botox in treating excessive armpit sweating?

5.24 Helping cocaine addicts. Cocaine addiction is hard to break. Addicts need cocaine to feel any pleasure, so perhaps giving them an antidepressant drug will help. An experiment assigned 72 chronic cocaine users to take either an antidepressant drug called desipramine, lithium, or a placebo. (Lithium is a standard drug to treat cocaine addiction. A placebo is a dummy drug, used so that the effect of being in the study but not taking any drug can be seen.) One-third of the subjects, chosen at random, received each drug. Here are the results after three years:[10]

	Desipramine	Lithium	Placebo
Relapse	10	18	20
No relapse	14	6	4
Total	24	24	24

(a) Compare the effectiveness of the three treatments in preventing relapse. Use percents and draw a bar graph.

(b) Do you think that this study gives good evidence that desipramine actually *causes* a reduction in relapses?

5.25 Fast food for baby. The Gerber company sponsored a large survey of the eating habits of American infants and toddlers. Among the many questions parents were asked was whether their child had eaten fried potatoes on one given day. Here are the data broken down by the children's age range:[11]

	Ate fried potatoes	Did not
9–11 months	61	618
15–18 months	62	246
19–24 months	82	234

(a) Calculate the conditional distribution of children who ate fried potatoes for each age range. Briefly describe your findings. (The study also found that fried potatoes were actually the most common cooked vegetable in the diet of the 19- to 24-month-old toddlers, way ahead of green beans and peas!)

(b) Do you think that the association between age and diet found by this study is evidence that age actually *causes* a change in diet?

5.26 Weight-lifting injuries. Resistance training is a popular form of conditioning aimed at enhancing sports performance and is widely used among high school, college, and professional athletes, although its use for younger athletes is controversial. A random sample of 4111 patients between the ages of 8 and 30 admitted to U.S. emergency rooms with the injury code "weightlifting" was obtained. These injuries were classified as "accidental" if caused by dropped weight or improper equipment use. The patients were also classified into four age categories. Here is a two-way table of the results:[12]

	Accidental	Not accidental
8–13 years	295	102
14–18 years	655	916
19–22 years	239	533
23–30 years	363	1008

Compare the distributions of ages for accidental and nonaccidental injuries. Use percents and draw a bar graph. What do you conclude?

5.27 Preventing neural-tube defect. A key study in the 1990s examined the effectiveness of prenatal vitamin and mineral supplements in preventing neural-tube defect and congenital malformations in newborns. A total of 4156 women planning a pregnancy were given either a vitamin and mineral supplement (including folic acid) or a mineral-only supplement to take daily for at least one month before conception. Here is a two-way table of the number of births with a congenital malformation, a neural-tube defect, or neither in both groups:[13]

	Vitamins and minerals	Minerals only
Congenital malformation	28	47
Neural-tube defect	0	6
Neither	2076	1999
Total	2104	2052

Find the conditional distribution of birth outcomes, given the type of prenatal supplement taken. Use your results to describe the effect of adding vitamins to prenatal mineral supplements in terms of newborn health.

5.28 HIV testing in the United States. The Centers for Disease Control and Prevention (CDC) recommend that HIV testing be a part of routine clinical care for all sexually active patients. But do people get tested for HIV regularly? The 2012 National Health Interview Survey gives the number of adults in the United States (in thousands) classified by age group and whether they have or have not been tested for HIV in the past 12 months. Here are the data:

	Tested	Not tested
15–19 years	2,331	18,964
20–24 years	4,866	15,893
25–29 years	5,023	16,270
30–34 years	3,377	15,039
35–39 years	3,249	17,694
40–44 years	2,425	18,753

Find the conditional distribution of HIV-testing status for each age group, and describe the state of HIV testing in the United States.

5.29 Type 1 and type 2 diabetes. Type 1 diabetes is usually first diagnosed in children or young adults and is due to the immune system interfering with the ability of the pancreas to make insulin. Type 2 diabetes is by far the more common form of diabetes and can develop at any age but is especially common among older individuals; it arises typically in response to an excessively rich diet and limited physical activity, inducing insulin resistance in the body. Records from a diabetes clinic in the United Kingdom show the status (still alive or dead) of long-term diabetic patients as a function of their diabetes type:[14]

Patients 40 or Younger

	Type 1	Type 2
Alive	129	15
Dead	1	0
Total	130	15

Patients Older than 40

	Type 1	Type 2
Alive	124	311
Dead	104	218
Total	228	529

(a) Compare the survival rates (percents still alive) for type 1 and for type 2 diabetes among the younger patients. Do the same for the older patients. What have you learned from these percents?

(b) Combine the data into a single two-way table of patient status (alive or dead) by diabetes type (1 or 2). Now calculate the survival rates for type 1 and for type 2 diabetes among all patients together. Which type of diabetes shows the higher survival rate?

(c) This is an example of Simpson's paradox. What is the lurking variable here? Explain in simple language how the paradox can happen.

5.30 Obesity and health. Recent studies have shown that earlier reports underestimated the health risks associated with being overweight. The error was due to overlooking lurking variables. In particular, smoking tends both to reduce weight and to lead to earlier death. Illustrate Simpson's paradox by a simplified version of this situation. That is, make up two-way tables of overweight (yes or no) by early death (yes or no) separately for smokers and nonsmokers such that

- overweight smokers and overweight nonsmokers both tend to die earlier than those not overweight,
- but when smokers and nonsmokers are combined into a two-way table of overweight by early death, persons who are not overweight tend to die earlier.

5.31 Python eggs. 4 STEP How is the hatching of water python eggs influenced by the temperature of the snake's nest? Researchers assigned newly laid eggs to one of three temperatures: hot, neutral, or cold. Hot duplicates the warmth provided by the mother python. Neutral and cold are cooler, as when the mother is absent. Here are the data on the number of eggs and the number that hatched:[15]

	Cold	Neutral	Hot
Number of eggs	27	56	104
Number hatched	16	38	75

(a) Notice that this is not a two-way table! Explain why and how you could use the data to create a two-way table.

(b) The researchers anticipated that eggs would hatch less well at cooler temperatures. Do the data support that anticipation? Follow the four-step process as in Example 5.3.

5.32 Do angry people have more heart disease? 4 STEP People who get angry easily tend to have more heart disease. That's the conclusion of a study that followed a random sample of 12,986 people for about four years. All subjects were free of heart disease at the beginning of the study. The subjects took the Spielberger Trait Anger Scale test, which measures how prone a person is to sudden anger. Here are data compiled at the end of the study for the 8474 people in the sample who had normal blood pressure.[16] CHD stands for "coronary heart disease." This includes people who had heart attacks and those who needed medical treatment for heart disease.

	Low anger	Moderate anger	High anger	Total
CHD	53	110	27	190
No CHD	3057	4621	606	8284
Total	3110	4731	633	8474

Do these data support the study's conclusion about the relationship between anger and heart disease? Follow the four-step process as in Example 5.3.

5.33 Characteristics of a forest. 4 STEP Forests are complex, evolving ecosystems. For instance, pioneer tree species can be displaced by successional species better adapted to the changing environment. Ecologists mapped a large Canadian forest plot dominated by Douglas fir with an understory of western hemlock and western red cedar. The two-way table below records all the trees in the plot by species and by life stage. The distinction between live and sapling trees is made for live trees taller or shorter than 1.3 meters, respectively.[17]

	Dead	Live	Sapling	Total
Western red cedar	48	214	154	416
Douglas fir	326	324	2	652
Western hemlock	474	420	88	982
Total	848	958	244	2050

What can you tell from these data about the current composition and the evolution of this forest? Which tree species appears to be taking over and becoming dominant? Follow the four-step process as in Example 5.3.

5.34 Complications of bariatric surgery. 4 STEP Bariatric surgery, or weight-loss surgery, includes a variety of procedures performed on people who are obese. Weight loss is achieved by reducing the size of the stomach with an implanted medical device (gastric banding), by removing a portion of the stomach (sleeve gastrectomy), or by resecting and rerouting the small intestines to a small stomach pouch (gastric bypass surgery). Because there can be complications using any of these methods, the National Institutes of Health recommend bariatric surgery for obese people with a body mass index (BMI) of at least 40 and for people with BMI 35 and serious coexisting medical conditions such as diabetes. Serious complications include potentially life-threatening, permanently disabling, and fatal outcomes. Here is a two-way table for data collected in Michigan over several years giving counts of non-life-threatening complications, serious complications, and no complications for these three types of surgeries:[18]

	Non-life-threatening complication	Serious complication	No complication	Total
Gastric banding	81	46	5253	5380
Sleeve gastrectomy	31	19	804	854
Gastric bypass	606	325	8110	9041

What do the data say about differences in complications for the three types of surgeries? Follow the four-step process.

5.35 Smoking cessation. 4 STEP A large randomized trial was conducted to assess the efficacy of Chantix for smoking cessation compared with bupropion (more commonly known as Wellbutrin or Zyban) and a placebo. Chantix is different from most other quit-smoking products in that it targets nicotine receptors in the brain, attaches to them, and blocks nicotine from reaching them, while bupropion is an antidepressant often used to help people stop smoking. Generally healthy smokers who smoked at least 10 cigarettes per day were assigned at random to take Chantix ($n = 352$), bupropion ($n = 329$), or a placebo ($n = 344$). The response measure is continuous cessation from smoking for Weeks 9 through 12 of the study. Here is a two-way table of the results:[19]

	Treatment		
	Chantix	Bupropion	Placebo
No smoking in Weeks 9–12	155	97	61
Smoked in Weeks 9–12	197	232	283

How does whether a subject smoked in Weeks 9 to 12 depend on the treatment received? Follow the four-step process.

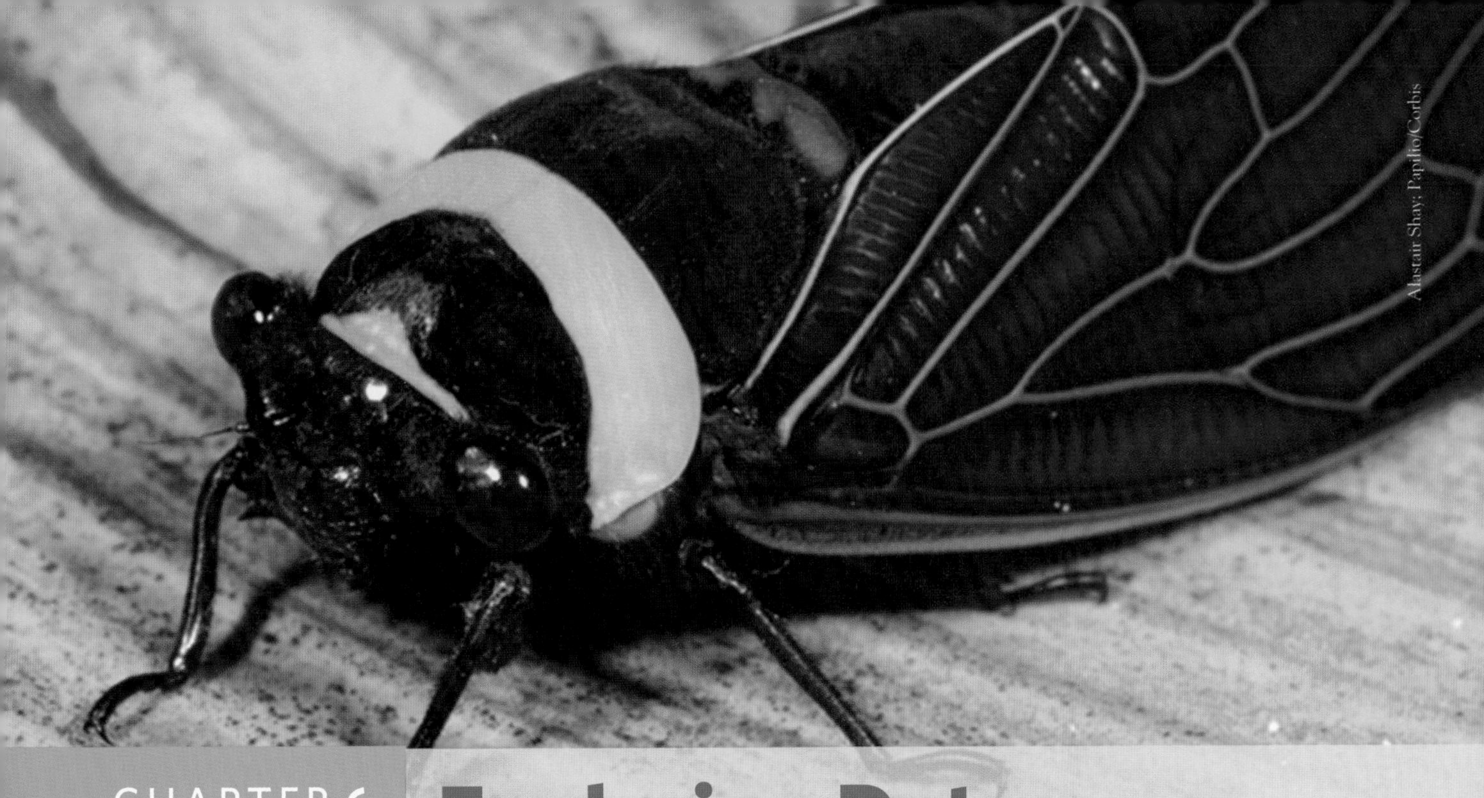

CHAPTER 6 Exploring Data: Part I Review

Data analysis is the art of describing data using graphs and numerical summaries. The purpose of data analysis is to help us see and understand the most important features of a set of data. Chapter 1 described graphs to display distributions: pie charts and bar graphs for categorical variables, histograms and stemplots for quantitative variables, and time plots to show how a quantitative variable changes over time. Chapter 2 presented numerical tools for describing the center and spread of the distribution of one variable.

The first STATISTICS IN SUMMARY figure on the next page organizes the big ideas for exploring a quantitative variable. Plot your data, then describe their center and spread using either the mean and standard deviation or the five-number summary. The question marks at the last stage remind us that the usefulness of numerical summaries depends on what we find when we examine graphs of our data. No short summary does justice to irregular shapes or to data with several distinct clusters.

Chapters 3 and 4 applied the same ideas to relationships between two quantitative variables. The second STATISTICS IN SUMMARY figure retraces the big ideas, with details that fit the new setting. Always begin by making graphs of your data. In the case of a scatterplot, we have learned a numerical summary only for data that show a roughly linear pattern on the scatterplot. The summary is then the means and standard deviations of the two variables and their correlation.

IN THIS CHAPTER WE COVER...

- Part I Summary
- Review Exercises
- Supplementary Exercises
- EESEE Case Studies

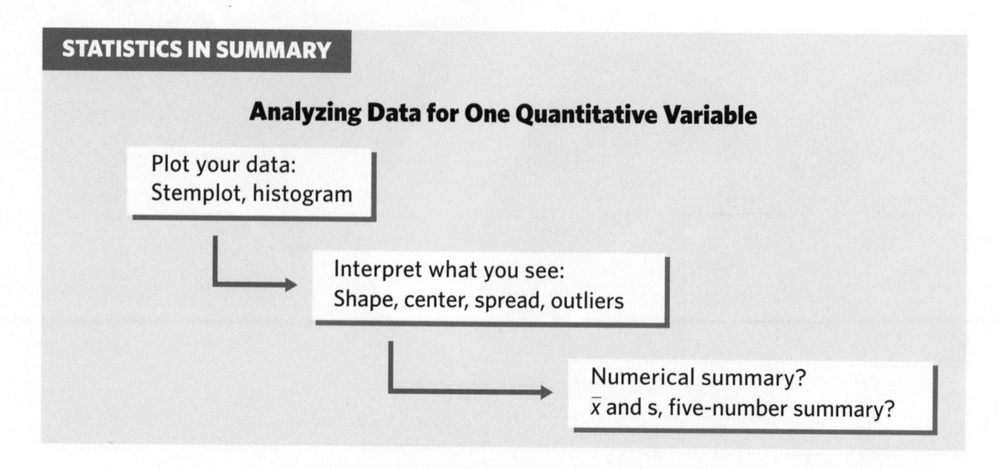

A regression line drawn on the plot gives a compact description of the overall pattern that we can use for prediction. The question marks at the last three stages remind us that correlation and regression describe only straight-line relationships. Chapter 5 shows how to understand relationships between two categorical variables; comparing well-chosen percents is the key.

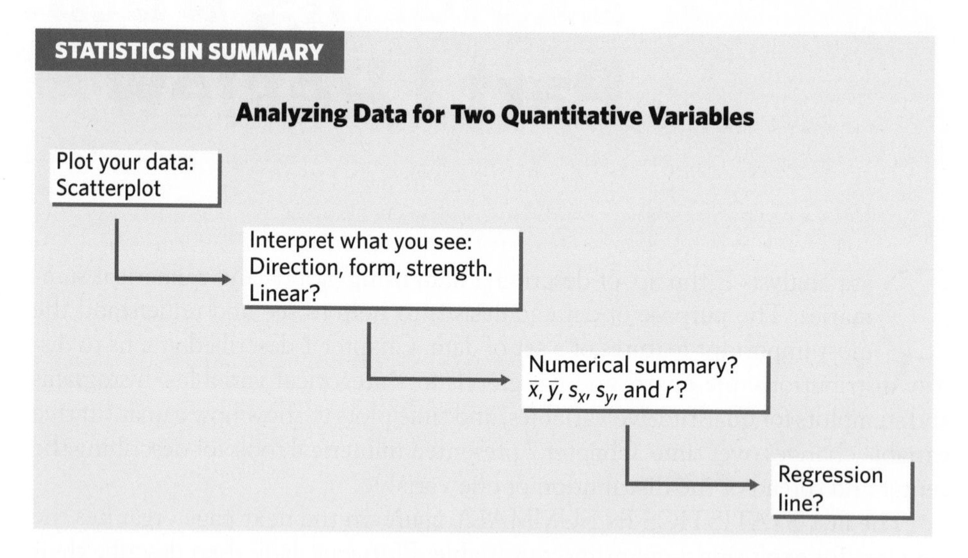

You can organize your work in any open-ended data analysis setting by following the four-step State, Plan, Solve, and Conclude process first introduced in Chapter 2. After we have mastered the extra background needed for statistical inference, this process will also guide practical work on inference later in the book.

PART I SUMMARY

Here are the most important skills you should have acquired from reading Chapters 1 to 5.

A. Data

1. Identify the individuals and variables in a set of data.

2. Identify each variable as categorical or quantitative. Identify the units in which each quantitative variable is measured.
3. Identify the explanatory (independent) and response (dependent) variables in situations where one variable explains or influences another.

B. Displaying Distributions

1. Recognize when a pie chart can and cannot be used.
2. Make a bar graph of the distribution of a categorical variable, or in general to compare related quantities.
3. Interpret pie charts and bar graphs.
4. Make a time plot of a quantitative variable recorded over time. Recognize patterns such as trends and cycles in time plots.
5. Make a histogram of the distribution of a quantitative variable.
6. Make a stemplot of the distribution of a small set of observations. Round (or truncate) leaves or split stems as needed to make an effective stemplot.
7. Make a dotplot of the distribution of a small set of observations.
8. Describe the shape, center, and spread of a quantitative distribution from a histogram, a stemplot, or a dotplot. Identify outliers of the distribution.

C. Describing Distributions (Quantitative Variable)

1. Look for the overall pattern and for major deviations from the pattern.
2. Assess from a histogram or stemplot whether the shape of a distribution is roughly symmetric, distinctly skewed, or neither. Assess whether the distribution has one or more major peaks.
3. Describe the overall pattern by giving numerical measures of center and spread in addition to a verbal description of shape.
4. Decide which measures of center and spread are more appropriate: the mean and standard deviation (especially for symmetric distributions) or the five-number summary (especially for skewed distributions).
5. Recognize outliers and give plausible explanations for them.

D. Numerical Summaries of Distributions

1. Find the median M and the quartiles Q_1 and Q_3 for a set of observations.
2. Find the five-number summary and draw a boxplot; assess center, spread, symmetry, and skewness from a boxplot.
3. (Optional) Use the $1.5 \times IQR$ rule to identify suspected outliers.
4. Find the mean $\overline{x}$ and the standard deviation s for a set of observations.
5. Understand that the median is more resistant than the mean. Recognize that skewness in a distribution moves the mean away from the median toward the long tail.

6. Know the basic properties of the standard deviation: $s \geq 0$ always; $s = 0$ only when all observations are identical and increases as the spread increases; s has the same units as the original measurements; s is pulled strongly up by outliers or skewness.

E. Scatterplots and Correlation

1. Make a scatterplot to display the relationship between two quantitative variables measured on the same subjects. Place the explanatory variable (if any) on the horizontal scale of the plot.
2. Add a categorical variable to a scatterplot by using a different plotting symbol or color.
3. Describe the direction, form, and strength of the overall pattern of a scatterplot. In particular, recognize positive or negative association and linear (straight-line) patterns. Recognize outliers in a scatterplot.
4. Judge whether it is appropriate to use correlation to describe the relationship between two quantitative variables. Find the correlation r.
5. Know the basic properties of correlation: r measures the direction and strength of only straight-line relationships; r is always a number between -1 and 1; $r = \pm 1$ only for perfect straight-line relationships; r moves away from 0 toward ± 1 as the straight-line relationship gets stronger.

F. Regression Lines

1. Understand that regression requires an explanatory variable and a response variable. Use a calculator or software to find the least-squares regression line of a response variable y on an explanatory variable x from data.
2. Explain what the slope b and the intercept a mean in the equation $\hat{y} = a + bx$ of a regression line.
3. Draw a graph of a regression line when you are given its equation.
4. Use a regression line to predict y for a given x. Recognize extrapolation and be aware of its dangers.
5. Find the slope and intercept of the least-squares regression line from the means and standard deviations of x and y and their correlation.
6. Use r^2, the square of the correlation, to describe how much of the variation in one variable can be accounted for by a straight-line relationship with another variable.
7. Recognize outliers and potentially influential observations from a scatterplot with the regression line drawn on it.

G. Cautions about Correlation and Regression

1. Understand that both r and the least-squares regression line can be strongly influenced by a few extreme observations.
2. Recognize possible lurking variables that may explain the observed association between two variables x and y.
3. Understand that even a strong correlation does not mean that there is a cause-and-effect relationship between x and y.

4. Give plausible explanations for an observed association between two variables: direct cause and effect, the influence of lurking variables, or both.

H. Relationships in Categorical Data (Optional)

1. From a two-way table of counts, find the marginal distributions of both variables by obtaining the row sums and column sums.
2. Express any distribution in percents by dividing the category counts by their total.
3. Describe the relationship between two categorical variables by computing and comparing percents. Often this involves comparing the conditional distributions of one variable for the different categories of the other variable.
4. Recognize Simpson's paradox and be able to explain it.

REVIEW EXERCISES

Review exercises help you solidify the basic ideas and skills presented in Chapters 1 to 5.

6.1 Describing colleges. Popular magazines rank colleges and universities on their "academic quality" in serving undergraduate students. Give one categorical variable and two quantitative variables that you would like to see measured for each college if you were choosing where to study.

6.2 Genetic engineering for cancer treatment. Here's a new idea for treating advanced melanoma, the most serious kind of skin cancer. Genetically engineer white blood cells to better recognize and destroy cancer cells, then infuse these cells into patients. The subjects in a small initial study were 11 patients whose melanoma had not responded to existing treatments. One question was how rapidly the new cells would multiply after infusion, as measured by the doubling time in days. Here are the doubling times:[1]

1.4 1.0 1.3 1.0 1.3 2.0 0.6 0.8 0.7 0.9 1.9

Another outcome was the increase in the presence of cells that trigger an immune response in the body and so may help fight cancer. Here are the increases, in counts of active cells per 100,000 cells:

27 7 0 215 20 700 13 510 34 86 108

Make stemplots of both distributions (use split stems and leave out any extreme outliers). Describe the overall shapes and any outliers. Then give the five-number summary for both. (We can't compare the summaries, because the two variables have different scales.)

6.3 Detecting outliers (optional). In the previous exercise you gave five-number summaries for the distributions of two responses to a new cancer treatment. Do the data contain any observations that are suspected outliers by the $1.5 \times IQR$ rule (page 48)?

6.4 Weights are skewed. The heights of people of the same sex and similar ages have a reasonably symmetric distribution. Weights, on the other hand, do not. The weights of women aged 20 to 29 have mean 141.7 pounds and median 133.2 pounds. The first and third quartiles are 118.3 and 157.3 pounds, respectively. What can you say about the shape of the weight distribution? Why?

6.5 Squirrels and their food supply. That animal species produce more offspring when their supply of food goes up isn't surprising. That some animals appear able to anticipate unusual food abundance is more surprising. Red squirrels eat seeds from pine cones, a food source that occasionally has very large crops (called seed masting). Here are data on an index of the abundance of pine cones and the average number of offspring per female over 16 years:[2]

Don Johnston/Alamy

Cone index x	0.00	2.02	0.25	3.22	4.68	0.31	3.37	3.09
Offspring y	1.49	1.10	1.29	2.71	4.07	1.29	3.36	2.41
Cone index x	2.44	4.81	1.88	0.31	1.61	1.88	0.91	1.04
Offspring y	1.97	3.41	1.49	2.02	3.34	2.41	2.15	2.12

Describe the relationship with both a graph and numerical measures, then summarize it in words. What is striking is that the offspring are conceived in the spring, *before* the cones mature in the fall to feed the new young squirrels through the winter.

6.6 Cicadas as fertilizer? 4 STEP Every 17 years, swarms of cicadas emerge from the ground in the eastern United States, live for about six weeks, then die. (There are several "broods," so we experience cicada eruptions more often than every 17 years.) There are so many cicadas that their dead bodies can serve as fertilizer and increase plant growth. In an experiment, a researcher added 10 cicadas under some plants in a natural plot of American bellflowers in a forest, leaving other plants undisturbed. One of the response variables was the size of seeds produced by the plants. Here are data (seed mass in milligrams) for 39 cicada plants and 33 undisturbed (control) plants:[3]

Alastair Shay; Papilio/Corbis

Cicada plants				Control plants			
0.237	0.277	0.241	0.142	0.212	0.188	0.263	0.253
0.109	0.209	0.238	0.277	0.261	0.265	0.135	0.170
0.261	0.227	0.171	0.235	0.203	0.241	0.257	0.155
0.276	0.234	0.255	0.296	0.215	0.285	0.198	0.266
0.239	0.266	0.296	0.217	0.178	0.244	0.190	0.212
0.238	0.210	0.295	0.193	0.290	0.253	0.249	0.253
0.218	0.263	0.305	0.257	0.268	0.190	0.196	0.220
0.351	0.245	0.226	0.276	0.246	0.145	0.247	0.140
0.317	0.310	0.223	0.229	0.241			
0.192	0.201	0.211					

Do the data support the idea that dead cicadas can serve as fertilizer? Follow the four-step process (page 57) in your work.

6.7 More about cicadas. Let's examine in more detail the distribution of seed mass for plants in the cicada group from the previous exercise.

(a) Make a stemplot. Is the overall shape roughly symmetric or clearly skewed? There are both low and high observations that we might call outliers.

(b) Find the mean and standard deviation of the seed masses. Then remove both the smallest and the largest masses and find the mean and standard deviation of the remaining 37 plants. Why does removing these two observations reduce s? Why does it have little effect on $\bar{x}$?

6.8 Outliers? (optional) In the previous exercise, you noticed that the smallest and largest observations might be called outliers. Are either of these observations suspected outliers by the $1.5 \times IQR$ rule (page 48)?

6.9 Balancing calories. The Centers for Disease Control and Prevention (CDC) estimated that, in 2010, about 70% of adults in the United States were either overweight or obese. A 2010 survey of 1024 American adults asked, "To what extent, if at all, do you make a conscious effort to monitor the balance between how many calories you consume and how many calories you burn/use per day?" Here are the answers:[4]

Answer choice	Count
Always	59
Sometimes	376
Rarely	305
Not at all	284

Make a pie chart to display these data. What percent of survey participants said they do not make an effort to balance calories in and calories out ("not at all" and "rarely")?

6.10 Comparing tropical flowers. Researchers studied the flower lengths of varieties of *Heliconia* on the island of Dominica to see how they relate to the beak lengths of the hummingbird species that fertilize them. Table 6.1 gives length measurements (in millimeters) for samples of three varieties of *Heliconia*, each fertilized by a different species of hummingbird.[5]

(a) Display the flower lengths in side-by-side dotplots. Do the three varieties display distinct distributions of length?

(b) Find the five-number summary for each data set and use boxplots to display the distributions. Are the boxplots an appropriate summary for these findings?

TABLE 6.1 Flower lengths (millimeters) for three *Heliconia* varieties

H. bihai	47.12	46.75	46.81	47.12	46.67	47.43	46.44	46.64
	48.07	48.34	48.15	50.26	50.12	46.34	46.94	48.36
H. caribaea red	41.90	42.01	41.93	43.09	41.47	41.69	39.78	40.57
	39.63	42.18	40.66	37.87	39.16	37.40	38.20	38.07
	38.10	37.97	38.79	38.23	38.87	37.78	38.01	
H. caribaea yellow	36.78	37.02	36.52	36.11	36.03	35.45	38.13	37.10
	35.17	36.82	36.66	35.68	36.03	34.57	34.63	

6.11 Paw preference in tree shrews. Tree shrews, *Tupaia belangeri*, are small omnivorous mammals phylogenetically related to primates. A research team examined paw preference among tree shrews born and raised in captivity. They computed a "pawedness index" (PI), a unitless continuous variable taking values between −1 and 1, depending on the relative use of the right and left paws during food-grabbing tasks. An animal always using its left paw would have a PI value of −1, but an animal using its right and left paws equally would have a PI value of 0. Here are the PI values for the 36 tree shrews, sorted for convenience:[6]

−1.00	−1.00	−1.00	−1.00	−1.00	−1.00	−1.00	−1.00	−1.00	−1.00
−0.96	−0.95	−0.93	−0.89	−0.89	−0.25	−0.11	−0.08	−0.06	−0.03
0.13	0.24	0.41	0.50	0.56	0.67	0.86	0.87	0.90	0.94
1.00	1.00	1.00	1.00	1.00	1.00				

(a) Display the PI values in a dotplot and use it to describe the distribution of PI values.

(b) Explain why it would be misleading to report the mean and standard deviation for this data set. The five-number summary would provide more information, but it would still not be an appropriate summary description of this data set. You should always plot your data before computing numerical summaries.

6.12 A big-toe problem. Hallux abducto valgus (call it HAV) is a deformation of the big toe that is not common in youth and often requires surgery. Doctors used X-rays to measure the angle (in degrees) of deformity in 38 consecutive patients under the age of 21 who came to a medical center for surgery to correct HAV.[7] The angle is a measure of the seriousness of the deformity. The data appear in Table 6.2 as "HAV angle." Make a graph and give a numerical description of this distribution. Are there any outliers? Write a brief discussion of the shape, center, and spread of the angle of deformity among young patients needing surgery for this condition.

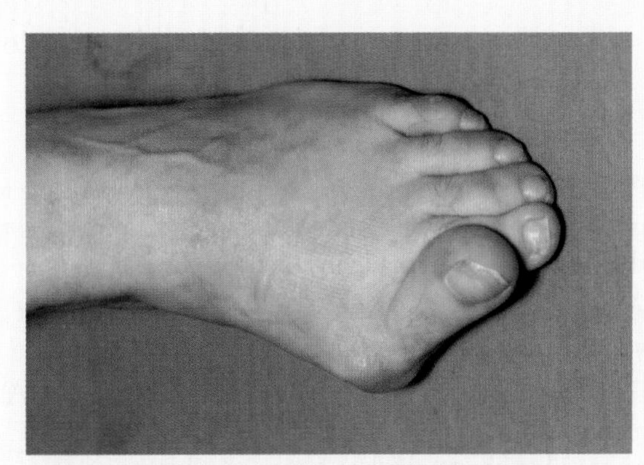

Wellcome Trust Medical Library/Custom Medical Stock Photo

6.13 More on a big-toe problem. The HAV angle data in the previous exercise contain one high outlier. Calculate the median, the mean, and the standard deviation for the full data set and also for the 37 observations remaining when you remove the outlier. How strongly does the outlier affect each of these measures?

6.14 Predicting foot problems. Metatarsus adductus (call it MA) is a turning in of the front part of the foot that is common in adolescents and usually corrects itself. Table 6.2 gives the severity of MA ("MA angle") as well. Doctors speculate that the severity of MA can help predict the severity of HAV.

(a) Make a scatterplot of the data. (Which is the explanatory variable?)

(b) Describe the form, direction, and strength of the relationship between MA angle and HAV angle. Are there any clear outliers in your graph?

(c) Do you think the data confirm the doctors' speculation? Why or why not?

TABLE 6.2 Angle of deformity (degrees) for two types of foot deformity

HAV angle	MA angle	HAV angle	MA angle	HAV angle	MA angle
28	18	21	15	16	10
32	16	17	16	30	12
25	22	16	10	30	10
34	17	21	7	20	10
38	33	23	11	50	12
26	10	14	15	25	25
25	18	32	12	26	30
18	13	25	16	28	22
30	19	21	16	31	24
26	10	22	18	38	20
28	17	20	10	32	37
13	14	18	15	21	23
20	20	26	16		

6.15 Predicting foot problems, continued.

(a) Find the equation of the least-squares regression line for predicting HAV angle from MA angle. Add this line to the scatterplot you made in the previous exercise.

(b) A new patient has an MA angle of 25 degrees. What do you predict this patient's HAV angle to be?

(c) Does knowing MA angle allow doctors to predict HAV angle accurately? Explain your answer using the scatterplot, then calculate a numerical measure to support your finding.

6.16 Sulfur, the ocean, and the sun. Sulfur in the atmosphere affects climate by influencing the formation of clouds. The main natural source of sulfur is dimethyl sulfide (DMS) produced by small organisms in the upper layers of the oceans. DMS production is in turn influenced by the amount of energy the upper ocean receives from sunlight. Here are monthly data on solar radiation dose (SRD, in watts per square meter) and surface DMS concentration (in nanomolars) for a region in the Mediterranean:[8]

SRD	12.55	12.91	14.34	19.72	21.52	22.41	37.65	48.41
DMS	0.796	0.692	1.744	1.062	0.682	1.517	0.736	0.720

SRD	74.41	94.14	109.38	157.79	262.67	268.96	289.23
DMS	1.820	1.099	2.692	5.134	8.038	7.280	8.872

(a) Make a scatterplot that shows how DMS responds to SRD.

(b) Describe the overall pattern of the data. Find the correlation r between DMS and SRD. Because SRD changes with the seasons of the year, the close relationship between SRD and DMS helps explain other seasonal patterns.

(c) Find the least-squares regression line predicting DMS from SRD. What percent of the observed variation in DMS can be explained by this regression model?

(d) Use your regression line to predict the DMS produced for a solar radiation dose of 100 watts per square meter.

6.17 Monkey calls. The usual way to study the brain's response to sounds is to have subjects listen to "pure tones." The response to recognizable sounds may differ. To compare responses, researchers anesthetized macaque monkeys. They fed pure tones and also monkey calls directly into their brains by inserting electrodes. Response to the stimulus was measured by the firing rate (electrical spikes per second) of neurons in various areas of the brain. Table 6.3 contains the responses for 37 neurons.[9]

(a) One important finding is that responses to monkey calls are generally stronger than responses to pure tones. For how many of the 37 neurons is this true?

(b) We might expect some neurons to have strong responses to any stimulus and others to have consistently weak responses. There would then be a strong relationship between tone response and call response. Make a scatterplot of monkey call response against pure-tone response (explanatory variable). Find the correlation r between tone and call responses. How strong is the linear relationship?

TABLE 6.3 Neuron response (spikes per second) to tones and monkey calls

Neuron	Tone	Call	Neuron	Tone	Call	Neuron	Tone	Call
1	474	500	14	145	42	26	71	134
2	256	138	15	141	241	27	68	65
3	241	485	16	129	194	28	59	182
4	226	338	17	113	123	29	59	97
5	185	194	18	112	182	30	57	318
6	174	159	19	102	141	31	56	201
7	176	341	20	100	118	32	47	279
8	168	85	21	74	62	33	46	62
9	161	303	22	72	112	34	41	84
10	150	208	23	20	193	35	26	203
11	19	66	24	21	129	36	28	192
12	20	54	25	26	135	37	31	70
13	35	103						

SUPPLEMENTARY EXERCISES

Supplementary exercises apply the skills you have learned in ways that require more thought or more elaborate use of technology.

6.18 Change in the Serengeti. 4 STEP Long-term records from the Serengeti National Park in Tanzania show interesting ecological relationships. When wildebeest are more abundant, they graze the grass more heavily, so there are fewer fires and more trees grow. Lions hunt more successfully when there are more trees, so the lion population increases. Here are data on one part of this cycle, wildebeest abundance (in thousands of animals) and the percent of the grass area that burned in the same year:[10]

Wildebeest (1000s)	Percent burned	Wildebeest (1000s)	Percent burned	Wildebeest (1000s)	Percent burned
396	56	360	88	1147	32
476	50	444	88	1173	31
698	25	524	75	1178	24
1049	16	622	60	1253	24
1178	7	600	56	1249	53
1200	5	902	45		
1302	7	1440	21		

To what extent do these data support the claim that more wildebeest reduce the percent of grasslands that burn? How rapidly does burned area decrease as the number of wildebeest increases? Include a graph and suitable calculations in your answer. Follow the four-step process (page 57).

6.19 Prey attract predators. 4 STEP Here is one way in which nature regulates the size of animal populations: High population density attracts predators, who remove a higher proportion of the population than when the density of the prey is low. One study looked at kelp perch and their common predator, the kelp bass. The researcher set up four large circular pens on sandy ocean bottom in southern California. He chose young perch at random from a large group and placed 10, 20, 40, and 60 perch in the four pens. Then he dropped the nets protecting the pens, allowing bass to swarm in, and counted the perch left after 2 hours. Here are data on the proportions of perch eaten in four repetitions of this setup:[11]

Perch	Proportion killed			
10	0.0	0.1	0.3	0.3
20	0.2	0.3	0.3	0.6
40	0.075	0.3	0.6	0.725
60	0.517	0.55	0.7	0.817

Do the data support the principle that "more prey attract more predators, who drive down the number of prey"? Follow the four-step process (page 57) in your answer.

6.20 Extrapolation. Your work in Exercise 6.18 no doubt included a regression line. Use the equation of this line to illustrate the danger of extrapolation, taking advantage of the fact that the percent of grasslands burned cannot be less than zero.

Falling through the ice. *The Nenana Ice Classic is an annual contest to guess the exact time in the spring thaw when a tripod erected on the frozen Tanana River near Nenana, Alaska, will fall through the ice. The 2013 jackpot was $318,500. The contest has been run since 1917. Table 6.4 gives simplified data that record only the date on which the tripod fell each year. The earliest date so far is April 20. To make the data easier to use, the table gives the date each year in days starting with April 20. That is, April 20 is 1, April 21 is 2, and so on. You will need software or a graphing calculator to analyze these data in Exercises 6.21 to 6.23.*[12]

6.21 When does the ice break up? We have 97 years of data on the date of ice breakup on the Tanana River. Describe the distribution of the breakup date with both a graph or graphs and appropriate numerical summaries. What is the median date (month and day) for ice breakup?

6.22 Global warming? Because of the high stakes, the falling of the tripod has been carefully observed for many years. If the date the tripod falls has been getting earlier, that may be evidence for the effects of global warming.

(a) Make a time plot of the date the tripod falls against year.

(b) There is a great deal of year-to-year variation. Fitting a regression line to the data may help us see the trend. Fit the least-squares line and add it to your time plot. What do you conclude?

(c) There is much variation about the line. Give a numerical description of how much of the year-to-year variation in ice breakup time is accounted for by the time trend represented by the regression line.

6.23 More on global warming. Side-by-side boxplots offer a different look at the data. Group the data into periods of roughly equal length: 1917 to 1939, 1940 to 1964, 1965 to 1989, and 1990 to 2013. Make boxplots to compare ice breakup dates in these four time periods. Write a brief description of what the plots show.

6.24 Saving the eagles. The pesticide DDT was especially threatening to bald eagles because they are at the top of their food chain. Below are data on the productivity of the eagle population in northwestern Ontario, Canada.[13] The eagles nest in an area free of DDT but migrate south and eat prey contaminated with the pesticide. DDT was banned at the end

TABLE 6.4 Days from April 20 for the Tanana River tripod to fall

Year	Day	Year	Day	Year	Day	Year	Day	Year	Day	Year	Day
1917	11	1934	11	1951	11	1968	19	1985	23	2002	18
1918	22	1935	26	1952	23	1969	9	1986	19	2003	10
1919	14	1936	11	1953	10	1970	15	1987	16	2004	5
1920	22	1937	23	1954	17	1971	19	1988	8	2005	9
1921	22	1938	17	1955	20	1972	21	1989	12	2006	13
1922	23	1939	10	1956	12	1973	15	1990	5	2007	8
1923	20	1940	1	1957	16	1974	17	1991	12	2008	17
1924	22	1941	14	1958	10	1975	21	1992	25	2009	12
1925	16	1942	11	1959	19	1976	13	1993	4	2010	10
1926	7	1943	9	1960	13	1977	17	1994	10	2011	15
1927	23	1944	15	1961	16	1978	11	1995	7	2012	4
1928	17	1945	27	1962	23	1979	11	1996	16	2013	31
1929	16	1946	16	1963	16	1980	10	1997	11		
1930	19	1947	14	1964	31	1981	11	1998	1		
1931	21	1948	24	1965	18	1982	21	1999	10		
1932	12	1949	25	1966	19	1983	10	2000	12		
1933	19	1950	17	1967	15	1984	20	2001	19		

of 1972. The researcher observed every nesting area he could reach every year between 1966 and 1981. He measured productivity by the count of young eagles per nesting area.

Year	Count	Year	Count	Year	Count	Year	Count
1966	1.26	1970	0.54	1974	0.46	1978	0.82
1967	0.73	1971	0.60	1975	0.77	1979	0.98
1968	0.89	1972	0.54	1976	0.86	1980	0.93
1969	0.84	1973	0.78	1977	0.96	1981	1.12

(a) Make a time plot of the data. Does the plot support the claim that banning DDT helped save the eagles?

(b) It appears that the overall pattern might be described by *two* straight lines. Find the least-squares line for 1966 to 1972 (pre-ban) and also the least-squares line for 1975 to 1981 (allowing a few years for DDT to leave the environment after the ban). Draw these lines on your plot. Would you use the second line to predict young per nesting area in the several years after 1981?

6.25 Thin monkeys, fat monkeys. Animals and people that take in more energy than they expend will get fatter. Here are data on 12 rhesus monkeys: 6 lean monkeys (4% to 9% body fat) and 6 obese monkeys (13% to 44% body fat). The data report the energy expended in 24 hours (kilojoules per minute) and the lean body mass (kilograms, leaving out fat) for each monkey.[14]

Lean		Obese	
Mass	Energy	Mass	Energy
6.6	1.17	7.9	0.93
7.8	1.02	9.4	1.39
8.9	1.46	10.7	1.19
9.8	1.68	12.2	1.49
9.7	1.06	12.1	1.29
9.3	1.16	10.8	1.31

(a) What is the mean lean body mass of the lean monkeys? Of the obese monkeys? Because animals with higher lean mass usually expend more energy, we can't directly compare energy expended between the two groups.

(b) Instead, look at how energy expended is related to body mass. Make a scatterplot of energy versus mass, using different plot symbols for lean and obese monkeys. Then add to the plot two regression lines, one for lean monkeys and one for obese monkeys. What do these lines suggest about the monkeys?

6.26 Weeds among the corn. Lamb's-quarter is a common weed that interferes with the growth of corn. An agriculture researcher planted corn at the same rate in 16 small plots of ground, then weeded the plots by hand to allow a fixed number of lamb's-quarter plants to grow in each meter of corn row. No other weeds were allowed to grow. Here are the yields of corn (bushels per acre) in each of the plots:[15]

Weeds per meter	Corn yield	Weeds per meter	Corn yield	Weeds per meter	Corn yield	Weeds per meter	Corn yield
0	166.7	1	166.2	3	158.6	9	162.8
0	172.2	1	157.3	3	176.4	9	142.4
0	165.0	1	166.7	3	153.1	9	162.8
0	176.9	1	161.1	3	156.0	9	162.4

(a) What are the explanatory and response variables in this experiment?

(b) Make side-by-side stemplots (after rounding to the nearest bushel) or dotplots of the yields. Give the median yield for each group (using the unrounded data). What do you conclude about the effect of this weed on corn yield?

6.27 Weeds among the corn, continued. We can also use regression to analyze the data on weeds and corn yield from the previous exercise. The advantage of regression over the side-by-side comparison in the previous exercise is that we can use the fitted line to draw conclusions for counts of weeds other than the ones the researcher actually used.

(a) Make a scatterplot of corn yield against weeds per meter. Find the least-squares regression line and add it to your plot. What does the slope of the fitted line tell us about the effect of lamb's-quarter on corn yield?

(b) Predict the yield for corn grown under these conditions with 6 lamb's-quarter plants per meter of row.

6.28 Manatee deaths: trends. Example 3.3 (page 67) describes a data set from Florida relating manatee deaths from collisions with powerboats and the number of powerboats registered in a given year. The data, available in Table 3.1, appear in a scatterplot in Figure 3.1. That scatterplot shows a clear, strong positive linear relationship between the number of manatee deaths from collisions with powerboats and the number of powerboats registered in a given year. We also want to study the trends over time for these two variables. Create and interpret the corresponding two time plots. What do you conclude?

6.29 Influence: monkey calls. Table 6.3 contains data on the response of 37 monkey neurons to pure tones and to monkey calls. You made a scatterplot of these data in Exercise 6.17.

(a) Find the least-squares line for predicting a neuron's call response from its pure-tone response. Add the line to your scatterplot. Mark on your plot the point (call it A) with the largest residual (furthest from the regression line, vertically) and also the point (call it B) that is an outlier in the x direction.

(b) How influential are each of these points for the correlation r?

(c) How influential are each of these points for the regression line?

6.30 Influence: bushmeat. Table 3.5 (page 87) gives data on fish catches in a region of West Africa and the percent change in the biomass (total weight) of 41 animals in nature reserves. It appears that years with smaller fish catches see greater declines in animals, probably because local people turn to "bushmeat" when other sources of protein are not available. The next year (1999) had a fish catch of 23.0 kilograms per person and animal biomass change of −22.9%.

(a) Make a scatterplot that shows how change in animal biomass depends on fish catch. Be sure to include the additional data point. Describe the overall pattern. The added point is a low outlier in the y direction.

(b) Find the correlation between fish catch and change in animal biomass both with and without the outlier. The outlier is influential for correlation. Explain from your plot why adding the outlier makes the correlation smaller.

(c) Find the least-squares line for predicting change in animal biomass from fish catch both with and without the additional data point for 1999. Add both lines to your scatterplot from (a). The outlier is not influential for the least-squares line. Explain from your plot why this is true.

6.31 Optional Chapter 5: Preventing urinary tract infections.  Cranberries were used by Native Americans to treat a number of ailments, particularly those related to the urinary tract. Could a daily intake of cranberry juice help prevent urinary tract infections (UTIs) in women? Researchers studied the recurrence of UTIs in 149 women just cured from a UTI with antibiotics. The women were asked to take one of three preventive treatments over a six-month period (cranberry juice daily, lactobacillus drink daily, or neither drink). The findings are summarized in the table below:[16]

	Outcome	
Treatment	**UTI recurrence**	**No UTI recurrence**
Cranberry juice	8	42
Lactobacillus drink	19	30
Neither drink	18	32

What do the data say about the relative effectiveness of the three treatments in preventing the recurrence of UTIs? Use the four-step process (page 57) in your answer.

6.32 Optional Chapter 5: Smoking on campus. In February 2013, SurveyUSA interviewed a random sample of 800 adults in the Tampa, Florida, area. Two of the questions asked were "Should smoking be allowed? Or banned on college campuses?" and "What is your smoking status?" The two-way table below summarizes the answers to these two questions.[17]

	Smoking Status		
Opinion	**Current smoker**	**Former smoker**	**Never a smoker**
Allowed	115	93	79
Banned	20	213	188
Not sure	23	39	30

What do the data say about differences in opinion about smoking on college campuses for individuals with different smoking habits? Use the four-step process (page 57) in your answer.

EESEE CASE STUDIES

The Electronic Encyclopedia of Statistical Examples and Exercises (EESEE) is available on the text CD and website. These more elaborate stories, with data, provide settings for longer case studies. Here are some suggestions for EESEE stories that apply the ideas you have learned in Chapters 1 to 5.

6.33 Checkmating and Reading Skills. Write a report based on Question 1 in this case study. (Describing a distribution.)

6.34 Keeping Balance as You Age. Write a report based on Questions 1 and 4 in this case study. (Describing and comparing distributions.)

6.35 Counting Calories. Respond to Questions 1, 4, and 6 for this case study. (Describing and comparing distributions.)

6.36 Habitat of the Spotted Owl. Write a report based on Questions 1 and 3. (Describing and comparing distributions, scatterplots, and correlation.)

6.37 Visibility of Highway Signs. Write a report based on Questions 2 and 5. (Describing and comparing distributions, scatterplots, and correlation.)

6.38 Mercury in Florida's Bass. Respond to Question 5. (Scatterplots, form of relationships. By the way, "homoscedastic" means that the scatter of points about the overall pattern is roughly the same from one side of the scatterplot to the other.)

6.39 Brain Size and Intelligence. Write a response to Question 3. (Scatterplots, correlation, and lurking variables.)

6.40 Acorn Size and Oak Tree Range. Write a report based on Questions 1 and 2. (Scatterplots, correlation, and regression.)

From Exploration to Inference

The purpose of statistics is to gain understanding from data. We have studied one approach to data, *exploratory data analysis*, in some detail. Now we move from data analysis toward *statistical inference*. Both types of reasoning are essential to effective work with data. Here is a brief sketch of the differences between them.

EXPLORATORY DATA ANALYSIS	STATISTICAL INFERENCE
Purpose is unrestricted exploration of the data, searching for interesting patterns.	Purpose is to answer specific questions, posed before the data were produced.
Conclusions apply only to the individuals and circumstances for which we have data in hand.	Conclusions apply to a larger group of individuals or a broader class of circumstances.
Conclusions are informal, based on what we see in the data.	Conclusions are formal, backed by a statement of our confidence in them.

Our journey toward inference begins in Chapters 7 and 8 with *data production*, statistical ideas for producing data to answer specific questions. Chapters 9, 11, and 13 (and the optional Chapters 10 and 12) concern *probability*, the language of formal statistical conclusions. Finally, Chapters 14 and 15 present the core concepts of *inference*. Chapter 16 reviews the material of Chapters 7 to 15 and provides more comprehensive exercises. In practice, data analysis and inference cooperate. Successful inference requires good data production, data analysis to ensure that the data are regular enough, and the language of probability to state conclusions.

PART II

PRODUCING DATA

PROBABILITY AND SAMPLING DISTRIBUTIONS

THE IDEA OF INFERENCE

Dr. Gary Gaugler/Science Source

iStockphoto

CHAPTER 7 Samples and Observational Studies

Statistics, the science of data, provides ideas and tools that we can use in many settings. Raw data should always be carefully examined for trends and deviations from these trends using the tools of exploratory data analysis. But how do we *collect data* that can help us answer interesting scientific questions?

Suppose our question is "What percent of adult Americans use complementary and alternative medicine (CAM)?" To answer the question, we ask individuals who are 20 and older whether they have used any form of CAM in the past 12 months. We can't realistically ask every adult in America, so we put the question to a *sample* chosen to represent the entire adult *population*. How shall we choose such a sample? In this chapter we examine how to choose appropriate samples for observational studies.

IN THIS CHAPTER WE COVER...

- Observation versus experiment
- Sampling
- Sampling designs
- Sample surveys
- Cohorts and case-control studies

Observation versus experiment

We collect data in order to understand phenomena or answer scientific questions. There are two approaches for collecting data. One is the *observational study*, in which the subject of our scientific inquiry is disturbed as little as possible by the act of gathering information. Observational studies can answer questions such as "What is the range of gestation times in the prairie dog?" or "What is the proportion of Americans who are overweight?" The other approach is that of the *experiment*. In doing an experiment, we actively impose some *treatment* or condition in order to observe the *response*. Experiments can answer questions such as "Does aspirin reduce the chance of a heart attack?" and "Does ginkgo biloba enhance memory?" Experiments and observational studies provide useful data only when properly designed. The distinction between experiments and observational studies is one of the most important ideas in statistics.

OBSERVATION VERSUS EXPERIMENT

An **observational study** observes individuals and measures variables of interest but does not attempt to influence the responses. The purpose of an observational study is to describe and compare existing groups or situations.

An **experiment,** on the other hand, deliberately imposes some treatment on individuals in order to observe their responses. The purpose of an experiment is to study whether the treatment causes a change in the response.

Observational studies are essential sources of data about topics ranging from the heights of adult Americans to the behavior of animals in the wild. But an observational study, even one based on a statistical sample, is a poor way to gauge the effect of an intervention. To see the response to a change, we must actually impose the change. When our goal is to understand cause and effect, experiments are the only source of fully convincing data.

EXAMPLE 7.1 The ups and downs of hormone replacement

Should women take hormones such as estrogen after menopause, when natural production of these hormones ends? In 1992, several major medical organizations said "yes." In particular, women who took hormones seemed to reduce their risk of a heart attack by 35% to 50%. The risks of taking hormones appeared small compared with the benefits.

The evidence in favor of hormone replacement came from a number of observational studies that compared women who were taking hormones on their own accord with others who were not. But women who elect to take hormones are most likely quite different from women who do not: They may be better informed and see doctors more often, and they may do many other things to maintain their health. So, is it surprising that they have fewer heart attacks?

Experiments don't let women decide what to do. The Women's Health Initiative (WHI) trial sponsored by the National Institutes of Health assigned women either to hormone replacement or to dummy pills that look and taste the same as the hormone pills. The assignment was done by a coin toss, so that all kinds of women were equally likely to get either treatment. By 2002, the WHI trial published its first results, which indicated that women who took hormones had a *higher* incidence of cardiovascular disease and breast cancer. Taking hormones after menopause quickly fell out of favor. This first WHI study, however, had focused on older women, with an average age of 63 years. A follow-up WHI trial of women in their 50s was published in 2007 and showed that younger women taking hormone therapy had lower levels of calcium deposits in their arteries, which may lower their risk of heart disease.[1] ■

When we simply observe women, the effects of actually taking hormones are *confounded* with (mixed up with) the characteristics of women who choose to take hormones.

CONFOUNDING

Two variables (explanatory variables or lurking variables) are **confounded** when their effects on a response variable cannot be distinguished from each other.

Observational studies of the effect of one variable on another often fail to demonstrate causality because the explanatory variable is confounded with lurking variables. We will see in the next chapter that well-designed experiments take steps to defeat confounding.

EXAMPLE 7.2 Wine, beer, or spirits?

Moderate use of alcohol is associated with better health. Observational studies suggest that drinking wine rather than beer or spirits confers added health benefits. But people who prefer wine are different from those who drink mainly beer or stronger stuff. Moderate wine drinkers as a group are richer and better educated. They eat more fruit and vegetables and less fried food. Their diets contain less fat and less cholesterol. They are less likely to smoke. The explanatory variable (What type of alcoholic beverage do you drink most often?) is confounded with many lurking variables (education, wealth, diet, and so on). A large study therefore concludes: "The apparent health benefits of wine compared with other alcoholic beverages, as described by others, may be a result of confounding by dietary habits and other lifestyle factors."[2] Figure 7.1 shows the confounding in picture form. ■

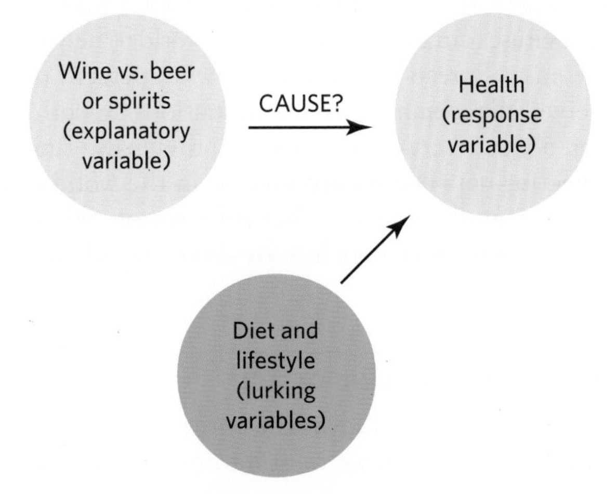

FIGURE 7.1 Confounding, for Example 7.2: We can't distinguish the effects of what people drink from the effects of their overall diet and lifestyle.

APPLY YOUR KNOWLEDGE

7.1 Brain asymmetry. A study aims to determine the percent of adults whose right cerebral hemisphere is larger than their left hemisphere. A group of individuals aged 20 and older representative of the American adult population is asked to come

iStockphoto

to a brain-imaging facility to have their brains scanned. Each individual's scan is then analyzed to identify whether the right or left cerebral hemisphere is larger. Is this an observational study or an experiment? Explain your answer.

7.2 Cell phones and brain cancer. A study of cell phones and the risk of brain cancer looked at a group of 469 people who have brain cancer. The investigators matched each cancer patient with a person of the same sex, age, and race who did not have brain cancer, then asked about use of cell phones. Result: "Our data suggest that use of handheld cellular telephones is not associated with risk of brain cancer."[3] Is this an observational study or an experiment? Why? What are the explanatory and response variables?

7.3 Global warming. Instrumental records show an increase in global mean surface air and ocean temperatures on the order of 1°F (0.5°C) over the 20th century. This trend is also evident in the reduced extent of snow cover, the accelerated rate of rise of sea level, and the increasingly earlier arrival and breeding times of migratory birds. This global warming coincides with dramatic changes in atmospheric gas composition due to human activity such as the burning of fossil fuels. The Committee on the Science of Climate Change states: "Because of the large and still uncertain level of natural variability inherent in the climate record . . ., a causal linkage between the buildup of greenhouse gases in the atmosphere and the observed climate changes during the 20th century cannot be unequivocally established."[4]

(a) Is the evidence for global warming experimental or observational?

(b) What explanatory variable is thought to influence climate change? What lurking variable is confounded with that explanatory variable?

Sampling

The Centers for Disease Control and Prevention (CDC) want to know the DNA sequences of the flu viruses circulating around the world to help prepare next year's flu vaccine. The Census Bureau wants to know what percent of children 16 and younger have received all recommended immunizations. In both cases, we want to gather information about a very large group of individuals. Time, cost, and practicality forbid collecting data about each flu case in the world or each American child's immunization records. So we gather information about only part of the group, a *sample*, in order to draw conclusions about the whole, the *population* of interest.

POPULATION, SAMPLE, SAMPLING DESIGN

The **population** in a statistical study is the entire group of individuals (not necessarily people) about which we want information.

A **sample** is the part of the population from which we actually collect information. We use a sample to draw conclusions about the entire population.

A **sampling design** describes exactly how to choose a sample from the population.

We often draw conclusions about a whole on the basis of a sample. Everyone has sipped a spoonful of soup and judged the entire bowl on the basis of that taste. When your doctor is concerned about your white blood cell count, you give a "blood sample" to see how low your white cell count really is. But the bowl of soup and your blood are quite homogeneous, so that the taste of a single spoonful represents the whole soup and a small vial of your blood represents your whole blood composition at the time of extraction. These are relatively easy sampling designs.

Choosing a representative sample from a large and varied population is not so simple. The first step in a proper sample design is to say exactly *what population* we want to describe. The second step is to say exactly *what we want to measure,* that is, to give exact definitions of our variables.

EXAMPLE 7.3 The Current Population Survey

The Current Population Survey (CPS) is an important sample survey conducted by the U.S. government. The CPS contacts about 60,000 households each month. It produces the monthly unemployment rate and various demographic characteristics. Supplemental questionnaires on a variety of topics are added at regular intervals. They are designed to update other government surveys such as the National Health Interview Survey conducted yearly.

One question of interest to the government is the proportion of current U.S. smokers who attempted to quit smoking within the previous 12 months. The first step in producing data for this question is to specify the population we want to describe. Which age groups will we include? Will we include illegal aliens or people in prisons? The CPS defines its population as all U.S. residents (whether citizens or not) who are at least 15 years of age and are civilians and are not in an institution such as a prison. Persons from all 50 states are considered, representing both genders and all ethnic groups.

The second step involves choosing what precisely we want to measure: What does it mean to be a "current smoker"? Should we include individuals who smoke only occasionally? What constitutes an "attempt to quit smoking"? The CPS defines current smokers as individuals who currently smoke either every day or some days. If you are a current smoker, the interviewer then goes on to ask about quitting attempts. A quitting attempt is defined as an attempt to quit smoking during the past year that lasted for 24 hours or more. ■

EXAMPLE 7.4 Animal studies

Krys Bailey/Alamy

Tree shrews, *Tupaia belangeri,* are small omnivorous mammals phylogenetically related to primates. Do tree shrews exhibit a paw preference when grasping food?[5] The first step in producing data for this question is to specify the population we want to describe. This would be the population of all tree shrews—a population naturally found in Asian tropical forests. What are researchers' options for obtaining a sample of live animals? They can capture tree shrews in the wild or obtain tree shrews from a dedicated breeding facility. The researchers in this study chose to obtain 36 tree shrews from a breeding facility run by the University of Veterinary Medicine, Hanover, Germany. The advantage is that their population of interest is similar to that studied by other researchers in their field. The disadvantage is that these animals are not representative of the population of wild tree shrews. The second step involves choosing what we want to measure: How do we assess "paw preference"? The researchers observed the tree shrews during food-grasping tasks and recorded which paw was used for grasping. They used this information to compute a quantitative "pawedness index" reflecting the relative use of the right and left paws during numerous grasping attempts; they then used this index value to label each animal as having a right, left, or ambidextrous paw preference.

Another study examined the biomechanics of jumping in the hedgehog flea, *Archaeopsyllus erinacei*.[6] The target population is all adult hedgehog fleas (only the adults jump), but how do you sample from this population? The researchers obtained 10 adult specimens from hedgehogs at St. Tiggywinkles Wildlife Hospital Trust in the United Kingdom. These fleas are representative of wild fleas in the area around the wildlife hospital, but there is no guarantee that they are truly representative of all hedgehog fleas everywhere. One aspect of jump examined in this study was takeoff angle. The researchers filmed the fleas in slow motion during spontaneous jumps and recorded the takeoff angle of each jump. ■

These two examples bring up an important issue. *Not all statistical studies use samples that are drawn directly from the entire population of interest*. Studies of animal behavior, for example, cannot realistically capture animals from their entire wildlife habitat. Whether the results may also apply to a larger population (the entire tree shrew breeding colony in Hanover or even the entire population of tree shrews) is a biological argument, not a statistical one. Experiments, in particular, rarely use samples from an entire population of interest, because of practical and ethical considerations. The consequence is that conclusions from the experiment cannot always be extended to the whole intended population.

EXAMPLE 7.5 Gender and clinical research

Researchers and drug manufacturers have historically avoided testing new drugs on women. Yet these same drugs have been routinely prescribed regardless of gender. Of course, an untested drug could potentially be harmful to the fetus if a woman were to become pregnant. However, this concern can easily be alleviated by enrolling only women willing to use contraceptives over the timeline of the clinical trial.

The major reasons for omitting women from clinical research are variations in a woman's menstrual cycle and hormonal changes over time with menopause or oral contraceptive use. Response to treatment might be influenced by these hormonal fluctuations, confounding the results. Or the treatment itself might interfere with concurrent use of oral contraceptives. Hormonal fluctuations during the menstrual cycle and the use of an oral contraceptive have indeed been shown to affect how drugs are metabolized. This is true, for example, of such common drugs as acetaminophen, aspirin, diazepam, and even caffeine. Some drugs, like the antibiotic rifampin, also reduce the effectiveness of oral contraceptives.[7] The question is: Is this a good reason to exclude women from clinical trials?

Absolutely not! We know that samples are representative of only the population from which they are taken. Clinical trials based entirely on men ensure drug effectiveness and safety for men alone. Not including women in clinical research does not protect them: It only delays the assessment of efficacy and side effects until the drug is made available to the public.

Since the mid-1990s, the Food and Drug Administration (FDA) requires that clinical trials for drugs targeting both men and women include both genders, in proportions reflecting the actual proportions of men and women in the target population. ■

APPLY YOUR KNOWLEDGE

7.4 **Sampling members.** A health care provider wants to rate its members' satisfaction with physical therapy. A questionnaire is mailed to 800 members of the health plan selected at random from the list of members who were prescribed physical therapy in the past 12 months. Only 212 questionnaires are returned.

(a) What is the population in this study? Be careful: What group does the health care provider *want information about?*

(b) What is the sample? Be careful: From what group does the provider *actually obtain information?*

7.5 **Live births.** Every year, the state of California publishes the state's vital statistics. Among these are the total live births based on actual birth certificates and hospital records. There were 502,023 live births in California in 2011.[8] Is this number derived from a sample of the California live-birth population, or does it represent the actual population?

Sampling designs

The purpose of a sample is to give us information about a larger population. The process of drawing conclusions about a population on the basis of sample data is called statistical **inference** because we *infer* information about the population from what we *know* about the sample. There are a variety of possible sampling designs, but not all are appropriate for statistical purposes.

inference

Poor sampling designs The easiest—but not the best—sampling design just chooses individuals close at hand. This is called a **convenience sample,** and it often produces unrepresentative data. If we are interested in finding out what percent of adults take vitamin and mineral supplements, for example, we might go to a shopping mall and ask people we meet. However, a shopping mall sample will almost surely overrepresent middle-class and retired people and underrepresent the poor. This will happen almost every time we take such a sample. That is, it is a systematic error caused by a bad sampling design, not just bad luck on one sample. This is *bias:* The outcomes of shopping mall surveys will repeatedly miss the truth about the population in the same ways.

convenience sample

BIAS

The design of a statistical study is **biased** if it systematically favors certain outcomes.

Another biased sampling design, the **voluntary response sample,** lets individuals choose whether to participate or not. Opt-in polls (write-in, call-in, or online quick votes) are examples of voluntary response samples. They are not scientific polls. These types of samples are biased because people with strong opinions are most likely to respond. The problem is that *people who take the trouble to respond to an open invitation are usually not representative of any clearly defined population.*

voluntary response sample

Unfortunately, convenience and voluntary response samples have become very popular with the media because they are cheap and often (misleadingly) sensational.[9]

EXAMPLE 7.6 Opt-in poll: genome sequencing

An article published on National Public Radio's website in the fall of 2012 discussed the recent advances in genome sequencing and offered an opt-in poll. The question asked was "Would you have your genome sequenced if you could afford it?" The poll stayed opened for two weeks, during which time 6627 clicks were received (5398 for yes, 656 for undecided, and 573 for no).[10] What can we conclude from this? Nothing beyond the fact that NPR.org received 5398 yes clicks, 656 undecided clicks, and 573 no clicks. We have no way of knowing what type of individuals clicked on the poll or even if they clicked only once. There is no larger, clearly defined population here about which to infer. ■

Simple random samples In a voluntary response sample, people choose whether to respond. In a convenience sample, the interviewer makes the choice. In both cases, personal choice produces bias. The statistician's remedy is to allow impersonal chance to choose the sample. A sample chosen by chance allows neither favoritism by the sampler nor self-selection by respondents. Choosing a sample by chance mitigates bias by giving all individuals an equal chance to be chosen. Rich and poor, young and old, male and female, all have the same chance to be in the sample.

probability sampling

A sample chosen by chance is called a probability sample. The most common **probability sampling** designs are simple random sampling, stratified random sampling, and multistage random sampling. The simplest way to use chance to select a sample is to place names in a hat (the population) and draw out a handful (the sample). This is the idea of *simple random sampling*.

SIMPLE RANDOM SAMPLE

A **simple random sample (SRS)** of size n consists of n individuals from the population chosen in such a way that every set of n individuals has an equal chance to be the sample actually selected.

An SRS not only gives each individual an equal chance to be chosen but also gives every possible sample an equal chance to be chosen. When you think of an SRS, picture drawing names from a hat to remind yourself that an SRS doesn't favor any part of the population. That's why an SRS is a better method of choosing samples than convenience or voluntary response sampling.

APPLET

table of random digits

But writing names on slips of paper and drawing them from a hat is slow and inconvenient, especially for large populations (for example, all American adults). In practice, most people use technology. The *Simple Random Sample* applet makes choosing an SRS very fast. If you don't use the applet or other software, you can use a **table of random digits** consisting of a long random sequence of the digits 0, 1, 2, 3, 4, 5, 6, 7, 8, 9.

Table A at the back of the book is a table of random digits. Using this method to select an SRS is antiquated in the computer era, but it helps us understand the concept of random sampling from random digits, a concept on which software and applets rely. Exercise 7.7 guides you through the process. Table A begins with the digits 19223950340575628713 (ignore spacing, introduced only for the sake of readability). To use Table A, label and select. **Step 1. Label:** Give each member of the population a unique numerical label of fixed length. **Step 2. Select:** Read from Table A successive groups of digits of the length you used as labels; your sample contains the individuals whose labels you find in the table.

We can trust that an SRS avoids selection bias, because it uses impersonal chance to select the individuals in the sample. Voluntary opt-in polls and mall interviews also produce samples. But we can't trust results from these samples, because they are chosen in ways that invite bias. *The first question to ask about any sample is whether it was chosen at random.*

EXAMPLE 7.7 Who gets the flu?

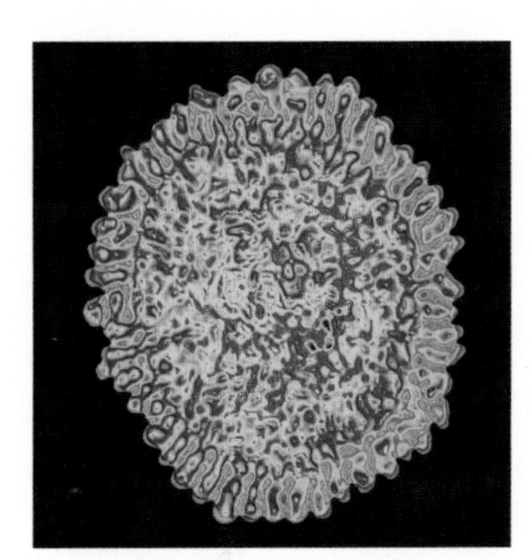
James Cavallini/Photo Researchers, Inc.

The Gallup-Healthways Well-Being Index survey asked a sample of people in October 2012 whether they were sick with the flu yesterday. The survey found that 2.1% of the respondents were. Should we trust that 2.1% to represent a larger population? Ask first how Gallup selected its sample. The page describing the findings states that the results are based on telephone interviews (both landline telephones and cell phones) conducted between October 1 and October 31, 2012, with 28,295 adults, aged 18 and older, living in all 50 U.S. states and the District of Columbia.[11]

Gallup tells us what population it has in mind: people at least 18 years old who live anywhere in the United States and can be contacted by phone. We know that Gallup called both landline telephones and cell phones to sample U.S. adults regardless of which type of telephone they use. We also know that the sample from this population was of size 28,295 and that it was chosen at random. ■

Other probability sampling designs The use of chance to select a sample is the essential principle of statistical sampling. Some probability sampling designs (such as an SRS) give each member of the population an equal chance to be selected. This may not be true in more elaborate sampling designs.

For example, it is common to sample important groups within the population separately and then to combine these samples. This is the idea of a **stratified** random sample. It can be useful when the population contains groups that are more rare or harder to reach than others. For example, a stratified random sample might select individuals of different ethnic groups separately in order to ensure inclusion of both majority and minority groups according to desired proportions.

stratified samples

Nationwide or statewide governmental surveys often use **multistage** samples for reasons of practicality. These samples typically involve choosing SRSs within SRSs. For example, an SRS of counties is first chosen; for each chosen county, an SRS of schools is chosen; then within each chosen school, an SRS of classrooms is drawn. This is a more practical and cost-effective method than directly selecting an SRS of students from all over the United States.

multistage samples

Most large-scale sample surveys use stratified, multistage samples that combine SRSs from each stage of the sampling process. Analysis of data from sampling designs more complex than an SRS takes us beyond basic statistics and will not be treated further in this book. But the SRS is the building block of more elaborate designs, and analysis of other probability designs differs more in complexity of detail than in fundamental concepts.

APPLY YOUR KNOWLEDGE

7.6 Sampling on campus: poor sampling. Your college wants to know students' perceptions of the nutritional quality of campus food. It isn't practical to contact all students.

(a) Give an example of a way to choose a sample of students that is poor practice because it depends on voluntary response.

(b) Give an example of a bad way to choose a sample of students that doesn't use voluntary response.

7.7 Sampling the forest: a simple random sample. To gather data on a 1200-acre pine forest in Louisiana, the U.S. Forest Service laid a grid of 1410 equally spaced circular plots over a map of the forest. A ground survey visited a sample of 10% of these plots.[12]

(a) How would you label the plots to select an SRS, using numerical labels starting with the number 1?

(b) Use software or Table A to choose the first 3 plots in an SRS of 141 plots. If you use Table A, begin at line 105. Make sure that all labels have the same length (think about how many digits are needed for the largest numerical label, in this case 1410). Ignore the spacing between digits (their only purpose is readability), and read the digits in groups of the same length as your numerical labels. The first three groups that match a plot label make up the beginning of your sample. Ignore any repeated labels, because you don't want to choose the same plot twice.

7.8 America's State of Mind Report: a stratified random sample. Medco Health Solutions Inc. is a prescription drug insurance provider with approximately 65 million members. Medco took a random sample of its members to analyze the utilization of mental health-related medications among the insured population. Sampling was stratified by age, sex, and region to match the demographics of Medco's members. Explain how a stratified sample is different from a simple random sample and what they have in common.

7.9 The National Health Interview Survey: a multistage random sample. The CDC conducts the yearly National Health Interview Survey (NHIS). Here is a description of its methodology: "To achieve sampling efficiency and to keep survey operations manageable, cost-effective, and timely, the NHIS survey planners used multistage sampling techniques to select the sample of persons and households for the NHIS." Explain how a multistage sample is different from a simple random sample and what they have in common.

Sample surveys

A sample survey is an observational study that relies on a random sample drawn from the entire population at one point in time. Sample surveys most often assess the characteristics or opinions of people, with a wide array of applications. The Current Population Survey, described in Example 7.3, is a comprehensive survey of households from all parts of the United States (about 100 million U.S. households in all). Opinion polls are sample surveys that cover all sorts of topics and typically use voter registries or telephone numbers to select their samples. In epidemiology, sample surveys are sometimes called **cross-sectional studies** and are used to establish the prevalence of various medical conditions and diseases (proportion of individuals in the population with the condition at one point in time). The Gallup-Healthways survey of individuals sick with the flu, described in Example 7.7, is one such epidemiological survey.

cross-sectional

Sample surveys may collect lots of data for each individual sampled. In addition to the main questions of interest, personal information about the respondents is often recorded, such as gender, age, and sociodemographics. This enables the comparison of results between subgroups of individuals at the data analysis stage. For example, the Gallup flu survey in Example 7.7 broke down its findings between men (2.0% sick with the flu) and women (2.2%) and also between smokers (2.7% sick with the flu) and nonsmokers (1.9%).

Random selection of the individuals surveyed eliminates bias in the choice of a sample from a list of the population. In practice, this process can be quite challenging. To begin, we need an accurate and complete list of the population. Because such a list is rarely available, most samples suffer from some degree of **undercoverage.** Undercoverage occurs when some groups in the population are left out of the process of choosing the sample. A sample survey of households, for example, will miss not only homeless people but also prison inmates and students in dormitories. An opinion poll conducted by calling landline telephone numbers will miss households that have only cell phones as well as households without a phone. The results of national sample surveys therefore have some bias if the people not covered differ from the rest of the population.

undercoverage

Computer-assisted interviewing The days of the interviewer with a clipboard are past. Interviewers now use computers to read questions and enter responses. The computer skips irrelevant items—once a woman says that she has no children, further questions about her children never appear. The computer can even present questions in random order to avoid bias due to always following the same order. The tedious process of transferring responses from paper to computer, once a source of errors, has disappeared.

EXAMPLE 7.8 Cell phones and undercoverage

A few national sample surveys, especially government surveys, interview some or all of their subjects in person. Most national surveys contact subjects by telephone using the random digit dialing (RDD) method. However, federal regulations forbid RDD technology for dialing cell phone numbers.[13] How much of a problem is this?

The number of cell-phone-only households is increasing rapidly. By 2008, 18% of American households had a cell phone but no landline phone, and by 2012 that number had increased to 38%.[14] Surveys reaching only landline numbers therefore suffer from considerable undercoverage. Pollsters can add cell phone numbers if they are hand-dialed, but the additional expense can be prohibitive. Additionally, a cell phone can be located anywhere, which makes breaking down a sample by location difficult. And a cell phone user may be driving or otherwise unable to talk safely.

People who screen calls and people who have only a cell phone tend to be younger than the general population. In fact, at the extremes, 58% of adults aged 25 to 29 live in households with only cell phone coverage, compared with only 8% of adults aged 65 and over. So surveys using RDD tend to oversample older individuals and undersample young adults.

Careful surveys weight their responses to reduce bias. For example, if a sample contains too few young adults, the responses of the young adults who do respond are given extra weight. Some polling organizations also include a minimum quota of cell phone users in their samples to minimize this bias. ■

nonresponse

Another potential source of bias in most sample surveys is **nonresponse,** which occurs when a selected individual cannot be contacted or refuses to cooperate. Nonresponse to sample surveys is often very large, even with careful planning and several callbacks. Because nonresponse is higher in urban areas, many sample surveys substitute other people in the same area to avoid favoring rural areas in the final sample. If the people contacted differ from those who are rarely at home or who refuse to answer questions, some bias remains.

New York, New York New York City, they say, is bigger, richer, faster, ruder. Maybe there's something to that. The sample survey firm Zogby International says that as a national average it takes 5 telephone calls to reach a live person. When calling to New York, it takes 12 calls. Survey firms assign their best interviewers to make calls to New York and often pay them bonuses to cope with the stress.

EXAMPLE 7.9 How bad is nonresponse?

The Census Bureau's American Community Survey (ACS) is a monthly survey of about 250,000 U.S. households. It has the lowest nonresponse rate of any poll we know: In 2011, only about 2.4% of the households selected failed to respond.[15] Participation in the ACS is mandatory by law, and the Census Bureau follows up by telephone and then in person if a household fails to return the mail questionnaire. Why so much effort? The Census Bureau states that the information collected helps to determine the allocation of more than $400 billion dollars of federal funding each year. An unexpected budget cut in 2004 prevented this kind of extensive follow-up during the month of January, resulting in an increase of about four percentage points in the nonresponse rate for the whole year.

The University of Chicago's General Social Survey (GSS) is the nation's most important social science survey. The GSS contacts its sample in person, and it is run by a university. Despite these advantages, its 2010 survey had a 40.2% rate of nonresponse.[16]

What about opinion polls by news media and opinion-polling firms? Few disclose their rates of nonresponse. Every year, the Pew Research Center conducts a survey using standard telephone polling methodology to assess issues such as nonresponse in commercial polls. The rate of nonresponse for the 2012 survey was 91%. When the same survey was conducted using additional polling efforts, such as an extended polling time period, monetary incentives for respondents, and letters to households that initially declined to be interviewed, the rate of nonresponse dropped to 78%.[17] ■

Do not confuse nonresponse in a probability sample and opt-in polls that use a voluntary response sample. Surveys using a probability sample may suffer bias from nonresponse, but this bias can be alleviated to some extent by weighting various groups to make the sample more closely representative of the target population. With opt-in polls relying on voluntary response, we never know who the respondents are and what kind of population they might represent—some individuals may even participate more than once (there are documented examples of thousands of online submissions coming from the same computer[18]). Nothing can be done with voluntary response samples to make the respondents more representative of the intended target population.

In addition to undercoverage and nonresponse, the behavior of the respondent or of the interviewer can cause **response bias** in sample results. Drinking a lot of alcohol and having many sexual partners, for example, typically have negative connotations, and many interviewees understate their response. Responses may also be influenced by whether the interviewer is male or female or from one ethnic group or another. Answers to questions that require recalling past events are often inaccurate because of faulty memory. For example, many people "telescope" events in the past, bringing them forward in memory to more recent time periods. "Have you visited a dentist in the last six months?" will often draw a "yes" from someone who last visited a dentist eight months ago.[19] Careful training of interviewers and careful supervision to avoid variation among the interviewers can reduce response bias. Good interviewing technique is another aspect of a well-done sample survey.

response bias

The **wording of questions** can strongly influence the answers given to a sample survey. Confusing or leading questions can introduce strong bias, and even minor changes in wording can change a survey's outcome. For example, only 13% of Americans surveyed think we are spending too much on "assistance to the poor," but 44% think we are spending too much on "welfare."[20] The order in which questions are asked can make a difference too.

wording effects

EXAMPLE 7.10 Are you happy?

Ask a sample of college students these two questions:

"How happy are you with your life in general?"

(Answers on a scale of 1 to 5)

"How many dates did you have last month?"

The correlation between answers is $r = -0.012$ when asked in this order. It appears that dating has little to do with happiness. Reverse the order of the questions, however, and $r = 0.66$. Asking a question that brings dating to mind makes dating success a big factor in happiness. ■

Don't trust the results of a sample survey until you have read the exact questions asked. The amount of nonresponse and the date of the survey are also important. A trustworthy survey must go beyond a good sampling design.

APPLY YOUR KNOWLEDGE

7.10 Question wording. Comment on each of the following as a potential sample survey question. Is the question clear? Is it slanted toward a desired response?

(a) "It is estimated that disposable diapers account for less than 2% of the trash in today's landfills. In contrast, beverage containers, third-class mail, and yard

He said, she said If you need data about body weight, should you ask the individuals in your sample or should you take actual measurements? When *asked* their weight, women tend to say they weigh less than they really do. Heavier men also underreport their weight—but lighter men claim to weigh more than the scale shows; and men tend to exaggerate their height. We leave you to ponder the psychology of the two sexes. Just remember that "say so" is no substitute for measuring.

wastes are estimated to account for about 21% of the trash in landfills. Given this, in your opinion, would it be fair to ban disposable diapers?"

(b) "Given the current trend of more home runs and more injuries in baseball today, do you think that steroid use should continue to be banned even though it is not enforced?"

(c) "In view of escalating environmental degradation and incipient resource depletion, would you favor economic incentives for recycling of resource-intensive consumer goods?"

7.11 A survey of 100,000 physicians. In 2010, the Physicians Foundation conducted a survey of physicians' attitudes about health care reform, calling the report "a survey of 100,000 physicians." The survey was sent to 100,000 randomly selected physicians practicing in the United States, 40,000 via post office mail and 60,000 via email. A total of 2379 completed surveys were received.[21]

(a) State carefully what population is sampled in this survey and what is the sample size. Could you draw conclusions from this study about all physicians practicing in the United States?

(b) What is the rate of nonresponse for this survey? How might this affect the credibility of the survey results?

(c) Why is it misleading to call the report "a survey of 100,000 physicians"?

Cohorts and case-control studies

Case-control studies Some biological traits are too rare to study effectively with a sample survey. For instance, a trait found in about 1 in 1000 individuals would require a random sample of 10,000 to observe maybe about 10 individuals with the trait of interest. It would be much simpler to select two random samples. In epidemiology, this design is often called *case-control:* A random sample of individuals with a condition (the cases) is compared with a random sample of individuals without the condition (the controls).

CASE-CONTROL STUDY

In a **case-control** observational study, case-subjects are selected based on a defined outcome, and a control group of subjects is selected separately to serve as a baseline with which the case group is compared.

EXAMPLE 7.11 Aflatoxicosis in Kenya

Aflatoxins are toxic compounds secreted by a fungus found in damaged crops. Aflatoxicosis is a rare but very severe poisoning that results from ingestion of aflatoxins in contaminated food. Kenya experienced an outbreak of aflatoxicosis in 2004, resulting in over three hundred cases of liver failure. The Kenya Ministry of Health suspected that improper maize (corn) storage was at least in part responsible for the outbreak. Forty case-patients and 80 healthy controls were asked how they had stored and prepared their maize.[22]

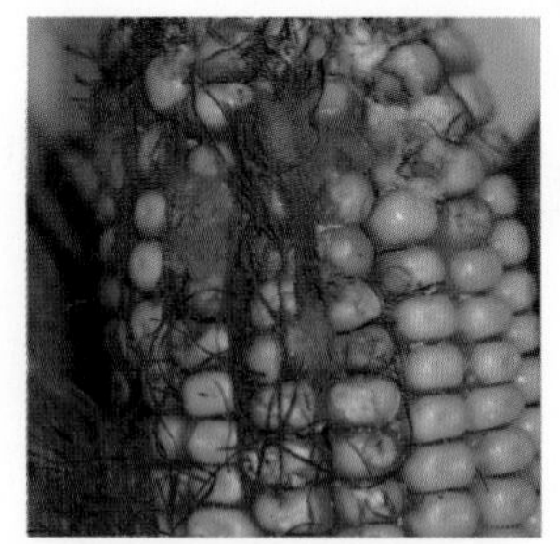

A. Robertson, Iowa State University

The case-patients were randomly selected from a list of individuals admitted to a hospital during the 2004 outbreak for unexplained acute jaundice. For each case-patient, the researchers selected at random two individuals from the patient's village who had no history of jaundice symptoms. Selecting individuals living in the same village ensured that the control subjects shared similar soil and microclimate conditions and farming practices with the case-patients. Because age or gender did not appear to be relevant to the epidemic, the controls were simply randomly selected from the village. ■

Case-control studies start with two samples of individuals selected for their different outcome and look for differences between the two groups. Most commonly, researchers look for exposure factors in the subjects' pasts that differ. Looking back into the past is called a **retrospective** approach. One conclusion from the Kenyan study was that improper maize storage before the epidemic was associated with a higher rate of aflatoxicosis but that the amount of maize previously consumed did not seem to matter. Memories are not always accurate, though, and recall bias may be a confounding factor. For instance, the individuals who were hospitalized for acute liver failure might have different recollections from the control subjects, possibly because the issue matters more to them.

retrospective

Case-control studies are efficient ways to approach rare outcomes and often give fast results, within the usual limitations of observational studies. The main challenge is in identifying a random sample of control individuals as similar to the case subjects as possible. The aflatoxicosis study had very carefully selected controls, allowing the researcher to find plausible causes for the aflatoxin epidemic.

Cohort studies Some observational studies aim to examine the emergence of a specific condition over time and how it relates to a number of variables. *Cohort* studies enlist a homogeneous group of fairly similar individuals and keep track of them over a long period of time. For the most part, these are **prospective** studies that record at regular intervals all sorts of new relevant information about the study participants.[23] After some time, individuals who have developed a condition are then compared with the remaining, unaffected individuals. Observational studies that monitor a sample of individuals repeatedly over time are sometimes called **longitudinal studies.**

prospective

longitudinal

Unlike case-control studies, cohort studies are very costly and better suited to investigating common outcomes such as type 2 diabetes or heart disease. By using the existing medical records of members of a large health care organization to form their cohort, some cohort studies compare members with different conditions in a retrospective approach to identify previous health events that might have been influential.

COHORT STUDY

In a **cohort study,** subjects sharing a common demographic characteristic are enrolled and observed at regular intervals over an extended period of time.

Tom Grill/Corbis

EXAMPLE 7.12 The Nurses' Health Study

The Nurses' Health Study is one of the largest prospective observational studies designed to examine factors that may affect major chronic diseases in women. Since 1976 the study has followed a cohort of over 100,000 registered nurses with the idea that nurses would be able to respond accurately to technically worded medical questionnaires. Every two years, enrolled nurses receive a follow-up questionnaire about diseases and health-related topics such as diet and lifestyle. The response rates to the questionnaires are about 90% for each two-year cycle.

In the 2007 newsletter, study investigators reported their findings on age-related memory loss. About 20,000 women aged 70 and older had completed telephone interviews every two years to assess their memory with a set of cognitive tests. One of the study findings was that the more women walked during their late 50s and 60s, the better their memory was at age 70 and older. The study was observational because the investigators did not randomly assign different walking regimens. Instead, they observed that women who had walked more during their late 50s and 60s ended up with the better memory scores at age 70 and older. ■

Cohort studies accumulate enormous amounts of detailed information and can examine the compounded effect of various factors over time. However, they also take a long time to complete and lose subjects over time, creating a potential confounding effect. This is especially true when studying older individuals, who may die before the end of the study. For example, it is possible that the women with the greatest memory loss also had poor overall health and died younger and were therefore not included in the analysis.

Overall, though, cohort studies are less prone to confounding than case-control studies because cohorts start with one homogeneous group. This has other important advantages. Cohorts can provide information about the relative health risks of different subgroups. They also support incidence calculations, unlike case-control studies, which select subjects based on an existing disease status.

However, like all observational studies, cohort designs cannot establish whether observed differences between groups can be attributed to the groups' differentiating feature or to confounding variables. For example, maybe the women with the better health were both more capable of walking and also less prone to memory loss. Therefore, we cannot unambiguously conclude that walking has a protective effect against memory loss.

Observational studies play an important role in the building of scientific knowledge. But they can easily be plagued by bias and confounding. Experiments, on the other hand, provide an opportunity for manipulating the environment and confounding variables. In the next chapter we study ways in which experiments can be designed to avoid bias and confounding.

APPLY YOUR KNOWLEDGE

7.12 Cell phones and brain cancer. Exercise 7.2 describes a study of the link between cell phone use and brain cancer. What type of observational study is this? What population or populations are represented in the study?

7.13 Antibiotic exposure during infancy. Early life corresponds to the first bacterial colonization of the digestive track, creating a microbiome that appears to play an important role in human development. What might be the effect of early exposure to antibiotics on children's development? One study followed 11,532 full-term babies all born in the county of Avon (United Kingdom) during 1991 and 1992. Exposures to antibiotics during infancy were recorded, along with recurring measures of body mass over several years. The study found that antibiotic exposure during the first six months of life was associated with increased body mass later on, a pattern not seen with antibiotic exposure occurring after the first six months.[24] What type of observational study is this? What population or populations are represented in the study? What are the explanatory and response variables?

CHAPTER 7 SUMMARY

- We can produce data intended to answer specific questions by **observational studies** or **experiments.** Experiments, unlike observational studies, actively impose some treatment on the subjects of the experiment.
- Observational studies often fail to show that changes in an explanatory variable actually cause changes in a response variable, because the explanatory variable is **confounded** with lurking variables. Variables are confounded when their effects on a response can't be distinguished from each other.
- An observational study selects a **sample** from the **population** of all individuals about which we desire information. We base conclusions about the population on data from the sample.
- The **design** of a sample describes the method used to select the sample from the population. **Probability sampling** designs use chance to select a sample.
- The basic probability sample is a **simple random sample (SRS).** An SRS gives every possible sample of a given size the same chance to be chosen. Software or a table of random digits can be used to select an SRS from a known, finite population.
- **Stratified random samples** and **multistage random samples** are more complex forms of probability sampling that are also commonly used.
- Failure to use probability sampling often results in **bias,** or systematic errors in the way the sample represents the population. **Voluntary response samples,** in which the respondents choose themselves, are particularly prone to large bias.
- **Sample surveys** are observational studies of a random sample drawn from the entire population. **Undercoverage** and **nonresponse** are two major challenges that can result in substantial bias. In addition, **response bias** and **poorly worded questions** can create misleading conclusions.
- **Case-control** studies select two random samples based on a known characteristic and typically examine **retrospectively** what might have led to

the difference. They are well suited to the study of rare conditions. However, finding controls that closely resemble the case-subjects is the most challenging part of this observational design.

- In contrast, **cohort** studies enroll subjects who share a common demographic factor at one point in time and typically follow them **prospectively** over time, with data collected at regular intervals. These observational studies are very comprehensive but also expensive and may not be adequate to study rare outcomes.

THIS CHAPTER IN CONTEXT

The methods of Chapters 1 to 5 can be used to describe data regardless of how the data were obtained. However, if we want to reason from data to give answers to specific questions or to draw conclusions about the larger population, then the method that was used to collect the data is important. Sampling is one way to collect data, but it does not guarantee that we can draw meaningful conclusions. Biased sampling methods, such as convenience sampling and voluntary response samples, produce data that can be misleading, resulting in incorrect conclusions. The use of impersonal chance to select the sample is key to a useful sampling process. Probability sampling avoids bias and produces data that can lead to valid conclusions regarding the population. Even with perfect sampling methods, there is still sample-to-sample variation; we will begin our study of the connection between sampling variation and drawing conclusions (statistical inference) in Chapter 13.

Even when we take a simple random sample, our conclusions can be weakened by practical difficulties in obtaining a representative sample, such as undercoverage and nonresponse, and difficulties in obtaining reliable data, such as poor wording of questions and response bias. Careful attention must be given to all aspects of the sampling process to ensure that the conclusions we make are valid. However, even well-designed observational studies cannot readily be used to determine causality from an observed association (as also discussed in Chapter 4). In Chapter 8 we will examine various types of experimental designs and what we can and cannot conclude from them.

CHECK YOUR SKILLS

7.14 The *New England Journal of Medicine* (*NEJM*) posted an opt-in poll on its website, next to an editorial about regulation of sugar-sweetened beverages. The poll asked, "Do you support government regulation of sugar-sweetened beverages?" You only needed to click on a response (yes or no) to become part of the sample. The poll stayed open for several weeks in October 2012. Of the 1290 votes cast, 864 were "yes" responses. You can conclude that

(a) about two-thirds of Americans support government regulation of sugar-sweetened beverages.

(b) about two-thirds of *NEJM* readers support government regulation of sugar-sweetened beverages.

(c) the poll uses voluntary response, so the results tell us little about any particular population.

7.15 What population is represented by the opt-in poll in the previous exercise?

(a) All Americans with access to the internet
(b) All internet users who visit the *NEJM* website
(c) Only those internet users who chose to participate in this particular opt-in poll

7.16 A National Public Radio, Robert Wood Johnson Foundation, Harvard School of Public Health survey was conducted in March 2012 via telephone, interviewing 1508 nationally representative adults aged 18 and over. Of these respondents, 291 had been hospitalized in the past 12 months. The sample in this setting is

(a) all 235 million adults living in the United States.
(b) the 1508 people interviewed.
(c) the 291 people interviewed who had been hospitalized in the past 12 months.

7.17 The population represented by the survey described in the previous exercise is

(a) all 235 million adults living in the United States.
(b) the 1508 people interviewed.
(c) the 291 people interviewed who had been hospitalized in the past 12 months.

7.18 In June 2012, SurveyUSA conducted a survey of 665 adults living in California, asking their opinions on the Supreme Court decision on the constitutionality of the Affordable Care Act. SurveyUSA broke down the findings for the 330 males and the 335 females in the sample. In the methodology section, we find out that the poll was conducted by telephone and that respondent households were selected at random, using random digit dialing (RDD). What type of sample was used for this survey?

(a) A simple random sample (SRS)
(b) A stratified random sample
(c) A multistage random sample

7.19 A sample of households in a community is selected at random from the telephone directory. In this community, 4% of households have no telephone and another 35% have unlisted telephone numbers. The sample will certainly suffer from

(a) nonresponse bias.
(b) undercoverage bias.
(c) response bias.

7.20 A Gallup poll of Americans' smoking behavior and attitudes was conducted in July 2012. A total of 1014 American adults were randomly selected and interviewed by phone. Of the 166 current smokers in the sample, only 1% said that they smoke more than one pack of cigarettes per day, a historical low. In its report Gallup states, "It is possible that the decline in reports of smoking is the result of respondents' awareness that smoking is socially undesirable. Therefore, respondents may aim to present themselves in the best possible light to the interviewer and underestimate the amount they truly smoke." To which type of bias is Gallup referring in this statement?

(a) nonresponse bias
(b) response bias
(c) interviewer bias

7.21 A study enrolled 517 children aged 2 to 5 years with a diagnosis of autism spectrum disorder and 315 control children of the same age group but without such diagnosis. Information was collected about metabolic conditions of the mothers at the time of their pregnancy, such as type 2 diabetes, hypertension, or obesity. Metabolic conditions during pregnancy tended to be more frequent among the mothers of children with autism spectrum disorder than among mothers of control children.[25] This is

(a) an observational study with a cohort design.
(b) an observational study with a case-control design.
(c) an experiment.

7.22 A study followed from 1986 to 2000 a large number of male health professionals aged 40 to 75. Each respondent filled out a lifestyle questionnaire every two years. The 22,086 subjects who had reported having good erectile function in 1986 were included in an analysis of risk factors for the development of erectile dysfunction. The analysis revealed that "obesity and smoking were positively associated, and physical activity was inversely associated, with the risk of erectile dysfunction developing."[26] This is

(a) an observational study with a cohort design.
(b) an observational study with a case-control design.
(c) an experiment.

7.23 How strong is the evidence cited in the previous exercise that physical activity may lower the risk of erectile dysfunction?

(a) Quite strong because it comes from an experiment
(b) Quite strong because it comes from a large random sample
(c) Weak because physical activity is confounded with many other variables

CHAPTER 7 EXERCISES

7.24 Safety of anesthetics. The National Halothane Study was a major investigation of the safety of anesthetics used in surgery. Records of over 850,000 operations performed in 34 major hospitals showed the following death rates for four common anesthetics:[27]

Anesthetic	A	B	C	D
Death rate	1.7%	1.7%	3.4%	1.9%

There is a clear association between the anesthetic used and the death rate of patients. Anesthetic C appears dangerous.

(a) Explain why we call the National Halothane Study an observational study rather than an experiment, even though it compared the results of using different anesthetics in actual surgery.

(b) When the study looked at other variables that are confounded with a doctor's choice of anesthetic, it found that Anesthetic C was not causing extra deaths. Suggest important lurking variables that are confounded with what anesthetic a patient receives.

7.25 Legalizing marijuana. A January 2010 SurveyUSA poll asked a random sample of adults living in the state of Washington, "State lawmakers are considering making marijuana possession legal. Do you think legalizing marijuana is a good idea or a bad idea?" Of the 500 adults interviewed, 280 said "good idea."

(a) What population is the poll targeting?

(b) What sampling design was used? What is the sample size?

7.26 More on legalizing marijuana. A January 2010 SurveyUSA poll asked a random sample of adults living in the San Diego, CA, area, "Do you think marijuana should? or should not? be legal when used for recreational purposes?" Of the 500 adults interviewed, 219 answered "should."

(a) What population is the poll targeting?

(b) Give at least two ways in which this poll differs from the poll described in the previous exercise.

(c) What percent of respondents seemed to support the idea of legalizing marijuana in each poll? Explain how the differences in poll design might explain this apparent difference of opinion.

7.27 Monitoring chimpanzee behavior. Jane Goodall dedicated years of her life to the study of chimpanzee behavior in their natural habitat of East Africa. Initially, the chimps would flee at the sight of her and she had to observe them from a distance with binoculars. But she persisted until the animals eventually became accustomed enough to her presence to ignore her.[28]

(a) What kind of bias was she trying to minimize?

(b) Explain why this lengthy step is particularly important when observing animal behavior.

7.28 Telephone area codes. There are approximately 371 active telephone area codes covering Canada, the United States, and some Caribbean regions. (More are created regularly.) You want to choose an SRS of 25 of these area codes for a study of available telephone numbers. Label the codes 001 to 371 (ignore the fact that area codes are numerical sequences themselves), and use the *Simple Random Sample* applet or other software to choose your sample. (If you use Table A, start at line 129 and choose only the first 5 codes in the sample.)

7.29 Sampling Amazon forests. Stratified samples are widely used to study large areas of forest. Based on satellite images, a forest area in the Amazon basin is divided into 14 types. Foresters studied the four most commercially valuable types: alluvial climax forests of quality levels 1, 2, and 3, and mature secondary forest. They divided the area of each type into large parcels, chose parcels of each type at random, and counted tree species in a 20- by 25-meter rectangle randomly placed within each parcel selected. Here is some detail:

Forest type	Total parcels	Sample size
Climax 1	36	4
Climax 2	72	7
Climax 3	31	3
Secondary	42	4

The researchers chose the stratified sample of 18 parcels described in the table. Explain how a stratified random sample is different from a simple random sample. Why do you think the researchers preferred to use a stratified design?

7.30 Random digit dialing. The list of individuals from which a sample is actually selected is called the *sampling frame*. Ideally, the frame should list every individual in the population, but in practice this is often difficult. A frame that leaves out part of the population is a common source of undercoverage.

(a) Suppose that a sample of households in a community is selected at random from the telephone directory. What households are omitted from this frame? What types of people do you think are likely to live in these households? These people will probably be underrepresented in the sample.

(b) It is usual in telephone surveys to use random digit dialing equipment that selects the last four digits of a telephone number at random after being given the exchange (the first three digits). Which of the households you mentioned in your answer to (a) will be included in the sampling frame by random digit dialing?

7.31 Systematic random samples. *Systematic random samples* go through a list of the population and choose individuals at fixed intervals from a randomly chosen starting point. For example, a study of exercise and diet among college students takes a systematic sample of 200 students at a university as follows: Start with a list of all 9000 students. Because $9000/200 = 45$, choose one of the first 45 names on the list at random and then every 45th name after that. For example, if the first name chosen is at position 23, the systematic sample consists of the names at positions 23, 68, 113, 158, and so on, up to 8978. Like an SRS, a systematic sample gives all individuals the same chance to be chosen. Explain why this is true, then explain carefully why a systematic sample is nonetheless *not* an SRS.

7.32 Seat belt use. A study in El Paso, Texas, looked at seat belt use by drivers. Drivers were observed at randomly chosen convenience stores. After they left their cars, they were invited to answer questions that included questions about seat belt use. In all, 75% said they always used seat belts, yet only 61.5% were wearing seat belts when they pulled into the store parking lots.[29] Explain the reason for the bias observed in responses to the survey. Do you expect bias in the same direction in most surveys about seat belt use?

7.33 Wording survey questions. Comment on each of the following as a potential sample survey question. Is the question clear? Is it slanted toward a desired response?

(a) "Some cell phone users have developed brain cancer. Should all cell phones come with a warning label explaining the danger of using cell phones?"

(b) "Do you agree that a national system of health insurance should be favored because it would provide health insurance for everyone and would reduce administrative costs?"

(c) "In view of the negative externalities in parent labor force participation and pediatric evidence associating increased group size with morbidity of children in day care, do you support government subsidies for day care programs?"

7.34 Protecting the environment. A 2013 Gallup poll reports that nearly half of respondents (47%) believe that the government is doing too little to protect the environment. The section on survey methods states, "Results for this Gallup poll are based on telephone interviews conducted March 7–10, 2013, with a random sample of 1,022 adults, aged 18 and older, living in all 50 U.S. states and the District of Columbia." The survey was conducted by phone in the following manner: "Landline telephone numbers are chosen at random among listed telephone numbers. Cellphone numbers are selected using random digit dial methods. Landline respondents are chosen at random within each household on the basis of which member had the most recent birthday."[30]

(a) What is the population that this survey aims to describe? What individuals from this population might not be represented in the survey and why?

(b) Explain how including both landline telephones and cell phones helps reach the target population.

(c) Landline phones reach a whole household. Why do you think that Gallup selects the adult with the most recent birthday rather than the first one to answer the phone?

7.35 Write your own bad questions. Write your own examples of bad sample survey questions.

(a) Write a biased question designed to get one answer rather than another. Then rewrite it to avoid that bias.

(b) Write a question that is confusing, so that it is hard to answer. Then rewrite it in an improved form.

7.36 Depression and income level. Gallup reports a survey of depression based on telephone interviews with 258,141 national adults conducted January through September 2009. The findings show that 17% of adult Americans have been told at some point by a physician or nurse that they have depression. However, when looking only at individuals in the lower income bracket (annual income of less than $24,000), that number goes up to 30%. Gallup concludes that lower-income Americans have a higher rate of clinical depression but cautions against attempts to see a causal relationship between income and depression. Here is what Gallup says:

It is important to remember that some variables that appear to be risk factors for depression can also be consequences of a major depressive episode. For instance, chronic depression may cause an inability to work, which may lead to poverty. Variables that appear to be risk factors can also be consequences of variables that occur even further up in a causal chain. For instance, serious health problems may simultaneously lead to both depression and poverty.

(a) What important limitation of surveys, and observational studies in general, does Gallup describe here?
(b) Summarize Gallup's argument in a diagram similar to that of Figure 7.1.

7.37 Protecting skiers and snowboarders. Most alpine skiers and snowboarders do not use helmets. Do helmets reduce the risk of head injuries? A study in Norway compared skiers and snowboarders who suffered head injuries with a control group of skiers and snowboarders who were not injured. Of 578 injured subjects, 96 had worn a helmet. Of the 2992 in the control group, 656 wore helmets.[31]

(a) What type of observational study is this? Explain your answer.
(b) Conclusions from observational studies are limited because of potential confounding effects. Give an example of a possible confounding variable in this study.

7.38 Bone loss by nursing mothers. Breast-feeding mothers secrete calcium into their milk. Some of the calcium may come from their bones, resulting in reduced bone mineral density. Researchers compared 47 breast-feeding women with 22 women of similar age who were neither pregnant nor lactating. They measured the percent change in the mineral content of the women's spines over three months.[32]

(a) What type of observational study is this? What two populations did the investigators want to compare?
(b) Give one possible confounding variable for the relationship between breast-feeding and change in bone mineral content.

7.39 Coffee and depression. A study examined coffee consumption between 1980 and 2004 for 50,739 American nurses from the Nurses' Health Study who were free of depressive symptoms at baseline in 1996. Ten years later, 2607 nurses had reported a diagnosis of depression and antidepressant use at some point. The study found that depression risk decreased with greater caffeinated coffee consumption (but not for decaf coffee consumption).[33]

(a) What type of observational study is this? Explain your answer.
(b) The investigators state that "this study cannot prove that caffeine or caffeinated coffee reduces the risk of depression but only suggests the possibility of such a protective effect." Explain why this is the case.
(c) What would be a possible confounding variable for this study?

7.40 Brain size and autism. Is autism marked by different brain growth patterns in early life? One study looked at the brain sizes of 30 autistic boys and 12 nonautistic boys who had all received an MRI scan while toddlers. The autistic boys had received the MRI as part of a diagnostic for behavioral concerns, and the nonautistic boys had been participants in a previous quantitative MRI study reporting age-related changes in head and brain features.[34]

(a) What type of observational study is this specifically? Explain your answer.
(b) The researcher honestly disclosed the limitations of the study, an important part of the ethical conduct of research. Suggest some potential confounding variables for this study.

7.41 Red wine and prostate cancer. In 1986, over 50,000 American male health professionals 40 to 75 years of age were enrolled in the Health Professionals Follow-up Study (HPFS). At the time of enrollment and again every two years, the subjects reported their average consumption of red wine, white wine, beer, and liquor, and prostate cancer diagnoses were recorded. Investigators found no clear evidence of association between amount of red wine consumption and risk of prostate cancer.[35]

(a) What type of observational study is this? Explain your answer.
(b) By 2002, the HPFS had recorded 3348 cases of prostate cancer. How is this information relevant when interpreting the study results?

7.42 Hormone therapy and thromboembolism. Oral estrogen-replacement therapy (ERT) activates blood coagulation and increases the risk of venous thromboembolism (a traveling blood clot, sometimes fatal) in postmenopausal women. A study compared a random sample of 155 postmenopausal women who had had a first thromboembolism and a control random sample of 381 postmenopausal women with no history of thromboembolism. One of the study findings was that 21% of the women in the thromboembolism group were currently using oral ERT, compared with only 7% of the women in the control group.[36]

(a) What type of observational study is this?
(b) Can you conclude that oral ERT causes the higher rate of thromboembolism? Explain your answer.

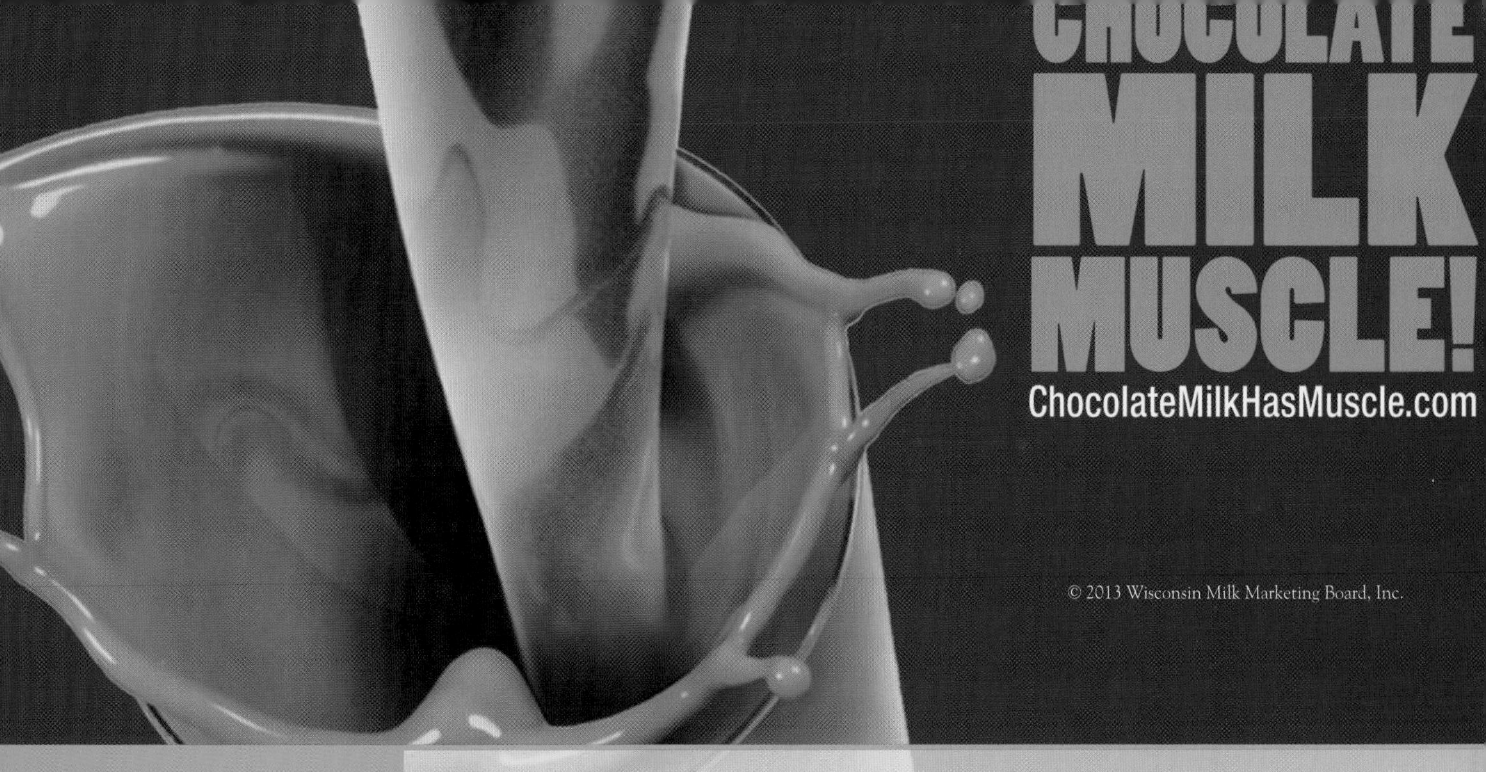

CHAPTER 8 Designing Experiments

How do we *produce data* that can help us answer interesting scientific questions? In the previous chapter we looked at ways to select samples from a population of interest and their applications for observational studies.

But what if we asked, "Is a new cholesterol-lowering drug more efficient than the standard-treatment drug?" To answer this question, we would give the new drug to a group of high-cholesterol patients and measure the decrease in cholesterol level after one month of treatment. However, that wouldn't be enough to answer the question completely. We would also need to give the standard drug to another group of high-cholesterol patients and compare their results with those of the patients given the new drug. In this chapter we study specifically how to design experiments such as this.

IN THIS CHAPTER WE COVER...

- Designing experiments
- Randomized comparative experiments
- Common experimental designs
- Cautions about experimentation
- Ethics in experimentation
- *Discussion: The Tuskegee syphilis study*

Designing experiments

SUBJECTS, FACTORS, TREATMENTS

The **individuals** (or units) studied in an experiment are often called **subjects** when they are people, and **experimental units** otherwise.

The explanatory variables in an experiment are often called **factors.**

A **treatment** is any specific experimental condition applied to the subjects. If an experiment has more than one factor, a treatment is a combination of specific values of each factor.

EXAMPLE 8.1 Protective effect of the Mediterranean diet

Mediterranean populations are known to suffer comparatively less from heart disease and from some types of cancer. Their diet high in natural fibers, vitamins, and oleic acid has been offered as an explanation for this phenomenon.

To test this hypothesis, an experiment assigned 605 survivors of a first heart attack to follow for 5 years either a Mediterranean-type diet or a diet close to the American Heart Association (AHA) "prudent" diet (30% or less of caloric intake from fats, low cholesterol intake). The patients were assessed regularly and a variety of outcomes were recorded, from plasma fatty acid levels, to occurrences of cancer or heart attack, to death (cardiac, cancer, or other). The assignment to one diet or the other was random, and the two groups did not differ substantially in demographics. Possible confounders such as smoking and activity levels were closely monitored and included in the statistical analysis, but they were not restricted or imposed.[1]

This experiment compares two *treatments:* Mediterranean diet and AHA prudent diet. There is a single *factor:* "diet." The *subjects* are survivors of a first heart attack and there are many *response variables* recorded over 5 years, such as cancer and heart attack. ■

Image Source/Getty Images

EXAMPLE 8.2 Effect of photoperiod and wavelength on flowering

Photoperiod, the relative lengths of light and dark periods in a 24-hour cycle, is a common environmental cue for flowering, allowing plants to bloom during an appropriate time of the year. Chrysanthemum (*Chrysanthemum morifolium*) is known as a short-day plant, blooming during times of longer nights. Plant physiologists grew chrysanthemum plants in controlled greenhouses under different combinations of photoperiods (short day, long day, continuous light, and interrupted night) and light wavelengths (blue light, red light, and blue + red light). The plants were kept in these conditions for 5 weeks and examined regularly to assess whether flowering had occurred during this time.[2]

This experiment has two *factors:* photoperiod, with 4 values (short day, long day, continuous light, and interrupted night), and light wavelength, with 3 values (blue, red, and blue + red). The 12 combinations of both factors form 12 *treatments*. Figure 8.1 shows the layout of the treatments. The *individuals* (or *experimental units*) are chrysanthemum plants, and the *response variable* is flowering, a categorical variable. ■

Examples 8.1 and 8.2 illustrate the advantages of experiments over observational studies. In an experiment, we can study the effects of the specific treatments in which we are interested. By assigning subjects to treatments, we can avoid

		Factor B Light wavelength		
		Blue	Red	Blue + red
Factor A Photoperiod	Short day	1	2	3
	Long day	4	5	6
	Continuous light	7	8	9
	Interrupted night	10	11	12

Plants assigned to Treatment 3 were exposed to blue + red light on a short-day lighting cycle.

FIGURE 8.1 The treatments in the experimental design of Example 8.2. Combinations of values of the two factors form 12 treatments.

confounding. If, for example, we simply compare heart attack survivors who decided to follow a Mediterranean diet with those who decided to follow an AHA prudent diet, we may find that one group is richer or more social or more physically active. Example 8.1 avoids that. Moreover, we can control the environment of the subjects to hold constant factors that are not directly of interest to us, such as soil, temperature, and humidity in the greenhouses where the chrysanthemums are grown in Example 8.2.

Another advantage of experiments is that we can study the combined effects of several factors simultaneously. The interaction of several factors can produce effects that could not be predicted from looking at the effect of each factor alone. Perhaps interrupting the night period inhibits flowering in chrysanthemums, but only if the light used during the interruption contains some red light. The two-factor experiment in Example 8.2 will help us find out.

Experiments are the preferred method for examining the effect of one variable on another. By imposing the specific treatment of interest and controlling other influences, we can pin down cause and effect. Statistical designs are often essential for effective experiments, just as they are for sample surveys and observational studies. To see why, let's start with an example of a bad design.

What's news? Randomized comparative experiments provide the best evidence for medical advances. Do newspapers care? Maybe not. University researchers looked at 1192 articles in medical journals, of which 7% were turned into stories by the two newspapers examined. Of the journal articles, 37% concerned observational studies and 25% described randomized experiments. Among the articles publicized by the newspapers, 58% were observational studies and only 6% were randomized experiments. Conclusion: the newspapers want exciting stories, whether or not the evidence is good.

EXAMPLE 8.3 An uncontrolled experiment

Gastric freezing was introduced in the 1960s by a prominent surgeon as a way to relieve ulcer pain. Patients would swallow a deflated balloon that was then filled with a refrigerated liquid. The idea was that the refrigerant would cool the stomach lining, reduce acid production, and reduce pain. The procedures performed showed that gastric freezing did reduce ulcer pain, and the treatment was recommended based on this evidence.[3]

This experiment has a very simple design. A group of subjects (the patients) were exposed to a treatment (gastric freezing), and the outcome (pain reduction) was observed. Here is the design:

$$\text{Subjects} \longrightarrow \text{Gastric freezing} \longrightarrow \text{Pain reduction}$$

Some critical pieces of information are missing from this experiment. What would have happened if we had left the patients alone? Would their pain have abated even without intervention? What if patients were inclined to say they were feeling better (maybe because they expected to get better, or to please the experimenter)? And what if patients genuinely felt better but did so because the care and attention they received made them feel better overall, not because of the cooling of their stomach lining? This is a real possibility, and similar psychological improvements have been documented in many human experiments. We call it the *placebo effect*.

None of these questions can be addressed by the simple experiment. The effect of gastric freezing is confounded with the effect of two lurking variables: spontaneous, natural improvement and psychological improvement due to medical care. Figure 8.2 shows the confounding in picture form. Because of confounding, subject improvement cannot be unequivocally attributed to the effect of gastric freezing. ■

Many laboratory experiments use a design like that of Example 8.3:

$$\text{Subjects} \longrightarrow \text{Treatment} \longrightarrow \text{Measure response}$$

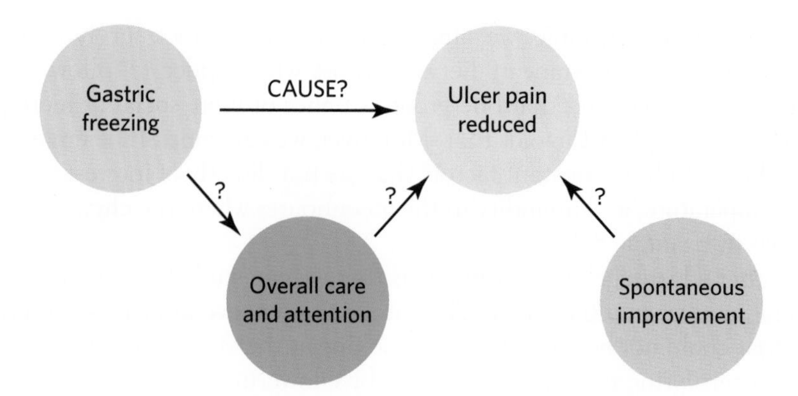

FIGURE 8.2 Confounding. We can't distinguish the effect of the treatment from the effects of lurking variables in Example 8.3.

In the controlled environment of the laboratory, simple designs may work well. However, *field experiments and experiments with human subjects are exposed to more variable conditions and deal with more variable individuals. A simple design often yields worthless results because of confounding with lurking variables.*

APPLY YOUR KNOWLEDGE

8.1 Ginkgo extract and the post-lunch dip. The post-lunch dip is the drop in mental alertness after a midday meal. Does an extract of the leaves of the ginkgo tree reduce the post-lunch dip? A study assigns healthy people aged 18 to 40 to take either ginkgo extract or a placebo pill. After lunch, they are asked to read seven pages of random letters and place an X over every *e*. We count the number of misses. What are the individuals, the treatments, and the response variable in this experiment?

8.2 Improving adolescents' health habits. Most American adolescents don't eat well and don't exercise enough. Can middle schools increase physical activity among their students? Can they persuade students to eat better? Investigators designed a "physical activity intervention" to increase activity in physical education classes and during leisure periods throughout the school day. They also designed a "nutrition intervention" that improved school lunches and offered ideas for healthy home-packed lunches. Each participating school was randomly assigned to one of the interventions, both interventions, or no intervention. The investigators observed physical activity and lunchtime consumption of fat. Identify the individuals, the factors, and the response variables in this experiment. Use a diagram like that in Figure 8.1 to display the treatments.

8.3 Quitting smoking and risk for type 2 diabetes. Researchers studied a group of 10,892 middle-aged adults over a period of nine years. They found that smokers who quit had a higher risk for diabetes within three years of quitting than either nonsmokers or continuing smokers.[4] Does this show that stopping smoking causes the short-term risk for diabetes to increase? (Weight gain has been shown to be a major risk factor for developing type 2 diabetes and is often a side effect of quitting smoking.) Based on this research, should you tell a middle-aged adult who smokes that stopping smoking can *cause* diabetes and advise him or her to continue smoking? Carefully explain your answers to both questions.

Randomized comparative experiments

The remedy for the confounding in Example 8.3 is to do a *comparative experiment* in which some patients undergo the gastric-freezing treatment and other, similar patients undergo a sham procedure or no procedure at all. Comparing patients undergoing gastric freezing with untreated patients allows us to control for the effect of natural improvement. However, that alone would not tell us if a lower pain rate in the *experimental group* (gastric freezing) is due to the benefits of the medical procedure itself or to the psychological impact of undergoing any kind of treatment—the *placebo effect*. Only by comparing the gastric-freezing procedure with a sham procedure would we be able to conclude that gastric freezing itself is responsible for any observed improvement in patient condition.

Most well-designed experiments compare two or more treatments. Part of the design of an experiment is a description of the factors (explanatory variables) and the layout of the treatments, with comparison as the leading principle.

EXPERIMENTAL GROUP, CONTROL, PLACEBO

An **experimental group** is a group of individuals receiving a treatment whose effect we seek to understand.

A **control** group serves as a baseline with which the experimental group is compared.

A **placebo** is a control treatment that is fake (for example, taking a sugar pill) but otherwise indistinguishable from the treatment in the experimental group.

EXAMPLE 8.4 A controlled experiment

The simple experiment outlined in Example 8.3 was followed up many years later with a randomized, controlled experiment in which patients undergoing gastric freezing were compared with similar patients undergoing the same procedure except that the liquid pumped was not refrigerated. This second group was a placebo group. The comparative study found that the condition of 28 of the 82 gastric-freezing patients improved, while 30 of the 78 patients in the placebo group improved.

This comparative experiment demonstrated that cooling the stomach lining was not more effective in reducing ulcer pain than a sham procedure. The treatment was then abandoned. ■

Placebo for baby The most famous and maybe most powerful placebo in the world is the one parents use on their little ones for every small misfortune. Different techniques exist but they all work equally well: a kiss, a hug, or even a Band-Aid. The effect gradually disappears, however, as children realize that sometimes they can get better without help and that at times more advanced treatment is provided.

The *placebo effect* is documented in human experimentation only, in contexts as diverse as the study of asthma or high blood pressure. The effect is particularly strong when the response variable is pain. The mechanisms of this mind-body effect are unknown but are not without biological foundation. A neural response to the placebo effect has been documented with neuroimaging techniques as early in the sensory process as the spinal chord.[5] We also know that the nervous system interacts with the immune system and that depression, for instance, suppresses the immune response.

Like experiments, observational studies often compare different groups and may even include a control group that does not exhibit the characteristic of interest. For example, a study could follow individuals over time and record diagnoses of lung cancer, comparing a sample of cigarette-smoking adults and a sample of nonsmoking adults. However, in observational studies the groups compared may differ by more characteristics than the ones used to distinguish the groups. For example, the smokers may also be poorer overall or more prone to other forms of addictions that render them more susceptible to lung cancer. Controls in observational studies are very informative, but they are not sufficient to conclude about a causal effect.

Thus, comparison alone isn't enough to produce strong conclusions. If the treatments are given to groups that differ markedly when the experiment begins, bias will result. In Example 8.4, if we select all the patients most likely to improve to make up the gastric-freezing treatment group, we will see better results in the treatment group even if the procedure itself makes no contribution to the improvement. Personal choice would bias our results in the same way that volunteers bias the results of online opinion polls. The solution to the problem of bias is the same for experiments and for survey samples: *randomization*. That is, use impersonal chance to select the groups.

RANDOMIZED COMPARATIVE EXPERIMENT

An experiment that uses both comparison of two or more treatments and chance assignment of subjects to treatments is a **randomized comparative experiment.**

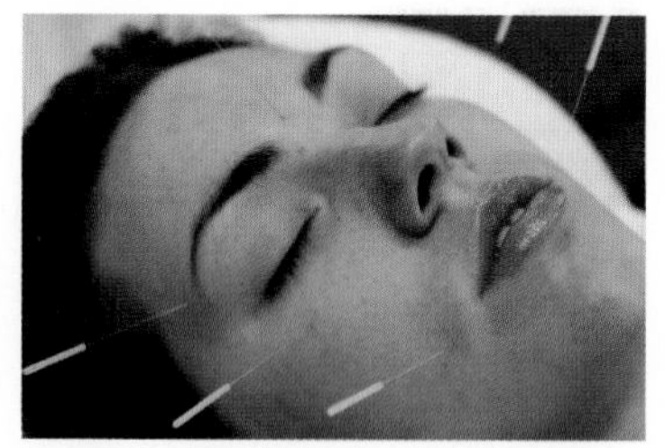

SuperStock/Alamy

EXAMPLE 8.5 Acupuncture for migraines

Acupuncture is an ancient traditional Chinese medicine that uses very thin needles inserted at specific body points to relieve pain and other conditions. Researchers investigated whether acupuncture can help prevent migraine attacks. Patients suffering from migraines were randomly assigned to either acupuncture, sham acupuncture (needles inserted at nontraditional anatomical points), or a waiting list for acupuncture. The sham acupuncture acted as a placebo. The waiting-list group acted as a control (no medical treatment). The patients were asked to record headache days in diaries for 12 weeks.[6] Figure 8.3 outlines the design in graphical form.

The selection procedure is exactly the same as it is for sampling: label and select. **Step 1. Label** the 302 patients 001 to 302. **Step 2. Select.** Go to a table of random digits, such as Table A at the back of the book (reading successive three-digit groups), or

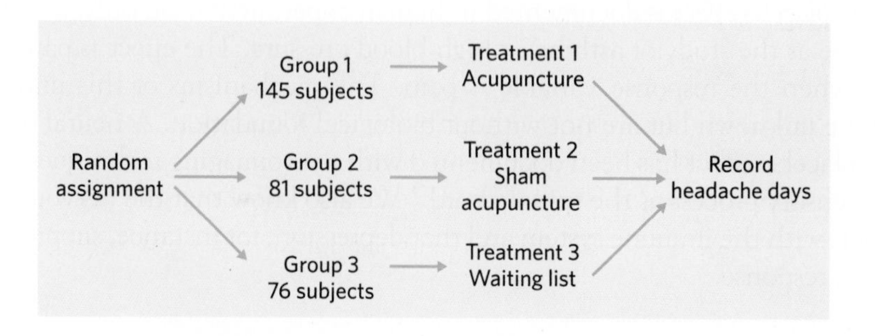

FIGURE 8.3 Outline of a randomized comparative experiment to compare the effect of acupuncture, sham acupuncture, and placebo on headaches, for Example 8.5.

use software, such as the *Simple Random Sample* applet. The first 145 labels encountered select the acupuncture group. Ignore repeated labels and groups of digits not used as labels. After the first group is selected, the next 81 labels encountered select the sham acupuncture group. The remaining 76 labels make up the waiting-list group. ■

The design in Figure 8.3 is *comparative* because it compares three treatments (acupuncture, sham acupuncture, and waiting list). It is *randomized* because the subjects are assigned to the treatments by chance. The "flowchart" outline presents all the essentials: randomization, the sizes of the groups and which treatment they receive, and the response variable. Here the researchers chose to place about half the patients in the acupuncture group and to split the other half between the two control treatments (sham and waiting list).

Randomized comparative experiments are designed to give good evidence that differences in the treatments actually *cause* the differences we see in the response. The logic is as follows:

- Random assignment of subjects forms groups that should be similar in all respects before the treatments are applied. Exercise 8.53 uses the *Simple Random Sample* applet to demonstrate this. The random assignment also allows us to use probability laws to conduct statistical inference on the data collected, as we will see in later chapters.

- Comparative design ensures that influences other than the experimental treatments operate equally on all groups.
- Therefore, differences in average response must be due either to the treatments or to the play of chance in the random assignment of subjects to the treatments.

This last statement deserves more thought. Even with an excellent experimental design, we can never say that *any* difference between two groups is entirely due to differences in the treatments applied. Variables, by definition, vary from individual to individual. We would not trust an experiment like that of Example 8.5 if it had just one patient in each group. The results would depend too much on which group got lucky and received a patient with less frequent headaches, for example. If we assign many subjects to each group, however, the effects of chance will average out, and there will be little difference in the *average* responses in the three groups unless the treatments themselves cause a difference. "Use enough subjects to reduce chance variation" is the third big idea of statistical design of experiments.

PRINCIPLES OF EXPERIMENTAL DESIGN

The basic principles of statistical design of experiments are

1. **Control** the effects of lurking variables on the response, most simply by comparing two or more treatments.
2. **Randomize**—use impersonal chance to assign subjects to treatments.
3. **Use enough subjects** in each group to reduce chance variation in the results.

We hope to see a difference in the responses so large that it is unlikely to happen just because of chance variation. We can use the laws of probability, which give a mathematical description of chance behavior, to learn if the treatment effects are larger than we would expect to see if only chance assignment to groups were operating. If they are, we call them *statistically significant*.

> **STATISTICAL SIGNIFICANCE**
>
> An observed effect so large that it would rarely occur by chance is called **statistically significant.**

If we observe statistically significant differences among the groups in a randomized comparative experiment, we have good evidence that the treatments actually caused these differences. You will often see the phrase "statistically significant" in scientific reports. The great advantage of randomized comparative experiments is that they can produce data that give good evidence for a cause-and-effect relationship between the explanatory and response variables. We know that in general a strong *observed* association does not imply causation. A statistically significant association in data from a well-designed *experiment* does imply causation.

APPLY YOUR KNOWLEDGE

8.4 Adolescent obesity. Adolescent obesity is a serious health risk affecting more than 5 million young people in the United States alone. Laparoscopic adjustable gastric banding has the potential to provide a safe and effective treatment. Fifty adolescents between 14 and 18 years old with a body mass index higher than 35 were recruited from the Melbourne, Australia, community for the study. Twenty-five were randomly selected to undergo gastric banding, and the remaining 25 were assigned to a supervised lifestyle intervention program involving diet, exercise, and behavior modification. All subjects were followed for two years and their weight loss was recorded.[7]

(a) Outline the design of this experiment, following the model of Figure 8.3. What is the response variable?

(b) Carry out the random assignment of 25 adolescents to the gastric-banding group, using the *Simple Random Sample* applet, other software, or Table A (starting at line 130, and starting your labels at 1). The remaining 25 subjects will make up the lifestyle intervention group.

8.5 More rain for California? The changing climate will probably bring more rain to California, but we don't know whether the additional rain will come during the winter wet season or extend into the long dry season in spring and summer. Kenwyn Suttle of the University of California at Berkeley and his coworkers carried out a randomized controlled experiment to study the effects of more rain in either season. They randomly assigned plots of open grassland to 3 treatments: added water equal to 20% of annual rainfall either during January to March (winter) or during April to June (spring), and no added water (control). Thirty-six circular plots of area 70 square meters were available (see the photo), of which 18 were used for this study. One response variable was total plant biomass, in grams per square meter, produced in a plot over a year.[8]

(a) Outline the design of the experiment, following the model of Figure 8.3.

(b) Number all 36 plots and choose 6 at random for each of the 3 treatments. Be sure to explain how you did the random selection.

8.6 Effectiveness of sport drinks. The Gatorade Sport Science Institute prepared a report on the scientific evidence supporting the benefits of sport drinks for soccer players. Here are some of the findings:

> *Creatine supplementation was tested in 14–19 year-old male soccer players from the 1st Yugoslav Junior League (Ostojic, 2004). Before and after 7 days of supplementation with either creatine or placebo, 10 players in each group performed a timed soccer-dribbling test around cones set 3 m apart, a sprint-power test that lasted about 3 s, a vertical-jump test, and a shuttle endurance run that lasted about 11 min. In contrast to two [other] studies of soccer skills, the authors correctly reported overall effects of creatine versus placebo, not simply pretest-posttest results within groups. The creatine group was significantly better than placebo for the dribbling test, the sprint test, and the vertical jump.*[9]

(a) A friend thinks that "significantly" in this article has its plain English meaning, roughly "I think this is important." Explain in simple language what "significantly better" tells us here.

(b) Your friend is puzzled by the comment on other studies not using a placebo. Explain what a placebo is and what using a placebo group in this study accomplishes.

Common experimental designs

Completely randomized designs The simplest experimental design assigns the individuals (subjects or experimental units) to the treatments completely at random. We call such designs *completely randomized*. Example 8.5, about acupuncture for migraines, describes an experiment with a completely randomized design, and Figure 8.3 illustrates the randomization process. *Note that it is not necessary for a completely randomized design to assign the same number of individuals to each treatment.* It might be advantageous or more practical to assign a larger number of individuals to one particular treatment. This is the case for the acupuncture group in Example 8.5.

CAUTION

> **COMPLETELY RANDOMIZED DESIGN**
>
> In a **completely randomized** experimental design, all the individuals are allocated at random among all the treatments.

Completely randomized designs can compare any number of treatments. They can also have more than one factor. The chrysanthemum-flowering experiment of Example 8.2 has two factors: photoperiod and light wavelength. Their combinations form the 12 treatments outlined in Figure 8.1. A completely randomized design assigns individual chrysanthemum plants at random to these 12 treatments.

Once the layout of treatments is set, the randomization needed for a completely randomized design is tedious but straightforward, as illustrated in Example 8.5.

Completely randomized designs are the simplest statistical designs for experiments. They illustrate clearly the principles of comparison, randomization, and adequate number of subjects. However, experiments sometimes need additional sophistication—for example, to deal with substantial individual-to-individual variability or potentially confounding variables.

Block designs One such design is the *block design*, in which individuals sharing the same characteristic are pooled, forming a *block*. The characteristic used to form the basis of the blocks is preexisting, representing a feature of the individuals rather than an imposed condition or treatment.

> **BLOCK DESIGN**
>
> A **block** is a group of individuals that are known before the experiment to be similar in some way that is expected to affect the response to the treatments.
>
> In a **block design,** the random assignment of individuals to treatments is carried out separately within each block.

Block designs can have blocks of any size. Blocks are another form of *control*. They control the effects of some outside variables by bringing those variables into the experiment to form the blocks. Here are some typical examples of block designs.

EXAMPLE 8.6 Smoking exposure and asthma care

An estimated 9.5% of children in the United States suffer from asthma. Children with persistent asthma should receive daily preventive anti-inflammatory medication. Unfortunately, this recommendation is not always followed. A study compared the effect of the usual at-home asthma care and a school-based asthma care program (where school nurses administer treatment) on asthma symptoms of elementary school children. Environmental tobacco smoke can trigger and worsen asthma symptoms, so the experiment used home exposure to tobacco smoke as a blocking variable.[10] Two separate randomizations were done, one assigning the children exposed to smoke to the two treatments, and the other assigning the children not exposed to smoke. Figure 8.4 outlines the design of this experiment. Note that there was no randomization involved in making up the blocks. They were groups of subjects who differed in some way (home smoke exposure in this case) that existed before the experiment began. ■

IIC/Axiom/Getty Images

EXAMPLE 8.7 Soil differences

The soil type and fertility of farmland differ by location. Because of this, experiments on plant growth often divide the available planting area into smaller fields making up the blocks. A test of the effect of tillage type (two types) and pesticide application (three application schedules) on soybean yields used small fields as blocks. Each block was further divided into six plots, and the six treatments were randomly assigned to plots separately within each block, as illustrated in Figure 8.5. Field location, used to create the blocks, was a preexisting feature of the experimental units (the plots). ■

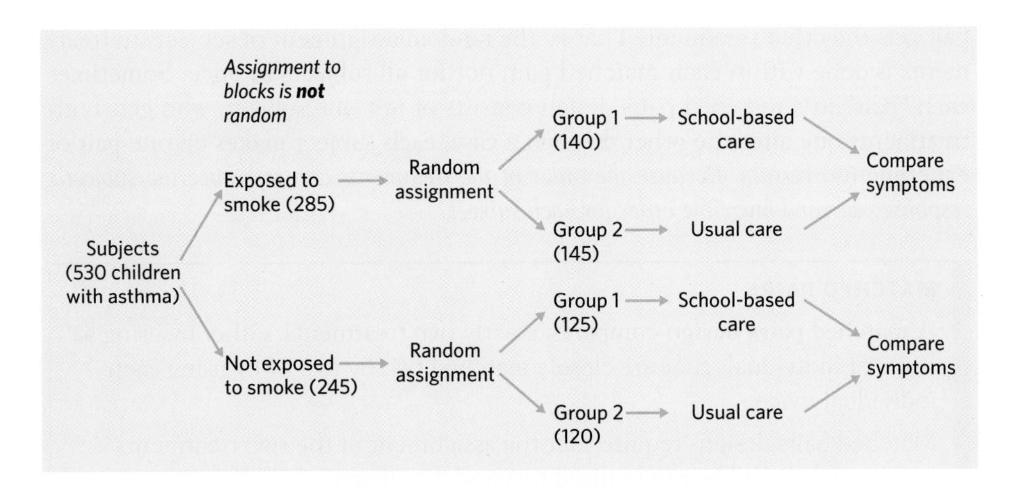

FIGURE 8.4 Outline of a block design comparing two asthma care options for asthmatic children who are or are not exposed to smoke at home, for Example 8.6.

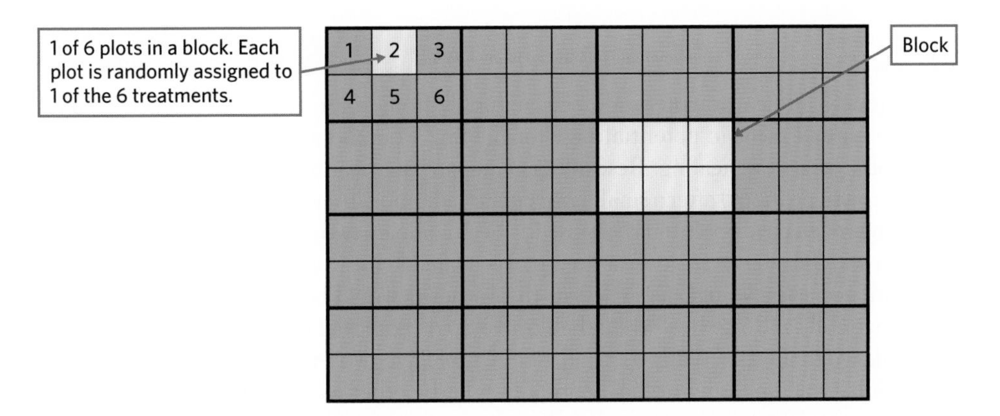

FIGURE 8.5 Outline of a block design, for Example 8.7. The blocks are land areas subdivided into six planting plots. The treatments are six combinations of tillage type and pesticide application.

In both examples, blocking allowed the experimenters to include in their analysis an additional variable (smoke exposure in Example 8.6 and field location in Example 8.7) that was not an imposed treatment in the experiment. Blocks allow us to draw separate conclusions about each block and to assess whether the response to the treatments (such as the two asthma care options in Example 8.6) may depend on the block variable.

The idea of blocking is an important additional principle of statistical design of experiments. A wise experimenter will form blocks based on the most important unavoidable sources of variability among the subjects or experimental units. Randomization will then average out the effects of the remaining variation and allow an unbiased comparison of the treatments.

Matched pairs designs Another experimental design is especially useful when substantial differences between individuals exist or are suspected. A *matched pairs design* uses a form of blocking to compare just two treatments. Choose pairs of subjects that are as closely matched as possible. These can be individuals of the same sex, age, weight, and stature, for example, or they can be genetically related individuals such as twins or animals born in the same litter. Use chance to decide which subject in a pair gets one specific treatment. The other subject in that pair gets the other treatment. That is, the random assignment of subjects to treatments is done within each matched pair, not for all subjects at once. Sometimes each "pair" in a matched pairs design consists of just one subject, who gets both treatments one after the other. In such a case, each subject makes up one pair of experimental results. *Because the order of the treatments can influence the subject's response, we randomize the order for each subject.*

MATCHED PAIRS

A **matched pairs design** compares exactly two treatments, either by using a series of individuals that are closely matched two by two or by using each individual twice.

Matched pairs designs require that the assignment of the two treatments within each "pair" be **randomized** to avoid a systematic bias.

EXAMPLE 8.8 Does nature heal best?

Differences of electric potential occur naturally from point to point on a body's skin. Is the natural electric potential best for helping wounds to heal? If so, changing the potential will slow healing. Researchers compare the healing rates (in micrometers of skin growth per hour) after a small razor cut is made and either it is left to heal naturally or a change of electric potential is imposed.[11] Let's compare two designs for this experiment. The researchers have 14 anesthetized newts available for this experiment.

In a *completely randomized design*, the 14 newts would be assigned at random, 7 to heal naturally and 7 to heal while the skin's electric potential is disturbed. In the *matched pairs design* that was actually used, all 14 newts receive two small razor cuts, one on each forelimb. One limb is left to heal naturally, and the other receives the electric potential disturbance. The side receiving the disturbance is assigned at random, in case healing differs on the right and left sides.

Some newts spontaneously heal faster than others. The completely randomized design relies on chance to distribute the faster-healing newts roughly evenly between the two groups. The matched pairs design compares each newt's healing rate with and without the electric potential disturbance. This makes it easier to see the effects of disturbing the skin's natural electric potential. ■

repeated measures

Fields that study mainly human subjects, such as psychology and sociology, often reserve the term "matched pairs" for actual pairs of similar but different individuals. When the same individuals are used for both treatments, the term **repeated measures** is often used instead. One added bonus to the term "repeated measures" is that it applies also to experiments in which each subject is exposed to more than two treatments.

© 2013 Wisconsin Milk Marketing Board, Inc.

EXAMPLE 8.9 Chocolate milk for athletes

Chocolate milk is gaining in popularity among athletes as a sport drink because it contains proteins in addition to the carbohydrates (sugar) and minerals found in commercial sport drinks like Gatorade. Researchers wanted to see if drinking chocolate milk would result in higher endurance among athletes than drinking Gatorade or mineral-enriched water.

Athletes vary substantially in their physical abilities. They vary so much that individual variations across athletes might be more substantial than the effect of one drink over another. So the researchers asked trained male cyclists to cycle to exhaustion on 3 different weeks: once drinking chocolate milk, once drinking Gatorade, and once drinking mineral-enriched water, in a randomized order. That is, each athlete was subjected to all 3 treatments, in a repeated-measures design.[12] ■

Matched pairs and repeated-measures designs use the principles of comparison of treatments and randomization. However, the randomization is not complete—we do not randomly assign all the subjects at once to the treatments. Instead, we randomize only within each matched pair or, when using the same individuals for different treatments, we randomize the order of the treatments for each individual. This differs from a completely randomized design which produces groups that are *independent:* That is, the individuals in one group are completely unrelated to the individuals in the other group(s). *Being able to distinguish a matched pairs design from a completely randomized design is critical because it impacts the statistical analysis and conclusions, as we will see in the later chapters on inference.*

Students sometimes confuse the case-control design described in Chapter 7 and the matched pairs design described in this chapter. *The case-control design applies to observational studies (no treatment is imposed), whereas the matched pairs design described in this chapter applies to experiments.* Also, the selection of control individuals in case-control studies does not necessarily follow a strict matching process and investigators may select more than one control for each case.

Like the design of samples, the design of experiments can get quite complex. There are both advantages and disadvantages to complex designs. In a complex design, the causal effect of one variable on another might be better isolated from influences of other factors. On the other hand, the data collection, analysis, and interpretation are likely to also become much more complex. In the rest of the book, we will focus on simple completely randomized, matched pairs, and block designs.

APPLY YOUR KNOWLEDGE

8.7 **Cricket mating songs.** Male crickets sing to attract females, but this also increases their risk of being found by a predator. What other important function could the singing serve? The singing may provide females with insight into the health of male crickets. Or the singing may indicate a male's age, demonstrating the male's ability to survive. A researcher wants to compare the female mating choices for healthy male crickets and for similar crickets that have been infected with intestinal parasites. Young male crickets 7 to 12 days old and older male crickets 15 days old or more are available for the experiment.

(a) Outline the design of this experiment, as in Figure 8.4. What type of design is it?

(b) Explain how you would carry out the randomization if you had 20 young crickets and 20 older crickets. (You need not actually do the randomization.)

8.8 **Comparing hand strength.** Is the right hand generally stronger than the left in right-handed people? You can crudely measure hand strength by placing a bathroom scale on a shelf with the end protruding, then squeezing the scale between the thumb below and the four fingers above it. The reading of the scale shows the force exerted.

(a) Describe the design of a matched pairs experiment to compare the strength of the right and left hands, using 10 right-handed people as subjects.

(b) Explain how you would carry out the randomization. (You need not actually do the randomization.)

8.9 **How long did I work?** A psychologist wants to know if the difficulty of a task influences our estimate of how long we spend working at it. She designs two sets of mazes that subjects can work through on a computer. One set has easy mazes and the other has hard mazes. Subjects work until told to stop (after six minutes, but subjects do not know this). They are then asked to estimate how long they worked. The psychologist has 30 students available to serve as subjects.

(a) Describe the design of a completely randomized experiment to learn the effect of difficulty on estimated time.

(b) Describe the design of a matched pairs experiment using the same 30 subjects.

(c) Describe the design of a block experiment considering men and women separately.

Cautions about experimentation

CAUTION

The logic of a randomized comparative experiment depends on our ability to treat all the subjects identically in every way except for the actual treatments being compared. Good experiments therefore require careful attention to details.

Statistics in Your World **Scratch my furry ears** Rats and rabbits, specially bred to be uniform in their inherited characteristics, are the subjects in many experiments. Animals, like people, are quite sensitive to how they are treated. This can create opportunities for hidden bias. For example, human affection can change the cholesterol level of rabbits. Choose some rabbits at random and regularly remove them from their cages to have their heads scratched by friendly people. Leave other rabbits unloved. All the rabbits eat the same diet, but the rabbits that receive affection have lower cholesterol.

EXAMPLE 8.10 Ginkgo biloba and tinnitus

Tinnitus is a ringing noise in the ears, a fairly common condition without an effective pharmaceutical treatment. A study asked whether the herbal supplement ginkgo biloba could help alleviate tinnitus.

A total of 978 healthy adults with tinnitus were matched by age and sex, making up 489 pairs. Within each pair, one individual was randomly selected to receive the ginkgo biloba treatment, while the other individual was given a placebo. The pairing was done to ensure that the two groups were similar in age and gender mix. The placebo condition was used to make sure that any improvement in the ginkgo biloba group was not simply reflecting a placebo effect.

In addition, the study was *double-blind:* The subjects didn't know whether they were getting ginkgo biloba or a placebo, and neither did the investigators. The placebo tablets were identical to the ginkgo biloba tablets in shape, size, color, and packaging. A third-party pharmaceutical company produced the tablets and mailed them to the subjects in coded bottles, revealing the code only after the experiment was completed and all the data were collected. ■

DOUBLE-BLIND EXPERIMENTS

In a **double-blind** experiment, neither the subjects nor the people who interact with them know which treatment each subject is receiving.

Double-blinding is obviously necessary whenever the investigator evaluates the experimental outcome. In clinical trials, how a doctor interacts with a patient can also influence the patient's perception of his or her prognosis and thus create or reinforce a placebo effect. Double-blinding controls the influence of these human interactions by preventing potential bias. In many medical studies, only the statistician who does the randomization knows which treatment each patient is receiving.

Blinding is a challenge if subjects are able to differentiate the treatments. Example 8.5 described a study of acupuncture to treat migraines in which real acupuncture was compared with a sham acupuncture treatment. But how would you compare acupuncture with medication while avoiding a placebo effect? One solution, described in the following example, involves a **double-dummy.**

double-dummy

EXAMPLE 8.11 Acupuncture versus medication

A study compared the effect of acupuncture and the established drug flunarizine in preventing migraines among adult subjects suffering from recurring migraines. Patients can obviously tell the difference between needles and pills, but they cannot tell the difference between real and sham acupuncture or between a flunarizine pill and a placebo pill. So all subjects received both types of treatment (needles and pills), in a double-dummy design. The subjects were randomly assigned to two groups: One group received real acupuncture and placebo pills for four weeks; the other group received sham acupuncture and actual flunarizine pills for four weeks. All subjects kept diaries to record migraine occurrences and intensity.[13] ■

In practice, there isn't always a solution for every experimental challenge. There are instances when the use of a placebo is unethical. Blinding might not always be practical. For instance, the experiment in Example 8.11 was **single-blind** rather than double-blind, because the medical practitioner performing the real and sham acupuncture *had* to know which treatment each subject received.

single-blind

Many—perhaps most—experiments have some weaknesses in detail. The environment of an experiment can influence the outcomes in unexpected ways. Although experiments are the gold standard for evidence of cause and effect, really convincing evidence usually requires that the study be **replicated** successfully by its investigators as well as by independent investigators in different locations. All researchers are required to keep detailed records of their experiments and must provide these records if other teams have been unsuccessful at replicating the published results.

replication

EXAMPLE 8.12 Environmental influence

To study genetic influence on behavior, experimenters "knock out" a gene in one group of mice and compare their behavior with that of a control group of normal mice. The results of these experiments often don't agree as well as hoped, so investigators did exactly the same experiment with the same genetic strain of mice in Oregon, Alberta (Canada), and New York. Many results were very different. It appears that small differences in the lab environments have big effects on the behavior of the mice. Remember this the next time you read that our genes control our behavior.[14] ■

Always be very cautious in your interpretation of an experiment if the likely sources of bias have not all been carefully controlled for.

lack of realism

The most serious potential weakness of experiments is **lack of realism.** The subjects or treatments or setting of an experiment may not realistically duplicate the conditions we really want to study. Carcinogenicity, for example, is often tested in extreme situations unlikely to reflect realistic human use.

EXAMPLE 8.13 Carcinogenicity of saccharin

The carcinogenicity of a product is its propensity to induce cancer when used, ingested, or simply present in the environment. Saccharin has been used as an artificial sweetener for almost a century. Because of experimental evidence of carcinogenicity in rats, in 1981 saccharin was listed on the U.S. *Report on Carcinogens* as reasonably anticipated to be a human carcinogen.

However, closer examination of the observed urinary bladder cancers in rats indicate that the biological mechanism of this cancer is specific to the rat urinary system and related to saccharin consumption at concentrations unlikely to represent realistic human consumption. Saccharin was officially delisted from the *Report on Carcinogens* in 2005 with the conclusion that "the factors thought to contribute to tumor induction by sodium saccharin in rats would not be expected to occur in humans."[15] ■

Lack of realism can limit our ability to apply the conclusions of an experiment to the settings of greatest interest. Most experimenters want to generalize their conclusions to some setting wider than that of the actual experiment. Example 7.5 (page 160) discussed how women had often been excluded from clinical trials. A disturbing consequence is that medications were approved for use in women without knowing whether they would work at the dosage prescribed or what complications might be expected in this half of the population. It has been only about 15 years since the Food and Drug Administration started requiring both male and female participation in clinical trials of drugs applicable to both sexes.

Statistical analysis of an experiment cannot tell us how far the results will generalize. The ability to extend conclusions beyond the experimental setting depends on a thorough discussion of the biological merits of the experiment and the representativeness of the subjects used. Nevertheless, a randomized comparative experiment, because of its ability to give convincing evidence for causation, is one of the most important ideas in statistics.

APPLY YOUR KNOWLEDGE

8.10 Testosterone for older men. As men age, their testosterone levels gradually decrease. This may cause a reduction in lean body mass, an increase in fat, and other undesirable changes. Do testosterone supplements reverse some of these effects? A study in the Netherlands assigned 237 men aged 60 to 80 with low or low-normal testosterone levels to either a testosterone supplement or a placebo. The report in the *Journal of the American Medical Association* described the study as a "double-blind, randomized, placebo-controlled trial."[16] Explain each of these terms to someone who knows no statistics.

8.11 Does meditation reduce anxiety? An experiment that claimed to show that meditation reduces anxiety proceeded as follows. The experimenter interviewed the subjects and rated their level of anxiety. Then the subjects were randomly assigned to two groups. The experimenter taught one group how to meditate and they meditated daily for a month. The other group was simply told to relax more. At the end of the month, the experimenter interviewed all the subjects again and rated their anxiety level. The meditation group now had less anxiety. Psychologists said that the results were suspect because the ratings were not blind. Explain what this means and how lack of blindness could bias the reported results.

Ethics in experimentation

We have dedicated the bulk of this chapter to describing what kinds of experiments *can* be done and discussing their respective scientific merits. We will now ask what kinds of experiments *should* or *should not* be done. This critically important aspect of experimental design is the domain of **ethics.**

ethics

Ethical issues are particularly important in biological experimentation because investigators deal with live forms. Whether the experimental units are humans, animals, or ecosystems, experimentation needs to be thought out not only to be statistically powerful and elegant but also to be *responsible*.

Experimental ethics is about all the questions we need to ask before, during, and after experimentation. But, unlike design, ethics doesn't always have clear-cut answers. Therefore, our objective here won't be to give you all the right answers but rather to help you ask all the right questions.

Human experimentation undoubtedly lends itself to particular ethical scrutiny. All experiments involving humans, even the least intrusive ones, raise ethical issues. Many experiments involving babies and young children have, for instance, helped us understand the nature and importance of attachment between a parent and a child. Yet we might legitimately ask whether the subjects themselves benefited from these experiments and whether they suffered any negative impacts. These same questions become paramount in the area of medical experimentation, where benefits and risks are clear and potentially great. Here are some questions we should ask ourselves:

- *Are all experiments acceptable in the name of science, knowledge, or the public's greater good?* Both medical ethics and international human rights standards say that "the interests of the subject must always prevail over the interests of science and society." The quoted words are from the 1964 Helsinki Declaration of the World Medical Association, the most respected international standard. The most outrageous examples of unethical experiments are those that ignore the interests of the subjects. Chief among them is the Tuskegee syphilis study, discussed further below.
- *Is it morally right to include a placebo in a study design when there are safe and efficient treatments already available?* Because "the interests of the subject must always prevail," medical treatments can be tested in clinical trials

only when there is reason to hope that they will help the patients who are subjects in the trials. Future benefits aren't enough to justify experiments with human subjects. So why is it ethical to give a control group of patients a placebo? Well, we know that placebos often work. What is more, placebos have no harmful side effects other than withholding the benefits of a proven treatment. If we knew which treatment was better, we would give it to everyone. When we don't know, it is ethical to try both and compare them.

- *Should we stop an experiment if we find that one of the treatments shows early signs of adverse effects? And what if one of the treatments shows early signs of clear superiority?* Again, "the interests of the subject must always prevail." Disregard of information gained in the course of experimentation or follow-up studies has been the basis of many private and class-action lawsuits against pharmaceutical companies and is fueling the current debate over the prescription of Vioxx, for instance.
- *Is it ever acceptable to leave subjects in the dark about the purpose and expected results of an experiment?* This question goes hand in hand with the first question on what experiments are morally acceptable. Subjects should be able to judge for themselves the acceptability of their participation. That's why all human experiments are now required to obtain the subject's *informed consent* in writing as well as obtaining the formal approval of an ethics committee. But can subjects truly grasp the purpose and consequences of an experiment? And is complete informed consent achievable without compromising the experiment? Some experiments, especially in psychology, must disguise their true purpose in order to obtain genuine responses from the participants.

EXAMPLE 8.14 Studying drinking behavior

A study wanted to examine the impact of the shape of the glass (straight or curved) on how fast adult social drinkers finish a beer or a soft drink. But informing participants of this objective would likely change their natural drinking behavior. So the researchers obtained special approval from their university's Faculty of Science Research Ethics Committee to disguise the purpose of their study. Instead, the researchers told participants that the study examined "the effects of alcohol consumption on wordsearch performance," and they implemented the deception by having the participants actually complete a wordsearch after they had finished their drinks.[17] ■

- *Should subjects be rewarded for participation?* This is a tricky question related to informed consent and the freedom to choose participation. The poor and uninsured could easily become primary targets for the riskiest experiments. In fact, it was once common to test new vaccines on prison inmates who gave their consent in return for good-behavior credit. The law now forbids medical experiments in prisons. And, of course, we must also learn to think globally because new treatments can be tested anywhere, including in parts of the world where subjects' rights have not yet fully matured.

By law, all studies funded by the federal government must obey the following principles:

- The organization that carries out the study must have an *institutional review board* that reviews all planned studies in advance in order to protect the subjects from possible harm.
- All individuals who are subjects in a study must give their *informed consent* in writing before data are collected. Subjects must be informed about the nature of a study and any risk of harm it may bring.
- All individual data must be kept *confidential*. Only statistical summaries for groups of subjects may be made public. Any breach of confidentiality is a serious violation of data ethics. The best practice is to separate the identity of the subjects from the rest of the data at once.

EXAMPLE 8.15 Who can consent?

Are there some subjects who can't give informed consent? It was once common, for example, to test new vaccines on prison inmates who gave their consent in return for good-behavior credit. Now we worry that prisoners are not really free to refuse, and the law forbids almost all medical research in prisons.

Children can't give fully informed consent, so the usual procedure is to ask their parents. A study of new ways to teach reading is about to start at a local elementary school, so the study team sends consent forms home to parents. Many parents don't return the forms. Can their children take part in the study because the parents did not say "no," or should we allow only children whose parents returned the form and said "yes"?

What about research into new medical treatments for people with mental disorders? What about studies of new ways to help emergency room patients who may be unconscious? In most cases, there is not time to get the consent of the family. Does the principle of informed consent bar realistic trials of new treatments for unconscious patients?

These are questions without clear answers. Reasonable people differ strongly on all of them. Not everything about informed consent is simple.[18] ■

While it is important to question the necessity and ethicality of human and animal experimentation, we also must consider the implications of not performing carefully designed experiments. There are many unfortunate examples of pharmaceutical or surgical methods that have been widely used without this important preliminary stage. The result is that the whole population becomes unsuspecting test subjects of an "experiment" that has no guidelines, no safeguards, and no way to reach informative conclusions about the benefits and dangers of the drug or procedure.

Lobotomy, for instance, which is the severance of the frontal lobes from the rest of the brain, became a popular method in the 1950s and 1960s for dealing with psychotic individuals. It was performed on consenting and somewhat-informed individuals or with the approval of family members wishing to stop undesirable behavior (even in children). After thousands of lobotomies in the United States alone, it became clear that the procedure did provide improvement for some patients, but others showed no improvement at all, and the rest suffered irreversible, lifelong psychological damage.[19]

This is not an isolated case. Surgical practice, especially, has evolved largely based on the ideas and attempts of individual surgeons. Because surgery is by definition invasive, there has historically been a reluctance to perform randomized, controlled experiments in this field out of concern for the subjects. However, with any new procedure, a placebo or sham operation may in fact prove as beneficial or more as the procedure tested and have many fewer or milder side effects. Gastric freezing for the treatment of stomach ulcers (described in Example 8.3) and mammary artery ligation for the treatment of angina (described in Exercise 15.2, page 367) are two medical procedures that were widely used before randomized comparative experiments showed that they performed no better than placebos and the procedures were abandoned. Likewise, the use of routine episiotomy to speed up child delivery and prevent natural tearing is now questioned by multiple studies,[20] and its use has been declining rapidly (as shown in Exercise 1.11, page 26). Fortunately, these drastic practice reversals have led to a greater awareness of the necessity for careful experimentation to legitimate the use of medical and surgical procedures in humans.

Courtesy of the National Archives at Atlanta

DISCUSSION The Tuskegee syphilis study

"For 40 years, the U.S. Public Health Service has conducted a study in which human guinea pigs, not given proper treatment, have died of syphilis and its side effects. The study was conducted to determine from autopsies what the disease does to the human body." The *Washington Evening Star* published these lines in a front-page article in the summer of 1972. The article would end what was perhaps the most controversial human experiment ever in America and result in drastic changes to federal regulations related to scientific and medical experimentation requiring human subjects.

The Public Health Service and the Tuskegee Institute began a study on syphilis in 1932. Nearly 400 poor black men with syphilis from Macon County, Alabama, were enrolled in the study. They were never told they had syphilis and were never treated for it even after penicillin became known as a highly effective cure by 1943 and widely available by 1947. Also, they were never told that they were part of a medical experiment.

Syphilis

Syphilis is a sexually transmitted disease caused by the spirochete bacterium *Treponema pallidum*. The pathology progresses in stages, starting with hard ulcers primarily around the genitalia and followed by a transient, widespread rash. Without treatment the third stage of syphilis may develop, damaging the nervous system, eyes, heart, blood vessels, liver, bones, and joints, causing blindness, madness, and paralysis.

Untreated men can transmit the disease to their female partners, who, in turn, can transmit it to future offspring. About 40% of pregnancies where the mother has untreated syphilis result in stillbirth (baby dead at birth). If the baby survives birth, it then has a 40% to 70% chance of being infected with syphilis. If left untreated, infected newborns may later become developmentally disabled, have seizures, or die because of the disease.

How could this happen?

How could a study designed to deny treatment to uninformed subjects have occurred? The historical context can help us understand the logical journey.

At the end of the 19th century, philanthropists dreamed of black economic development, and the Tuskegee Institute was founded to help develop local schools, businesses, and agriculture. It was later expanded to fight the devastating impact of a syphilis pandemic that affected nearly a third of the African American population in the South. At that time, diagnosis and the currently known treatment were offered to participants.

When the stock market crashed in 1929, funds for the endeavor were cut. In 1932, the Tuskegee philanthropic effort was recycled into a study of the impact of untreated syphilis. In exchange for participating in the study, the men were given free medical exams, free meals, and free burial insurance. The study was originally intended to run for six months, but it lasted 40 years, until a whistle-blower turned to the press.

Acknowledging misconduct

The Tuskegee study was well known in the scientific community, and several papers were published in peer-reviewed journals. The first concerns over ethical aspects of the study were raised in 1968, but the study was terminated only after the outrage generated by the 1972 press release. It then took 25 more years for the government to voice a formal apology. Here is an excerpt from President Clinton's 1997 formal apology on behalf of the nation (the full transcript can be read at `http://clinton4.nara.gov/textonly/New/Remarks/Fri/19970516-898.html`):

> *The United States government did something that was wrong—deeply, profoundly, morally wrong. It was an outrage to our commitment to integrity and equality for all our citizens.*
>
> *To the survivors, to the wives and family members, the children and the grandchildren, I say what you know: No power on Earth can give you back the lives lost, the pain suffered, the years of internal torment and anguish. What was done cannot be undone. But we can end the silence. We can stop turning our heads away. We can look at you in the eye and finally say on behalf of the American people, what the United States government did was shameful, and I am sorry.*

How Tuskegee changed the ethics of human experimentation

The revelation of the Tuskegee syphilis study was the spark that initiated profound and long-lasting reforms of the ethical code of conduct for research requiring the use of human subjects in America. In 1974, the National Research Act was signed into law, creating the National Commission for the Protection of Human Subjects of Biomedical and Behavioral Research.

As a consequence, federally supported studies using human subjects now must be reviewed by institutional review boards, which decide whether study protocols meet ethical standards, and researchers must get voluntary informed consent from all study participants.

Complementing this approach, emphasis has also been placed on training researchers to be aware of the ethical issues related to human experimentation. For instance, the National Institutes of Health now require that all prebaccalaureate, predoctoral, and postdoctoral trainees supported by institutional training grants receive instruction in the responsible conduct of research.

Experimentation on animals, ecosystems, cells, and embryos also raises many ethics questions. But standards here are much more variable. Federal agencies also regulate the use of animals for laboratory experiments, but the guidelines and scrutiny applied are widely different for different species. For example, primate experiments in the United States are closely monitored, in drastic contrast to research on insects. Rats are the typical animal model used for testing suspected carcinogenicity. Such experiments obviously do not have the animals' well-being in mind. But rats are considered pests routinely exterminated by farmers and homeowners, and we consider the information gained from carcinogenicity studies important for the well-being and safety of entire human populations. Rat experiments used to test beauty products, on the other hand, have been much more controversial. Animal-rights activists advocate giving all animals (or at least species with some level of self-consciousness) the same ethical consideration as humans. The debate over animal experimentation is still far from being settled, and this is exemplified by the various laws, regulations, and practices enforced in different countries.

An alternative to actual experimentation that has tremendous potential is to run "experiments" on the computer instead. With computers getting so powerful and fast and with our knowledge of the world ever deepening, we can *simulate* everything we know about a system and the forces that regulate it. The system can then be modified and observed, just as we would apply a treatment and observe the outcome in the real world.

Finally, even the status of cells has been a subject of ethical controversy when these cells have the potential to become human beings. Stem cell research and its applications have been debated all over the world, with varying results. Within the United States, the debate has changed over time. In 2001 President Bush changed the U.S. policy on human embryonic stem cell research by allowing federal funds to be used for research only on existing stem cell lines.[21] In 2004 California passed a stem cell initiative allowing for $3 billion in funding over 10 years for stem cell research and research facilities in California.[22] In 2009 President Obama lifted the 2001 restrictions on stem cell research.[23]

The answers to questions of ethics in experimentation are rarely easy. But perhaps as important as the decisions we make is keeping the debate alive and open to public discussion.

APPLY YOUR KNOWLEDGE

8.12 Informed consent. Long ago, doctors drew a blood specimen from you as part of treating minor anemia. Unknown to you, the sample was stored. Now researchers plan to used stored samples from you and many other people to look for genetic factors that may influence anemia. It is no longer possible to ask your consent. Modern technology can read your entire genetic makeup from the blood sample.

(a) Do you think it violates the principle of informed consent to use your blood sample if your name is on it but you were not told that it might be saved and studied later?

(b) Suppose that your identity is not attached. The blood sample is known only to come from (say) "a 20-year-old white female being treated for anemia." Is it now OK to use the sample for research?

(c) Perhaps we should use biological materials such as blood samples only from patients who have agreed to allow the material to be stored for later use in research. It isn't possible to say in advance what kind of research, so this falls short of the usual standard for informed consent. Is it nonetheless acceptable, given complete confidentiality and the fact that using the sample can't physically harm the patient?

8.13 What is ethical? Effective drugs for treating AIDS are very expensive, so most African nations cannot afford to give them to large numbers of people. Yet AIDS is more common in parts of Africa than anywhere else. Several clinical trials are looking at ways to prevent pregnant mothers infected with HIV from passing the infection to their unborn children, a major source of HIV infections in Africa. Some people say these trials are unethical because they do not give effective AIDS drugs to their subjects, as would be required in rich nations. Others reply that the trials are looking for treatments that can work in the real world in Africa and that they promise benefits at least to the children of their subjects. What do you think?

CHAPTER 8 SUMMARY

- In an experiment, we impose one or more **treatments** on individuals, often called **subjects** or **experimental units.** Each treatment is a combination of values of the explanatory variables, which we call **factors.**
- The **design** of an experiment describes the choice of treatments and the manner in which the subjects are assigned to the treatments.
- The basic principles of statistical design of experiments are **control** and **randomization** to combat bias and **using enough subjects** to reduce chance variation.
- The simplest form of control is **comparison.** Experiments should compare two or more treatments in order to avoid **confounding** of the effect of a treatment with other influences, such as lurking variables.

- **Randomization** uses chance to assign subjects to the treatments. Randomization creates treatment groups that are similar (except for chance variation) before the treatments are applied. Randomization and comparison together prevent **bias,** or systematic favoritism, in experiments. Randomization also allows the use of statistical analyses based on laws of probability.
- You can carry out randomization by using software or by giving numerical labels to the subjects and using a table of random digits to choose treatment groups.
- Applying each treatment to many subjects reduces the role of chance variation and makes the experiment more sensitive to differences among the treatments.
- Good experiments require attention to detail as well as good statistical design. Many behavioral and medical experiments are **double-blind.** Some give a **placebo** to a control group or introduce a **double-dummy** when the treatments compared are too visibly different for a simple placebo. **Replicating** an experiment helps ensure that the results were not merely due to unusual conditions. **Lack of realism** in an experiment can prevent us from generalizing its results.
- In addition to comparison, a second form of control is to restrict randomization by forming **blocks** of individuals that are similar in some way that is important to the response. Randomization is then carried out separately within each block.
- **Matched pairs** are a common form of blocking for comparing just two treatments. In some matched pairs designs, the subjects are matched in pairs as closely as possible, and each subject in a pair receives one of the treatments. In others, each subject receives both treatments in a random order, a design also called **repeated measures.**
- In the end, **ethics** must be the overarching principle guiding experimentation. No matter what experiments could be done and what we might gain from them, we need to ask ourselves if the end justifies the means and be open to a continuing debate.

THIS CHAPTER IN CONTEXT

Observational studies and experiments are two methods for producing data. Observational studies are useful when the conclusion involves describing a group or situation without disturbing the scene we observe. In contrast, experiments are used when the situation calls for a conclusion about whether a treatment *causes* a change in a response. The distinction between observational studies and experiments will be important when stating your conclusions in later chapters.

Only well-designed experiments provide a sound basis for concluding that there is a cause-and-effect relationship. In the simplest comparative experiment, two treatments are imposed on two groups of individuals. Reaching the conclusion that a resulting difference between the groups is caused by the treatments rather than lurking variables requires that the two groups of individuals be similar at the outset. A randomized comparative experiment is used to create groups that are similar. If there is a sufficiently large difference between the groups after imposing the treatments, we can say that the results are statistically significant and conclude that the differences in the response were most likely *caused* by the treatments. In Parts III and IV, the specific statistical procedures for reaching these conclusions will be described. We will also see that which statistical procedure should be used depends in part on the type of experimental design used to collect the data.

As with sampling, when conducting an experiment, attention to detail is important because our conclusions can be weakened by several factors. A lack of blinding can result in the expectations of the researcher influencing the results, while the placebo effect can confound the comparison between a treatment and a control group. In many instances, a more complex design is required to overcome difficulties and can produce more precise results.

CHECK YOUR SKILLS

8.14 What electrical changes occur in muscles as they get tired? Student subjects hold their arms above their shoulders until they drop. Meanwhile, the electrical activity in their arm muscles is measured. This is

(a) an observational study.
(b) an uncontrolled experiment.
(c) a randomized comparative experiment.

8.15 Can changing diet reduce high blood pressure? Vegetarian diets and low-salt diets are both promising. Men with high blood pressure are assigned at random to four diets: (1) normal diet with unrestricted salt, (2) vegetarian with unrestricted salt, (3) normal with restricted salt, and (4) vegetarian with restricted salt. This experiment has

(a) one factor, the choice of diet.
(b) two factors, normal/vegetarian diet and unrestricted/restricted salt.
(c) four factors, the four diets being compared.

8.16 The response variable in the experiment described in Exercise 8.15 is

(a) the amount of salt in the subject's diet.
(b) which of the four diets a subject is assigned to.
(c) change in blood pressure after 8 weeks on the assigned diet.

8.17 In the experiment described in Exercise 8.15, the 240 subjects are labeled 001 to 240. Software assigns an SRS of 60 subjects to Diet 1, then an SRS of 60 of the remaining 180 to Diet 2, then an SRS of 60 of the remaining 120 to Diet 3. The 60 who are left get Diet 4. This is a

(a) completely randomized design.
(b) block design, with four blocks.
(c) matched pairs design.

8.18 A study of the relationship between dengue infection and miscarriage enrolled a group of women who had just had a miscarriage. The investigators matched each woman who had suffered a miscarriage to three women with viable pregnancies at a similar gestational stage, then asked them if they had had a recent dengue infection.[24] This is

(a) an observational study with a case-control design.
(b) an experiment with a block design.
(c) an experiment with a matched pairs design.

8.19 A medical experiment compares an antidepressant with a placebo for relief of chronic headaches. There are 38 headache patients available to serve as subjects: 28 women and 10 men. The investigators randomly select 16 female patients to receive the antidepressant and they also randomly select 5 men to receive the antidepressant. The remaining patients will receive the placebo. This experiment has

(a) a completely randomized design.
(b) a matched pairs design.
(c) a block design.

8.20 The Community Intervention Trial for Smoking Cessation asked whether a community-wide advertising campaign would reduce smoking. The researchers located 11 pairs of communities that were similar in location, size, economic status, and so on. One community in each pair participated in the advertising campaign and the other did not. This is

(a) an observational study.
(b) a matched pairs experiment.
(c) a completely randomized experiment.

8.21 To decide which community in each pair in the previous exercise should get the advertising campaign, it is best to

(a) toss a coin.
(b) choose the community that will help pay for the campaign.
(c) choose the community with a mayor who will participate.

8.22 The Food and Drug Administration should be reluctant to use the results of the experiment described in Exercise 8.15 for its guideline on diets for Americans with high blood pressure because

(a) the study used a matched pairs design instead of a completely randomized design.
(b) results from men with high blood pressure may not generalize to the population of all Americans with high blood pressure.
(c) the study did not use a placebo and was not double-blind.

8.23 An experimental design randomly assigns volunteer arthritis patients suffering from chronic pain to take either a new pain medication, aspirin (the traditional nonsteroidal anti-inflammatory treatment), or a placebo. An institutional review board discusses whether the experiment should include a placebo. This is

(a) an ethical issue.
(b) an issue of lack of replication.
(c) an issue of lack of realism.

CHAPTER 8 EXERCISES

In all exercises that require randomization, you may use Table A, the Simple Random Sample applet, or other software.

8.24 Wine, beer, or spirits? Example 7.2 (page 157) describes a study that compared three groups of people. The first group drinks mostly wine, the second drinks mostly beer, and the third drinks mostly spirits. This study is comparative, but it is not an experiment. Why not?

8.25 Treating breast cancer. The most common treatment for breast cancer discovered in its early stages was once removal of the breast. It is now usual to remove only the tumor and nearby lymph nodes, followed by radiation. To study whether these treatments differ in their effectiveness, a medical team examines the records of 25 large hospitals and compares the survival times after surgery of all women who have had either treatment.

(a) What are the explanatory and response variables?
(b) Explain carefully why this study is not an experiment.
(c) Explain why confounding will prevent this study from discovering which treatment is more effective. (The current treatment was in fact recommended after several large randomized comparative experiments.)

8.26 Dental bonding. A dentistry study evaluated the effect of tooth etch time on resin bonding strength. A total of 78 undamaged, recently extracted first molars (baby teeth) were randomly assigned to be etched with phosphoric acid gel for either 15, 30, or 60 seconds. Composite resin cylinders of identical size were then bonded to the tooth enamel. The researchers examined the bond strength after 24 hours by finding the failure load (in megapascals) for each bond.

(a) What are the explanatory and response variables?
(b) Outline the design of this experiment.

8.27 Bird songs. Bird songs have been hypothesized to be a secondary sexual character signaling an individual's health status. Researchers designed an experiment in which they randomly assigned male collared flycatchers (*Ficedula albicollis*) to two groups. One group received an immune challenge in the form of an injection of sheep red blood cells, and the other group received a placebo injection. The researchers then recorded the changes in song rate (in strophes per minute) after the injection for the 15 males in the immune challenge group and the 12 males in the placebo group.

(a) Outline the design of this experiment.
(b) Label the birds from 1 through 27. Use software or Table A at line 131 to carry out the randomization your design requires.

8.28 Growing in the shade. Ability to grow in shade may help pines in the dry forests of Arizona resist drought. How well do these pines grow in shade? Investigators planted 45 pine seedlings in a greenhouse in either full light, light reduced to

25% of normal by shade cloth, or light reduced to 5% of normal. At the end of the study, they dried the young trees and weighed them.

(a) What are the individuals, the treatments, and the response variable in this experiment?

(b) Outline the design of this experiment. Use software or Table A to randomly assign seedlings to the treatment groups.

8.29 Exercise and heart attacks. Does regular exercise reduce the risk of a heart attack? Here are two ways to study this question. Explain clearly why the second design will produce more trustworthy data.

1. A researcher finds 2000 men over 40 who exercise regularly and have not had heart attacks. She matches each with a similar man who does not exercise regularly, and she follows both groups for five years.
2. Another researcher finds 4000 men over 40 who have not had heart attacks and are willing to participate in a study. She assigns 2000 of the men to a regular program of supervised exercise. The other 2000 continue their usual habits. The researcher follows both groups for five years.

8.30 Improving adolescents' health habits. APPLET Twenty-four public middle schools agree to participate in the experiment described in Exercise 8.2. Use a diagram to outline a completely randomized design for this experiment. Do the randomization required to assign schools to treatments. If you use the *Simple Random Sample* applet or other software, choose all four treatment groups. If you use Table A, start at line 105 and choose only the first two groups.

8.31 Relieving headaches. Doctors identify "chronic tension-type headaches" as headaches that occur almost daily for at least six months. Can antidepressant medications or stress management training reduce the number and severity of these headaches? Are both together more effective than either alone?

(a) Use a diagram like Figure 8.1 to display the treatments in a design with two factors: medication, yes or no; and stress management, yes or no. Then outline the design of a completely randomized experiment to compare these treatments.

(b) APPLET The headache sufferers named below have agreed to participate in the study. Randomly assign the subjects to the treatments. If you use the *Simple Random Sample* applet or other software, assign all the subjects. If you use Table A, start at line 130 and assign subjects to only the first treatment group.

Abbott	Decker	Herrera	Lucero	Richter
Abdalla	Devlin	Hersch	Masters	Riley
Alawi	Engel	Hurwitz	Morgan	Samuels
Broden	Fuentes	Irwin	Nelson	Smith
Chai	Garrett	Jiang	Nho	Suarez
Chuang	Gill	Kelley	Ortiz	Upasani
Cordoba	Glover	Kim	Ramdas	Wilson
Custer	Hammond	Landers	Reed	Xiang

8.32 Oligofructose and calcium absorption. Nondigestible oligosaccharides are known to stimulate calcium absorption in rats. A double-blind, randomized experiment investigated whether the consumption of oligofructose similarly stimulates calcium absorption in 11 healthy male adolescents 14 to 16 years old. The subjects took a pill for nine days and had their calcium absorption tested on the last day. The experiment was repeated three weeks later. Some subjects received the oligofructose pill in the first round and then a pill containing sucrose (serving as a control) in the second round. The order was switched for the remaining subjects.

(a) What are the explanatory and response variables here?

(b) What design was used in this experiment? Why was the order of treatment switched for some subjects?

(c) The study was double-blind. What does this mean?

8.33 Cell phone use and brain metabolism. A study examined the impact of cell phone use on brain metabolism in 47 healthy adults. Here is a quote from the published abstract: "Cell phones were placed on the left and right ears and positron emission tomography with (^{18}F)fluorodeoxyglucose injection was used to measure brain glucose metabolism twice, once with the right cell phone activated (sound muted) for 50 minutes (ON condition) and once with both cell phones deactivated (OFF condition)."[25]

(a) This experiment has a matched pairs design. How can you tell?

(b) What is the advantage of using a matched pairs design rather than a completely randomized design in this context?

(c) The abstract reports that "metabolism in the region closest to the antenna (orbitofrontal cortex and temporal pole) was significantly higher for ON than OFF conditions." Explain in simple terms what "significantly higher" means in this context.

8.34 Athletes taking oxygen. We often see players on the sidelines of a football game inhaling oxygen. Their coaches think this will speed their recovery. We might measure recovery from intense exercise as follows. Have a football player run 100 yards three times in quick succession. Then allow three minutes to rest before running 100 yards again. Time the final run. Because players vary greatly in speed, you plan a matched pairs experiment using 25 football players as subjects.

(a) Discuss the design of such an experiment to investigate the effect of inhaling oxygen during the rest period. What is the name for this experimental design?

(b) Use a diagram to outline the design.

8.35 Protecting ultramarathon runners. An ultramarathon, as you might guess, is a footrace longer than the 26.2 miles of a marathon. Runners commonly develop respiratory infections after an ultramarathon. Will taking 600 milligrams of vitamin C daily reduce these infections? Researchers randomly assigned ultramarathon runners to receive either vitamin C or a placebo. Separately, they also randomly assigned these treatments to a group of nonrunners the same age as the runners. All subjects were watched for 14 days after the big race to see if infections developed.

(a) What is the name for this experimental design?

(b) Use a diagram to outline the design.

8.36 Reducing spine fractures. Fractures of the spine are common and serious among women with advanced osteoporosis (low mineral density in the bones). Can taking strontium renelate help? A large medical trial assigned 1649 women to take either strontium renelate or a placebo each day. All the subjects had osteoporosis and had suffered at least one fracture. All were taking calcium supplements and receiving standard medical care. The response variables were measurements of bone density and counts of new fractures over three years. The subjects were treated at 10 medical centers, each in a different country. Outline a block design for this experiment, with the medical centers as blocks. Explain why this is the proper design.

8.37 Does ginkgo improve memory? The law allows marketers of herbs and other natural substances to make health claims that are not supported by evidence. Brands of ginkgo extract claim to "improve memory and concentration." A randomized comparative experiment found no evidence for such effects. The subjects were 230 healthy people over 60 years old. They were randomly assigned to ginkgo or a placebo pill. All the subjects took a battery of tests for learning and memory before treatment started and again after six weeks.

(a) Following the model of Figure 8.3, outline the design of this experiment.

(b) Use the *Simple Random Sample* applet, other software, or Table A to assign half the subjects to the ginkgo group. If you use software, report the first 20 members of the ginkgo group (in the applet's "Sample bin") and the first 20 members of the placebo group (those left in the "Population hopper"). If you use Table A, start at line 103 and choose only the first 5 members of the ginkgo group.

APPLET

8.38 Wine, beer, or spirits? Example 7.2 (page 157) describes an observational study of the health benefits of drinking (in moderation) mostly wine, mostly beer, or mostly spirits. Imagine that you have 120 women and 180 men volunteers for an experiment comparing the effect of these drinks on heart disease. Outline a block design for comparing the 3 treatments (wine, beer, and spirits), using women and men as blocks. Be sure to say how many subjects you would put in each group in your design. (Such an experiment might be unethical, because the subjects would risk developing alcohol addiction.)

8.39 "Doctor Offers Women Flashy New Treatment." Here is an excerpt from a 2007 NBC news report:

> *Millions of women suffer from severe mood swings and hot flashes associated with menopause and other causes of estrogen depletion. . . . Lipov is treating menopausal women by using an injection which he believes can reset the brain's thermometer to reduce or eliminate hot flashes. . . . Lipov is the first and only doctor in the U.S. to target hot flashes. To make the procedure more safe, he first injects a dye into the targeted area. Complications, like seizures, are rare and so far Lipov's patients have had no problems. The hot flash relief lasts from weeks to months, Lipov said, and almost all of the 22 women he's used the technique on have gotten some benefit.*[26]

(a) What is the design of this study?

(b) Can we trust the results of this study? Explain why or why not.

8.40 Quick randomizing? Here's a quick and easy way to randomize. You have 100 subjects, 50 women and 50 men. Toss a coin. If it's heads, assign the men to the treatment and the women to the control group. If the coin comes up tails, assign the women to treatment and the men to control. This gives every individual subject a 50-50 chance of being assigned to treatment or control. Why isn't this a good way to randomly assign subjects to treatment groups?

8.41 Explaining medical research. Observational studies had suggested that vitamin E reduces the risk of heart disease. Careful experiments showed that vitamin E has no effect, at least for women. According to a commentary in the *Journal of the American Medical Association*:

> *Thus, vitamin E enters the category of therapies that were promising in epidemiologic and observational studies but failed to deliver in adequately powered randomized controlled trials. As in other studies, the "healthy user" bias must be considered, ie, the healthy lifestyle behaviors that characterize individuals who care enough about their health to take various supplements are actually responsible for the better health, but this is minimized with the rigorous trial design.*[27]

A friend who knows no statistics asks you to explain this.

(a) What is the difference between observational studies and experiments?

(b) What is a "randomized controlled trial"? (We'll discuss "adequately powered" in Chapter 15.)

(c) How does "healthy user bias" explain how people who take vitamin E supplements have better health in observational studies but not in controlled experiments?

8.42 Treating sinus infections. Sinus infections are common, and doctors often treat them with antibiotics. Another treatment is to spray a steroid solution into the nose. A well-designed clinical trial found that these treatments, alone or in combination, do not reduce the severity or the length of sinus infections. The clinical trial was a completely randomized experiment that assigned 240 patients at random among 4 treatments as follows:

	Antibiotic pill	Placebo pill
Steroid spray	53	64
Placebo spray	60	63

(a) What are the factors and treatments in this experiment?

(b) Outline the design of the experiment.

(c) Explain briefly how you would do the random assignment of patients to treatments.

8.43 Treating sinus infections, continued. If the random assignment of patients to treatments in the previous exercise did a good job of eliminating bias, possible lurking variables such as smoking history, asthma, and hay fever should be similar in all 4 groups. After recording and comparing many such variables, the investigators said that "all showed no significant difference between groups." Explain to someone who knows no statistics what "no significant difference" means. Does it mean that the presence of all these variables was exactly the same in all four treatment groups?

8.44 Growing tomato plants. An experiment measures the amounts of various nutrients in tomato plants subjected to three levels of nitrogen fertilization. To avoid confounding by the soil properties, the planting area was divided into 12 blocks within which subplots were randomly assigned to the 3 treatments. Draw an outline of this block design as in Figure 8.5.

8.45 An herb for depression? Does the herb Saint-John's-wort relieve major depression? A study examined this issue. Here are some excerpts from the published report:

> *Design: Randomized, double-blind, placebo-controlled clinical trial. . . . Participants . . . were randomly assigned to receive either Saint-John's-wort extract (n = 98) or placebo (n = 102). . . . The primary outcome measure was the rate of change in the Hamilton Rating Scale for Depression over the treatment period.*[28]

(a) Use the information provided to draw a diagram outlining the design of this study.

(b) The report concluded that Saint-John's-wort is no more effective than a placebo at relieving symptoms of depression. Based on the information quoted, would you be inclined to trust this conclusion? Explain your answer.

8.46 Calcium and blood pressure. Some medical researchers suspect that added calcium in the diet reduces blood pressure. You have available 40 men with high blood pressure who are willing to serve as subjects.

(a) Outline an appropriate design for the experiment, taking the placebo effect into account.

(b) The researchers conclude that "the blood pressure of the calcium group was significantly lower than that of the placebo group." "Significant" in this conclusion means statistically significant. Explain what "statistically significant" means in the context of this experiment, as if you were speaking to a doctor who knows no statistics.

8.47 Experimentation for emergency treatments. How do you deliver oxygen to cells in emergency situations when the lungs have failed? A team of researchers devised a new method that encapsulates oxygen within microparticles of fat so that oxygen can be injected directly into the bloodstream without risking an embolism. The method was tested experimentally in rabbits and shown to be successful. But how do you then test it in humans?

(a) One step would involve giving the oxygen injections to healthy volunteers to test its safety. What are some ethical challenges to using healthy volunteers to test an unknown medical procedure?

(b) Another step would require giving the oxygen injections during real-life emergencies involving oxygen deprivation. Obtaining informed consent from the patient is likely to be impossible, and there may not be enough time to obtain informed consent from family members. Would it be ethical to test the procedure anyway?

8.48 Institutional review boards. Your college or university has an institutional review board that screens all studies that use human subjects. Get a copy of the document that describes this board (you can probably find it online).

(a) According to this document, what are the duties of the board?

(b) How are members of the board chosen? How many members are not scientists? How many members are not employees of the college? Do these members have some special expertise, or are they simply members of the "general public"?

8.49 World ethics. Researchers from Yale, working with medical teams in Tanzania, wanted to know how common infection with the AIDS virus is among pregnant women in that African country. To do this, they planned to test blood samples drawn from pregnant women. Yale's institutional review board insisted that the researchers get the informed consent of each woman and tell her the results of the test. This is the usual procedure in developed nations. The Tanzanian government did not want to tell the women why blood was drawn or tell them the test results. The government feared panic if many people turned out to have an incurable disease for which the country's medical system could not provide care. The study was canceled. Do you think that Yale was right to apply its usual standards for protecting subjects?

8.50 What is minimal risk? Federal regulations say that "minimal risk" means the risks are no greater than "those ordinarily encountered in daily life or during the performance of routine physical or psychological examinations or tests." You are a member of your college's institutional review board. You must decide whether several research proposals qualify for lighter review because they involve only minimal risk to subjects. Which of these do you think qualifies as "minimal risk"?

(a) Draw a drop of blood by pricking a finger to measure blood sugar.

(b) Draw blood from the arm for a full set of blood tests.

(c) Insert a tube that remains in the arm, so blood can be drawn regularly.

8.51 The Willowbrook hepatitis studies. In the 1960s, children entering the Willowbrook State School, an institution for the developmentally disabled, were deliberately infected with hepatitis. The researchers argued that almost all children in the institution quickly became infected anyway. The studies showed for the first time that two strains of hepatitis existed. This finding contributed to the development of effective vaccines. Despite these valuable results, the Willowbrook studies are now considered an example of unethical research. Explain why, according to current ethical standards, useful results are not enough to allow a study.

8.52 AIDS trials in Africa. The drug programs that treat AIDS in rich countries are very expensive, so some African nations cannot afford to give them to large numbers of people. Yet AIDS is more common in parts of Africa than anywhere else. "Short-course" drug programs that are much less expensive might help, for example, in preventing infected pregnant women from passing the infection to their unborn children. Is it ethical to compare a short-course program with a placebo in a clinical trial? Some say "no": This is a double standard, because in rich countries the full drug program would be the control treatment. Others say "yes": The intent is to find treatments that are practical in Africa, and the trial does not withhold any treatment that subjects would otherwise receive. What do you think?

8.53 Randomization avoids bias. APPLET Suppose that you have 50 subjects available, 25 women and 25 men, to test reaction time to a visual cue in the presence or absence of a noise distraction. Let's label the women from 1 to 25 and the men from 26 to 50. We hope that randomization will distribute these subjects roughly equally between the noise and no-noise groups. Use the *Simple Random Sample* applet to take 20 samples of size 25 from the 50 subjects. (Be sure to click "Reset" after each sample.) Record the counts of women (labels 1 to 25) in each of your 20 samples. You see that there is considerable chance variation but no systematic bias in favor of one or the other group in assigning the women. Larger samples from a larger population will, on average, do an even better job of creating two similar groups.

CHAPTER 9 Introducing Probability

Why is probability, the mathematics of chance behavior, needed to understand statistics, the science of data? Let's look at a typical sample survey.

IN THIS CHAPTER WE COVER ...

- The idea of probability
- Probability models
- Probability rules
- Discrete probability models
- Continuous probability models
- Random variables
- Personal probability*
- Risk and odds*

EXAMPLE 9.1 Who gets the flu?

Example 7.7 (page 163) described a Gallup-Healthways survey seeking to find out what proportion of American adults were sick with the flu in October 2012. In that month, Gallup took a national random sample of 28,295 adults and found that 594 of the people in the sample said they were sick with the flu the previous day. The proportion who were sick with the flu was

$$\text{sample proportion} = \frac{594}{28{,}295} = 0.021 \quad \text{(that is, 2.1\%)}$$

Because all adults had the same chance to be among the chosen 28,295, it seems reasonable to use this 2.1% as an estimate of the unknown proportion in the population. It's a *fact* that 2.1% of the sample were sick with the flu—we know because Gallup asked them. We don't know what percent of all adults in the United States were sick with the flu in October 2012, but we *estimate* that about 2.1% were. This is a basic move in statistics: Use a result from a sample to estimate something about a population. ■

What if Gallup took a second random sample of 28,295 adults and asked them the same question? The new sample would have different people in it. It is almost certain that there would not be exactly 594 "sick with the flu" responses. That is,

Gallup's estimate of the proportion of adults who were sick with the flu in October 2012 will vary from sample to sample. Could it happen that one random sample finds that 2.1% of adults were sick with the flu and a second random sample finds that 5.3% were? *Random samples eliminate bias from the act of choosing a sample, but they can still be wrong because of the variability that results when we choose at random*. If the variation when we take repeat samples from the same population is too great, we can't trust the results of any one sample.

This is where we need facts about probability to make progress in statistics. Because Gallup uses chance to choose its samples, the laws of probability govern the behavior of the samples. We will address this in greater detail in Chapter 13 on sampling distributions. Our purpose in this chapter is to understand the language of probability, but without going into the mathematics of probability theory.

The idea of probability

To understand why we can trust random samples and randomized comparative experiments, we must look closely at chance behavior. The big fact that emerges is this: **Chance behavior is unpredictable in the short run but has a regular and predictable pattern in the long run.**

Toss a coin or choose an SRS. The result can't be predicted in advance, because the result will vary when you toss the coin or choose the sample repeatedly. But there is still a regular pattern in the results, a pattern that emerges clearly only after many repetitions. This remarkable fact is the basis for the idea of probability.

EXAMPLE 9.2 Coin tossing

When you toss a coin, there are only two possible outcomes, heads or tails. Figure 9.1 shows the results of tossing a coin 5000 times on two separate occasions. For each number of tosses from 1 to 5000, we have plotted the proportion of those tosses that gave a head. Trial A (solid line) begins tail, head, tail, tail. You can see that the proportion of heads for Trial A starts at 0 on the first toss, rises to 0.5 when the second toss gives a head, then falls to 0.33 and 0.25 as we get two more tails. Trial B (dotted line), on the other hand, starts with five straight heads, so the proportion of heads is 1 until the sixth toss.

The proportion of tosses that produce heads is quite variable at first. Trial A starts low and Trial B starts high. As we make more and more tosses, however, the proportion of heads for both trials gets close to 0.5 and stays there. If we made yet a third trial at tossing the coin a great many times, the proportion of heads would again settle down to 0.5 in the long run. This is the intuitive idea of probability. Probability 0.5 means "occurs half the time in a very large number of trials." The probability 0.5 appears as a horizontal line on the graph. ■

We might suspect that a coin has probability 0.5 of coming up heads just because the coin has two sides. But we can't be sure. The coin might be unbalanced. In fact, spinning a penny or nickel on a flat surface, rather than tossing the coin, doesn't give heads probability 0.5. The idea of probability is empirical. That is, it is based on observation rather than theorizing. Probability describes what happens in very many trials, and we must actually observe many trials to pin down a probability. In the case of tossing a coin, some diligent people have in fact made thousands of tosses.

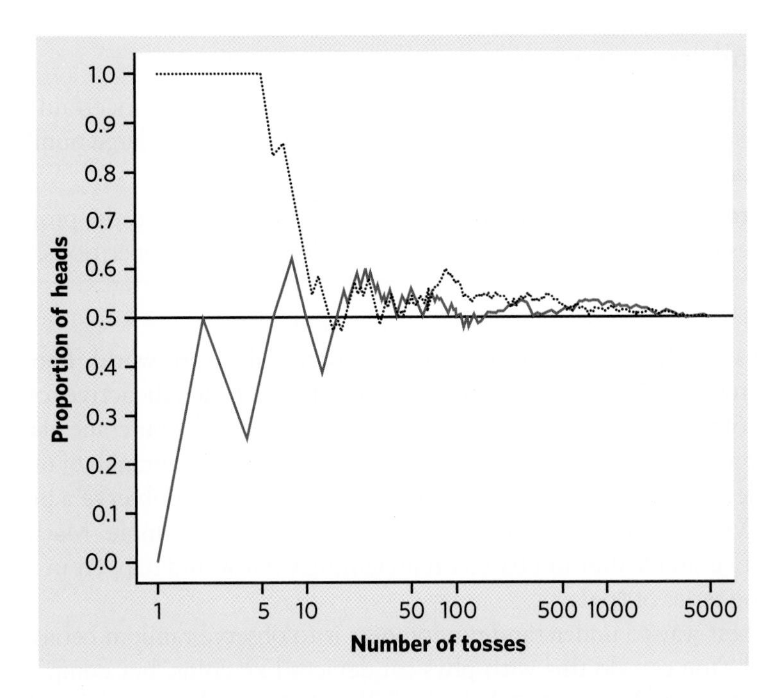

FIGURE 9.1 The proportion of tosses of a coin that give a head changes as we make more tosses. Eventually, however, the proportion approaches 0.5, the probability of a head. This figure shows the results of two trials of 5000 tosses each.

EXAMPLE 9.3 Theory and practice

How does practice agree with theory? When it comes to coin tossing, we have a few historical examples of very large trial numbers. The French naturalist Count Buffon (1707–1788) tossed a coin 4040 times. Around 1900, the English statistician Karl Pearson heroically tossed a coin 24,000 times. While imprisoned by the Germans during World War II, the South African mathematician John Kerrich tossed a coin 10,000 times. Here are their results:

	Buffon	Pearson	Kerrich
Total tosses	4,040	24,000	10,000
Number of heads	2,048	12,012	5,067
Proportion of heads	0.5069	0.5005	0.5067

Genetic theory predicts an equal proportion over the long term of male and female newborns because each spermatozoid carries either an X or a Y chromosome. However, many factors affect the dispersion and success rate of gametes, and birth certificates indicate a slight departure from the equal-proportions model. For example, the U.S. National Center for Health Statistics reports the number of births (in thousands) by gender in the United States for the following years:[1]

	1990	2000	2002	2008
Males	2129	2077	2058	2173
Females	2029	1982	1964	2074
Proportion of males	0.5120	0.5117	0.5117	0.5117

■

> **RANDOMNESS AND PROBABILITY**
>
> We call a phenomenon **random** if individual outcomes are uncertain but there is nonetheless a regular distribution of outcomes in a large number of repetitions.
>
> The **probability** of any outcome of a random phenomenon is the proportion of times the outcome would occur in a very long series of repetitions.

That some things are random is an observed fact about the world. The outcome of a coin toss, the time between emissions of particles by a radioactive source, and the sexes of the next litter of lab rats are all random. So is the outcome of a random sample or a randomized experiment. Probability theory is the branch of mathematics that describes random behavior. Of course, we can never observe a probability exactly. We could always continue tossing the coin, for example. Mathematical probability is an idealization based on imagining what would happen in an indefinitely long series of trials.

The best way to understand randomness is to observe random behavior, as in Figure 9.1. You can do this with physical devices like coins, but computer simulations (imitations) of random behavior allow faster exploration. The *Probability* applet is a computer simulation that animates Figure 9.1. It allows you to choose the probability of a head and simulate any number of tosses of a coin with that probability. Experience shows that the proportion of heads gradually settles down close to the chosen probability. Equally important, it also shows that *the proportion in a small or moderate number of tosses can be far from the probability. Probability describes only what happens in the long run*. This is called a **frequentist** approach to defining probabilities, because we rely on the relative frequency (proportion) of one particular outcome among very many observations of the random phenomenon. The optional section on personal probabilities later in this chapter discusses another approach to obtaining probabilities.

frequentist

Computer simulations like the *Probability* applet start with given probabilities and imitate random behavior, but we can estimate a real-world probability only by actually observing many trials. Nonetheless, computer simulations are very useful because we need long runs of trials. In situations such as coin tossing, the proportion of an outcome often requires several hundred trials to settle down to the probability of that outcome. Shorter runs give only rough estimates of a probability.

What looks random? Toss a coin six times and record heads (H) or tails (T) on each toss. Which of these outcomes is more probable: HTHTTH or TTTHHH? Almost everyone says that HTHTTH is more probable, because TTTHHH does not "look random." In fact, both are equally probable. That heads has probability 0.5 says that about half of a very long sequence of tosses will be heads. It doesn't say that heads and tails must come close to alternating in the short run. The coin doesn't know what past outcomes were, and it can't try to create a balanced sequence.

APPLY YOUR KNOWLEDGE

9.1 **Hemophilia.** Hemophilia refers to a group of rare hereditary disorders of blood coagulation. Because the disorder is caused by defective genes on the X chromosome, hemophilia affects primarily men. According to the Centers for Disease Control and Prevention, the prevalence of hemophilia (the number in the population with hemophilia at any given time) among American males is 13 in 100,000. Explain carefully what this means. In particular, explain why it does *not* mean that if you obtain the medical records of 100,000 males, exactly 13 will be diagnosed with hemophilia.

9.2 Random digits. The table of random digits (Table A) was produced by a random mechanism that gives each digit probability 0.1 of being a 0.

(a) What proportion of the first 50 digits in the table are 0s? This proportion is an estimate, based on 50 repetitions, of the true probability, which in this case is known to be 0.1.

(b) The *Probability* applet can imitate random digits. Set the probability of heads in the applet to 0.1. Check "Show true probability" to show this value on the graph. A head stands for a 0 in the random digits table, and a tail stands for any other digit. Simulate 200 digits. If you kept going forever, presumably you would get 10% heads. What was the result of your 200 tosses?

9.3 Probability says ... Probability is a measure of how likely an event is to occur. Match one of the probabilities that follow with each statement of likelihood given. (The probability is usually a more exact measure of likelihood than is the verbal statement.)

(a) This event is impossible. It can never occur.

(b) This event is certain. It will occur on every trial.

(c) This event is very unlikely, but it will occur once in a while in a long sequence of trials.

(d) This event will occur more often than not.

Probability models

Gamblers have known for centuries that the fall of coins, cards, and dice displays clear patterns in the long run. After all, the first formal studies of probabilities were aimed at understanding various games of chance. The idea of probability rests on the observed fact that the average result of many thousands of chance outcomes can be known with near certainty. How can we give a mathematical description of long-run regularity?

To see how to proceed, think first about a very simple random phenomenon, the birth of one child. When the child is conceived, we cannot know the outcome in advance. What do we know? We are willing to say that the outcome will be either male or female. We believe that each of these outcomes has probability 1/2. This description of a child's birth has two parts:

- A list of possible outcomes
- A probability for each outcome

Such a description is the basis for all probability models. Here is the basic vocabulary we use.

What is the chance of rain? We have all checked the weather forecast to see whether rain will spoil our weekend plans. A report says that the chance of rain in your city tomorrow is 20%. What does that mean and how is this probability calculated? The National Weather Service keeps a historical database of daily weather conditions such as temperature, pressure, and humidity. The chance of rain on a given day is calculated as the percent of days in the database with similar weather conditions that had rain. So, a 20% chance of rain tomorrow means that it has rained in only 20% of days with similar weather conditions.

PROBABILITY MODELS

The **sample space S** of a random phenomenon is the set of all possible outcomes.

An **event** is an outcome or a set of outcomes of a random phenomenon. That is, an event is a subset of the sample space.

A **probability model** is a mathematical description of a random phenomenon consisting of two parts: a sample space S and a way of assigning probabilities to events.

A sample space S can be very simple or very complex. When one child is born, there are only two outcomes, male and female. The sample space is $S = \{M, F\}$. When the National Health Survey records the body weights in pounds of a random sample of adults, the sample space contains all possible adult weights over a realistic interval.

CDC/Janice Haney Carr

EXAMPLE 9.4 Blood types

Your blood type greatly impacts, for instance, the kind of blood transfusion or organ transplant you can safely get. There are 8 different blood types based on the presence or absence of certain molecules on the surface of red blood cells. A person's blood type is given as a combination of a group (O, A, B, or AB) and a Rhesus factor (+ or −). They make up the sample space S:

$$S = \{O+, O-, A+, A-, B+, B-, AB+, AB-\}$$

How can we assign probabilities to this sample space? First of all, these 8 blood types are represented differently in different ethnic groups. Within a given ethnic group, we can use the blood types' frequencies in that group to assign their respective probabilities. The American Red Cross reports that, among Asian Americans, there are 39% blood type O+, 1% O−, 27% A+, 0.5% A−, 25% B+, 0.4% B−, 7% AB+, 0.1% AB−.[2] Because 39% of all Asian Americans have blood type O+, the probability that a randomly chosen Asian American has blood type O+ is 39%, or 0.39. We can thus construct the complete probability model for blood types among Asian Americans:

Blood type	O+	O−	A+	A−	B+	B−	AB+	AB−
Probability	0.39	0.01	0.27	0.005	0.25	0.004	0.07	0.001

What if we were interested only in the person's Rhesus factor? For any randomly selected Asian American, the Rhesus factor can only be positive or negative. Therefore, the sample space for this new question is:

$$S = \{Rh+, Rh-\}$$

Based on the known population percents, the probability model for Rhesus factor is:

Rh factor	Rh+	Rh−
Probability	0.98	0.02

■

In Example 9.4, we used the known frequencies of blood types among Asian Americans to construct the probability models. In some cases, we can use known properties of the random phenomenon (for example, physical properties, genetic laws) to compute the probabilities of the outcomes in the sample space. Here is an example.

EXAMPLE 9.5 A boy or a girl?

Young couples often discuss how many children they would like to have. One couple wants to have two children. There are four possible outcomes when we examine the gender of the two children in order (first child, second child). Figure 9.2 displays these four outcomes.

		First child	
		Girl	Boy
Second child	Girl	GG	BG
	Boy	GB	BB

FIGURE 9.2 The four possible outcomes in gender sequence for couples having two children, for Example 9.5. If we assume that male and female newborns are equally likely, all four of these outcomes have the same probability.

Parents often care more about how many boys or girls they could have than about the particular order. The sample space for the number of girls a couple with two children could have is

$$S = \{0, 1, 2\}$$

What are the probabilities for this sample space? For each newborn, the probability that it will be a girl and the probability that it will be a boy are approximately equal. We also know that the gender of the first child does not influence the gender of the second child. Therefore, all four outcomes in Figure 9.2 will be *equally likely*. That is, each of the four outcomes will, in the long run, come up in one-fourth of all couples with two children. So each outcome in Figure 9.2 has probability 1/4. However, the three possible outcomes in our sample space are *not* equally likely, because there are two ways to have exactly one girl and only one way to have no girl at all. So "no girl" has probability 1/4 but "one girl" has probability 2/4 (2 outcomes from Figure 9.2). Here is the complete probability model:

Number of girls	0	1	2
Probability	1/4	2/4	1/4

■

We built the probability model in Example 9.5 by assuming equal probability of both genders at birth. This model is reasonably accurate. However, we have seen in Example 9.3 that national data point to a slight imbalance between the genders at birth. So, in reality, all four outcomes in Figure 9.2 are not exactly equally likely.

APPLY YOUR KNOWLEDGE

9.4 Sample space. Choose a student at random from a large statistics class. Describe a sample space S for each of the following. (In some cases, you may have some freedom in specifying S.)

(a) Ask whether the student is male or female.

(b) Ask how tall the student is.

(c) Ask what the student's blood type is.

(d) Ask how many times a day the student brushes his or her teeth.

(e) Ask how long since the student's last flu or cold.

9.5 More on boys and girls. A couple wants to have three children. Assume that the probabilities of a newborn being male or being female are the same and that the gender of one child does not influence the gender of another child.

(a) There are 8 possible arrangements of girls and boys. What is the sample space for having three children (gender of the first, second, and third child)? All 8 arrangements are (approximately) equally likely.

(b) The future parents are wondering how many boys they might get if they have three children. Give a probability model (sample space and probabilities of outcomes) for the number of boys. Follow the method of Example 9.5.

Probability rules

Examples 9.4 and 9.5 describe pretty simple random phenomena. However, we don't always have a probability model available to answer our questions. We can make progress by listing some facts that must be true for *any* assignment of probabilities. These facts follow from the idea of probability as "the long-run proportion of repetitions on which an event occurs."

1. **Any probability is a number between 0 and 1, inclusively.** Any proportion is a number between 0 and 1, so any probability is also a number between 0 and 1. An event with probability 0 never occurs, and an event with probability 1 occurs on every trial. An event with probability 0.5 occurs in half the trials in the long run.
2. **All possible outcomes together must have probability 1.** Because some outcome must occur on every trial, the sum of the probabilities for all possible outcomes must be exactly 1.
3. **If two events have no outcomes in common, the probability that one or the other occurs is the sum of their individual probabilities.** If one event occurs in 40% of all trials, a different event occurs in 25% of all trials, and the two can never occur together, then one or the other occurs on 65% of all trials because 40% + 25% = 65%.
4. **The probability that an event does not occur is 1 minus the probability that the event does occur.** If an event occurs in (say) 70% of all trials, it fails to occur in the other 30%. The probability that an event occurs and the probability that it does not occur always add to 100%, or 1.

We can use mathematical notation to state Facts 1 to 4 more concisely. Capital letters near the beginning of the alphabet denote events. If A is any event, we write its probability as $P(A)$. Here are our probability facts in formal language. As you apply these rules, remember that they are just another form of intuitively true facts about long-run proportions.

PROBABILITY RULES

Rule 1. The probability $P(A)$ of any event A satisfies $0 \leq P(A) \leq 1$.

Rule 2. If S is the sample space in a probability model, then $P(S) = 1$.

Rule 3. Two events A and B are **disjoint** (mutually exclusive) if they have no outcomes in common and so can never occur together. If A and B are disjoint,

$$P(A \text{ or } B) = P(A) + P(B)$$

This is the **addition rule for disjoint events.**

Rule 4. For any event A,

$$P(A \text{ does not occur}) = 1 - P(A)$$

The addition rule extends to more than two events that are disjoint in the sense that no two have any outcomes in common. If events A, B, and C are disjoint, the probability that one of these events occurs is $P(A) + P(B) + P(C)$.

EXAMPLE 9.6 Using the probability rules

We already used the addition rule, without calling it by that name, to find the probabilities in Example 9.5. The event "one girl" contains the two disjoint outcomes displayed in Figure 9.2, so the addition rule (Rule 3) says that its probability is

$$\begin{aligned} P(\text{one girl}) &= P(\text{GB}) + P(\text{BG}) \\ &= \frac{1}{4} + \frac{1}{4} \\ &= \frac{2}{4} = 0.5 \end{aligned}$$

Check that the probabilities in Example 9.5, found using the addition rule, are all between 0 and 1 and add to exactly 1. That is, this probability model obeys Rules 1 and 2.

What is the probability that a couple with two children would not have two girls? By Rule 4,

$$\begin{aligned} P(\text{couple does not have two girls}) &= 1 - P(\text{has two girls}) \\ &= 1 - 0.25 = 0.75 \end{aligned}$$

■

APPLY YOUR KNOWLEDGE

9.6 **What's your weight?** According to the U.S. Census Bureau, 29% of American males 18 years of age or older are obese, 41% are overweight, 29% have a healthy weight, and 1% are underweight.[3]

(a) Does this assignment of probabilities to adult American males satisfy Rules 1 and 2?

(b) What percent of adult American males have a weight higher than what is considered healthy?

(c) The Census Bureau reports that the percent of adult American females with a weight higher than what is considered healthy is 56%. Does the assignment of probabilities for males and females who are over a healthy weight satisfy Rules 1 and 2?

Gregory G. Dimijian, M.D./Science Source

9.7 **Rabies in Florida.** Rabies is a viral disease of mammals transmitted through the bite of a rabid animal. The virus infects the central nervous system, causing encephalopathy and ultimately death. The Florida Department of Health reports the distribution of documented cases of rabies for all of 2011:[4]

Species	Raccoon	Bat	Fox	Other
Probability	0.57	0.15	0.11	?

(a) What probability should replace "?" in the distribution?

(b) What is the probability that a reported case of rabies is not a raccoon?

(c) What is the probability that a reported case of rabies is either a bat or a fox?

Discrete probability models

Examples 9.4, 9.5, and 9.6 illustrate one way to assign probabilities to events: Assign a probability to every individual outcome, then add these probabilities to find the probability of any event. This idea works well when there are only a finite (fixed and limited) number of outcomes.

DISCRETE PROBABILITY MODEL

A probability model with a sample space made up of a list of individual outcomes[5] is called **discrete.**

To assign probabilities in a discrete model, list the probabilities of all the individual outcomes. These probabilities must be numbers between 0 and 1 and must have sum 1. The probability of any event is the sum of the probabilities of the outcomes making up the event.

EXAMPLE 9.7 Hearing impairment in dalmatians

Pure dog breeds are often highly inbred, leading to high numbers of congenital defects. A study examined hearing impairment in 5333 dalmatians.[6] Call the number of ears impaired (deaf) in a randomly chosen dalmatian X for short. The researchers found the following probability model for X:

X	0	1	2
Probability	0.70	0.22	0.08

Check that the probabilities of the outcomes sum to exactly 1. This is therefore a legitimate discrete probability model.

The probability that a randomly chosen dalmatian has some hearing impairment is the probability that X is equal to or greater than 1:

$$P(X \geq 1) = P(X = 1) + P(X = 2)$$
$$= 0.22 + 0.08 = 0.30$$

Almost a third of dalmatians are deaf in one or both ears. This is a very high proportion that may be explained in part by the fact that breeders cannot detect partial deafness behaviorally. The study suggested giving the dogs a hearing test before considering them for breeding.

Note that the probability that X is greater than or equal to 1 is not the same as the probability that X is strictly greater than 1. The latter probability here is

$$P(X > 1) = P(X = 2) = 0.08$$

The outcome $X = 1$ is included in "greater than or equal to" and is not included in "strictly greater than." ■

APPLY YOUR KNOWLEDGE

9.8 Soda consumption. A survey by Gallup asked a random sample of American adults about their soda consumption.[7] Let's call X the number of glasses of soda consumed on a typical day. Gallup found the following probability model for X:

X	0	1	2	3	4+
Probability	0.52	0.28	0.09	0.04	0.07

age fotostock/SuperStock

Consider the events

$$A = \{\text{number of glasses of soda is 1 or greater}\}$$
$$B = \{\text{number of glasses of soda is 2 or less}\}$$

(a) What outcomes make up the event A? What is $P(A)$?

(b) What outcomes make up the event B? What is $P(B)$?

(c) What outcomes make up the event "A or B"? What is $P(A \text{ or } B)$? Why is this probability not equal to $P(A) + P(B)$?

9.9 Physically active high schoolers. The 2011 National Youth Risk Behavior Survey provides insight on the physical activity of high school students in the United States. Over 15,000 high schoolers were asked, "During the past 7 days, on how many days were you physically active for a total of at least 60 minutes per day?" Physical activity was defined as any activity that increased heart rate. Call the response X for short. The survey results give the following probability model for X:[8]

Days	0	1	2	3	4	5	6	7
Probability	0.15	0.08	0.10	0.11	0.10	0.12	0.07	0.27

(a) Verify that this is a legitimate discrete probability model.

(b) Describe the event $X < 7$ in words. What is $P(X < 7)$?

(c) Express the event "physically active at least one day" in terms of X. What is the probability of this event?

Continuous probability models

When we use the table of random digits to select a digit between 0 and 9, the discrete probability model assigns probability 1/10 to each of the 10 possible outcomes. Suppose that we want to choose a number at random between 0 and 1, allowing *any* number between 0 and 1 as the outcome. Software random number generators will do this. For example, here is the result of asking software to produce five random numbers between 0 and 1:

0.2893511 0.3213787 0.5816462 0.9787920 0.4475373

The sample space is now an entire interval of numbers:

$$S = \{\text{all numbers between 0 and 1}\}$$

Call the outcome of the random number generator Y for short. How can we assign probabilities to such events as $\{0.3 \leq Y \leq 0.7\}$? As in the case of selecting a random digit, we would like all possible outcomes to be equally likely. But we cannot assign probabilities to each individual value of Y and then add them, because there are infinitely many possible values.

We use a new way of assigning probabilities directly to events—as *areas under a density curve*. Density curves are models for continuous distributions. Any density curve has area exactly 1 underneath it, corresponding to total probability 1.

DENSITY CURVE

A **density curve** is a curve that

- is always on or above the horizontal axis, and
- has area exactly 1 underneath it.

A density curve describes the overall pattern of a distribution. The area under the curve and above any range of values on the horizontal axis is the proportion of all observations that fall in that range.

In Chapter 1 we described continuous distributions with histograms. Sometimes the overall pattern of a large number of observations is so regular that we can describe it by a smooth curve.

EXAMPLE 9.8 From histogram to density curve

Figure 9.3 is a histogram of the heights of a large government sample survey of women 40 to 49 years of age.[9] Overall, the distribution of heights is quite regular. The histogram is symmetric, and both tails fall off smoothly from a single center peak. There are no

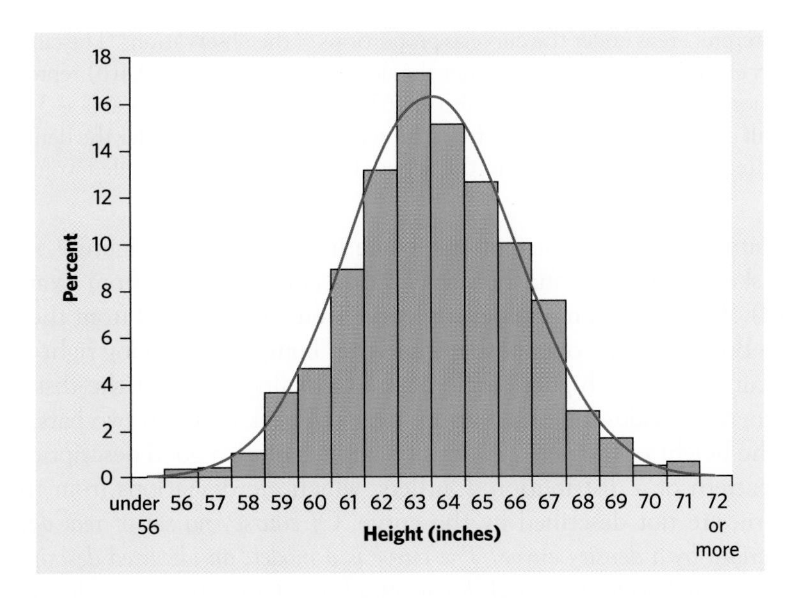

FIGURE 9.3 Histogram of the heights in inches of women aged 40 to 49 in the United States, for Example 9.8. The smooth curve shows the overall shape of the distribution.

large gaps or obvious outliers. The smooth curve drawn over the histogram is a good description of the overall pattern of the data. Our eyes respond to the *areas* of the bars in a histogram. The bar areas represent the percents of the observations. Figure 9.4(a) is a copy of Figure 9.3 with the leftmost bars shaded. The area of the shaded bars represents the women 62 inches tall or less. They make up 31.9% of all women in the sample—this is the cumulative percent, the sum of all bars for 62 inches and below.

Now look at the curve drawn through the bars. In Figure 9.4(b), the area under the curve to the left of 62 inches is shaded. The smooth curve we use to model the histogram distribution is chosen with the specific constraint that *the total area under the curve is exactly 1*. The total area represents 100%, that is, all the observations. We can

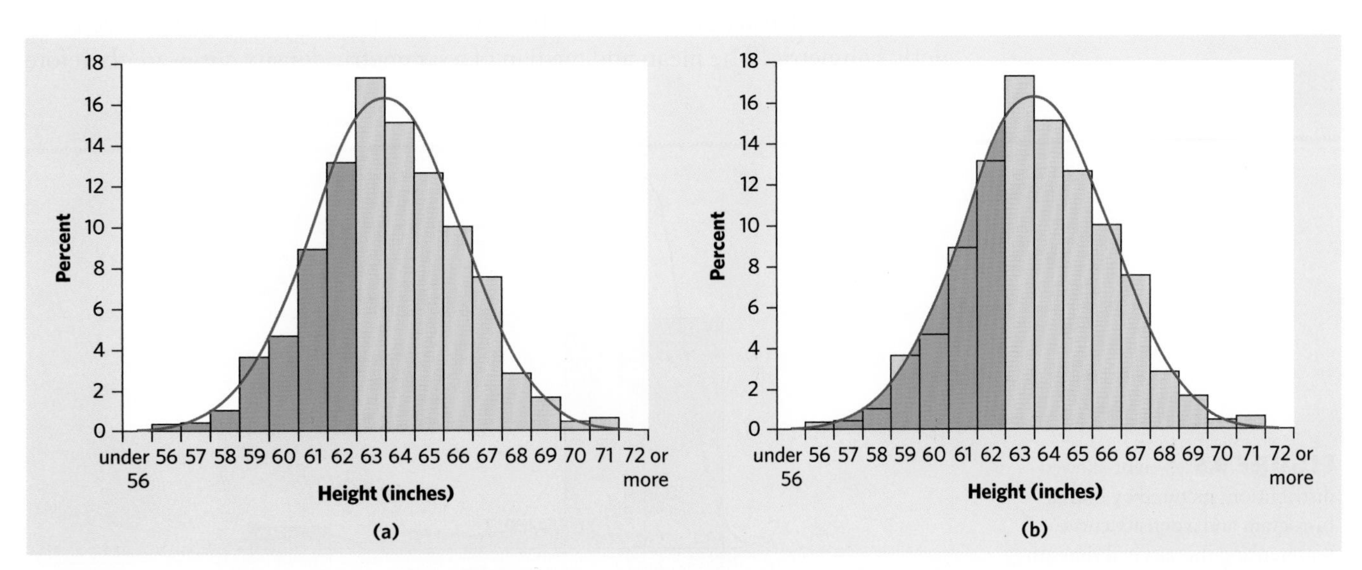

FIGURE 9.4 (a) The proportion of women 62 inches tall or less in the sample is 0.319. (b) The proportion of women 62 inches tall or less calculated from the density curve is 0.316. The density curve is a good approximation to the distribution of the data.

then interpret areas under the curve as proportions of the observations. The curve is now a *density curve*. The shaded area under the density curve in Figure 9.4(b) represents the proportion of women in their 40s who are 62 inches or shorter. This area is 31.6%, less than half a percentage point away from the actual 31.9%. Areas under the density curve give quite good approximations to the actual distribution of the sampled women. ■

Density curves, like distributions, come in many shapes. Figure 9.5 shows a strongly skewed distribution: the survival times of guinea pigs from Example 1.7 (page 18). The histogram and density curve were both created from the data by software. Both show the overall shape and the "bumps" in the long right tail. The density curve shows a higher single peak as a main feature of the distribution. The histogram divides the observations near the peak between two bars, thus reducing the height of the peak. A density curve is often a good description of the overall pattern of a distribution. Outliers, which are deviations from the overall pattern, are not described by the curve. *Of course, no set of real data is exactly described by a density curve. The curve is a model, an idealized description that is easy to use and accurate enough for practical use.* Conceptually, a density curve is similar to a regression line: We use a least-squares regression line to model an observed linear trend and to make predictions about similar individuals in the population.

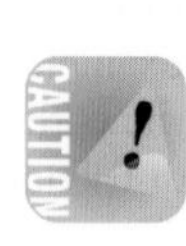

Measures of center and spread apply to density curves as well as to actual sets of observations. Areas under a density curve represent proportions of the total number of observations. The median is the point with half the observations on either side. So *the median of a density curve is the equal-areas point.* The mean of a set of observations is their arithmetic average. If we think of observations as a series of different weights strung out along a thin rod, the mean is the point at which the rod would balance. This fact is also true of density curves. *The mean is the point at which the curve would balance if made of solid material of varying weight.*

Because density curves are idealized patterns, a symmetric density curve is exactly symmetric. The mean and median of a symmetric density curve are therefore

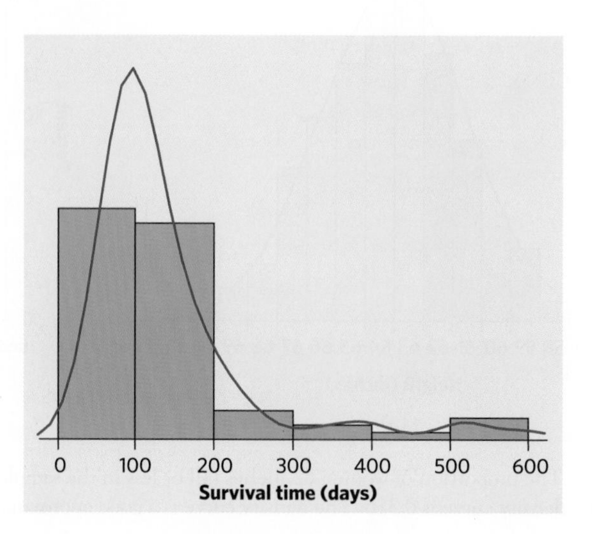

FIGURE 9.5 A right-skewed distribution, pictured by both a histogram and a density curve, representing the survival times in days of guinea pigs infected with a pathogen.

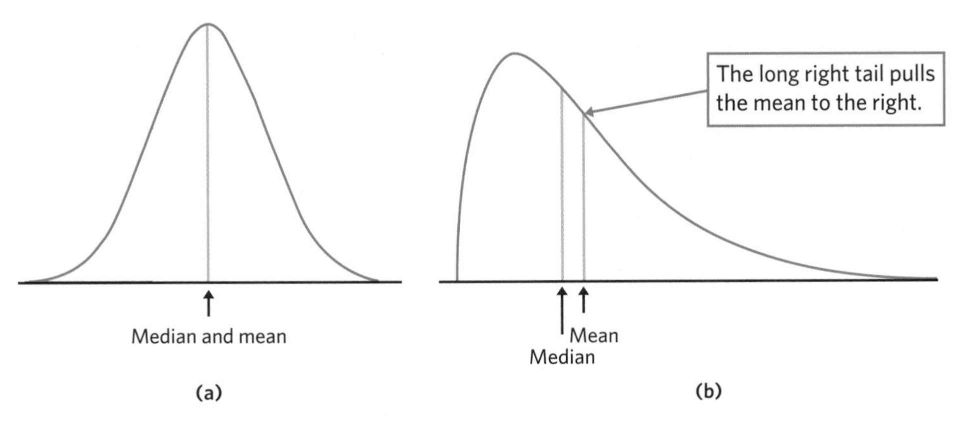

FIGURE 9.6 The mean and median of (a) a symmetric density curve and (b) a skewed density curve.

equal, as shown in Figure 9.6(a). The mean of a skewed distribution is pulled toward the long tail more than is the median, as shown in Figure 9.6(b) (finding the mean and standard deviation of a density curve is beyond the scope of this textbook). Density curves represent the whole population. Their mean and standard deviation are expressed with Greek letters to distinguish them from the mean $\overline{x}$ and standard deviation s computed from actual sample observations. The usual notation for the **mean of a density curve** is μ (the Greek letter mu). We write the **standard deviation of a density curve** as σ (the Greek letter sigma).

mean μ
standard deviation σ

How do we use density curves to assign probabilities over continuous intervals? There is a direct relationship between the representation (or proportion) of a given type of individual in a population and the probability that one individual randomly selected from the population will be of that given type. Just as with discrete probabilities, we define probabilities over continuous intervals by the relative frequency of relevant individuals in the population—in this case, all individuals from the population that belong to the desired interval.

EXAMPLE 9.9 From population distribution to probability distribution

In Example 9.8 we used a density curve to estimate that the proportion of women in their 40s who are 62 inches or shorter is 31.6%. This is the area under the density curve for heights of 62 inches or less, as shown in Figure 9.4(b).

Let's now ask: What is the probability that a randomly chosen woman in her 40s has a height of 62 inches or less? Because the selection is random, this probability depends on the relative frequency of women in the population who are 62 inches tall or less, and that relative frequency is 31.6%. Therefore, the probability that a randomly selected woman in her 40s would measure 62 inches or less is 0.316. This is the area under the density curve for heights 62 inches or less. ■

Blend Images/Getty Images

CONTINUOUS PROBABILITY MODEL

A **continuous probability model** assigns probabilities as areas under a density curve. The area under the curve and above any range of values on the horizontal axis is the probability of an outcome in that range.

EXAMPLE 9.10 Random numbers

The random number generator will spread its output uniformly across the entire interval from 0 to 1 as we allow it to generate a long sequence of numbers. The results of many trials are represented by the uniform density curve shown in Figure 9.7. This density curve has height 1 over the interval from 0 to 1. The area under the curve is 1, and the probability of any event is the area under the curve and above the event in question.

As Figure 9.7(a) illustrates, the probability that the random number generator produces a number between 0.3 and 0.7 is

$$P(0.3 \leq Y \leq 0.7) = 0.4$$

because the area under the density curve and above the interval from 0.3 to 0.7 is 0.4. The height of the curve is 1 and the area of a rectangle is the product of height and length, so the probability of any interval of outcomes is just the length of the interval.

Similarly,

$$P(Y \leq 0.5) = 0.5$$
$$P(Y > 0.8) = 0.2$$
$$P(Y \leq 0.5 \text{ or } Y > 0.8) = 0.7$$

The last event consists of two nonoverlapping intervals, so the total area above the event is found by adding two areas, as illustrated by Figure 9.7(b). This assignment of probabilities obeys all our rules for probability. ■

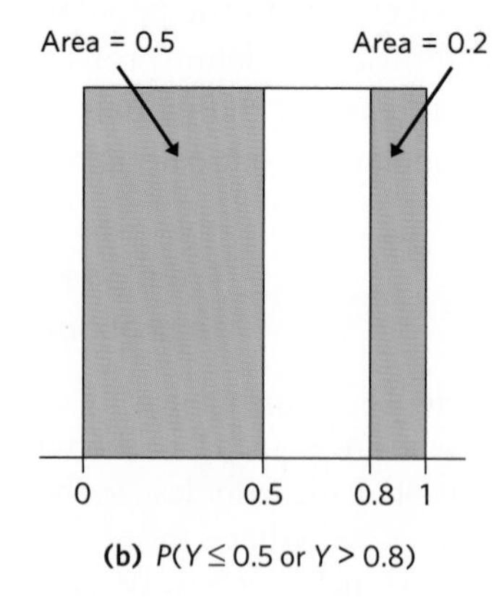

FIGURE 9.7 Probability as area under a density curve, for Example 9.10. This uniform density curve spreads probability evenly between 0 and 1.

The probability model for a continuous random variable assigns probabilities to intervals of outcomes rather than to individual outcomes. In fact, *all continuous probability models assign probability 0 to any individual outcome*. *Only intervals of values have positive (non-zero) probabilities*. To see that this is true, consider a specific outcome such as $P(Y = 0.8)$ in Example 9.10. The probability of any interval is the same as its length. The point 0.8 has no length, so its probability is 0.

We can use any density curve to assign probabilities. In Chapter 11 we will examine *Normal curves*, a particularly useful family of density curves in probability and statistics.

APPLY YOUR KNOWLEDGE

9.10 Sketch density curves. Sketch density curves that describe distributions with the following shapes:

(a) symmetric, but with two peaks (that is, two strong clusters of observations)

(b) single peak and skewed to the left

9.11 Random numbers. Let X be a random number between 0 and 1 produced by the random number generator described in Example 9.10 and Figure 9.7. Find the following probabilities:

(a) $P(X \leq 0.4)$

(b) $P(X < 0.4)$

(c) $P(0.3 \leq X \leq 0.5)$

(d) $P(X < 0.3 \text{ or } X > 0.5)$

9.12 TV viewing among high school students. Some argue that TV viewing promotes obesity because of physical inactivity and high exposure to commercials for foods with high fat and sugar content. In the two examples provided here, decide whether the probability model is discrete or continuous. Explain your reasoning.

(a) According to the National Center for Health Statistics, there is a 38% chance that a high school student watches at least 3 hours of television on an average school day.[10] This nationwide government survey recorded the amount of time high school students watch TV on an average school day.

(b) A large sample survey of school-age children by the Kaiser Family Foundation found a 42% chance that the TV is on most of the time at home.[11] The survey asked whether, at home, the TV was on most of the time, some of the time, a little bit of the time, or never.

Random variables

Examples 9.7 and 9.10 use a notation that is often convenient. It is especially useful when using mathematical functions to describe probability distributions, as we will see in Chapters 11 and 12. In Example 9.7 we let X stand for the result of choosing a dalmatian at random and assessing its hearing impairment. We know that X may take a different value if we make another random choice. Because its value changes from one random choice to another, we call the hearing impairment X a *random variable*.

RANDOM VARIABLE

A **random variable** is a variable whose value is a numerical outcome of a random phenomenon.

The **probability distribution** of a random variable X tells us what values X can take and how to assign probabilities to those values.

There are two main types of random variables, corresponding to two types of probability models: *discrete* and *continuous*.

EXAMPLE 9.11 Discrete and continuous random variables

The hearing impairment X in Example 9.7 is a random variable whose possible values are the whole numbers {0, 1, 2}. The distribution of X assigns a probability to each of these outcomes. Random variables that have a countable (typically finite) list of possible outcomes are called **discrete.**

discrete random variable

Compare this with the value Y obtained with the random number generator in Example 9.10. The values of Y fill the entire interval of numbers between 0 and 1. The probability distribution of Y is given by its density curve, shown in Figure 9.7. Random variables that can take on any value in an interval, with probabilities given as areas under a density curve, are called **continuous.** ■

continuous random variable

Be sure to consider closely the true nature of your random variable. Example 9.8 described the process of assigning a density curve to model women's heights in inches. While most people report their heights in whole numbers of inches, they do not grow one inch at a time. All possible values of height within a realistic interval can actually exist, for example, 65.125 inches. So the random variable height is truly a continuous random variable.

APPLY YOUR KNOWLEDGE

9.13 Discrete or continuous random variable? Indicate in the following examples whether the random variable X is discrete or continuous. Explain your reasoning.

(a) X is the number of days last week that a randomly chosen child exercised for at least one hour.

(b) X is the amount of time in hours that a randomly chosen child spends watching television today.

(c) X is the number of television sets in a randomly chosen household.

9.14 Discrete or continuous random variable? Indicate in the following examples whether the random variable X is discrete or continuous. Explain your reasoning.

(a) X is the number of petals on a randomly chosen daisy.

(b) X is the stem length in centimeters of a randomly chosen daisy.

(c) X is the number of daisies found in a randomly chosen grassy area 1 square meter in size.

(d) X is the average number of petals per daisy computed from all the daisies found in a randomly chosen grassy area 1 square meter in size.

Personal probability*

We began our discussion of probability with one idea: The probability of an outcome of a random phenomenon is the proportion of times that outcome would occur in a very long series of repetitions. This idea ties probability to actual outcomes. It allows us, for example, to estimate probabilities by simulating random phenomena. Yet we often meet another, quite different, idea of probability.

*This section is optional.

EXAMPLE 9.12 Intelligent life and the universe

Joe reads an article discussing the Search for Extraterrestrial Intelligence (SETI) project. We ask Joe, "What's the chance that we will find evidence of extraterrestrial intelligence in this century?" Joe responds, "Oh, about 1%."

Does Joe assign probability 0.01 to humans finding extraterrestrial intelligence this century? The outcome of our search is certainly unpredictable, but we can't reasonably ask what would happen in many repetitions. This century will happen only once and will differ from all other centuries in many ways, especially in terms of technology. If probability measures "what would happen if we did this many times," Joe's 0.01 is not a probability. The frequentist definition of probability is based on data from many repetitions of the same random phenomenon. Joe is giving us something else, his personal judgment. ■

Although Joe's 0.01 isn't a probability in the frequentist sense, it gives useful information about Joe's opinion. Closer to home, a government asking, "How likely is it that building a new nuclear power plant will pay off within five years?" can't employ an idea of probability based on many repetitions of the same thing. The opinions of science and business advisers are nonetheless useful information, and these opinions can be expressed in the language of probability. These are *personal probabilities*.

PERSONAL PROBABILITY

A **personal probability** of an outcome is a number between 0 and 1 that expresses an individual's judgment of how likely the outcome is.

Rachel's opinion about finding extraterrestrial intelligence may differ from Joe's, and the opinions of several advisers about the new power plant may differ. Personal probabilities are indeed personal: They vary from person to person. Moreover, a personal probability can't be called right or wrong. If we say, "In the long run, this coin will come up heads 60% of the time," we can find out if we are right by actually tossing the coin several thousand times. If Joe says, "I think there is a 1% chance of finding extraterrestrial intelligence this century," that's just Joe's opinion.

Why think of personal probabilities as probabilities? Because any set of personal probabilities that makes sense obeys the same basic Rules 1 to 4 that describe any legitimate assignment of probabilities to events. If Joe thinks there's a 1% chance that we find extraterrestrial intelligence this century, he must also think that there's a 99% chance that we won't. There is just one set of rules of probability, even though we now have two interpretations of what probability means.

APPLY YOUR KNOWLEDGE

9.15 Will you have an accident? The probability that a randomly chosen adult will be involved in a car accident in the next year is about 0.2. This is based on the proportion of millions of drivers who have accidents. "Accident" includes things like crumpling a fender in your own driveway, not just highway accidents.

(a) What do you think is your own probability of being in an accident in the next year? This is a personal probability.

(b) Give some reasons why your personal probability might be a more accurate prediction of your "true chance" of having an accident than the probability for a random driver.

(c) Almost everyone says that their personal probability is lower than the random driver probability. Why do you think this is true?

Risk and odds*

Random events can be described in a variety of ways. The most common descriptors of random events in the life sciences are probability, risk, and odds. So far in this chapter we have introduced the idea of probability and fundamental probability rules. We now briefly describe what risk and odds are and how they relate to the notion of probability. Chapter 20 will further discuss the applications of risk and odds in clinical research when comparing two groups.

risk

Risk means different things in different fields. In statistics, **risk** corresponds to the probability of an undesirable event such as death, disease, or side effects. This term is particularly common in the health sciences, which aim to assess risk and find ways to reduce it. The risk of a given adverse event, like its probability, is defined by the frequency of that adverse event in a population or sample of interest.

odds

Odds are a somewhat less intuitive concept, with a foundation in gambling (although mathematical odds are not to be confused with betting odds, which typically reflect payoffs offered by bookies to winning bets). An **odds** is a ratio of two probabilities where the numerator represents the probability of an event and the denominator represents the complementary probability of that event not occurring. Therefore, odds can take any positive value, including values greater than 1. The odds of an event can be expressed as the numerical value of the ratio or as a ratio of two integers with no common denominator.

> **RISK AND ODDS**
>
> The **risk** of an *undesirable* outcome of a random phenomenon is the probability of that undesirable outcome.
>
> The **odds** of any outcome of a random phenomenon is the ratio of the probability of that outcome over the probability of that outcome not occurring.
>
> That is, if an outcome A has probability p of occurring, then
>
> $$\text{risk}(A) = p$$
> $$\text{odds}(A) = p/(1-p)$$

---

*This section is optional.

EXAMPLE 9.13 Blood clots in immobilized patients

Patients immobilized for a substantial amount of time can develop deep vein thrombosis (DVT), a blood clot in a leg or pelvis vein. DVT can have serious adverse health effects and can be difficult to diagnose. On its website, drug manufacturer Pfizer reports the outcome of a study looking at the effectiveness of the drug Fragmin (dalteparin) in preventing DVT in immobilized patients. Of the 1518 randomly chosen immobilized patients given Fragmin, 42 experienced a complication from DVT (the remaining 1476 patients did not).[12]

The proportion of patients experiencing DVT complications is $42/1518 = 0.0277$, or 2.77%. We can use this information to compute the risk and odds of experiencing DVT complications for immobilized patients treated with Fragmin:

$$\text{risk} = 0.0277, \text{ or } 2.77\%$$

$$\text{odds} = 0.0277/(1 - 0.0277) = 42/1476 = 0.0285$$

The odds of experiencing DVT complications among immobilized patients given Fragmin are 42:1476, or about 1:35. That is, for every such patient experiencing a DVT complication, about 35 do not experience a DVT complication. ■

The numerical values for the risk and odds in this example are very close, 0.277 and 0.0285. In general, when the sample size is very large and the undesirable event not very frequent, risk and odds give similar numerical values. In other situations, risk and odds can be very different.

EXAMPLE 9.14 Sickle-cell anemia

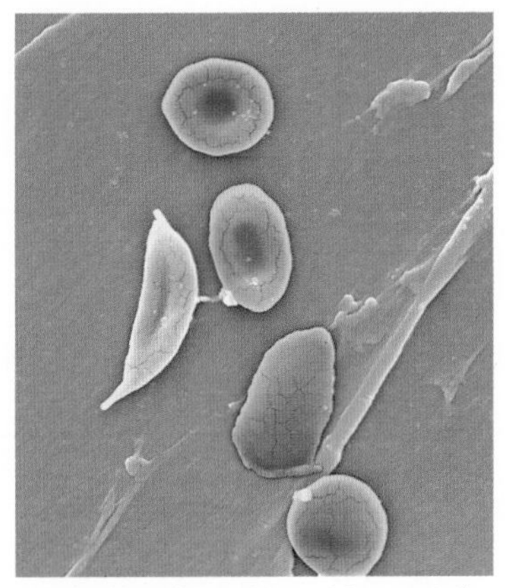

CDC/Sickle Cell Foundation of Georgia: Jackie George, Beverly Sinclair

Sickle-cell anemia is a serious, inherited blood disease affecting the shape of red blood cells. Individuals with both genes causing the defect suffer pain from blocked arteries and can have their life shortened from organ damage. Individuals carrying only one copy of the defective gene ("sickle-cell trait") are generally healthy but may pass on the gene to their offspring. An estimated two million Americans carry the sickle-cell trait.

If a couple learns from blood tests that they both carry the sickle-cell trait, genetic laws of inheritance tell us that there is a 25% chance that they could conceive a child suffering from sickle-cell anemia. That is, the risk of conceiving a child who will suffer from sickle-cell anemia is 0.25, or 25%. The odds of this are

$$\text{odds} = 0.25/(1 - 0.25) = 0.333, \text{ or } 1:3$$

■

In this second example, the risk and the odds of conceiving a child suffering from sickle-cell anemia have quite different numerical values, 0.25 and 0.33. Always make sure that you understand how risk and odds relate to probability when you read reports about these concepts.

APPLY YOUR KNOWLEDGE

9.16 Blood clots in immobilized patients, continued. The Fragmin study from Example 9.13 compared patients treated with Fragmin with patients given a placebo in a randomized, double-blind design. Of the 1473 immobilized patients given a placebo, 73 experienced a complication from DVT.

(a) Compute the proportion of patients given a placebo who experienced a complication from DVT. What are the risk and the odds of experiencing a complication from DVT when an immobilized patient is given a placebo? How do these values compare?

(b) Compare your results with those of Example 9.13. What do you conclude? We will see in Chapter 20 how to formally compare risks or odds for two groups.

9.17 **HPV infections in women.** Human papillomavirus (HPV) infection is the most common sexually transmitted infection. Certain types of HPV can cause genital warts in both men and women and cervical cancer in women. The U.S. National Health and Nutrition Examination Survey (NHANES) contacted a representative sample of 1921 women between the ages of 14 and 59 years and asked them to provide a self-collected vaginal swab specimen. Of these, 515 tested positive for HPV, indicating a current HPV infection.[13]

(a) Give the probability, risk, and odds that a randomly selected American woman between the ages of 14 and 59 years has a current HPV infection.

(b) The survey broke down the data by age group:

Age group (years)	14–19	20–24	25–29	30–39	40–49	50–59
Percent HPV positive	24.5	44.8	27.4	27.5	25.2	19.6

Give the risk and the odds of being HPV positive for women in each age group. Which age group is the most at risk (has the highest odds) of testing HPV positive?

CHAPTER 9 SUMMARY

- A **random phenomenon** has outcomes that we cannot predict but nonetheless has a regular distribution of outcomes in very many repetitions.
- The **probability** of an event is the proportion of times the event occurs in many repeated trials of a random phenomenon.
- A **probability model** for a random phenomenon consists of a sample space S and an assignment of probabilities P.
- The **sample space** S is the set of all possible outcomes of the random phenomenon. Sets of outcomes are called **events.** P assigns a number $P(A)$ to an event A as its probability.
- Any assignment of probability must obey the rules that state the basic properties of probability:

  1. $0 \leq P(A) \leq 1$ for any event A.
  2. $P(S) = 1$.
  3. **Addition rule:** Events A and B are **disjoint** (or mutually exclusive) if they have no outcomes in common. If A and B are disjoint, then $P(A \text{ or } B) = P(A) + P(B)$.
  4. For any event A, $P(A \text{ does not occur}) = 1 - P(A)$.

- When a sample space S contains a finite number of possible values, a **discrete probability model** assigns each of these values a probability between 0 and 1 such that the sum of all the probabilities is exactly 1. The probability of any event is the sum of the probabilities of all the values that make up the event.
- A sample space can contain all values in some interval of numbers. A **continuous probability model** assigns probabilities as areas under a density curve. A **density curve** has total area 1 underneath it. The probability of any event is the area under the curve and above the values on the horizontal axis that make up the event.
- A density curve is an idealized description of the overall pattern of a distribution that smooths out the irregularities in the actual data. We write the **mean of a density curve** as μ and the **standard deviation of a density curve** as σ to distinguish them from the mean $\bar{x}$ and standard deviation s of actual sample data.
- A **random variable** is a variable taking numerical values determined by the outcome of a random phenomenon. The **probability distribution** of a random variable X tells us what the possible values of X are and how probabilities are assigned to those values.
- A random variable X and its distribution can be **discrete** or **continuous.** A **discrete random variable** has distinct values that can be listed. Its distribution gives the probability of each value. A **continuous random variable** takes all values in some numerical interval. A density curve describes the probability distribution of a continuous random variable.
- **Personal probabilities** also follow the rules of probability, although they represent an individual's judgment of how likely an outcome is rather than its frequency in a long run of trials.
- A probability can also be described as a **risk** when it applies to an undesirable outcome. An **odds** is the ratio of the probability of an outcome over the probability of that outcome not occurring.

THIS CHAPTER IN CONTEXT

This chapter begins our study of probability. The important fact is that random phenomena are unpredictable in the short run but have a regular and predictable behavior in the long run. Probability rules and probability models provide the tools for describing and predicting the long-run behavior of random phenomena.

Chapter 10 will allow us to explore probability rules in greater depth, including the use of conditional probabilities for medical screening. In Chapters 11 and 12 we will study some important random variables with a wide range of practical and theoretical applications: Normal, binomial, and Poisson.

Probability helps us understand why we can trust random samples and randomized comparative experiments, the subjects of Chapters 7 and 8. It is the key to generalizing what we learn from data produced by random samples and randomized comparative experiments to some wider universe or population. Chapter 13 will focus specifically on the probability properties linking random samples and populations. This will be the founding principle for statistical inference, which we will address in Chapters 14 and on.

CHECK YOUR SKILLS

9.18 You read that in native Hawaiians, the probability of having blood type AB is 1/100 (1 in 100). What does this mean?

(a) If you pick 100 Hawaiians randomly, the fraction of them having blood type AB will be very close to 1/100.
(b) If you pick 100 Hawaiians randomly, exactly 1 of them will have blood type AB.
(c) If you pick 10,000 Hawaiians randomly, exactly 100 of them will have blood type AB.

9.19 A cat is about to have 6 kittens. The sample space for counting the number of female kittens she has is

(a) S = any number between 0 and 1.
(b) S = whole numbers 0 to 6.
(c) S = all sequences of 6 males or females by order of birth, such as FMMFFF.

Here is the probability model for the blood type of a randomly chosen person of Hispanic ethnicity in the United States, according to the Red Cross. Exercises 9.20 to 9.22 use this information.

Blood type	O	A	B	AB
Probability	0.57	0.31	0.10	?

9.20 This probability model is

(a) continuous.
(b) discrete.
(c) discretely continuous.

9.21 The probability that a randomly chosen American of Hispanic ethnicity has type AB blood must be

(a) any number between 0 and 1.
(b) 0.02.
(c) 0.2.

9.22 Individuals with type B blood can safely receive blood transfusions from people with blood types O and B. What is the probability that a randomly chosen American of Hispanic ethnicity can donate blood to someone with type B blood?

(a) 0.10
(b) 0.57
(c) 0.67

9.23 According to the *National Vital Statistics Reports*, in all of 2010, 1.0% of live vaginal births were delivered using forceps and 4.4% were delivered using vacuum extraction. The probability that a live baby was delivered vaginally without the use of either forceps or vacuum extraction is

(a) 0.054.
(b) 0.550.
(c) 0.946.

9.24 In a table of random digits such as Table A, each digit is equally likely to be any of 0, 1, 2, 3, 4, 5, 6, 7, 8, or 9. What is the probability that a digit in the table is a 0?

(a) 1/9
(b) 1/10
(c) 9/10

9.25 In a table of random digits such as Table A, each digit is equally likely to be any of 0, 1, 2, 3, 4, 5, 6, 7, 8, or 9. What is the probability that a digit in the table is 7 or greater?

(a) 7/10
(b) 4/10
(c) 3/10

9.26 A study of freely forming groups in bars throughout Europe examined the number of individuals found in groups that were laughing together. Let X be the number of individual in laughing groups. Here is the probability model for X:[14]

Number of individuals X	2	3	4	5	6
Probability	0.51	0.34	0.10	0.04	0.01

What percent of laughing groups had more than 4 individuals?

(a) 5%
(b) 10%
(c) 15%

9.27 In the previous exercise, if we picked a laughing group at random, the probability that X is equal to 3 or less is

(a) 0.34.
(b) 0.51.
(c) 0.85.

CHAPTER 9 EXERCISES

9.28 Taking your pulse. The throbbing of your arteries is a consequence of the heartbeat. A healthy resting pulse rate for an adult ranges from 60 to 90 beats per minute (BPM). While at rest, take your pulse over 30 seconds (for example, at the neck or at the wrist) and record whether your pulse was higher than 75 BPM or not. Repeat the experiment until you have 20 measurements. How many times did you get a pulse above 75 BPM? What is the approximate probability of a pulse above 75 BPM?

9.29 Manatee deaths. Manatees are an endangered species of herbivorous, aquatic mammals found primarily in the rivers and estuaries of Florida. As part of its conservation efforts, the Florida Fish and Wildlife Commission records the cause of death for every recovered manatee carcass in its waters. Of the 392 recorded manatee deaths in 2012, 68 were perinatal and 81 were caused by collision with a watercraft.[15]

(a) What is the probability that the death of a randomly selected manatee was due to collision with a watercraft?

(b) What is the probability that the death was not due to collision with a watercraft?

(c) What is the probability that the cause of death was due to perinatal problems or collision with a watercraft? What is the probability that it was due to some other cause?

9.30 Sample space. A randomly chosen subject arrives for a study of exercise and fitness. Describe a sample space for each of the following. (In some cases, you may have some freedom in your choice of S.)

(a) You record the gender of the subject.

(b) After 10 minutes on an exercise bicycle, you ask the subject to rate his or her effort on the Rate of Perceived Exertion (RPE) scale. RPE ranges in whole-number steps from 6 (no exertion at all) to 20 (maximal exertion).

(c) You measure VO_2, the maximum volume of oxygen consumed per minute during exercise. VO_2 is generally between 2.5 and 6 liters per minute.

(d) You measure the maximum heart rate (beats per minute).

9.31 More sample spaces. In each of the following situations, describe a sample space S for the random phenomenon. In some cases, you have some freedom in your choice of S.

(a) A seed is planted in the ground. It either germinates or fails to grow.

(b) A patient with a usually fatal form of cancer is given a new treatment. The response variable is the length of time that the patient lives after treatment.

(c) A nutrition researcher feeds a new diet to a young male white rat. The response variable is the weight (in grams) that the rat gains in eight weeks.

(d) Buy a hot dog and record how many calories it has.

9.32 Probability models? In each of the following situations, state whether or not the given assignment of probabilities to individual outcomes is legitimate, that is, satisfies the rules of probability. If not, give specific reasons for your answer.

(a) Among Navajo Native Americans, the distribution of blood type is[16] $P(\text{A}) = 0.27$, $P(\text{B}) = 0.00$, $P(\text{AB}) = 0.00$, $P(\text{O}) = 0.73$.

(b) Choose a cat at random from an animal shelter and record sex and hair type:
$P(\text{female short-haired}) = 0.56$,
$P(\text{female long-haired}) = 0.24$,
$P(\text{male short-haired}) = 0.44$,
$P(\text{male long-haired}) = 0.16$.

(c) A hospital tests every newborn for phenylketonuria (PKU). PKU is one of the most common genetic disorders, easily treated but resulting in brain damage if undiagnosed. The incidence of PKU varies all over the world, as follows:[17] $P(\text{U.S.}) = 1/10,000$, $P(\text{Ireland}) = 1/4,500$, $P(\text{Finland}) = 1/100,000$, $P(\text{rest of the world}) = 1/15,000$.

9.33 Causes of death among young people. Government data assign a single cause for each death that occurs in the United States. The data show that, among persons aged 15 to 24 years, the probability is 0.41 that a randomly chosen death was an accident, 0.16 that it was a homicide, and 0.15 that it was a suicide.

(a) What is the sample space for the probability model of major causes of death in this age group based on the information you are given?

(b) What is the probability that a death was either an accident, a homicide, or a suicide? What is the probability that the death was due to some other cause?

9.34 Osteoporosis. Choose a postmenopausal woman at random. The probability is 0.40 that the woman chosen has osteopenia (low bone density) and 0.07 that she has osteoporosis (pathologically low bone density). Otherwise, her bone density is considered healthy.[18]

(a) What must be the probability that a randomly chosen postmenopausal woman has a healthy bone density?

(b) What is the probability that a randomly chosen postmenopausal woman has some problem with low bone density (osteopenia or osteoporosis)?

9.35 Hepatitis C. The CDC provides the breakdown of the sources of infection leading to hepatitis C in Americans. Here are the probabilities of each infection source for a randomly chosen individual with hepatitis C:[19]

Source of infection	Probability
Intravenous drug use	0.60
Unprotected sex	0.15
Transfusion (before screening)	0.10
Unknown or other	0.11
Occupational	?

(a) What is the probability that a person with hepatitis C was infected in the course of his or her professional occupation?

(b) What is the probability that a person with hepatitis C was infected through a known risky behavior (intravenous drugs or unprotected sex)?

9.36 Birth order. The National Vital Statistics database reports all births in 2011 by the reported birth order of the child (first child of the mother, second, etc.). Here is how the 3,923,381 documented births break down:

Birth order	Percent
First child	40.2
Second child	31.6
Third child	16.5
Fourth child and over	11.7

(a) What is the probability that a randomly selected baby born in 2011 wasn't a woman's first child?

(b) What is the probability that a randomly selected woman giving birth in 2011 had given birth to exactly 1 child before? Think carefully about the birth order of her latest baby.

(c) What is the probability that a randomly selected woman giving birth in 2011 had given birth to fewer than 2 children before?

9.37 Taste buds. Two wine tasters rate each wine they taste on a scale of 1 to 5. From data on their ratings of a large number of wines, we obtain the following probabilities for both tasters' ratings of a randomly chosen wine:

	Taster B				
Taster A	1	2	3	4	5
1	0.03	0.02	0.01	0.00	0.00
2	0.02	0.08	0.05	0.02	0.01
3	0.01	0.05	0.25	0.05	0.01
4	0.00	0.02	0.05	0.20	0.02
5	0.00	0.01	0.01	0.02	0.06

(a) Why is this a legitimate assignment of probabilities to outcomes?

(b) What is the probability that the tasters agree when rating a wine?

(c) What is the probability that Taster A rates a wine higher than 3? What is the probability that Taster B rates a wine higher than 3?

9.38 Overweight? Choose an American adult at random. Define two events:

$A =$ the person chosen is obese

$B =$ the person chosen is overweight but not obese

According to the 2011 National Health Interview Survey, $P(A) = 0.28$ and $P(B) = 0.35$.

(a) Explain why events A and B are disjoint.

(b) Say in plain language what the event "A or B" is. What is $P(A \text{ or } B)$?

(c) If C is the event that the person chosen has normal weight or less, what is $P(C)$?

9.39 Dentist appointment. A large governmental survey of dental care among children 2 to 17 years old found the following distribution of times since the last visit to the dentist:[20]

Time since last visit	Probability
6 months or less	0.57
More than 6 months but no more than 1 year	0.18
More than 1 year but no more than 2 years	0.08
More than 2 years but no more than 5 years	0.03
More than 5 years	?

(a) What must be the probability that a child has not seen a dentist in more than 5 years?

(b) What is the probability that a randomly chosen child has not seen a dentist within the last 6 months?

(c) What is the probability that the child has seen a dentist within the last year?

9.40 Eye color in hawks. An observational study examined iris coloration in a population of North American Cooper's hawks. Iris colors range from yellow to red and are numerically coded to reflect pigmentation intensity.[21] Among female hawks, the coded eye color X has the following distribution:

Eye color X	Yellow 1	Light orange 2	Orange 3	Dark orange 4	Red 5
Probability	0.05	0.29	0.47	0.18	0.01

(a) As defined here, is the random variable X discrete or continuous? Why?

(b) Write the event "eye color at least 4" in terms of X. What is the probability of this event?

(c) Describe the event $X \leq 2$ in words. What is its probability? What is the probability that $X < 2$?

9.41 Eye color in hawks again. The study in the previous exercise also gave the distribution of eye color X among male hawks:

Eye color X	Yellow 1	Light orange 2	Orange 3	Dark orange 4	Red 5
Probability	0.00	0.11	0.40	0.32	0.17

(a) Write the event "eye color less than 2 or more than 4" in terms of X. What is the probability of this event?

(b) Describe the event $2 < X \leq 4$ in words. What is its probability?

9.42 Understanding density curves. Remember that it is areas under a density curve, not the height of the curve, that give proportions in a distribution. To illustrate this, sketch a density curve that has its peak at 0 on the horizontal axis but has greater area within 0.25 on either side of 1 than within 0.25 on either side of 0.

9.43 Random numbers. Many random number generators allow users to specify the range of the random numbers to be produced. Suppose that you specify that the random number Y can take any value between 0 and 2. Then the density curve of the outcomes has constant height between 0 and 2, and height 0 elsewhere.

(a) Is the random variable Y discrete or continuous? Why?

(b) What is the height of the density curve between 0 and 2? Draw a graph of the density curve.

(c) Use your graph from (b) and the fact that probability is area under the curve to find $P(Y \leq 1)$.

9.44 More random numbers. Find these probabilities as areas under the density curve you sketched in the previous exercise.

(a) $P(0.5 < Y < 1.3)$

(b) $P(Y \geq 0.8)$

9.45 Anorexia. Exercise 1.8 (page 20) introduced a Canadian study of the age at onset of anorexia nervosa—an eating disorder—for 691 diagnosed adolescent girls. Figure 9.8 shows the density curve (solid line) summarizing the histogram of the data (dotted lines). The density curve has been chopped into three parts shown in different shades.

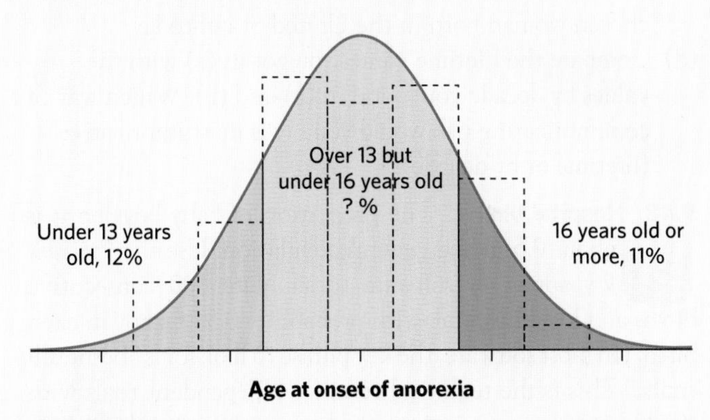

FIGURE 9.8 Distribution of age of onset of anorexia nervosa, pictured by both a histogram and a density curve, for Exercise 9.45.

(a) If we call X the age at onset of anorexia, describe the event $13 \leq X < 16$ in words. What is its probability?

(b) What is the probability that the age at onset for a randomly chosen adolescent girl is sometime after the 13th birthday? Write this event in terms of X.

9.46 Osteoporosis, continued (optional). Go back to Exercise 9.34.

(a) What are the risk and the odds of osteopenia in postmenopausal women?

(b) What are the risk and the odds of osteoporosis in postmenopausal women?

9.47 Breast cancer (optional). The National Cancer Institute (NCI) compiles U.S. epidemiology data for a number of different cancers (the Surveillance, Epidemiology, and End Results Program, or SEER). Among women, breast cancer is one of the most common and deadliest types of cancer.

(a) Based on SEER data, the NCI estimates that a woman in her 30s has a 0.43% chance of being diagnosed with breast cancer.[22] Compute the risk and the odds of being diagnosed with breast cancer for a randomly chosen woman in her 30s.

(b) The risk of breast cancer increases with age. The NCI estimates that a woman in her 60s has a 3.65% chance of being diagnosed with breast cancer. Compute the risk and the odds of being diagnosed with breast cancer for a randomly chosen woman in her 60s.

(c) The NCI also reports a lifetime risk of breast cancer. This is the probability that a woman born in the United States today will develop breast cancer at some time in her life, based on current rates of breast cancer. The most recent estimate is 12.7%. Give the probability, risk, and odds of developing breast cancer over a lifetime for a randomly chosen woman born in the United States today.

(d) Compare the lifetime values you got in (c) with the values by decade you found in (a) and (b). Which way of communicating risk would you say is most informative (lifetime or by decade)?

9.48 Hospital births. APPLET The proportion of baby boys born is about 0.5 in the general population. Use the *Probability* applet or software to simulate 100 consecutive births at a local hospital, with probability 0.5 of a boy in each birth. (In most software, the key phrase to look for is "Bernoulli trials." This is the technical term for independent trials with Yes/No outcomes. Our outcomes here are "Boy" and "Girl.")

(a) What percent of the 100 births are boys?

(b) Examine the sequence of boys and girls. How long was the longest run of boys? Of girls? (Sequences of random outcomes often show runs longer than our intuition thinks likely.)

9.49 Simulating a survey. APPLET The culinary herb cilantro is very polarizing: Some people love it and others hate it. A large survey of American adults of European ancestry found that 26% disliked the taste of cilantro.[23] Suppose that this holds exactly true for all American adults of European ancestry. Choosing such an individual at random then has probability 0.26 of getting one who dislikes cilantro. Use the *Probability* applet or your statistical software to simulate choosing many people at random. (In most software, the key phrase to look for is "Bernoulli trials." This is the technical term for independent trials with Yes/No outcomes. Our outcomes here are "dislikes cilantro" or not.)

(a) Simulate drawing 25 people, then 100 people, then 400 people (click "TOSS" twice with 200 selected). What proportion, in each simulated sample, dislike cilantro? We expect (but because of chance variation we can't be sure) that the proportion will be closer to 0.26 in larger samples.

(b) Simulate drawing 25 people 10 times (click "RESET" after each simulation) and record the percents in each simulated sample who dislike cilantro. Then simulate drawing 400 people 10 times and again record the 10 percents. Which set of 10 results is less variable? We expect the results of 400 drawings to be more predictable (less variable) than the results of 25 drawings. That is "long-run regularity" showing itself.

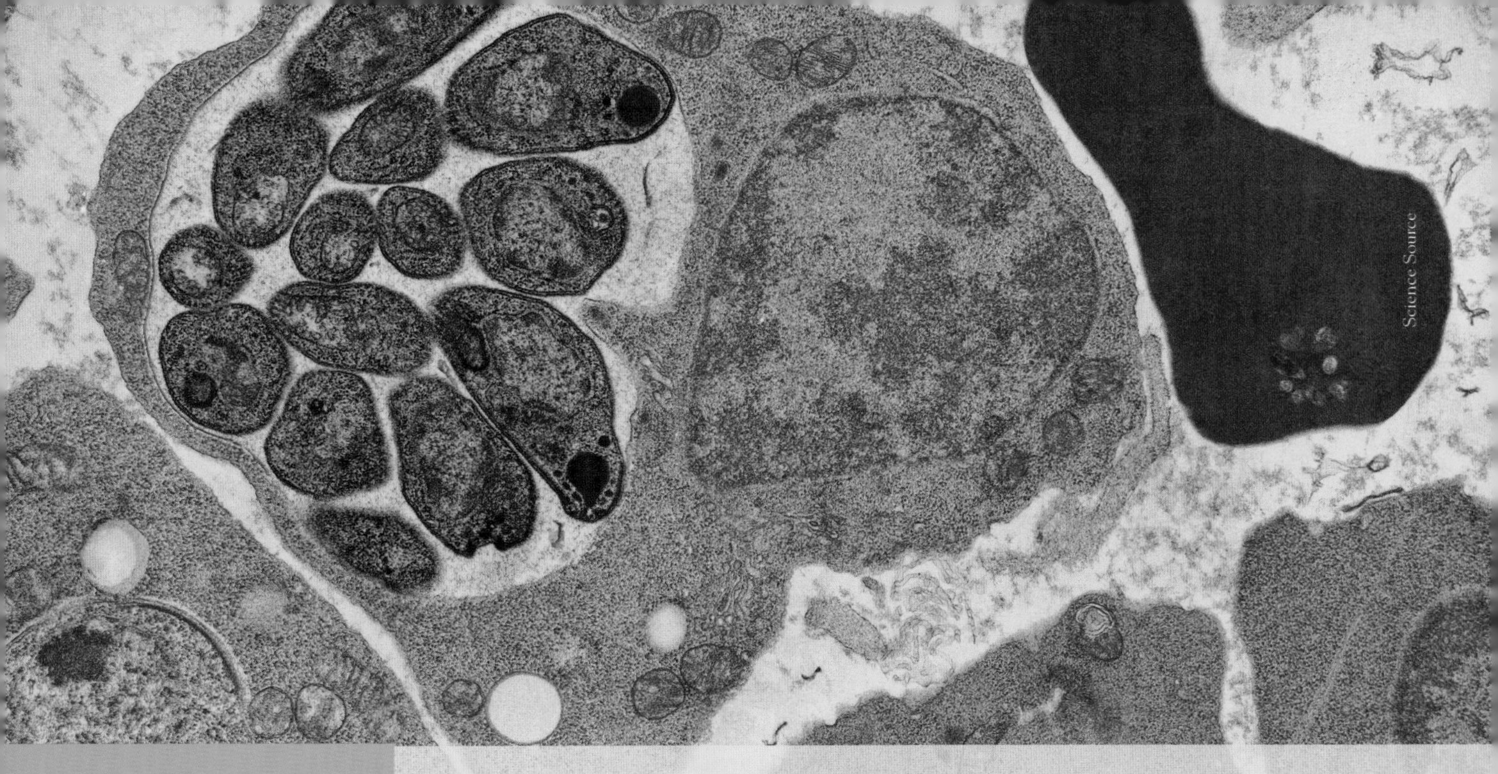

CHAPTER 10 General Rules of Probability*

The mathematics of probability can provide models to describe the genetic makeup of populations, the spread of epidemics, and risk factors for a given disorder. Although we are interested in probability because of its usefulness in statistics, the mathematics of chance is important in many fields of study. Our study of probability in Chapter 9 concentrated on basic ideas and facts. Now we look at some details. With more probability at our command, we can model more complex random phenomena.

IN THIS CHAPTER WE COVER...

- Relationships among several events
- Conditional probability
- General probability rules
- Tree diagrams
- Bayes's theorem
- *Discussion: Making sense of conditional probabilities in diagnostic tests*

Relationships among several events

In the previous chapter we described the probability that *one or the other* of two events A and B occurs in the special situation when A and B cannot occur together. Now we will describe the probability that *both* events A and B occur.

You may find it helpful to draw a picture to display relationships among several events. A picture like Figure 10.1 that shows the sample space S as a rectangular area and events as areas within S is called a **Venn diagram.** The events A and

Venn diagram

*This chapter gives more details about probability that are not needed to read the rest of the book. These concepts, however, are particularly important in medicine and clinical research.

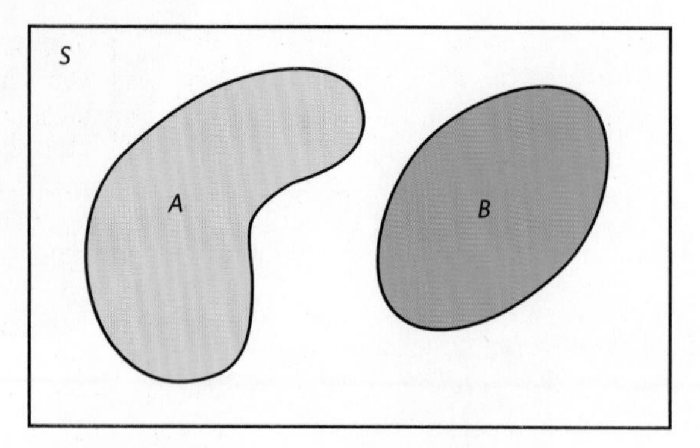

FIGURE 10.1 Venn diagram showing disjoint events A and B.

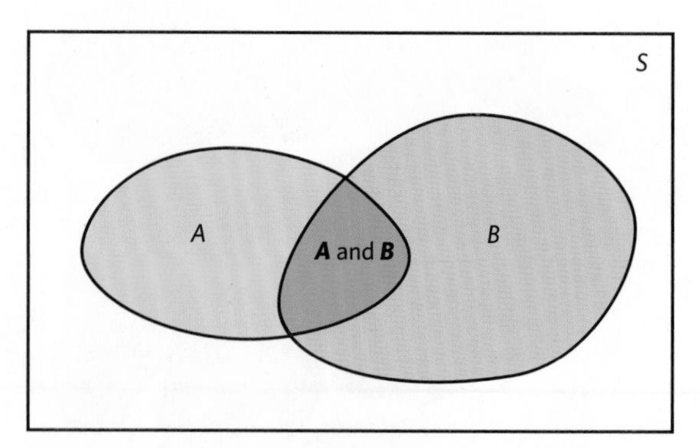

FIGURE 10.2 Venn diagram showing events A and B that are not disjoint. The event $\{A \text{ and } B\}$ consists of outcomes common to A and B.

B in Figure 10.1 are disjoint because they do not overlap. The Venn diagram in Figure 10.2 illustrates two events that are not disjoint. The event $\{A \text{ and } B\}$ appears as the overlapping area that is common to both A and B.

Suppose that you want to study the next two single births at a local hospital. You are counting girls, so two events of interest are

$$A = \text{first baby is a girl}$$
$$B = \text{second baby is a girl}$$

The events A and B are not disjoint (they are not mutually exclusive). They occur together whenever the next two single births at the hospital are baby girls. We want to find the probability of the event $\{A \text{ and } B\}$ that *both* babies are girls.

Genetic laws and the U.S. statistics described at the beginning of Chapter 9 make us willing to assign probability 1/2 to a girl when a baby is born. So

$$P(A) = 0.5$$
$$P(B) = 0.5$$

What is $P(A \text{ and } B)$? Common sense says that it is 1/4. The first baby will be a girl half the time. And when the first baby is a girl, the second baby will be a girl half the time. So both babies will be girls on $1/2 \times 1/2 = 1/4$ of all births in the long run (half of half). This reasoning assumes that the second baby still has probability 1/2 of being a girl after the first one born was a girl. This is true—the birth outcome of one couple is not influenced by the birth outcome of another couple. We say that the events "girl on the first birth" and "girl on the second birth" are independent. **Independence** means that the outcome of the first event cannot influence the outcome of the second event.

independence

EXAMPLE 10.1 Independent or not?

Because a coin has no memory and most coin tossers cannot influence the fall of the coin, it is safe to assume that successive coin tosses are independent. For a balanced coin this means that after we see the outcome of the first toss, we still assign probability 1/2 to heads on the second toss.

On the other hand, the colors of successive cards dealt from the same deck are not independent. A standard 52-card deck contains 26 red and 26 black cards. For the first card dealt from a shuffled deck, the probability of a red card is $26/52 = 0.50$ (equally likely outcomes). Once we see that the first card is red, we know that there are only 25 reds among the remaining 51 cards. The probability that the second card is red is therefore only $25/51 = 0.49$. Knowing the outcome of the first card dealt changes the probability for the second.

If a nurse measures your height twice, it is reasonable to assume that the two results are independent observations. Each records your actual height plus a measurement error, and the size of the error in the first result does not influence the instrument that makes the second reading. But if you take an IQ test or other mental test twice in succession, the two test scores are not independent. The learning that occurs on the first attempt influences your second attempt. ■

We want a boy Misunderstanding independence can be disastrous. "Dear Abby" once published a letter from a mother of eight girls. She and her husband had planned a family of four children. When all four were girls, they kept trying for a boy. After seven girls, her doctor assured her that "the law of averages was in our favor 100 to 1." Unfortunately, having children is like tossing coins. Having eight girls in a row is highly unlikely, but once you have seven girls it is not at all unlikely that the next child will be a girl—and she was.

MULTIPLICATION RULE FOR INDEPENDENT EVENTS

Two events A and B are **independent** if knowing that one occurs does not change the probability that the other occurs. If A and B are independent,

$$P(A \text{ and } B) = P(A)P(B)$$

Conversely, if this condition is not satisfied, then events A and B are **dependent.**

The multiplication rule also extends to collections of more than two events, provided that all are independent. Independence of events A, B, and C means that no information about any one or any two can change the probability of the remaining events.

If two events A and B are independent, the event that A does not occur is also independent of B, and so on. For example, 45% of all Americans have blood type O.[1] If the Census Bureau interviews two individuals chosen independently, the probability that the first is type O and the second is not type O is $(0.45)(0.55) = 0.2475$.

Independence is often assumed in setting up a probability model when the events we are describing seem to have no logical connection. *However, assuming independence by default is a common mistake and results in faulty computations when the assumption is incorrect.* Do not assume that two events are independent unless you are told that they are or unless the events have no logical connection (for example, when two individuals are independently selected from a large population).

EXAMPLE 10.2 Rapid HIV testing

STATE: Many people who come to clinics to be tested for HIV, the virus that causes AIDS, don't come back to learn the test results. Clinics now use "rapid HIV tests" that give a result while the client waits. In a clinic in Malawi, for example, use of rapid tests increased the percent of clients who learned their test results from 69% to 99.7%.

The trade-off for fast results is that rapid tests are less accurate than slower laboratory tests. Applied to people who have no HIV antibodies, one rapid test has a probability of about 0.004 of producing a false-positive (that is, of falsely indicating that antibodies are present).[2] If a clinic tests 200 people who are free of HIV antibodies, what is the chance that at least one false-positive will occur?

PLAN: It is reasonable to assume that the test results for different individuals are independent. We have 200 independent events, each with probability 0.004. What is the probability that at least one of these events occurs?

SOLVE: "At least one" combines many outcomes. It is much easier to use the fact that

$$P(\text{at least one positive}) = 1 - P(\text{no positives})$$

and find $P(\text{no positives})$ first.

The probability of a negative result for any one person is $1 - 0.004 = 0.996$. To find the probability that all 200 people tested have negative results, use the multiplication rule:

$$\begin{aligned} P(\text{no positives}) &= P(\text{all 200 negative}) \\ &= (0.996)(0.996)\cdots(0.996) \\ &= 0.996^{200} = 0.4486 \end{aligned}$$

The probability we want is therefore

$$P(\text{at least one positive}) = 1 - 0.4486 = 0.5514$$

CONCLUDE: The probability is greater than 1/2 that at least one of the 200 people will test positive for HIV even though none of them has the virus. ■

The multiplication rule $P(A \text{ and } B) = P(A)P(B)$ holds if A and B are independent, but not otherwise. The addition rule $P(A \text{ or } B) = P(A) + P(B)$ discussed in Chapter 9 holds if A and B are disjoint, but not otherwise. Resist the temptation to use these simple rules when the circumstances that justify them are not present. *You must also be careful not to confuse disjointness and independence.* If A and B are disjoint, then the fact that A occurs tells us that B cannot occur—look again at Figure 10.1. So disjoint events are not independent; disjoint events are dependent. Unlike disjointness, we cannot "see" independence in a Venn diagram, because it involves the probabilities of the events rather than just the outcomes that make up the events.

APPLY YOUR KNOWLEDGE

10.1 Unintended pregnancies. Pharmaceutical companies advertise for the birth control pill an annual efficacy of 99.5% in preventing pregnancy. However, under typical use the real efficacy is only about 95%. That is, 5% of women taking the pill for a year will experience an unplanned pregnancy that year. The difference between these two rates is that the real world is not perfect: For example, a woman might get sick or forget to take the pill one day, or she might be prescribed antibiotics, which interfere with hormonal metabolism.[3] If a sexually active woman takes the pill for the four years she is in college, what is the chance that she will become pregnant at least once? Give your answer using first the theoretical efficacy of the pill and then the real efficacy of the pill. Compare both answers.

10.2 More on unintended pregnancies. Condoms have a failure rate of about 14%. That is, if a woman uses condoms as a means of contraception for one year, she has a 14%

chance of becoming pregnant that year. Condom failure is independent of failure of the pill in preventing pregnancy.[4]

(a) If a woman chooses to use both the pill and condoms as a means of contraception for a year, what is the chance that she will become pregnant that year? Use the pill's real efficacy (95%), not its theoretical efficacy.

(b) If a woman uses both condoms and the pill for the four years she is in college, what is the chance that she will become pregnant at least once?

10.3 Spinal cord injuries. The state of Florida reports that 75% of all patients first diagnosed with a spinal cord injury are males.[5]

(a) What is the probability that the next 2 patients diagnosed with a spinal cord injury in Florida will be males?

(b) Of the next 5 patients diagnosed with a spinal cord injury in Florida, what is the probability that at least 1 will be female? What is the probability that at least 1 will be male?

10.4 Chronic diseases. A Gallup-Healthways survey of over 350,000 American adults in 2011 found that 30.0% had been told by a physician they have high blood pressure and that 26.2% had been told by a physician that they have high cholesterol. Nonetheless, we can't conclude that because $(0.300)(0.262) = 0.0786$, about 7.9% of American adults have been told by a physician that they have high blood pressure and that they have high cholesterol. Why not?

Conditional probability

When 2 events A and B are independent, the outcome of the first event cannot influence the outcome of the second event. This is true of 2 independent coin tosses or 2 independent measurements of your height. However, not all events are independent. Sometimes the probability we assign to an event can change if we know that some other event is true or has occurred. This idea is the key to many applications of probability.

EXAMPLE 10.3 Characteristics of a forest

Forests are complex, evolving ecosystems. For instance, pioneer tree species can be displaced by successional species better adapted to the changing environment. Ecologists mapped a large Canadian forest plot dominated by pioneer Douglas fir with an understory of the invading successional species western hemlock and western red cedar. The two-way table below records the distribution of all 2050 trees in the plot by species and by life stage. The distinction between live and sapling trees is made for live trees taller or shorter than 1.3 meters, respectively.[6]

	Dead	Live	Sapling	Total
Western red cedar (RC)	0.02	0.10	0.08	0.20
Douglas fir (DF)	0.16	0.16	0.00	0.32
Western hemlock (WH)	0.23	0.21	0.04	0.48
Total	0.41	0.47	0.12	

The "Total" row and column are obtained from the probabilities in the body of the table by the addition rule. For example, the probability that a randomly selected tree is a western hemlock (WH) is

$$\begin{aligned} P(\text{WH}) &= P(\text{WH and dead}) + P(\text{WH and live}) + P(\text{WH and sapling}) \\ &= 0.23 + 0.21 + 0.04 = 0.48 \end{aligned}$$

Now we are told that the tree selected is a sapling. That is, it is one of the 12% in the "Sapling" column of the table. The probability that a tree is a western hemlock, *given the information that the tree is a sapling,* is the proportion of western hemlocks in the "Sapling" column,

$$P(\text{WH} \mid \text{sapling}) = \frac{0.04}{0.12} = 0.33$$

This is a conditional probability. You can read the bar | as "given the information that." ■

Although 48% of the trees in this forest are western hemlock, only 33% of sapling trees are western hemlock. It's common sense that knowing that one event (the tree is a sapling) occurs often influences the probability of another event (the tree is a western hemlock). The example also shows how we should define *conditional probability*. The idea of a conditional probability $P(B \mid A)$ of one event B, given that another event A occurs, is the proportion *of all occurrences of* A for which B also occurs.

CONDITIONAL PROBABILITY

When $P(A) > 0$, the **conditional probability** of B, given A, is

$$P(B \mid A) = \frac{P(A \text{ and } B)}{P(A)}$$

The conditional probability $P(B \mid A)$ makes no sense if the event A can never occur, so we require that $P(A) > 0$ whenever we talk about $P(B \mid A)$. *Be sure to keep in mind the distinct roles of the events A and B in $P(B \mid A)$.* Event A represents the information we are given, and B is the event whose probability we are calculating. Here is an example that emphasizes this distinction.

EXAMPLE 10.4 Characteristics of a forest

What is the conditional probability that a randomly chosen tree is a sapling, *given the information that it is a western hemlock (WH)?* Using the definition of conditional probability,

$$\begin{aligned} P(\text{sapling} \mid \text{WH}) &= \frac{P(\text{sapling and WH})}{P(\text{WH})} \\ &= \frac{0.04}{0.48} = 0.08 \end{aligned}$$

Only 8% of western hemlock trees are saplings.

Be careful not to confuse the two different conditional probabilities

$$P(\text{WH} \mid \text{sapling}) = 0.33$$
$$P(\text{sapling} \mid \text{WH}) = 0.08$$

The first answers the question "What proportion of sapling trees are western hemlocks?" The second answers "What proportion of western hemlock trees are saplings?" ■

If you have covered optional Chapter 5 on two-way tables, you will immediately notice how similar conditional probabilities are to the conditional distributions in two-way tables.

Personal probability, defined in Chapter 9, is a special case of conditional probability. Different persons may assign different personal probabilities to a given event because their knowledge and beliefs are different. We also often update the probability we give to a particular event when new information is provided to us (or at least we should; failure to do so is known as "stubbornness"!).

APPLY YOUR KNOWLEDGE

10.5 **Composition of a forest.** What do the data in Example 10.3 say about the current composition and the evolution of this Canadian forest? Compute the conditional probabilities that a randomly chosen tree is dead, given information about its species, for each of the three tree species in this forest. What do these probabilities say about the past of this forest?

10.6 **Composition of a forest, continued.** Using the data provided in Example 10.3, compute the conditional probabilities that a randomly chosen tree is a sapling, given information about its species, for each of the three tree species. What does this say about the likely future of this forest? Which tree species appears to be taking over and becoming dominant?

10.7 **Diabetes and chronic kidney disease.** Diabetes and chronic kidney disease (CKD) are two diseases that increasingly burden the senior population in the United States. A large national sample of healthcare visits to the Veteran's Affairs Health System indicates that 9.1% of veterans in their 60s have stage 3–5 CKD. However, this rate is 14.1% among veterans in their 60s diagnosed with diabetes and only 6.4% among veterans in their 60s without diabetes.[7] Express these 3 percents as probabilities for a randomly selected veteran 60 to 69 years of age. Use proper probability notation as in Example 10.3.

10.8 **Alcohol dependency.** A study of the U.S. clinical population found that 22.5% are diagnosed with a mental disorder, 13.5% are diagnosed with an alcohol-related disorder, and 5% are diagnosed with both disorders.[8]

(a) What is the probability that someone from the clinical population is diagnosed with a mental disorder, given that the person is diagnosed with an alcohol-related disorder?

(b) What is the probability that someone from the clinical population is diagnosed with an alcohol-related disorder, given that the person is diagnosed with a mental disorder?

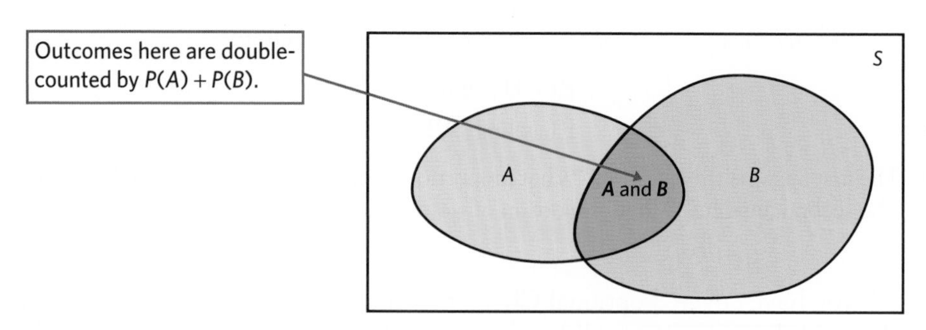

FIGURE 10.3 The general addition rule: $P(A \text{ or } B) = P(A) + P(B) - P(A \text{ and } B)$ for any two events A and B.

General probability rules

General addition rule We know that if A and B are disjoint events, then $P(A \text{ or } B) = P(A) + P(B)$. If events A and B are *not* disjoint, they can occur together. The probability that one or the other occurs is then *less* than the sum of their probabilities. As Figure 10.3 illustrates, outcomes common to both are counted twice when we add probabilities, so we must subtract this probability once. Here is the addition rule for any two events, disjoint or not.

GENERAL ADDITION RULE FOR ANY TWO EVENTS

For any two events A and B,

$$P(A \text{ or } B) = P(A) + P(B) - P(A \text{ and } B)$$

If A and B are disjoint, the event $\{A \text{ and } B\}$ that both occur contains no outcomes and therefore has probability 0. So the general addition rule includes Rule 3, the addition rule for disjoint events.

Petra Wegner/Alamy

EXAMPLE 10.5 Hearing impairment in dalmatians

Congenital sensorineural deafness is the most common form of deafness in dogs and is often associated with congenital pigmentation deficiencies. A study of hearing impairment in dogs examined over five thousand dalmatians for both hearing impairment and iris color. "Impaired" was defined as deafness in either one or both ears. Dogs with one or both irises blue (a trait due to low iris pigmentation) were labeled "blue."

The study found that 28% of the dalmatians were hearing impaired, 11% were blue eyed, and 5% were hearing impaired and blue eyed.[9] Choose a dalmatian at random. Then

$$P(\text{blue or impaired}) = P(\text{blue}) + P(\text{impaired}) - P(\text{blue impaired})$$
$$= 0.11 + 0.28 - 0.05 = 0.34$$

That is, 34% of dalmatians were either blue eyed or hearing impaired. A dalmatian is an unimpaired, brown-eyed dog if it is *neither* hearing impaired nor blue eyed. So

$$P(\text{neither}) = P(\text{unimpaired brown}) = 1 - 0.34 = 0.66$$ ■

Venn diagrams are a great help in finding such probabilities because you can just think of adding and subtracting areas. Figure 10.4 shows two Venn diagrams based on the information from Example 10.5. Both diagrams show some events

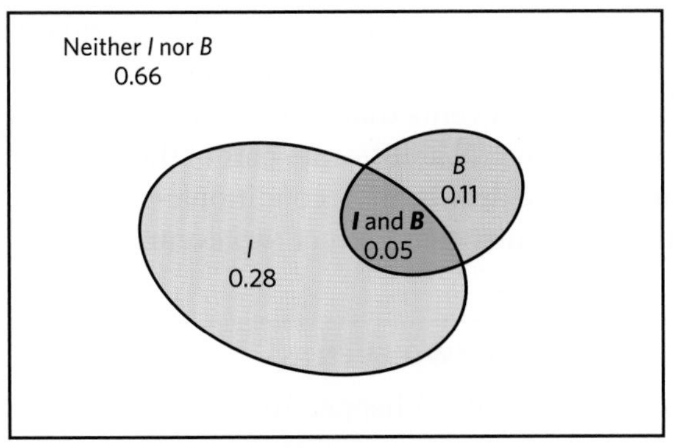

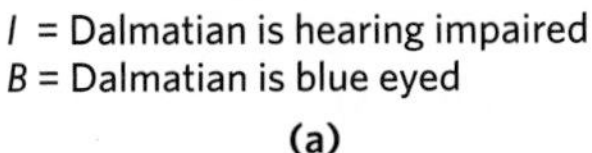

FIGURE 10.4 Venn diagrams and probabilities for hearing impairment and eye color in Dalmatian dogs, for Example 10.5. (a) This Venn diagram displays the simple probabilities of I and B. (b) This Venn diagram displays the probabilities for all possible combinations of I and B. Notice that, in both diagrams, the probability values shown match their corresponding labels.

and their corresponding probabilities. Venn diagram (a) displays the information we were provided: the labels B and I represent all blue-eyed dalmatians and all hearing-impaired dalmatians, respectively. Because B and I are not disjoint events, the probabilities displayed in (a) do not add to 1. In contrast, Venn diagram (b) labels the four disjoint events that make up the sample space, and therefore, the corresponding four probabilities do add to 1 (but you need to compute most of these values before you can complete the diagram). Be sure to always check carefully the labels in a Venn diagram so that you do not confuse what the probabilities displayed represent.

What is the probability that a randomly chosen dalmatian is a hearing-impaired, brown-eyed dog? The Venn diagram in Figure 10.4(b) shows the answer directly (I and not B, 0.23). Figure 10.4(a) shows that this is the probability that the dog is hearing-impaired minus the probability that it is an hearing-impaired, blue-eyed dog, $0.28 - 0.05 = 0.23$.

Alternatively, you could display the probabilities given in Example 10.5 using a two-way table like the one in Example 10.3. Start with the information you were provided (the values shown in bold), and complete the rest of the table using the addition rule. Here is what we get:

	B	Not B	Total
I	**0.05**	0.23	**0.28**
Not I	0.06	0.66	0.72
Total	**0.11**	0.89	

Some students have a strong preference for either Venn diagrams or two-way tables. You should be able to understand and use both formats, but if given the choice, use the one that you find more natural and helpful.

General multiplication rule The definition of conditional probability reminds us that in principle all probabilities, including conditional probabilities, can be found from the assignment of probabilities to events that make up a random phenomenon. More often, however, conditional probabilities are part of the information given to us in a probability model. The definition of conditional probability then turns into a rule for finding the probability that both of two events occur.

GENERAL MULTIPLICATION RULE FOR ANY TWO EVENTS

The probability that both of two events A and B happen together can be found by

$$P(A \text{ and } B) = P(A)P(B \mid A)$$

Here $P(B \mid A)$ is the conditional probability that B occurs, given the information that A occurs.

In words, you could think of this rule as saying that for both of two events to occur, one must occur and, given that this event has occurred, the other must occur. This is often just common sense expressed in the language of probability, as the following example illustrates.

EXAMPLE 10.6 Soda consumption and weight

A 2012 Gallup survey of a random sample of American adults found that 52% say they do not drink any soda on a typical day. Of those who say they do not drink soda on a typical day, 39% describe themselves as overweight. Let's consider this sample survey to be representative of the U.S. adult population. What percent of American adults drink no soda *and* are overweight?

Use the general multiplication rule:

$$\begin{aligned} P(\text{no soda}) &= 0.52 \\ P(\text{overweight} \mid \text{no soda}) &= 0.39 \\ P(\text{no soda and overweight}) &= P(\text{no soda}) \times P(\text{overweight} \mid \text{no soda}) \\ &= (0.52)(0.39) = 0.2028 \end{aligned}$$

That is, about 20% of all American adults are overweight and drink no soda on typical days.

You should think your way through this: If 52% of Americans drink no soda and 39% *of these* are overweight, then 39% of 52% are overweight and drink no soda. ■

The conditional probability $P(B \mid A)$ is generally not equal to the unconditional probability $P(B)$. That's because the occurrence of event A generally gives us some additional information about whether or not event B occurs. If knowing that A occurs gives no additional information about B, then A and B are independent events. The precise definition of independence is expressed in terms of conditional probability.

INDEPENDENT EVENTS

Two events A and B that both have positive probability are **independent** if

$$P(B \mid A) = P(B)$$

Because of the multiplication rule, this also implies that

$$P(A \text{ and } B) = P(A)P(B)$$

You can use either formula to verify independence.

We now see that the multiplication rule for independent events, $P(A \text{ and } B) = P(A)P(B)$, is a special case of the general multiplication rule, $P(A \text{ and } B) = P(A)P(B \mid A)$, just as the addition rule for disjoint events is a special case of the general addition rule.

The multiplication rule extends to the probability that all of several events occur. The key is to condition each event on the occurrence of *all* the preceding events. For example, we have for three events A, B, and C that

$$P(A \text{ and } B \text{ and } C) = P(A)P(B \mid A)P(C \mid A \text{ and } B)$$

The special multiplication rule $P(A \text{ and } B) = P(A)P(B)$ *applies only to independent events; you cannot use it if events are not independent.* Here is a distressing example of misuse of the multiplication rule.

EXAMPLE 10.7 Sudden infant death syndrome

Sudden infant death syndrome (SIDS) causes babies to die suddenly (often in their cribs) with no explanation. Deaths from SIDS have been greatly reduced by placing babies on their backs, but as yet no cause is known.

When more than one SIDS death occurs in a family, the parents are sometimes accused. One "expert witness" (a pediatrician) popular with prosecutors in England told juries that there is only a 1 in 73 million chance that two children in the same family could have died naturally. Here's his calculation: The rate of SIDS in a nonsmoking middle-class family is 1 in 8500. So the probability of two deaths is

$$\frac{1}{8500} \times \frac{1}{8500} = \frac{1}{72{,}250{,}000}$$

Several women were convicted of murder on this basis, without any direct evidence that they harmed their children.

The Royal Statistical Society described this reasoning as nonsense. The argument assumes that SIDS deaths in the same family are independent events. However, the cause of SIDS is unknown: "There may well be unknown genetic or environmental factors that predispose families to SIDS, so that a second case within the family becomes much more likely."[10] The British government decided, years later, to review the cases of 258 parents convicted of murdering their babies when no direct proof was available. ■

Independence is often part of the information given to us in a probability model. However, it can also be an important property that we want to identify.

Lack of independence in biology can suggest, for example, a shared mechanism of action or transmission, genetic linkage, or interaction between two factors.

EXAMPLE 10.8 Dalmatians again

Example 10.5 described the frequencies of hearing impairment and eye color in dalmatians. We know that there is a 28% chance that a randomly chosen dalmatian is hearing impaired, an 11% chance that it is blue eyed, and a 5% chance that it is hearing impaired *and* blue eyed. We want to know if the two traits, hearing impairment and eye color, are independent or if they are linked (dependent) in dalmatians.

We know that

$$P(\text{blue and impaired}) = 0.05$$

If both events are independent, we should find the same value by using the multiplication rule for independent events:

$$P(\text{blue}) \times P(\text{impaired}) = 0.11 \times 0.28 = 0.03$$

The true frequency of hearing-impaired, blue-eyed dalmatians (0.05) is larger than we would predict if the two traits were independent (0.03). Therefore, hearing impairment and blue eye pigmentation in dalmatians are not independent; they are dependent. Eye color is a well-known genetic trait. The lack of independence suggests two possible biological interpretations: Hearing impairment may also be genetically inherited in some dogs (microbial infections are known to sometimes cause hearing loss); or the lack of pigmentation may, in some indirect way, render the dogs more susceptible to hearing loss. ■

APPLY YOUR KNOWLEDGE

10.9 At the gym. Suppose that 10% of adults belong to health clubs, and 40% of these health club members go to the club at least twice a week. What percent of all adults go to a health club at least twice a week? Write the information given in terms of probabilities and use the general multiplication rule.

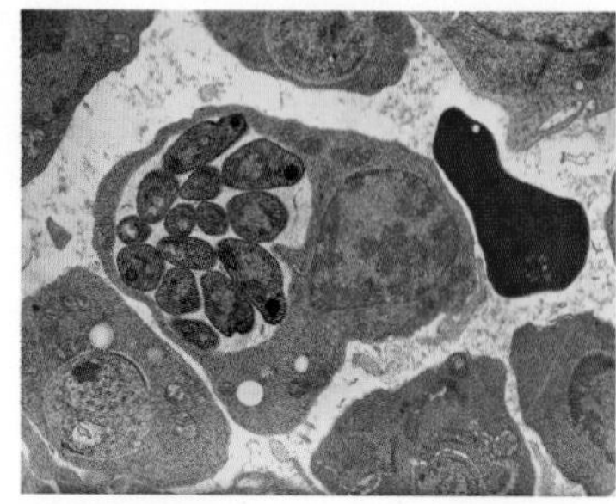

Science Source

10.10 Sickle-cell trait and malaria. Sickle-cell anemia is a hereditary medical condition that affects red blood cells and is thought to protect against malaria, a debilitating parasitic infection of the liver and blood. That would explain why the sickle-cell trait is found in people who originally came from Africa, where malaria is widespread. A landmark study in Africa tested 543 children for the sickle-cell trait and also for malaria infection. In all, 25% of the children had the sickle-cell trait, and 6.6% of the children had both the sickle-cell trait and malaria. Overall, 34.6% of the children had malaria.[11]

(a) Make a Venn diagram with the information provided.

(b) What is the probability that a given child has neither malaria nor the sickle-cell trait? What is the probability that a given child has either malaria or the sickle-cell trait?

10.11 Sickle-cell trait and malaria, continued. Write the information in the previous exercise in terms of probabilities and use them to answer the following questions.

(a) What is the probability that a child has malaria, given that the child has the sickle-cell trait? What is the probability that a child has malaria, given that the child doesn't have the sickle-cell trait?

(b) Are the events sickle-cell trait and malaria independent? What might that tell you about the relationship between the sickle-cell trait and malaria?

10.12 Alcohol dependency, continued. Go back to Exercise 10.8 and make a Venn diagram representing the information given. Use your diagram to answer the following questions.

(a) What percent of the U.S. clinical population is diagnosed with a mental disorder but not with an alcohol-related disorder?

(b) What percent has neither a mental disorder nor an alcohol-related disorder?

Tree diagrams

People often find it difficult to understand formal probability calculations, especially when they involve conditional probabilities. The discussion (page 255) on interpreting medical test results illustrates this problem. There are, fortunately, some graphical representations of probabilities that can greatly help guide calculations.

If you have covered optional Chapter 5 on two-way tables, you will have noticed the similarity between conditional probabilities and conditional distributions calculated from a two-way table. It can indeed be easier to arrange a problem in a two-way table (such as those used for Examples 10.3 and 10.5) before calculating conditional probabilities.

However, some problems can have too many stages to be easily represented in a two-way table, or the information provided may contain conditional probabilities that will not fit directly inside the table. We can still display these problems graphically by using *tree diagrams*. Here is an example.

EXAMPLE 10.9 Skin cancer in men and women

STATE: Intense or repetitive sun exposure can lead to skin cancer. Patterns of sun exposure, however, differ between men and women. A study of the presence of cutaneous malignant melanoma at a single body location among the Italian population found that 15% of skin cancers are located on the head and neck area, another 41% on the trunk, and the remaining 44% on the limbs. Moreover, 44% of the individuals with a skin cancer on the head are men, as are 63% of those with a skin cancer on the trunk and only 20% of those with a skin cancer on the limbs.[12] What percent of all individuals with skin cancer are women?

PLAN: To use the tools of probability, restate the percents as probabilities. If we choose an individual with skin cancer at random,

$$P(\text{head}) = 0.15$$
$$P(\text{trunk}) = 0.41$$
$$P(\text{limbs}) = 0.44$$

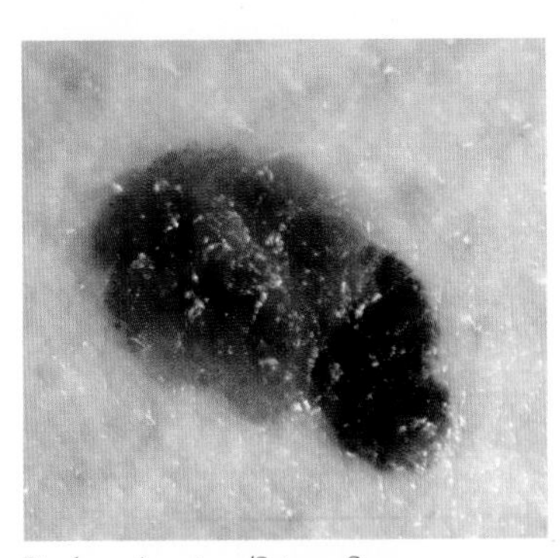

Biophoto Associates/Science Source

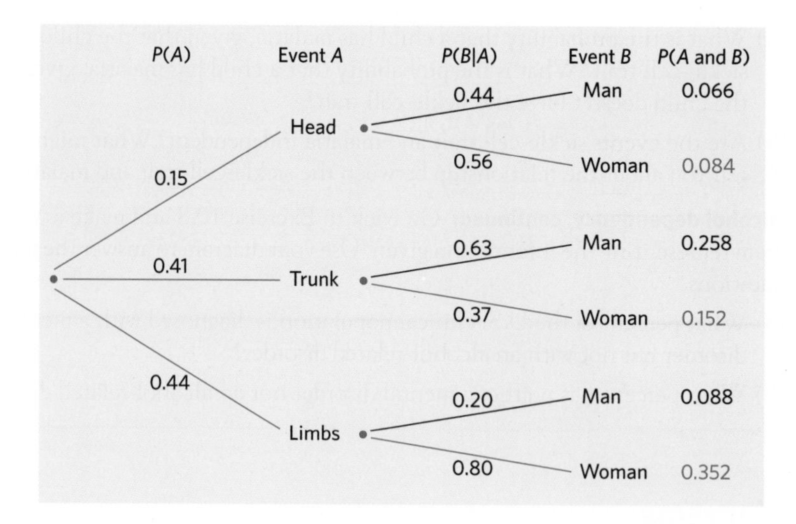

FIGURE 10.5 Tree diagram of skin cancer, for Example 10.9. The two stages are the location on the body of the cancer and the gender of the cancer patient.

These three probabilities add to 1 because every individual with skin cancer is in one of these three groups. The percents in these groups that are men are *conditional* probabilities:

$$P(\text{man} \mid \text{head}) = 0.44$$
$$P(\text{man} \mid \text{trunk}) = 0.63$$
$$P(\text{man} \mid \text{limbs}) = 0.20$$

We want to find the unconditional probability $P(\text{woman})$.

tree diagram

SOLVE: The **tree diagram** in Figure 10.5 organizes this information. Each segment in the tree is one stage of the problem. Each complete branch shows a path through the two stages. The probability written on each segment is the conditional probability that an individual with skin cancer will follow that segment, given that he or she has reached the node from which it branches.

Starting at the left, an individual with skin cancer falls into one of the three cancer location groups. The probabilities of these groups mark the leftmost branches in the tree. Look at the limbs group, the bottom branch. The two segments going out from the "limbs" branch node carry the conditional probabilities

$$P(\text{man} \mid \text{limbs}) = 0.20$$
$$P(\text{woman} \mid \text{limbs}) = 0.80$$

The full tree shows the probabilities for all three cancer location groups.

Now use the multiplication rule. The probability that a randomly chosen individual with skin cancer is a woman with cancer on the limbs is

$$\begin{aligned} P(\text{limbs and woman}) &= P(\text{limbs})P(\text{woman} \mid \text{limbs}) \\ &= (0.44)(0.80) = 0.352 \end{aligned}$$

This probability appears at the end of the bottom branch. You see that the probability of any complete branch in the tree is the product of the probabilities of the segments of that branch.

There are three disjoint paths to "woman." These paths are colored red in Figure 10.5. Because the three paths are disjoint, the probability that an individual with skin cancer is a woman is the sum of their probabilities:

$$P(\text{woman}) = (0.15)(0.56) + (0.41)(0.37) + (0.44)(0.80)$$
$$= 0.084 + 0.152 + 0.352 = 0.588$$

CONCLUDE: About 59% of all individuals with skin cancer are women. ■

It takes longer to explain a tree diagram than it does to use it. Once you have understood a problem well enough to draw the tree, the rest is easy. Here is another question about skin cancer that the tree diagram helps us answer.

EXAMPLE 10.10 Skin cancer in men and women

STATE: What percent of women with skin cancer have the cancer on the limbs?

PLAN: In probability language, we want the conditional probability

$$P(\text{limbs} \mid \text{woman}) = \frac{P(\text{limbs and woman})}{P(\text{woman})}$$

SOLVE: Look again at the tree diagram. $P(\text{woman})$ is the overall outcome. $P(\text{limbs and woman})$ is the result of following one branch of the tree diagram. So

$$P(\text{limbs} \mid \text{woman}) = \frac{0.352}{0.588} = 0.599$$

CONCLUDE: About 60% of skin cancers in women are located on the limbs. Compare this conditional probability with the original information (unconditional) that 44% of skin cancers are located on the limbs. Knowing that a person is a woman increases the probability that her skin cancer is located on the limbs. ■

Notice how $P(\text{limbs} \mid \text{woman})$ and $P(\text{woman} \mid \text{limbs})$ are two very different probabilities. The conditional probability $P(\text{woman} \mid \text{limbs})$ is displayed on the tree and is complementary to $P(\text{man} \mid \text{limbs})$—that is, they add up to 1. The conditional probability $P(\text{limbs} \mid \text{woman})$, on the other hand, is calculated from other probabilities and is complementary to $P(\text{trunk} \mid \text{woman})$ and $P(\text{head} \mid \text{woman})$.

Examples 10.9 and 10.10 illustrate a common setting for tree diagrams. Some outcome (such as the person's gender) has several sources (such as the three cancer location groups). Starting from

- the probability of each source and
- the conditional probability of the outcome, given each source,

the tree diagram leads to the overall probability of the outcome. Example 10.9 does this. You can then use the probability of the outcome and the definition of conditional probability to find the conditional probability of one of the sources, given that the outcome occurred. Example 10.10 shows how.

APPLY YOUR KNOWLEDGE

10.13 Eye color, hair color, and freckles. A large study of children of Caucasian descent in Germany looked at the effect of eye color, hair color, and freckles on the reported extent of burning from sun exposure.[13] The population's distribution of hair color,

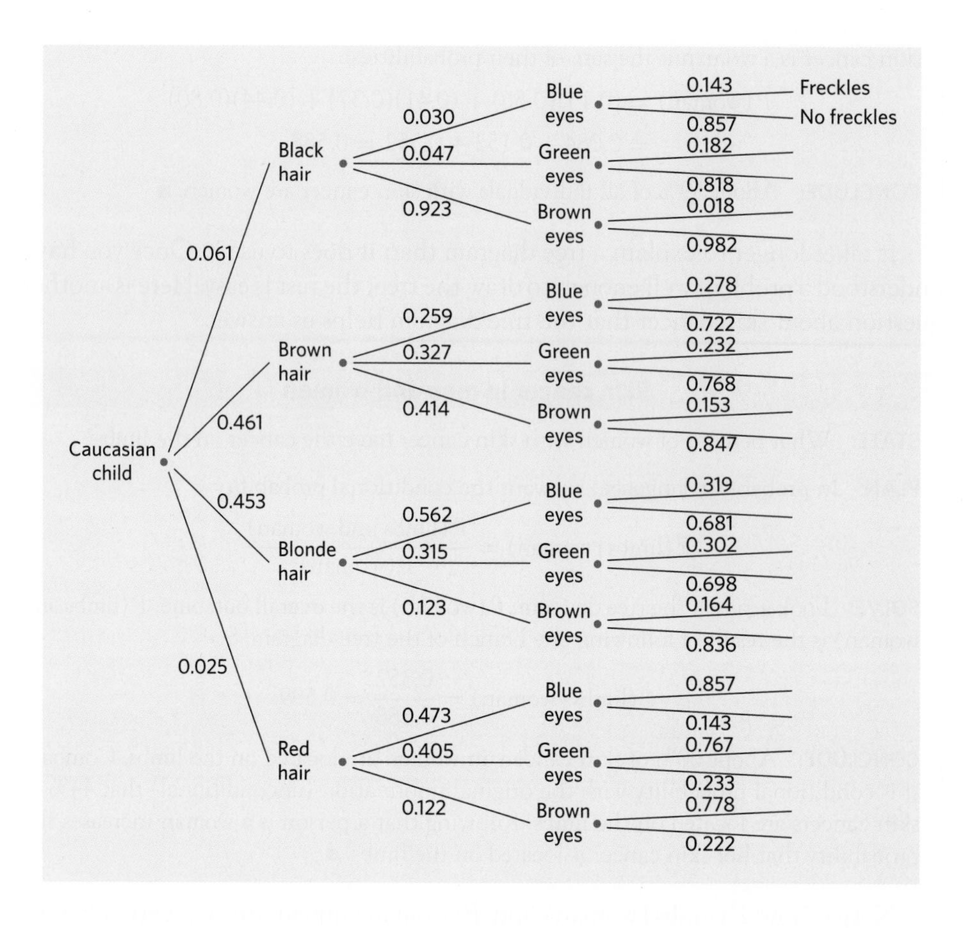

FIGURE 10.6 Tree diagram of the features of children of Caucasian descent in Germany, for Exercise 10.13. The three stages are hair color, eye color, and whether or not the child has freckles.

eye color, and freckles was as shown in the tree diagram in Figure 10.6. Find the following probabilities and describe them in plain English:

(a) P(blue eyes | red hair), P(blue eyes and red hair)

(b) P(freckles | red hair and blue eyes), P(freckles and red hair and blue eyes)

10.14 **Testing for HIV.** Enzyme immunoassay tests are used to screen blood specimens for the presence of antibodies to HIV, the virus that causes AIDS. Antibodies indicate the presence of the virus. The test is quite accurate but is not always correct. Here are approximate probabilities of positive and negative test results when the blood tested does and does not actually contain antibodies to HIV:[14]

	Test Result	
	+	−
When antibodies are present	0.9985	0.0015
When antibodies are absent	0.0060	0.9940

Suppose that 1% of a large population carries antibodies to HIV in its blood.

(a) Draw a tree diagram representing the HIV status of a person from this population (outcomes: antibodies present or absent) and the blood test result (outcomes: test positive or negative). Notice that the probabilities provided for the test results are conditional probabilities.

(b) What is the probability that the test is positive for a randomly chosen person from this population?

10.15 Eye color, hair color, and freckles, continued. Continue your work from Exercise 10.13 using the tree diagram in Figure 10.6. Find the following probabilities and describe them in plain English:

(a) P(red hair), P(blue eyes), P(freckles)

(b) P(freckles and red hair), P(freckles | red hair)

Bayes's theorem

In some situations, we know the conditional probability $P(B \mid A)$ but are more interested in $P(A \mid B)$. Here is an important example.

EXAMPLE 10.11 Diagnostic tests in medicine

The performance of a diagnostic test for a given disease can be defined by two properties. The first one is the test's ability to appropriately give a positive result when a person tested has the disease. This is the test's **sensitivity,** or, in probability language, P(positive test | disease). The second property is the test's ability to come up negative when a person tested doesn't have the disease. This is the test's **specificity,** P(negative test | no disease). These two values are determined experimentally.

sensitivity

specificity

Information about the sensitivity of a medical test can easily be found in the Internet age. However, when a patient takes a diagnostic test, what we really want to know is the probability that a positive test result truly reflects the presence of the disease. This is the conditional probability of having the disease, knowing that you received a positive test result: P(disease | positive test). It is called the **positive predictive value,** or PPV, of a test. ■

positive predictive value

Many patients—and indeed even some physicians[15]—assume that if you get a positive test result, then you must have the disease. That's not true. No test is perfectly accurate, and human error in interpreting or communicating the results can also occur. Figure 10.7 shows all possible outcomes of a diagnostic test, displayed in a two-way table and in a tree diagram. The information contained in both displays

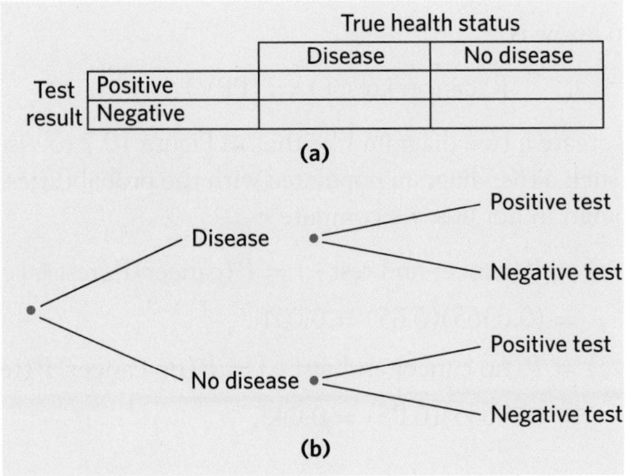

		True health status	
		Disease	No disease
Test result	Positive		
	Negative		

(a)

Disease: Positive test, Negative test

No disease: Positive test, Negative test

(b)

FIGURE 10.7 All possible outcomes of a diagnostic test, represented with (a) a two-way table and (b) a tree diagram.

is the same, but you might feel that one is more intuitive for you than the other. It is clear in both displays that

- there are 4 possible outcomes to a diagnostic test, and
- a positive test may be obtained for subjects who have the disease (these results are called "true-positives") and for subjects who do not have the disease ("false-positives").

Therefore, the probability that a subject whose test result is positive indeed has the disease, $P(\text{disease} \mid \text{positive test})$, depends on what proportion of all positive tests (both true-positives and false-positives) can be expected to be true positives. And this depends on both the properties of the diagnostic test (the test's sensitivity and specificity) and how common or rare the disease is in the population of interest (estimated from large-scale epidemiological studies).

EXAMPLE 10.12 Mammography screening

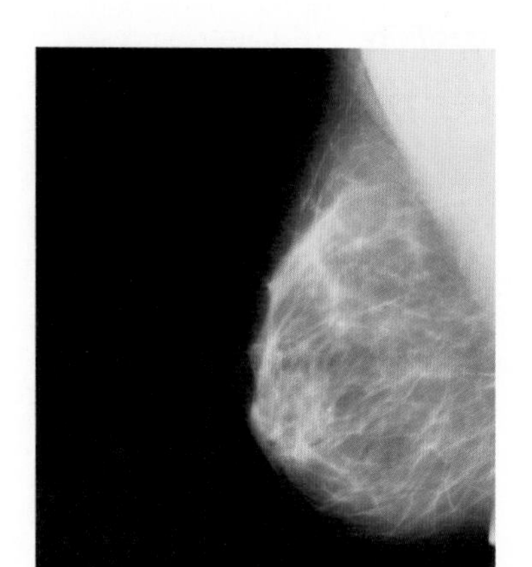

Jim Cummins/Getty Images

STATE: Breast cancer occurs most frequently among older women. Of all age groups, women in their 60s have the highest rate of breast cancer. The National Cancer Institute (NCI) compiles U.S. epidemiology data for a number of different cancers. The NCI estimates that 3.65% of women in their 60s get breast cancer.[16]

Mammograms are X-ray images of the breast used to detect breast cancer. A mammogram can typically identify correctly 85% of cancer cases (sensitivity = 85%) and 95% of cases without cancer (specificity = 95%).[17] If a woman in her 60s gets a positive mammogram, what is the probability that she indeed has breast cancer?

PLAN: We know the following probabilities:

$$P(\text{cancer}) = 0.0365 \text{ (cancer rate)}$$
$$P(\text{no cancer}) = 0.9635$$

$$P(\text{test}+ \mid \text{cancer}) = 0.85 \text{ (sensitivity)}$$
$$P(\text{test}- \mid \text{cancer}) = 0.15$$

$$P(\text{test}+ \mid \text{no cancer}) = 0.05$$
$$P(\text{test}- \mid \text{no cancer}) = 0.95 \text{ (specificity)}$$

What we want to know is

$$P(\text{cancer} \mid \text{test}+) = ? \text{ (PPV)}$$

SOLVE: We can create a tree diagram like that in Figure 10.7 to visualize the problem. Figure 10.8(a) is such a tree diagram populated with the probabilities from this example. For a random woman in her 60s, we compute

$$\begin{aligned} P(\text{true-positive}) &= P(\text{cancer and test}+) = P(\text{cancer})P(\text{test}+ \mid \text{cancer}) \\ &= (0.0365)(0.85) = 0.031 \\ P(\text{false-positive}) &= P(\text{no cancer and test}+) = P(\text{no cancer})P(\text{test}+ \mid \text{no cancer}) \\ &= (0.9635)(0.05) = 0.048 \end{aligned}$$

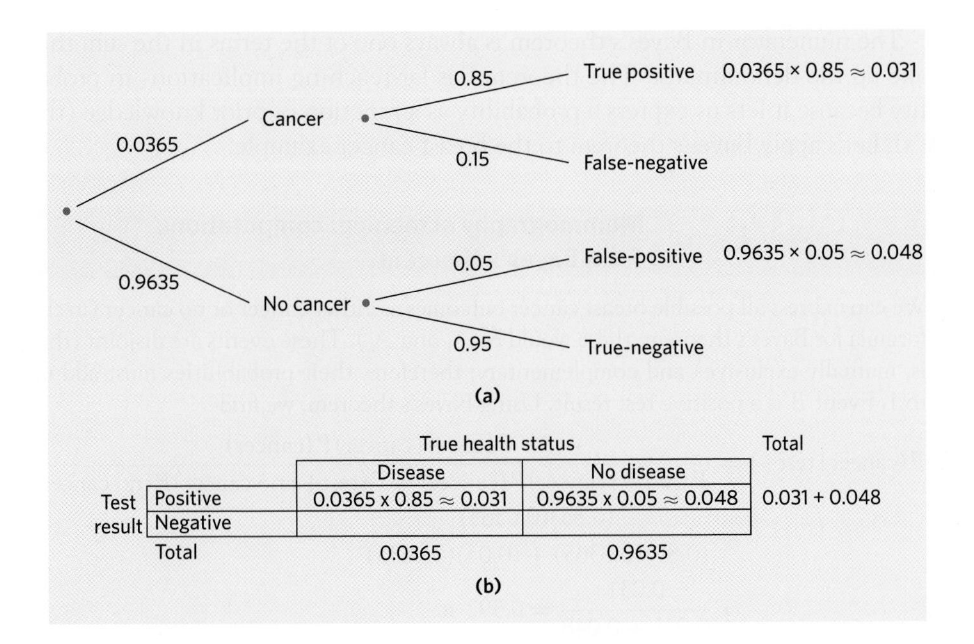

		True health status		Total
		Disease	No disease	
Test result	Positive	$0.0365 \times 0.85 \approx 0.031$	$0.9635 \times 0.05 \approx 0.048$	$0.031 + 0.048$
	Negative			
	Total	0.0365	0.9635	

(b)

FIGURE 10.8 (a) Tree diagram for mammograms given to women in their 60s, for Example 10.12. (b) Corresponding two-way table.

The positive predictive value (PPV) is simply the proportion of true-positives among all positive test results for women in their 60s:

$$\text{PPV} = P(\text{cancer} \mid \text{test+}) = \frac{P(\text{true-positive})}{P(\text{true-positive}) + P(\text{false-positive})}$$

$$= \frac{0.031}{0.031 + 0.048} = 0.392$$

CONCLUDE: If a randomly selected woman in her 60s gets a positive mammogram, there is a 39% chance that she indeed has breast cancer. It also means that there is a 61% chance that she does *not* have breast cancer. This important information should be clear to the physician and communicated clearly to the patient. ■

The trial of John Hinckley John Hinckley attempted to assassinate President Reagan in 1981. At the trial, the defense described how Hinckley's brain scan showed brain atrophy, which could be evidence of schizophrenia. About 1.5% of Americans suffer from schizophrenia. Brain atrophy is observed in 30% of schizophrenics, but in only 2% of normal individuals. How convincing is this argument? The answer lies in the probability that Hinckley has schizophrenia, given the evidence of brain atrophy. This is $P(\text{schizophrenia} \mid \text{atrophy}) = (0.3 \times 0.015)/(0.3 \times 0.015 + 0.02 \times 0.985) = 0.186$.

What we have done in Example 10.12 is state that the probability we want (PPV) represents the frequency of one outcome (the true-positives) in a series of complementary and mutually exclusive outcomes (all positive test results: true-positives and false-positives). Using a tree diagram made it easy visually to understand why. The same information was presented in a two-way table for students who prefer working with this format.

Our approach can be summarized formally in an equation known as Bayes's theorem:

BAYES'S THEOREM

Suppose that $A_1, A_2, \ldots, A_k$ are disjoint events whose probabilities are not 0 and add to exactly 1. That is, any outcome has to be exactly in one of these events. Then if B is any other event whose probability is not 0 or 1,

$$P(A_i \mid B) = \frac{P(B \mid A_i)P(A_i)}{P(B \mid A_1)P(A_1) + P(B \mid A_2)P(A_2) + \cdots + P(A_k)P(B \mid A_k)}$$

The numerator in Bayes's theorem is always one of the terms in the sum that make up the denominator. The theorem has far-reaching implications in probability because it lets us express a probability as a function of prior knowledge (the A_i's). Let's apply Bayes's theorem to the breast cancer example.

EXAMPLE 10.13 Mammography screening: computations from Bayes's theorem

We can express all possible breast cancer outcomes as either cancer or no cancer (in the formula for Bayes's theorem, these would be A_1 and A_2). These events are disjoint (that is, mutually exclusive) and complementary; therefore, their probabilities must add up to 1. Event B is a positive test result. Using Bayes's theorem, we find

$$P(\text{cancer} \mid \text{test+}) = \frac{P(\text{test+} \mid \text{cancer})P(\text{cancer})}{P(\text{test+} \mid \text{cancer})P(\text{cancer}) + P(\text{test+} \mid \text{no cancer})P(\text{no cancer})}$$

$$= \frac{(0.85)(0.0365)}{(0.85)(0.0365) + (0.05)(0.9635)}$$

$$= \frac{0.031}{0.031 + 0.048} = 0.392$$ ■

Bayes's theorem is extremely powerful and extensively used in more advanced statistics. The importance of Bayes's theorem justifies its inclusion in an introductory textbook. However, many struggle with its formal use. It is far better to think your way through such problems rather than memorize these formal expressions. The discussion that follows illustrates this in the context of medical diagnostic tests.

APPLY YOUR KNOWLEDGE

10.16 False HIV positives. Continue your work from Exercise 10.14.

(a) What is the probability that a person has the antibody, given that the test is positive? Explain in your own words what this means.

(b) Identify the test's sensitivity, specificity, and positive predictive value.

10.17 Assessing the new OraQuick HIV test. In 2012, the U.S. Food and Drug Administration approved the first over-the-counter HIV testing kit, the OraQuick In-Home HIV Test. Example 10.11 told us that the sensitivity and specificity of a diagnostic test are determined experimentally. The OraQuick test was approved based on a study that examined the test results of a sample of 4410 adults from a high-risk population who did not yet know their HIV status. Each person's actual HIV status was later confirmed via several other medical examinations. Here are the findings:[18]

105 true-positives (HIV and positive test)	1 false-positive (no HIV and positive test)
8 false-negatives (HIV and negative test)	4296 true-negatives (no HIV and negative test)

(a) What is the sensitivity of OraQuick, $P(\text{positive test} \mid \text{HIV})$? This represents the fraction of all HIV cases who received a positive test result. Use the counts in the table above to obtain your answer.

HIV screening In 2006, the Centers for Disease Control and Prevention (CDC) announced a new recommendation that HIV screening be offered to all patients aged 13 to 64 as part of routine medical care. This is a sharp departure from previous guidelines recommending that only high-risk individuals be tested. Why such a drastic change? The CDC estimates that about 25% of all HIV-positive Americans are not aware of their serologic status. This is thought to be one of the primary reasons why new HIV infections remain high in the United States, with about 40,000 new cases every year.

(b) What is the specificity of OraQuick, P(negative test | no HIV)? This represents the fraction of all individuals with no HIV who received a negative test result.

(c) Now use the study findings to find the approximate rate of HIV in this high-risk population. What is the positive predictive value of OraQuick, P(HIV | positive test), in this high-risk population?

(d) Explain in your own words why the positive predictive value of this test would be different for a population with a different rate of HIV, whereas the test's sensitivity and specificity would remain the same.

DISCUSSION Making sense of conditional probabilities in diagnostic tests

Conditional probabilities are often grossly misinterpreted. Yet they are very important, particularly in medicine, where understanding test results requires understanding these probabilities. Many studies have documented that most physicians understand and communicate the true meaning of test results incorrectly. Here is a revealing quote from one such study:

For instance, doctors with an average of 14 years of professional experience were asked to imagine using the Haemoccult test to screen for colorectal cancer. The prevalence [rate] of cancer was 0.3%, the sensitivity of the test was 50%, and the false positive rate was 3%. The doctors were asked: what is the probability that someone who tests positive actually has colorectal cancer? The correct answer is about 5%. However, the doctors' answers ranged from 1% to 99%, with about half of them estimating the probability as 50% (the sensitivity) or 47% (sensitivity minus false positive rate). If patients knew about this degree of variability and statistical innumeracy they would be justly alarmed.[19]

The positive predictive value, PPV, describes the probability that a person who receives a positive test result actually has a given disease, P(disease | positive test). You can check that PPV in this study is indeed only

$$\frac{(0.003)(0.5)}{(0.003)(0.5)+(0.997)(0.03)} = 0.048, \text{ or about } 5\%$$

The discrepancy between a perceived (incorrect) PPV and its actual value is particularly large for screening tests because they are typically administered to a large population with a relatively low rate of disease. When a disease is rare in the population screened, the percent of false-positives tends to be large, even when sensitivity and specificity are high. The consequence is a low PPV in the context of *indiscriminate* screening. (Note that these testing procedures have a much higher PPV when they are used to screen higher-risk individuals or to diagnose patients with a suspected condition based on existing symptoms.)

The potential merit of screening is that early detection may play a key role in patient prognosis. However, the emotional, medical, and financial consequences of misinterpreted test results cannot be ignored. For instance, up until the summer of 2012, the U.S. Food and Drug Administration had refused approval to over-the-counter HIV screening because of concerns with public reaction to false-positives.[20] Inversely, the 2009 recommendation by the U.S. Preventive Services Task Force against routine mammography screening for women in their 40s caused an uproar,[21] despite the fact that the PPV in this age group is in the 10% to 20% range.

Beyond debating the pros and cons of various diagnostic and screening tests, we should ask how such misunderstanding can happen. Physicians are highly trained professionals, and studies show that most do know the definitions of "sensitivity" and "specificity." The challenge seems to be interpreting conditional probabilities, particularly confusing $P(A \mid B)$ with $P(B \mid A)$. In the study cited, many physicians confused the PPV, $P(\text{disease} \mid \text{positive test})$, with the test's sensitivity, $P(\text{positive test} \mid \text{disease})$. Probability concepts tend to be more abstract than natural frequencies. Studies show that reframing information in a simpler, more natural format using a concrete example increases dramatically the proportion of physicians who correctly identify the PPV. The colorectal cancer case, for example, can be reframed as follows:

> Ten thousand patients take the Haemoccult test. Of these 10,000 patients, we expect that about 30 actually have colorectal cancer and the remaining 9970 do not. Of the 30 patients with colorectal cancer, about 15 (50%) can be expected to receive a positive test result. Of the 9970 patients without colorectal cancer, about 299 (3%) can also be expected to receive a positive test result. What proportion of patients who receive a positive test result can be expected to actually have colorectal cancer?

The PPV seems much more intuitive now as the proportion of positive tests that are true-positives: PPV $= 15/(15 + 299) = 0.048$, or about 5%. Researchers indeed advocate providing information about diagnostic and screening tests in terms of natural frequencies in addition to giving the usual sensitivity and specificity conditional probabilities. Rather than being expected to do elaborate probability calculations on their own, physicians and patients should be given clear information about the predictive value of a test. This would improve medical practices and support informed patient decisions. Figure 10.9 for Exercise 10.44 on the PSA test illustrates a different and effective way to communicate probabilities for such tests.

Until the general public understands these statistical concepts, we will continue to pay a high medical and financial price. Here is a most unusual and interesting illustration. In 2006, a Texas judge uncovered an extensive legal scam in which screening tests were used to fuel massive class action asbestos and silicone lawsuits. Instead of representing workers who sought them out because of existing health problems, some lawyers had advertised widely to the general public and indiscriminately tested very large numbers of people, including some with existing diagnoses for unrelated lung diseases. This resulted, obviously, in large numbers of false-positives... and lots of money for the lawyers and screening companies. The many workers who tested positive received a meager check in the mail and were never informed that a positive test could be a false-positive. Interestingly, the scheme was uncovered by a judge who had been a nurse earlier in her career. Judge Janis Jack was the first person, in years of such litigation, to ever ask how the plaintiffs had been screened.[22]

CHAPTER 10 SUMMARY

- The **conditional probability** $P(B \mid A)$ of an event B, given an event A, is defined by

$$P(B \mid A) = \frac{P(A \text{ and } B)}{P(A)}$$

  when $P(A) > 0$. In practice, we sometimes find conditional probabilities from directly available information rather than from the definition.
- Events A and B are **independent** if knowing that one event occurs does not change the probability we would assign to the other event; that is, $P(B \mid A) = P(B)$. In that case, $P(A \text{ and } B) = P(A)P(B)$.
- Remember from Chapter 9 that events A and B are **disjoint** (mutually exclusive) if they have no outcomes in common.
- Any assignment of probability obeys these rules:

  Addition rule for disjoint events: If events $A, B, C, \ldots$ are all disjoint in pairs, then

$$P(\text{at least one of these events occurs}) = P(A) + P(B) + P(C) + \cdots$$

  Multiplication rule for independent events: If events $A, B, C, \ldots$ are independent, then

$$P(\text{all the events occur}) = P(A)P(B)P(C)\cdots$$

  General addition rule: For any two events A and B,

$$P(A \text{ or } B) = P(A) + P(B) - P(A \text{ and } B)$$

  General multiplication rule: For any two events A and B,

$$P(A \text{ and } B) = P(A)P(B \mid A)$$

- Bayes's theorem states that, if a series of disjoint and complementary events A_1 through A_k have nonzero probabilities, we can express the conditional probability of event A_i, knowing that event B occurred (when event B has probability not equal to 0 or 1), as

$$P(A_i \mid B) = \frac{P(B \mid A_i)P(A_i)}{P(B \mid A_1)P(A_1) + P(B \mid A_2)P(A_2) + \cdots + P(A_k)P(B \mid A_k)}$$

THIS CHAPTER IN CONTEXT

Probability models provide the important connection between the observed data and the process that generated the data. Probability is also the foundation of statistical inference and provides the language with which we answer questions and draw conclusions from our data. For these reasons, it is important that we have at least a basic understanding of what we mean by probability and of some of the rules and properties of probabilities.

Chapter 9 introduced basic ideas and facts about probability, and in this chapter we have considered some further details. The conditional distributions discussed in optional Chapter 5 are an example of the computation of conditional probabilities. Conditional probabilities play a central role when we study the relationship between two categorical variables, and we will return to them in Chapter 22 when examining the chi-square test for two-way tables. In terms of later chapters, the most important idea of this chapter is independence. This is an assumption that many statistical procedures make about the observations in a data set, and therefore, it is an important assumption to fully understand.

CHECK YOUR SKILLS

10.18 The weather forecast for the weekend is a 50% chance of rain for Saturday and a 50% chance of rain for Sunday. If we assume that consecutive days are independent events, the probability that it rains over the weekend (either Saturday or Sunday) is

(a) 0.75.
(b) 1.0.
(c) unknown, given the information provided.

10.19 In fact, weather on consecutive days depends on similar atmospheric conditions and therefore consecutive days are not independent events. With the weather forecast of the previous exercise, the probability that it rains over the weekend (either Saturday or Sunday) is

(a) 0.75.
(b) 1.0.
(c) unknown, given the information provided.

10.20 An athlete suspected of having used steroids is given two tests that operate independently of each other. Test A has probability 0.9 of being positive if steroids have been used. Test B has probability 0.8 of being positive if steroids have been used. What is the probability that *neither* test is positive if steroids have been used?

(a) 0.72 (b) 0.38 (c) 0.02

The culinary herb cilantro is very polarizing: Some people love it and others hate it. A large survey of 12,087 American adults of European ancestry asked whether they like or dislike the taste of cilantro. Here are the numbers of men and women in the study for each answer:[23]

	Number of men	Number of women
Dislikes cilantro	1632	1549
Likes cilantro	5165	3741

Exercises 10.21 to 10.24 are based on this table.

10.21 The conditional probability that such an adult likes cilantro, given that the adult is male, is about

(a) 0.580. (b) 0.737. (c) 0.760.

10.22 The conditional probability that such an adult is male, given that the adult likes cilantro, is about

(a) 0.500. (b) 0.580. (c) 0.760.

10.23 Choose an American adult of European ancestry at random. The probability that this person is male and likes cilantro is about

(a) 0.414. (b) 0.427. (c) 0.580.

10.24 Let L be the event that an American adult of European ancestry likes cilantro, and M the event that this person is male. The proportion of individuals who like cilantro among males is expressed in probability notation as

(a) $P(L \mid M)$. (b) $P(M \mid L)$. (c) $P(L \text{ and } M)$.

10.25 The Centers for Disease Control and Prevention states on its website that "in the United States, cigarette smoking causes about 90% of lung cancers."[24] If S is the event "smokes cigarettes" and L is the event "has lung cancer," then the probability 0.90 is expressed in probability notation as
(a) $P(S \text{ and } L)$. (b) $P(S \mid L)$. (c) $P(L \mid S)$.

10.26 Of people who died in the United States in a recent year, 86% were white, 12% were black, and 2% were Asian. (This ignores a small number of deaths among other races.) Diabetes caused 2.8% of deaths among whites, 4.4% among blacks, and 3.5% among Asians. The probability that a randomly chosen death is a white who died of diabetes is about
(a) 0.107. (b) 0.030. (c) 0.024.

10.27 Using the information in the previous exercise, the probability that a randomly chosen death was due to diabetes is about
(a) 0.107. (b) 0.030. (c) 0.024.

CHAPTER 10 EXERCISES

10.28 Blood types. All human blood can be "ABO-typed" as O, A, B, or AB, but the distribution of the types varies a bit among groups of people. Here are the distributions of blood types for a randomly chosen person in China and in the United States:

	O	A	B	AB
China probability	0.35	0.27	0.26	0.12
U.S. probability	0.45	0.40	0.11	0.04

Choose an American and a Chinese at random, independently of each other. What is the probability that both have type O blood? What is the probability that both have the same blood type?

10.29 Universal blood donors. People with type O-negative blood are universal donors. That is, any patient can receive a transfusion of O-negative blood. Only 7.2% of the American population has O-negative blood. If 10 people appear at random to give blood, what is the probability that at least 1 of them is a universal donor?

10.30 The Rhesus factor. Human blood is typed as O, A, B, or AB and also as Rh-positive or Rh-negative. ABO type and Rh-factor type are independent because they are governed by different genes on different chromosomes. In the American population, 84% of people are Rh-positive. Use the information about ABO type in Exercise 10.28 to give the probability distribution of blood type (ABO and Rh) for a randomly chosen American.

10.31 Tasting phenylthiocarbamide. Phenylthiocarbamide, or PTC, is a molecule that tastes very bitter to those with the ability to taste it. However, in about 30% of the world population a recessive allele causes a chemosensory deficit for PTC.[25] A researcher administers a simple PTC tasting test to 10 randomly chosen subjects. What is the probability that none of these 10 subjects is PTC chemodeficient? What is the probability that at least 1 subject is chemodeficient for PTC?

10.32 Polydactyly. Polydactyly is a fairly common congenital abnormality in which a baby is born with one or more extra fingers or toes. It is reported in about 1 child in every 500.[26] A young obstetrician celebrates her first 100 deliveries. What is the probability that she has delivered no child with polydactyly? What is the probability that she has delivered at least 1 child with polydactyly?

10.33 Hair color and freckles. A large study of German children of Caucasian descent found that 46% have brown hair and that 25% have freckles.[27] Nonetheless, we can't conclude that because $(0.46)(0.25) = 0.115$, about 11.5% of German children of Caucasian descent are brown-haired with freckles. Why not?

10.34 Cancer-detecting dogs. Research has shown that specific biochemical markers are found exclusively in the breath of patients with lung cancer. However, no lab test can currently distinguish the breath of lung cancer patients from that of other subjects. Could dogs be trained to identify these markers in specimens of human breath, as they can be to detect illegal substances or to follow a person's scent? An experiment trained dogs to distinguish breath specimens of lung cancer patients from breath specimens of control individuals by using a food-reward training method. After the training was complete, the dogs were tested on new breath specimens without any reward or clue using a double-blind, completely randomized design. Here are the results for a random sample of 1286 breath specimens:[28]

Dog test result	Breath specimen from a Control subject	Cancer subject	Total
Negative	708	10	718
Positive	4	564	568
Total	712	574	1286

(a) The sensitivity of a diagnostic test is its ability to correctly give a positive result when a person tested has the disease, or P(positive test | disease). Find the sensitivity of the dog cancer-detection test for lung cancer.

(b) The specificity of a diagnostic test is the conditional probability that the test comes up negative, given that the subject tested doesn't have the disease. Find the specificity of the dog cancer-detection test for lung cancer.

10.35 Tendon surgery. You have torn a tendon and are facing surgery to repair it. The surgeon explains the risks to you: Infection occurs in 3% of such operations, the repair fails in 14%, and infection and failure occur together in 1%. What percent of these operations succeed and are free of infection?

10.36 Lactose intolerance. Lactose intolerance causes difficulty digesting dairy products that contain lactose (milk sugar). It is particularly common among people of African and Asian ancestry. In the United States (ignoring other groups and people who consider themselves to belong to more than one race), 82% of the population is white, 14% is black, and 4% is Asian. Moreover, 15% of whites, 70% of blacks, and 90% of Asians are lactose intolerant.[29]

(a) What percent of the entire population is lactose intolerant?

(b) What percent of people who are lactose intolerant are Asian?

10.37 Deer and pine seedlings. As suburban gardeners know, deer will eat almost anything green. In a study of 871 pine seedlings at an environmental center in Ohio, researchers noted how deer damage varied with how much of the seedling was covered by thorny undergrowth:[30]

Thorny Cover	Deer Damage Yes	No
None	60	151
< 1/3	76	158
1/3 to 2/3	44	177
> 2/3	29	176

(a) What is the probability that a randomly selected seedling was damaged by deer?

(b) What are the conditional probabilities that a randomly selected seedling was damaged, given each level of cover?

(c) Does knowing about the amount of thorny cover on a seedling change the probability of deer damage? If so, cover and damage are not independent.

10.38 Deer and pine seedlings, continued. In the setting of Exercise 10.37, what percent of the seedlings that were damaged by deer were less than 1/3 covered by thorny plants?

10.39 Deer and pine seedlings, continued. In the setting of Exercise 10.37, what percent of the seedlings that were not damaged by deer were more than 2/3 covered by thorny plants?

10.40 Continuous probability. Choose a point at random in a square with sides $0 \leq x \leq 1$ and $0 \leq y \leq 1$. This means that the probability that the point falls in any region within the square is equal to the area of that region. Let X be the x coordinate and Y the y coordinate of the point chosen. Find the conditional probability $P(Y < 1/2 \mid Y > X)$. (*Hint:* Draw a diagram of the square and the events $Y < 1/2$ and $Y > X$.)

10.41 The geometric distributions. Osteoporosis is a disease defined as very low bone density resulting in a high risk of fracture, hospitalization, and immobilization. Seven percent of postmenopausal women suffer from osteoporosis.[31] A physician administers routine checkups. Visits by postmenopausal women are independent. We are interested in how long we must wait to get the first postmenopausal woman with osteoporosis.

(a) Considering only patient visits by postmenopausal women, the probability of a patient with osteoporosis on the first visit is 7%. What is the probability that the first patient does not have osteoporosis and the second patient does?

(b) What is the probability that the first two patients do not have osteoporosis and the third one does? This is the probability that the first patient with osteoporosis occurs on the third visit.

(c) Now you see the pattern. What is the probability that the first patient with osteoporosis occurs on the fourth visit? On the fifth visit? Give the general result: What is the probability that the first patient with osteoporosis occurs on the kth visit?

Comment: The distribution of the number of trials to the first "success" (in our example, a patient with osteoporosis) is called a **geometric distribution.** In this problem you have found geometric distribution probabilities when the probability of a success on each trial is 7%. The same idea works for any probability of success.

10.42 Hepatitis C and HIV coinfection. According to the 2003 report of the U.S. Health Resources and Services Administration, coinfection of HIV and hepatitis C (HCV) is on the rise. Only about 2% of Americans have HCV, but 25% of Americans with HIV have HCV. And about 10% of Americans with HCV also have HIV.[32]

(a) Write the information provided in terms of probabilities. Are the two viral infections independent? Explain your reasoning.
(b) What is the probability that a randomly chosen American has both HIV and HCV?

10.43 Cystic fibrosis. Cystic fibrosis (CF) is a hereditary lung disorder that often results in death. It can be inherited only if both parents are carriers of an abnormal gene. In 1989, the CF gene that is abnormal in carriers of cystic fibrosis was identified. The probability that a randomly chosen person of European ancestry carries an abnormal CF gene is 1/25. (The probability is less in other ethnic groups.) The CF20m test detects most but not all harmful mutations of the CF gene. The test is positive for 90% of people who are carriers. It is (ignoring human error) never positive for people who are not carriers.[33] Jason tests positive. What is the probability that he is a carrier?

10.44 Should I have a PSA test? That's the question addressed in a Kaiser Permanente pamphlet on prostate cancer. The prostate-specific antigen (PSA) test is a blood test that is used to screen for prostate cancer. Elevated PSA levels are often associated with prostate cancer, but they can also reflect other conditions. Thus, a positive PSA test does not necessarily indicate prostate cancer. And not all cases of prostate cancer can be detected with a PSA test. The pamphlet uses a clever diagram to represent the possible outcomes of a PSA test with their respective frequencies.[34]

(a) The Kaiser diagram is reproduced in Figure 10.9. Use this diagram to find the probabilities of a true-positive, a false-positive, a false-negative, and a true-negative.
(b) What is the positive predictive value (PPV) of the test? That is, what is the conditional probability that a randomly screened patient has prostate cancer, given that he receives a positive PSA test result?
(c) In addressing the question "Should I have a PSA test?" Kaiser discusses the relatively low PPV of the test. They also mention that treatment for prostate cancer can cause serious side effects and that most men diagnosed with prostate cancer die of other causes. (Some prostate cancers are aggressive and spread, but most don't.) The last two facts are unrelated to the reliability of the test as assessed by its PPV. Explain why they are nonetheless important considerations in deciding whether or not to include a PSA test as part of routine medical checkups for men over the age of 50.

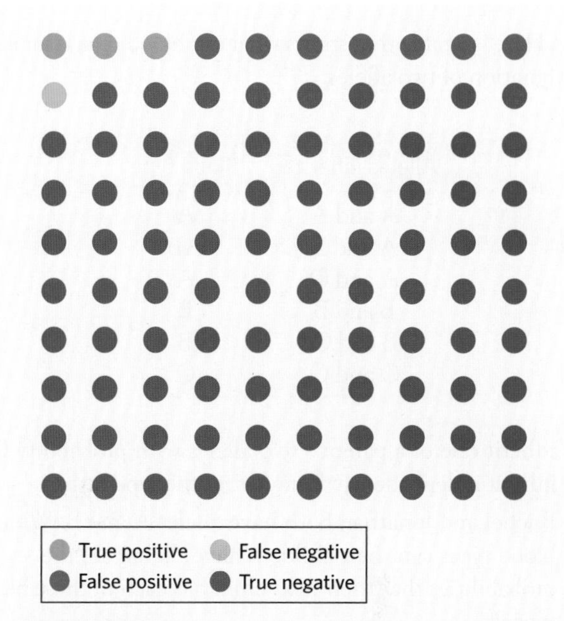

FIGURE 10.9 Kaiser Permanente's outcome diagram of 100 PSA tests, for Exercise 10.44. Each dot represents the result of one PSA test.

10.45 Peanut allergies among children. A recent survey of over 40,000 children found that 2% were allergic to peanuts.[35] Choose 5 children at random, and let the random variable X be the number in this sample who are allergic to peanuts. The possible values X can take are 0, 1, 2, 3, 4, and 5. Make a five-stage tree diagram of the outcomes (allergic or not allergic) for the 5 individuals, and use it to find the probability distribution of X. *(We will see in Chapter 12 that X follows a binomial distribution.)*

10.46 Mendelian inheritance. Some traits of plants and animals depend on inheritance of a single gene. This is called Mendelian inheritance, after Gregor Mendel (1822–1884). Each of us has an ABO blood type, which describes whether two characteristics called A and B are present. Every human being has two blood-type alleles (gene forms), one inherited from our mother and one from our father. Each of these alleles can be A, B, or O. Which two we inherit determines our blood type. Here is a table that shows what our blood type is for each combination of two alleles:

Alleles inherited	Blood type
A and A	A
A and B	AB
A and O	A
B and B	B
B and O	B
O and O	O

We inherit one of a parent's two alleles with probability 0.5. We inherit independently from our mother and father.

(a) Rachel and Jonathan both have alleles A and B. What blood types can their children have? What are the probabilities that their next child has each of these blood types?

(b) Sarah and David both have alleles B and O. What blood types can their children have? What are the probabilities that their next child has each of these blood types?

(c) Jasmine has alleles A and O. Tyrone has alleles B and O. What is the probability that a child of these parents has blood type O? If Jasmine and Tyrone have three children, what is the probability that all three have blood type O? What is the probability that the first child has blood type O and the next two do not?

10.47 Tay-Sachs disease. Tay-Sachs (TS) disease is a fatal, recessive genetic disorder in which harmful quantities of a fatty waste substance accumulate in the brain, slowly progressing to death by age five. TS carriers are common among individuals of Eastern European Jewish descent, with a carrier frequency of about 1 in 27. In the general population, on the other hand, the carrier rate is about 1 in 250.[36]

(a) Give the probability that both parents are carriers in each of the following conditions: both mom and dad are of European Jewish descent; mom is of European Jewish descent but dad isn't; both mom and dad are not of European Jewish descent.

(b) If both parents are carriers, the probability is 0.25 that their child has TS. Otherwise, the probability is zero. Draw the tree diagram for the children of parents who are both of European Jewish descent. Use the tree to find the probability that a child has TS, given that both parents are of European Jewish descent.

(c) What would be the probability that a child has TS, given that neither parent is of European Jewish descent?

10.48 Albinism. The gene for albinism in humans is recessive. That is, carriers of this gene have probability 1/2 of passing it to a child, and the child is albino only if both parents pass the albinism gene. Parents pass their genes independently of each other. If both parents carry the albinism gene but are not albino, what is the probability that their first child is albino? If they have two children (who inherit independently of each other), what is the probability that both are albino? That neither is albino?

10.49 Asking sensitive questions. When asked personal questions of a sensitive nature, survey respondents are not always truthful. We saw in Chapter 7 that this is an issue of response bias. One way to make it easier for respondents to answer truthfully is to give them an opportunity to preserve the anonymity of their answer. For example, one randomized-response technique lets respondents flip a coin without showing anyone. If the result is a head, the participant answers the sensitive question truthfully ("yes" or "no"); otherwise, the participant just answers "yes" regardless of the truth. Therefore, no one but the respondent can tell if a "yes" was actually an answer to the sensitive question. However, when all the answers are compiled, researchers can use the data to figure out what proportion of individuals answered "yes" to the sensitive question. Here is how.

(a) Let's call p the proportion of individuals who answered "yes" to the sensitive question. Make a tree diagram representing all possible outcomes for a survey respondent, starting with the coin flip. Be sure to place the correct probabilities on your tree.

(b) Use your tree diagram to compute the probability that a survey respondent would answer "yes." Note that this information, $P(\text{yes})$, is readily available from the complete survey data.

(c) Assume that the sensitive question was "Have you used illegal drugs in the past six months?" and that 59% of the participants answered "yes" in this randomized-response design. Using your work in part (b), find the proportion p of individuals who used illegal drugs in the past six months.

CHAPTER 11 The Normal Distributions

IN THIS CHAPTER WE COVER...

- Normal distributions
- The 68-95-99.7 rule
- The standard Normal distribution
- Finding Normal probabilities
- Using the standard Normal table*
- Finding a value, given a probability or proportion
- Normal quantile plots*

Of all the continuous probability distributions, none is more widely used than the family of Normal distributions. Normal distributions are extremely important in statistical inference, as we will see starting with Chapter 14. They are also a good mathematical model for many (but certainly not all) biological variables, such as blood pressure, bone mineral density, and the height of plants or animals.

Normal distributions

Chapter 9 introduced the idea of density curves to describe continuous probability distributions. One particularly important class of density curves has already appeared in Figures 9.4 and 9.8. These density curves are symmetric, single-peaked, and bell-shaped. They are called *Normal curves*, and they describe *Normal distributions*. Normal distributions play a large role in statistics, but they are rather special and not at all "normal" in the sense of being usual or average. We capitalize "Normal" to remind you that these curves are special.

All Normal distributions have the same overall shape. The exact equation describing a Normal distribution is beyond the scope of this textbook, but from this equation we know that *the exact density curve for a particular Normal distribution is described by giving its mean μ and its standard deviation σ*. The mean is located at the

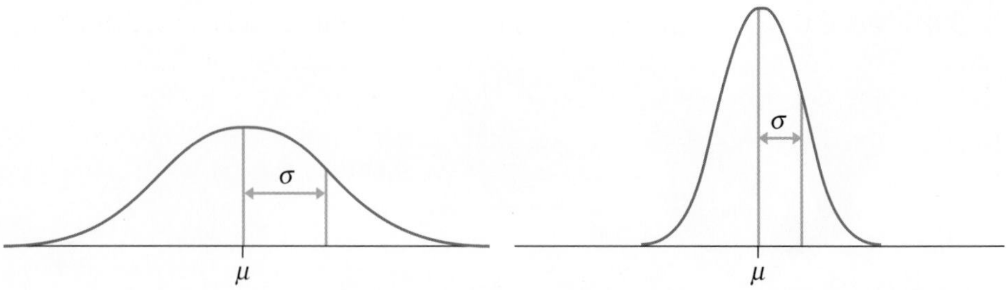

FIGURE 11.1 Two Normal curves, showing the mean μ and standard deviation σ.

center of the symmetric curve and is the same as the median. Changing μ without changing σ moves the Normal curve along the horizontal axis without changing its spread. The standard deviation σ controls the spread of a Normal curve. Figure 11.1 shows two Normal curves with different values of σ. The curve with the larger standard deviation is more spread out. Remember that density curves, by definition, are scaled so that the total area under the curve equals 1 (the probability of the entire sample space). This is why the Normal curve on the left with the larger standard deviation σ is not only wider but also shorter than the Normal curve on the right with the smaller σ.

The standard deviation σ is the natural measure of spread for Normal distributions. Not only do μ and σ completely determine the shape of a Normal curve, but we can locate σ by eye on the curve. Here's how. Imagine that you are skiing down a mountain that has the shape of a Normal curve. At first, you descend at an ever-steeper angle as you go out from the peak:

Fortunately, before you find yourself going straight down, the slope begins to grow flatter rather than steeper as you go out and down:

The points at which this change of curvature takes place are located at distance σ on either side of the mean μ. You can feel the change as you run a pencil along a Normal curve, and so find the standard deviation. Remember that *μ and σ alone do not specify the shape of most distributions*, and that the shape of density curves in general does not reveal σ. These are special properties of Normal distributions.

NORMAL DISTRIBUTIONS

A **Normal distribution** is described by a Normal density curve. Any particular Normal distribution is completely specified by two numbers: its mean and standard deviation.

The mean of a Normal distribution is at the center of the symmetric Normal curve. The standard deviation is the distance from the center to the change-of-curvature points on either side.

Why are the Normal distributions important in statistics? Here are three reasons. First, Normal distributions are good descriptions for some distributions of *real data*. Distributions that are often close to Normal include scores on tests taken by many people (such as IQ tests and SAT exams), repeated careful measurements of the same quantity, and characteristics of biological populations (such as lengths of crickets and yields of corn). Second, Normal distributions are good approximations to the results of many kinds of *chance outcomes*, such as the proportion of boys in many hospital births. Third, we will see that many *statistical inference* procedures based on Normal distributions work well for other, roughly symmetric distributions.

Students and professionals often make the error of assuming that their variable is Normally distributed without first verifying this assumption by plotting the data. However, many data sets do not follow a Normal distribution. Distributions of body weights or survival times, for example, are skewed to the right and so are not Normal. Figure 9.5 (page 220), showing the distribution of guinea pig survival times after infection, is a clear example of non-Normal data. Keep in mind that non-Normal data are actually quite common, and so be sure to always plot your data.

The 68-95-99.7 rule

Although there are many Normal curves (with different values of μ and σ), they all have common properties. In particular, all Normal distributions obey the following rule.

THE 68-95-99.7 RULE

In the Normal distribution with mean μ and standard deviation σ:

- Approximately **68%** of the observations fall within σ of the mean μ.
- Approximately **95%** of the observations fall within 2σ of μ.
- Approximately **99.7%** of the observations fall within 3σ of μ.

Expressed in probability terms, this means that if the random variable X has a Normal distribution with mean μ and standard deviation σ, then

$$P(\mu - \sigma < X < \mu + \sigma) \approx 0.68$$
$$P(\mu - 2\sigma < X < \mu + 2\sigma) \approx 0.95$$
$$P(\mu - 3\sigma < X < \mu + 3\sigma) \approx 0.997$$

Figure 11.2 illustrates the 68–95–99.7 rule. By remembering these three numbers, you can think about Normal distributions without constantly making detailed calculations. This rule can also be very useful to quickly check that your detailed calculations are not obviously wrong.

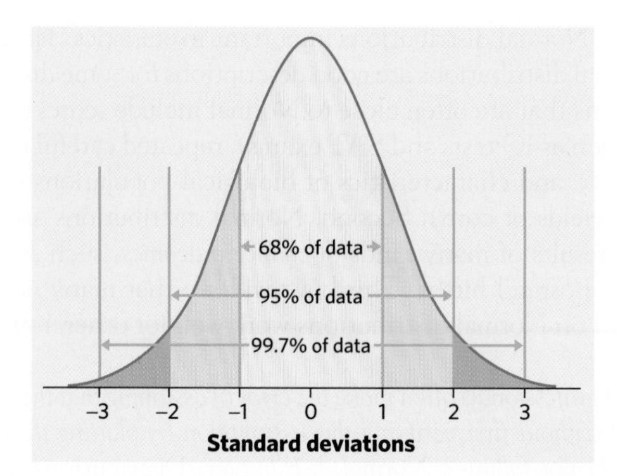

FIGURE 11.2 The 68–95–99.7 rule for Normal distributions.

EXAMPLE 11.1 Heights of young women

The distribution of heights of young women aged 18 to 24 is approximately Normal with mean $\mu = 64.5$ inches and standard deviation $\sigma = 2.5$ inches. Figure 11.3 applies the 68–95–99.7 rule to this distribution.

The 95 part of the 68–95–99.7 rule says that the heights of the middle 95% of young women are approximately between

$$\mu - 2\sigma = 64.5 - (2)(2.5) = 64.5 - 5 = 59.5$$

and

$$\mu + 2\sigma = 64.5 + (2)(2.5) = 64.5 + 5 = 69.5$$

that is, these women are between 59.5 and 69.5 inches tall.

The other 5% of young women have heights outside this range. Because the Normal distributions are symmetric, half of these women are on the tall side. So the tallest 2.5% of young women are taller than 69.5 inches. Rephrased in probability terms, this means that there is probability 0.025 approximately that a randomly selected woman is taller than 69.5 inches. ■

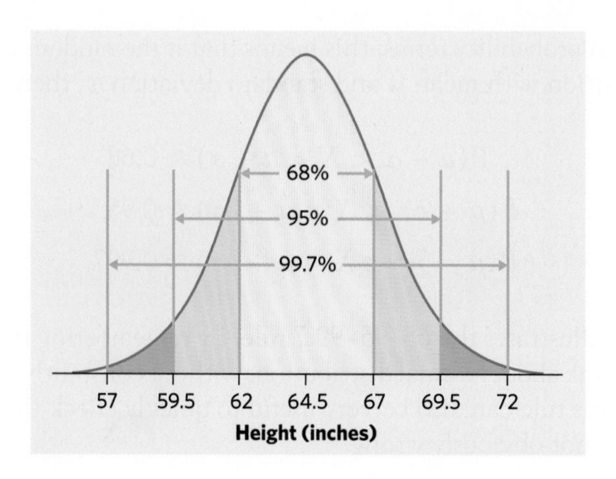

FIGURE 11.3 The 68–95–99.7 rule applied to the distribution of heights of young women aged 18 to 24 years, with $\mu = 64.5$ inches and $\sigma = 2.5$ inches.

EXAMPLE 11.2 Heights of young women

Look again at Figure 11.3. A height of 62 inches is one standard deviation below the mean. What is the probability that a woman is taller than 62 inches? Find the answer by adding areas in the figure. Here is the calculation in pictures:

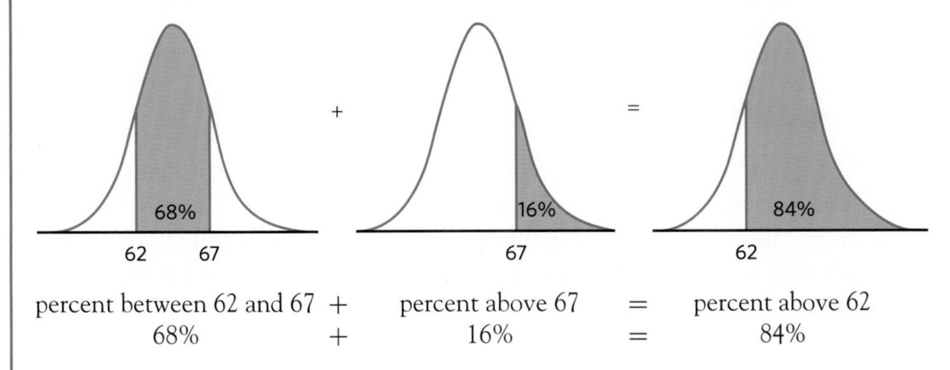

percent between 62 and 67	+	percent above 67	=	percent above 62
68%	+	16%	=	84%

Be sure you see where the 16% came from: 32% of heights are outside the range 62 to 67 inches, and half of these are above 67 inches. ■

The 68–95–99.7 rule describes distributions that are exactly Normal. Real data are never "exactly" Normal, however. Look back at Example 9.8 (page 218) on the heights of women in their 40s. The example and Figure 9.4 show that calculations using the histogram of raw data are quite close to calculations using a Normal density curve. Differences are due, in part, to the fact that someone's height is usually reported only to the nearest quarter of an inch. A height is reported as 59.0 or 59.25, but not 59.1436. We can use a Normal distribution because it's a good approximation and because we know that heights are actually continuous—people do not grow by steps of a quarter of an inch!

Because we will mention Normal distributions often, a short notation is helpful. We abbreviate the Normal distribution with mean μ and standard deviation σ as $N(\mu, \sigma)$. For example, the distribution of young women's heights is approximately $N(64.5, 2.5)$.

APPLY YOUR KNOWLEDGE

11.1 Men's bladders. The distribution of bladder volume in men is approximately Normal with mean 550 milliliters (ml) and standard deviation 100 ml.[1] Draw a Normal curve on which this mean and standard deviation are correctly located. (*Hint:* Draw the curve first, locate the points where the curvature changes, then mark the horizontal axis.)

11.2 Men's bladders. The distribution of bladder volume in men is approximately Normal with mean 550 ml and standard deviation 100 ml. Use the 68–95–99.7 rule to answer the following questions. (Use the sketch you made in the previous exercise.)

(a) Between what volumes do the middle 95% of men's bladders fall?

(b) What percent of men's bladders have a volume larger than 650 ml?

11.3 Women's bladders. The distribution of bladder volume in women is approximately Normal with mean 400 ml and standard deviation 75 ml. Use the 68–95–99.7 rule to answer the following questions.

(a) Between what values do almost all (99.7%) of women's bladder volumes fall?

(b) How small are the smallest 2.5% of all bladders among women?

The standard Normal distribution

As the 68–95–99.7 rule suggests, all Normal distributions share many common properties. In fact, all Normal distributions are the same if we measure in units of size σ about the mean μ as center. Changing to these units is called *standardizing*. To standardize a value, subtract the mean of the distribution and then divide by the standard deviation.

STANDARDIZING AND z-SCORES

If x is an observation from a distribution that has mean μ and standard deviation σ, the **standardized value** of x is

$$z = \frac{x - \mu}{\sigma}$$

A standardized value is often called a **z-score.**

A z-score tells us how many standard deviations the original observation falls away from the mean, and in which direction. Observations larger than the mean are positive when standardized, and observations smaller than the mean are negative.

EXAMPLE 11.3 Standardizing women's heights

The heights of young women aged 18 to 24 are approximately Normal with $\mu = 64.5$ inches and $\sigma = 2.5$ inches. The standardized height is

$$z = \frac{\text{height} - 64.5}{2.5}$$

A woman's standardized height is the number of standard deviations by which her height differs from the mean height of all young women. A woman 70 inches tall, for example, has standardized height

$$z = \frac{70 - 64.5}{2.5} = 2.2$$

or 2.2 standard deviations above the mean. Similarly, a woman 5 feet (60 inches) tall has standardized height

$$z = \frac{60 - 64.5}{2.5} = -1.8$$

or 1.8 standard deviations less than the mean height. ■

We often standardize observations from symmetric distributions to express them in a common scale. We might, for example, measure the height of a boy at two different ages by calculating each z-score. The standardized heights tell us where the child stands in the distribution for his age group and whether he might be growing more slowly than other children of his age.

If the variable we standardize has a Normal distribution, standardizing does more than give a common scale. It makes all Normal distributions into a single distribution, and this distribution is still Normal. Standardizing a variable that has any Normal distribution produces a new variable that has the *standard Normal distribution*.

STANDARD NORMAL DISTRIBUTION

The **standard Normal distribution** is the Normal distribution $N(0, 1)$ with mean 0 and standard deviation 1.

If a variable x has any Normal distribution $N(\mu, \sigma)$ with mean μ and standard deviation σ, then the standardized variable

$$z = \frac{x - \mu}{\sigma}$$

has the standard Normal distribution.

Keep in mind that *standardizing a variable x by subtracting the mean and dividing by the standard deviation does not automatically make the resulting z-score Normal*. The resulting z-score will follow the standard Normal distribution $N(0, 1)$ only if the original x variable had a Normal distribution $N(\mu, \sigma)$.

APPLY YOUR KNOWLEDGE

11.4 Men's and women's heights. The heights of women aged 20 to 29 are approximately Normal with mean 64 inches and standard deviation 2.7 inches. Men the same age have mean height 69.3 inches with standard deviation 2.8 inches. What are the z-scores for a woman 6 feet tall and for a man 6 feet tall? Say in simple language what information the z-scores give that the actual heights do not.

11.5 Growth chart. Boys' growth charts indicate that the heights of five-year-old boys are approximately Normal with mean 43 inches and standard deviation 1.8 inches, whereas those of seven-year-old boys are approximately Normal with mean 48 inches and standard deviation 2.1 inches. John measured 44 inches on his fifth birthday, and now, on his seventh birthday, he measures 49 inches. Find John's standardized scores for both ages and compare them. Has he been growing faster or more slowly than the general population of boys his age?

Finding Normal probabilities

Areas under a Normal curve represent proportions (frequencies) of observations from that Normal distribution. Therefore, they also represent probabilities of randomly selecting an individual from that Normal distribution. There is no direct formula for areas under a Normal curve. Calculations use either software that calculates areas or a table of areas. Tables and most software calculate one kind of area, *cumulative probabilities*.

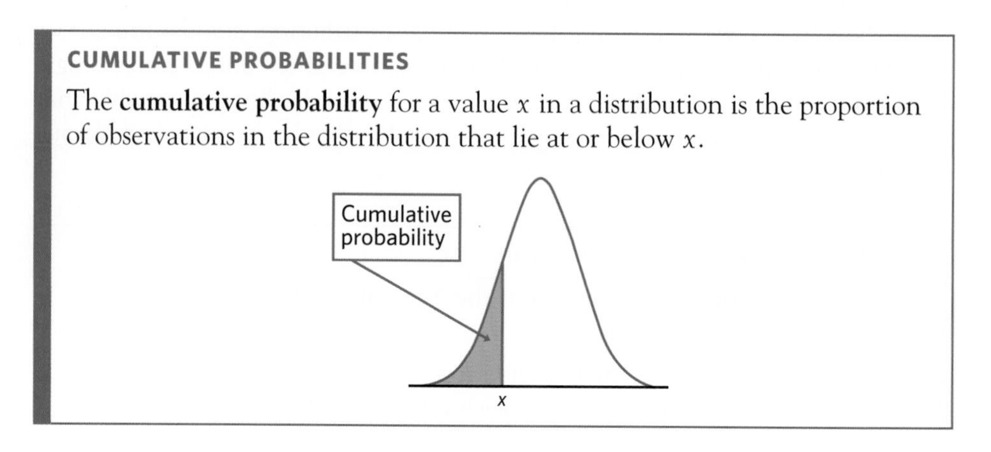

CUMULATIVE PROBABILITIES

The **cumulative probability** for a value x in a distribution is the proportion of observations in the distribution that lie at or below x.

The key to calculating Normal probabilities is to match the area you want with areas that represent cumulative probabilities (or cumulative proportions). *If you make a sketch of the area you want, you will almost never go wrong.* Find areas for cumulative proportions either from software or (with an extra step) from a table. The following example shows the method in a picture.

Brooke Fasani/Corbis

EXAMPLE 11.4 Length of human pregnancies

In the United States, the length of human pregnancies from conception to birth varies according to a distribution that is approximately Normal with mean 266 days and standard deviation 16 days. Babies born substantially before term must be given special medical attention. What percent of babies are born after 8 months (240 days) or more of gestation from conception?

Here is the calculation in a picture: The proportion of babies born with 240 days of gestation or more is the area under the curve to the right of 240. That's the total area under the curve (which is always 1) minus the cumulative proportion up to 240.

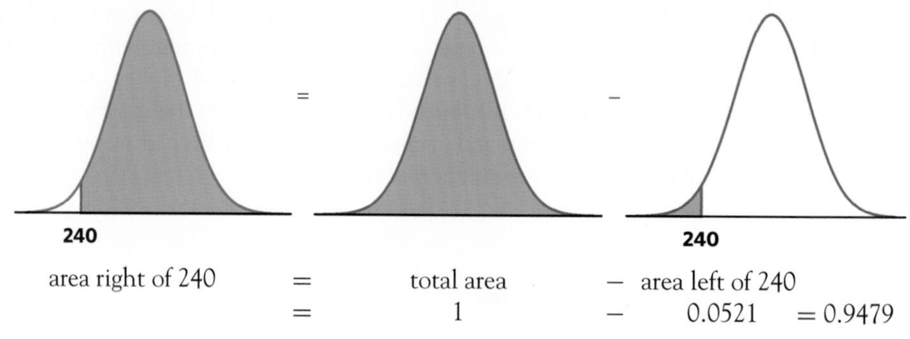

area right of 240 = total area − area left of 240
= 1 − 0.0521 = 0.9479

About 95% of all babies born in the United States have 240 days or more of gestation since conception. ■

There is *no* area under a smooth curve and exactly over the point 240. Consequently, the area to the right of 240 (the proportion of babies born at > 240 days) is the same as the area at or to the right of this point (the proportion of babies born at ≥ 240 days).

Actual birth records may show babies born 240 days after conception. However, a child born "exactly 239.8 days" after conception would be labeled as born 240 days after conception because of rounding and lack of precision in estimating time of conception. That the proportion of babies born exactly 240 days after conception is 0 for a Normal distribution is a consequence of the idealized smoothing of Normal distributions for data. Under a Normal distribution, the proportion of babies born exactly 239.8 days after conception is also 0. *Be sure to distinguish the everyday language we use to describe a variable—such as pregnancy length—and a mathematical description like the Normal distribution.*

To find the numerical value 0.0521 of the cumulative proportion in Example 11.4 using software, plug in mean 266 and standard deviation 16 and ask for the cumulative proportion for 240. Software often uses terms such as "cumulative distribution" or "cumulative probability." Figure 11.4 shows examples of software outputs, using Minitab, Excel, and R. The P in the Minitab output stands for "probability," but we can read it as "proportion of the observations" too. Minitab also displays the mean and standard deviation in its output, whereas Excel and R list them only within their command functions: `norm.dist(x,mean,standard_dev,cumulative)` for Excel, and `pnorm(x,mean=,sd=)` for R.

CrunchIt! and the *Normal Density Curve* applet are even handier because they draw pictures as well as finding areas. If you are not using software, you can find cumulative proportions for Normal curves from a table. That requires an extra step.

Minitab

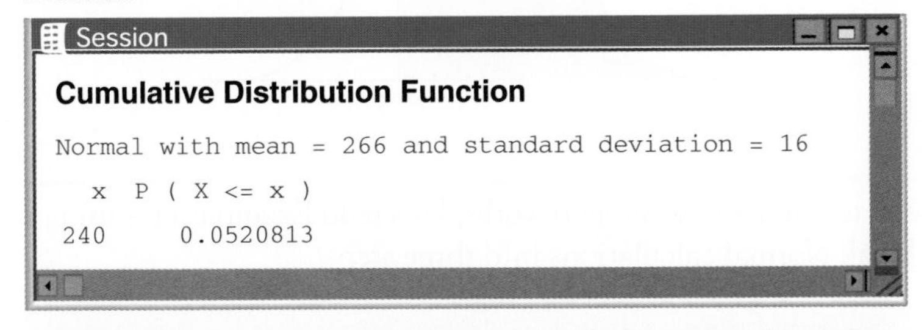

Session

Cumulative Distribution Function

```
Normal with mean = 266 and standard deviation = 16

  x  P ( X <= x )

240     0.0520813
```

Microsoft Excel

	A	B	C	D
1	0.052081	=NORM.DIST(240,266,16,TRUE)		
2				

R

```
> pnorm(240, mean=266, sd=16)
[1] 0.05208128
```

FIGURE 11.4 Software output for Example 11.4.

Using the standard Normal table*

To use a table to find cumulative proportions we must first standardize to express the problem in the standard scale of z-scores. This allows us to get by with just one table, a table of *standard Normal probabilities*.

Table B in the back of the book gives cumulative proportions (probabilities) for the standard Normal distribution. Table B also appears on the inside front cover. The pictures at the top of the table remind us that the entries are cumulative proportions, areas under the curve to the left of a value z.

EXAMPLE 11.5 The standard Normal table

What proportion of observations on a standard Normal variable z take values less than 1.47?

Solution: To find the area to the left of 1.47, locate 1.4 in the left-hand column of Table B, then locate the remaining digit 7 as .07 in the top row. The entry opposite 1.4 and under .07 is 0.9292. This is the cumulative proportion we seek. Figure 11.5 illustrates this area. ■

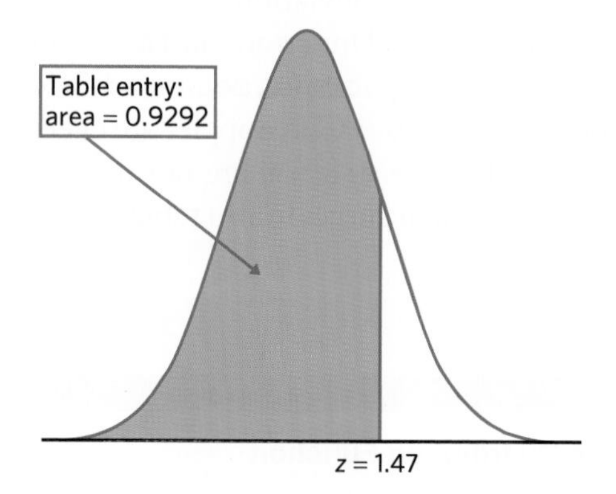

FIGURE 11.5 The area under a standard Normal curve to the left of the point $z = 1.47$ is 0.9292, for Example 11.5. Table B gives areas under the standard Normal curve.

Now that you see how Table B works, let's redo Example 11.4 using the table. We can break Normal calculations into three steps.

EXAMPLE 11.6 Length of human pregnancies

The lengths of human pregnancies from conception to birth follow the Normal distribution with mean $\mu = 266$ days and standard deviation $\sigma = 16$ days. What percent of pregnancies are at least 240 days long?

Step 1. Draw a picture. The picture is exactly as in Example 11.4.

*This section is unnecessary if you will always use software for Normal distribution calculations.

Step 2. **Standardize.** Call the pregnancy length x. Subtract the mean, then divide by the standard deviation, to transform the problem about x into a problem about a standard Normal z:

$$x \geq 240$$
$$\frac{x - 266}{16} \geq \frac{240 - 266}{16}$$
$$z \geq -1.625$$

Step 3. **Use the table.** The picture says that we want the cumulative proportion for $x = 240$. Step 2 says this is the same as the cumulative proportion for $z = -1.625$, or approximately $z = -1.63$. The Table B entry for $z = -1.63$ says that this cumulative proportion is 0.0516. The area to the right of -1.63 is therefore $1 - 0.0516 = 0.9484$, or about 95%. ■

The area from the table in Example 11.6 (0.9484) is slightly less accurate than the area from software in Example 11.4 (0.9479) because we must round z to two decimal places when we use Table B. The difference is rarely important in practice. Here's the method in outline form.

FINDING NORMAL PROBABILITIES WITH TABLE B

1. State the problem in terms of the observed variable x. **Draw a picture** that shows the proportion you want in terms of cumulative proportions.
2. **Standardize** x to restate the problem in terms of a standard Normal variable z.
3. **Use Table B** and the fact that the total area under the curve is 1 to find the required area under the standard Normal curve.

EXAMPLE 11.7 Heights of young women

The heights of young women aged 18 to 24 are approximately Normal with $\mu = 64.5$ inches and $\sigma = 2.5$ inches. What is the probability that a randomly selected woman from this age group measures between 60 and 68 inches?

Step 1. **State the problem and draw a picture.** Call the height x. The variable x has the $N(64.5, 2.5)$ distribution. The probability that a randomly selected young woman measures between 60 and 68 inches reflects the proportion of women measuring between 60 and 68 inches in the population of these young women. What is this proportion? Here is the picture:

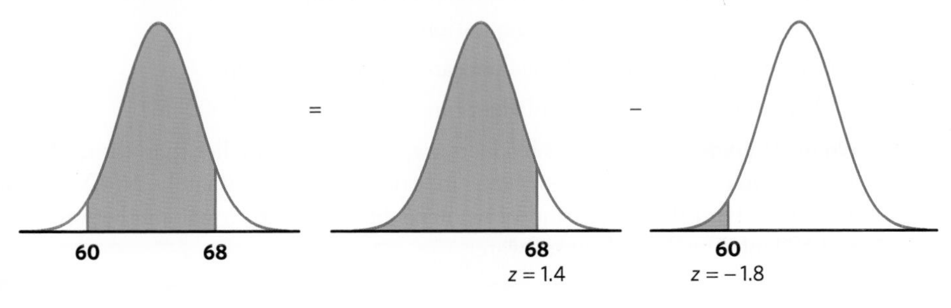

Step 2. **Standardize.** Subtract the mean, then divide by the standard deviation, to turn x into a standard Normal z:

$$60 \leq x < 68$$
$$\frac{60 - 64.5}{2.5} \leq \frac{x - 64.5}{2.5} < \frac{68 - 64.5}{2.5}$$
$$-1.8 \leq z < 1.4$$

Step 3. **Use the table.** Follow the picture (we added the z-scores to the picture label to help you):

$$\text{area between } -1.8 \text{ and } 1.4 = (\text{area left of } 1.4) - (\text{area left of } -1.8)$$
$$= 0.9192 - 0.0359 = 0.8833$$

The probability that a randomly selected young woman measures between 60 and 68 inches is about 0.88, or 88%. ■

Using the R software, we get the value 0.883313 when typing `pnorm(68, mean=64.5, sd=2.5) – pnorm(60, mean=64.5, sd=2.5)`. We also get 0.883313 by typing `pnorm(1.4, mean=0, sd=1) – pnorm(−1.8, mean=0, sd=1)`. The two methods are equivalent, which is why we can always compute a z-score and use Table B whether we are studying heights in inches or pregnancy lengths in days.

Sometimes we encounter a value of z more extreme than those appearing in Table B. For example, the area to the left of $z = -4$ is not given directly in the table. The z-values in Table B leave only an area of 0.0002 in each tail unaccounted for. Therefore, there is an area of 0.0004 outside the range of Table B. For practical purposes, we can act as if there is zero area outside the range of Table B.

APPLY YOUR KNOWLEDGE

11.6 **Finding a proportion.** Use Table B or software to find the proportion of observations from a standard Normal distribution that satisfies each of the following statements. In each case, sketch a standard Normal curve and shade the area under the curve that is the answer to the question.

(a) $z < 2.85$

(b) $z > 2.85$

(c) $z > -1.66$

(d) $-1.66 < z < 2.85$

11.7 **Men's bladders.** The distribution of bladder volume in men is approximately Normal with mean $\mu = 550$ ml and standard deviation $\sigma = 100$ ml. Use Table B or software to answer the following questions.

(a) What proportion of male bladders are larger than 500 ml?

(b) What proportion of male bladders are between 500 and 600 ml?

11.8 **Women's bladders.** Women's bladders are, on average, smaller than those of men. The distribution of bladder volume in women is approximately $N(400, 75)$ in milliliters. Find the proportion of women's bladders with a volume between 500 and 600 ml. Use Table B or software.

Finding a value, given a probability or a proportion

Examples 11.4 and 11.6 illustrate the use of software or Table B to find what proportion of the observations satisfies some condition, such as "pregnancies of 240 days or longer." We may instead want to find the observed value with a given proportion of the observations above or below it. Statistical software will do this directly.

EXAMPLE 11.8 Find the top 10% using software

The lengths of human pregnancies in days from conception to birth follow approximately the $N(266, 16)$ distribution. How long are the longest 10% of pregnancies?

We want to find the pregnancy length x with area 0.1 to its *right* under the Normal curve with mean $\mu = 266$ and standard deviation $\sigma = 16$. That's the same as finding the pregnancy length x with area 0.9 to its *left*, because the value x we are looking for separates the shortest 90% of pregnancies from the longest 10%.

Most software will tell you x when you plug in mean 266, standard deviation 16, and cumulative probability 0.9. Figure 11.6 displays the output from the TI-83 calculator and Minitab. Both software programs give approximately $x = 286.505$. So pregnancies longer than 286 days (287 days or longer) are in the top 10%. (Round up because pregnancy durations are expressed in whole numbers.) That's three weeks or more longer than the mean pregnancy length of 266 days. ■

TI-83

```
invNorm(0.9,266,
16)
        286.5048251
```

Minitab

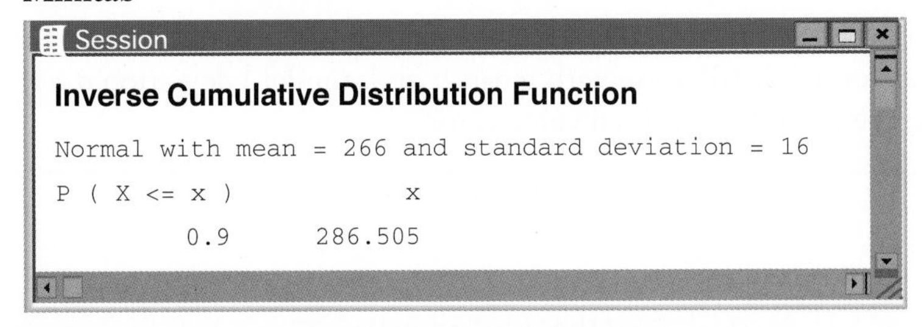

FIGURE 11.6 Software output for Example 11.8.

Without software, use Table B backward. Find the given proportion in the body of the table and then read the corresponding z from the left column and top row. There are again three steps.

EXAMPLE 11.9 Find the top 10% using Table B

The lengths of human pregnancies in days from conception to birth follow approximately the $N(266, 16)$ distribution. How long are the longest 10% of pregnancies?

Step 1. State the problem and draw a picture. This step is exactly as in Example 11.8. The picture is shown on next page.

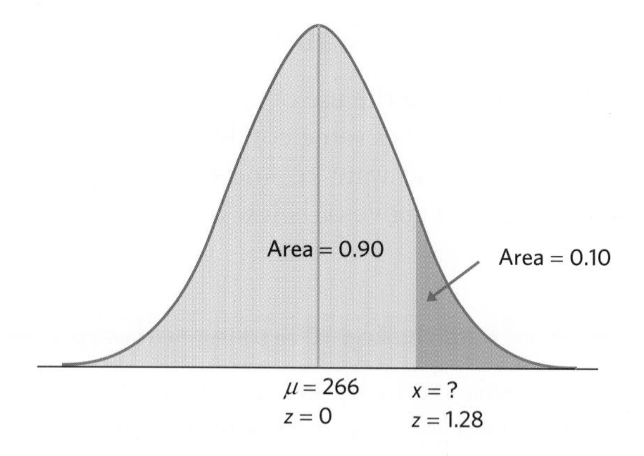

Step 2. **Use the table.** Look in the body of Table B for the entry closest to 0.9. It is 0.8997. This is the entry corresponding to $z = 1.28$. So $z = 1.28$ is the standardized value with approximately area 0.9 to its left.

Step 3. **Unstandardize** to transform z back to the original x scale. We know that the standardized value of the unknown x is $z = 1.28$. So x itself satisfies

$$\frac{x - 266}{16} = 1.28$$

Solving this equation for x gives

$$x = 266 + (1.28)(16) = 286.48$$

This equation should make sense: It says that x lies 1.28 standard deviations above the mean on this particular Normal curve. That is the "unstandardized" meaning of $z = 1.28$. Pregnancies longer than 286 days are among the 10% longest pregnancies. ■

Here is the general formula for unstandardizing a z-score. To find the value x from the Normal distribution with mean μ and standard deviation σ corresponding to a given standard Normal value z, use

$$x = \mu + z\sigma$$

Glow Images/SuperStock

EXAMPLE 11.10 Find the third quartile

The hatching weights of commercial chickens can be modeled accurately using a Normal distribution with mean $\mu = 45$ grams (g) and standard deviation $\sigma = 4$ g.[2] What is the third quartile of the distribution of hatching weights?

Step 1. **State the problem and draw a picture.** Call the hatching weight x. The variable x has the $N(45, 4)$ distribution. The third quartile is the value with 75% of the distribution to its left. Figure 11.7 is the picture.

Step 2. **Use the table.** Look in the body of Table B for the entry closest to 0.75. It is 0.7486. This is the entry corresponding to $z = 0.67$. So $z = 0.67$ is the standardized value with area 0.75 to its left.

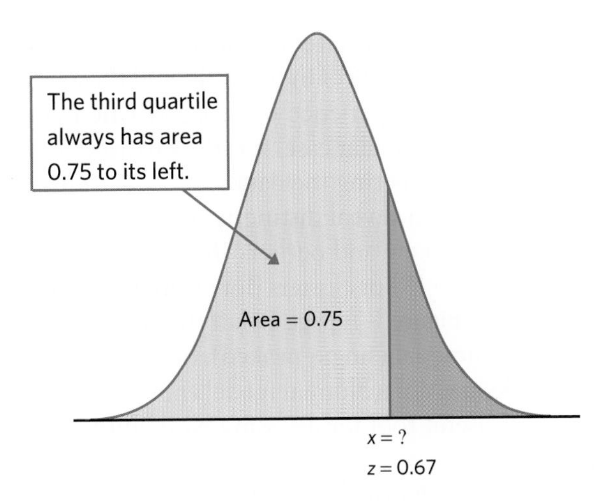

FIGURE 11.7 Locating the third quartile of a Normal curve, for Example 11.10.

Step 3. Unstandardize. The hatching weight corresponding to $z = 0.67$ is

$$x = \mu + z\sigma$$
$$= 45 + (0.67)(4) = 47.68$$

We could combine Steps 2 and 3 using technology. For example, typing `invNorm(0.75,45,4)` in the TI-83 calculator gives a hatching weight, in grams, of 47.697959.

The third quartile of hatching weights of commercial chickens is about 47.7 g. ■

APPLY YOUR KNOWLEDGE

11.9 Finding an unknown value. Use Table B or software to find the value z of a standard Normal variable that satisfies each of the following conditions. (If using Table B, select the value of z in the table that comes closest to satisfying the condition.) In each case, sketch a standard Normal curve with your value of z marked on the axis.

(a) The point z with 25% of the observations falling below it.

(b) The point z with 40% of the observations falling above it.

11.10 IQ test scores. Scores on the Wechsler Adult Intelligence Scale are approximately Normally distributed with $\mu = 100$ and $\sigma = 15$. Use Table B or software to answer the following questions.

(a) What scores fall in the lowest 25% of the distribution?

(b) How high a score is needed to be in the highest 5%?

The bell curve? Does the distribution of human intelligence follow the "bell curve" of a Normal distribution? Scores on IQ tests do roughly follow a Normal distribution. That is because a test score is calculated from a person's answers in a way that is designed to produce a Normal distribution in the population. To conclude that *intelligence* follows a bell curve, we must agree that the test scores directly measure intelligence. Many psychologists don't think there is one human characteristic that we can call "intelligence" and that can be measured by a single test score.

Normal quantile plots*

The Normal distributions provide good descriptions of some distributions of real data, such as IQ tests or individual heights. The distributions of some other common variables are skewed and therefore distinctly non-Normal. Examples include

*This section is optional for a quarter-based course.

the survival times of cancer patients after treatment, the amount of a given toxin in the blood among a random sample of individuals, and the average size or weight of mammalian species. While experience can suggest whether or not a Normal distribution is plausible in a particular case, it is risky to assume that a distribution is Normal without actually inspecting the data.

A histogram or dotplot can reveal distinctly non-Normal features of a distribution, such as pronounced skew and outliers (for example, the survival time of guinea pigs in Figure 1.6, page 19) or clusters (for example, the age of patients diagnosed with Lyme disease in Figure 1.7, page 19). If the dotplot or histogram appears roughly symmetric and unimodal (single-peaked), however, we need a more sensitive way to judge the adequacy of a Normal model (not all symmetric distributions are Normal). The most useful tool for assessing Normality is another graph, the **Normal quantile plot.**

Normal quantile plot

Here is the basic idea of a Normal quantile plot. The graphs produced by software use more sophisticated versions of this idea. It is not practical to make Normal quantile plots by hand.

1. Arrange the observed data values from smallest to largest. Record what percentile of the data each value occupies. For example, the smallest observation in a set of 20 is at the 5% point, the second smallest is at the 10% point, and so on.
2. Do Normal distribution calculations to find the z-scores at these same percentiles. For example, $z = -1.645$ is the 5% point of the standard Normal distribution, and $z = -1.282$ is the 10% point.
3. Plot each data point x against the corresponding z. If the data distribution is close to standard Normal, the plotted points will lie close to the 45-degree line $x = z$. If the data distribution is close to any Normal distribution, the plotted points will lie close to some straight line.

Any Normally distributed data set will produce a straight line on a Normal quantile plot, because the data distribution and the z distribution are both Normal and their relationship is thus linear. If the data are not Normally distributed, the data and the z distribution are unrelated, and a Normal quantile plot will not be linear.

USE OF NORMAL QUANTILE PLOTS

If the points on a **Normal quantile plot** lie close to a straight line, the plot indicates that the data are Normal. Systematic deviations from a straight line indicate a non-Normal distribution. Outliers appear as points that are far away from the overall pattern of the plot.

EXAMPLE 11.11 Shark body lengths

Figure 11.8 is a Normal quantile plot of the body lengths of sharks from Example 1.5 (page 14). Lay a transparent straightedge over the center of the plot to see that most

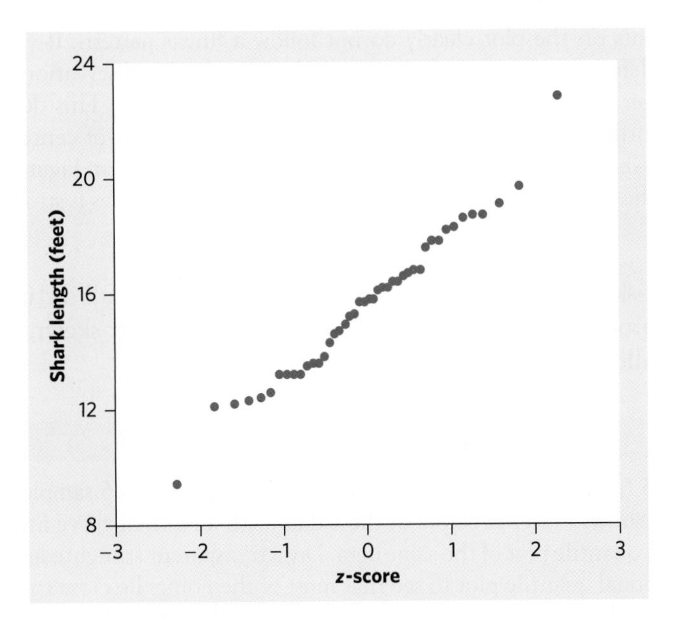

Mark Conlin/Oxford Scientific/Getty Images

FIGURE 11.8 Normal quantile plot of shark lengths, for Example 11.11. This distribution is approximately Normal except for a low outlier and a high outlier.

of the points lie close to a straight line. A Normal distribution describes these points quite well. The low and high outliers at 9.4 and 22.8 feet lie, respectively, below and above the line formed by the center of the data. Compare Figure 11.8 with the histogram of these data in Figure 1.5(b) (page 15). ■

EXAMPLE 11.12 Guinea pig survival times

Figure 11.9 is a Normal quantile plot of the guinea pig survival times from Example 1.7 (page 18). The data x are plotted vertically against the corresponding standard Normal z-score plotted horizontally. The z-score scale extends from -3 to 3 because almost all of a standard Normal curve lies between these values.

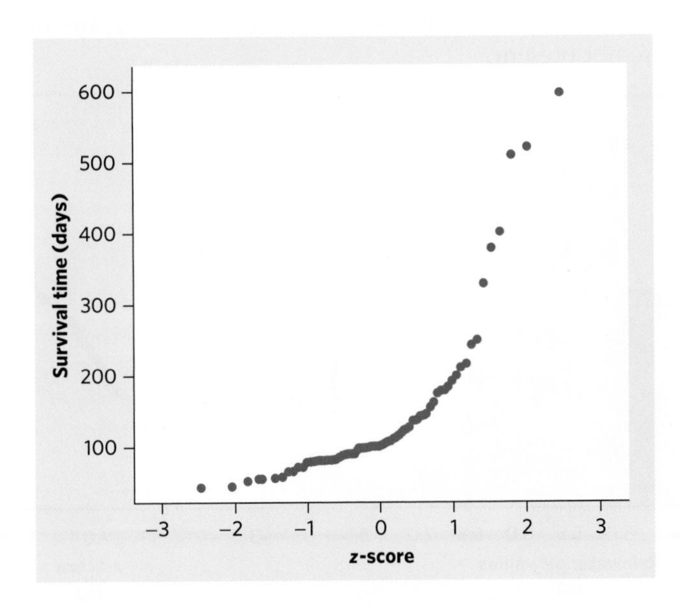

FIGURE 11.9 Normal quantile plot of guinea pig survival times, for Example 11.12. This distribution is skewed to the right.

The points on the plot clearly do not follow a linear pattern. If you draw a line through the leftmost points, which correspond to the smaller observations, you will find that the larger observations fall systematically above this line. This deviation in the Normal quantile plot indicates a right-skew. That is, the right-of-center observations have larger values than in a Normal distribution. The histogram in Figure 1.6 (page 19) also clearly shows that the distribution of guinea pig survival times is strongly skewed to the right. ■

In a right-skewed distribution, the largest observations fall distinctly above a line drawn through the main body of points. Similarly, left-skewness is evident when the smallest observations fall below the line.

EXAMPLE 11.13 Acid rain

Exercise 1.33 (page 34) provided data on the acidity (pH) of 105 samples of rainwater. Figure 11.10 shows (a) a histogram of these data with a Normal curve fitted over it and (b) a Normal quantile plot of the same data. Lay a transparent straightedge over the center of the Normal quantile plot to see that most of the points lie close to a straight line. A Normal distribution describes these points quite well. Notice also that the majority of the points on the Normal quantile plot have z-scores between -1 and 1 and few have z-scores close to 3 or -3, reflecting the 68–95–99.7 property of Normal distributions.

As Exercise 1.33 illustrated, histograms don't settle the question of approximate Normality, because their particular shape depends on the choice of classes. The Normal quantile plot makes it clear that a Normal distribution is a good description for these data—there are only minor wiggles in a generally straight-line pattern. ■

As Figure 11.10 illustrates, real data almost always show some departure from the theoretical Normal model. *When you examine a Normal quantile plot, look for shapes that show clear departures from Normality. Don't overreact to minor wiggles in the plot.* When we discuss statistical methods that are based on the Normal model, we will pay attention to the sensitivity of each method to departures from Normality. Many common methods work well as long as the data are approximately Normal and outliers are not present.

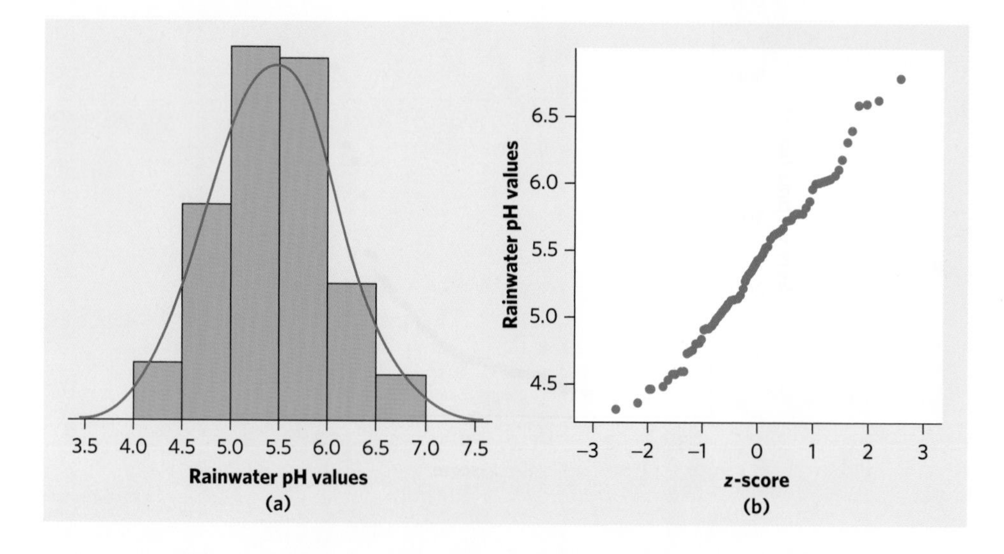

FIGURE 11.10 (a) Histogram with superimposed Normal curve and (b) Normal quantile plot of rainwater pH values, for Example 11.13. This distribution is roughly Normal.

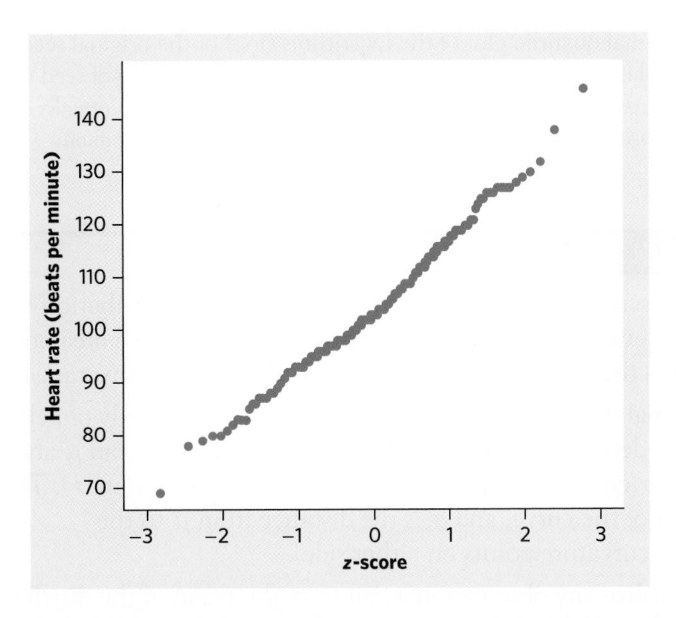

FIGURE 11.11 Normal quantile plot of the heart rates of 200 male runners, for Exercise 11.11.

APPLY YOUR KNOWLEDGE

11.11 Heart rate. Figure 11.11 is a Normal quantile plot of the heart rates of 200 male runners after a 6-minute treadmill run.[3] The distribution is close to Normal. How can you see this? Describe the nature of the small deviations from Normality that are visible in the plot.

11.12 Transforming data: counting seeds. Table 3.4 (page 86) gives data on the mean number of seeds produced in a year by several common tree species. (Some species appear twice because their seeds were counted in two locations.)

(a) Figure 11.12(a) is a Normal quantile plot of these seed count values. In what way is the distribution of seed counts non-Normal? Is it right-skewed or left-skewed? Look carefully at Figure 11.12(a): Are there more points on the graph corresponding to small seed counts or to large seed counts?

(b) Biologists know that, when dealing with sizes and counts, the logarithms of the original data are often a judicious choice of variable. Figure 11.12(b) is a

i4images horticulture/Alamy

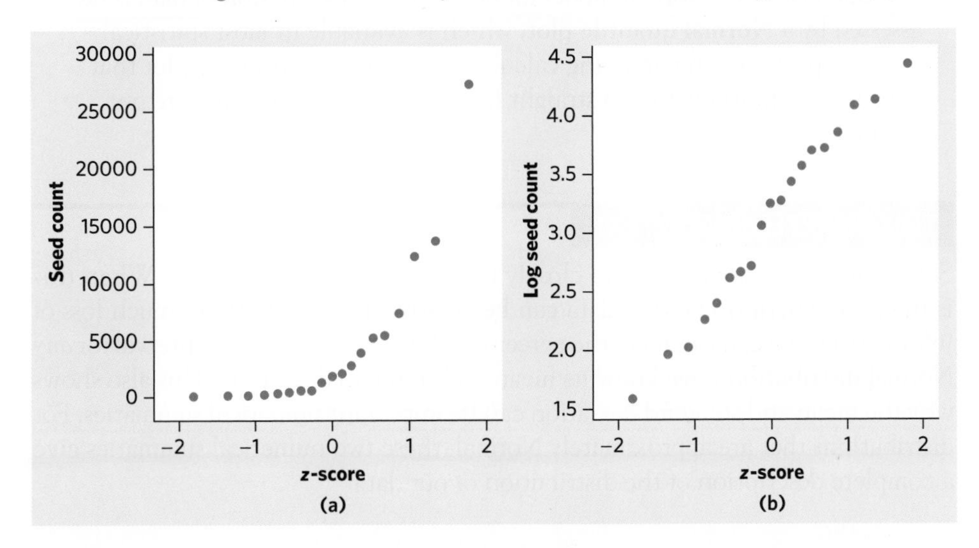

FIGURE 11.12 Normal quantile plots of (a) seed count for 19 tree species and (b) logarithm of seed count for these tree species, for Exercise 11.12.

Normal quantile plot of the logarithms (log) of the original seed count values. Explain why we can say that the distribution of the log of seed counts is approximately Normal. *Note: When a variable x is not Normally distributed but its logarithm log(x) is, we say that x has a lognormal distribution.*

CHAPTER 11 SUMMARY

- We can describe the overall pattern of a continuous distribution by a **density curve.** A density curve has total area 1 underneath it. An area under a density curve gives the proportion of observations that fall within a range of values.
- The **Normal distributions** are described by a special family of bell-shaped, symmetric density curves, called **Normal curves.** The mean μ and standard deviation σ completely specify a Normal distribution $N(\mu, \sigma)$. The mean is the center of the curve, and σ is the distance from μ to the change-of-curvature points on either side.
- To **standardize** any observation x, subtract the mean of the distribution and then divide by the standard deviation. The resulting **z-score**

$$z = \frac{x - \mu}{\sigma}$$

  says how many standard deviations x lies away from the distribution mean.
- All Normal distributions are the same when measurements are transformed to the standardized scale. In particular, all Normal distributions satisfy the **68–95–99.7 rule,** which describes what percent of observations lie within one, two, and three standard deviations of the mean.
- If x has the $N(\mu, \sigma)$ distribution, then the **standardized variable** $z = (x - \mu)/\sigma$ has the **standard Normal distribution** $N(0, 1)$ with mean 0 and standard deviation 1. Table B gives the **cumulative proportions** of standard Normal observations that are less than z for many values of z. By standardizing, we can use Table B for any Normal distribution.
- The adequacy of a Normal model for describing a distribution of data is best assessed by a **Normal quantile plot,** which is available in most statistical software packages and graphing calculators. A pattern on such a plot that deviates substantially from a straight line indicates that the data are not Normal.

THIS CHAPTER IN CONTEXT

Some data sets can be shown to closely follow a Normal distribution. When this is true, the description of the data can be greatly simplified without much loss of information. We can calculate the percent of the distribution in an interval for *any* Normal distribution if we know its mean and standard deviation. This also shows why the mean and standard deviation can be important numerical summaries. For distributions that are approximately Normal, these two numerical summaries give a complete description of the distribution of our data.

It is important to remember that not all distributions can be well approximated by a Normal curve. In particular, the distribution of data from a random sample is likely to reflect the true distribution of the population. When the distribution of the variable in the population is unknown, we can plot sample data using a histogram or, better yet, a Normal quantile plot to determine whether the sample data are roughly Normal; if they are, this suggests that, in the population from which we sampled, the variable is likely to also be roughly Normal. This approach will be key to checking whether the assumptions of many statistical inference procedures in Part III of this book are met.

Normal distributions are also good approximations to many kinds of chance outcomes, and we will use them to describe sampling distributions in Chapter 13. And when we can use a Normal distribution to model the sampling distribution of a statistic, we will be able to use Normal calculations like those done in this chapter to obtain confidence intervals and P-values for hypothesis tests. We will examine this closely in Chapters 14 and 15.

CHECK YOUR SKILLS

11.13 Which of these variables is most likely to have a Normal distribution?

(a) Lengths of newly hatched pythons
(b) Weight category (underweight, normal weight, overweight, obese) of adults
(c) Number of heart attacks experienced by men before the age of 65

11.14 To completely specify the shape of a Normal distribution, you must give

(a) the mean and the standard deviation.
(b) the five-number summary.
(c) the mean and the median.

The respiratory rate per minute in newborns varies according to a distribution that is approximately Normal with mean 50 and standard deviation 5. Exercises 11.15 to 11.17 use this information.[4]

11.15 Approximately 95% of all newborn respiratory rates per minute are between

(a) 45 and 55.
(b) 40 and 60.
(c) 35 and 65.

11.16 The probability that a randomly chosen newborn has a respiratory rate of 55 per minute or more is approximately

(a) 0.16.
(b) 0.68.
(c) 0.84.

11.17 The probability that a randomly chosen newborn has a respiratory rate per minute between 40 and 55 is approximately

(a) 0.498. (b) 0.815. (c) 0.997.

11.18 The proportion of observations from a standard Normal distribution that take values less than 1.15 is about

(a) 0.1251. (b) 0.8531. (c) 0.8749.

11.19 The proportion of observations from a standard Normal distribution that take values larger than −0.75 is about

(a) 0.2266. (b) 0.7734. (c) 0.8023.

The common fruit fly, Drosophila melanogaster, *is the most studied organism in genetic research because it is small, easy to grow, and reproduces rapidly. The length of the thorax (where the wings and legs attach) in a population of male fruit flies is approximately Normal with mean 0.800 millimeters (mm) and standard deviation 0.078 mm. Exercises 11.20 to 11.22 use this information.*

11.20 Choose a male fruit fly at random. The probability that the fly you choose has a thorax longer than 1 mm is about

(a) 0.995. (b) 0.5. (c) 0.005.

11.21 What proportion of male fruit flies have thorax length between 0.9 and 1 mm?

(a) 0.0254. (b) 0.0947. (c) 0.5398.

11.22 About 16% of male fruit flies have thorax length shorter than

(a) 0.644 mm. (b) 0.722 mm. (c) 0.878 mm.

CHAPTER 11 EXERCISES

11.23 IQ test scores. The Wechsler Adult Intelligence Scale (WAIS) is the most common "IQ test." The scale of scores is set separately for each age group and is approximately Normal with mean 100 and standard deviation 15. Use the 68–95–99.7 rule to answer the following questions.

(a) What percent of people have WAIS scores above 100? Above 145? Below 85?

(b) The organization MENSA, which calls itself "the high IQ society," requires an IQ score of 130 or higher for membership. What percent of adults would qualify for membership?

(c) People with WAIS scores below 70 are considered mentally retarded when, for example, applying for Social Security disability benefits. About what percent of adults are mentally retarded by this criterion?

11.24 Pregnancy length in horses. Bigger mammals tend to carry their young longer before birth. The length of horse pregnancies from conception to birth varies according to a roughly Normal distribution with mean 336 days and standard deviation 3 days. Use the 68–95–99.7 rule to answer the following questions.

(a) Almost all (99.7%) horse pregnancies fall within what range of lengths?

(b) What percent of horse pregnancies are longer than 339 days?

11.25 The standard Normal curve. Use Table B or software to find the proportion of observations from a standard Normal distribution that fall in each of the following regions. In each case, sketch a standard Normal curve and shade the area representing the region.

(a) $z \leq -2.25$

(b) $z \geq -2.25$

(c) $z > 1.77$

(d) $-2.25 < z < 1.77$

11.26 The standard Normal curve. Use Table B or software to find the proportion of observations from a standard Normal distribution that satisfy each of the following statements. In each case, sketch a standard Normal curve and shade the area under the curve that is the answer to the question.

(a) $z < 1.85$

(b) $z > 1.85$

(c) $z > -0.66$

(d) $-0.66 < z < 1.85$

11.27 The standard Normal curve. Use Table B or software to find the standardized value z that satisfies each of the following conditions (report the value of z that comes closest to satisfying the condition).

(a) The probability is 0.8 that a randomly selected observation falls below z.

(b) The probability is 0.35 that a randomly selected observation falls above z.

11.28 The standard Normal curve. Use Table B or software to find the z-score that satisfies each of the following conditions (report the value of z that comes closest to satisfying the condition). In each case, sketch a standard Normal curve with your value of z marked on the axis.

(a) 20% of the observations fall below z.

(b) 30% of the observations fall above z.

(c) 30% of the observations fall below z.

11.29 Grading on the curve. Grades in large classes are often approximately Normally distributed, and therefore, many large college classes "grade on a bell curve." A common practice is to give 16% A grades, 34% B grades, 34% C grades, and 16% D and F grades. Assuming a Normal distribution of grades, what are the ranges of z-scores for each of these letter grades?

11.30 Hemoglobin. The distribution of hemoglobin in grams per deciliter of blood is approximately $N(14, 1)$ in women and approximately $N(16, 1)$ in men.[5]

(a) What percent of women have more hemoglobin than the men's mean hemoglobin?

(b) What percent of men have less hemoglobin than the women's mean hemoglobin?

(c) What percent of men have more hemoglobin than the women's mean hemoglobin?

11.31 Birth weights. The distribution of baby weights at birth is left-skewed because of premies who have particularly low birth weights. However, within a close range of gestation times, birth weights are approximately Normally distributed. For babies born at full term (37 to 39 completed weeks of gestation), for instance, the distribution of birth weight (in grams) is approximately $N(3350, 440)$.[6] Low-birth-weight babies (weighing less than 2500 grams, or about 5 pounds 8 ounces) are at an increased risk of serious health problems. Among those, very-low-birth-weight babies (weighing less than 1500 grams, or about 3 pounds 4 ounces) have the highest risk for health problems.

(a) What proportion of babies born full term are low-birth-weight babies?
(b) What proportion of babies born full term are very-low-birth-weight babies?

11.32 Birth weights, continued. How much difference do a couple of weeks make for birth weight? Late-preterm babies are born with 34 to 36 completed weeks of gestation. The distribution of birth weights (in grams) for late-preterm babies is approximately $N(2750, 560)$.

(a) What is the probability that a randomly chosen late-preterm baby would have a low birth weight (less than 2500 grams)?
(b) What is the probability that a randomly chosen late-preterm baby would have a very low birth weight (less than 1500 grams)?
(c) How do these results compare with your findings in the previous exercise?

11.33 Osteoporosis. Osteoporosis is a condition in which the bones become brittle due to loss of minerals. To diagnose osteoporosis, an elaborate apparatus measures bone mineral density (BMD). BMD is usually reported in standardized form. The standardization is based on a population of healthy young adults. The World Health Organization (WHO) criterion for osteoporosis is a BMD 2.5 or more standard deviations *below* the mean for young adults. BMD measurements in a population of people who are similar in age and sex roughly follow a Normal distribution.[7]

(a) What percent of healthy young adults have osteoporosis by the WHO criterion?
(b) Women aged 70 to 79 are, of course, not young adults. The mean BMD in this age is about −2 on the standard scale for young adults. Suppose that the standard deviation is the same as for young adults. What percent of this older population has osteoporosis?

11.34 Osteopenia. The previous exercise described the criterion for diagnosing osteoporosis. Likewise, osteopenia is low bone mineral density, defined by the WHO as a BMD between 1 and 2.5 standard deviations *below* the mean of young adults.

(a) What percent of healthy young adults have osteopenia by the WHO criterion?
(b) The mean BMD among women aged 70 to 79 is about −2 on the standard scale for young adults. Suppose that the standard deviation is the same as for young adults. What percent of this older population has osteopenia?

11.35 Foot length. A middle school in Irvine, California, had its seventh-graders measure their right foot length (in centimeters, cm) at the beginning of the school year. A Normal quantile plot indicates that a Normal curve is an excellent approximation. We will model foot length in this population using the Normal distribution with mean 23.4 cm and standard deviation 1.7 cm.[8]

(a) What percent of seventh-graders in this school had a foot length above 20.5 cm at the beginning of the school year? What percent had a foot length above 25.5 cm?
(b) Find the first and third quartiles of foot lengths in this population. Follow Example 11.10 in your approach.

11.36 Cholesterol in middle-aged men. The blood cholesterol levels of men aged 55 to 64 are approximately Normal with mean 222 milligrams per deciliter (mg/dl) and standard deviation 37 mg/dl. What percent of these men have high cholesterol (levels above 240 mg/dl)? What percent have elevated cholesterol (between 200 and 240 mg/dl)?

11.37 Pregnancy length in humans. The length of human pregnancies from conception to birth varies according to a distribution that is approximately Normal with mean 266 days and standard deviation 16 days.

(a) What percent of pregnancies last less than 270 days (about 9 months)?
(b) What percent of pregnancies last between 240 and 270 days (roughly between 8 and 9 months)?
(c) How long do the longest 20% of pregnancies last?

11.38 Pregnancy length in humans, continued. The quartiles of any distribution are the values with cumulative proportions 0.25 and 0.75.

(a) What are the quartiles of the standard Normal distribution?
(b) Using your numerical values from (a), write an equation that gives the quartiles of the $N(\mu, \sigma)$ distribution in terms of μ and σ.
(c) The length of human pregnancies from conception to birth varies according to a distribution that is approximately Normal with mean 266 days and standard deviation 16 days. Apply your result from (b): What are the quartiles of the distribution of lengths of human pregnancies?

11.39 Body mass index. Your body mass index (BMI) is your weight in kilograms divided by the square of your height in meters. Many online BMI calculators allow you to enter weight in pounds and height in inches. High BMI is a common but controversial indicator of overweight or obesity. A study by the National Center for Health Statistics found that the BMI of American young women (ages 20 to 29) is approximately Normal with mean 26.8 and standard deviation 7.4.[9]

(a) People with BMI less than 18.5 are often classed as "underweight." What percent of young women are underweight by this criterion?
(b) People with BMI over 30 are often classed as "obese." What percent of young women are obese by this criterion?

11.40 Arsenic in blood. Arsenic is a compound that occurs naturally in very low concentrations. Arsenic blood concentrations in healthy adults are Normally distributed with mean $\mu = 3.2$ micrograms per deciliter (μg/dl) and standard deviation $\sigma = 1.5$ μg/dl.[10] What is the range of arsenic blood concentrations corresponding to the middle 90% of healthy adults?

11.41 Body temperature. An 1868 paper by German physician Carl Wunderlich reported, based on over a million body temperature readings, that healthy-adult body temperatures are approximately Normal with mean $\mu = 98.6$ degrees Fahrenheit (°F) and standard deviation $\sigma = 0.6$°F. This is still the most widely quoted result for human temperature.
(a) According to this study, what is the range of body temperatures that can be found in 95% of healthy adults? (We are looking for the middle 95% of the adult population.)
(b) A more recent study suggests that healthy-adult body temperatures are better described by an $N(98.2, 0.7)$ distribution. Based on this later study, what is the range of values representing the 95% most common body temperatures in healthy adults? Compare this range with the one you obtained in (a).

The Normal Density Curve applet allows you to do Normal calculations quickly. It is somewhat limited by the number of pixels available for use, so it can't hit every value exactly. In Exercises 11.42 and 11.43, use the closest available values. In each case, make a sketch of the curve from the applet marked with the values you used to answer the questions.

11.42 How accurate is 68–95–99.7? APPLET The 68–95–99.7 rule for Normal distributions is a useful approximation. To see how accurate the rule is, drag one flag across the other so that the applet shows the area under the curve between the two flags.
(a) Place the flags one standard deviation on either side of the mean. What is the area between these two values? What does the 68–95–99.7 rule say this area is?
(b) Repeat for locations two and three standard deviations on either side of the mean. Again compare the 68–95–99.7 rule with the area given by the applet.

11.43 Where are the quartiles? APPLET How many standard deviations above and below the mean do the quartiles of any Normal distribution lie? (Use the standard Normal distribution to answer this question.)

11.44 Exploring models for climate change. Can the recent noticeable increase in the number of extreme weather events, such as heat waves, be explained by global warming? We can use Normal curves to understand how various climate change scenarios would affect the probability of observing such extreme weather events.[11]
(a) We start by describing historical temperature variations with a model using a Normal curve such that cold weather and hot weather occupy, respectively, the tail areas below and above two standard deviations from the historical mean. Record-cold and record-hot temperatures occupy the tail areas below and above three standard deviations from the historical mean. Draw such a Normal curve, indicating on your sketch the areas corresponding to record-cold, cold, hot, and record-hot weather events.
(b) Using the same scale, draw underneath your first Normal curve another Normal curve with the same standard deviation but a somewhat higher (warmer) mean temperature (roughly one standard deviation greater than the historical mean). Using your first Normal curve as a reference, indicate on your sketch the areas corresponding to record-cold, cold, hot, and record-hot weather events in this new model. Would we observe more or fewer record-cold, cold, hot, and record-hot weather events?
(c) Underneath these, draw a third Normal curve with the same mean as that of your second curve but a somewhat larger standard deviation (aim for roughly one and a quarter to one and a half times as large). Using your first Normal curve as a reference, indicate on your sketch the areas corresponding to record-cold, cold, hot, and record-hot weather events in this new model. Would we observe more or fewer record-cold, cold, hot, and record-hot weather events?

Exercises 11.45 to 11.49 make use of the optional material on Normal quantile plots.

11.45 Acorn sizes. Figure 11.13 shows the Normal quantile plot for the volumes of 28 acorns from the Atlantic area.[12] Examine the plot for deviations from Normality and explain your conclusions.

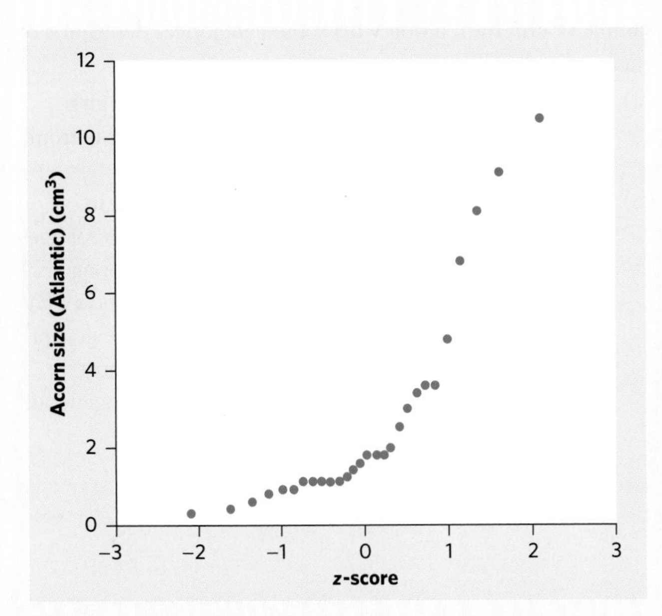

FIGURE 11.13 Normal quantile plot of the volumes of 28 acorns from the Atlantic region, for Exercise 11.45.

11.46 Onset of anorexia nervosa. Figure 11.14 shows the Normal quantile plot for data on the age at onset of anorexia nervosa in 691 teenage girls from Exercise 1.8. Examine the Normal quantile plot for deviations from Normality. Explain your conclusions and compare this plot with the histogram of the same data shown in Figure 1.8 (page 20).

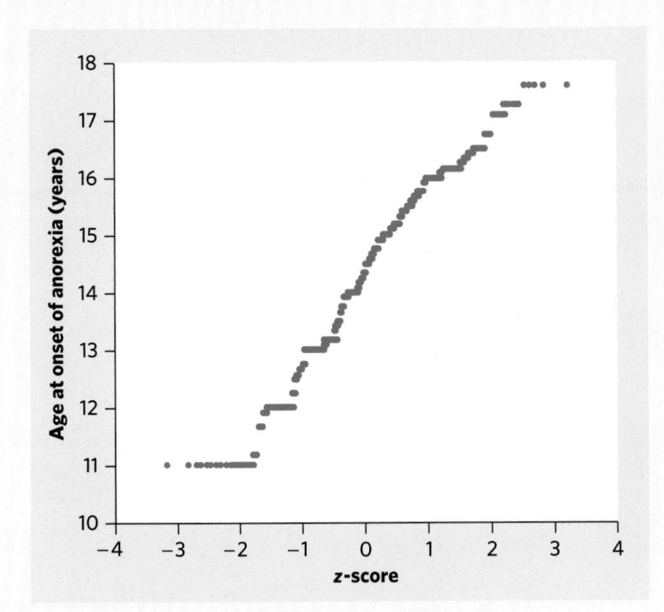

FIGURE 11.14 Normal quantile plot of age at onset of anorexia nervosa in 691 teenage girls, for Exercise 11.46.

11.47 Fruit flies. Figure 11.15 shows the Normal quantile plot for data on the thorax lengths of 49 fruit flies.[13] Examine the plot for deviations from Normality and explain your conclusions. Because of limited precision in measurements, many observations happen to have the same numerical value. This gives a steplike appearance to the Normal quantile plot. This phenomenon is called **granularity.** It is not an important feature and can be ignored in your interpretation.

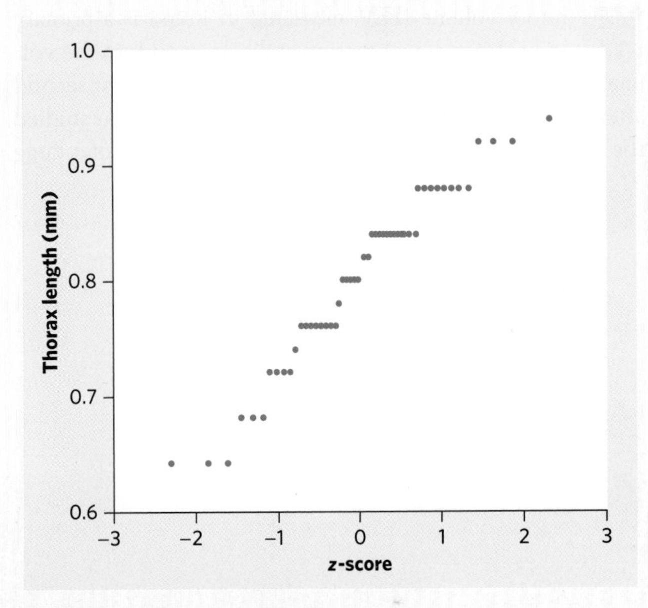

FIGURE 11.15 Normal quantile plot of the thorax lengths of 49 male fruit flies, for Exercise 11.47.

11.48 Logging in the rain forest. Exercise 2.39 (page 63) described a study of the effects of logging on tree counts in the Borneo rain forest.

(a) Make your own Normal quantile plot of the data in Group 3 for plots of forest that were logged 8 years earlier: First, sort and rank the data from smallest to largest. Second, calculate z-scores for the ranks. Third, make a scatterplot of number of trees per plot against their calculated z-scores.

(b) Examine your Normal quantile plot for deviations from Normality and explain your conclusions.

11.49 Logging in the rain forest, continued. Go back to the previous exercise. Use software to make Normal quantile plots for each of the three groups of forest plots. (If you completed the previous exercise, compare your own Normal quantile plot with the corresponding one produced by software.) Are the three distributions roughly Normal? What are the most prominent deviations from Normality that you see?

11.50 Estimating body fat in men (optional). The data file *Large.Bodyfat* contains data on a number of body measurements for 252 adult men. In Exercise 1.45 (page 37) you made histograms for the distributions of the men's body weights (in pounds) and heights (in inches). Make a Normal quantile plot for each variable and use them to complement your description of these two distributions.

LARGE DATA SET

11.51 Lung capacity of children (optional). The forced expiratory volume (FEV, measured in liters) is a primary indicator of lung function and corresponds to the volume of air that can forcibly be blown out in the first second after full inspiration. In Exercise 1.44 (page 37) you studied the *Large.FEV* data file containing the FEV values of a large sample of children, along with some categorical descriptor of each individual.

LARGE DATA SET

(a) Make separate Normal quantile plots of FEV for girls ($sex = 0$) and for boys ($sex = 1$). What can you say from these two graphs?

(b) Now use the age group variable that you created in Exercise 1.44 (*preschool* for ages 3 to 5, *elementary* for ages 6 to 10, *middle* for ages 11 to 13, and *highschool* for ages 14 and above). Make separate Normal quantile plots of FEV for each of the eight combinations of sex and age group. What can you conclude now? Explain why we must sometimes break down a data set to isolate homogeneous subgroups.

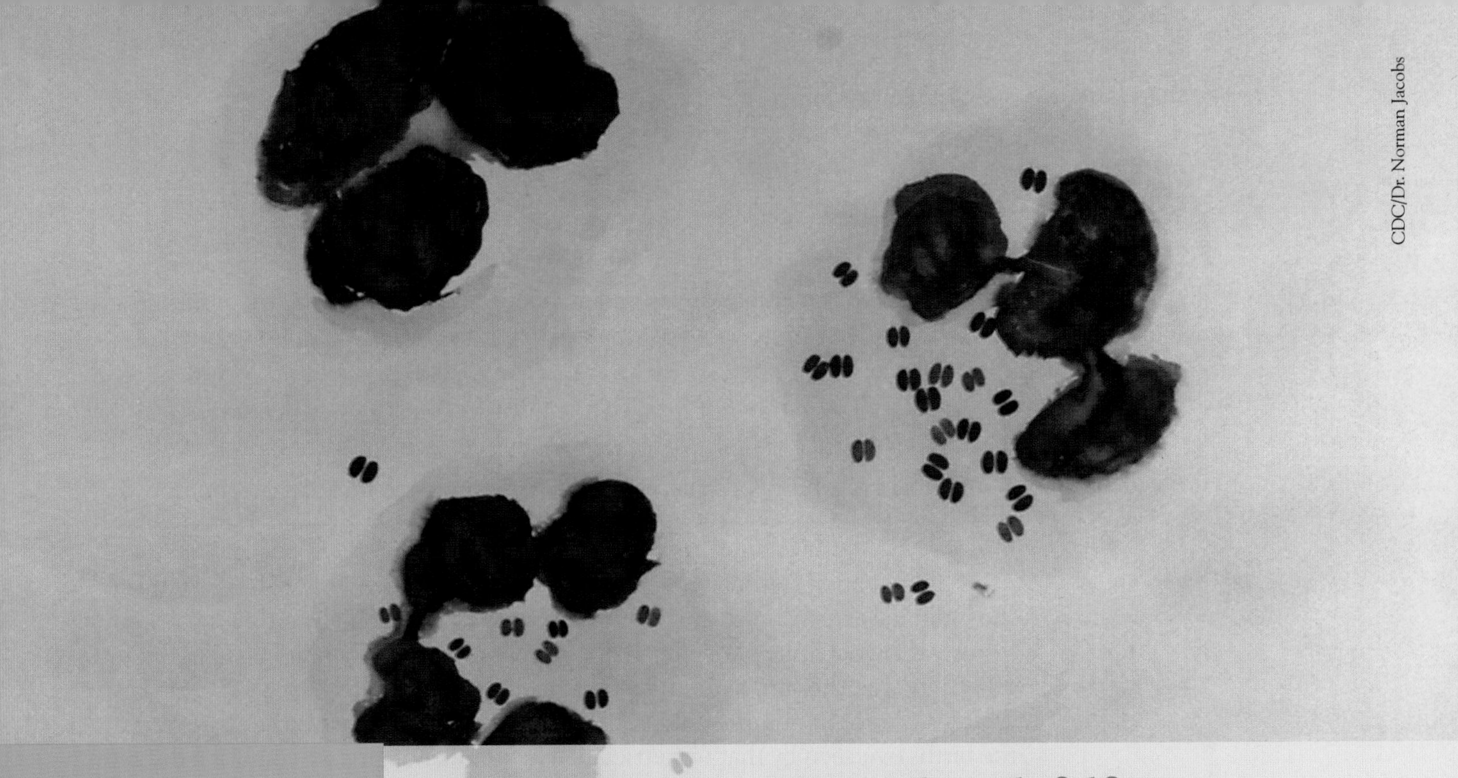
CDC/Dr. Norman Jacobs

CHAPTER 12 Discrete Probability Distributions*

An elementary school administers eye exams to 800 students. How many students have perfect vision? A new treatment for pancreatic cancer is tried on 250 patients. How many survive for five years? You plant 10 dogwood trees. How many live through the winter? In all these situations, we want a probability model for a *count* of successful outcomes.

IN THIS CHAPTER WE COVER...

- The binomial setting and binomial distributions
- Binomial distributions in statistical sampling
- Binomial probabilities
- Using technology
- Binomial mean and standard deviation
- The Normal approximation to binomial distributions
- The Poisson distributions
- Using technology

The binomial setting and binomial distributions

The distribution of a count depends on how the data are produced. Here is a common situation.

*This more advanced chapter concerns a special topic in probability. It is not needed to read the rest of the book.

THE BINOMIAL SETTING

1. There are a fixed number n of observations.
2. The n observations are all **independent.** That is, knowing the result of one observation does not change the probabilities we assign to other observations.
3. Each observation falls into one of just two categories, which for convenience we call "success" and "failure."
4. The probability of a success, call it p, is the same for each observation.

Think of an obstetrician overseeing n single-birth deliveries as an example of the binomial setting. Each single-birth delivery is either a baby girl or a baby boy. Knowing the outcome of one birth doesn't change the probability of a girl on any other birth, so the births are independent. If we call a girl a "success" (this is an arbitrary choice), then p is the probability of a girl and remains the same over all the births overseen by that obstetrician. The number of girls we count is a discrete random variable X. The distribution of X is called a *binomial distribution.*

BINOMIAL DISTRIBUTION

The count X of successes in the binomial setting has the **binomial distribution** with parameters n and p. The parameter n is the number of observations, and p is the probability of a success on any one observation. The possible values of X are the whole numbers from 0 to n.

The binomial distributions are an important class of probability distributions. *Pay attention to the binomial setting, because not all counts have binomial distributions.*

EXAMPLE 12.1 Inheriting blood type

Genetics says that children receive genes from their parents independently. Each child of one particular pair of parents has probability 0.25 of having blood type O. If these parents have 5 children, the number who have blood type O is the count X of successes in 5 independent observations with probability 0.25 of a success on each observation. So X has the binomial distribution with $n = 5$ and $p = 0.25$. ■

EXAMPLE 12.2 Selecting volunteer subjects

A researcher has access to 40 healthy volunteers, 20 men and 20 women. The researcher selects 10 of them at random to participate in an experiment on the effect of sport drinks taken during endurance exercise. There are 10 observations, and each is either a man or a woman. But the observations are *not* independent. If the first volunteer selected is a man, the second one is now more likely to be a woman because there are more women (20) than men (19) left in the pool of volunteers. The count X of men randomly selected in this situation does *not* have a binomial distribution. ■

Binomial distributions in statistical sampling

The binomial distributions are important in statistics when we wish to make inferences about the proportion p of "successes" in a population. Here is a typical example.

EXAMPLE 12.3 Selecting an SRS

A pharmaceutical company inspects an SRS of 10 empty plastic containers from a shipment of 10,000. The containers are examined for traces of benzene, a common chemical solvent but also a known human carcinogen. Suppose that (unknown to the company) 10% of the containers in the shipment contain traces of benzene. Count the number X of containers contaminated with benzene in the sample.

This is not quite a binomial setting. Just as selecting 1 man in Example 12.2 changes the makeup of the remaining pool of volunteers, removing 1 plastic container changes the proportion of contaminated containers remaining in the shipment. So the probability that the second container chosen is contaminated changes when we know that the first is contaminated. But removing 1 container from a shipment of 10,000 changes the makeup of the remaining 9999 very little. In practice, the distribution of X is very close to the binomial distribution with $n = 10$ and $p = 0.1$. ■

Allan Shoemake/Getty Images

Example 12.3 shows how we can use the binomial distributions in the statistical setting of selecting an SRS. When the population is much larger than the sample, a count of successes in an SRS of size n has approximately the binomial distribution with n equal to the sample size and p equal to the proportion of successes in the population.

SAMPLING DISTRIBUTION OF A COUNT

Choose an SRS of size n from a population with proportion p of successes. When the population is much larger than the sample, the count X of successes in the sample has approximately the binomial distribution with parameters n and p.

APPLY YOUR KNOWLEDGE

In each of Exercises 12.1 to 12.3, X is a count. Does X have a binomial distribution? Give your reasons in each case.

12.1 **Random digit dialing.** When an opinion poll calls residential telephone numbers at random, only 20% of the calls reach a live person. You watch the random dialing machine make 15 calls. X is the number that reach a live person.

12.2 **A boy or a girl?** A couple really wants to have a boy. They decide that they will keep having children until they have their first boy. Successive births are independent and each has about probability 0.5 of being a boy or a girl. X is the number of the pregnancy that finally succeeds in bringing a baby boy to the couple.

12.3 **High school students and sex.** A study of the sexual health of high school students in New York State indicates that 44% have ever had sexual intercourse.[1] X is the number who have ever had intercourse in an SRS of 100 high school students from New York State.

Binomial probabilities

We can find a formula for the probability that a binomial random variable takes any value by adding probabilities for the different ways of getting exactly that many successes in n observations. Here is an example that illustrates the idea.

EXAMPLE 12.4 Inheriting blood type

Each child born to a particular set of parents has probability 0.25 of having blood type O. If these parents have 5 children, what is the probability that exactly 2 of them have type O blood?

The count of children with type O blood is a binomial random variable X with $n = 5$ tries and probability $p = 0.25$ of a success on each try (as explained in Example 12.1). We want $P(X = 2)$. ■

Because the method doesn't depend on the specific example, let's use "S" for success and "F" for failure for short. Do the work in two steps.

Step 1. Find the probability that a specific 2 of the 5 tries—say the first and the third—give successes. This is the outcome SFSFF. Because tries are independent (the blood type of one child doesn't change the probability that the next child has blood type O), the multiplication rule for independent events applies (see optional Chapter 10). The probability we want is

$$\begin{aligned} P(\text{SFSFF}) &= P(\text{S})P(\text{F})P(\text{S})P(\text{F})P(\text{F}) \\ &= (0.25)(0.75)(0.25)(0.75)(0.75) \\ &= (0.25)^2(0.75)^3 \end{aligned}$$

Unusual performances When a baseball player hits .300, people applaud or suspect performance-enhancing drugs. A .300 hitter gets a hit in 30% of times at bat. Could a .300 year just be luck? Typical major leaguers bat about 500 times a season and hit about .260. A hitter's successive tries seem to be independent, so we have a binomial setting. From this model, the probability of hitting .300 is about 0.025. Out of 100 run-of-the-mill major league hitters, two or three each year will bat .300 because they were lucky.

Step 2. Observe that *any one arrangement* of 2 S's and 3 F's has this same probability. This is true because we multiply together 0.25 twice and 0.75 three times whenever we have 2 S's and 3 F's. The probability that $X = 2$ is the probability of getting 2 S's and 3 F's in any arrangement whatsoever. Here are all the possible arrangements:

SSFFF SFSFF SFFSF SFFFS FSSFF
FSFSF FSFFS FFSSF FFSFS FFFSS

There are 10 of them, all with the same probability. The overall probability of 2 successes is therefore

$$P(X = 2) = 10(0.25)^2(0.75)^3 = 0.2637$$

The pattern of this calculation works for any binomial probability. To use it, we must count the number of arrangements of k successes in n observations. We use the following fact to do the counting without actually listing all the arrangements.

BINOMIAL COEFFICIENT

The number of ways of arranging k successes among n observations is given by the **binomial coefficient**

$$\binom{n}{k} = \frac{n!}{k!\,(n-k)!}$$

for $k = 0, 1, 2, \ldots, n$.

The formula for binomial coefficients uses the **factorial** notation. For any positive whole number n, its factorial $n!$ is

factorial

$$n! = n \times (n-1) \times (n-2) \times \cdots \times 3 \times 2 \times 1$$

Also, $0! = 1$.

The larger of the two factorials in the denominator of a binomial coefficient will cancel much of the $n!$ in the numerator. For example, the binomial coefficient we need for Example 12.4 is

$$\binom{5}{2} = \frac{5!}{2!\,3!}$$

$$= \frac{(5)(4)(3)(2)(1)}{(2)(1) \times (3)(2)(1)}$$

$$= \frac{(5)(4)}{(2)(1)} = \frac{20}{2} = 10$$

The notation $\binom{n}{k}$ *is not related to the fraction* $\frac{n}{k}$. A helpful way to remember its meaning is to read it as "binomial coefficient n choose k." Binomial coefficients have many uses, but we are interested in them only as an aid to finding binomial probabilities. The binomial coefficient $\binom{n}{k}$ counts the number of different ways in which k successes can be arranged among n observations. The binomial probability $P(X = k)$ is this count multiplied by the probability of any one specific arrangement of the k successes. Here is the result we seek.

BINOMIAL PROBABILITY

If X has the binomial distribution with n observations and probability p of success on each observation, the possible values of X are $0, 1, 2, \ldots, n$. If k is any one of these values,

$$P(X = k) = \binom{n}{k} p^k (1-p)^{n-k}$$

EXAMPLE 12.5 Inspecting pharmaceutical containers

The number X of containers contaminated with benzene in Example 12.3 has approximately the binomial distribution with $n = 10$ and $p = 0.1$.

According to the addition rule for disjoint events, the probability that the sample contains no more than 1 contaminated container is

$$\begin{aligned} P(X \leq 1) &= P(X = 1) + P(X = 0) \\ &= \binom{10}{1}(0.1)^1(0.9)^9 + \binom{10}{0}(0.1)^0(0.9)^{10} \\ &= \frac{10!}{1!\,9!}(0.1)(0.3874) + \frac{10!}{0!\,10!}(1)(0.3487) \\ &= (10)(0.1)(0.3874) + (1)(1)(0.3487) \\ &= 0.3874 + 0.3487 = 0.7361 \end{aligned}$$

This calculation uses the facts that $0! = 1$ and that $a^0 = 1$ for any number a other than 0. We see that about 74% of all samples will contain no more than 1 contaminated container. In fact, 35% of the samples will contain no contaminated container. A sample of size 10 cannot be trusted to alert the pharmaceutical company to the presence of unacceptable containers in the shipment. ■

Using technology

The binomial probability formula is awkward to use, particularly for the probabilities of events that contain many outcomes. You can find tables of binomial probabilities $P(X = k)$ and cumulative probabilities $P(X \leq k)$ for selected values of n and p.

The most efficient way to do binomial calculations, however, is to use technology. Figure 12.1 shows output for the calculation in Example 12.5 from a graphing calculator, three statistical programs, and a spreadsheet program. We asked all five to give cumulative probabilities. The TI-83, CrunchIt!, SPSS, and Minitab have menu entries for binomial cumulative probabilities. Excel has no menu entry, but the worksheet function BINOMDIST is available. All the outputs agree with the result 0.7361 of Example 12.5.

APPLY YOUR KNOWLEDGE

12.4 Random digit dialing. When a survey calls residential telephone numbers at random, only 20% of the calls reach a live person. You watch the random dialing machine make 15 calls.

(a) What is the probability that exactly 3 calls reach a person?

(b) What is the probability that 3 or fewer calls reach a person?

12.5 Antibiotic resistance. Antibiotic resistance occurs when disease-causing microbes become resistant to antibiotic drug therapy. Because this resistance is typically genetic and transferred to the next generations of microbes, it is a very serious public

CrunchIt!

Binomial Distribution Calculator

Parameters

n: 10

p: 0.1

Probability

P(X ≤ 1) = 0.7360989290999997

Help Cancel Calculate

TI-83

```
binomcdf (10,0.1,
1)
                .7361
```

Minitab

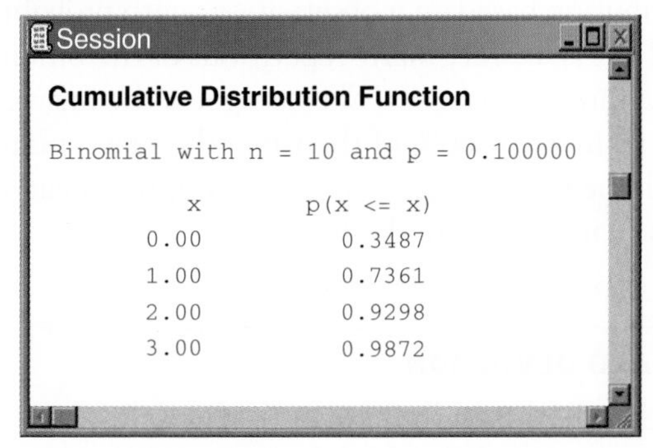

Session
Cumulative Distribution Function
Binomial with n = 10 and p = 0.100000
x p(x <= x)
0.00 0.3487
1.00 0.7361
2.00 0.9298
3.00 0.9872

SPSS

Untitled - SPSS Data Editor

	X	CumBinomial
1	0	.3487
2	1	.7361
3	2	.9298
4	3	.9872
5	4	.9984

Microsoft Excel

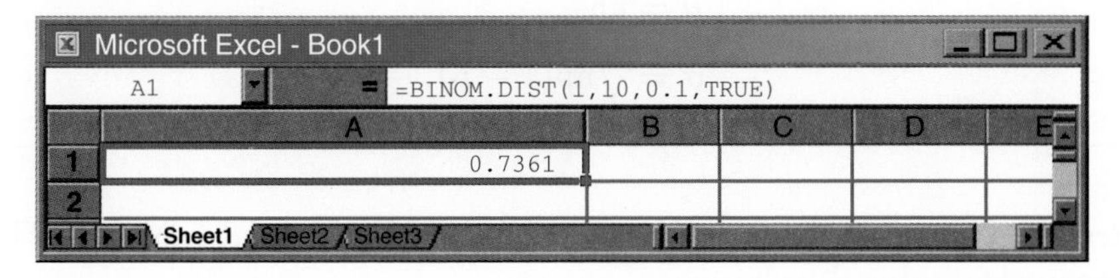

FIGURE 12.1 The binomial probability $P(X \leq 1)$, for Example 12.5: output from a graphing calculator, three statistical programs, and a spreadsheet program.

health problem. According to the Centers for Disease Control and Prevention (CDC), 27% of gonorrhea cases tested in 2010 were resistant to at least one of the three major antibiotics commonly used to treat gonorrhea.[2] A physician treated 10 cases of gonorrhea during one week of 2010.

(a) What is the distribution of the cases resistant to at least one of the three major antibiotics?

(b) What is the probability that exactly 1 out of the 10 cases was resistant to at least one of the three major antibiotics? What is that probability for exactly 2 out of 10?

(c) Also find the probability that 1 or more out of the 10 were resistant to at least one of the three major antibiotics. (*Hint:* It is easier to first find the probability that exactly 0 of the 10 cases were resistant.)

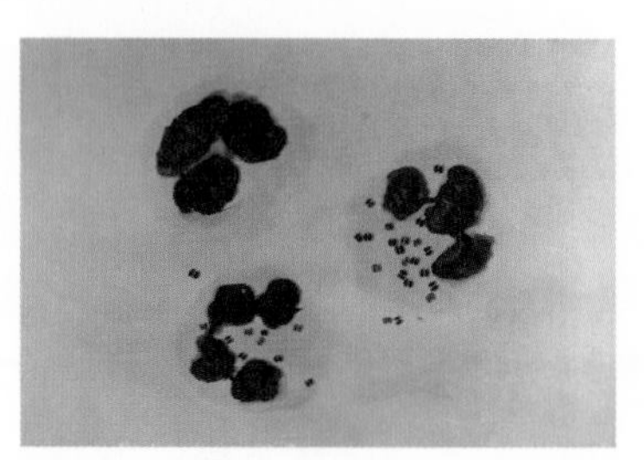

CDC/Dr. Norman Jacobs

12.6 Antibiotic resistance, continued. According to the CDC, 7% of gonorrhea cases tested in 2010 were resistant to all three major antibiotics commonly used to treat gonorrhea. Let's go back to the physician who treated 10 cases of gonorrhea during one week of 2010.

(a) What is the distribution of the cases resistant to all three antibiotics?

(b) What is the probability that exactly 1 out of the 10 cases was resistant to all three antibiotics? What is that probability for exactly 2 out of 10?

(c) Also find the probability that 1 or more out of the 10 were resistant to all three antibiotics. (*Hint:* It is easier to first find the probability that exactly 0 of the 10 cases were resistant.)

Binomial mean and standard deviation

If a count X has the binomial distribution based on n observations with probability p of success, what is its mean μ? That is, in very many repetitions of the binomial setting, what will be the average count of successes? We can guess the answer. If a psychic correctly guesses a pattern in a test 60% of the time, the mean number of correct guesses in 10 tries should be 60% of 10, or 6. In general, the mean of a binomial distribution should be $\mu = np$. Here are the facts.

BINOMIAL MEAN AND STANDARD DEVIATION

If a count X has the binomial distribution with number of observations n and probability of success p, the **mean** and **standard deviation** of X are

$$\mu = np$$
$$\sigma = \sqrt{np(1-p)}$$

Remember that these short formulas are good only for binomial distributions. They can't be used for other distributions.

EXAMPLE 12.6 Inspecting pharmaceutical containers

Continuing Example 12.5, the count X of containers contaminated with benzene is binomial with $n = 10$ and $p = 0.1$. The histogram in Figure 12.2 displays this probability distribution. (Because probabilities are long-run proportions, using probabilities as the heights of the bars shows what the distribution of X would be in very many repetitions.) The distribution is strongly skewed. Although X can take any whole-number value from 0 to 10, the probabilities of values larger than 5 are so small that they do not appear in the histogram.

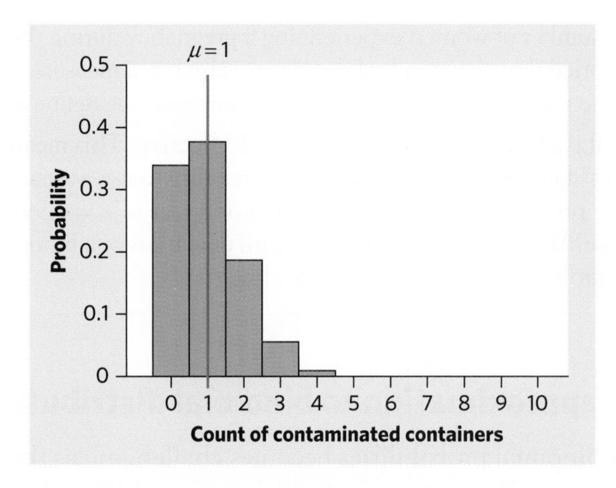

FIGURE 12.2 Probability histogram for the binomial distribution with $n = 10$ and $p = 0.1$, for Example 12.6.

The mean and standard deviation of the binomial distribution in Figure 12.2 are

$$\begin{aligned}\mu &= np \\ &= (10)(0.1) = 1 \\ \sigma &= \sqrt{np(1-p)} \\ &= \sqrt{(10)(0.1)(0.9)} = \sqrt{0.9} = 0.9487\end{aligned}$$

The mean is marked on the probability histogram in Figure 12.2. ■

APPLY YOUR KNOWLEDGE

12.7 Antibiotic resistance, continued.

(a) Out of 10 gonorrhea cases in the year 2010, what is the mean number of cases that are resistant to at least one of the three major antibiotics, as described in Exercise 12.5?

(b) What is the standard deviation σ of the count of cases showing resistance to at least one of the three major antibiotics?

(c) In the year 2010, the probability that a gonorrhea case was resistant to all three major antibiotics was only $p = 0.07$, as described in Exercise 12.6. How does this new p affect the standard deviation? What is the standard deviation for $p = 0.07$? What does your work show about the behavior of the standard deviation of a binomial distribution as the probability of a success gets closer to 0?

12.8 Pregnancy and the pill. The actual rate of unintended pregnancy for women taking the pill is 5%. That is, if a doctor prescribes the pill to 100 new female users, 5 of them will have become pregnant within one year while taking the pill. Therefore, the pill's true efficacy under typical usage conditions is 95%.[3]

(a) For 20 new female users, what is the mean number of unintended pregnancies? What is the mean number of women not experiencing any pregnancy within the first year of usage? You see that these two means must add to 20, the total number of women prescribed the pill.

(b) What is the standard deviation σ of the number of women not experiencing a pregnancy during the first year on the pill? What is the standard deviation σ

of the number of women experiencing a pregnancy during the first year on the pill? Notice that the standard deviation, which is a measure of variability, is the same regardless of which of the two outcomes we define as "success."

(c) The pill is often advertised as being 99% effective. This means that under "perfect" laboratory conditions, the probability that a woman would not become pregnant over a usage period of one year is $p = 0.99$. What is σ in this case? What happens to the standard deviation of a binomial distribution as the probability of a "success" gets close to 1?

The Normal approximation to binomial distributions

The formula for binomial probabilities becomes challenging as the number of observations n increases. You can use software or a graphing calculator to handle many problems for which the formula is not practical. If technology does not rescue you, there is another alternative: *As the number of observations n gets larger, the binomial distribution gets close to a Normal distribution.* When n is large, we can use Normal probability calculations to approximate binomial probabilities. We will see in Chapter 13 that this approximation has useful applications beyond the simplifying of binomial computations.

Tips Images/SuperStock

EXAMPLE 12.7 Overweight Americans

According to the 2011 Health Interview Survey, 62% of American adults are either overweight or obese, based on their self-reported body mass index (BMI). Suppose that we take a random sample of 2000 adults. What is the probability that 1250 or more of the sample are overweight or obese? ■

Because there are over 200 million adults in the United States, we can take the BMI values of 2000 randomly chosen adults to be independent. So the number in our sample who are either overweight or obese is a random variable X having the binomial distribution with $n = 2000$ and $p = 0.62$. To find the probability that at least 1250 of the people in the sample are overweight or obese, we must add the binomial probabilities of all outcomes from $X = 1250$ to $X = 2000$. This isn't practical. Here are two ways to do this problem:

1. Use technology. For $P(X \geq 1250)$, the software R gives us 0.3313682 if we type "1 − pbinom(1249, size=2000, prob=0.62)." This answer is exactly correct to seven decimal places. Note that some programs give only cumulative probabilities. So remember that $P(X \geq 1250) = 1 - P(X \leq 1249)$. Other programs may not even perform the intensive calculations required for such a large n.
2. Use the Normal approximation to the binomial distribution. Figure 12.3 shows the actual binomial probability distribution for X, with events $X \geq 1250$ in color. The shape of this distribution closely resembles the Normal distribution with the same mean and standard deviation as the

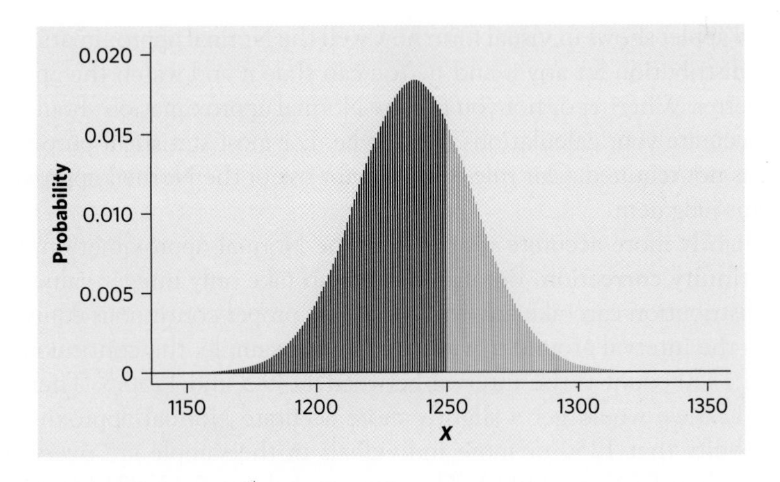

FIGURE 12.3 Probability distribution for the binomial model $n = 2000$, $p = 0.62$, displayed graphically. The height of each bar represents the probability for X when it takes a value on the horizontal axis. The green bars represent probabilities for values of $X \geq 1250$. Notice how the shape of this binomial probability distribution closely resembles a Normal curve.

binomial variable X:

$$\mu = np = (2000)(0.62) = 1240$$
$$\sigma = \sqrt{np(1-p)} = \sqrt{(2000)(0.62)(0.38)} = 21.707$$

EXAMPLE 12.8 Normal calculation of a binomial probability

If we act as though the count X has the $N(1240, 21.707)$ distribution, the software R gives us $P(X \geq 1250) = 0.3225138$ if we type "1 − pnorm(1250,mean=1240,sd=21.707)." The Normal approximation differs from the binomial result 0.3313682 by only 0.009.

Here is $P(X \geq 1250)$ with the Normal approximation, using Table B instead:

$$\begin{aligned} P(X \geq 1250) &= P\left(\frac{X - 1240}{21.707} \geq \frac{1250 - 1240}{21.707}\right) \\ &= P(Z \geq 0.46) \\ &= 1 - 0.6772 = 0.3228 \end{aligned}$$ ■

NORMAL APPROXIMATION FOR BINOMIAL DISTRIBUTIONS

Suppose that a count X has the binomial distribution with n observations and success probability p. When n is large, the distribution of X is approximately Normal, $N(np, \sqrt{np(1-p)})$.

As a rule of thumb, we will use the Normal approximation when n is so large that $np \geq 10$ and $n(1-p) \geq 10$.

The Normal approximation is easy to remember because it says that X is Normal with its binomial mean and standard deviation. The accuracy of the Normal approximation improves as the sample size n increases. It is most accurate for any fixed n when p is close to 0.5 and least accurate when p is near 0 or 1. This is why the rule of thumb in the box depends on p as well as n. The *Normal Approximation*

to Binomial applet shows in visual form how well the Normal approximation fits the binomial distribution for any n and p. You can slide n and watch the approximation get better. Whether or not you use the Normal approximation should depend on how accurate your calculations need to be. For most statistical purposes great accuracy is not required. Our rule of thumb for use of the Normal approximation reflects this judgment.

continuity correction

For slightly more accurate results with the Normal approximation, you can use a **continuity correction**. Because counts can take only integer values but the Normal distribution can take any real value, the proper continuous equivalent to a count is the interval around it with size 1. For example, the continuous equivalent to a 1250 count is the interval between 1249.5 and 1250.5. Therefore, in Example 12.8 we would get a slightly more accurate Normal approximation of the probability that 1250 or more individuals in the sample are overweight or obese by finding $P(X \geq 1249.5)$. This gives probability 0.3308212 (typing "1 − pnorm(1249.5,mean=1240,sd=21.707)" in R), which is closer still to the binomial software output. The continuity correction is especially helpful when the sample size is small.

APPLY YOUR KNOWLEDGE

Solid bone matrix — Weakened bone matrix

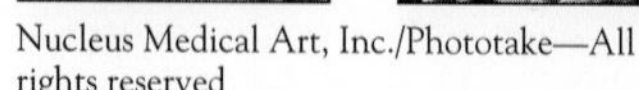
Nucleus Medical Art, Inc./Phototake—All rights reserved

12.9 **Bone density.** Forty percent of postmenopausal women have low bone density (osteopenia), placing them at risk for osteoporosis with ensuing spontaneous fractures. Osteoporosis is estimated to cost $14 billion per year in medical expenses alone, and yet it can be prevented if treated early enough. A physicians group has 248 postmenopausal primary patients and has diagnosed 82 with low bone density.

(a) What is the probability that only 82 or fewer of a set of 248 patients have low bone density? Use the Normal approximation to find this probability.

(b) If you have access to technology, find the probability in (a) using the binomial probability model. How does it compare with the Normal approximation?

(c) Based on your answer, what would you suggest at the next physicians meeting?

12.10 **Checking for survey errors.** One way of checking the effect of undercoverage, nonresponse, and other sources of error in a sample survey is to compare the sample with known facts about the population. About 12% of American adults are black. The number X of blacks in a random sample of 1500 adults should therefore vary with the binomial ($n = 1500$, $p = 0.12$) distribution.

(a) What are the mean and standard deviation of X?

(b) Use the Normal approximation to find the probability that the sample will contain between 165 and 195 blacks. Be sure to check that you can safely use the approximation.

The Poisson distributions

Binomial distributions are not the only type of discrete distribution for whole-number counts. Exercises 10.41 and 12.49 show examples of geometric distributions, and Exercise 12.51 walks you through an example of a multinomial distribution. These distributions are beyond the scope of an introductory textbook and will not be discussed beyond these examples.

Poisson distributions are yet another class of discrete distributions and have important applications in the life sciences. They describe the count of events that occur at random over a chosen unit of space or time. For example, a Poisson distribution would be a good model for the count of chocolate chips in cookies made from the same well-mixed batch or for the number of times you walk up the stairs on any workday. Note that, unlike counts in a binomial distribution, counts in a Poisson distribution are open-ended: That is, they can, in theory, go on to infinity.

POISSON DISTRIBUTION

A **Poisson distribution** describes the count X of occurrences of a defined event in fixed, finite intervals of time or space when

1. occurrences are all **independent** (that is, knowing that one event has occurred does not change the probability that another event may occur), and
2. the probability of an occurrence is the same over all possible intervals of the same size.

EXAMPLE 12.9 Counting daisies

On a lazy summer afternoon, you lie in a field adorned with daisies growing here and there. In an inquisitive spirit, you divide the field into quadrants of 1 square foot and count the number of daisies in each quadrant. The number of daisies per quadrant is the random variable X. If the daisies present in each quadrant are random, independent events, then the distribution of X follows approximately a Poisson distribution.

Figure 12.4 illustrates this concept via a simulation. The large rectangle represents a field divided into 50 quadrants of equal size. Each pink circle is a daisy on the simulated

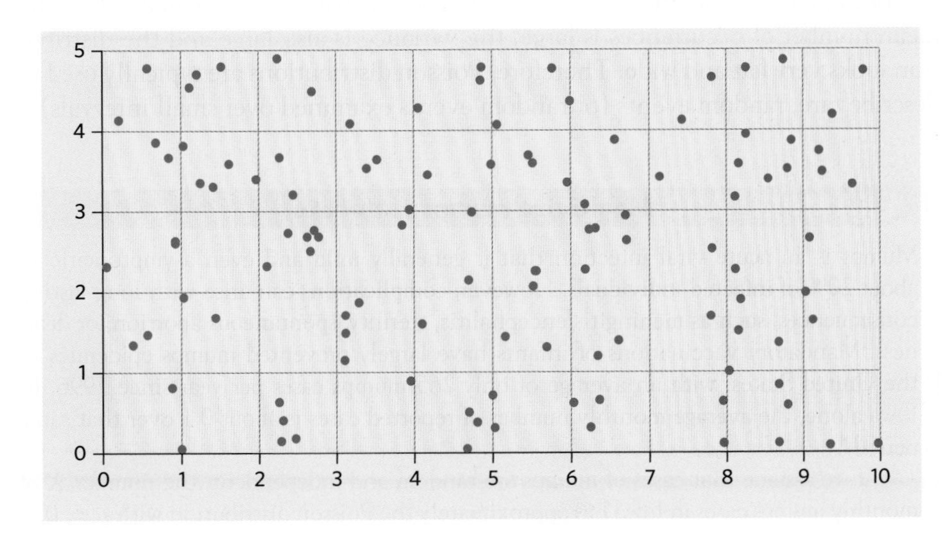

FIGURE 12.4 Graphical display of a simulated field planted with daisies (pink circles), for Example 12.9. Daisy location on the rectangular field follows a Poisson distribution with mean 2.1 daisies per quadrant.

field. The daisies were randomly "planted" according to a Poisson distribution, giving an overall mean of 2.1 daisies per quadrant. Starting from the left, the first ten quadrants on the upper row have, respectively, 2, 3, 2, 1, 3, 3, 0, 2, 2, and 2 daisies each. Notice how the overall pattern of daisies on the field appears to indeed be entirely random. ■

The Poisson distributions can be used to predict what counts we would expect for a random event occurring within an interval of fixed size (in time or space). For instance, a medical group can model the count of non-life-threatening emergency room visits on weekend days and decide whether to open a walk-in clinic and how to staff it.

Because Poisson distributions describe random, independent events, it can also be interesting to find out that a random variable substantially deviates from that model. For instance, when the count of meningitis cases in a county exceeds what would reasonably be expected under a Poisson distribution, we can conclude that the meningitis cases are not randomly and independently distributed, suggesting instead an epidemic outbreak.

POISSON PROBABILITY

If X has the Poisson distribution with mean number of occurrences per interval μ, the possible values of X are 0, 1, 2, ... If k is any one of these values,

$$P(X = k) = \frac{e^{-\mu}\mu^k}{k!}$$

The mean and variance of the Poisson distribution are both equal to μ, the mean number of occurrences per interval. The distribution's standard deviation σ is equal to $\sqrt{\mu}$.

Because the mean and variance of a Poisson distribution are equal, when the mean number of occurrences is large, the variance is also large and the distribution looks very flat and wide. Therefore, Poisson distributions are typically used to describe rare, random events (or random events examined over small intervals).

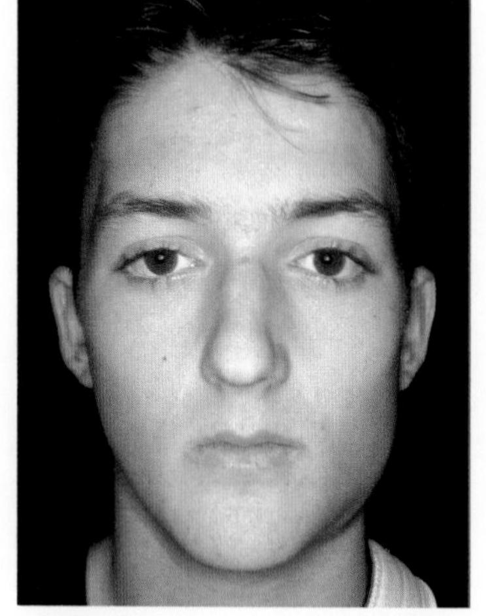
ISM/Phototake

EXAMPLE 12.10 Monitoring mumps outbreaks

Mumps is an acute viral infection that is generally mild and even asymptomatic in about 20% of infected individuals. However, complications can arise with long-lasting consequences, such as meningitis/encephalitis, sterility, spontaneous abortion, or deafness. Mandatory vaccinations of infants have largely prevented mumps epidemics in the United States, with an average of only 265 mumps cases per year since 1996. In Iowa alone, the average monthly number of reported cases is about 0.1 over that same period.[4]

If we assume that cases of mumps are random and independent, the number X of monthly mumps cases in Iowa has approximately the Poisson distribution with $\mu = 0.1$.

The probability that in a given month there is no more than 1 mumps case is

$$
\begin{aligned}
P(X \le 1) &= P(X = 0) + P(X = 1) \\
&= \frac{e^{-0.1} \times 0.1^0}{0!} + \frac{e^{-0.1} \times 0.1^1}{1!} \\
&= \frac{0.9048 \times 1}{1} + \frac{0.9048 \times 0.1}{1} \\
&= 0.9048 + 0.0905 = 0.9953
\end{aligned}
$$

We would expect no more than 1 case of mumps per month with probability 99.53%. That is, we would consider seeing two or more mumps cases in a month a very unlikely event (probability 1 minus 0.9953) if cases were random and independent. ■

Mumps in perspective According to the CDC, before approval of the mumps vaccine in 1967, nearly everyone in the United States experienced mumps, with about 200,000 new cases reported every year. The mumps vaccine has a reported effectiveness of about 95%. That is, some individuals will not be helped by the vaccine because they lack the genetic ability to produce antibodies against that particular viral strain.

Using technology

The Poisson probability formula is simple to use for any value of X but can be more cumbersome for cumulative probabilities. Again, technology is very useful. Figure 12.5 shows output for the calculation in Example 12.10 from the TI-83 graphing calculator and from the statistical programs Minitab and R. All three outputs agree with the result 0.9953 of Example 12.10.

Minitab

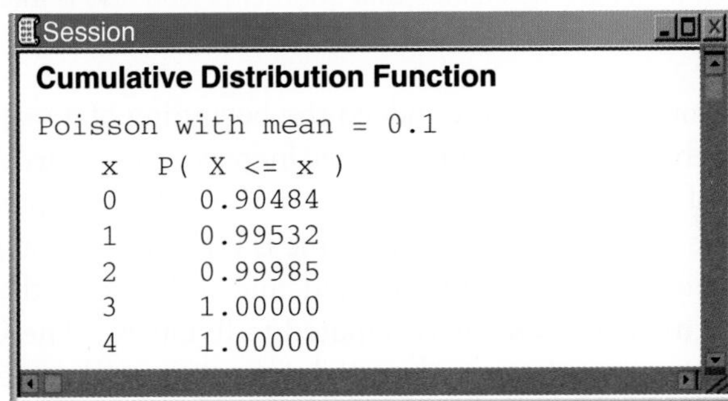

Cumulative Distribution Function

Poisson with mean = 0.1

x	P(X <= x)
0	0.90484
1	0.99532
2	0.99985
3	1.00000
4	1.00000

TI-83

```
Poissoncdf (0.1,1
)
        .995321159
```

R

```
> ppois (1, lambda=0.1)
[1] 0.9953212
```

FIGURE 12.5 The Poisson probability $P(X \le 1)$, for Example 12.10: output from two statistical programs and a graphing calculator.

EXAMPLE 12.11 Monitoring mumps outbreaks

Mumps cases, like all cases of rare diseases, must be reported to the CDC and are published in the *Morbidity and Mortality Weekly Report*. In January 2006, Iowa reported 4 cases of mumps. What is the probability of getting 4 or more cases of mumps under the model used in Example 12.10?

Microsoft Excel

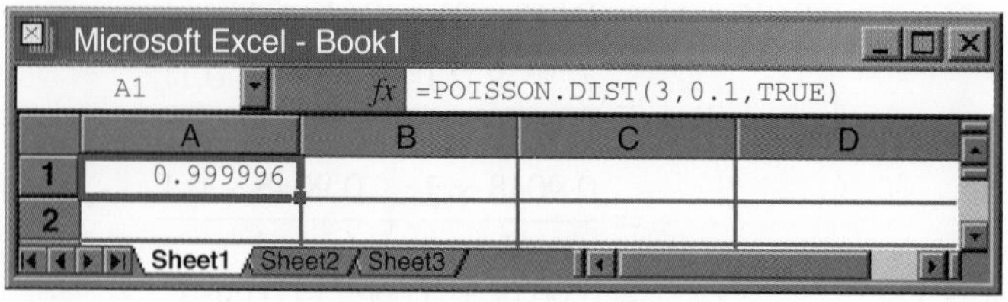

Microsoft Excel - Book1

A1 | fx =POISSON.DIST(3,0.1,TRUE)

	A	B	C	D
1	0.999996			
2				

Sheet1 / Sheet2 / Sheet3

FIGURE 12.6 The Poisson probability $P(X \leq 3)$, for Example 12.11: output from a spreadsheet program.

The probability that in a given month there are 4 or more mumps cases is

$$P(X \geq 4) = 1 - P(X \leq 3)$$
$$= 1 - 0.999996 = 0.000004$$

We found $P(X \leq 3)$ with the spreadsheet program Excel, using the cumulative Poisson distribution with parameter $\mu = 0.1$. Excel's function and output are displayed in Figure 12.6.

Assuming that cases of mumps are random and independent, we would expect to almost never see 4 or more cases of mumps in Iowa in any given month. The unusually high count of mumps cases points to a substantial departure from the Poisson model, for instance, because of a contagious outbreak. In a contagion model, knowing that one person is infected increases the chance that another person also is infected. Therefore, contagion events are not independent. ■

These first 4 mumps cases were indeed the beginning of a major mumps outbreak in Iowa, with the number of new cases increasing every week, totaling 219 cases by the end of March and 2597 by the end of April 2006, spread over several states. The mumps strain in the Iowa epidemic is known as genotype G, the same strain that has been responsible for other epidemic outbreaks in the United States and the United Kingdom, even in vaccinated individuals.[5] The CDC tracks all reports of rare diseases in order to understand their origin and prevent their spread and future recurrence.

Plant and animal distributions can also be compared with a Poisson distribution to determine whether the events are indeed random and independent. In Example 12.9 you looked at the distribution of daisies in field quadrants. Plants disseminated by animals or by wind are expected to have geographical distributions that can be approximated by a Poisson model. *Remember that biological data should not be expected to exactly match a mathematical model. This is a common misconception. Rather, people have found mathematical models that are reasonable representations for various biological features*. Some plants disseminate primarily asexually, for example, through underground rhizomes capable of sprouting new clones (like the lily of the valley). Such plants are expected to be more clustered than would be predicted by a Poisson model.

APPLY YOUR KNOWLEDGE

12.11 Polydactyly. In the United States, 1 in every 500 babies is born with polydactyly, and cases are random and independent. This is a fairly common congenital abnormality in which a baby is born with one or more extra fingers or toes. A hospital delivers an average of 268 children per month. Therefore, we would expect to see, on average, 0.536 babies born with polydactyly at that hospital per month. Let X be the count of babies born with polydactyly in a month at that hospital.

(a) What values could X take? Notice how there is no clear limit to this range.

(b) What distribution does X follow, approximately?

(c) Give the mean and standard deviation of X.

12.12 Typhoid fever. The CDC receives reports of 7.7 cases of typhoid fever per week, on average, from all over the United States, although most cases were acquired while traveling internationally. Typhoid fever is a life-threatening bacterial illness caused by *Salmonella typhi* and is transmitted by ingestion of contaminated food or drinks.[6]

(a) What distribution would best represent the distribution of weekly counts of typhoid fever?

(b) What are the mean and standard deviation of this distribution?

12.13 Polydactyly, continued. In the setting of Exercise 12.11,

(a) what is the probability that no child is born with polydactyly in a given month at that hospital?

(b) what is the probability that exactly 1 child is born with polydactyly in a given month?

(c) what is the probability that more than 1 child is born with polydactyly in a given month?

12.14 Typhoid fever, continued. In the setting of Exercise 12.12,

(a) use technology to find the cumulative probabilities for X equals 0, 1, 2, 3, to 15.

(b) what is the probability that the CDC receives reports of more than 15 cases next month? If this were to happen, what would you conclude?

CHAPTER 12 SUMMARY

- A count X of successes has a **binomial distribution** in the **binomial setting:** There are n observations, the observations are independent of each other, each observation results in a success or a failure, and each observation has the same probability p of a success.
- The binomial distribution with n observations and probability p of success gives a good approximation to the sampling distribution of the count of successes in an SRS of size n from a large population containing proportion p of successes.
- If X has the binomial distribution with parameters n and p, the possible values of X are the whole numbers 0, 1, 2, ..., n. The **binomial probability** that X takes any value is

$$P(X = k) = \binom{n}{k} p^k (1 - p)^{n-k}$$

In practice, binomial probabilities are best found using software.

- The **binomial coefficient**

$$\binom{n}{k} = \frac{n!}{k!\,(n-k)!}$$

counts the number of ways k successes can be arranged among n observations. Here the **factorial $n!$** is

$$n! = n \times (n-1) \times (n-2) \times \cdots \times 3 \times 2 \times 1$$

for positive whole numbers n, and $0! = 1$.

- The **mean** and **standard deviation** of a binomial count X are

$$\mu = np$$
$$\sigma = \sqrt{np(1-p)}$$

- The **Normal approximation** to the binomial distribution says that if X is a count having the binomial distribution with parameters n and p, then when n is large, the distribution of X is approximately $N(np, \sqrt{np(1-p)}\,)$. Use this approximation only when $np \geq 10$ and $n(1-p) \geq 10$.
- The **Poisson distribution** with mean μ describes the count X of occurrences of random, independent events in fixed intervals of time or space when the probability of an event occurring is the same over all possible intervals.
- If X has the Poisson distribution with mean number of occurrences per interval μ, the possible values of X are 0, 1, 2, ... The **Poisson probability** that X takes any value is

$$P(X = k) = \frac{e^{-\mu}\mu^k}{k!}$$

- The **mean** and **standard deviation** of a Poisson count X are

$$\mu = \text{mean number of occurrences per interval}$$
$$\sigma = \sqrt{\mu}$$

THIS CHAPTER IN CONTEXT

The binomial distribution is used to compute probabilities for the count of successes among n observations that are produced under the binomial setting. An important situation for which the binomial setting can be used is when we choose an SRS from a large population with a proportion p of successes. Calculations can be performed using the binomial formula or, when n and p are such that both the mean number of successes and the mean number of failures are large enough, the Normal approximation can be used instead.

Another important application of the binomial distribution is in making inferences about the unknown *proportion* p of some outcome in a population. The proportion of some outcome in an SRS is simply the count of such outcomes divided by the total number of observations n. The methodology used for inference about a proportion is a direct consequence of the Normal approximation for the binomial distribution, as we will see in Chapter 13 when discussing sampling distributions. In Chapter 19 we will examine how to use the sampling distribution of a proportion to estimate the unknown population parameter (the true proportion of an outcome of interest in a given population) or to test a hypothesis about it.

The Poisson distribution is used to compute probabilities for the count of rare random events in fixed intervals of time or space when the events are independent and occur with constant mean rate μ. Poisson probabilities are a good model for completely random events. We will see in Chapter 21 an example of statistical inference using the chi-square test for goodness of fit to test the hypothesis that a random event has a Poisson distribution.

CHECK YOUR SKILLS

Tay-Sachs disease is a fatal, recessive genetic disorder in which harmful quantities of a fatty waste substance accumulate in the brain, slowly progressing to death by age five. When both parents are carriers of the Tay-Sachs mutation, the probability is 0.25 that their child will have Tay-Sachs disease. Successive children inherit genetic material from their parents independently. Use this information to answer Exercises 12.15 and 12.16.

12.15 For a couple in which both parents carry the Tay-Sachs mutation, the number of children with Tay-Sachs disease that they could have if they want 3 children (ignore the possibility of identical twins) has the distribution

(a) binomial with $n = 4$ and $p = 1/4$.
(b) binomial with $n = 3$ and $p = 1/4$.
(c) binomial with $n = 3$ and $p = 1/3$.

12.16 The probability that at least 1 of their 3 children will have Tay-Sachs disease is

(a) 0.68. (b) 0.58. (c) 0.30.

12.17 In a group of 10 college students, 6 have never smoked tobacco regularly. You choose 3 of the 10 students at random and ask about their smoking status. The distribution of the number you choose who have never smoked regularly is

(a) binomial with $n = 10$ and $p = 0.6$.
(b) binomial with $n = 3$ and $p = 0.6$.
(c) not binomial.

About 5% of emergency room visits in South Carolina are for respiratory and chest symptoms, making up the top reason for emergency room visits.[7] *Assume that visits to the emergency room are independent. Exercises 12.18 to 12.20 use this setting.*

12.18 If 2 of 10 emergency room visits are because of respiratory and chest symptoms, in how many ways can you arrange the sequence of emergency room visits due to respiratory and chest symptoms and due to other symptoms?

(a) $\binom{10}{2} = 45$ (b) $\binom{10}{2} = 5$ (c) $\binom{8}{2} = 28$

12.19 Of 10 emergency room visits in South Carolina, the probability that the first 2 are because of respiratory and chest symptoms but not the remaining 8 is about

(a) 0.2. (b) 0.05. (c) 0.0017.

12.20 Of 10 emergency room visits in South Carolina, the probability that 2 are because of respiratory and chest symptoms is about

(a) 0.2. (b) 0.0746. (c) 0.0017.

12.21 Each entry in a table of random digits like Table A has probability 0.1 of being a 0, and digits are independent of each other. Ten lines in the table contain 400 digits. The count of 0s in these lines is approximately Normal with

(a) mean 40 and standard deviation 36.
(b) mean 40 and standard deviation 6.
(c) mean 36 and standard deviation 6.

12.22 The mean number of daily surgeries at a local hospital is 6.2. If we assume that surgeries are random, independent events, the count of daily surgeries follows approximately

(a) a binomial distribution with mean 6.2 and standard deviation 3.8.
(b) a Poisson distribution with mean 6.2 and standard deviation 6.2.
(c) a Poisson distribution with mean 6.2 and standard deviation 2.49.

12.23 The mean number of daily surgeries at a local hospital is 6.2. If we assume that surgeries are random, independent events, the probability that there would be only 2 or fewer surgeries in a given day is approximately

(a) 0.015. (b) 0.039. (c) 0.054.

12.24 Which of the following would you expect to follow approximately a Poisson distribution?

(a) The weekly count of toddlers with a runny nose at a day care.

(b) The count of adults with a disability visiting the post office daily.

(c) The count of children with color blindness in a classroom of 20.

CHAPTER 12 EXERCISES

12.25 Binomial setting? In each situation below, is it reasonable to use a binomial distribution for the random variable X? Give reasons for your answer in each case.

(a) A quality controller for medical catheters randomly chooses one catheter each hour during an 8-hour production shift for a detailed quality inspection. One variable recorded is the count X of catheters with an unacceptable diameter (too small or too large).

(b) A sample survey asks 100 persons chosen at random from the adult residents of a large city whether they oppose the legalization of medical marijuana; X is the number who say "yes."

(c) A pediatrician sees 24 unrelated children on one winter day; X is the number of patients who came because of cold or flu symptoms.

12.26 Binomial setting? Your pulse reflects the rate at which your heart beats. A healthy resting pulse rate is between 60 and 100 beats per minute in adults. In which of these two settings is a binomial distribution more likely to be at least approximately correct? Explain your answer.

(a) You select at random 10 times in the next 24-hour period to take your pulse. You then go about your usual activities, including exercising and sleep. The random variable X is the number of measurements that are below 100 beats per minute.

(b) A nurse instructs an adult patient to take his resting pulse 10 times over the next few days. A resting pulse requires resting for at least 10 minutes before taking the pulse. The random variable X is the number of measurements that are below 100 beats per minute.

12.27 Testing ESP. In a test for ESP (extrasensory perception), a subject is told that cards only the experimenter can see contain either a star, a circle, a wave, or a square. As the experimenter looks at each of 20 cards in turn, the subject names the shape on the card. A subject who is just guessing has probability 0.25 of guessing correctly on each card.

(a) The count of correct guesses in 20 cards has a binomial distribution. What are n and p?

(b) What is the mean number of correct guesses in 20 cards for subjects who are just guessing?

(c) What is the probability of exactly 5 correct guesses in 20 cards if a subject is just guessing?

12.28 Malaria epidemics. Malaria is a debilitating and potentially deadly parasitic infection that is mostly preventable and curable, yet many developing countries lack the necessary resources. The West African country of Guinea has the highest rate of malaria in the world, with 75% of its population infected.[8] A physician from Doctors Without Borders sees 6 patients in a given hour. Assuming that patient visits are independent, what is the probability that all of them are infected? What is the probability that all but 1 of these 6 patients are infected with malaria?

12.29 Malaria epidemics, continued. In the previous exercise, X is the count of patients who are infected with malaria among the 6 patients the doctor saw.

(a) The count X has a binomial distribution. What are n and p?

(b) What are the possible values that X can take?

(c) Find the probability of each value of X. Draw a probability histogram for the distribution of X. (See Figure 12.2 for an example of a probability histogram.)

(d) What are the mean and standard deviation of this distribution? Mark the location of the mean on your histogram.

12.30 False-positives in testing for HIV. A rapid test for the presence in the blood of antibodies to HIV, the virus that causes AIDS, gives a positive result with probability about 0.004 when a person who is free of HIV antibodies is tested. A clinic tests 1000 people who are all free of HIV antibodies.

(a) What is the distribution of the number of positive tests?

(b) What is the mean number of positive tests?

(c) You cannot safely use the Normal approximation for this distribution. Explain why.

12.31 Universal donors. People with type O-negative blood are universal donors whose blood can be safely given to anyone. Only 7.2% of the population have O-negative blood. A blood center is visited by 20 donors in an afternoon. What is the probability that there are at least 2 universal donors among them?

Whooping cough (pertussis) is a highly contagious bacterial infection that was a major cause of childhood deaths before the development of vaccines. About 80% of unvaccinated children who are exposed to whooping cough will develop the infection, as opposed to only about 5% of vaccinated children. Exercises 12.32 to 12.35 are based on this information.

12.32 Vaccination at work. A group of 20 children at a nursery school are exposed to whooping cough by playing with an infected child.

(a) If all 20 have been vaccinated, what is the mean number of new infections? What is the probability that no more than 2 of the 20 children develop infections?
(b) If none of the 20 have been vaccinated, what is the mean number of new infections? What is the probability that 18 or more of the 20 children develop infections?

12.33 A whooping cough outbreak. In 2007, Bob Jones University ended its fall semester a week early because of a whooping cough outbreak; 158 students were isolated and another 1200 given antibiotics as a precaution.[9] Authorities react strongly to whooping cough outbreaks because the disease is so contagious. Because the effect of childhood vaccination often wears off by late adolescence, treat the Bob Jones students as if they were unvaccinated. It appears that about 1400 students were exposed. What is the probability that at least 75% of these students develop infections if not treated? (Fortunately, whooping cough is much less serious after infancy.)

12.34 A mixed group: means. A group of 20 children at a nursery school are exposed to whooping cough by playing with an infected child. Of these children, 17 have been vaccinated and 3 have not.

(a) What is the distribution of the number of new infections among the 17 vaccinated children? What is the mean number of new infections?
(b) What is the distribution of the number of new infections among the 3 unvaccinated children? What is the mean number of new infections?
(c) Add your means from parts (a) and (b). This is the mean number of new infections among all 20 exposed children.

12.35 A mixed group: probabilities. We would like to find the probability that exactly 2 of the 20 exposed children in the previous exercise develop whooping cough.

(a) One way to get 2 infections is to get 1 among the 17 vaccinated children and 1 among the 3 unvaccinated children. Find the probability of exactly 1 infection among the 17 vaccinated children. Find the probability of exactly 1 infection among the 3 unvaccinated children. These events are independent: What is the probability of exactly 1 infection in each group?
(b) Write down all the ways in which 2 infections can be divided between the two groups of children. Follow the pattern of part (a) to find the probability of each of these possibilities. Add all your results (including the result of part (a)) to obtain the probability of exactly 2 infections among the 20 children.

12.36 Genetics. According to genetic theory, the blossom color in the second generation of a certain cross of sweet peas should be red or white in a 3:1 ratio. That is, each plant has probability 3/4 of having red blossoms, and the blossom colors of separate plants are independent.

(a) What is the probability that exactly 6 out of 8 of these plants have red blossoms?
(b) What is the mean number of red-blossomed plants when 80 plants of this type are grown from seeds?
(c) What is the probability of obtaining at least 50 red-blossomed plants when 80 plants are grown from seeds?

12.37 Inheriting blood type. If the parents in Example 12.4 have 5 children, the number who have type O blood is a random variable X that has the binomial distribution with $n = 5$ and $p = 0.25$.

(a) What are the possible values of X?
(b) Find the probability of each value of X. Draw a histogram to display this distribution. (Because probabilities are long-run proportions, a histogram with the probabilities as the heights of the bars shows what the distribution of X would be in very many repetitions.)

12.38 Inheriting blood type, continued. What are the mean and standard deviation of the number of children with type O blood in the previous exercise? Mark the location of the mean on the probability histogram you made in that exercise.

12.39 Sex ratio at birth. Example 9.3 (page 209) gave the counts of male and female babies born in the United States in 2008. There were 2,074,000 baby girls and 2,173,000 baby boys born. We often assume that the probability of having a boy or a girl is 0.5 because of the chromosomal determination of sex in humans. Is there reason to think that the sex ratio in the United States favors more baby boys? To answer this question, find the probability that an equal chance of getting a boy would give 2,173,000 boys or more in 4,247,000 births. What do you conclude?

12.40 The continuity correction. One reason why the Normal approximation may fail to give accurate estimates of binomial probabilities is that the binomial distributions are discrete and the Normal distributions are continuous. That is, counts take only whole-number values, but Normal variables can take any value. We can improve the Normal approximation by treating each whole-number count as if it occupied the interval from 0.5 below the number to 0.5 above the number. For example, you can approximate a binomial probability $P(X \geq 10)$ by finding the Normal probability $P(X \geq 9.5)$. Be careful: Binomial $P(X > 10)$ is approximated by Normal $P(X \geq 10.5)$.

In Exercise 12.3 we learned that 44% of high school students in New York State have ever had sexual intercourse. The exact binomial probability that 50 or more of an SRS of 100 high school students from New York State have ever had sexual intercourse is 0.1341.

(a) Show that this setting satisfies the rule of thumb for use of the Normal approximation.
(b) What is the Normal approximation to $P(X \geq 50)$?
(c) What is the Normal approximation using the continuity correction? That's much closer to the true binomial probability.

12.41 Monitoring rabies in Florida. Rabies is an often-fatal disease typically transmitted through the bite of an infected animal. The state of Florida has been recording all known cases of rabies for the past 20 years, with data indicating an average rate of 3.6 cases of rabies per week.[10] Let X be the count of rabies cases reported in a given week.

(a) What distribution does X follow, approximately? Give the mean and standard deviation of X.
(b) What is the probability that there would be no case of rabies reported in a given week? What is the probability that there would be at least 1 case reported in a given week?

12.42 Pediatric appointments. A health clinic gets an average of 4.5 pediatric appointments per day for routine vaccination. The appointments are unrelated and so we can assume that they are random, independent events. Let X be the count of pediatric appointments per day for routine vaccination. What distribution does X follow, approximately? Give the mean and standard deviation of X.

12.43 Pediatric appointments, continued. In the setting of the previous exercise,

(a) what is the probability that the health clinic gets no pediatric appointments for routine vaccination on a given day?
(b) what is the probability that the health clinic gets exactly 1 such appointment on a given day?
(c) what is the probability that the health clinic gets 2 or more such appointments on a given day?

12.44 Blood plasma donations. Blood type AB is somewhat rare in the United States (4% of the general population) and is particularly useful for plasma transfusions because this plasma can be accepted by individuals with all blood types. The plasma is the liquid part of the blood once all cells are removed. You count the number X of AB donors that come to a fixed Red Cross location per day. This count has the Poisson distribution with parameter μ being the mean number of AB donors per day.

(a) This donation center sees an average of 2.1 AB donors per day. Find the Poisson probabilities for $X = 0, 1, 2, 3, 4, 5$.
(b) What is the probability that this particular location would get more than 5 AB donors in one day?

12.45 Planning for hospital admissions. A private hospital gets an average of 3.2 hospital admissions a day from its emergency room. X is the count of hospital admissions from the emergency room in a day. To plan for the number of beds that should be kept available, you compute the probabilities $P(X \geq 5)$, $P(X \geq 6)$, and $P(X \geq 7)$. What do you conclude?

12.46 Salmonellosis outbreak. Improperly handled or undercooked poultry and eggs are the most frequent cause of salmonella food poisoning. Salmonellosis can be deadly, although most individuals recover on their own. Cases must be reported to the CDC, where they are closely monitored. The average number of salmonellosis cases per month in South Dakota is 1.67.

(a) If X is the monthly count of salmonellosis cases in South Dakota, what distribution does X follow, approximately? Give the mean and standard deviation of X.
(b) Give the probabilities $P(X = 0)$, $P(X \leq 1)$, $P(X \leq 2)$, $P(X \leq 3)$, and $P(X \leq 4)$. What is the probability that

there would be more than 4 cases of salmonellosis in South Dakota in a given month?

(c) In September 1994, 14 cases of salmonellosis in South Dakota were reported to the CDC.[11] What is the probability that 14 or more cases would arise in one month in South Dakota? What does this unusual report suggest? If you worked at the CDC at that time, what do you think would be your next action?

12.47 Salmonellosis outbreak, continued. Refer to the previous exercise. The CDC reports that the average number of salmonellosis cases per month in Wisconsin is 15.58.

(a) If X is the monthly count of salmonellosis cases in Wisconsin, what distribution does X follow, approximately? Give the mean and standard deviation of X.

(b) Use technology to find the probabilities $P(X = 0)$, $P(X \leq 5)$, $P(X \leq 15)$, and $P(X \leq 25)$. What is the probability that there would be more than 25 cases of salmonellosis in Wisconsin in a given month?

(c) As with South Dakota, September 1994 had an unusually high number of salmonellosis cases in Wisconsin with 48 cases that month. What is the probability that 48 or more cases would arise in one month in Wisconsin? What do you make of the fact that a salmonellosis outbreak occurred in two states at the same time? In fact, the salmonellosis outbreak was traced back to nationally distributed ice cream products from one factory. The factory was shut down until cleared, and its contaminated production recalled, thus preventing many more salmonellosis cases.

12.48 A new mumps outbreak. For the week of February 6, 2010, the CDC reported a total of 68 cases of mumps, whereas the historical weekly average was only 11 cases, based on data from the past five years.[12]

(a) Assuming that cases of mumps are random and independent, the number X of weekly mumps cases in the United States has approximately what distribution? Given this assumption, what are the mean and standard deviation of X?

(b) Based on this model, use technology to find the probabilities $P(X \leq 15)$, $P(X \leq 30)$, and $P(X \leq 67)$. What is the probability that there would be 68 or more cases of mumps in one week in the United States?

(c) Do these data from February 2010 support the idea that mumps cases are random and independent with a mean of 11 cases per week? Explain your answer. In fact, this was part of the largest mumps outbreak in the United States since the 2006 outbreak described in Example 12.11, with over 1500 cases of mumps diagnosed as of February 28, 2010. The outbreak began in June 2009 with the return of an 11-year-old boy from a trip to the United Kingdom while it was experiencing a local mumps epidemic.

The last three exercises explore other well-known discrete probability distributions not described in the core of this chapter. They are optional material.

12.49 Osteoporosis. Osteoporosis is a disease defined as very low bone density resulting in a high risk of fracture, hospitalization, and immobilization. Seven percent of postmenopausal women suffer from osteoporosis.[13] A physician administers routine check-ups. Visits by postmenopausal women are independent. We are interested in how long we must wait to get the first postmenopausal woman with osteoporosis.

(a) Considering only patient visits by postmenopausal women, the probability of a patient with osteoporosis on the first visit is 7%. What is the probability that the first patient does not have osteoporosis and the second patient does?

(b) What is the probability that the first two patients do not have osteoporosis and the third one does? This is the probability that the first patient with osteoporosis occurs on the third visit.

(c) Now you see the pattern. What is the probability that the first patient with osteoporosis occurs on the fourth visit? On the fifth visit? Give the general result: what is the probability that the first patient with osteoporosis occurs on the kth visit?

Comment: The distribution of the number of trials to the first "success" (in our example, a patient with osteoporosis) is called a **geometric distribution.** In this problem you have found geometric distribution probabilities when the probability of a success on each trial is 7%. The same idea works for any probability of success.

12.50 Sex selection. A couple decides to continue to have children until their first girl is born; X is the total number of children the couple has. Does X have a binomial distribution?

12.51 Overweight Americans. The National Center for Health Statistics estimates that 3% of adult American women are underweight, 42% have a healthy weight, and 55% are overweight or obese.[14] A gynecologist sees 3 patients for a routine exam.

(a) What is the probability that the first one has a healthy weight and the next two are overweight?
(b) What is the probability that the second patient has a healthy weight and the other two are overweight? What is the probability that the first two are overweight and the last one has a healthy weight?
(c) Using the results from (a) and (b), give the probability that out of any three patients, one has a healthy weight and two are overweight. This is a **multinomial probability** with three possible outcomes. Multinomial probabilities can represent any number of possible outcomes, generalizing the binomial setting (their mathematical description is beyond the scope of this introductory textbook). Binomial probabilities are a special case of the multinomial model when only two outcomes are possible.

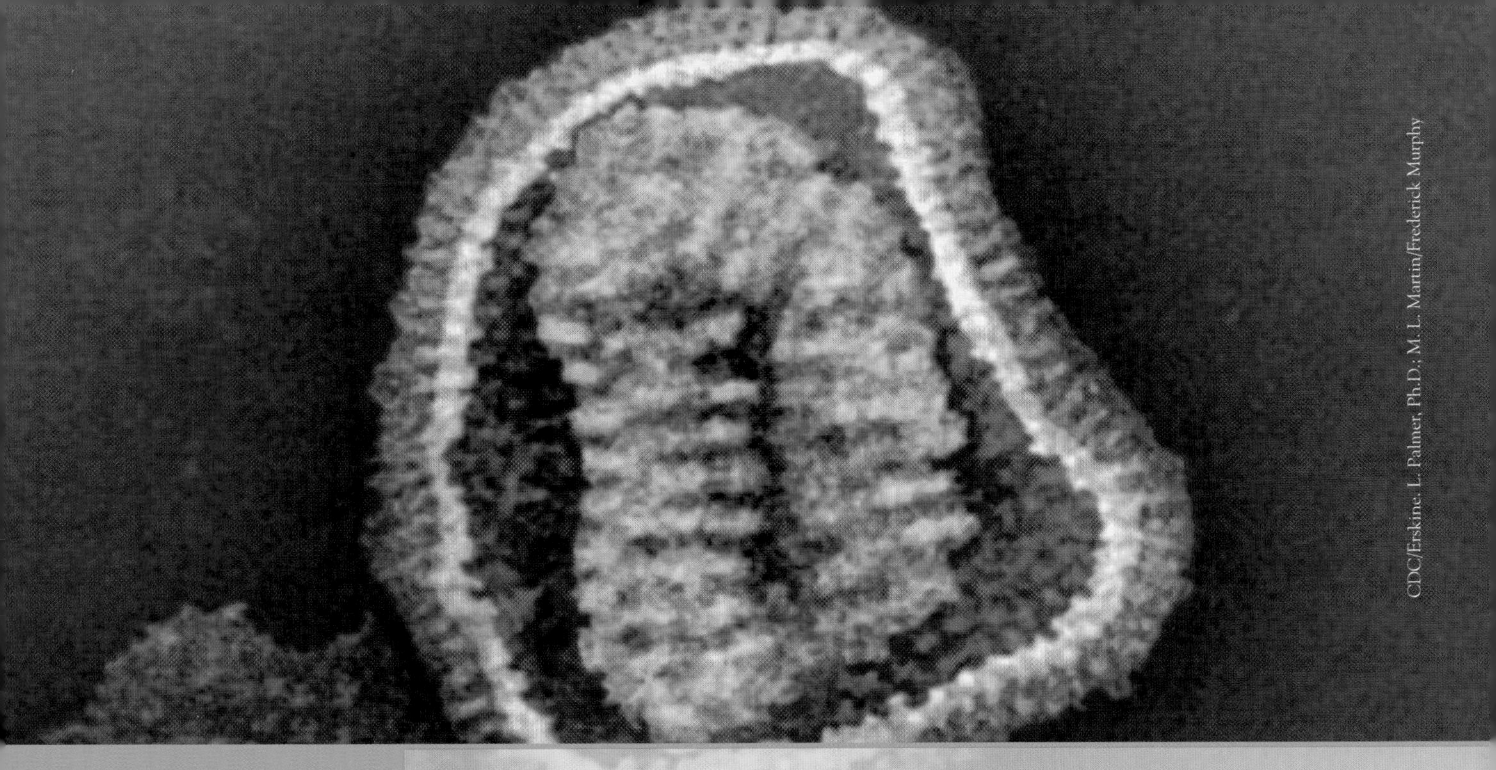

CDC/Erskine. L. Palmer, Ph.D.; M. L. Martin/Frederick Murphy

CHAPTER 13 Sampling Distributions

Obesity has become the unrelenting focus of health debates. But how do we know that it is a problem and that it is worsening? Gallup has been asking a representative sample of American adults since 1990. In 1990, the average weight of American adult males over the age of 18 was $\overline{x} = 180$ pounds based on a sample of roughly 500 individuals. Two decades later, a 2012 Gallup random sample of 515 adult men produced an average weight of $\overline{x} = 196$ pounds. The values 180 and 196 describe the samples taken in 1990 and 2012, respectively, but we use them to estimate the mean weight of all American adult males in 1990 and in 2012. This is an example of statistical inference: We use information from a sample to infer something about a wider population.

Because the results of random samples and randomized comparative experiments include an element of chance, we can't guarantee that our inferences are correct. What we can guarantee is that our methods usually give correct answers. We will see that the reasoning of statistical inference rests on asking, "How often would this method give a correct answer if I used it very many times?" If our data come from random sampling or randomized comparative experiments, the laws of probability answer the question "What would happen if we did this many times?" This chapter presents some facts about probability that help answer this question.

IN THIS CHAPTER WE COVER...

- Parameters and statistics
- Statistical estimation and sampling distributions
- The sampling distribution of $\overline{x}$
- The central limit theorem
- The sampling distribution of $\hat{p}$
- The law of large numbers*

Parameters and statistics

As we begin to use sample data to draw conclusions about a wider population, we must take care to keep straight whether a number describes a sample or a population. Here is the vocabulary we use.

PARAMETER, STATISTIC

A **parameter** is a number that describes the population. In statistical practice, the value of a parameter is not known, because we cannot examine the entire population.

A **statistic** is a number that can be computed from the sample data without making use of any unknown parameters. In practice, we often use a statistic to estimate an unknown parameter.

EXAMPLE 13.1 Obesity

The mean weight of the sample of American adult males contacted by the 2012 Gallup survey was $\bar{x} = 196$ pounds. The number 196 is a *statistic*, because it describes this one Gallup sample. The population that the survey wants to draw conclusions about is all American adult males over the age of 18. The *parameter* of interest is the mean weight of all these individuals. We don't know the value of this parameter. ■

Remember that **s**tatistics come from **s**amples, and **p**arameters come from **p**opulations. The distinction between population and sample is essential, and the notation we use must reflect this distinction.

population mean μ

sample mean $\bar{x}$

When dealing with quantitative data, we write μ (the Greek letter mu) for the **mean of a population.** This is a fixed parameter that is unknown when we use a sample for inference. The **mean of the sample** is the familiar $\bar{x}$, the average of the observations in the sample. This is a statistic that would almost certainly take a different value if we chose another sample from the same population. The sample mean $\bar{x}$ from a sample or an experiment is an estimate of the mean μ of the underlying population.

population proportion p

sample proportion $\hat{p}$

Likewise, when dealing with categorical data, we use the notation p to indicate the **population proportion** of individuals of a given type, a parameter often of unknown value. The **sample proportion**, $\hat{p}$, is the proportion of individuals of a given type in a sample or an experiment. It is a statistic that would almost certainly take a different value if we chose another sample from the same population. The sample proportion $\hat{p}$ is an estimate of the population proportion p.

APPLY YOUR KNOWLEDGE

13.1 Nerve conduction in rats. Extensive scientific research on nerve conduction has led to the conclusion that the mean refractory period in a healthy rat is **1.3** millisecond (ms). An experiment designed to study the effect of the pesticide DDT on nerve conduction finds that the mean refractory period of 4 rats poisoned with DDT is **1.75** ms. Is each of the boldface numbers a parameter or a statistic?

13.2 Disease surveillance. According to the Centers for Disease Control and Prevention (CDC), there were 6153 reported cases of animal rabies in the United States in all of 2010. Raccoons were the most common rabid animals, making up **36.5%** of all reported cases that year. Gallup contacted by phone a random sample of 27,300 American adults in December 2012 and found that **3.2%** had been sick with the flu. Is each of the boldface numbers a parameter or a statistic?

Statistical estimation and sampling distributions

Statistical inference uses sample data to draw conclusions about the entire population. *Because good samples are chosen randomly, statistics such as $\bar{x}$ and $\hat{p}$ are random variables*. We can describe the behavior of a sample statistic by a probability model that answers the question "What would happen if we did this many times?" Here is an example that will lead us toward the probability ideas most important for statistical inference.

EXAMPLE 13.2 Does this wine smell bad?

Sulfur compounds such as dimethyl sulfide (DMS) are sometimes present in wine. DMS causes "off-odors" in wine, so winemakers want to know the odor threshold, the lowest concentration of DMS that the human nose can detect. Different people have different thresholds, so we start by asking about the mean threshold μ in the population of all adults. The number μ is a parameter that describes this population.

Enigma/Alamy

To estimate μ, we present tasters with both natural wine and the same wine spiked with DMS at different concentrations to find the lowest concentration at which they identify the spiked wine. Here are the odor thresholds (measured in micrograms of DMS per liter of wine) for 10 randomly chosen subjects:

28 40 28 33 20 31 29 27 17 21

The mean threshold for these subjects is $\bar{x} = 27.4$. It seems reasonable to use the sample result $\bar{x} = 27.4$ to estimate the unknown μ. An SRS should fairly represent the population, so the mean $\bar{x}$ of the sample should be somewhere near the mean μ of the population. Of course, we don't expect $\bar{x}$ to be exactly equal to μ. We realize that if we choose another SRS, the luck of the draw will probably produce a different $\bar{x}$. ■

If $\bar{x}$ is rarely exactly right and varies from sample to sample, why is it nonetheless a reasonable estimate of the population mean μ? What can we say about $\bar{x}$ from 10 subjects as an estimate of μ? We ask: "What would happen if we took many samples of 10 subjects from this population?" Here's how to answer this question theoretically:

- Take a large number of samples of size 10 from the population.
- Calculate the sample mean $\bar{x}$ for each sample.

- Make a histogram of the values of $\overline{x}$.
- Examine the distribution displayed in the histogram for shape, center, and spread, as well as for outliers or other deviations.

EXAMPLE 13.3 What would happen in many samples?

Extensive studies have found that the DMS odor threshold of adults follows roughly a Normal distribution with mean $\mu = 25$ micrograms per liter and standard deviation $\sigma = 7$ micrograms per liter. With this information, we can use software to create many repetitions of Example 13.2 with different subjects drawn at random from the population. Using software to imitate chance behavior is called **simulation.**

simulation

Figure 13.1 illustrates the process of choosing many samples and finding the mean threshold $\overline{x}$ for each sample. Follow the flow of the figure from the population at the left, to choosing an SRS and finding the $\overline{x}$ for this sample, to collecting together the $\overline{x}$'s from many samples. The first sample has $\overline{x} = 26.42$. The second sample contains a different 10 people, with $\overline{x} = 24.28$, and so on. The histogram at the right of the figure shows the distribution of the values of $\overline{x}$ from 1000 separate SRSs of size 10. This histogram displays the *sampling distribution* of the statistic $\overline{x}$. ■

SAMPLING DISTRIBUTION

The **sampling distribution** of a statistic is the distribution of values taken by the statistic in all possible samples of the same size from the same population.

Strictly speaking, the sampling distribution is the ideal pattern that would emerge if we looked at all possible samples of size 10 from our population. A distribution obtained from a fixed number of trials, like the 1000 trials in Figure 13.1, is

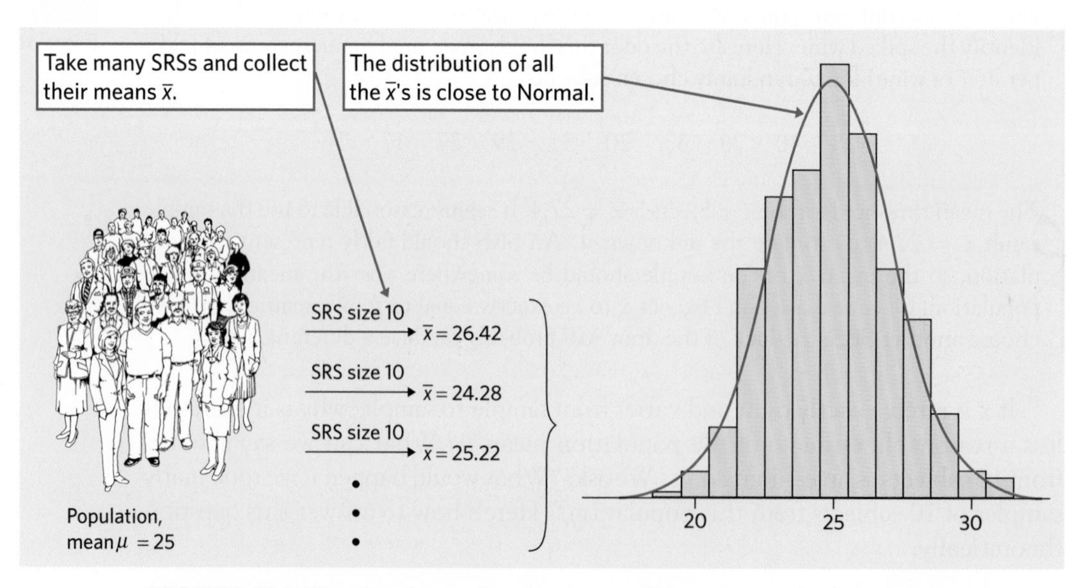

FIGURE 13.1 The idea of a sampling distribution: Take many samples from the same population, collect the $\overline{x}$'s from all the samples, and display the distribution of $\overline{x}$. The histogram shows the results of 1000 samples.

only an approximation to the sampling distribution. One of the uses of probability theory in statistics is to obtain exact sampling distributions without simulation. The interpretation of a sampling distribution is the same, however, whether we obtain it by simulation or by the mathematics of probability.

We can use the tools of data analysis to describe any distribution. Let's apply those tools to Figure 13.1. What can we say about the shape, center, and spread of this distribution?

- **Shape:** It looks Normal! Detailed examination confirms that the distribution of $\overline{x}$ from many samples does have a distribution that is very close to Normal.
- **Center:** The mean of the 1000 $\overline{x}$'s is 24.95. That is, the distribution is centered very close to the population mean $\mu = 25$.
- **Spread:** The standard deviation of the 1000 $\overline{x}$'s is 2.217, notably smaller than the standard deviation $\sigma = 7$ of the population of individual subjects.

Although these results describe just one simulation of a sampling distribution, they reflect facts that are true whenever we use random sampling. We shall describe these properties next.

Students are sometimes tempted to take several samples to build a partial sampling distribution. However, the sampling distribution of a statistic describes theoretically all possible samples of a given size that could be taken from a given population. Sampling distributions are important mathematical concepts that we will use for the purpose of statistical estimation.

APPLY YOUR KNOWLEDGE

13.3 Generating a sampling distribution for pedagogical purposes. Let's illustrate the idea of a sampling distribution in the case of a very small sample from a very small population. The population is the 10 female students in a class:

Student	0	1	2	3	4	5	6	7	8	9
Weight	136	99	118	129	125	170	130	128	120	147

The parameter of interest is the mean weight in pounds μ in this population. The sample is an SRS of size $n = 4$ drawn from the population. Because the students are labeled 0 to 9, a single random digit from Table A (at the back of the book) chooses one student for the sample.

(a) Find the mean of the 10 weights in the population. This is the population mean μ.

(b) Use the first 4 digits in row 116 of Table A to draw an SRS of size 4 from this population. What are the four weights in your sample? What is their mean $\overline{x}$? This statistic is an estimate of μ.

(c) Repeat this process 9 more times, using the first four digits in rows 117 to 125 of Table A. Make a histogram of the 10 values of $\overline{x}$. You are constructing the sampling distribution of $\overline{x}$. Is the center of your histogram close to μ?

(d) This is a lengthy process, even for such a small population. Explain why it would be unrealistic to try building a sampling distribution by hand for a large population.

13.4 A sampling distribution. We can use the *Simple Random Sample* applet to help us grasp the idea of a sampling distribution (for pedagogical purposes). Form a population labeled 1 to 100. We will choose an SRS of 10 of these numbers. That is, in this exercise, the numbers themselves are the population, not just labels for 100 individuals. The mean of the whole numbers 1 to 100 is $\mu = 50.5$. This is the population mean.

(a) Use the applet to choose an SRS of size 10. Which 10 numbers were chosen? What is their mean? This is the sample mean $\bar{x}$.

(b) Although the population and its mean $\mu = 50.5$ remain fixed, the sample mean changes as we take more samples. Take another SRS of size 10. (Use the "Reset" button to return to the original population before taking the second sample.) What are the 10 numbers in your sample? What is their mean? This is another value of $\bar{x}$.

(c) Take 8 more SRSs from this same population and record their means. You now have 10 values of the sample mean $\bar{x}$ from 10 SRSs of the same size from the same population. Make a histogram of the 10 values and mark the population mean $\mu = 50.5$ on the horizontal axis. Are your 10 sample values roughly centered at the population value μ? (If you kept going forever, your $\bar{x}$-values would form the sampling distribution of the sample mean; the population mean μ would indeed be the center of this distribution.)

(d) Even with software, explain some of the obstacles to trying to build a sampling distribution through repeated samples.

The sampling distribution of $\bar{x}$

Figure 13.1 suggests that when we choose many SRSs from a population, the sampling distribution of the sample means is centered at the mean of the original population and less spread out than the distribution of individual observations. Although we won't elaborate, this can be demonstrated mathematically. Here are simply the facts.

MEAN AND STANDARD DEVIATION OF A SAMPLE MEAN

Suppose that $\bar{x}$ is the mean of an SRS of size n drawn from a large population with mean μ and standard deviation σ. Then the sampling distribution of $\bar{x}$ has **mean** μ and **standard deviation** $\sigma/\sqrt{n}$.

These facts about the mean and the standard deviation of the sampling distribution of $\bar{x}$ are true for *any* population, provided that the population is much larger than the sample (say, at least 20 times larger).[1] Both facts have important implications for statistical inference.

- The mean of the statistic $\overline{x}$ is always equal to the mean μ of the population. That is, the sampling distribution of $\overline{x}$ is centered at μ. In repeated sampling, $\overline{x}$ will sometimes fall above the true value of the parameter μ and sometimes below it, but there is no systematic tendency to overestimate or underestimate the parameter. This makes the idea of lack of bias in the sense of "no favoritism" more precise. Because the mean of $\overline{x}$ is equal to μ, we say that the statistic $\overline{x}$ is an **unbiased estimator** of the parameter μ.

unbiased estimator

- An unbiased estimator is "correct on the average" over many samples. How close the estimator falls to the parameter in most samples is determined by the spread of the sampling distribution. If individual observations have standard deviation σ, then sample means $\overline{x}$ from samples of size n have standard deviation $\sigma/\sqrt{n}$. That is, **averages are less variable than individual observations.**

We have described the center and spread of the sampling distribution of a sample mean $\overline{x}$, but not its shape. The shape of the distribution of $\overline{x}$ depends on the shape of the population. Here is one important case: If measurements in the population follow a Normal distribution, then so does the sample mean.

SAMPLING DISTRIBUTION OF A SAMPLE MEAN FOR A NORMAL POPULATION

If individual observations have the $N(\mu, \sigma)$ distribution, then the sample mean $\overline{x}$ of an SRS of size n has the $N(\mu, \sigma/\sqrt{n})$ distribution.

EXAMPLE 13.4 Population distribution versus sampling distribution

If we measure the DMS odor thresholds of individual adults, the values follow the Normal distribution with mean $\mu = 25$ micrograms per liter and standard deviation $\sigma = 7$ micrograms per liter. We call this the **population distribution** because it shows how measurements vary within the population.

population distribution

Take many SRSs of size 10 from this population and find the sample mean $\overline{x}$ for each sample, as in Figure 13.1. The *sampling distribution* describes how the values of $\overline{x}$ vary among samples. That sampling distribution is also Normal, with mean $\mu = 25$ and standard deviation

$$\frac{\sigma}{\sqrt{n}} = \frac{7}{\sqrt{10}} = 2.2136$$

Figure 13.2 contrasts these two Normal distributions. Both are centered at the population mean, but sample means are much less variable than individual observations. ■

Not only is the standard deviation of the distribution of $\overline{x}$ smaller than the standard deviation of individual observations, but it gets smaller as we take larger samples. **The results of large random samples are less variable than the results of small samples.** If n is large, the standard deviation of $\overline{x}$ is small, and almost all samples will give values of $\overline{x}$ that lie very close to the true parameter μ. That is, the

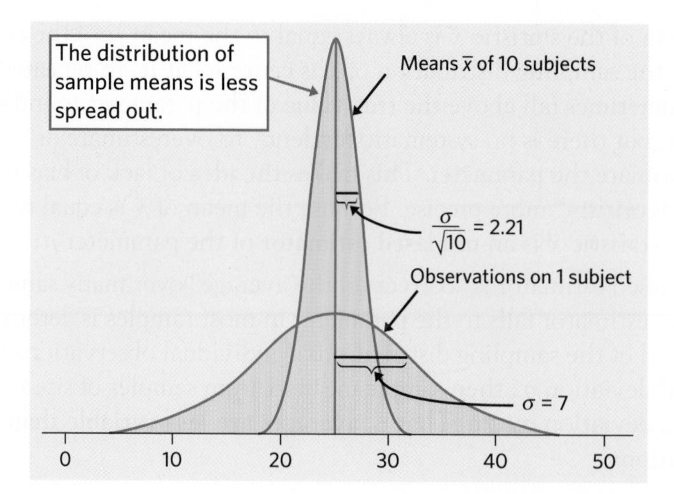

FIGURE 13.2 The distribution of single observations compared with the distribution of the means $\bar{x}$ of 10 observations. Averages are less variable than individual observations.

sample mean from a large sample can be trusted to estimate the population mean accurately. This property of sampling distributions has many important, direct applications in statistics. *However, the standard deviation of the sampling distribution gets smaller only at the rate* $\sqrt{n}$. *To cut the standard deviation of* $\bar{x}$ *by 10, we must take 100 times as many observations, not just 10 times as many, and large sample sizes are not always an option.*

APPLY YOUR KNOWLEDGE

13.5 A sample of young men. A government sample survey plans to measure the blood cholesterol levels of an SRS of men aged 20 to 34 years. The researchers will report the mean $\bar{x}$ from their sample as an estimate of the mean cholesterol level μ in this population.

(a) Explain to someone who knows no statistics what it means to say that $\bar{x}$ is an "unbiased" estimator of μ.

(b) The sample result $\bar{x}$ is an unbiased estimator of the true population μ no matter what size SRS the study uses. Explain to someone who knows no statistics why a large sample gives more trustworthy results than a small sample.

Spencer Grant/PhotoEdit

13.6 Measurements in the lab. Juan makes a measurement in a chemistry laboratory and records the result in his lab report. The standard deviation of lab measurements made by students is $\sigma = 10$ milligrams. Juan repeats the measurement 3 times and records the mean $\bar{x}$ of his 3 measurements.

(a) What is the standard deviation of Juan's mean result? (That is, if Juan kept on making 3 measurements and averaging them, what would be the standard deviation of all his $\bar{x}$'s?)

(b) How many times must Juan repeat the measurement to reduce the standard deviation of $\bar{x}$ to 5? Explain to someone who knows no statistics the advantage of reporting the average of several measurements rather than the result of a single measurement.

13.7 A sample of young men, continued. Suppose that in fact the blood cholesterol levels of all men aged 20 to 34 years follows the Normal distribution with mean $\mu = 188$ milligrams per deciliter (mg/dl) and standard deviation $\sigma = 41$ mg/dl.

(a) Choose an SRS of 100 men from this population. What is the sampling distribution of $\overline{x}$? What is the probability that $\overline{x}$ takes a value between 185 and 191 mg/dl? This is the probability that $\overline{x}$ estimates μ within ± 3 mg/dl.

(b) Choose an SRS of 1000 men from this population. Now what is the probability that $\overline{x}$ falls within ± 3 mg/dl of μ? The larger sample is much more likely to give an accurate estimate of μ.

The central limit theorem

The facts about the mean and standard deviation of $\overline{x}$ are true no matter what the shape of the population distribution may be. But what about the *shape of the sampling distribution*? We saw in Figure 13.2 from Example 13.4 that the shape of the sampling distribution is Normal when the population distribution is Normal. But what if the population distribution is not Normal? It is a remarkable fact that as the sample size increases, the distribution of $\overline{x}$ changes shape: It looks less like that of the population and more like a Normal distribution. When the sample is large enough, the distribution of $\overline{x}$ is very close to Normal. This is true no matter what shape the population distribution has, as long as the population has a finite standard deviation σ. This famous fact of probability theory is called the *central limit theorem* and is extremely useful because of its wide applications.

CENTRAL LIMIT THEOREM

Draw an SRS of size n from any population with mean μ and finite standard deviation σ. When n is large, the sampling distribution of the sample mean $\overline{x}$ is approximately Normal:

$$\overline{x} \text{ is approximately } N\left(\mu, \frac{\sigma}{\sqrt{n}}\right)$$

The central limit theorem allows us to use Normal probability calculations to answer questions about sample means from many observations (questions relying on the sampling distribution of the sample mean) even when the population distribution is not Normal.

EXAMPLE 13.5 The central limit theorem in action

Let's examine sampling distributions from a population that is definitely not Normal. Figure 13.3(a) shows the density curve of a strongly skewed distribution called an *exponential distribution*. Exponential distributions are good models, for example, of radioactive decay and the time the body takes to metabolize and clear away drugs. The mean μ of the exponential distribution displayed is 1, and its standard deviation σ is also 1.

Figures 13.3(b), (c), and (d) show, respectively, the density curves of the sample means of 2, 10, and 25 observations from this population. The shape of the sampling

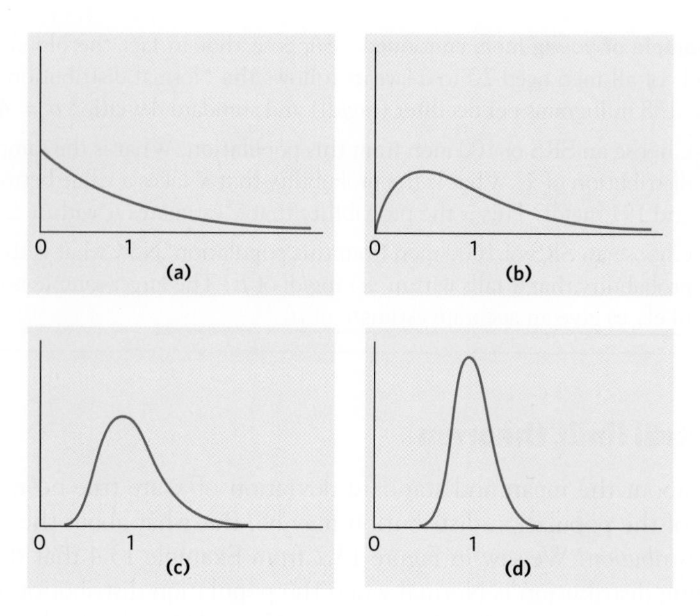

FIGURE 13.3 The central limit theorem in action: The distribution of sample means $\overline{x}$ from a strongly non-Normal population is more Normal for larger sample sizes. (a) The distribution of single observations. (This is the population distribution.) (b) The sampling distribution of $\overline{x}$ for samples of 2 observations. (c) The sampling distribution of $\overline{x}$ for samples of 10 observations. (d) The sampling distribution of $\overline{x}$ for samples of 25 observations.

distribution of $\overline{x}$ is more Normal for the larger values of n. The distribution mean remains at $\mu = 1$, and the standard deviation decreases, taking the value $1/\sqrt{n}$. The density curve for samples of 10 observations is still somewhat skewed to the right but already resembles a Normal curve having $\mu = 1$ and $\sigma = 1/\sqrt{10} = 0.32$. The density curve of the distribution of $\overline{x}$ when $n = 25$ is yet more Normal. The contrast between the shape of the population distribution and the shapes of the sampling distributions of the mean of $\overline{x}$ for samples of 10 or 25 observations is striking.

The *Central Limit Theorem* applet on the companion website allows you to watch the central limit theorem in action, using computer simulations. Select the exponential distribution on the right and choose a sample size n. The applet will simulate taking 10,000 random samples of size n and plot the resulting 10,000 sample means. If you select sample sizes of 1 (individual observations from the population), 2, 10, and 25, you should obtain distributions with patterns similar to those of Figure 13.3. ■

The central limit theorem applies to sampling distributions, not to the distribution (histogram) of the data from one random sample of n observations. Many students mistakenly believe that larger sample sizes yield more Normal sample histograms. This is not the case. If a population is skewed, for instance, chances are that a histogram of a random sample from that population will be skewed too. The central limit theorem describes only what happens to the distribution of the averages from repeated samples of a given size.

How large a sample size n is needed for $\overline{x}$ to be close to Normal depends on the population distribution. More observations are required if the shape of the population distribution is far from Normal. The following is always true, though.

THINKING ABOUT SAMPLE MEANS

Means of random samples are **less variable** than individual observations.

Means of random samples are **more Normal** than individual observations.

Mathematically proving the central limit theorem is beyond the scope of an introductory statistics textbook. We will instead focus on its practical uses.

EXAMPLE 13.6 Practical use of the central limit theorem

Figure 13.4 shows the histograms for three real data sets. By observing the overall shape of a histogram we can guess whether the population from which the data were taken might be approximately Normal, somewhat Normal, or definitely not Normal. From this and the size of the sample shown we can deduce whether the corresponding sampling distribution of $\bar{x}$ might be approximately Normal or not.

Figure 13.4(a) shows the angle of big-toe deformations in 38 patients. The histogram is nearly Normal except for one mild outlier. Because the outlier is not extreme and the sample size is reasonably large, the sampling distribution of $\bar{x}$ should be approximately Normal.

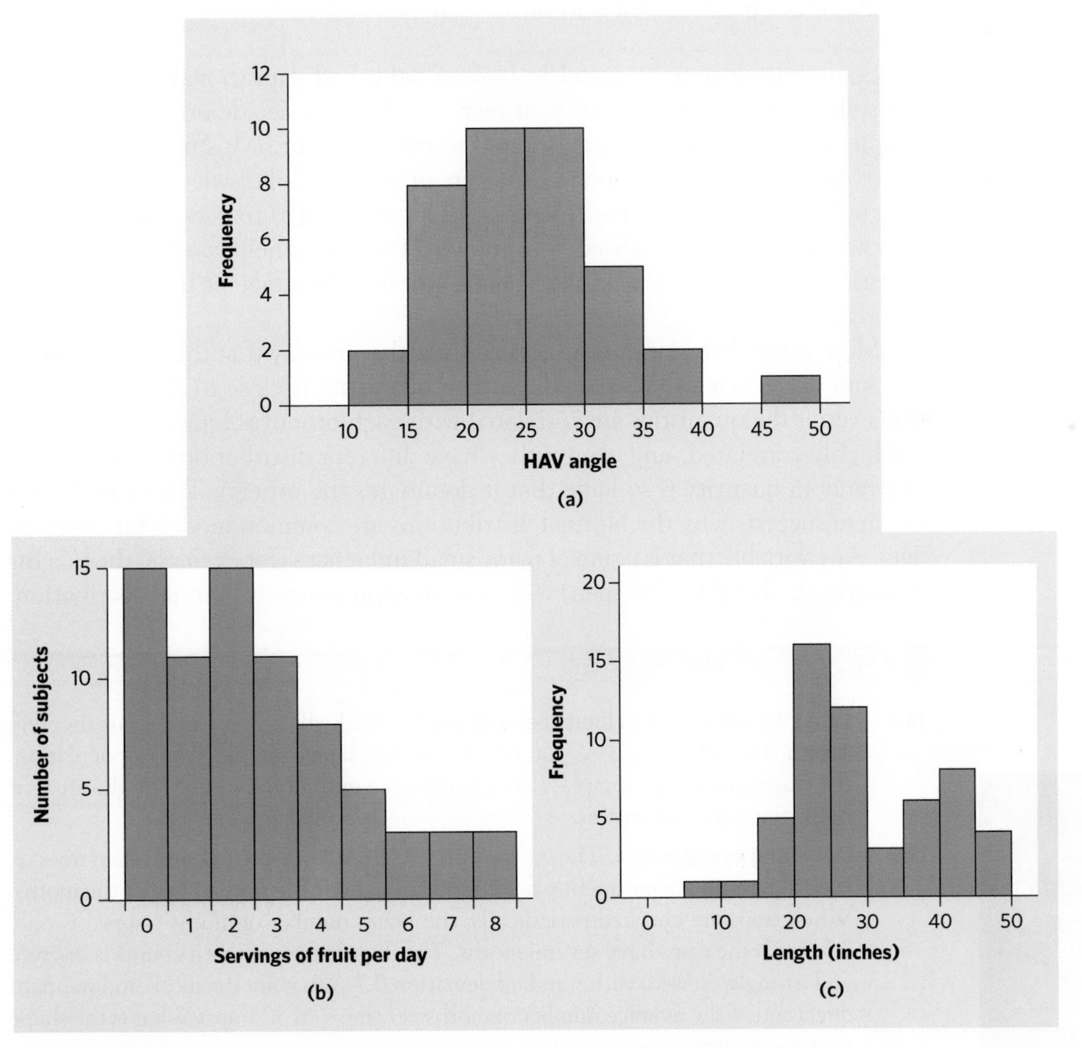

FIGURE 13.4 Histograms of three data sets for Example 13.6. (a) The angle of big-toe deformations (hallux abducto valgus, HAV) in 38 patients. (b) The number of daily fruit servings for 74 adolescent girls. (c) The lengths of 56 perch from a Swedish lake.

Figure 13.4(b) shows the number of daily fruit servings for 74 adolescent girls. The histogram is strongly right-skewed, suggesting that the population is not Normal but skewed. The sampling distribution for this variable should be Normal, nonetheless, because the sample size is very large.

Figure 13.4(c) shows the lengths of 56 perch from a Swedish lake. The histogram is not particularly skewed and does not have any outliers. With a sample of 56 fish, the central limit theorem guarantees that the sampling distribution would be approximately Normal. However, it is clear from the histogram that this sample is not homogeneous and, instead, seems to be made up of two types of fish: some smaller fish with a peak at 20 to 25 inches, and some larger fish with a peak at 40 to 45 inches. The histogram does not reveal what might be the reason for this bimodal distribution, and we would have to examine each fish closely to find out. What is clear is that the sampling distribution would be centered on a value "μ" that would reflect neither of these two subpopulations of fish and, therefore, would be misleading. Always keep in mind that poorly designed studies create problems that mathematics cannot remedy. ■

Sometimes it may be useful to look at published data to make an educated guess about the distribution of your own variable and decide on an appropriate sample size. For example, if you wanted to conduct your own study on the number of daily fruit servings among teenagers in your school district, you could rely on a publication showing the histogram of Figure 13.4(b) to guess that your own data would probably be skewed, too; knowing this would help you to decide to use at least 40 teenagers in your own sample to take advantage of the central limit theorem.

More general versions of the central limit theorem say that the distribution of any sum or average of many small random quantities is close to Normal. This is true even if the quantities are correlated with each other (as long as they are not too highly correlated) and even if they have different distributions (as long as no one random quantity is so large that it dominates the others). The central limit theorem suggests why the Normal distributions are common models for observed data. Any variable that is a sum of many small influences (for example, the heights of men or the heights of women) will have an approximately Normal distribution.

APPLY YOUR KNOWLEDGE

13.8 What does the central limit theorem say? Asked what the central limit theorem says, a student replies, "As you take larger and larger samples from a population, the histogram of the sample values looks more and more Normal." Is the student right? Explain your answer.

Bruce Coleman/Alamy

13.9 Detecting gypsy moths. The gypsy moth is a serious threat to oak and aspen trees. A state agriculture department places traps throughout the state to detect the moths. When traps are checked periodically, the mean number of moths trapped is only 0.5, but some traps have several moths. The distribution of moth counts is discrete and strongly skewed, with standard deviation 0.7. What are the mean and standard deviation of the average number of moths per trap $\bar{x}$ in 50 traps? What is the shape of this sampling distribution?

13.10 Insurance. An insurance company knows that, in the entire population of millions of insured workers, the mean annual cost of workers' compensation claims is

$\mu = \$439$ per insured worker, and the standard deviation is $\sigma = \$20,000$. The distribution of losses is strongly right-skewed: Most policies have no loss, but a few have large losses, up to millions of dollars.

(a) If the company sells 40,000 policies, what is the shape, mean, and standard deviation of the sampling distribution of the mean claim loss? Consider these 40,000 policies a random sample of all workers' compensation insurance policies.

(b) If the company sells 40,000 policies, what is the probability that the mean claim loss would be no greater than \$500? No greater than \$700?

The sampling distribution of $\hat{p}$

An experiment finds that 6 of 20 birds exposed to an avian flu strain develop flu symptoms. The number of birds that develop flu symptoms is a random variable X. X is a **count** of the occurrences of some categorical outcome in a fixed number of observations. If the number of observations is n, then the **sample proportion** is

count

sample proportion

$$\hat{p} = \frac{\text{count of successes in sample}}{\text{size of sample}} = \frac{X}{n}$$

Like the sample average $\overline{x}$ when studying quantitative variables, sample counts and sample proportions are common statistics when dealing with categorical data. This section describes their sampling distributions.

EXAMPLE 13.7 Who had the flu?

The 2012-13 flu season struck relatively early and hard. What proportion of the adult population got the flu in December 2012? The Gallup-Healthways Well-Being Index survey asked a random sample of 27,300 American adults in December 2012 if they were "sick with the flu yesterday." Of these, 874 answered "yes."

The value 874 is the count X of adults in the sample who were sick with the flu on the day before the interview. The proportion of adults among the 27,300 in the sample who were sick with the flu is the sample proportion $\hat{p} = 874/27{,}300 = 3.2\%$. The statistic 3.2% is representative of the population proportion p of all American adults who were sick with the flu on any given day of December 2012. ■

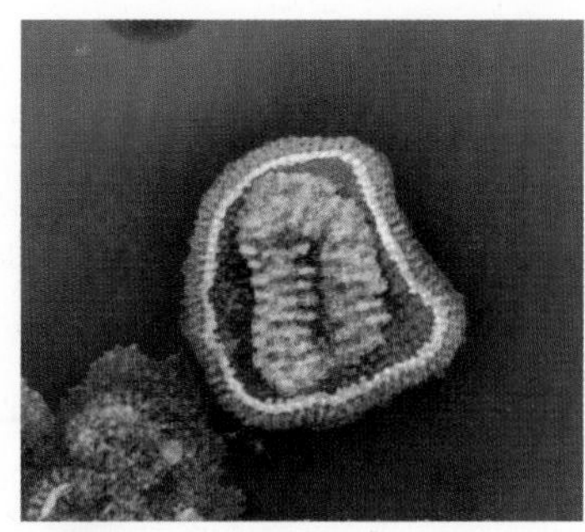

CDC/Erskine. L. Palmer, Ph.D.; M. L. Martin/Frederick Murphy

Categorical variables can take any of a finite number of possible outcomes. We may choose to call one such possible outcome a "success" and define all other possible outcomes as nonsuccesses, or failures. (This is an arbitrary decision, not a moral judgment.) If you have covered Chapter 12, you know that the distribution of the count X of a random "success" event occurring within n observations is binomial when the probability p of that event is constant over all observations and successive observations are independent. You also learned that the binomial distribution of X can be approximated by the Normal distribution $N(np, \sqrt{np(1-p)})$ when n is so large that the count of successes and the complementary count of failures in the sample are each at least 10.

A count of successes in itself is meaningful only in the context of the total number of observations and is therefore of limited use when comparing different studies. That's why we typically favor a more informative statistic, the sample proportion $\hat{p}$ of successes. How good is the statistic $\hat{p}$ as an estimate of the parameter p, the proportion of successes in the population? To find out, we ask, "What would happen if we took many samples?" The sampling distribution of $\hat{p}$ answers this question. Here are the facts.

SAMPLING DISTRIBUTION OF A SAMPLE PROPORTION

Choose an SRS of size n from a large population that contains population proportion p of successes. Let $\hat{p}$ be the **sample proportion** of successes,

$$\hat{p} = \frac{\text{count of successes in the sample}}{n}$$

Then:

- The **mean** of the sampling distribution is p.
- The **standard deviation** of the sampling distribution is $\sqrt{p(1-p)/n}$.
- As the sample size increases, the sampling distribution of $\hat{p}$ becomes **approximately Normal.**

Figure 13.5 summarizes these facts in a form that helps you recall the big idea of a sampling distribution. The behavior of sample proportions $\hat{p}$ is similar to the behavior of sample means $\bar{x}$. When the sample size n is large, the sampling distribution is approximately Normal. The larger the sample, the more nearly Normal the distribution is. The mean of the sampling distribution is the true value of the population proportion p. That is, $\hat{p}$ is an unbiased estimator of p. The standard deviation of $\hat{p}$ gets smaller as the sample size n gets larger, so that estimation is likely to be more accurate when the sample is larger. As is the case for $\bar{x}$, the standard deviation gets smaller only at the rate $\sqrt{n}$. We need 100 times as many observations to cut the standard deviation by 10.

You should not use the Normal approximation to the distribution of $\hat{p}$ when the sample size n is small. What is more, the formula given for the standard deviation of $\hat{p}$ is

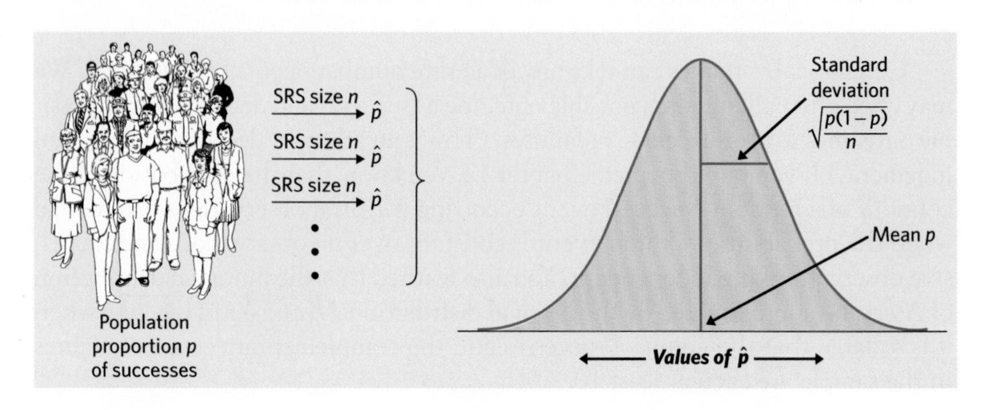

FIGURE 13.5 Select a large SRS from a population in which a proportion p are successes. The sampling distribution of the proportion $\hat{p}$ of successes in the sample is approximately Normal. The mean is p, and the standard deviation is $\sqrt{p(1-p)/n}$.

not accurate unless the population is much larger than the sample—say, at least 20 times larger.[2] We will give guidelines to help you decide when methods for inference based on this sampling distribution are trustworthy.

EXAMPLE 13.8 Blood types

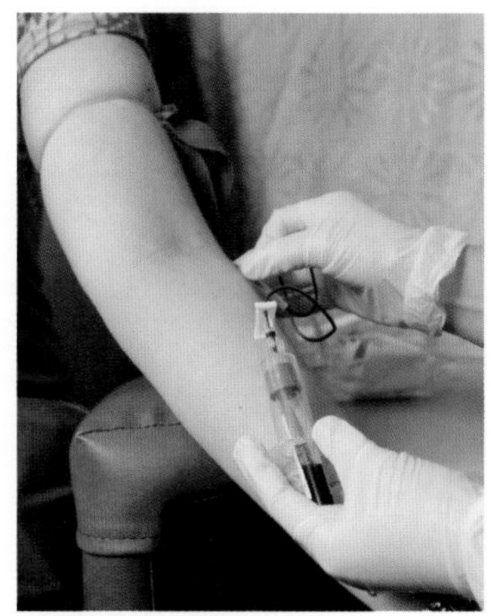
CDC/Amanda Mills

The American Red Cross reports that 45% of Americans have blood type O. We plan to take a random sample of 1000 medical records. What is the probability that no more than 42% of the individuals in our sample would have blood type O?

If the sample size is $n = 1000$ and the population proportion is $p = 0.45$, the sample proportion $\hat{p}$ has mean 0.45 and standard deviation

$$\sqrt{\frac{p(1-p)}{n}} = \sqrt{\frac{(0.45)(0.55)}{1000}}$$
$$= 0.01573$$

We want the probability that $\hat{p}$ is 0.42 or less.

Using the software R, the function pnorm(0.42, mean=0.45, sd=0.01573) gives us a probability for $P(\hat{p} \leq 0.42)$ of 0.0282. Alternatively, we can use Table B at the back of the book. First, standardize $\hat{p}$ by subtracting the mean 0.45 and dividing by the standard deviation 0.01573. This produces a new statistic that has approximately the standard Normal distribution. As usual, we call this statistic z:

$$z = \frac{\hat{p} - 0.45}{0.01573}$$

Figure 13.6 shows the probability we want as an area under the standard Normal curve.

$$P(\hat{p} \leq 0.42) = P\left(\frac{\hat{p} - 0.45}{0.01573} \leq \frac{0.42 - 0.45}{0.01573}\right)$$
$$= P(Z \leq -1.91)$$
$$= 0.0281$$

If we took many, many random samples of 1000 medical records, only 2.8% of all these samples would contain fewer than 42% of records from individuals with blood type O. ■

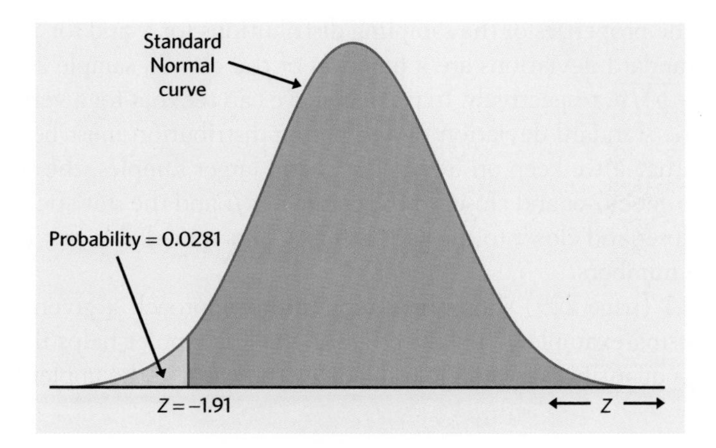

FIGURE 13.6 Probability in Example 13.8 as area under the standard Normal curve.

The Normal approximation for the sampling distribution of $\hat{p}$ is least accurate when p is close to 0 or 1. You can see that if $p = 0$, any sample must contain only failures. That is, $\hat{p} = 0$ every time and there is no Normal distribution in sight. In just the same way, the approximation works poorly when p is close to 1. In practice, this means that we need larger n for values of p near 0 or 1.

APPLY YOUR KNOWLEDGE

13.11 Aging American population. The 2010 U.S. census shows that **6%** of the American population are at least 75 years of age. To test a random digit dialing device, you use the device to call randomly chosen residential telephones in your county. Of the 150 members of the households contacted, **4%** are 75 years or older.

(a) Is each of the boldface numbers a parameter or a statistic?

(b) Assume that your county's population is similar to the American population in age distribution. What are the mean and standard deviation of the proportion $\hat{p}$ who are at least 75 years of age in samples of 150 respondents?

13.12 Baby boys. We saw in Example 9.3 (page 209) that, historically, about 51% of the babies born in the United States are boys. That is, we know that $p = 0.51$. A survey plans to study 650 randomly selected newborns.

(a) What are the mean and standard deviation of the statistic $\hat{p}$, the sample proportion who are boys?

(b) Use the Normal approximation to find the probability that $\hat{p}$ would be less than 0.50.

13.13 Aging American population, continued. Based on the information presented in Exercise 13.11, what is the probability that a random sample of 150 respondents would contain less than 4% of individuals at least 75 years of age?

13.14 Baby boys, continued. Exercise 13.12 asks you to find the probability that the sample proportion $\hat{p}$ is less than 0.5 in a sample of 650 newborns. Find this probability for SRSs of sizes 1000, 4000, and 16,000. What general fact do your results illustrate?

The law of large numbers*

Looking at the properties of the sampling distributions for $\overline{x}$ and for $\hat{p}$, we notice that their standard deviations are a function of the chosen sample size n: $\sigma/\sqrt{n}$ and $\sqrt{p(1-p)/n}$, respectively. In particular, we can see that for a very large sample size n, the standard deviation of a sampling distribution must be very small. This means that if we keep on taking larger and larger samples, the statistic $\overline{x}$ is *guaranteed* to get closer and closer to the parameter μ and the statistic $\hat{p}$ is *guaranteed* to get closer and closer to the parameter p. This remarkable fact is called the **law of large numbers.**

law of large numbers

Figure 9.1 (page 209) shows how proportions approach a given probability in a coin-tossing example. The *Law of Large Numbers* applet helps illustrate this concept for a quantitative variable using dice. You can use the applet to watch $\overline{x}$

*This section is optional.

change as you average more observations until it eventually settles down at the mean μ.

The law of large numbers is usually impractical for direct applications in the life sciences. Indeed, it is only *in the very long run* that the mean outcome is predictable, and studies in the life sciences rarely deal with such large numbers. However, the law of large numbers is the foundation of such businesses as gambling casinos and insurance companies. The winnings (or losses) of a gambler on a few plays are uncertain—that's why, unfortunately, gambling can be addictive. The house plays tens of thousands of times. So the house, unlike individual gamblers, can count on the long-run regularity described by the law of large numbers. The average winnings of the house on tens of thousands of plays will be very close to the mean of the distribution of winnings, guaranteeing the house a profit. For other scenarios, however, the law of large numbers is impractical.

The probability of dying We can't predict whether a specific person will die next year. But if we observe millions of people, deaths are random. Life insurance is based on this fact. Each year, the proportion of men aged 25 to 34 who die is about 0.0021. We can use this as the probability that a given young man will die next year. For women in that age group, the probability of death is about 0.0007. An insurance company that sells many policies to people aged 25 to 34 will have to pay off on about 0.21% of the policies sold to men and about 0.07% of the policies sold to women.

APPLY YOUR KNOWLEDGE

13.15 More on insurance. The idea of insurance is that we all face risks that are unlikely but carry high cost. Workers' compensation insurance provides for the cost of medical care, rehabilitation, and other related expenses for injured workers. In a given year, most workers will not have any work injury, some will have minor injuries, and a few will die or become permanently disabled. Insurance spreads the risk: We all pay a small amount, and the insurance policy pays a large amount to these few extreme claims. The National Academy of Social Insurance reports that the mean workers' compensation claim cost in a year is $\mu = \$439$ per covered worker. An insurance company plans to sell workers' compensation insurance for \$439 plus enough to cover its administrative costs and profit. Explain clearly why it would be unwise to sell only 12 policies. Then explain why selling thousands of such policies is a safe business.

13.16 Means in action. Use the *Law of Large Numbers* applet. The applet simulates successive rolls of a die and recomputes the average number of spots on the up-face after each roll. Over the long run, the mean number of spots is 3.5.

(a) Type 20 in the "Rolls" window. Then click on the "Roll dice" button. What do you observe? Reset the applet and repeat this simulation three more times. How are these four simulations different? What do they have in common? You can expect a lot of variability for such small sample sizes.

(b) Now type 100 in the "Rolls" window and click on the "Roll dice" button three times so that you have a total of 300 rolls. What do you observe? Reset the applet and repeat this simulation one more time. How are these two simulations different? What do they have in common? Three hundred is not a huge number, but you can get an idea of how the law of large numbers works.

CHAPTER 13 SUMMARY

- A **parameter** is a number that describes a population, whereas a **statistic** is a number that describes a sample.
- When we want information about the **population mean** μ for some quantitative variable, we often take an SRS and use the **sample mean** $\bar{x}$ to estimate the unknown parameter μ.

- The **sampling distribution** of $\overline{x}$ describes how the statistic $\overline{x}$ varies in all possible SRSs of the same size from the same population.
- The **mean** of the sampling distribution of $\overline{x}$ is the same as the population mean μ, and therefore $\overline{x}$ is an **unbiased estimator** of μ.
- The **standard deviation** of the sampling distribution of $\overline{x}$ is $\sigma/\sqrt{n}$ for an SRS of size n if the population has standard deviation σ. That is, averages are less variable than individual observations.
- If the population has a Normal distribution, so does the sampling distribution of $\overline{x}$.
- The **central limit theorem** states that for large n the sampling distribution of $\overline{x}$ is approximately Normal for any population with finite standard deviation σ. That is, averages are more Normal than individual observations. We can use the $N(\mu, \sigma/\sqrt{n})$ distribution to calculate approximate probabilities for events involving $\overline{x}$.
- When we want information about the **population proportion** p for some categorical variable, we often take an SRS and use the **sample proportion** $\hat{p}$ to estimate the unknown parameter p.
- The **sampling distribution of the proportion** $\hat{p}$ **of successes** in an SRS of size n from a large population containing proportion p of successes has mean p and standard deviation $\sqrt{p(1-p)/n}$.
- The **mean** of the sampling distribution of $\hat{p}$ is p, and therefore $\hat{p}$ is an **unbiased estimator** of p.
- When the sample size n is large, when both np and $n(1-p)$ are large enough, and when the population is at least 20 times larger than the sample, the sampling distribution of $\hat{p}$ has approximately the $N(p, \sqrt{np(1-p)/n})$ distribution. We can use this distribution to calculate approximate probabilities for events involving $\hat{p}$.
- The **law of large numbers** states that the observed mean outcome $\overline{x}$ must approach the mean μ of the population as the number of observations increases. Likewise, the observed proportion $\hat{p}$ of a categorical outcome must approach the proportion of this outcome in the population as the number of observations increases.

THIS CHAPTER IN CONTEXT

In this chapter we discussed the concept of sampling distributions and how they relate to random samples and populations. It is essential to understand the distinction between these three concepts. We first examine the case of a quantitative variable, x.

Population. A population contains all the individuals making up a group of interest, along with their associated numerical value for the variable x. We don't usually have information about each individual in the population (too many to examine). Instead, we use a density curve to model the probability distribution of the variable x studied in the population, as we have seen in Chapters 9 and 11.

Normal curves, described in Chapter 11, are one type of density curve that we can sometimes use to model a population distribution. The population mean is labeled μ, and the population standard deviation is labeled σ. (Sometimes we will be given the values of μ and σ, but in most cases we will not know them.)

Random sample. A random sample is a subset of n individuals from a population of interest, along with their associated numerical value x. The sample can be obtained by randomly selecting the individuals from the population or by randomly assigning individuals to experimental conditions, as seen in Chapters 7 and 8. We do have data for these n individuals, and we can plot the data in a histogram or a dotplot (Chapter 1). In Chapter 2, we saw how we can compute the mean $\bar{x}$ and the standard deviation s of sample data.

Sampling distribution. A sampling distribution contains all the means $\bar{x}$ for all possible random samples of size n from a population of interest. We do not usually have information about each sample mean (too many to examine). Instead, we use a density curve to model the probability distribution of the variable $\bar{x}$ for samples of size n from this population. The mean of the sampling distribution is equal to the population mean μ, and the standard deviation of the sampling distribution is equal to $\sigma/\sqrt{n}$ (the population standard deviation σ divided by the square root of the sample size n). We can use a Normal curve to model the sampling distribution of $\bar{x}$ when the population is Normally distributed. Because of the central limit theorem, we can also use a Normal curve to model the sampling distribution of $\bar{x}$ when the sample size n is large enough. The central limit theorem does not apply to one random sample—it applies only to the shape of a sampling distribution. In the next two chapters, we will see that the sampling distribution of $\bar{x}$ is interesting because it allows us to perform statistical inference about the population parameter μ.

When dealing with a categorical variable, we can define one particular outcome of interest as a "success" and any other outcome as a "failure." This approach should be familiar if you have read optional Chapter 12. The population contains all the individuals making up a group of interest, along with their associated outcome (success or failure). The population parameter is the proportion of successes p in the population. (The value of this parameter is sometimes given, but most often it is unknown.) For a random sample, we can compute the sample proportion $\hat{p}$ of successes. The sampling distribution of $\hat{p}$ has a mean equal to p and a standard deviation equal to $\sqrt{np(1-p)/n}$. We can use a Normal distribution to model the sampling distribution of $\hat{p}$ when the sample size n is large enough. We will see in Chapter 19 that the sampling distribution of $\hat{p}$ is interesting because it allows us to perform statistical inference about the population parameter p.

CHECK YOUR SKILLS

13.17 The National Center for Health Statistics interviewed all adult household members of a nationally representative sample of the civilian noninstitutionalized household population; **62%** of the interviewees had used some form of complementary or alternative medicine, including prayer, during the past 12 months. The boldface number is a

(a) sampling distribution.

(b) parameter.

(c) statistic.

13.18 A newborn baby has extremely low birth weight (ELBW) if it weighs less than 1000 grams. A study of the health of such children in later years examined a random sample of 219 children. Their mean weight at birth was $\bar{x} =$ 810 grams. This sample mean is an *unbiased estimator* of the mean weight μ in the population of all ELBW babies. This means that

(a) in many samples from this population, the mean of the many values of $\bar{x}$ will be equal to μ.
(b) as we take larger and larger samples from this population, $\bar{x}$ will get closer and closer to μ.
(c) in many samples from this population, the many values of $\bar{x}$ will have a distribution that is close to Normal.

Cholesterol levels among fourteen-year-old boys are roughly Normal with mean 170 and standard deviation 30 milligrams per deciliter (mg/dl). Use this information for Exercises 13.19 to 13.21.

13.19 You choose an SRS of 4 fourteen-year-old boys and average their cholesterol levels. If you do this many times, the mean of the average cholesterol levels you get will be close to
(a) 170. (b) $170/4 = 42.5$. (c) $170/\sqrt{4} = 85$.

13.20 You choose an SRS of 4 fourteen-year-old boys and average their cholesterol levels. If you do this many times, the standard deviation of the average cholesterol levels you get will be close to
(a) 30. (b) $4/\sqrt{30} = 0.73$. (c) $30/\sqrt{4} = 15$.

13.21 In an SRS of 4 fourteen-year-old boys, the probability that the average cholesterol level is 200 mg/dl or more is close to
(a) 0.023. (b) 0.159. (c) 0.977.

13.22 The survival times of guinea pigs inoculated with an infectious viral strain vary from animal to animal. The distribution of survival times is strongly skewed to the right. The central limit theorem says that

(a) as we study more and more infected guinea pigs, their average survival time gets ever closer to the mean μ for all infected guinea pigs.
(b) the average survival time of a large number of infected guinea pigs has a distribution of the same shape (strongly skewed) as the distribution for individual infected guinea pigs.
(c) the average survival time of a large number of infected guinea pigs has a distribution that is close to Normal.

According to the CDC, the proportion of all gonorrhea cases tested in 2010 that were resistant to a major antibiotic was 0.27.[3] *A researcher takes an SRS of 50 medical records of patients diagnosed with gonorrhea in 2010 and records the proportion that had shown resistance to a major antibiotic. Use this information for Exercises 13.23 to 13.25.*

13.23 The number 0.27 is a
(a) sampling distribution.
(b) parameter.
(c) statistic.

13.24 If you take samples of 50 medical records many times, the standard deviation of the proportion of gonorrhea cases that were resistant to a major antibiotic will be close to
(a) 0.27.
(b) $0.27 \times 0.73/\sqrt{50} = 0.028$.
(c) $\sqrt{0.27 \times 0.73/50} = 0.063$.

13.25 In an SRS of 50 medical cases, the probability of getting 25% or more cases that are resistant to a major antibiotic is close to
(a) 0.530. (b) 0.625. (c) 0.762

13.26 (Optional) Medical care and compensation costs for workers injured on the job vary a lot and are strongly skewed as a whole. The National Academy of Social Insurance reports that the mean workers' compensation claim cost in a year is $\mu = \$439$ per insured worker, with standard deviation $\sigma = \$20,000$. The law of large numbers says that

(a) an insurance company can get an average workers' compensation claim cost lower than the mean $439 by insuring a large number of workers.
(b) as a company insures more and more workers chosen at random, the average claim cost gets ever closer to $439.
(c) if a company insures a large number of workers chosen at random, that company's average claim cost will have approximately a Normal distribution.

CHAPTER 13 EXERCISES

13.27 Women's heights. A random sample of female college students has a mean height of **65** inches, which is greater than the **64**-inch mean height of all young women. Is each of the bold numbers a parameter or a statistic? Explain your answer.

13.28 Lightning strikes. The number of lightning strikes on a square kilometer of open ground in a year has mean 6 and standard deviation 2.4. (These values are typical of much of the United States.) The National Lightning Detection

Network uses automatic sensors to watch for lightning in a sample of 25 randomly chosen square kilometers. What are the mean and standard deviation of $\bar{x}$, the mean number of strikes per square kilometer?

13.29 Heights of male students. Suppose that the distribution of heights of all male students on your campus is Normal with mean 70 inches and standard deviation 2.8 inches.

(a) If you choose one student at random, what is the probability that he is between 69 and 71 inches tall?

(b) What is the standard deviation of $\bar{x}$, the mean height for a sample of n male students? How large an SRS must you take to reduce the standard deviation of the sample mean to 0.5 inch?

(c) What is the probability that the mean height of the sample in (b) is between 69 and 71 inches?

13.30 Lightning strikes, continued. The number of lightning strikes on a square kilometer of open ground in a year has mean 6 and standard deviation 2.4. What is the probability that the average number of lightning strikes per square kilometer per year for 25 randomly chosen square kilometers is 4.8 or less?

13.31 Heights of male students, continued. Suppose that the distribution of heights of all male students on your campus is Normal with mean 70 inches and standard deviation 2.8 inches.

(a) What standard deviation must $\bar{x}$ have so that 99.7% of all samples give an $\bar{x}$ within 0.5 inch of μ? (Use the 68–95–99.7 rule.)

(b) How large an SRS do you need to reduce the standard deviation of $\bar{x}$ to the value you found in (a)?

13.32 Glucose testing. Shelia's doctor is concerned that she may suffer from gestational diabetes (high blood glucose levels during pregnancy). There is variation both in the actual glucose level and in the blood test that measures the level. A patient is classified as having gestational diabetes if her glucose level is above 140 milligrams per deciliter (mg/dl) one hour after ingesting a sugary drink. Shelia's measured glucose level one hour after the sugary drink varies according to the Normal distribution with $\mu = 125$ mg/dl and $\sigma = 10$ mg/dl.

(a) If a single glucose measurement is made, what is the probability that Shelia is diagnosed as having gestational diabetes?

(b) If measurements are made on 4 separate days and the mean result is compared with the criterion 140 mg/dl, what is the probability that Shelia is diagnosed as having gestational diabetes?

13.33 Blood potassium level. Judy's doctor is concerned that she may suffer from hypokalemia (low potassium in the blood, measured in millimoles per liter, mmol/l). There is variation both in the actual potassium level and in the blood test that measures the level. Judy's measured potassium level varies according to the Normal distribution with $\mu = 3.8$ and $\sigma = 0.2$ mmol/l. A patient is classified as hypokalemic if the potassium level is below 3.5 mmol/l.

(a) If a single potassium measurement is made, what is the probability that Judy is diagnosed as hypokalemic?

(b) If measurements are made instead on 4 separate days and the mean result is compared with the criterion 3.5, what is the probability that Judy is diagnosed as hypokalemic?

13.34 Glucose testing, continued. Shelia's measured glucose level one hour after ingesting a sugary drink varies according to the Normal distribution with $\mu = 125$ mg/dl and $\sigma = 10$ mg/dl. Let's consider what could happen if we took 4 separate measurements from Shelia. What is the blood glucose level L such that the probability is only 0.05 that the average of 4 measurements is larger than L? (*Hint:* This requires a backward Normal calculation.)

13.35 Blood potassium level, continued. Judy's measured potassium level varies according to the Normal distribution with $\mu = 3.8$ and $\sigma = 0.2$ mmol/l. Let's consider what could happen if we took 4 separate measurements from Judy. What is the blood potassium level L such that the probability is only 0.05 that the average of 4 measurements is less than L? (*Hint:* This requires a backward Normal calculation.)

13.36 Pregnancy lengths. The length of human pregnancies from conception to birth varies according to a distribution that is approximately Normal with mean 266 days and standard deviation 16 days. A study enrolls a random sample of 16 pregnant women.

(a) What are the mean and standard deviation of the sampling distribution of $\bar{x}$?

(b) What is the probability that the average pregnancy length exceeds 270 days?

13.37 Deer mice. Deer mice (*Peromyscus maniculatus*) are small rodents native of North America. Their adult body lengths (excluding tail) are known to vary approximately Normally, with mean $\mu = 86$ millimeters (mm) and standard deviation $\sigma = 8$ mm.[4] A researcher uses a random sample of 14 deer mice.

(a) What are the mean and standard deviation of the sampling distribution of $\bar{x}$?

(b) What is the probability that the sample mean body length would be less than 80 mm?

13.38 Blood alcohol content. The distribution of blood alcohol content in evening drivers is very skewed: Most drivers don't drink and drive, and some limit their drinking, while a few are intoxicated. If you wanted to make Normal calculations using the sampling distribution of the average blood alcohol content of evening drivers, would a random sample of 10 drivers be sufficient? 50 drivers? 200 drivers? Explain your answers.

13.39 Smoking and pregnancy. Cotinine, the metabolized form of nicotine, crosses the protective placental barrier and actually reaches the fetus. Traces can be found in a newborn's first stool at birth, also called meconium. You want to study meconium cotinine levels, but you know from a review of the scientific literature that they are typically very strongly right-skewed. Your statistical analysis would benefit from an approximately Normal sampling distribution. What minimum sample size would be advisable for your study? Explain your reasoning.

13.40 Extrasensory perception. In tests for extrasensory perception (ESP), an experimenter looks at cards that are hidden from the subject. Each card contains 1 of 5 symbols, and the subject must name the symbol on each card presented. A subject with no suspected psychic ability has, by chance alone, a 20% chance of correctly guessing the symbol on each card presented. Because the cards are independent, the overall proportion of success can thus be expected to be **20%**. You have no psychic ability and score **28%** correct guesses out of a 25-card deck. Is each of the bold values a parameter or a statistic? Explain your answer.

13.41 Sampling bias. One way of checking the effect of undercoverage, nonresponse, and other sources of error in a sample survey is to compare the sample with known demographic facts about the population. The 2010 census found that 13.9%, or 32,576,000, of the 235,016,000 adults (aged 18 and over) in the United States identified themselves as of Hispanic origin. Is the value 13.9% a parameter or a statistic? Explain your answer.

13.42 Students on diets. A sample survey interviews an SRS of 267 college women. Suppose (as is roughly true) that 70% of all college women have been on a diet within the past 12 months. What are the mean and standard deviation of the sampling distribution for samples of 267 women?

13.43 Sampling bias, continued. The 2010 census found that 13.9% of adults in the United States identified themselves as of Hispanic origin.

(a) An opinion poll plans to interview 1500 adults at random. What are the mean and standard deviation of the sampling distribution of the proportion of individuals of Hispanic origin for samples of this size?

(b) Use a Normal approximation to find the probability that such a sample will contain 12% or fewer individuals of Hispanic origin.

13.44 Students on diets, continued. A sample survey interviews an SRS of 267 college women. Suppose (as is roughly true) that 70% of all college women have been on a diet within the past 12 months. Use a Normal approximation to find the probability that 75% or more of the women in the sample have been on a diet.

13.45 Contraception and unintended pregnancies. Pharmaceutical companies advertise that the birth control pill has an efficacy of 99.5% in preventing pregnancy. However, under typical use the real efficacy is only about 95%. That is, 5% of women taking the pill for a year will experience an unplanned pregnancy that year.[5] A gynecologist looks back at a random sample of medical records of patients who were prescribed the pill one year ago.

(a) What are the mean and standard deviation of the distribution of sample proportions of women experiencing unplanned pregnancies?

(b) The gynecologist takes an SRS of 200 records and finds that 14 women had become pregnant within one year while taking the pill. How surprising is this finding? Give the probability of finding 7% or more pregnant women in the sample.

(c) How surprising would it be if the gynecologist had found 16 pregnant women (8%)? 20 pregnant women (10%)?

CDC/Amanda Mills

CHAPTER 14 Introduction to Inference

After we have selected a sample, we know the responses of the individuals in the sample. The usual reason for taking a sample is not to learn about the individuals in the sample but to *infer* from the sample data some conclusion about the wider population that the sample represents.

IN THIS CHAPTER WE COVER...

- The reasoning of statistical estimation
- Margin of error and confidence level
- Confidence intervals for the mean μ
- The reasoning of tests of significance
- Stating hypotheses
- P-value and statistical significance
- Tests for a population mean
- Tests from confidence intervals

> **STATISTICAL INFERENCE**
>
> **Statistical inference** provides methods for drawing conclusions about a population from sample data.

Because a different sample might lead to different conclusions, we can't be certain that our conclusions are correct. Statistical inference uses the language of probability to say how trustworthy our conclusions are. This chapter introduces the two most common types of inference, *confidence intervals* for estimating the value of a population parameter and *tests of significance* to assess the evidence for or against a claim about a population. Both types of inference are based on the sampling distributions of statistics. That is, both report probabilities that state what would happen if we used the inference method many times.

This chapter presents the basic reasoning of statistical inference. To make the reasoning as clear as possible, we start with a setting that is too simple to be realistic, especially in the life sciences. Here is the setting for our work in this chapter.

SIMPLE CONDITIONS FOR INFERENCE ABOUT A MEAN

1. We have an SRS from the population of interest. There is no nonresponse or other practical difficulty.
2. The variable we measure has a perfectly Normal distribution $N(\mu, \sigma)$ in the population.
3. We don't know the population mean μ. But we do know the population standard deviation σ.

The conditions that we have a perfect SRS, that the population is perfectly Normal, and that we know the population σ are all unrealistic. We will discuss some important practical issues when using inference in Chapter 15, and later chapters will show how inference deals with more realistic settings.

The reasoning of statistical estimation

The National Health and Nutrition Examination Survey (NHANES) reports the weights and heights of Americans every few years. Each survey is based on a nationwide probability sample of the U.S. civilian noninstitutionalized population. What can we learn from the survey about the height of eight-year-old boys in America?

Nick White/Photodisc/Getty Images

EXAMPLE 14.1 How tall are children these days?

The most recent NHANES reports that the mean height of a sample of 217 eight-year-old boys was $\bar{x} = 132.5$ centimeters (that's 52.2 inches). On the basis of this sample, we want to estimate the mean height μ in the population of over a million American eight-year-old boys.

To match the "simple conditions," we will treat the NHANES sample as a perfect SRS of all American eight-year-old boys and the height in this population as having an exactly Normal distribution with standard deviation $\sigma = 10$ cm.[1] ■

Here is the reasoning of statistical estimation in a nutshell.

1. To estimate the unknown population mean μ, use the mean $\bar{x} = 132.5$ of the random sample. We don't expect $\bar{x}$ to be exactly equal to μ, so we want to say how accurate this estimate is.
2. We know the sampling distribution of $\bar{x}$. In repeated samples, $\bar{x}$ has the Normal distribution with mean μ and standard deviation $\sigma/\sqrt{n}$. So the sampling distribution of the average height $\bar{x}$ of 217 eight-year-old American boys has standard deviation

$$\frac{\sigma}{\sqrt{n}} = \frac{10}{\sqrt{217}} = 0.7 \text{ cm (rounded off)}$$

3. The 95 part of the 68–95–99.7 rule for Normal distributions says that $\bar{x}$ is within 1.4 cm (that's two standard deviations) of the mean μ in 95% of all

samples. That is, for 95% of all samples, 1.4 cm is the maximum distance separating $\overline{x}$ and μ. So if we estimate that μ lies somewhere in the interval from $\overline{x} - 1.4$ to $\overline{x} + 1.4$, we'll be right 95% of the times we take a sample. For this particular NHANES sample, this interval is

$$\overline{x} - 1.4 = 132.5 - 1.4 = 131.1$$

to

$$\overline{x} + 1.4 = 132.5 + 1.4 = 133.9$$

Because we got the interval 131.1 to 133.9 cm from a method that captures the population mean μ 95% of the time, we say that we are 95% *confident* that the mean height of all eight-year-old boys in the United States is some value in that interval, maybe as low as 131.1 cm or as high as 133.9 cm.

Ranges are for statistics? Many people like to think that statistical estimates are exact. The Nobel prize-winning economist Daniel McFadden tells a story of his time on the Council of Economic Advisers. Presented with a range of forecasts for economic growth, President Lyndon Johnson replied: "Ranges are for cattle; give me one number."

The big idea is that the sampling distribution of $\overline{x}$ tells us how close to μ the sample mean $\overline{x}$ is likely to be. A confidence interval just turns that information around to say how close to $\overline{x}$ the unknown population mean μ is likely to be.

EXAMPLE 14.2 Statistical estimation in pictures

Figure 14.1 summarizes the idea of the sampling distribution using what we learned in Chapter 13. Starting with the population, imagine taking many SRSs of 217 eight-year-old American boys. The first sample has mean height $\overline{x} = 132.5$ cm, the second has mean $\overline{x} = 134.2$ cm, the third has mean $\overline{x} = 131.8$ cm, and so on. If we collect all these sample means and display their distribution, we get the Normal distribution with mean equal to the unknown value of the parameter μ and standard deviation 0.7 cm.

The 68–95–99.7 rule for Normal calculations says that in 95% of all samples, the sample mean $\overline{x}$ lies within 1.4 cm (two standard deviations) of the population mean μ. Whenever this happens, the interval $\overline{x} \pm 1.4$ captures μ.

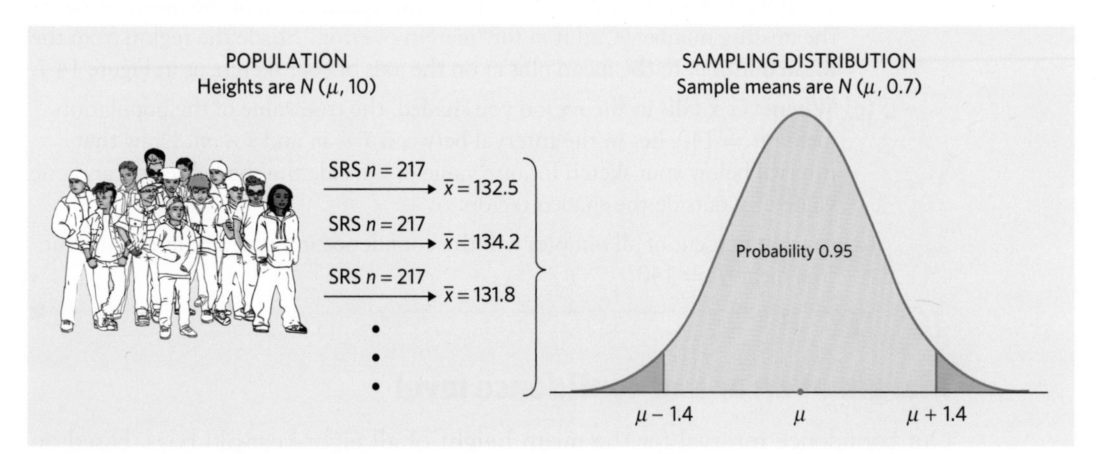

FIGURE 14.1 The sampling distribution of the mean height $\overline{x}$ of an SRS of 217 eight-year-old boys, for Example 14.2. In 95% of all samples, $\overline{x}$ lies within ± 1.4 of the unknown population mean μ.

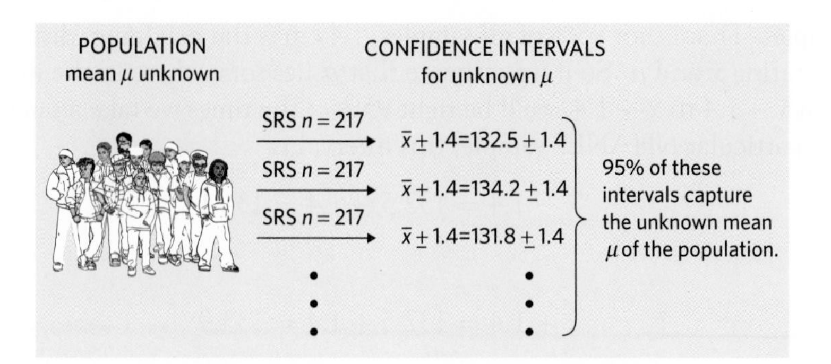

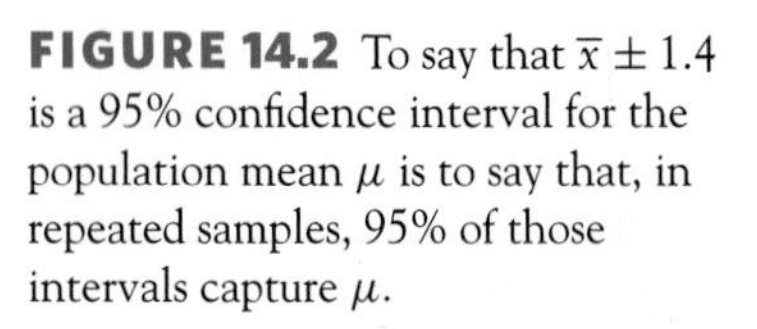

FIGURE 14.2 To say that $\bar{x} \pm 1.4$ is a 95% confidence interval for the population mean μ is to say that, in repeated samples, 95% of those intervals capture μ.

Figure 14.2 illustrates this idea graphically. Starting with the population, imagine taking many SRSs of 217 eight-year-old American boys. The formula $\bar{x} \pm 1.4$ gives an interval based on each sample; 95% *of these intervals capture the unknown population mean* μ. That is, a statistical method using the interval $\bar{x} \pm 1.4$ has a 95% *success rate* in capturing within that interval the mean height μ of all eight-year-old American boys. ■

In practice, we don't take many SRSs to estimate μ. Because the method has a 95% success rate, all we need is one SRS of 217 eight-year-old American boys to compute just one interval $\bar{x} \pm 1.4$. This interval of numbers between the values $\bar{x} - 1.4$ and $\bar{x} + 1.4$ is called a 95% *confidence interval* for μ.

APPLY YOUR KNOWLEDGE

14.1 Girls' heights. Suppose that you measure the heights of an SRS of 400 eight-year-old American girls, a population with mean $\mu = 140$ cm and standard deviation $\sigma = 8$ cm. The mean $\bar{x}$ of the 400 heights will vary if you take repeated samples.

(a) The sampling distribution of $\bar{x}$ is approximately Normal. It has mean $\mu = 140$ cm. What is its standard deviation?

(b) Sketch the Normal curve that describes how $\bar{x}$ varies in many samples from this population. Mark the mean $\mu = 140$. According to the 68–95–99.7 rule, about 95% of all the values of $\bar{x}$ fall within ______ cm of the mean. What is the missing number? Call it m for "margin of error." Shade the region from the mean minus m to the mean plus m on the axis of your sketch, as in Figure 14.1.

(c) Whenever $\bar{x}$ falls in the region you shaded, the true value of the population mean, $\mu = 140$, lies in the interval between $\bar{x} - m$ and $\bar{x} + m$. Draw that interval below your sketch for one value of $\bar{x}$ inside the shaded region and one value of $\bar{x}$ outside the shaded region.

(d) In what percent of all samples will the confidence interval $\bar{x} \pm m$ capture the true mean $\mu = 140$?

Margin of error and confidence level

Our confidence interval for the mean height of all eight-year-old boys, based on the NHANES sample, is 132.5 ± 1.4. Most confidence intervals take the form

$$\text{estimate} \pm \text{margin of error}$$

The estimate ($\overline{x} = 132.5$ in our example) is the center of the interval. It is our unbiased guess for the value of the unknown parameter, based on the sample data. The *margin of error* (in our example, ± 1.4) shows how accurate we believe our guess is, based on the variability of the estimate. We have a 95% confidence interval because the interval catches the unknown value of the parameter (μ in this case) with 95% probability based on the overall success rate of the method. That is, we are 95% confident that the unknown parameter is one value within the confidence interval.

CONFIDENCE INTERVAL

A **level C confidence interval** for a parameter has two parts:

- An interval calculated from the data, usually of the form

$$\text{estimate} \pm \text{margin of error}$$

  where the estimate is a sample statistic and the **margin of error** represents the accuracy of our guess for the parameter.
- A **confidence level** C, which gives the probability that the interval will capture the true parameter value in repeated samples. That is, the confidence level is the success rate for the method.

Users can choose the confidence level, usually 90% or higher because we want to be quite sure of our conclusions. It may be disconcerting at first to rely on a method that is not 100% reliable. However, we must keep in mind that the only way to be 100% confident about the unknown value of a parameter would be to record the entire population; this is rarely an option, especially in the life sciences.

INTERPRETING A CONFIDENCE INTERVAL

The confidence level is the success rate of the method that produces the interval. We don't know whether the 95% confidence interval from a particular sample is one of the 95% that capture μ or one of the unlucky 5% that miss.

To say that we are **95% confident** that the unknown value of μ lies between 131.1 and 133.9 cm is shorthand for **"We got these numbers using a method that gives correct results 95% of the time."**

Figure 14.2 is one way to picture the idea of a 95% confidence interval. Figure 14.3 illustrates the idea in a different form. Study these figures carefully. If you understand what they say, you have mastered one of the big ideas of statistics.

Figure 14.3 shows the result of simulating a number of SRSs from the same population and calculating a 95% confidence interval from each sample. Each SRS in this simulation consists of $n = 15$ observations randomly selected from the population. Both the population distribution and the sampling distribution of $\overline{x}$ appear

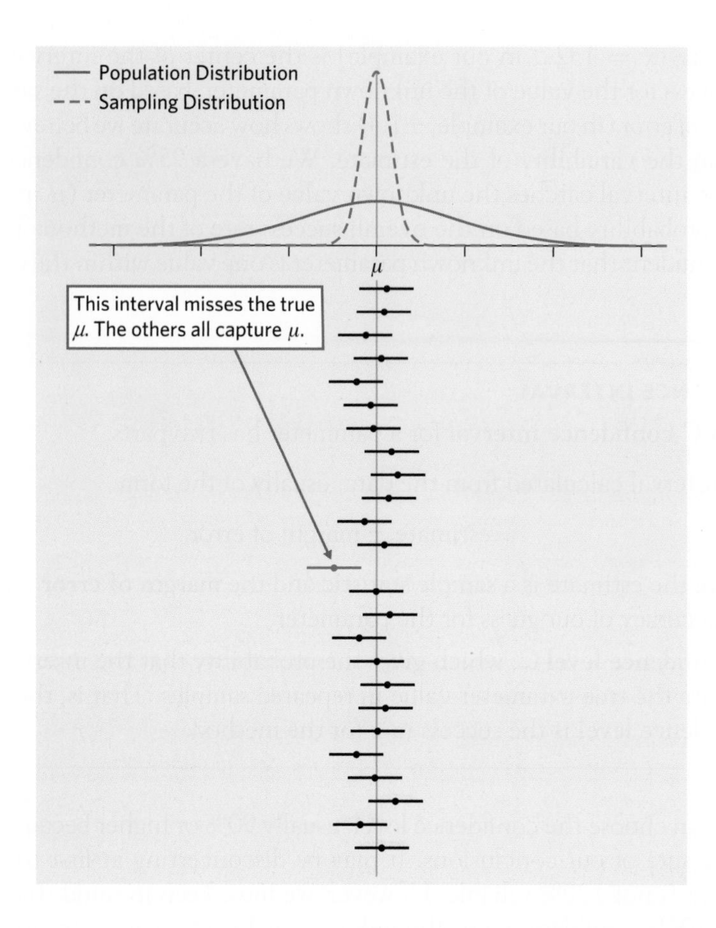

FIGURE 14.3 Twenty-five samples from the same population gave these 95% confidence intervals. In the long run, 95% of all samples give an interval that contains the population mean μ.

at the top of the figure (in solid and dotted lines, respectively). The population mean μ is at the center of both distributions. The 95% confidence intervals from 25 different SRSs appear underneath. The center of each interval is at $\bar{x}$ (marked by a dot) and therefore varies from sample to sample. The arrows on either side of the dot span the confidence interval. All except 1 of these 25 intervals contain the true value of μ. In a very large number of samples, 95% of the confidence intervals would contain μ.

The *Confidence Intervals* applet lets you perform such simulations and watch confidence intervals from one sample after another capture or fail to capture the true parameter. Notice that two things influence the size of the simulated confidence intervals: the confidence level C and the sample size n. The applet also allows you to display the n individual observations making up an SRS by clicking on the interval.

APPLY YOUR KNOWLEDGE

14.2 Confidence intervals in action. The idea of an 80% confidence interval is that the interval captures the true parameter value in 80% of all samples. That's not high enough confidence for practical use, but 80% hits and 20% misses make it easy to see how a confidence interval behaves in repeated samples from the same population. Go to the *Confidence Intervals* applet.

(a) Pick a sample size, say, $n = 15$. Set the confidence level to 80%. Click "Sample" to choose an SRS and calculate the confidence interval. Do this 10 times to simulate 10 SRSs with their 10 confidence intervals. How many of the 10 intervals captured the true mean μ? How many missed?

(b) You see that we can't predict whether the next sample will hit or miss. The confidence level, however, tells us what percent will hit in the long run. Reset the applet and click "Sample 25" to get the confidence intervals from 25 SRSs. How many hit? Keep clicking "Sample 25" and record the percent of hits among 100, 500, and 1000 SRSs. Even 1000 samples is not truly "the long run," but we expect the percent of hits in 1000 samples to be fairly close to the confidence level, 80%.

14.3 Spending on food. A 2012 Gallup survey of a random sample of 1014 American adults indicates that American families spend on average \$151 per week on food. The report further states that, with 95% confidence, this estimate has a margin of error of ±\$7.

(a) This confidence interval is expressed in the following form: "estimate ± margin of error." What is the range of values (lower bound, upper bound) that corresponds to this confidence interval?

(b) What is the parameter captured by this confidence interval? What does it mean to say that we have "95% confidence" in this interval?

Confidence intervals for the mean μ

To find a 95% confidence interval for the mean height of eight-year-old American boys, we first caught the central 95% of the Normal sampling distribution by going out two standard deviations in both directions from the mean. To find a level C confidence interval, we first catch the central area C under the Normal sampling distribution. Because all Normal distributions are the same in the standard scale, we can obtain everything we need from the standard Normal curve.

Figure 14.4 shows how the central area C under a standard Normal curve is marked off by two points z^* and $-z^*$. Numbers like z^* that mark off specified areas are called **critical values** of the standard Normal distribution. Values of z^* for many choices of C appear in the bottom row of Table C in the back of the book. This

critical value

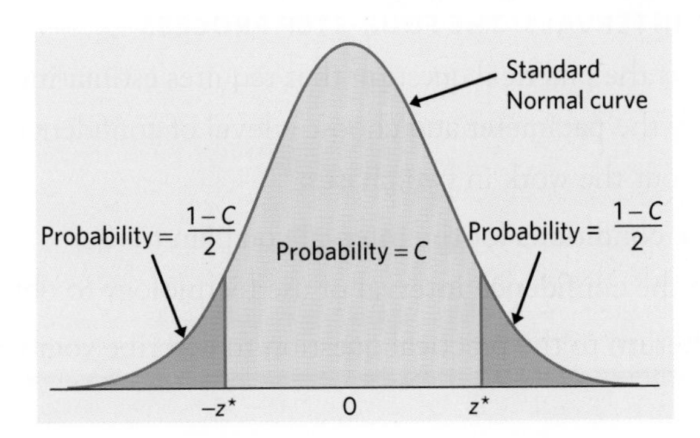

FIGURE 14.4 The critical value z^* is the number that catches central probability C under a standard Normal curve between $-z^*$ and z^*.

row is labeled z^*. Here are the entries for the most common confidence levels:

Confidence level C	90%	95%	99%
Critical value z^*	1.645	1.960	2.576

Notice that for C = 95% the table gives $z^* = 1.960$. This is a bit more precise than the approximate value $z^* = 2$ based on the 68–95–99.7 rule. You can of course use software to find critical values z^*, as well as the entire confidence interval.

Figure 14.4 shows that there is area C under the standard Normal curve between $-z^*$ and z^*. If we start at the sample mean $\bar{x}$ and go out z^* standard deviations, we get an interval that contains the population mean μ in a proportion C of all samples. This interval is

$$\text{from} \quad \bar{x} - z^* \frac{\sigma}{\sqrt{n}} \quad \text{to} \quad \bar{x} + z^* \frac{\sigma}{\sqrt{n}}$$

or

$$\bar{x} \pm z^* \frac{\sigma}{\sqrt{n}}$$

It is a level C confidence interval for μ.

CONFIDENCE INTERVAL FOR THE MEAN OF A NORMAL POPULATION

Draw an SRS of size n from a Normal population having unknown mean μ and known standard deviation σ. A level C **confidence interval for μ** is

$$\bar{x} \pm z^* \frac{\sigma}{\sqrt{n}}$$

The critical value z^* is illustrated in Figure 14.4 and found in Table C or using technology. Technology can also be used to obtain the confidence interval directly.

The steps in finding a confidence interval mirror the overall four-step process for organizing statistical problems.

CONFIDENCE INTERVALS: THE FOUR-STEP PROCESS

STATE: What is the practical question that requires estimating a parameter?

PLAN: Identify the parameter and choose a level of confidence.

SOLVE: Carry out the work in two phases:

1. **Check the conditions** for the interval you plan to use.
2. Calculate the **confidence interval** or use technology to obtain it.

CONCLUDE: Return to the practical question to describe your results in this setting.

EXAMPLE 14.3 IQ scores

STATE: We are interested in the mean IQ test score of seventh-grade girls in a Midwest school district. Here are the scores for 31 randomly selected seventh-grade girls in the district:[2]

114	100	104	89	102	91	114	114	103	105	
108	130	120	132	111	128	118	119	86	72	
111	103	74	112	107	103	98	96	112	112	93

PLAN: We will estimate the mean IQ score μ for all seventh-grade girls in this Midwest school district by giving a 95% confidence interval.

SOLVE: We should start by checking the conditions for inference. For this first example, however, we will find the interval, then discuss how statistical practice deals with conditions that are never perfectly satisfied.

The mean of the sample is $\bar{x} = 105.84$. IQ scores are typically Normally distributed with standard deviation $\sigma = 15$. As part of the "simple conditions," suppose that it is also true of the population of seventh-grade girls in this school district. For 95% confidence, the critical value is $z^* = 1.960$. A 95% confidence interval for μ is therefore

$$\bar{x} \pm z^* \frac{\sigma}{\sqrt{n}} = 105.84 \pm 1.960 \frac{15}{\sqrt{31}}$$

$$= 105.84 \pm 5.28$$

$$= 100.56 \text{ to } 111.12$$

CONCLUDE: We are 95% confident that the mean IQ score for all seventh-grade girls in this Midwest school district is between 100.6 and 111.1. ■

7	2 4
8	6 9
9	1 3 6 8
10	0 2 3 3 3 4 5 7 8
11	1 1 2 2 2 4 4 4 8 9
12	0 8
13	0 2

FIGURE 14.5 Stemplot of the IQ test scores in Example 14.3.

In practice, the first part of the *Solve* step is to check the conditions for inference. Here are the "simple conditions:"

1. **SRS:** The girls are an SRS from the population of all seventh-grade girls in the school district.
2. **Normal distribution:** IQ scores are typically rounded to the nearest integer when provided. For inference's sake, we treat IQ scores as a continuous random variable. We expect from experience that IQ scores in a homogeneous population with similar developmental age will follow approximately a Normal distribution. We can't look at the population, but we can examine the sample. Figure 14.5 is a stemplot of the data. The shape is slightly left-skewed, but there are no extreme outliers. So we have no reason to doubt that the population distribution is close to Normal. Additionally, the central limit theorem tells us that a sample size of 31 is large enough to overcome moderate departures from Normality for performing inference.
3. **Known σ:** Most IQ tests are designed so that $\sigma = 15$, so it is not too unrealistic to assume that we know that $\sigma = 15$ for our calculations.

In this introductory chapter we work with examples for which the "simple conditions" are likely to be satisfied. However, inference methods are often used when conditions like SRS and Normal population are not exactly met. In addition, knowing σ is the exception rather than the rule, and so it is not generally realistic to suppose that we know σ. You will see in Chapter 17 that it is easy to do away with the need to know σ. Chapter 15 and the later chapters on each inference topic will give you a basis to judge whether it is safe to do inference in specific instances when the simple conditions are not met.

APPLY YOUR KNOWLEDGE

14.4 Find a critical value. The critical value z^* for confidence level 97.5% is not in Table C. Use software or Table B of standard Normal probabilities to find z^*. Include in your answer a copy of Figure 14.4 with $C = 0.975$ that shows how much area is left in each tail when the central area is 0.975. (Note that the purpose of this exercise is to help you understand how confidence intervals are obtained. However, it is unlikely that you will encounter a 97.5% confidence interval in an actual publication.)

14.5 Pharmaceutical production. A manufacturer of pharmaceutical products analyzes each batch of a product to verify the concentration of the active ingredient. The chemical analysis is not perfectly precise. In fact, repeated measurements follow a Normal distribution with mean μ equal to the true concentration and with standard deviation $\sigma = 0.0068$ grams per liter (g/l). Three analyses of one batch give concentrations of 0.8403, 0.8363, and 0.8447 g/l. To estimate the true concentration, give a 95% confidence interval for μ. Follow the four-step process as illustrated in Example 14.3.

14.6 Executives' blood pressures. The National Center for Health Statistics reports that the systolic blood pressure for males 35 to 44 years of age has standard deviation 15. The medical director of a very large company looks at the medical records of 72 randomly selected male executives in this age group and finds that the mean systolic blood pressure in this sample is $\bar{x} = 126.07$. Estimate the mean systolic blood pressure for all executive males 35 to 44 years of age, using a 90% confidence interval. Follow the four-step process as illustrated in Example 14.3.

The reasoning of tests of significance

Confidence intervals are one of the two most common types of statistical inference. Use a confidence interval when your goal is to estimate a population parameter. The second common type of inference, called *tests of significance*, has a different goal: to assess the evidence provided by data about some claim concerning a population. Here is the reasoning of statistical tests in a nutshell.

EXAMPLE 14.4 Running a mile in 6 minutes

A high school student claims that he ran every day over the summer and that his average running mile time was 6 minutes. However, the coach sees that the student's average mile time on 5 different days during the first month of school was 7.1 minutes. You think, "Someone who runs a mile in 6 minutes on average would almost never get an average mile time of 7.1 minutes in 5 separate runs." So you don't believe his claim.

Your reasoning is based on asking what would happen if the student's claim was true and we repeated the sample of 5 runs many times—the student would almost never get an average mile time as high as 7.1 minutes. This outcome is so unlikely that it gives strong evidence that the student's claim is not true (perhaps the student's daily run in the summer was shorter than a whole mile or perhaps the student exaggerated).

You can say how strong the evidence against the student's claim is by giving the probability that his average mile time from 5 runs would be at least 7.1 minutes if his overall mean mile time really was 6 minutes. This probability (using a standard deviation of 0.5 minutes) is nearly zero. The small probability convinces you that the student's claim is false because *an outcome that would rarely happen if a claim was true is good evidence that the claim is not true*. ■

The reasoning of statistical tests, like that of confidence intervals, is based on asking what would happen if we repeated the sample or experiment many times. We will act as if the "simple conditions" listed on page 336 are true: We have a perfect SRS from an exactly Normal population with standard deviation σ known to us. Here is an example we will explore.

EXAMPLE 14.5 Inorganic phosphorus levels in the elderly

CDC/Amanda Mills

Phosphorus is a mineral that is essential to a whole range of metabolic processes as well as bone stiffening. Levels of inorganic phosphorus in the blood are known to vary among adults Normally with mean 1.2 and standard deviation 0.1 millimoles per liter (mmol/l). A study was conceived to examine inorganic phosphorus blood levels in older individuals to see if it decreases with age. Here are data from a retrospective chart review of 12 men and women between the ages of 75 and 79 years:[3]

1.26	1.39	1.00	1.00	1.00	1.10
0.87	1.23	1.19	1.29	1.03	1.18

Some of these values are above the adult population mean and some are below. The average inorganic phosphorus level is given by the sample mean $\bar{x} = 1.128$ mmol/l. Are these data good evidence that, on average, inorganic phosphorus levels among people aged 75 to 79 are lower than in the whole adult population? ■

The reasoning is the same as in Example 14.4. We make a claim and ask if the data give evidence *against* it. We seek evidence that the mean blood level of inorganic phosphorus in people aged 75 to 79 *is less than* 1.2 mmol/l, so the claim we test is that the mean for people aged 75 to 79 *is* 1.2 mmol/l. In that case, the mean blood level of inorganic phosphorus for the population of individuals between the ages of 75 and 79 would be $\mu = 1.2$.

If the claim that $\mu = 1.2$ is true, the sampling distribution of $\bar{x}$ from 12 individuals aged 75 to 79 is Normal with mean $\mu = 1.2$ and standard deviation

$$\frac{\sigma}{\sqrt{n}} = \frac{0.1}{\sqrt{12}} = 0.0289$$

Figure 14.6 shows this sampling distribution. We can judge whether any observed $\bar{x}$ is surprising by locating it on this distribution.

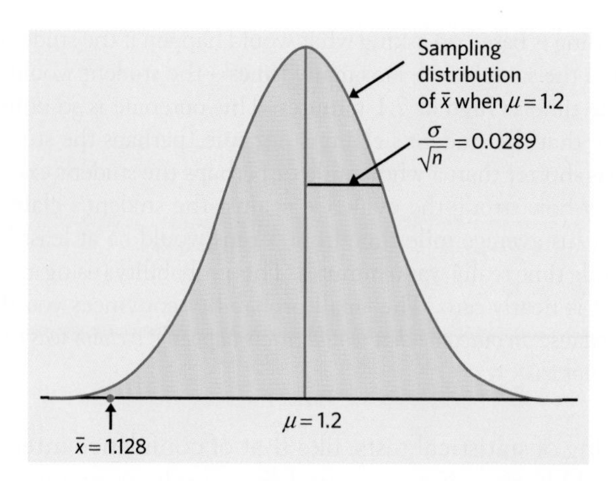

FIGURE 14.6 If people aged 75 to 79 do not differ from the whole adult population in their blood levels of inorganic phosphorus, the mean inorganic phosphorus level $\bar{x}$ for a random sample of 12 individuals aged 75 to 79 will have this sampling distribution. The actual result for this group is $\bar{x} = 1.128$. That's so far out on the Normal curve that it is good evidence that people aged 75 to 79 do have inorganic phosphorus levels lower than 1.2 mmol/l.

A sample mean close to the population mean, within the central bulk of the sampling distribution, would indicate that such a sample mean could easily occur just by chance when the population mean is $\mu = 1.2$. On the other hand, a sample mean located on the edges of the sampling distribution would indicate that it is somewhat unlikely to occur just by chance when the population mean is $\mu = 1.2$.

In our case, the sample of 12 individuals aged 75 to 79 produced $\bar{x} = 1.128$ mmol/l. That's far out on the Normal curve in Figure 14.6—so far out that *an observed value this small would rarely occur just by chance if the true μ were 1.2 mmol/l.* This observed value is good evidence that the true μ is in fact less than 1.2, that is, that elderly individuals between the ages of 75 and 79 years have a lower inorganic phosphorus blood level than the general population.

APPLY YOUR KNOWLEDGE

14.7 Anemia. Hemoglobin is a protein in red blood cells that carries oxygen from the lungs to body tissues. People with less than 12 grams of hemoglobin per deciliter of blood (g/dl) are anemic. A public health official in Jordan suspects that the mean μ for all children in Jordan is less than 12. He measures a sample of 50 children. Suppose that the "simple conditions" hold: The 50 children are an SRS from all Jordanian children, and the hemoglobin level in this population follows a Normal distribution with standard deviation $\sigma = 1.6$ g/dl.

(a) We seek evidence *against* the claim that $\mu = 12$. What is the sampling distribution of $\bar{x}$ in many samples of size 50 if in fact $\mu = 12$? Make a sketch of the Normal curve for this distribution. (Sketch a Normal curve, then mark the axis using what you know about locating the mean and standard deviation on a Normal curve.)

(b) The sample mean was $\bar{x} = 11.3$ g/dl. Mark this outcome on the sampling distribution. Also mark the outcome $\bar{x} = 11.8$ g/dl from a different study of 50 children from another country. Explain carefully from your sketch why one of these outcomes is good evidence that μ is lower than 12 and also why the other outcome is not good evidence for this conclusion.

14.8 Arsenic contamination. Arsenic is a compound naturally occurring in very low concentrations. Arsenic blood concentrations in healthy individuals are Normally distributed with mean $\mu = 3.2$ micrograms per deciliter (μg/dl) and standard deviation $\sigma = 1.5\ \mu g/dl$.[4] Some areas are known to have naturally elevated concentrations of arsenic in the ground and water supplies. We take two SRSs of 25 adults residing in two different high-arsenic areas.

(a) We seek evidence *against* the claim that $\mu = 3.2$. What is the sampling distribution of the mean blood arsenic concentration $\bar{x}$ in many samples of 25 adults if the claim is true? Sketch the density curve of this distribution. (Sketch a Normal curve, then mark the axis using what you know about locating the mean and standard deviation on a Normal curve.)

(b) Suppose that the data from the first sample give $\bar{x} = 3.35\ \mu g/dl$. Mark this point on the axis of your sketch. Suppose that the second sample gives $\bar{x} = 3.75\ \mu g/dl$. Mark this point on your sketch. Using your sketch, explain in simple language why one result is good evidence that the mean blood arsenic concentration of all adults in one high-arsenic area is greater than 3.2 $\mu g/dl$ and why the outcome for the other high-arsenic area is not.

Stating hypotheses

A statistical test starts with a careful statement of the claims we want to compare. In Example 14.5 we saw that the inorganic phosphorus blood data are not very plausible if people aged 75 to 79 years have mean level 1.2 mmol/l. Because the reasoning of tests looks for evidence *against* a claim, we start with the claim we seek evidence against, such as "no difference from the adult mean of 1.2 mmol/l."

> **NULL AND ALTERNATIVE HYPOTHESES**
>
> The claim tested by a statistical test is called the **null hypothesis.** The test is designed to assess the strength of the evidence *against* the null hypothesis. Usually the null hypothesis is a statement of "no effect" or "no difference."
>
> The claim about the population that we are trying to find evidence *for* is the **alternative hypothesis.** The alternative hypothesis is **one-sided** if it states that a parameter is *larger than* or that it is *smaller than* the null hypothesis value. It is **two-sided** if it states that the parameter is *different from* the null value (it could be either smaller or larger).

We abbreviate the null hypothesis H_0 and the alternative hypothesis H_a. *Hypotheses always refer to a population, not to a particular outcome. Be sure to state H_0 and H_a in terms of population parameters.* Because H_a most often expresses the effect that we hope to find evidence *for*, it may be easier to begin by stating H_a and then set up H_0 as the statement that the hoped-for effect is not present.

In Example 14.5 we are seeking evidence *for* a lower inorganic phosphorus level. The null hypothesis says that "there is no difference from the adult mean of 1.2 mmol/l" on average in a large population of elderly individuals between the ages of 75 and 79 years. The alternative hypothesis says that "their mean is lower than 1.2 mmol/l." So the hypotheses are

$$H_0\colon \mu = 1.2$$
$$H_a\colon \mu < 1.2$$

The alternative hypothesis is *one-sided* because we are interested only in whether the elderly have a *lower* inorganic phosphorus level, on average.

Getty Images/Glowimages

EXAMPLE 14.6 Manufacturing aspirin tablets

A pharmaceutical company manufactures aspirin tablets that are sold with the label "active ingredient: aspirin 325 mg." No production, even machine-made, is ever perfect, and individual tablets do vary a little bit in actual aspirin content. A small amount of variation is acceptable provided that, on average, the whole production has mean $\mu = 325$ mg.

The parameter of interest is the mean aspirin content per tablet, μ. The null hypothesis says that the population of all aspirin tablets produced by this manufacturer has mean $\mu = 325$ mg, that is,

$$H_0: \mu = 325 \text{ mg}$$

Proper dosage of prescription medicine is important, and deviations from the expected dose distribution in either direction (too much or too little aspirin per tablet) would need to be identified and then, of course, remedied. The alternative hypothesis is therefore *two-sided:*

$$H_a: \mu \neq 325 \text{ mg} \quad ■$$

The hypotheses should express the hopes or suspicions we have **before** *we see the data. It is cheating to first look at the data and then frame hypotheses to fit what the data show.* Thus, the fact that a random sample of aspirin tablets in Example 14.6 gave a sample average $\bar{x}$ larger than the population mean μ should not influence our choice of H_a. If you do not have a specific direction firmly in mind in advance, you must use a two-sided alternative.

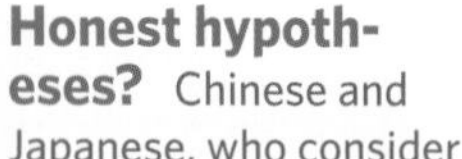

Honest hypotheses? Chinese and Japanese, who consider the number 4 unlucky, die more often on the fourth day of the month than on other days. The authors of a study did a statistical test of the claim that the fourth day has more deaths than other days and found good evidence in favor of this claim. Can we trust this? Not if the authors looked at all days, picked the one with the most deaths, then made "this day is different" the claim to be tested. A critic raised that issue, and the authors replied: "No, we had day 4 in mind in advance, so our test was legitimate."

APPLY YOUR KNOWLEDGE

14.9 Anemia. State the null and alternative hypotheses for the anemia study described in Exercise 14.7.

14.10 Arsenic contamination. State the null and alternative hypotheses for the study of blood arsenic levels described in Exercise 14.8.

14.11 Women's heights. Young American women aged 18 to 24 years have an average height of 64.5 inches. You wonder whether the mean height of female students at your university is different from the national average. You find that the mean height of a sample of 78 female students on campus is $\bar{x} = 63.1$ inches. What are your null and alternative hypotheses?

14.12 Stating hypotheses. In planning a study of the birth weights of babies whose mothers did not see a doctor before delivery, a researcher states the hypotheses as

$$H_0: \bar{x} = 1000 \text{ grams}$$
$$H_a: \bar{x} < 1000 \text{ grams}$$

What's wrong with this?

P-value and statistical significance

The idea of making a claim (H_0) in order to evaluate evidence *against* it may seem odd at first. The approach is not unique to tests of significance, though. Think of a criminal trial. The defendant is "innocent until proven guilty." That is, the null hypothesis is innocence, and the prosecution must try to provide convincing evidence against this hypothesis. That's exactly how statistical tests work, though in statistics we deal with evidence provided by data and use a probability to say how strong the evidence is.

The probability that measures the strength of the evidence against a null hypothesis is called a *P-value*. Statistical tests generally work as follows.

TEST STATISTIC AND *P*-VALUE

A **test statistic** calculated from the sample data measures how far the data diverge from the null hypothesis H_0. Large values of the statistic show that the data are far from what we would expect if H_0 was true.

The probability, computed assuming that H_0 is true, that the test statistic would take a value at least as extreme (in the direction of H_a) as that actually observed is called the **P-value** of the test. The smaller the P-value, the stronger the evidence against H_0 provided by the data.

Small P-values are evidence against H_0, because they say that the observed result would be unlikely to occur if H_0 was true. Large P-values fail to provide evidence against H_0. Statistical software will give you the P-value of a test when you enter your null and alternative hypotheses and your data. So your most important task is to understand what a P-value says.

EXAMPLE 14.7 Inorganic phosphorus: one-sided *P*-value

The study of inorganic phosphorus levels in the elderly in Example 14.5 tests the hypotheses

$$H_0: \mu = 1.2$$
$$H_a: \mu < 1.2$$

Because the alternative hypothesis says that $\mu < 1.2$, values of $\bar{x}$ less than 1.2 favor H_a over H_0. The 12 individuals aged 75 to 79 had a mean inorganic phosphorus level of $\bar{x} = 1.128$. *The P-value is the probability of getting an $\bar{x}$ as small as 1.128 or smaller when the null hypothesis is really true.*

The *P-Value of a Test of Significance* applet automates the work of finding P-values for samples of size 250 or smaller under the simple inference conditions. We enter the information for Example 14.5 into the applet—hypotheses, σ, n, and $\bar{x}$—then we click "Update." Figure 14.7 shows the applet output. It displays the P-value as an area under a Normal curve and its corresponding value $P = 0.0063$. We would rarely observe a mean inorganic phosphorus level of 1.128 or lower if H_0 was true. The small P-value provides strong evidence against H_0 and in favor of the alternative H_a: $\mu < 1.2$. ■

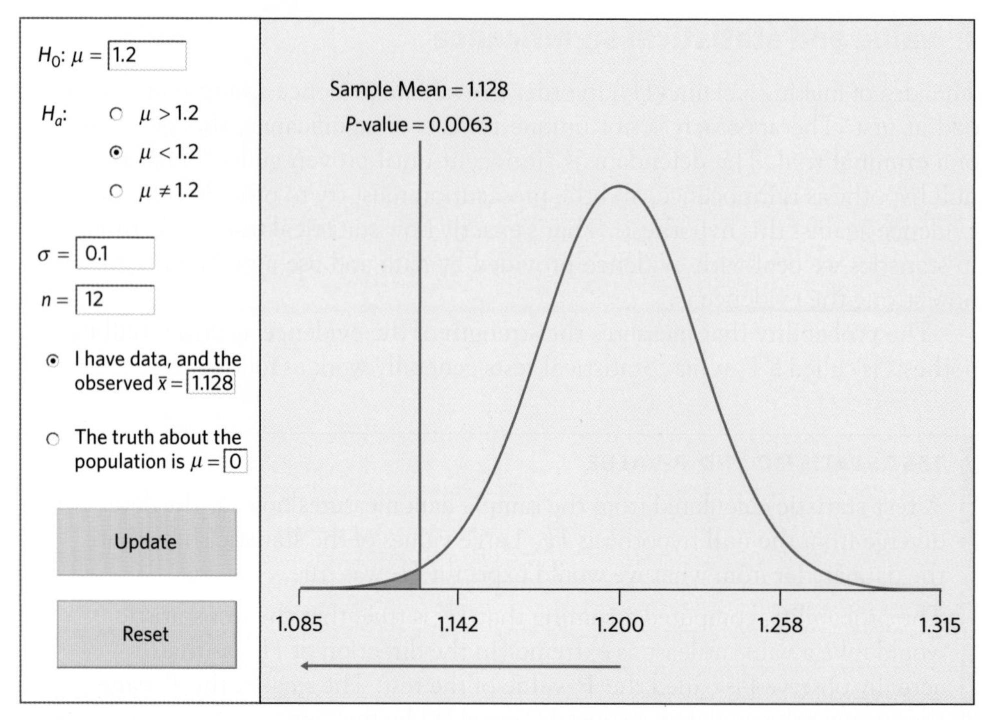

FIGURE 14.7 Output from the *P-Value of a Test of Significance* applet for Example 14.7. The *P*-value for the one-sided test, 0.0063, is the area under the Normal curve to the left of $\bar{x} = 1.128$.

The alternative hypothesis sets the direction that counts as evidence against H_0. In Example 14.7 only small values count because the alternative is one-sided on the low side. If the alternative is two-sided, both directions count.

EXAMPLE 14.8 Aspirin tablets: two-sided *P*-value

Suppose that we know that the aspirin content of aspirin tablets in Example 14.6 follows a Normal distribution with standard deviation $\sigma = 5$ mg. If the manufacturing process is well calibrated, the mean aspirin content is $\mu = 325$ mg. This is our null hypothesis. The alternative hypothesis says simply that "the mean is not 325 mg."

$$H_0: \mu = 325$$
$$H_a: \mu \neq 325$$

Data from a random sample of 10 aspirin tablets gives $\bar{x} = 326.9$ mg. *Because the alternative is two-sided, the P-value is the probability of getting an $\bar{x}$ at least as far from $\mu = 325$ in either direction as the observed $\bar{x} = 326.9$.*

Enter the information for this example into the *P-Value of a Test of Significance* applet and click "Update." Figure 14.8 shows the applet output as well as the information we entered. The *P*-value is the sum of the two shaded areas under the Normal curve. It is $P = 0.2295$. Values as far from 325 as $\bar{x} = 326.9$ (in either direction) would happen 23% of the time if the true population mean was $\mu = 325$. An outcome that would occur so often if H_0 was true is not good evidence against H_0. ■

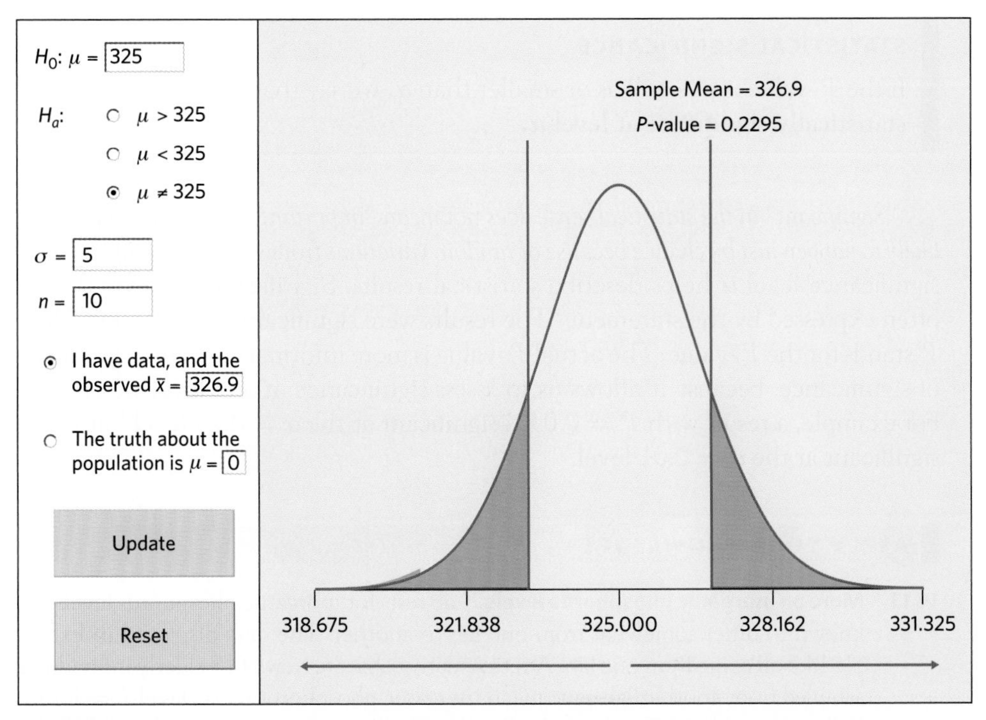

FIGURE 14.8 Output from the *P-Value of a Test of Significance* applet for Example 14.8. The *P*-value for the two-sided test, 0.2295, is twice the value of the area under the Normal curve to the right of $\bar{x} = 326.9$.

The conclusion of Example 14.8 is *not* that H_0 is true. The study looked for evidence against H_0: $\mu = 325$ and failed to find strong evidence. That is all we can say. Tests of significance assess the evidence *against* H_0. If the evidence is strong, we can confidently reject H_0 in favor of the alternative. *Failing to find evidence against H_0 means only that the data are consistent with H_0, not that we have clear evidence that H_0 is true*. The mean μ for the population of all aspirin tablets could, for instance, be a value very close but not exactly equal to 325 mg.

In Examples 14.7 and 14.8 we decided that *P*-value $P = 0.0063$ was strong evidence against the null hypothesis and that *P*-value $P = 0.2295$ did not give convincing evidence. There is no rule for how small a *P*-value we should require to reject H_0—it's a matter of judgment and depends on the specific circumstances.

Nonetheless, we can compare a *P*-value with some fixed values that are in common use as standards for evidence against H_0. The most common fixed values are 0.05 and 0.01. If $P \leq 0.05$, there is no more than 1 chance in 20 that a sample would give evidence this strong just by chance when H_0 is actually true. If $P \leq 0.01$, we have a result that in the long run would happen no more than once per 100 samples when H_0 is true. These fixed standards for *P*-values are called **significance levels.** We use α, the Greek letter alpha, to stand for a significance level.

significance level

> **STATISTICAL SIGNIFICANCE**
>
> If the P-value is as small as or smaller than α, we say that the data are **statistically significant at level α.**

"Significant" in the statistical sense does not mean "important." It means simply "not likely to happen just by chance because of random variations from sample to sample." The significance level α helps describe statistical results. Significance at level 0.01 is often expressed by the statement "The results were significant ($P < 0.01$)." Here P stands for the P-value. The actual P-value is more informative than a statement of significance, because it allows us to assess significance at any level we choose. For example, a result with $P = 0.03$ is significant at the $\alpha = 0.05$ level but is not significant at the $\alpha = 0.01$ level.

APPLY YOUR KNOWLEDGE

14.13 More on inorganic phosphorus levels. Lab tests for inorganic phosphorus levels are known to differ somewhat from one lab to another. The data displayed in Example 14.5 all came from one lab. A retrospective chart review of 18 elderly individuals sampled from another lab gave mean inorganic phosphorus blood level $\bar{x} = 1.152$. Follow Example 14.7 and use the *P-Value of a Test of Significance* applet to find the test P-value when $\bar{x} = 1.152$ and $n = 18$ using the same null and alternative hypotheses as in Example 14.7. Interpret your result.

Exactostock/SuperStock

14.14 Mechanical properties of spider webs. Researchers examined the mechanical properties of the sticky and the nonsticky silk threads used in *Nephila edulis* webs. Here is an excerpt from their published report:

> *The fibre diameter of the non-sticky spiral was significantly smaller than the diameter of radial fibres ($P < 0.001$), while no significant differences were found for fibre stiffness ($P = 0.07$).*[5]

(a) Explain in specific terms what $P < 0.001$ and $P = 0.07$ mean.

(b) The report also states that "a significance level of 0.05 was used for all tests." Using this significance level of $\alpha = 0.05$, can we conclude that the difference in fiber diameter is statistically significant? Can we conclude that the difference in fiber stiffness is statistically significant? Explain your answers.

14.15 Anemia. Go back to the anemia study of Exercise 14.7.

(a) Enter the hypotheses, n, σ, and $\bar{x}$ for the Jordan study into the *P-Value of a Test of Significance* applet. What is the P-value for this study? Is this outcome statistically significant at the $\alpha = 0.05$ level? At the $\alpha = 0.01$ level?

(b) Now enter the values for the other study of anemia. What is the P-value for that study? Is this outcome statistically significant at the $\alpha = 0.05$ level? At the $\alpha = 0.01$ level?

(c) Explain briefly why these P-values tell us that one outcome is strong evidence against the null hypothesis and that the other outcome is not.

14.16 Arsenic contamination. Go back to the study of arsenic contamination in two high-arsenic areas in Exercise 14.8.

(a) Enter the hypotheses, n, σ, and $\overline{x}$ for the first study in the *P-Value of a Test of Significance* applet. What is the P-value for this study? Is this outcome statistically significant at the $\alpha = 0.05$ level? At the $\alpha = 0.01$ level?

(b) Now enter into the applet the mean blood arsenic level for the residents of the second high-arsenic area. What is the P-value for this study? Is it statistically significant at the $\alpha = 0.05$ level? At the $\alpha = 0.01$ level?

(c) Explain briefly why these P-values tell us that one outcome is strong evidence against the null hypothesis and that the other outcome is not.

Tests for a population mean

We have used tests for hypotheses about the mean μ of a population, under the "simple conditions," to introduce statistical tests. The test compares the sample mean $\overline{x}$ with the claimed population mean stated by the null hypothesis H_0. The P-value says how unlikely an $\overline{x}$ this extreme is if H_0 is true. Now we can reduce this test to a rule.

z TEST FOR A POPULATION MEAN

Draw an SRS of size n from a Normal population that has unknown mean μ and known standard deviation σ. To **test the null hypothesis that μ has a specified value**

$$H_0: \mu = \mu_0$$

calculate the **one-sample z test statistic**

$$z = \frac{\overline{x} - \mu_0}{\sigma/\sqrt{n}}$$

In terms of a variable Z having the standard Normal distribution, the P-value for a test of H_0 against

$H_a: \mu > \mu_0$ is $P(Z \geq z)$

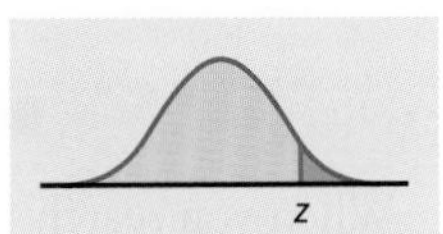

$H_a: \mu < \mu_0$ is $P(Z \leq z)$

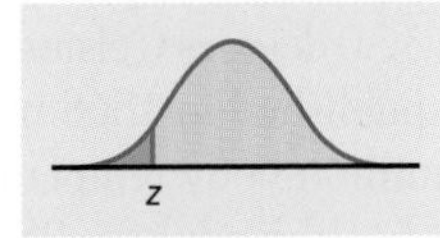

$H_a: \mu \neq \mu_0$ is $2P(Z \geq |z|)$

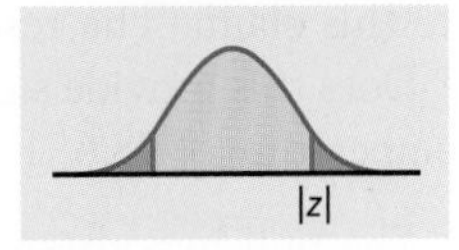

Software will find both z and its P-value for you.

The test statistic z measures how far the observed sample mean $\bar{x}$ deviates from the hypothesized population value μ_0. The measurement is in the standard scale obtained by dividing by the standard deviation of $\bar{x}$. When H_0 is true, z has the standard Normal distribution. The pictures that illustrate the P-value look just like Figures 14.7 and 14.8 except that they are in the standard scale.

one-tailed
two-tailed

Know that many statistical software packages call a P-value obtained for a one-sided H_a a "one-sided" or **"one-tailed"** P-value. A P-value obtained for a two-sided H_a is then called a "two-sided" or **"two-tailed"** P-value. The logic for this naming convention is apparent in Figures 14.7 and 14.8: A P-value computed for a two-sided H_a includes both "tails" of the Normal sampling distribution, whereas a P-value computed for a one-sided H_a includes only one.

The steps for carrying out a test of significance mirror the overall four-step process for organizing realistic statistical problems.

TESTS OF SIGNIFICANCE: THE FOUR-STEP PROCESS

STATE: What is the practical question that requires a statistical test?

PLAN: Identify the parameter, state null and alternative hypotheses, and choose the type of test that fits your situation.

SOLVE: Carry out the test in three phases:

1. **Check the conditions** for the test you plan to use.
2. Calculate the **test statistic.**
3. Find the **P-value** using a table of Normal probabilities or technology.

CONCLUDE: Return to the practical question to describe your results in this setting.

If you do not have access to technology, sketch the Normal sampling distribution and the area or areas under it representing the P-value. Then use Table B or Table C to obtain an approximate P-value from the calculated z value.

EXAMPLE 14.9 Body temperature

STATE: When you search the internet for what constitutes "normal" or healthy body temperature, you will probably find that an oral temperature of 98.6 degrees Fahrenheit (°F), or 37.0 degrees Celsius, is considered normal. This widely quoted value comes from a paper published in 1868 by German physician Carl Wunderlich in which he reported over a million body temperature readings. In it, he stated that the mean body temperature of healthy adults is 98.6 °F. More than a century later, one study was designed to evaluate this claim. The result is a mean oral temperature of $\bar{x} = 98.25$ °F from 130 adults.[6] Does this provide significant evidence that Wunderlich's claim of a mean adult body temperature of 98.6 °F is not correct?

PLAN: The null hypothesis is "no difference" from the accepted mean $\mu_0 = 98.6$ °F. The alternative is two-sided because the study did not have a particular direction in mind before examining the data. So the hypotheses about the unknown mean μ of the

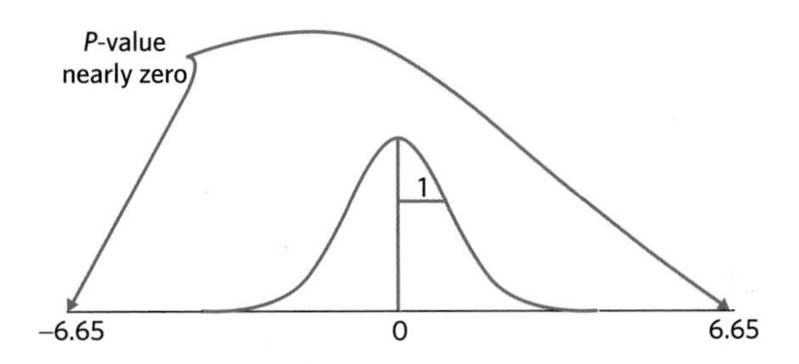

FIGURE 14.9 The P-value for the two-sided test in Example 14.9. The observed value of the test statistics is $z = -6.65$.

population of healthy adults are

$$H_0: \mu = 98.6$$
$$H_a: \mu \neq 98.6$$

SOLVE: As part of the simple conditions, suppose we know that individual body temperatures follow a Normal distribution with standard deviation $\sigma = 0.6$ °F. The one-sample z **test statistic** is

$$z = \frac{\bar{x} - \mu_0}{\sigma/\sqrt{n}} = \frac{98.25 - 98.6}{0.6/\sqrt{130}}$$
$$= -6.65$$

To help find a **P-value,** sketch the standard Normal curve and mark on it the observed value of z. Figure 14.9 shows that the P-value is the probability that a standard Normal variable Z takes a value at least 6.65 away from zero. From Table B or software, this probability is $P = 2P(Z \leq -6.65)$, which is nearly zero.

CONCLUDE: If $\mu = 98.6$ °F, it would be nearly impossible to obtain an SRS of size 130 from the healthy adult population with sample mean at least as far from 98.6 °F as that obtained here. The observed $\bar{x} = 98.25$ °F is therefore very strong evidence that the true mean body temperature of healthy adults is not 98.6 °F. ■

In this chapter we are acting as if the "simple conditions" stated on page 336 are true. In practice, you must verify these conditions.

1. **SRS:** The most important condition is that the 130 healthy adults in the sample are an SRS from the population of all healthy adults. We should check this requirement by asking how the data were produced. The researchers obtained the baseline oral temperature of volunteers participating in various vaccine clinical trials conducted at the University of Maryland. If these individuals enrolled in the clinical trial because of a specific medical concern, for example, the data may not be representative of the whole healthy-adult population and could be biased. The researchers assure us that the volunteers were all healthy. Because the subjects were volunteers, the sample is a random sample but not a true SRS. This is very common in the life sciences, especially when dealing with human subjects. We discuss this issue further in Chapter 15.
2. **Normal distribution:** We should also examine the distribution of the 130 observations to look for signs that the population distribution is not Normal. A histogram in the published report indicates that the data were roughly Normal.

3. **Known σ:** It really is unrealistic to suppose that we know that $\sigma = 0.6$ °F. This value is based on Wunderlich's original report of a very large number of body temperatures. We will see in Chapter 17 that it is easy to do away with the need to know σ.

APPLY YOUR KNOWLEDGE

14.17 Water quality. An environmentalist group collects a liter of water from each of 45 random locations along a stream and measures the amount of dissolved oxygen in each specimen. The mean is 4.62 milligrams (mg). Is this strong evidence that the stream has a mean oxygen content of less than 5 mg per liter? (Suppose that we know that dissolved oxygen varies among locations according to a Normal distribution with $\sigma = 0.92$ mg.) Follow the four-step process as illustrated in Example 14.9.

14.18 Hepatitis C and bone density. Does chronic hepatitis C infection impact bone mineral density? In the general population of healthy young adults, bone mineral density is Normally distributed, and density scores are standardized to have a mean μ of 0 and a standard deviation σ of 1. A study examined the bone densities of a random sample of 28 adult men suffering from chronic hepatitis C infection.[7] Their mean standardized bone density was −0.61. Does this provide good evidence that the mean standardized bone density of adult men suffering from chronic hepatitis C infection is less than 0? Follow the four-step process as illustrated in Example 14.9.

Tests from confidence intervals

Both tests and confidence intervals for a population mean μ start by using the sample mean $\bar{x}$ to estimate μ. Both rely on probabilities calculated from a Normal sampling distribution. In fact, a two-sided test at significance level α can be carried out from a confidence interval with confidence level $C = 1 - \alpha$.

CONFIDENCE INTERVALS AND TWO-SIDED TESTS

A level α two-sided significance test rejects a hypothesis H_0: $\mu = \mu_0$ exactly when the value μ_0 falls outside a level $1 - \alpha$ confidence interval for μ.

EXAMPLE 14.10 Tests from a confidence interval

In Example 14.9 a study found that the mean body temperature for a random sample of 130 healthy adults was $\bar{x} = 98.25$ °F. Is this value significantly different from the textbook value $\mu_0 = 98.6$ °F at the 5% significance level?

We can answer this question directly by a two-sided test or indirectly from a 95% confidence interval. The confidence interval is

$$\begin{aligned} \bar{x} \pm z^* \frac{\sigma}{\sqrt{n}} &= 98.25 \pm 1.96 \frac{0.6}{\sqrt{130}} \\ &= 98.25 \pm 0.10 \\ &= 98.15 \text{ to } 98.35 \end{aligned}$$

The hypothesized value $\mu_0 = 98.6$ falls *outside* this confidence interval, so we *can* reject

$$H_0: \mu = 98.6$$

at the 5% significance level. ■

Let's examine a different situation, purely for the sake of argument. If the mean temperature quoted in textbooks had been 98.3 °F, the value tested in H_0 would have fallen *inside* the confidence interval, and we would not have been able to reject the claim

$$H_0: \mu = 98.3$$

at the 5% level.

Statisticians argue over the relative merits of computing a confidence interval or a P-value. A test of significance is obviously particularly well suited when you have a clearly defined set of hypotheses. The P-value will say how strong the evidence against the null hypothesis is. However, if you reject H_0, you are left not knowing what kind of value the parameter might have. And if you fail to reject H_0, there are still other likely values for the parameter besides that defined by H_0. That is, the P-value gives no information about effect size (we will discuss this further in the next chapter). All things being equal, a confidence interval will provide more useful information than a P-value alone.

APPLY YOUR KNOWLEDGE

14.19 Test and confidence interval. The P-value for a two-sided test of the null hypothesis H_0: $\mu = 10$ is 0.06.

(a) Does the 95% confidence interval include the value 10? Why?

(b) Does the 90% confidence interval include the value 10? Why?

14.20 Body temperature. Examples 14.9 and 14.10 computed a P-value and a confidence interval from the data about body temperatures of healthy adults. What can be concluded from each method? What can be concluded from the confidence interval but not from the test of significance? What can be concluded from the test of significance but not from the confidence interval?

14.21 Confidence interval and test. A 95% confidence interval for a population mean is 31.5 ± 3.5.

(a) With a two-sided alternative, can you reject the null hypothesis that $\mu = 34$ at the 5% significance level? Why?

(b) With a two-sided alternative, can you reject the null hypothesis that $\mu = 36$ at the 5% significance level? Why?

CHAPTER 14 SUMMARY

- A **confidence interval** uses sample data to estimate an unknown population parameter with an indication of how accurate the estimate is and of how confident we are that the result is correct.

- Any confidence interval has two parts: an interval calculated from the data and a confidence level C. The **interval** often has the form

$$\text{estimate} \pm \text{margin of error}$$

- The **confidence level** is the success rate of the method that produces the interval. That is, C is the probability that the method will give a correct answer. If you use 95% confidence intervals often, in the long run 95% of your intervals will contain the true parameter value. You do not know whether a 95% confidence interval calculated from a particular set of data contains the true parameter value.
- A level C **confidence interval for the mean μ** of a Normal population with known standard deviation σ, based on an SRS of size n, is given by

$$\bar{x} \pm z^* \frac{\sigma}{\sqrt{n}}$$

The **critical value** z^* is chosen so that the standard Normal curve has area C between $-z^*$ and z^*.
- A **test of significance** assesses the evidence provided by data against a **null hypothesis** H_0 and in favor of an **alternative hypothesis** H_a.
- Hypotheses are always stated in terms of population parameters. Usually, H_0 is a statement that no effect is present, and H_a says that a parameter differs from its null value in a specific direction (**one-sided alternative**) or in either direction (**two-sided alternative**).
- The essential reasoning of a significance test is as follows. Suppose for the sake of argument that the null hypothesis is true. If we repeated our data production many times, would we often get data as inconsistent with H_0 as the data we actually have? If the data are unlikely when H_0 is true, they provide evidence against H_0.
- A test is based on a **test statistic** that measures how far the sample outcome is from the value stated by H_0.
- The **P-value** of a test is the probability, computed supposing H_0 to be true, that the test statistic will take a value at least as extreme as that actually observed. Small P-values indicate strong evidence against H_0. To calculate a P-value we must know the sampling distribution of the test statistic when H_0 is true.
- If the P-value is as small as or smaller than a specified value α, the data are **statistically significant** at significance level α.
- **Significance tests for the null hypothesis H_0: $\mu = \mu_0$** concerning the unknown mean μ of a population are based on the **one-sample z test statistic**

$$z = \frac{\bar{x} - \mu_0}{\sigma/\sqrt{n}}$$

The z test assumes an SRS of size n from a Normal population with known population standard deviation σ. P-values can be obtained either with computations from the standard Normal distribution or by using technology (applet or software).

THIS CHAPTER IN CONTEXT

The reason we collect data is not to learn about the individuals that we examined but to infer from the data to some wider population that the individuals represent. In this chapter we discuss the basic reasoning of statistical inference—confidence intervals and significance tests—with emphasis on inference for a population mean under simple conditions.

The computations described in this chapter build on Normal calculations you learned in Chapter 11 and on the properties of the sampling distribution of $\bar{x}$ introduced in Chapter 13. We examine other methods for computing confidence intervals in Chapters 17 through 20 and Chapter 26. All inference chapters (17 to 28) in this book describe some specific computational method for testing hypotheses and obtaining P-values. While the mathematical approach varies across these chapters, the interpretation of confidence intervals and significance tests remains fundamentally the same.

In the next chapter we discuss some limitations to obtaining and interpreting confidence intervals and significance tests. As with Chapter 14, the essence of Chapter 15 is directly applicable to all the inference chapters in the remainder of the book. In the optional sections of Chapter 15, we also consider some more mathematically sophisticated approaches relevant to study planning.

CHECK YOUR SKILLS

14.22 To give a 98% confidence interval for a population mean μ, you would use the critical value

(a) 1.960. (b) 2.054. (c) 2.326.

14.23 A study reports the mean change in HDL (high-density lipoprotein, or "good" cholesterol) of adults eating raw garlic six days a week for six months. The margin of error for a 95% confidence interval is given as plus or minus 6 milligrams per deciliter of blood (mg/dl).[8] This means that

(a) we can be certain that the study result is within 6 mg/dl of the truth about the population.
(b) we could be certain that the study result is within 6 mg/dl of the truth about the population if the conditions for inferences were satisfied.
(c) the study used a method that gives a result within 6 mg/dl of the truth about the population in 95% of all samples.

Use the following information for Exercises 14.24 through 14.26. A laboratory scale is known to have a standard deviation of $\sigma = 0.001$ gram in repeated weighings. Scale readings in repeated weighings are Normally distributed, with mean equal to the true weight of the specimen.

14.24 Three weighings of a specimen on this scale give 3.412, 3.416, and 3.414 grams. A 95% confidence interval for the true weight is

(a) 3.414 ± 0.00113.
(b) 3.414 ± 0.00065.
(c) 3.414 ± 0.00196.

14.25 Suppose that we needed a 99% confidence level for this specimen. The margin of error would be

(a) about the same.
(b) larger.
(c) smaller.

14.26 Another specimen is weighed 8 times on this scale. The average weight is 4.1602 grams. A 99% confidence interval for the true weight of this specimen is

(a) 4.1602 ± 0.00032.
(b) 4.1602 ± 0.00069.
(c) 4.1602 ± 0.00091.

Use the following information for Exercises 14.27 through 14.29. The average human gestation time is 266 days from conception. A researcher suspects that proper nutrition plays an important role and that poor women with inadequate food intake would have shorter gestation times even when given vitamin supplements. A random sample of 20 poor women given vitamin supplements throughout the pregnancy has a mean gestation time from conception of $\bar{x} = 256$ days.

14.27 The null hypothesis for the researcher's test is

(a) H_0: $\mu = 266$.
(b) H_0: $\mu = 256$.
(c) H_0: $\mu < 266$.

14.28 The researcher's alternative hypothesis for the test is

(a) H_a: $\mu \neq 256$. (b) H_a: $\mu < 266$. (c) H_a: $\mu < 256$.

14.29 Human gestation times are approximately Normal with standard deviation $\sigma = 16$ days. The P-value for the researcher's test is

(a) more than 0.1.
(b) less than 0.01.
(c) less than 0.001.

14.30 You use software to run a test of significance. The program tells you that the P-value is 0.031. This result is

(a) not significant at the 5% level.
(b) significant at the 5% level but not at the 1% level.
(c) significant at the 1% level.

14.31 Average human gestation time is 266 days, when counted from conception. A hospital gives a 90% confidence interval for the mean gestation time from conception among its patients. That interval is 264 ± 5 days. Is the mean gestation time in that hospital significantly different from 266 days?

(a) It is not significantly different at the 10% level and therefore is also not significantly different at the 5% level.
(b) It is not significantly different at the 10% level but might be significantly different at the 5% level.
(c) It is significantly different at the 10% level.

CHAPTER 14 EXERCISES

14.32 Explaining confidence. You calculate a 95% confidence interval of 27 ± 2 centimeters (cm) for the mean needle length of Torrey pine trees. You ask a friend to explain this result. He believes it means that "95% of all Torrey pine needles have lengths between 25 and 29 cm." Is he right? Explain your answer.

14.33 Explaining confidence. Here is an explanation from the Associated Press concerning one of its opinion polls. Explain briefly but clearly in what way this explanation is incorrect.

> *For a poll of 1,600 adults, the variation due to sampling error is no more than three percentage points either way. The error margin is said to be valid at the 95 percent confidence level. This means that, if the same questions were repeated in 20 polls, the results of at least 19 surveys would be within three percentage points of the results of this survey.*

14.34 Explaining confidence. You ask another friend to explain your finding that a 95% confidence interval for the mean needle length of Torrey pines is 27 ± 2 cm. She thinks it means that "we can be 95% confident that the true mean needle length of Torrey pine trees is 27 cm." Is she right? Explain your answer.

14.35 Deer mice. Deer mice (*Peromyscus maniculatus*) are small rodents native to North America. Their adult body lengths (excluding tail) are known to vary approximately Normally, with mean $\mu = 86$ millimeters (mm) and standard deviation $\sigma = 8$ mm.[9] Deer mice are found in diverse habitats and exhibit different adaptations to their environment. A random sample of 14 deer mice in a rich forest habitat gives an average body length of $\bar{x} = 91.1$ mm. Assume that the standard deviation σ of all deer mice in this area is also 8 mm.

(a) What is the standard deviation of the mean length $\bar{x}$?
(b) What critical value do you need to use in order to compute a 95% confidence interval for the mean μ?
(c) Give a 95% confidence interval for the mean body length of all deer mice in the forest habitat.

14.36 More about deer mice. The 14 deer mice described in the previous exercise had an average body length of $\bar{x} = 91.1$ mm. Assume that the standard deviation of body lengths in the population of all deer mice in the rich forest habitat is the same as the $\sigma = 8$ mm for the general deer mouse population.

(a) Following your approach in the previous exercise, now give a 90% confidence interval for the mean body length of all deer mice in the forest habitat.
(b) This confidence interval is shorter than your interval in the previous exercise, even though the intervals come from the same sample. Why does the second interval have a smaller margin of error?

14.37 Avandia and cholesterol. Drug maker GlaxoSmithKline investigated the potential impact of its oral antidiabetic drug Avandia (rosiglitazone maleate) on the blood lipid levels of adult diabetics who might benefit from taking Avandia. The mean baseline LDL (low-density lipoprotein, or "bad" cholesterol) level for a random sample of 964 such patients was $\bar{x} = 125.6$ milligrams per deciliter (mg/dl). The distribution of LDL levels in the adult population is known to be close to

TABLE 14.1 Heights in inches of female undergraduate students at UC-Irvine

54	54	56	57	58	58	58	59	60	60	60	60	60	60	60	60	61	61	61
61	61	61	61	62	62	62	62	62	62	62	62	62	62	62	62	62	62	62
63	63	63	63	63	63	63	63	63	63	63	63	63	63	63	63	63	63	63
63	63	64	64	64	64	64	64	64	64	64	64	64	64	64	64	64	64	65
65	65	65	65	65	65	65	66	66	66	67	67	67	67	68	68	70		

Normal with standard deviation $\sigma = 30$ mg/dl.[10] Assuming that σ is the same as in the general population, give a 90% confidence interval for the mean baseline LDL level μ among adult diabetics who might benefit from taking Avandia.

14.38 Student heights. In Example 11.1 (page 266) we saw that the distribution of heights of young women aged 18 to 24 is approximately Normal with mean $\mu = 64.5$ inches and standard deviation $\sigma = 2.5$ inches. A random sample of 93 female undergraduate students at the University of California at Irvine produced a sample mean height of 62.7 inches. The individual heights are displayed in Table 14.1.[11]

4 STEP

(a) Assume that the standard deviation of young women heights is the same at UC–Irvine as in the rest of the country ($\sigma = 2.5$ inches). Briefly discuss the other two simple conditions, using a histogram or a dotplot to verify that the distribution is roughly symmetric with no outliers.

(b) Following the four-step process, give a 95% confidence interval for the mean height of all female undergraduate students at UC–Irvine.

In all exercises that call for P-values, give the actual value if you use software or the P-value applet. Otherwise, calculate the test statistic and use Table B to obtain an approximate P-value.

14.39 Sodium intake. The Institute of Medicine sets the tolerable upper level of sodium intake for adults and teenagers at 2300 milligrams (mg) per day. A random sample of 3727 adolescents representative of the U.S. adolescent population found a mean sodium intake of 3486 mg per day. Does this finding suggest that U.S. adolescents consume sodium in excess of the tolerable upper level? State H_0 and H_a for a test to answer this question. State carefully what the parameter μ in your hypotheses is.

14.40 Cartons of eggs. According to the U.S. Department of Agriculture, cartons of 12 large eggs should have a minimum weight of 24 ounces (oz); that's an average egg weight of 2 oz or more.[12] You suspect that a certain producer sells eggs of inferior caliber (lower weight) in its cartons labeled "large eggs." State H_0 and H_a for a test to assess your suspicion. State what the parameter μ in your hypotheses is.

14.41 The wrong alternative. One of your friends is testing the effect of drinking coffee on the duration of cold symptoms. The common cold lasts, on average, 6 days. She starts with no expectations as to whether drinking coffee will have any effect at all on cold duration. After seeing the results of her experiment, in which the average cold duration was less than 6 days, she tests a one-sided alternative about the population mean cold duration when drinking coffee,

$$H_0\colon \mu_{coffee} = 6$$
$$H_a\colon \mu_{coffee} < 6$$

She finds $z = -1.68$ with one-sided P-value $P = 0.0465$.

(a) Explain why your friend should have used the two-sided alternative hypothesis.

(b) What is the correct two-sided P-value for $z = -1.68$?

14.42 Is this what *P* means? When asked to explain the meaning of "the P-value was 0.03," a student says, "This means there is only probability 0.03 that the null hypothesis is true." Is this an essentially correct explanation? Explain your answer.

14.43 Is this what significance means? Another student, when asked why statistical significance appears so often in research reports, says, "Because saying that results are significant tells us that they cannot easily be explained by chance variation alone." Do you think that this statement is essentially correct? Explain your answer.

14.44 Cicadas as fertilizer? Every 17 years, swarms of cicadas emerge from the ground in the eastern United States, live for about six weeks, and then die. There are so many cicadas that their dead bodies can serve as fertilizer. In an experiment, a researcher added cicadas under some plants in a natural plot of bellflowers on the forest floor, leaving other plants undisturbed. "In this experiment, cicada-supplemented bellflowers from a natural field population produced foliage with 12% greater nitrogen content relative to controls ($P = 0.031$)."[13]

A colleague who knows no statistics says that an increase of 12% isn't a lot—maybe it's just an accident due to natural variation among the plants. Explain in simple language how "$P = 0.031$" answers this objection.

14.45 Forests and windstorms. Does the destruction of large trees in a windstorm change forests in any important way? Here is the conclusion of a study that found that the answer is no:

> *We found surprisingly little divergence between treefall areas and adjacent control areas in the richness of woody plants ($P = 0.62$), in total stem densities ($P = 0.98$), or in population size or structure for any individual shrub or tree species.*[14]

The two P-values refer to null hypotheses that say "no difference" in measurements between treefall and control areas. Explain clearly why these values provide no evidence of a difference.

14.46 The wrong *P*. The report of a study of seat belt use by drivers says, "Hispanic drivers were not significantly more likely than White/non-Hispanic drivers to overreport safety belt use (27.4 vs. 21.1%, respectively; $z = 1.33$, $P > 1.0$)."[15] How do you know that the P-value given is incorrect? What is the correct one-sided P-value for test statistic $z = 1.33$?

14.47 Significance level: 5% versus 1%. Sketch the standard Normal curve for the z test statistic and mark off areas under the curve to show why a value of z that is significant at the 1% level in a one-sided test is always significant at the 5% level. If z is significant at the 5% level, what can you say about its significance at the 1% level?

14.48 Student heights. 4 STEP In Example 11.1 (page 266) we saw that the distribution of heights of young women aged 18 to 24 is approximately Normal with mean $\mu = 64.5$ inches and standard deviation $\sigma = 2.5$ inches. A random sample of 93 female undergraduate students at the University of California at Irvine produced a sample mean height of 62.7 inches. The individual heights are displayed in Table 14.1. Is there evidence that the mean height of all female undergraduate students at UC–Irvine is different from the mean height of all young women in the United States? Follow the four-step process, as illustrated in Example 14.9, in your answer.

14.49 Deer mice. 4 STEP The sample of 14 deer mice in Exercise 14.35 had an average body length of $\bar{x} = 91.1$ mm. Do forest deer mice on the average differ significantly in body length from the general deer mouse population (which is known to have mean $\mu = 86$ mm)? Assume that the standard deviation σ of all deer mice in this area is also 8 mm. Follow the four-step process, as illustrated in Example 14.9, in your answer.

14.50 IQ test scores. 4 STEP Example 14.3 gives the IQ test scores of 31 seventh-grade girls in a Midwest school district. IQ scores follow a Normal distribution with standard deviation $\sigma = 15$. Treat these 31 girls as an SRS of all seventh-grade girls in this district. IQ scores in a broad population are supposed to have mean $\mu = 100$. Is there evidence that the mean in this district differs from 100? Follow the four-step process, as illustrated in Example 14.9, in your answer.

14.51 Significance tests and confidence intervals. In Exercise 14.35 you obtained a 95% confidence interval for the mean body length μ of deer mice in a forest habitat. In Exercise 14.49 you used the same data to test the hypothesis that deer mice from this habitat have significantly different body lengths from the general population of deer mice. Compare the two results and explain why the confidence interval is more informative than the test P-value.

14.52 Significance tests and confidence intervals. In Example 14.3 we calculated a 95% confidence interval for the mean IQ μ of seventh-grade girls in a Midwest school district. In Exercise 14.50 you used the same data to test the hypothesis that this population has a mean IQ significantly different from 100. Compare your two results. What did you learn from the confidence interval? What did you learn from the test P-value? Describe the relative merits of both methods.

14.53 Significance tests and confidence intervals. To assess the accuracy of a laboratory scale, a reference weight known to weigh exactly 10 grams (g) is weighed repeatedly. The scale readings are Normally distributed with standard deviation $\sigma = 0.0002$ g. The reference weight is weighed 5 times on that scale. The mean result is 10.0023 g.

(a) Do the 5 weighings give good evidence that the scale is not well calibrated (that is, its mean μ for weighing this weight is not 10 g)?

(b) Give a 95% confidence interval for the mean weight on this scale for all possible measurements of the reference weight. What do you conclude about the calibration of this scale?

(c) Compare your results in parts (a) and (b). Explain why the confidence interval is more informative than the test result.

CHAPTER 15 Inference in Practice

IN THIS CHAPTER WE COVER...

- Conditions for inference in practice
- How confidence intervals behave
- How significance tests behave
- *Discussion: The scientific approach*
- Planning studies: sample size for confidence intervals*
- Planning studies: the power of a statistical test*

To this point, we have learned just two procedures for statistical inference. Both concern inference about the mean μ of a population when the "simple conditions" (page 336) are true: *The data are from an SRS, the population has a Normal distribution, and we know the standard deviation σ of the population.* Under these conditions, a confidence interval for the mean μ is

$$\bar{x} \pm z^* \frac{\sigma}{\sqrt{n}}$$

To test a hypothesis H_0: $\mu = \mu_0$, we use the one-sample z statistic:

$$z = \frac{\bar{x} - \mu_0}{\sigma/\sqrt{n}}$$

We call these **z procedures** because they both start with the one-sample z statistic and use the standard Normal distribution.

***z* procedures**

In later chapters we will modify these procedures for inference about a population mean to make them useful in practice. We will also introduce procedures for confidence intervals and tests in most of the settings we encountered in learning to explore data. There are libraries—both of books and of software—full of more elaborate statistical techniques. The reasoning of confidence intervals and tests is the same, no matter how elaborate the details of the procedure are.

There is a saying among statisticians that "mathematical theorems are true; statistical methods are effective when used with judgment." That the one-sample

z statistic has the standard Normal distribution when the null hypothesis is true is a mathematical theorem. Effective use of statistical methods requires more than knowing such facts.

This chapter begins the process of helping you develop the judgment needed to use statistics in practice. That process will continue in examples and exercises through the rest of this book.

Conditions for inference in practice

Any confidence interval or significance test can be trusted only under specific conditions. It's up to you to understand these conditions and judge whether they fit your problem. With that in mind, let's look back at the "simple conditions" for the z confidence interval and test.

The final "simple condition," that we know the standard deviation σ of the population, is rarely satisfied in practice. The z procedures are therefore of little practical use, especially in the life sciences. Fortunately, it's easy to do without the "known σ" condition. Chapter 17 shows how. The first two "simple conditions" (SRS, Normal population) are harder to escape. In fact, they are typical of the conditions needed if we are to trust the results of statistical inference. As you plan inference, you should always ask, "Where did the data come from?" and you must often also ask, "What is the shape of the population distribution?" This is the point where knowing mathematical facts gives way to the need for judgment.

Where did the data come from? *The most important requirement for any inference procedure is that the data come from a process to which the laws of probability apply*. Inference is most reliable when the data come from a probability sample or a randomized comparative experiment. Probability samples use chance to choose respondents. Randomized comparative experiments use chance to assign subjects to treatments. The deliberate use of chance ensures that the laws of probability apply to the outcomes, and this in turn ensures that statistical inference makes sense.

> **WHERE THE DATA COME FROM MATTERS**
>
> When you use statistical inference, you are acting as if your data are a probability sample or come from a randomized experiment.

If your data don't come from a probability sample or a randomized comparative experiment, your conclusions may be challenged. To answer the challenge, you must usually rely on subject-matter knowledge, not on statistics. It is common to apply statistics to data that are not produced by random selection. When you see such a study, ask whether the data can be trusted as a basis for the conclusions of the study.

EXAMPLE 15.1 The neurobiologist and the sociologist

A neurobiologist is interested in how our visual perception can be fooled by optical illusions. Her subjects are students in Neuroscience 101 at her university. Most neurobiologists would agree that it's safe to treat the students as an SRS of all people with normal vision. There is nothing unique about being a student that changes the neurobiology of visual perception.

A sociologist at the same university uses students in Sociology 101 to examine attitudes toward the use of human subjects in science. Students as a group are younger than the adult population as a whole. Even among young people, students as a group tend to come from more prosperous and better-educated homes. Even among students, this university isn't typical of all campuses. Even on this campus, students in a sociology course may have opinions that are quite different from those of engineering students or biology students. The sociologist can't reasonably act as if these students are a random sample from a population of interest. ■

Our first examples of inference, using the z procedures, act as if each data set is an SRS from the population of interest. Let's look back at the examples in Chapter 14.

EXAMPLE 15.2 Is it really an SRS?

The NHANES that produced the height data for Example 14.1 uses a complex multistage sample design, so it's a bit of an oversimplification to treat the height data as coming from an SRS from the population of eight-year-old boys.[1] The overall effect of the NHANES sample is close to an SRS, but professional statisticians would use more complex inference procedures to match the more complex design of the sample.

The 31 seventh-grade girls in the IQ scores study in Example 14.3 were randomly chosen from the population of seventh-grade girls in a Midwest school district. We can treat these girls as an SRS from the population of interest.

The study of inorganic phosphorus in Example 14.5 uses existing medical records from a random sample of 12 elderly men and women between the ages of 75 and 79 years. However, all these medical records came from the archives of a single laboratory in Ohio. This is a major problem for two reasons. First, individuals who have their blood tested in Ohio may not be representative of the whole population of Americans aged 74 to 79 years and of that population's overall susceptibility to various medical conditions. Second, different laboratories are known to produce slightly different blood test results. In fact, the secondary purpose of the study was to investigate just how different results from different laboratories can be (see Exercise 14.13, page 352). Collecting all 12 charts from the same laboratory means that our sample in Example 14.5 could be biased. These two issues should be clearly discussed when presenting the results of the statistical analysis because they severely limit our ability to generalize the findings. ■

These examples are fairly typical. Only one is an actual SRS of the population of interest. One represents a situation in which we assume an SRS might be used for a quick analysis of data from a more complex probability sample. The last one is an SRS taken from a small portion of the actual population of interest and may

not be appropriate to answer the study's question. *There is no simple rule for deciding when you can act as if a sample is an SRS. Pay attention to these cautions:*

- *Practical problems such as nonresponse in samples or dropouts from an experiment can hinder inference from even a well-designed study.* NHANES has about an 80% response rate. This is much higher than opinion polls and most other national surveys, so by realistic standards NHANES data are quite trustworthy. (NHANES uses advanced methods to try to correct for nonresponse, but these methods work a lot better when response is high to start with.)
- *Different methods are needed for different designs.* The z procedures aren't correct for probability samples more complex than an SRS. Later chapters give methods for some other designs, but we won't discuss inference for really complex designs like that used by NHANES. Always be sure that you (or your statistical consultant) know how to carry out the inference your design calls for.
- *There is no cure for fundamental flaws like voluntary response surveys, uncontrolled experiments, or biased samples.* Look back at the bad examples in Chapters 7 and 8 and steel yourself to just ignore data from such studies.

What is the shape of the population distribution? Most statistical inference procedures require some conditions on the shape of the population distribution. Many of the most basic methods of inference are designed for Normal populations. That's the case for the z procedures and also for the more practical procedures for inference about means that we will meet in Chapters 17 and 18. Fortunately, this condition is less essential than where the data come from.

This is true because the z procedures and many other procedures designed for Normal distributions are based on Normality of the sample mean $\bar{x}$, not Normality of individual observations. The central limit theorem tells us that $\bar{x}$ is more Normal than the individual observations and that $\bar{x}$ becomes more Normal as the size of the sample increases. In practice, the z procedures are reasonably accurate for any roughly symmetric distribution for samples of even moderate size. If the sample is large, $\bar{x}$ will be close to Normal even if individual measurements are strongly skewed, as Figure 13.3 (page 322) illustrates. Later chapters give practical guidelines for specific inference procedures.

There is one important exception to the principle that the shape of the population is less critical than how the data were produced. Outliers can distort the results of inference. *Any inference procedure based on sample statistics like the sample mean $\bar{x}$ that are not resistant to outliers can be strongly influenced by a few extreme observations.*

We rarely know the shape of the population distribution. In practice we rely on previous studies and on data analysis. Sometimes long experience suggests that our data are likely to come from a roughly Normal distribution, or not. For example, heights of people of the same sex and similar ages are close to Normal, but weights are not. Always explore your data before doing inference. When the data are chosen at random from a population, the shape of the data distribution mirrors the shape of the population distribution. Make a stemplot or a histogram of your data and look to see whether the shape is roughly Normal, or make a Normal quantile plot if you have access to technology. Remember that small samples have a lot of chance variation, so that Normality is hard to judge from just a few observations. Always look for outliers and try to correct them or justify their removal before performing the z procedures or other inference based on statistics like $\bar{x}$ that are not resistant. Read the discussion "Dealing with Outliers" in Chapter 2 (page 53) for detailed examples.

When outliers are present or the data suggest that the population is strongly non-Normal, consider alternative methods that don't require Normality and are not sensitive to outliers. Some of these methods appear in Chapter 27 (on the companion website).

Where are the details? Papers reporting scientific research must be short. Unfortunately, brevity allows researchers to hide details about their data. Did they choose their subjects in a biased way? Did they report data on only some of their subjects? Did they try several statistical analyses and report the one that looked best? The statistician John Bailar screened more than 4000 papers for the *New England Journal of Medicine.* He says, "When it came to the statistical review, it was often clear that critical information was lacking, and the gaps nearly always had the practical effect of making the authors' conclusions look stronger than they should have."

APPLY YOUR KNOWLEDGE

15.1 A medical journal takes a poll. The *New England Journal of Medicine* posts its peer-reviewed articles and editorials on its website. An opt-in poll was featured next to an editorial about the regulation of sugar-sweetened beverages. The poll asked, "Do you support government regulation of sugar-sweetened beverages?" You only needed to click on a response (yes or no) to become part of the sample. The poll stayed open for several weeks in October 2012. Of the 1290 votes cast, 864 were "yes" responses.[2]

(a) Would it be reasonable to calculate from these data a confidence interval for the percent answering "yes" in the American population? Explain your answer.

(b) Would it be reasonable to calculate from these data a confidence interval for the percent answering "yes" among online readers of the *New England Journal of Medicine*? Explain your answer.

15.2 Mammary artery ligation. Angina is the severe pain caused by inadequate blood supply to the heart. Perhaps we can relieve angina by tying off the mammary arteries to force the body to develop other routes to supply blood to the heart. Surgeons tried this procedure, called "mammary artery ligation." Patients reported a statistically significant reduction in angina pain. A randomized comparative experiment later showed that ligation was no more effective than a sham operation (placebo). Surgeons abandoned the procedure at once.[3] Explain how a study can have statistically significant results and nonetheless reach uninformative or even misleading conclusions.

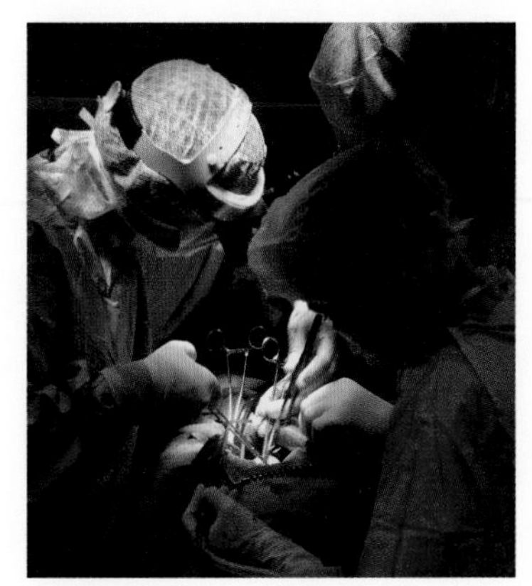
BSIP/Phototake

15.3 Sampling medical records. A retrospective study reported on complications from prostate needle biopsies by examining the medical records of 1000 consecutive cases of patients who underwent the procedure at one academic medical center between September 2001 and August 2010.[4] Why is it risky to regard these 1000 patients as an SRS from the population of all patients who have a prostate needle biopsy? Name some factors that might make these 1000 consecutive patient records unrepresentative of all prostate needle biopsy outcomes.

15.4 Exploratory data analysis. Tree shrews, *Tupaia belangeri*, are small omnivorous mammals phylogenetically related to primates. A research team examined paw preference during grasping tasks among 36 tree shrews born and raised in captivity.

Bruce Coleman Inc./Alamy

A "pawedness index" was computed as a unitless variable taking values between −1 and 1, depending on the relative use of the right and left paws, so that an animal always using its left paw would have a value of −1 but an animal using its right and left paws equally would have a value of 0. The individual pawedness index values are displayed below.[5]

−1.00	−1.00	−1.00	−1.00	−1.00	−0.96	−0.95	−0.89	−0.06
0.86	0.90	1.00	1.00	−1.00	−1.00	−1.00	−0.89	−0.25
0.87	0.94	1.00	1.00	−0.93	−0.03	−1.00	−1.00	−0.08
0.13	0.56	0.67	−0.11	0.24	0.50	1.00	1.00	0.41

Display these data in a dotplot and describe the distribution of pawedness index values. Explain why a confidence interval for the population mean pawedness index of tree shrews raised in captivity would not be very informative (assuming that we know the population standard deviation σ) and could easily be misinterpreted. Note that it is always a good idea to look carefully at your data before obtaining summary statistics and performing statistical inference procedures.

15.5 Shape of the population. In Exercise 14.6 (page 344) you calculated a 90% confidence interval for the mean blood pressure of executives at a large company based on an SRS of 72 executives from that company, assuming that blood pressures follow a Normal distribution with standard deviation $\sigma = 15$. Would you be willing to calculate a 90% confidence interval for μ if these blood pressures were strongly skewed? Explain your answer.

How confidence intervals behave

The z confidence interval $\bar{x} \pm z^*\sigma/\sqrt{n}$ for the mean of a Normal population illustrates several important properties that are shared by all confidence intervals in common use. The user chooses the confidence level, and the margin of error follows from this choice. We would like high confidence and also a small margin of error. High confidence says that our method almost always gives correct answers. A small margin of error says that we have pinned down the parameter quite precisely. The factors that influence the margin of error of the z confidence interval are typical of most confidence intervals.

How do we get a small margin of error? The margin of error for the z confidence interval is

$$\text{margin of error} = z^* \frac{\sigma}{\sqrt{n}}$$

This expression has z^* and σ in the numerator and $\sqrt{n}$ in the denominator. Therefore, the margin of error gets smaller when

- z^* gets smaller. A smaller z^* is the same as a lower confidence level C (look at Figure 14.4, page 341, again). *There is a trade-off between the confidence level and the margin of error. To obtain a smaller margin of error from the same data, you must be willing to accept lower confidence.*

- σ is smaller. The standard deviation σ measures the variation in the population. You can think of the variation among individuals in the population as noise that obscures the average value μ. It is easier to pin down μ when σ is small.
- n gets larger. Increasing the sample size n reduces the margin of error for any confidence level. Larger samples thus allow more precise estimates. However, *because n appears under a square root sign, we must take four times as many observations in order to cut the margin of error in half.*

EXAMPLE 15.3 Confidence level and margin of error

In Example 14.3 (page 343) we examined the IQ scores of 31 seventh-grade girls in a Midwest school district. The data gave $\bar{x} = 105.84$, and we know that $\sigma = 15$. The 95% confidence interval for the mean IQ score for all seventh-grade girls in this district is

$$\bar{x} \pm z^* \frac{\sigma}{\sqrt{n}} = 105.84 \pm 1.960 \frac{15}{\sqrt{31}}$$
$$= 105.84 \pm 5.28$$

The 90% confidence interval based on the same data replaces the 95% critical value $z^* = 1.960$ with the 90% critical value $z^* = 1.645$. This interval is

$$\bar{x} \pm z^* \frac{\sigma}{\sqrt{n}} = 105.84 \pm 1.645 \frac{15}{\sqrt{31}}$$
$$= 105.84 \pm 4.43$$

Lower confidence results in a smaller margin of error, ± 4.43 in place of ± 5.28. In the same way, you can calculate that the margin of error for 99% confidence is larger, ± 6.94. Figure 15.1 compares these three confidence intervals.

You can check that if we had a sample of 62 girls, the margin of error for 95% confidence would decrease from ± 5.28 to ± 3.73. Doubling the sample size does *not* cut the margin of error in half, because the sample size n appears under a square root sign. Also keep in mind that a sample of a different size would most likely produce a different sample average and thus change both the center and the width of the 95% confidence interval. ■

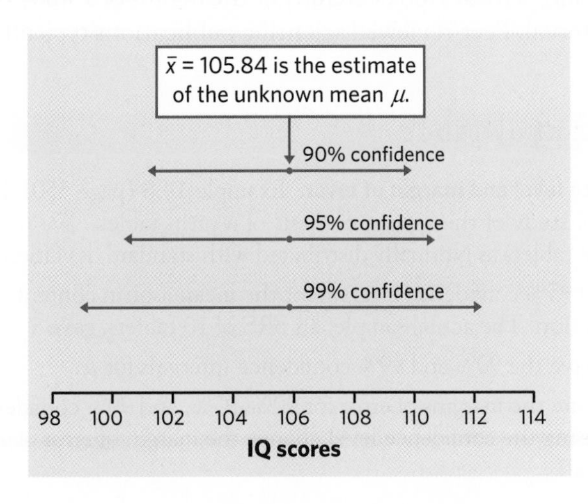

FIGURE 15.1 The lengths of three confidence intervals, for Example 15.3. All three are centered at the estimate $\bar{x} = 105.84$. When the data and the sample size remain the same, higher confidence requires a larger margin of error.

The margin of error accounts only for sampling error The most important caution about confidence intervals in general is a consequence of the use of a sampling distribution. A sampling distribution shows how a statistic such as $\overline{x}$ varies in repeated random sampling. This variation causes "random sampling error" because the statistic misses the true parameter by a random amount. No other source of variation or bias in the sample data influences the sampling distribution. So *the margin of error in a confidence interval ignores everything except the sample-to-sample variation due to choosing the sample randomly*.

> **THE MARGIN OF ERROR DOESN'T COVER ALL ERRORS**
>
> The margin of error in a confidence interval covers only random sampling errors.
>
> Practical difficulties such as undercoverage and nonresponse are often more serious than random sampling error. The margin of error does not take such difficulties into account.

Recall from Chapter 7 that national opinion polls often have response rates much less than 50% and that even small changes in the wording of questions can strongly influence results. In such cases, the announced margin of error is probably unrealistically small. The Gallup Organization always includes the following statement in its detailed survey results: "In addition to sampling error, question wording and practical difficulties in conducting surveys can introduce error or bias into the findings of public opinion polls." Unfortunately, this important cautionary statement is routinely omitted by news organizations.

Experimental studies also have their set of challenges, from producing random samples to problems with attrition. Clinical trials, for instance, can enroll only volunteers and, therefore, do not truly sample from the whole population of interest. In addition, subjects may move away and decide to drop out of the study, or they may be removed because of health concerns. Any systematic bias in the attrition process would not be taken into account in the margin of error of a simple inference procedure. Always look carefully at the details of a study before you trust a confidence interval. Peer-reviewed scientific publications typically address these issues in detail.

APPLY YOUR KNOWLEDGE

15.6 Confidence level and margin of error. Example 14.8 (page 350) described a quality control study of the aspirin content of aspirin tablets. We treated the aspirin content of tablets as Normally distributed with standard deviation $\sigma = 5$ mg.

(a) Give a 95% confidence interval for the mean aspirin content μ in this population. The actual sample, an SRS of 10 tablets, gave $\overline{x} = 326.9$ mg.

(b) Now give the 90% and 99% confidence intervals for μ.

(c) What are the margins of error for 90%, 95%, and 99% confidence? How does increasing the confidence level change the margin of error of a confidence interval?

15.7 Sample size and margin of error. In the previous exercise you calculated a 95% confidence interval for the population mean aspirin content μ based on the sample from Example 14.8.

(a) Suppose that the SRS had only 4 tablets. Obviously, the sample average would be specific to that particular sample. What would be the margin of error of a 95% confidence interval for the population mean μ in this case?

(b) Suppose that the SRS had 25 tablets. What would be the margin of error of a 95% confidence interval for μ?

(c) Compare the margins of error for samples of size 4, 10, and 25. How does increasing the sample size change the margin of error of a confidence interval?

15.8 Spending on food. A 2012 Gallup survey based on a random sample of 1014 American adults indicates that American families spend, on average, \$151 per week on food. The report further states that, with 95% confidence, this estimate has a margin of error of ±\$7. Which of the following sources of error are included in the margin of error?

(a) The poll dialed telephone numbers at random and so missed all people without phones.

(b) Nonresponse—some people whose phone numbers were chosen never answered the phone in several calls or answered but refused to participate in the poll.

(c) There is chance variation in the random selection of telephone numbers.

How significance tests behave

Significance tests are widely used in reporting the results of research in many fields of applied science and in industry. New pharmaceutical products require significant evidence of effectiveness and safety. Courts inquire about statistical significance in hearing class action lawsuits (many of which concern pharmaceutical products or biohazards in the workplace). Marketers want to know whether a new ad campaign significantly outperforms the old one, and medical researchers want to know whether a new therapy performs significantly better. In all these uses, statistical significance is valued because it points to an effect that is unlikely to occur simply by chance. Here are some points to keep in mind when using or interpreting significance tests.

Should tests be banned? Significance tests don't tell us how large or how important an effect is. Research in psychology has emphasized tests, so much so that some think their weaknesses should ban them from use. The American Psychological Association asked a group of experts. They said, "Use anything that sheds light on your study. Use more data analysis and confidence intervals." But: "The task force does not support any action that could be interpreted as banning the use of null hypothesis significance testing or *P*-values in psychological research and publication."

How small a *P* is convincing? The purpose of a test of significance is to describe the degree of evidence provided by the sample against the null hypothesis. The *P*-value does this. But how small a *P*-value is convincing evidence against the null hypothesis? This depends mainly on two circumstances:

- *How plausible is H_0?* If H_0 represents an assumption that the people you must convince have believed for years, strong evidence (small *P*) will be needed to persuade them.
- *What are the consequences of rejecting H_0?* If rejecting H_0 in favor of H_a means making an expensive changeover from one therapy to another, you need

strong evidence that the new treatment will save or improve lives. On the other hand, preliminary studies do not require taking any action beyond deciding whether or not to pursue the topic. It makes sense to continue work in light of even weakly significant evidence.

These criteria are a bit subjective. Different people will often insist on different levels of significance. Users of statistics have often emphasized standard levels of significance such as 5% and 1%. However, *there is no sharp border between "significant" and "insignificant," only increasingly strong evidence as the P-value decreases. It makes no sense to treat $P \leq 0.05$ as a universal rule for what is statistically significant.* Giving the P-value allows each of us to decide individually if the evidence is sufficiently strong.

Significance depends on the alternative hypothesis You may have noticed that the P-value for a one-sided test is one-half the P-value for the two-sided test of the same null hypothesis based on the same data. The two-sided P-value combines two equal areas, one in each tail of a Normal curve. The one-sided P-value is just one of these areas, in the direction specified by the alternative hypothesis. It makes sense that the evidence against H_0 is stronger when the alternative is one-sided, because the evidence is based on the data *plus* information about the direction of possible deviations from H_0. If you lack this added information, always use a two-sided alternative hypothesis.

Significance depends on sample size A sample survey shows that significantly fewer students are heavy drinkers at colleges that ban alcohol on campus. "Significantly fewer" is not enough information to decide whether there is an important difference in drinking behavior at schools that ban alcohol. *How important an effect is depends on the size of the effect as well as on its statistical significance.* If the number of heavy drinkers is only 1 percentage point less at colleges that ban alcohol than at other colleges, this is not an important effect even if it is statistically significant. In fact, the sample survey found that 38% of students at colleges that ban alcohol are "heavy episodic drinkers," compared with 48% at other colleges.[6] That difference is large enough to be important. (Of course, this observational study doesn't prove that an alcohol ban directly reduces drinking; it may be that colleges that ban alcohol attract more students who don't want to drink heavily.)

Such examples remind us to always look at the size of an effect (like 38% versus 48%) as well as its statistical significance. They also raise a question: Can a tiny effect really be highly significant? Yes. To see why, think of the z test statistic as follows:

$$z = \frac{\bar{x} - \mu_0}{\sigma/\sqrt{n}} = \frac{\text{size of the observed effect}}{\text{size of chance variation}}$$

The numerator $\bar{x} - \mu_0$ shows how far the data diverge from the null hypothesis. How significant this effect is depends on the size of the chance variation from sample to sample, measured by the standard deviation of $\bar{x}$, which is the denominator of z.

SAMPLE SIZE AFFECTS STATISTICAL SIGNIFICANCE

Because large random samples have small chance variation, very small population effects can be highly significant if the sample is large.

Because small random samples have a lot of chance variation, even large population effects can fail to be statistically significant if the sample is small.

Statistical significance does not tell us whether an effect is large enough to be important. That is, **statistical significance is not the same thing as practical significance.**

Keep in mind that statistical significance means "the sample showed an effect larger than would often occur just by chance." Significance says nothing about how large the effect actually is. *Always look at the actual size of an effect as well as at its statistical significance*. A confidence interval is often helpful, as is a good understanding of the subject matter.

EXAMPLE 15.4 Coaching and SAT scores

On average, high school juniors who retake the SAT exam as seniors improve their Mathematics score by 13 points. Professor Gamma has designed a rigorous coaching program that she thinks will produce a larger increase. A group of juniors goes through this program and retakes the SAT as seniors. Their scores improve by an average of $\bar{x} = 15$ points. Is this significant evidence that the coaching program raises scores by more than 13 points on the average?

We'll assume the simple conditions: The students in the program are an SRS of the population of all juniors who take the SAT, and the distribution of Math score gains in the entire population is Normal with unknown mean μ and known standard deviation $\sigma = 30$. For testing

$$H_0: \mu = 13$$
$$H_a: \mu > 13$$

the z test statistic is

$$z = \frac{\bar{x} - \mu_0}{\sigma/\sqrt{n}} = \frac{15 - 13}{30/\sqrt{n}}$$

The gain of 2 points is not very impressive. But how statistically significant it is depends on the sample size n:

Sample size	z statistic	P-value
50	0.4714	0.3187
100	0.6667	0.2525
500	1.4907	0.0680
1000	2.1082	0.0175
5000	4.7140	0.000001

Even such a very small effect can be highly significant if the sample size is very large. ■

Beware of multiple analyses Statistical significance ought to mean that you have found an effect that you were looking for. The reasoning behind statistical significance works well if you decide what effect you are seeking, design a study to search for it, and use a test of significance to weigh the evidence you get. In other settings, significance may have little meaning.

W. G. Murray/age fotostock

EXAMPLE 15.5 Cell phones and brain cancer

Might the radiation from cell phones be harmful to users? Many studies have found little or no connection between using cell phones and various illnesses. Here is part of a news account of one study:

> *A hospital study that compared brain cancer patients and a similar group without brain cancer found no statistically significant association between cell phone use and a group of brain cancers known as gliomas. But when 20 types of glioma were considered separately an association was found between phone use and one rare form. Puzzlingly, however, this risk appeared to decrease rather than increase with greater mobile phone use.*[7]

Think for a moment: Suppose that the 20 null hypotheses (no association) for these 20 significance tests are all true. Then each test has a 5% chance of being significant at the 5% level. That's what $\alpha = 0.05$ means: Results this extreme occur 5% of the time just by chance when the null hypothesis is true. Because 5% is 1/20, we expect about 1 of 20 tests to give a significant result just by chance. That's what the study observed. ■

Running one test and reaching the 5% level of significance is reasonably good evidence that you have found something. Running 20 tests and reaching that level only once is not. A related mistake happens when an experiment is performed under a number of different conditions but the experimenter decides to run a test of significance only on the most extreme results. This is still equivalent to performing multiple analyses.

The caution about multiple analyses applies to confidence intervals as well. A single 95% confidence interval has probability 0.95 of capturing the true parameter each time you use it. The probability that all of 20 confidence intervals will capture their parameters is much less than 95%. If you think that multiple tests or intervals may have discovered an important effect, you need to gather new data to do inference about that specific effect.

APPLY YOUR KNOWLEDGE

15.9 Detecting acid rain. Emissions of sulfur dioxide by industry set off chemical changes in the atmosphere that result in "acid rain." The acidity of liquids is measured by pH on a scale of 0 to 14. Distilled water has pH 7.0, and lower pH values indicate acidity. Typical rain is somewhat acidic, so acid rain is defined as rainfall with a pH below 5.0.

Suppose that pH measurements of rainfall on different days in a Canadian forest follow a Normal distribution with standard deviation $\sigma = 0.5$. A sample of

n days finds that the mean pH is $\bar{x} = 4.8$. Is this good evidence that the mean pH μ for all rainy days is less than 5.0? The answer depends on the size of the sample. Use the *P-Value of a Test of Significance* applet or technology for your computations. (If not, do the calculations by hand. Your results may differ slightly from those obtained with technology due to rounding error.)

(a) Enter H_0: $\mu = 5$, H_a: $\mu < 5$, $\sigma = 0.5$, and $\bar{x} = 4.8$. Then enter $n = 5$, $n = 15$, and $n = 40$ one after the other, clicking "UPDATE" each time to get the three P-values. What are they?

(b) Sketch the three Normal curves displayed by the applet, with $\bar{x} = 4.8$ marked on each curve. Explain why the P-value of the same result $\bar{x} = 4.8$ is smaller (more statistically significant) for larger sample sizes.

15.10 Is it statistically significant? A study examined dairy cows during pregnancy and after giving birth ("calving") and reported the following:

The effect of time was almost significant at the 5% level ($P = 0.058$), with the mean locomotion scores tending to be higher after calving than in late pregnancy.[8]

(a) Why is it more informative to report the actual P-value than to simply reject or fail to reject H_0 at a chosen significance level?

(b) Explain what "almost significant at the 5% level" means.

15.11 Confidence intervals help. Give a 95% confidence interval for the mean pH μ for each value of n used in Exercise 15.9. The intervals, unlike the P-values, give a clear picture of what mean pH values are plausible for each sample.

15.12 Treating shortness of breath. Dyspnea, or shortness of breath, is a common complaint in patients with chronic obstructive pulmonary disease (COPD). It is often assessed by an FEV_1 test measuring the forced expiratory volume in the first second. A study examined the effect of various treatments on perceived dyspnea in patients with advanced COPD. The researchers reported that, for the 23 patients who had received 6 weeks of therapy with a long-acting bronchodilator, "there was a small, statistically insignificant, increase in FEV_1."[9]

(a) What does "statistically insignificant" mean?

(b) Why is it important to know that the effect was small in size as well as insignificant?

15.13 Searching for ESP. A researcher looking for evidence of extrasensory perception (ESP) tests 500 subjects. Four of these subjects do significantly better ($P < 0.01$) than random guessing.

(a) Is it proper to conclude that these 4 people have ESP? Explain your answer.

(b) What should the researcher now do to test whether any of these 4 subjects have ESP?

15.14 More on cell phones and brain cancer. The INTERPHONE study is an international case-control study of individuals with brain cancer (2708 with glioma and 2409 with meningioma) and matched controls without brain cancer. The researchers compared the rates of each of these 2 types of brain cancer among subgroups of individuals, based on a number of factors: amount and history of regular cell phone use (8 levels), cell phone type (2 levels), and brain location of the cancer (5 levels). In all, the same inference procedure was therefore repeated 160 times. The statistical analysis showed a small significant "protective" effect of cell phone use in 7 groups, and a moderate significant "harmful" effect of cell phone use in 2 groups (the researchers also noted that some individuals in these two groups provided unrealistically high reports of cell phone use).[10] Explain why, using a significance level of 0.05, it shouldn't be surprising to obtain statistical significance in some of the 160 identical inference procedures. How does this affect your interpretation of the study findings?

DISCUSSION The scientific approach: hypotheses, study design, data collection, and statistical analysis

Citing a P-value is not the ultimate goal and focus point of a research study. In fact, a P-value obtained from a poor design means little whether it is statistically significant or not. A test of significance is only part of the whole scientific approach. Good scientific methodology follows a four-pronged approach: hypotheses, study design, data collection, and statistical analysis. Statistical analysis can lead to valuable conclusions only when the rest of the approach is sound.

So, how do we get there? Hypotheses arise in complicated ways from the results of past studies, preliminary data, theory, or even simply an educated guess. They must be specific enough to allow studies to be designed to confirm or refute them. Under appropriate circumstances, such as random sample selection or large enough samples, statistical inference can then be applied and sound conclusions reached. Here is an outstanding example of scientific method.

Uncovering the cause of peptic ulcers[11]

It was once commonplace to assume that stress, an anxious personality, and even spicy food caused peptic ulcers, painful lifelong lesions of the stomach or duodenum for which only temporary relief was available. Several reports had been made of unusual, curved bacteria found in biopsies of patients suffering from peptic ulcers. These were undeniable observations, but at the time, most researchers and physicians believed that nothing could survive very long in the stomach because of its extremely acidic environment (pH between 2 and 4). Thus, the reports had been generally ignored—except by Australian researchers Robin Warren and Barry Marshall. They thought that it was worth exploring a possible link between the curved bacteria and peptic ulcers, however counterintuitive it might be. They had a working hypothesis.

How can you test the hypothesis that a given bacterium causes peptic ulcers? It would obviously not be ethical to infect people with the bacteria and see if they developed a peptic ulcer. (Interestingly, that's exactly what Marshall did to himself later on, after he and Warren had developed a treatment against the bacteria.) The only real research option at this stage, though, was to conduct an observational study. They examined biopsies of the stomach lining from a random sample of patients given an endoscopy referral for any number of problems along the upper digestive tract. Statistical analysis showed that the bacteria were found significantly more often in patients with peptic ulcers than in patients with other digestive disorders. This clearly established a link between the curved bacteria and peptic ulcers, an important preliminary step in determining a possible causal effect.

Establishing causality would need direct experimentation, however, and ethically this would require offering possible treatments. If the bacteria caused the peptic ulcers, then a treatment killing them or inhibiting their growth should relieve the symptoms. This logic gave Warren and Marshall a second hypothesis to test. They therefore designed a double-blind experiment in which patients with an established diagnosis of peptic ulcer received one of several treatments or a placebo. They found that patients whose treatment completely eradicated the bacteria did significantly better over both the short term and the long term, with very few relapses even many years later. The experiment showed that the presence of the bacteria was indeed responsible, at least in part, for the peptic ulcers. It also showed that the majority of peptic ulcers could be successfully treated, thus preventing a lifetime of suffering in these patients.

How these curved bacteria, *Helicobacter pylori*, could survive in the acidic environment of the stomach and duodenum was still a mystery that needed to be resolved to undeniably establish the bacterial role in peptic ulcers. Studying *H. pylori* showed that it produces urease, an enzyme that had already been found in the gastric mucosa of animals. Urease breaks down urea into carbon dioxide and ammonia, resulting in a higher (more basic) pH in solution. Marshall and colleagues hypothesized that *H. pylori* might use urease to survive in the extremely acidic gastric environment by neutralizing the pH locally. To test this hypothesis, they compared the survival, in environments with various pH levels, of urease-producing *H. pylori* and two other types of bacteria with no or limited capacity to produce urease. In the absence of urea, none of the three bacteria types survived in significant numbers at pH below 3. When urea was added to the solution, however, *H. pylori* survived in significant numbers at pH levels as low as 1.5, whereas the other two types still couldn't survive below pH 3. This experiment showed that *H. pylori* possessed a mechanism enabling it to survive in the stomach long enough to have a lasting impact. It also paved the way for a new, fast, and noninvasive method to assess the presence of *H. pylori* in patients with gastric problems simply by having patients swallow a solution of carbon-labeled urea and monitoring signs of labeled carbon dioxide in their breath shortly afterward.

These are only three of the many studies Warren and Marshall conducted on the link between *H. pylori* and peptic ulcers over two decades. In 2005, they were awarded the Nobel Prize in Medicine "for their discovery of the bacterium *Helicobacter pylori* and its role in gastritis and peptic ulcer disease." More about this topic can be found on the Nobel Prize website at `nobelprize.org/nobel_prizes/medicine/laureates/2005`.

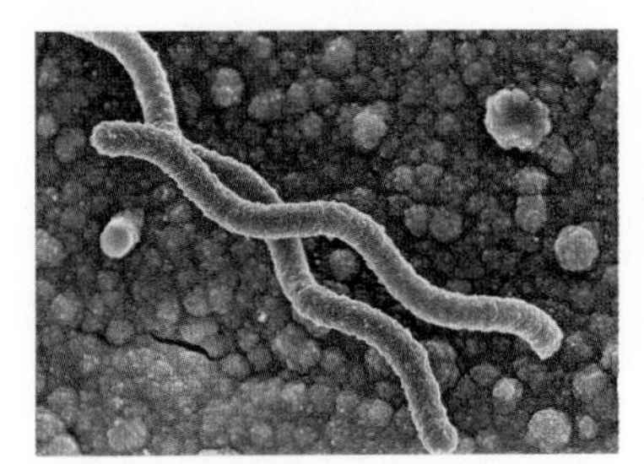

Dr. Gary Gaugler/Science Source

Planning studies: sample size for confidence intervals*

A wise user of statistics never plans a sample or an experiment without at the same time planning the inference. The number of observations is a critical part of planning a study. Larger samples give smaller margins of error in confidence intervals and make significance tests better able to detect effects in the population. But taking observations costs both time and money. How many observations are enough? We will look at this question first for confidence intervals and then for tests.

You can arrange to have both high confidence and a small margin of error by taking enough observations. The margin of error of the confidence interval for the mean of a Normally distributed population is $m = z^*\sigma/\sqrt{n}$. To obtain a desired margin of error m, put in the value of z^* for your desired confidence level, and solve for the sample size n. Here is the result.

> **SAMPLE SIZE FOR DESIRED MARGIN OF ERROR**
>
> The confidence interval for the mean of a Normal population will have a specified margin of error m when the sample size is
>
> $$n = \left(\frac{z^*\sigma}{m}\right)^2$$

Notice that it is the size of the sample that determines the margin of error. The size of the population does not influence the sample size we need. (This is true as long as the population is much larger than the sample.)

EXAMPLE 15.6 How many observations?

Example 14.10 (page 356) reports a study of the mean body temperature of healthy adults. We know that the population standard deviation is $\sigma = 0.6$ degrees Fahrenheit (°F). We want to estimate the mean body temperature μ for healthy adults within ± 0.05°F with 95% confidence. How many healthy adults must we measure?

The desired margin of error is $m = 0.05$°F. For 95% confidence, Table C gives $z^* = 1.96$. We know that $\sigma = 0.6$°F. Therefore,

$$n = \left(\frac{z^*\sigma}{m}\right)^2 = \left(\frac{1.96 \times 0.6}{0.05}\right)^2 = 553.2$$

Because 553 healthy adults will give a slightly larger margin of error than desired, and 554 healthy adults a slightly smaller margin of error, we must recruit 554 healthy adults. *Always round up to the next higher whole number when finding n.* ■

*The remainder of this chapter presents more advanced material that is not needed to read the rest of the book. The idea of planning studies is, however, very important in practice.

APPLY YOUR KNOWLEDGE

15.15 How tall are children these days? Example 14.1 (page 336) assumed that the heights of all American eight-year-old boys are Normally distributed with standard deviation $\sigma = 10$ cm. How large a sample would be needed to estimate the mean height μ in this population to within ± 1 cm with 95% confidence?

15.16 Estimating mean IQ. IQ scores are typically Normally distributed within a homogeneous population. How large a sample of schoolgirls in Example 14.3 (page 343) would be needed to estimate the mean IQ score μ within ± 5 points with 99% confidence, assuming that $\sigma = 15$ in this population?

15.17 Pharmaceutical production. Exercise 14.5 (page 344) described the manufacturing process of a pharmaceutical product. Repeated measurements follow a Normal distribution with mean μ equal to the true product concentration and standard deviation $\sigma = 0.0068$ g/l.

(a) How many measurements would be needed to estimate the true concentration within ± 0.001 g/l with 95% confidence?

(b) How many measurements would be needed to estimate the true concentration within ± 0.001 g/l with 99% confidence? What is the implication of choosing a higher confidence level?

15.18 Arsenic blood level. Arsenic is a poisonous compound naturally present in soil and water at very low levels. Arsenic blood concentrations in healthy individuals are Normally distributed with standard deviation $\sigma = 1.5$ micrograms per deciliter (μg/dl).[12]

(a) To assess the mean arsenic blood level μ of a population living in a defined geographical area, how many individuals should have their blood tested so that we can estimate μ to within ± 0.5 μg/dl with 95% confidence?

(b) How many would be needed to estimate μ to within ± 0.1 μg/dl with 95% confidence? What is the implication of requiring a smaller margin of error?

Planning studies: the power of a statistical test*

How large a sample should we take when we plan to carry out a test of significance? We know that if our sample is too small, even large effects in the population will often fail to give statistically significant results. Here are the questions we must answer to decide how large a sample we must take:

> **Significance level.** How much protection do we want against getting a statistically significant result from our sample when there really is no effect in the population?
>
> **Effect size.** How large an effect in the population is important in practice?
>
> **Power.** How confident do we want to be that our study will detect an effect of the size we think is important?

*This section is optional.

The three boldface terms are statistical shorthand for three pieces of information. *Power* is a new idea.

Bloomberg via Getty Images

EXAMPLE 15.7 Sweetening colas: planning a study

A quality control engineer plans a study to assess the impact of storage on the sweetness of a new cola. Ten trained tasters are available to rate the cola's sweetness on a 10-point scale before and after storage, so that each taster's judgment of loss of sweetness can be assessed. Industry records indicate that sweetness loss scores vary from taster to taster according to a Normal distribution with standard deviation about $\sigma = 1$. To see if the taste test gives reason to think that the cola does lose sweetness, the engineer will test

$$H_0: \mu = 0$$

$$H_a: \mu > 0$$

Are 10 tasters enough, or should more be used?

Significance level. Requiring significance at the 5% level is enough protection against declaring there is a loss in sweetness when in fact no change would be found if we could look at the entire population. This means that when there is no change in sweetness in the population, 1 out of 20 samples of tasters will wrongly find a statistically significant loss.

Effect size. A mean sweetness loss of 0.8 point on the 10-point scale will be noticed by consumers and so is important in practice.

Power. The engineer needs to be 90% confident that the test will detect a mean loss of 0.8 point in the population of all tasters. A significance level of 5% is standard in the food industry for detecting an effect. So the objective is to have probability at least 0.9 that a test at the $\alpha = 0.05$ level will reject the null hypothesis $H_0: \mu = 0$ when the true population mean is $\mu = 0.8$. ■

The probability that the test successfully detects a sweetness loss of the specified size is the *power* of the test. You can think of tests with high power as being highly sensitive to deviations from the null hypothesis. In Example 15.7, the engineer decided on power 90% when the truth about the population is $\mu = 0.8$.

POWER

The **power** of a test against a specific alternative is the probability that the test will reject H_0 at a chosen significance level α when the specified alternative value of the parameter is true.

For most statistical tests, calculating power is a job for comprehensive statistical software. Finding it for the z test is easier, but we will nonetheless skip the details. The two following examples illustrate two approaches: (1) an applet that shows the meaning of power and (2) statistical software. Exercise 15.22 walks you through the details of the calculations.

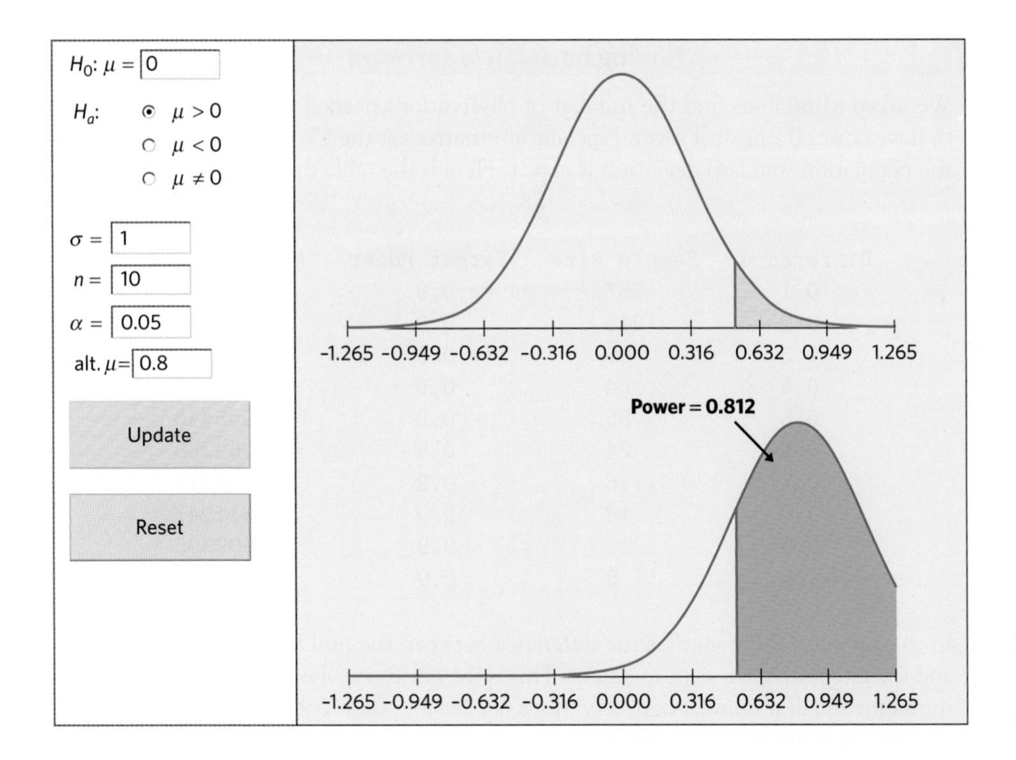

FIGURE 15.2 Output from the *Statistical Power* applet, for Example 15.8. The power of the test H_0: $\mu = 0$ against the specific alternative H_a: $\mu = 0.8$ is 0.812.

EXAMPLE 15.8 Finding power: use an applet

Finding the power of the z test is less challenging than most other power calculations because it requires only a Normal distribution probability calculation. The *Statistical Power* applet does this and illustrates the calculation with Normal curves. Enter the information from Example 15.7 into the applet: hypotheses, population standard deviation $\sigma = 1$, sample size $n = 10$, significance level $\alpha = 0.05$, and alternative $\mu = 0.8$. Click "UPDATE." The applet output appears in Figure 15.2.

The power of the test against the specific alternative $\mu = 0.8$ is 0.812. That is, the test will reject H_0 about 81% of the time when this alternative is true. So 10 observations are too few to give power 90%. ■

The two Normal curves in Figure 15.2 show the sampling distribution of $\bar{x}$ under the null hypothesis $\mu = 0$ (top) and also under the specific alternative $\mu = 0.8$ (bottom). The curves have the same shape because $\sigma/\sqrt{n}$ does not change. The top curve is centered at $\mu = 0$, and the bottom curve at $\mu = 0.8$. The shaded region at the right of the top curve has area 0.05. It marks off values of $\bar{x}$ that are statistically significant at the $\alpha = 0.05$ level. The lower curve shows the probability of these same values when $\mu = 0.8$. This area is the power, 0.812.

The applet will find the power for any given sample size. It's more helpful in practice to turn the process around and learn what sample size we need to achieve a given power. Statistical software will do this, but it usually doesn't show the helpful Normal curves that are part of the applet's output.

EXAMPLE 15.9 Finding power: use software

We asked Minitab to find the number of observations needed for the one-sided z test to have power 0.9 against several specific alternatives at the 5% significance level when the population standard deviation is $\sigma = 1$. Here is the table that results:

Difference	Sample Size	Target Power	Actual Power
0.1	857	0.9	0.900184
0.2	215	0.9	0.901079
0.3	96	0.9	0.902259
0.4	54	0.9	0.902259
0.5	35	0.9	0.905440
0.6	24	0.9	0.902259
0.7	18	0.9	0.907414
0.8	14	0.9	0.911247
0.9	11	0.9	0.909895
1.0	9	0.9	0.912315

In this output, "Difference" is the difference between the null hypothesis value $\mu = 0$ and the alternative we want to detect. This is the effect size. The "Sample Size" column shows the smallest number of observations needed for power 0.9 against each effect size.

We see again that a sample of 10 tasters is not large enough to be 90% confident of detecting (at the 5% significance level) an effect of size 0.8. To reach at least 90% power against effect size 0.8, at least 14 tasters are needed. The actual power with 14 tasters is 0.911247.

Statistical software, unlike the applet, will do power calculations for most of the tests in this book. ■

The table in Example 15.9 makes it clear that smaller effects require larger samples to reach 90% power. Here is an overview of influences on the sample size needed.

- If you insist on a smaller significance level (such as 1% rather than 5%), you will need a larger sample. A smaller significance level requires stronger evidence to reject the null hypothesis.
- If you insist on higher power (such as 99% rather than 90%), you will need a larger sample. Higher power gives a better chance of detecting an effect when it is really there.
- At any significance level and desired power, a two-sided alternative requires a larger sample than a one-sided alternative.
- At any significance level and desired power, detecting a small effect requires a larger sample than detecting a large effect.

Fish, fishermen, and power Are the stocks of cod in the ocean off eastern Canada declining? Studies over many years failed to find significant evidence of a decline. These studies had low power—that is, they might fail to find a decline even if one was present. When it became clear that the cod were vanishing, quotas on fishing ravaged the economy in parts of Canada. If the earlier studies had had high power, they would likely have seen the decline. Quick action might have reduced the economic and environmental costs.

Planning a serious statistical study always requires an answer to the question "How large a sample do I need?" If you intend to test the hypothesis $H_0\colon \mu = \mu_0$ about the mean μ of a population, you need at least a rough idea of the size of the population standard deviation σ and of how big a deviation $\mu - \mu_0$ of the population mean from its hypothesized value you want to be able to detect. More elaborate settings, such as comparing the mean effects of several treatments, require more elaborate advance information. You can leave the details to experts, but you should understand the idea of power and the factors that influence how large a sample you need.

To calculate the power of a test, we act as if we are interested in a fixed level of significance such as $\alpha = 0.05$. That's essential to do a power calculation, but remember that in practice we think in terms of P-values rather than a fixed level α. To effectively plan a new study you should find the power for a range of sample sizes and effect sizes to get a full picture of how the test will behave.

Type I and Type II errors in significance tests. We can assess the performance of a test by giving two probabilities: the significance level α and the power for an alternative that we want to be able to detect. The significance level of a test is the probability of making the *wrong* decision when the null hypothesis is true. The power for a specific alternative is the probability of making the *right* decision when that alternative is true. We can just as well describe the test by giving the probability of a *wrong* decision under both conditions.

TYPE I AND TYPE II ERRORS

If we reject H_0 when in fact H_0 is true, this is a **Type I error.**

If we fail to reject H_0 when in fact H_a is true, this is a **Type II error.** The probability of a Type II error is called β.

The **significance level** α of any fixed level test is the probability of a Type I error.

The **power** of a test against any alternative is 1 minus the probability of a Type II error for that alternative: power $= 1 - \beta$.

The possibilities are summed up in Figure 15.3. If H_0 is true, our decision is correct if we fail to reject H_0 and is a Type I error if we reject H_0. If H_a is true, our decision is either correct or a Type II error. Only one error is possible at one time.

		Truth about the population	
		H_0 true	H_a true
Decision based on sample	Reject H_0	Type I error (probability, α)	Correct decision
	Fail to reject H_0	Correct decision	Type II error (probability, β)

FIGURE 15.3 The two types of error in testing hypotheses.

EXAMPLE 15.10 Calculating error probabilities

Because the probabilities of the two types of error are just a rewording of significance level and power, we can see from Figure 15.2 what the error probabilities are for the test in Example 15.7.

$$\begin{aligned} P(\text{Type I error}) &= P(\text{reject } H_0 \text{ when in fact } \mu = 0) \\ &= \text{significance level } \alpha = 0.05 \\ P(\text{Type II error}) &= P(\text{fail to reject } H_0 \text{ when in fact } \mu = 0.8) \\ &= 1 - \text{power} = 1 - 0.812 = 0.188 \end{aligned}$$

The two Normal curves in Figure 15.2 are used to find the probabilities of a Type I error (top curve, $\mu = 0$) and of a Type II error (bottom curve, $\mu = 0.8$). ■

APPLY YOUR KNOWLEDGE

15.19 What is power? Exercise 14.17 (page 356) described a study of water quality that tested the following hypotheses about the mean oxygen content of a stream:

$$H_0\colon \mu = 5 \text{ mg/l versus } H_a\colon \mu < 5 \text{ mg/l}$$

Forty-five water specimens were taken from randomly chosen locations. We assume that dissolved oxygen varies among sampling locations according to a Normal distribution with $\sigma = 0.92$ mg/l. A statistician tells you that the power of the z test with $\alpha = 0.05$ against the alternative that the true mean oxygen content is $\mu = 4.75$ mg/l is 0.57. Explain in simple language what "power = 0.57" means.

15.20 Thinking about power. Answer these questions in the setting of the previous exercise about the oxygen content of a stream.

(a) To get higher power against the same alternative with the same α, what must we do?

(b) If we decide to use $\alpha = 0.10$ in place of $\alpha = 0.05$, does the power increase or decrease?

(c) If we shift our interest to the alternative $\mu = 4.5$ mg/l and change nothing else, does the power increase or decrease?

15.21 Detecting acid rain: power. Exercise 15.9 concerned detecting acid rain (rainfall with pH less than 5) from measurements made on a sample of n days for several sample sizes n. That exercise shows how the P-value for an observed sample mean $\bar{x}$ changes with n. It would be wise to do power calculations before deciding on the sample size. Suppose that pH measurements follow a Normal distribution with standard deviation $\sigma = 0.5$. You plan to test the hypotheses

$$H_0\colon \mu = 5$$
$$H_a\colon \mu < 5$$

at the 5% level of significance. You want to use a test that will almost always reject H_0 when the true mean pH is 4.7. Use the *Power of a Test* applet to find the power against the alternative $\mu = 4.7$ for samples of size $n = 5$, $n = 15$, and $n = 40$. What happens to the power as the size of the sample increases? Which of these sample sizes are adequate for use in this setting?

15.22 Detecting acid rain: power by hand. Even though software is used in practice to calculate power, doing the work by hand in a few examples builds your understanding. Find the power of the test in the previous exercise for a sample of size $n = 15$ by following these steps.

(a) Write the z test statistic for a sample of size 15. What values of z lead to rejecting H_0 at the 5% significance level?

(b) Starting from your result in (a), what values of $\bar{x}$ lead to rejecting H_0?

(c) What is the probability of rejecting H_0 when $\mu = 4.7$? This probability is the power against this alternative.

15.23 Two types of error. Your company markets a computerized medical diagnostic program used to evaluate thousands of people. The program scans the results of routine medical tests (pulse rate, blood tests, etc.) and refers the case to a doctor if there is evidence of a medical problem. The program makes a decision about each person.

(a) What are the two hypotheses and the two types of error that the program can make? Describe the two types of error in terms of "false-positive" and "false-negative" test results.

(b) The program can be adjusted to decrease one error probability, at the cost of an increase in the other error probability. Which error probability would you choose to make smaller, and why? (This is a matter of judgment. There is no single correct answer.)

CHAPTER 15 SUMMARY

- A specific confidence interval or test is correct only under specific conditions. The most important conditions concern the method used to produce the data. Other factors such as the shape of the population distribution may also be important.
- Whenever you use statistical inference, you are acting as if your data are a probability sample or come from a randomized comparative experiment.
- Always do data analysis before inference to detect outliers or other problems that would make inference untrustworthy.
- Other things being equal, the **margin of error** of a confidence interval gets smaller as
  - the confidence level C decreases,
  - the population standard deviation σ decreases, and
  - the sample size n increases.
- The margin of error in a confidence interval accounts for only the chance variation due to random sampling. In practice, errors due to nonresponse or undercoverage are often more serious.
- There is no universal rule for how small a P-value is convincing. Beware of placing too much weight on traditional significance levels, such as $\alpha = 0.05$.
- Very small effects can be highly significant (small P) when a test is based on a large sample. A statistically significant effect need not be practically

important. Plot the data to display the effect you are seeking, and use confidence intervals to estimate the actual values of parameters.

- On the other hand, lack of significance does not imply that H_0 is true. Even a large effect can fail to be statistically significant when a test is based on a small sample.
- Many tests run at once will probably produce some statistically significant results by chance alone, even if all the null hypotheses are true.
- Smart experimenters plan their studies with inference in mind. In particular, it is good practice to find what sample size will allow successful inference.
- The sample size required to obtain a confidence interval with specified margin of error m for a Normal mean is

$$n = \left(\frac{z^*\sigma}{m}\right)^2$$

  where z^* is the critical value for the desired level of confidence. Always round n up when you use this formula.
- The **power** of a significance test measures its ability to detect when a specific alternative hypothesis is true. The power of a test against a specific alternative is the probability that the test will reject H_0 when that alternative is true.
- Increasing the size of the sample increases the power of a significance test. Using statistical applets or software, you can find the sample size required to achieve a desired test power.
- We can describe the performance of a test at fixed level α by giving the probabilities of two types of error. A **Type I error** occurs if we reject H_0 when it is in fact true. A **Type II error** occurs if we fail to reject H_0 when in fact H_a is true.
- In a fixed level α significance test, the significance level α is the probability of a Type I error, and the power against a specific alternative is 1 minus the probability of a Type II error for that alternative.

THIS CHAPTER IN CONTEXT

This chapter is an extension of Chapter 14 and is designed to address the implications and limitations of statistical inference, again with emphasis on inference for a population mean under simple conditions. As with Chapter 14, what you learn here is also broadly relevant to all the chapters on statistical inference in the remainder of this book, even if the specific computations and applications vary with each inference procedure.

The limitations discussed here with respect to interpreting inference results draw heavily on everything you learned in Chapters 7 and 8 about the challenges of data collection. Concerns about how the data were collected should lead you to question the applicability or even the validity of the conclusions from any statistical inference based on these data.

The techniques of descriptive statistics you learned in Chapters 1 and 2 are also helpful in determining whether some data are adequate for the purpose of statistical inference. You will find that these techniques are even more critically important once we step away from the simple conditions considered in Chapters 14 and 15.

CHECK YOUR SKILLS

15.24 The most important condition for sound conclusions from statistical inference is usually

(a) that the data can be thought of as a random sample from the population of interest.
(b) that the population distribution is exactly Normal.
(c) that no calculation errors are made in the confidence interval or test statistic.

15.25 The coach of a college basketball team records the resting pulse rates of the team members. A confidence interval for the mean resting pulse rate of all college-age adults based on these data is of little use because

(a) the number of team members is small, so the margin of error will be large.
(b) many of the students in the team will probably refuse to respond.
(c) the college students in the basketball team can't be considered a random sample from the population.

15.26 You turn your Web browser to CNN's Quick Vote feature, which allows site visitors to choose an answer to the question of the day. You view yesterday's poll results, which are based on 26,494 responses. You should refuse to calculate any 95% confidence interval based on this sample because

(a) yesterday's responses are meaningless today.
(b) inference from a voluntary response sample can't be trusted.
(c) the sample is too large.

15.27 Many sample surveys use well-designed random samples, but half or more of the original sample can't be contacted or refuse to take part. Any errors due to this nonresponse

(a) have no effect on the accuracy of confidence intervals.
(b) are included in the announced margin of error.
(c) are in addition to the random variation accounted for by the announced margin of error.

15.28 Which of the following questions does a test of significance answer?

(a) Is the sample or experiment properly designed?
(b) Is the observed effect due to chance?
(c) Is the observed effect important?

15.29 Here's a quotation from a medical journal: "An uncontrolled experiment in 17 women found a significantly improved mean clinical symptom score after treatment. Methodologic flaws make it difficult to interpret the results of this study." The authors of this paper are skeptical about the significant improvement because

(a) there is no control group, so the improvement might be due to the placebo effect or to the fact that many medical conditions improve over time.
(b) the P-value given was $P = 0.03$, which is too large to be convincing.
(c) the response variable might not have an exactly Normal distribution in the population.

15.30 Vigorous exercise helps people live several years longer (on the average). Whether mild activities like slow walking extend life is not clear. Suppose that the added life expectancy from regular slow walking is just 2 months. A statistical test is more likely to find a significant increase in mean life if

(a) it is based on a very large random sample.
(b) it is based on a very small random sample.
(c) The size of the sample doesn't have any effect on the significance of the test.

15.31 (Optional) A laboratory scale is known to have a standard deviation of $\sigma = 0.001$ gram in repeated weighings. Scale readings in repeated weighings are Normally distributed, with mean equal to the true weight of the specimen. How many times must you weigh a specimen on this scale in order to get a margin of error no larger than ± 0.0005 with 95% confidence?

(a) 4 times (b) 15 times (c) 16 times

15.32 (Optional) A medical experiment compared the herb echinacea with a placebo for preventing colds. One response variable was "volume of nasal secretions" (if you have a cold, you blow your nose a lot). Take the average volume of nasal secretions in people without colds to be $\mu = 1$. An increase to $\mu = 3$ indicates a cold. The significance level of a test of H_0: $\mu = 1$ versus H_a: $\mu > 1$ is

(a) the probability that the test rejects H_0 when $\mu = 1$ is true.
(b) the probability that the test rejects H_0 when $\mu = 3$ is true.
(c) the probability that the test fails to reject H_0 when $\mu = 3$ is true.

15.33 (Optional) The power of the test in the previous exercise against the specific alternative $\mu = 3$ is

(a) the probability that the test rejects H_0 when $\mu = 1$ is true.
(b) the probability that the test rejects H_0 when $\mu = 3$ is true.
(c) the probability that the test fails to reject H_0 when $\mu = 3$ is true.

CHAPTER 15 EXERCISES

15.34 Deer mice. In Exercise 14.35 (page 360) you gave a confidence interval based on the body lengths of 14 deer mice (*Peromyscus maniculatus*) from a rich forest habitat. Before you can trust your results, you would like more information about the data. What facts would you most like to know?

15.35 Sensitive questions. The National AIDS Behavioral Surveys found that 170 individuals in its random sample of 2673 adult heterosexuals said they had multiple sexual partners in the past year. That's 6.36% of the sample. Why is this estimate likely to be biased? Does the margin of error of a 95% confidence interval for the proportion of all adults with multiple partners allow for this bias?

15.36 Asking about smoking habits. A 2012 Gallup survey of 1014 American adults asked participants about their smoking habits. Gallup has been conducting this survey yearly since 1944. The report was titled "In U.S., Smokers Light Up Less Than Ever," reflecting the fact that only 1% of smokers in this latest survey stated that they smoke more than a pack a day. Gallup added the following comment to its report:

> *It is possible that the decline in reports of smoking is the result of respondents' awareness that smoking is socially undesirable. Therefore, respondents may aim to present themselves in the best possible light to the interviewer and underestimate the amount they truly smoke.*

What type of bias is Gallup describing? And would it be included in the margin of error of a 95% confidence interval for the proportion of all adult smokers who smoke more than a pack a day?

15.37 Who is asking? A Harris Interactive Poll asked the same questions to two samples of adults: One group answered the survey on the phone with an interviewer and the other completed the survey online by themselves. When asked if they exercised regularly, 58% of those interviewed on the phone said "yes," compared with 35% of those who completed the survey online.

(a) What kind of answer variability or bias is most likely to produce such different results?
(b) Computing a 95% confidence interval for the true proportion of adult Americans who exercise regularly gives very different results depending on which sample is used. Is this kind of variability what the margin of error quantifies? Briefly explain your answer.

15.38 Saccharin no longer listed as a possible human carcinogen. In its 11th edition, the U.S. *Report on Carcinogens* deleted saccharin from the list of possible human carcinogens. Earlier carcinogenicity experiments had found rare instances of bladder tumors among male rats under very specific circumstances. Female rats never exhibited this sensitivity to saccharin, and neither did male or female mice. The toxicity was later explained by the unique metabolic pathway of certain male rats only. Years of observational studies in humans failed to show any sign of carcinogenicity from the regular consumption of saccharin.[13]

(a) What was the assumption made when saccharin was first placed on the list of possible human carcinogens based on the rat data? Why do you think this assumption was made?
(b) What simple condition is not met when applying animal research to humans? Explain why confirmation of the unique physiology of certain male rats led to the removal of saccharin from the list of possible human carcinogens.

15.39 When to use pacemakers. A medical panel prepared guidelines for when cardiac pacemakers should be implanted in patients with heart problems. The panel reviewed a large number of medical studies to judge the strength of the evidence supporting each recommendation. For each recommendation, they ranked the evidence as level A (strongest), B, or C (weakest). Here, in scrambled order, are the panel's descriptions of the three levels of evidence.[14] Which is A, which B, and which C? Explain your ranking.

Evidence was ranked as level ____ when data were derived from a limited number of trials involving comparatively small numbers of patients or from well-designed data analysis of nonrandomized studies or observational data registries.
Evidence was ranked as level ____ if the data were derived from multiple randomized clinical trials involving a large number of individuals.
Evidence was ranked as level ____ when consensus of expert opinion was the primary source of recommendation.

15.40 What distinguishes schizophrenics? A group of psychologists once measured 77 variables on a sample of schizophrenic people and a sample of people who were not schizophrenic. They compared the two samples using 77 separate significance tests. Two of these tests were statistically significant at the 5% level. Suppose that there is in fact no difference in any of the variables between people who are and people who are not schizophrenic, so that all 77 null hypotheses are true.

(a) What is the probability that 1 specific test shows a difference significant at the 5% level?
(b) Why is it not surprising that 2 of the 77 tests were statistically significant at the 5% level?

15.41 Why are larger samples better? Statisticians prefer reasonably large samples. Describe briefly the effect of increasing the size of a sample (or the number of subjects in an experiment), if all facts about the population remain unchanged, on each of the following:

(a) The margin of error of a 95% confidence interval
(b) The P-value of a test, when H_0 is false
(c) The impact of an outlier in the sample data on a confidence interval or a P-value

15.42 Legalizing marijuana. In January 2010, SurveyUSA conducted two polls on opinions about the legalization of marijuana. One poll was conducted in the state of Washington on January 13 and asked, "State lawmakers are considering making marijuana possession legal. Do you think legalizing marijuana is a good idea? Or a bad idea?" Of the 500 adults interviewed, 280 (56%) answered, "Good idea." The other poll was conducted in San Diego on January 12 and asked, "Do you think Marijuana should? or should not? be legal when used for recreational purposes?" Of the 500 adults interviewed, 219 (43.8%) answered, "Should." Both polls were conducted by telephone in the voice of a professional announcer. Respondent households were selected at random, using random digit dialing. Both polls reported a margin of sampling error of approximately 4.5 percentage points. The reports also stated that "there are other possible sources of error in all surveys that may be more serious than theoretical calculations of sampling error."[15]

(a) Which population or populations do these two surveys represent? Explain how this might impact the survey findings.
(b) What other major difference between these two surveys might explain the difference in support for marijuana legalization? Would it be covered by the stated margin of sampling error or would it fall under the category of "other possible sources of error"?

15.43 Adding fluoride to tap water. The widespread addition of fluoride to tap water in the United States and other countries has been credited with a drastic decrease in tooth decay. However, there is always some concern over possible harmful effects of any kind of treatment. Most studies so far have failed to show a negative impact of water fluoridation on health, but one study did report a significant decrease in IQ among individuals exposed to extra fluoride as children compared with individuals not exposed. Commenting on this study in a news article on National Public Radio, Dr. Myron Allukian of the Harvard School of Dental Medicine said that the effect on IQ was small, accounting for only about a half-point difference. He further explained that a half-point difference in IQ is meaningless, just like a difference in adult heights of half a millimeter.[16]

(a) Why is it important to know the size of an effect and not just that the effect was statistically significant?
(b) How can an effect be both small and statistically significant?

15.44 Physicians and health care reform. The Physicians Foundation's 2010 Physicians and Health Reform survey reports that "physicians' assessment of health reform in its early stages is predominantly negative, perhaps in part because they do not believe they had sufficient input into the new law." The survey was mailed to a random sample of 40,000 physicians and emailed to another random sample of 60,000 physicians, all actively practicing in the United States, in order to assess physicians' attitudes toward the Affordable Care Act within months of its passage. The published report also includes the following statements about the survey methodology:

A total of 2,379 completed surveys were received by August 23, 2010, for an overall response rate of 2.4%. (. . .) The overall margin of error for the entire survey is

$\pm$ 1.93%, indicating a low sampling error for a survey of this type (i.e., less than 2% error).[17]

What aspect of this survey makes you question the validity of the survey results? Is it covered by the margin of error cited? Briefly explain your answer.

The following exercises concern the optional material on planning studies for successful inference, choosing a sample size, and power calculations.

15.45 Aspirin tablets: how large a sample? Suppose we know that the aspirin content of aspirin tablets in Example 14.6 (page 348) follows a Normal distribution with standard deviation $\sigma = 5$ mg.

(a) You must verify the aspirin content of tablets produced in a day to within ± 1 mg with 99% confidence. How large a sample of aspirin tablets from the daily production do you need?

(b) If you needed to produce only a 95% confidence interval with a margin of error no larger than 1 mg, how large a sample of aspirin tablets from the daily production would you need instead?

15.46 Deer mice. Deer mice (*Peromyscus maniculatus*) are small rodents native to North America. Their body lengths (excluding tail) are known to vary approximately Normally, with mean $\mu = 86$ mm and standard deviation $\sigma = 8$ mm.[18] You decide to study the effect of vitamin supplements on deer mouse body length, but first, you want to figure out how many deer mice you would need for successful inference about the mean μ in the population of all deer mice that would ever receive vitamin supplements. How many deer mice would be needed to estimate μ to within ± 2 mm with 95% confidence?

15.47 Neural mechanism of Valium. Valium (diazepam) is a common antidepressant and sedative. A study investigated how Valium works by comparing its effect on sleep in 7 knock-in mice in which the $_{\alpha 2}GABA_A$ receptor is insensitive to Valium and in 8 wild-type control mice. The study found that Valium reduced sleep latency in both groups with no significant difference between the groups. The authors say that this lack of significance "is related to the large inter-individual variability that is also reflected in the low power (20%) of the test."[19]

(a) Explain in simple language why tests having low power often fail to give evidence against a null hypothesis even when the null hypothesis is really false.

(b) Which aspects of this experiment most likely contributed to a low test power?

15.48 Deer mice, continued. Refer to Exercise 15.46. You are thinking further about the design of your experiment.

(a) How many deer mice would be needed to estimate μ to within ± 3 mm with 95% confidence? When planning your study, what would be the most important factor in deciding that you should aim for a 2 mm versus a 3 mm margin of error?

(b) How should you design your study so that you can make valid conclusions about the effect of vitamin supplements on mouse body length?

15.49 Treating knee pain. Arthroscopic surgery is a minimally invasive surgical procedure on a joint, most commonly performed for persistent knee pain. Several uncontrolled, retrospective studies of the procedure reported pain relief among patients following the surgery.

(a) What can you conclude from these studies? What do they say about the benefits of arthroscopic surgery?

(b) A recent double-blind, randomized, placebo-controlled experiment found that outcomes after arthroscopic surgery are no better than after a sham (placebo) procedure.[20] How do you interpret this result? What does it say about the benefits of arthroscopic surgery?

(c) The controlled experiment was "designed to have 90 percent power, with a two-sided type I error of 0.04, to detect a moderate effect size (0.55) between the placebo group and the combined arthroscopic-treatment groups." How is this discussion of power important in interpreting the results of the controlled experiment? What do you conclude about the benefits of arthroscopic surgery in treating knee pain?

15.50 Error probabilities. You read that a statistical test at significance level $\alpha = 0.05$ has power 0.78. What are the probabilities of Type I and Type II errors for this test?

15.51 Power. You read that a statistical test at the $\alpha = 0.01$ level has probability 0.14 of making a Type II error when a specific alternative is true. What is the power of the test against this alternative?

15.52 Sweetening colas: calculating power. The cola maker of Example 15.7 wants to test at the 5% significance level the following hypotheses:

$$H_0: \mu = 0 \text{ versus } H_a: \mu > 0$$

Ten taste scores were used for the significance test. The distribution of taste scores is assumed to be roughly Normal with standard deviation $\sigma = 1$. We want to calculate the power of this test when the true mean sweetness loss is $\mu = 0.8$.

(a) What values of the z statistic would lead us to reject H_0? To what values of $\bar{x}$ do they correspond? Use the inverse Normal calculations to figure this out.

(b) When the true mean sweetness loss is $\mu = 0.8$, how often would we reject H_0? That is, what is the probability of obtaining sample averages like the ones defined in (a) when $\mu = 0.8$? Use Table A to calculate this probability. This probability is the power of your test against the alternative $\mu = 0.8$, the probability of rejecting H_0 when the alternative $\mu = 0.8$ is true.

15.53 Sweetening colas: power for various sample sizes. In the previous exercise you calculated by hand the power of the cola sweetness test when the true mean sweetness loss is $\mu = 0.8$.

(a) Use the *Statistical Power* applet to find the power against the alternative $\mu = 0.8$ as in the previous exercise. Check that your answers are similar.

(b) Now use the applet to calculate the power of the test against the alternative $\mu = 0.8$ for samples of size $n = 5$, $n = 15$, and $n = 40$. How does sample size affect power?

(c) Using a sample size of $n = 10$, find with the applet the power of the test against the alternatives $\mu = 0.5$, $\mu = 1.0$, and $\mu = 1.5$. How does effect size affect power?

15.54 Power of a two-sided test. The National Center for Health Statistics reports that the systolic blood pressure for males 35 to 44 years of age has mean 128 and standard deviation 15. The medical director of a large company looks at the medical records of 72 executives in this age group and finds that the mean systolic blood pressure in this sample is $\bar{x} = 126.07$. At a 5% significance level, the data fail to show significant evidence that the mean blood pressure of a population of executives differ from the national mean $\mu = 128$. The medical director now wonders if the test used would detect an important difference if one were present.

Power calculations for two-sided tests follow the same outline as for one-sided tests. The hypotheses tested here are H_0: $\mu = 128$ versus H_a: $\mu \neq 128$. What would be the power of this test against the alternative $\mu = 134$?

(a) This test rejects H_0 when $|z| \geq 1.96$. The test statistic z is

$$z = \frac{\bar{x} - 128}{15/\sqrt{72}}$$

Write the rule for rejecting H_0 in terms of the values of $\bar{x}$. (Because the test is two-sided, it rejects H_0 when $\bar{x}$ is either too large or too small.)

(b) Now find the probability that $\bar{x}$ takes values that lead to rejecting H_0 if the true mean is $\mu = 134$. This probability is the power.

(c) What is the probability that this test makes a Type II error when $\mu = 134$?

15.55 Power of a two-sided test, continued. Let's now use technology to perform the power calculations for the two-sided test of the previous exercise.

(a) Use the *Statistical Power* applet to find the power of the two-sided test against the alternative $\mu = 134$. Make sure that it is similar to what you calculated in the previous exercise.

(b) Use the applet to calculate the power of the test against the alternative $\mu = 122$. Can the test be relied on to detect a mean that differs from 128 by 6?

(c) If the alternative were farther from H_0, say, $\mu = 136$, would the power be higher or lower than the values calculated in (a) and (b)?

CHAPTER 16 From Exploration to Inference: Part II Review

IN THIS CHAPTER WE COVER...

Designs for producing data are essential parts of statistics in practice. Figures 16.1 and 16.2 display the big ideas visually. Random sampling and randomized comparative experiments are perhaps the most important statistical inventions of the twentieth century. Both were slow to gain acceptance, and you will still see many voluntary response samples and uncontrolled experiments. You should now understand good techniques for producing data and also why bad techniques often produce worthless data. The deliberate use of chance in producing data is a central idea in statistics. It not only reduces bias but also allows use of the laws of probability to analyze data. Fortunately, we need only some basic facts about probability in order to understand statistical inference.

Statistical inference draws conclusions about a population on the basis of sample data and uses probability to indicate how reliable the conclusions are. A confidence interval estimates an unknown parameter. A significance test shows how strong the evidence is for some claim about a parameter.

The probabilities in both confidence intervals and tests tell us what would happen if we used the method for the interval or test very many times.

- A confidence level is the success rate of the method for a confidence interval. This is the probability that the method actually produces an interval that

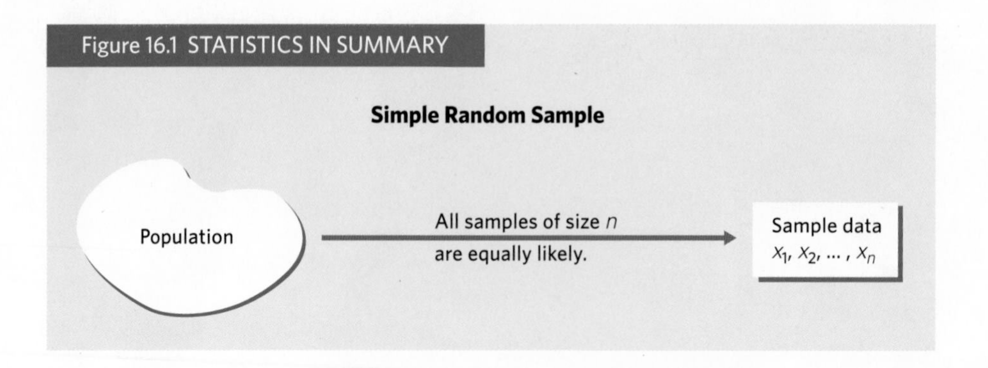

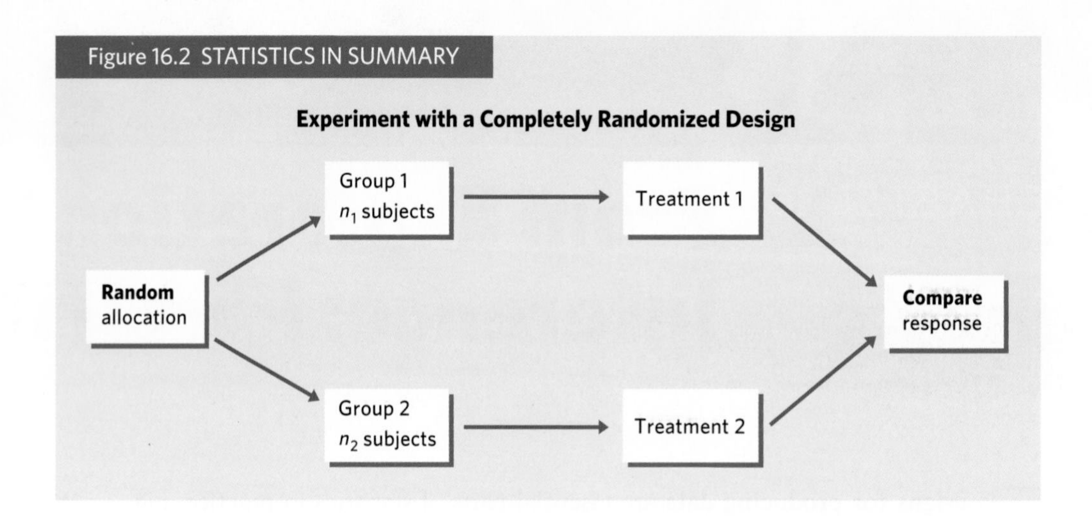

contains the unknown parameter. A 95% confidence interval gives a correct result 95% of the time when we use it repeatedly. That is, when we produce *one* 95% confidence interval, we do not know whether it gave a correct result or not, but we know that there is a 95% chance that it did.

- A P-value tells us how surprising the observed outcome would be if the null hypothesis was true. That is, P is the probability that the test would produce a result at least as extreme as the observed result if the null hypothesis really was true. Very surprising outcomes (small P-values) are good evidence that the null hypothesis is not true.

Figures 16.3 and 16.4 use the z procedures introduced in Chapter 14 to present in picture form the big ideas of confidence intervals and significance tests. These ideas are the foundation for the rest of this book. We will have much to say about many statistical methods and their use in practice. In every case, the basic reasoning of confidence intervals and significance tests remains the same.

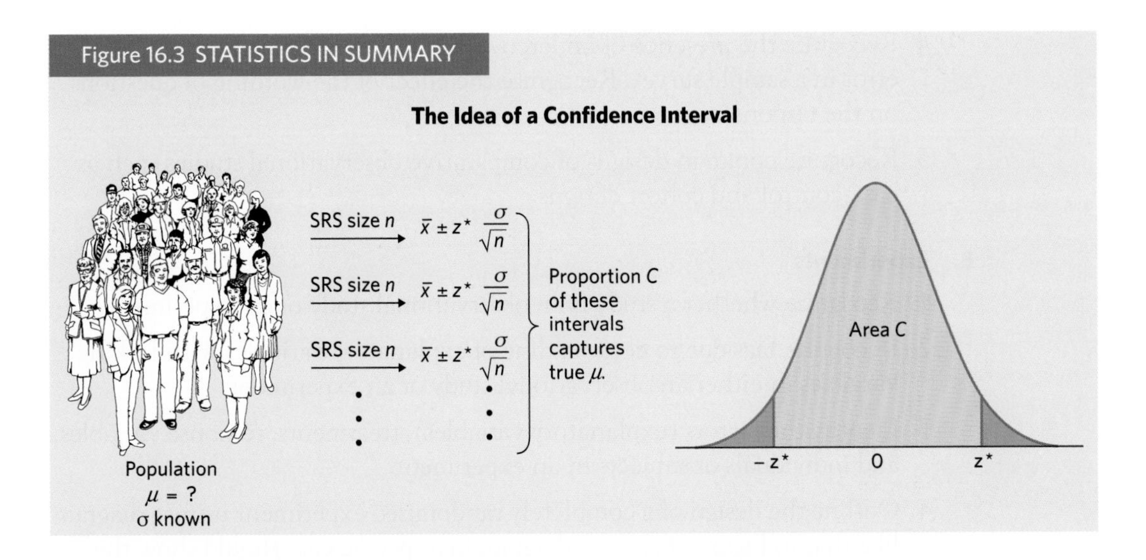

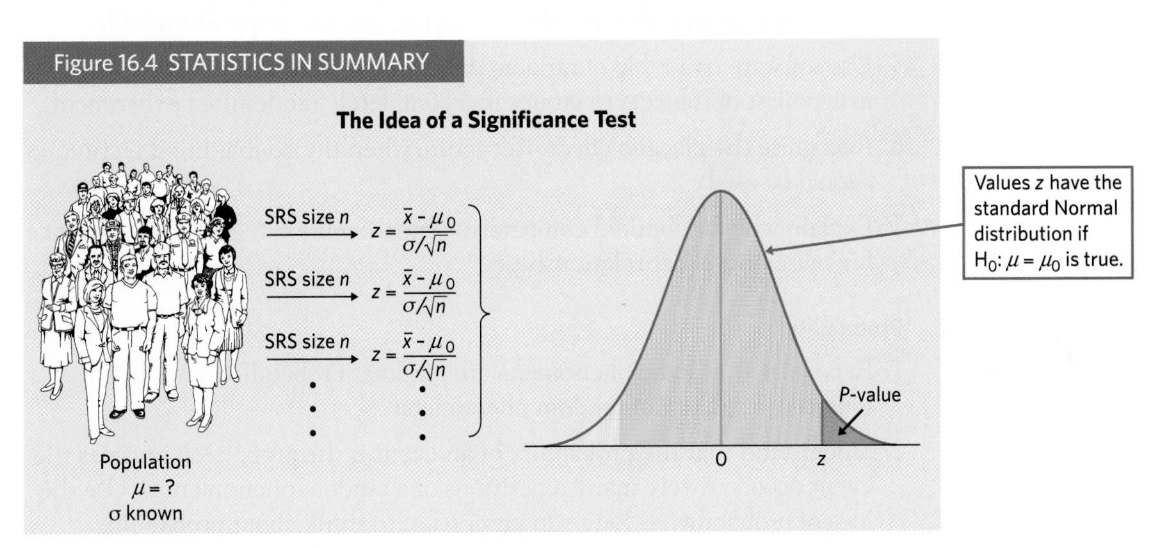

PART II SUMMARY

Here are the most important skills you should have acquired from reading Chapters 7 through 15.

A. Sampling and Observational Studies

1. Identify the population in a sampling situation.
2. Recognize bias due to voluntary response samples and other inferior sampling methods.
3. Use software or a table of random digits to select a simple random sample (SRS) from a population or to select a stratified random sample from a population when the strata are identified.

4. Recognize the presence of undercoverage and nonresponse as sources of error in a sample survey. Recognize the effect of the wording of questions on the responses.
5. Recognize common designs of comparative observational studies such as case-control and cohort designs.

B. Experiments

1. Recognize whether a study is an observational study or an experiment.
2. Recognize bias due to confounding of explanatory variables with lurking variables in either an observational study or an experiment.
3. Identify the factors (explanatory variables), treatments, response variables, and individuals or subjects in an experiment.
4. Outline the design of a completely randomized experiment using a diagram like that in Figure 16.2. The diagram in a specific case should show the sizes of the groups, the specific treatments, and the response variable.
5. Use software or a table of random digits to carry out the random assignment of subjects to groups in a completely randomized experiment.
6. Recognize the placebo effect. Recognize when the double-blind technique should be used.
7. Explain why randomized comparative experiments can give good evidence for cause-and-effect relationships.

C. Probability

1. Recognize that some phenomena are random. Probability describes the long-run regularity of random phenomena.
2. Understand that the probability of an event is the proportion of times the event occurs in very many repetitions of a random phenomenon. Use the idea of probability as long-run proportion to think about probability.
3. Use basic probability rules to detect illegitimate assignments of probability: Any probability must be a number between 0 and 1, and the total probability assigned to all possible outcomes must be 1.
4. Use basic probability rules to find the probabilities of events that are formed from other events. The probability that an event does not occur is 1 minus its probability. If two events are disjoint, the probability that one or the other occurs is the sum of their individual probabilities.
5. Find probabilities in a discrete probability model by adding the probabilities of their outcomes. Find probabilities in a continuous probability model as areas under a density curve.
6. Use the notation of random variables to make compact statements about random outcomes, such as $P(\overline{x} \leq 4) = 0.3$. Be able to interpret such statements.

D. General Rules of Probability (Optional)

1. Use Venn diagrams to picture relationships among several events.
2. Use the general addition rule to find probabilities that involve overlapping events.
3. Understand the idea of independence. Judge when it is reasonable to assume independence as part of a probability model.
4. Use the multiplication rule for independent events to find the probability that all of several independent events occur.
5. Use the multiplication rule for independent events in combination with other probability rules to find the probabilities of complex events.
6. Understand the idea of conditional probability. Find conditional probabilities for individuals or outcomes, chosen at random, from a table of counts of possible outcomes.
7. Use the general multiplication rule to find $P(A \text{ and } B)$ from $P(A)$ and the conditional probability $P(B \mid A)$.
8. Use two-way tables or tree diagrams to organize several-stage probability models.
9. Understand the idea of Bayes's theorem. Use the formula or a tree diagram to find a conditional probability based on other, known probabilities.

E. Density Curves and Normal Distributions

1. Know that density curves represent probability distributions for continuous variables. The total area under a density curve is 1.
2. Know that any given area under a density curve represents both the proportion of all corresponding observations in a population and the probability of randomly selecting such observations from the population.
3. Approximately locate the median (equal-areas point) and the mean (balance point) on a density curve.
4. Recognize the shape of Normal curves and estimate by eye both the mean and the standard deviation from such a curve.
5. Use the 68–95–99.7 rule and symmetry to state what percent of the observations from a Normal distribution fall between two points when both points lie at one, two, or three standard deviations on either side of the mean.
6. Find the standardized value (z-score) of an observation. Interpret z-scores and understand that any Normal distribution becomes standard Normal $N(0, 1)$ when standardized.
7. Given that a variable has a Normal distribution with a stated mean μ and standard deviation σ, calculate the proportion/probability of values above a stated number, below a stated number, or between two stated numbers.

8. Given that a variable has a Normal distribution with a stated mean μ and standard deviation σ, calculate the point having a stated proportion/probability above it or below it.

F. Discrete Distributions (Optional)

1. Recognize the binomial setting: a fixed number n of independent success-failure trials with the same probability p of success on each trial.
2. Recognize and use the binomial distribution of the count of successes in a binomial setting.
3. Use the binomial probability formula or software to find probabilities of events involving the count X of successes in a binomial setting for small values of n.
4. Find the mean and standard deviation of a binomial count X.
5. Recognize when you can use the Normal approximation to a binomial distribution. Use the Normal approximation to calculate probabilities that concern a binomial count X.
6. Recognize and use the Poisson distribution of the count of occurrences of a defined event in fixed, finite intervals of time or space.
7. Find the mean and standard deviation of a Poisson distribution.
8. Use the Poisson probability formula or software to find probabilities of events involving the count X of occurrences over a finite interval.

G. Sampling Distributions

1. Identify parameters and statistics in a statistical study.
2. Recognize the fact of sampling variability: A statistic will take different values when you repeat a sample or experiment.
3. Interpret a sampling distribution as describing the values taken by a statistic in all possible repetitions of a sample or experiment under the same conditions.
4. Interpret the sampling distribution of a statistic as describing the probabilities of its possible values.

H. The Sampling Distribution of a Sample Mean

1. Recognize when a problem involves the mean $\bar{x}$ of a sample. Understand that $\bar{x}$ estimates the mean μ of the population from which the sample is drawn.
2. Find the mean and standard deviation of a sample mean $\bar{x}$ from an SRS of size n when the mean μ and standard deviation σ of the population are known.
3. Understand that $\bar{x}$ is an unbiased estimator of μ and that the variability of $\bar{x}$ about its mean μ gets smaller as the sample size increases.
4. Understand that $\bar{x}$ has approximately a Normal distribution when the sample is large (central limit theorem). Use this Normal distribution to calculate probabilities that concern $\bar{x}$.

5. (Optional) Use the law of large numbers to describe the behavior of $\overline{x}$ as the size of the sample increases.

I. The Sampling Distribution of a Sample Proportion

1. Recognize when a problem involves the proportion $\hat{p}$ of successes in a sample. Understand that $\hat{p}$ estimates the proportion p of successes in the population from which the sample is drawn.
2. Find the mean and standard deviation of a sample proportion $\hat{p}$ from an SRS of size n when the population proportion p is known.
3. Understand that $\hat{p}$ is an unbiased estimator of p and that the variability of $\hat{p}$ about its mean p gets smaller as the sample size increases.
4. Understand that $\hat{p}$ has approximately a Normal distribution when the sample is large. Use this Normal distribution to calculate probabilities that concern $\hat{p}$.
5. (Optional) Use the law of large numbers to describe the behavior of $\hat{p}$ as the size of the sample increases.

J. Confidence Intervals

1. State in nontechnical language what is meant by "95% confidence" or other statements of confidence in statistical reports.
2. Know the four-step process (page 342) for any confidence interval.

3. Calculate a confidence interval for the mean μ of a Normal population with known standard deviation σ, using the formula $\overline{x} \pm z^*\sigma/\sqrt{n}$.
4. Understand how the margin of error of a confidence interval changes with the sample size and the level of confidence C.
5. (Optional) Find the sample size required to obtain a confidence interval of specified margin of error m when the confidence level and other information are given.
6. Identify sources of error in a study that are *not* included in the margin of error of a confidence interval, such as undercoverage or nonresponse.

K. Significance Tests

1. State the null and alternative hypotheses in a testing situation when the parameter in question is a population mean μ.
2. Explain in nontechnical language the meaning of the P-value when you are given the numerical value of P for a test.
3. Know the four-step process (page 354) for any significance test.

4. Calculate the one-sample z test statistic and the P-value for both one-sided and two-sided tests about the mean μ of a Normal population.
5. Assess statistical significance at standard levels α, either by comparing P with α or by comparing z with standard Normal critical values.
6. Recognize that significance testing does not measure the size or importance of an effect. Explain why a small effect can be significant in a

large sample and why a large effect can fail to be significant in a small sample.

7. Recognize that any inference procedure acts as if the data were properly produced. The z confidence interval and test require that the data be an SRS from the population.
8. (Optional) Recognize that the decision to reject or fail to reject H_0 may be the wrong decision. Identify the two types of error that can be made: A Type I error, which occurs if we reject H_0 when it is in fact true, and a Type II error, which occurs if we fail to reject H_0 when in fact H_a is true.
9. (Optional) Find the probability of making a Type I error for a test with a fixed significance level α. Obtain the probability of making a Type II error for a given alternative value of the parameter. Know that the power of a test against a specific alternative is 1 minus the probability of a Type II error.

REVIEW EXERCISES

Review exercises help you solidify the basic ideas and skills in Chapters 7 through 15.

16.1 Marijuana and driving. Questioning a sample of young people in New Zealand revealed a positive association between use of marijuana (cannabis) and traffic accidents caused by the members of the sample. Both cannabis use and accidents were measured by interviewing the young people themselves. The study report says, "It is unlikely that self reports of cannabis use and accident rates will be perfectly accurate."[1] Is the response bias likely to make the reported association stronger or weaker than the true association? Why?

16.2 Effects of binge drinking. A common definition of "binge drinking" is having five or more drinks at one sitting for men, and four or more for women. An observational study finds that students who binge have lower average GPAs than those who don't. Suggest some lurking variables that may be confounded with binge drinking. The possibility of confounding means that we can't conclude that binge drinking *causes* lower GPAs.

16.3 Elephants and bees. Elephants sometimes damage crops in Africa. It turns out that elephants dislike bees. They recognize beehives in areas where they are common and avoid them. Can this aversion be used to keep elephants away from trees? A group in Kenya placed active beehives in some trees and empty beehives in others. Will elephant damage be less in trees with active hives? Or will even empty hives keep elephants away?[2]

(a) Outline the design of an experiment to answer these questions using 72 acacia trees (be sure to also include a control group).

(b) Use software or the *Simple Random Sample* applet to choose the trees for the active-hive group, or use Table A at line 137 to choose the first 4 trees in that group.

(c) What is the response variable in this experiment?

16.4 Treating obese monkeys. The experimental drug adipotide is designed to limit the blood supply to adipose tissue and thus lead to gradual loss of body fat. A study tested the efficacy of adipotide in obese macaque monkeys. The experiment assigned 15 obese monkeys to two groups: 10 to the treatment group receiving daily injections of adipotide for four weeks, and 5 to a control group receiving daily injections of saline water for four weeks. The percent body fat of each monkey was assessed with magnetic resonance imaging at the end of the study.[3]

(a) Outline the design of this experiment.

(b) Use software or the *Simple Random Sample* applet to choose the 5 members of the control group, or use Table A at line 110.

(c) Explain why, for this particular experiment, a group receiving daily injections of saline water provides a better control than a group simply left untouched.

16.5 Cell phones and driving. Does talking on a hands-free cell phone distract drivers? Undergraduate students "drove" in a high-fidelity driving simulator equipped with a hands-free cell phone. The car ahead brakes: How quickly does the subject react? There are 40 student subjects available.

(a) Describe a completely randomized design for this experiment.

(b) Describe a matched pairs design for this experiment.

(c) The study actually used a matched pairs design. How is this design helpful in answering the study question?

16.6 Deceiving subjects. Healthy adult volunteers who drink socially but are not alcoholics signed up to be subjects in an experiment. They were told that the objective of the study was to examine the effects of alcohol consumption on word search performance. Participants were given a 12-ounce beverage and asked to consume it at their own pace while watching a nature documentary alone in a room. They were then given a word search task to complete.

The researchers had actually randomly assigned the participants to one of four conditions based on drink type (beer or soft drink) and glass type (straight or curved) and placed hidden cameras in the room to determine the time it took each participant to finish his or her drink. The findings showed that individuals drinking beer in a straight glass took a significantly longer time to finish their drinks than individuals in the other three groups, suggesting that the shape of a glass can influence the pace at which alcoholic beverages are consumed.[4]

(a) Why do you think the researchers chose to disguise the true objective of the study?

(b) Do you think this study is ethical?

16.7 Studying your blood. Long ago, doctors drew a blood specimen from you as part of treating minor anemia. Unknown to you, the sample was stored. Now researchers plan to use stored samples from you and many other people to look for genetic factors that may influence anemia. It is no longer possible to ask your consent. Modern technology can read your entire genetic makeup from the blood sample.

(a) Do you think it violates the principle of informed consent to use your blood sample if your name is on it but you were not told that it might be saved and studied later?

(b) Suppose that your identity is not attached. The blood sample is known only to come from (say) "a 20-year-old white female being treated for anemia." Is it now OK to use the sample for research?

(c) Perhaps we should use biological materials such as blood samples only from patients who have agreed to allow the material to be stored for later use in research. It isn't possible to say in advance what kind of research, so this falls short of the usual standard for informed consent. Is it nonetheless acceptable, given complete confidentiality and the fact that using the sample can't physically harm the patient?

16.8 Chronic diseases. A recent Gallup-Healthways Well-Being Index study of chronic diseases in the United States asked roughly 353,000 randomly selected American adults if they had ever been told by a doctor or a nurse that they have a chronic condition. Gallup reported that 30.0% of American adults have had a diagnosis of high blood pressure and that 26.2% have had a diagnosis of high cholesterol. Can we conclude that there is a 56.2% probability that a randomly selected American adult has had either a diagnosis of high blood pressure or a diagnosis of high cholesterol? Briefly explain your answer.

16.9 Birth order. The 2012 National Vital Statistics Report provides information on how the nearly four million live births that occurred in 2010 broke down by birth order. Choose at random a baby born in America in 2010:[5]

Baby's birth order	1	2	3	4	5	6	7
Probability	0.404	0.315	0.165	0.069	0.026	0.011	0.010

(The few babies who were 8th or more in birth order are included in the "7" birth order group.)

(a) Check that this distribution satisfies the two requirements for a legitimate discrete probability model.

(b) Describe in words the event $P(X \leq 2)$. What is the probability of this event?

(c) What is $P(X < 2)$?

(d) Write the event "a woman gives birth in 2010 to her 4th or more child" in terms of values of X. What is the probability of this event?

16.10 The addition rule. The addition rule for probabilities, $P(A \text{ or } B) = P(A) + P(B)$, is true only in some circumstances. Give (in words) an example of real-world events A and B for which this rule is not true.

16.11 Accidents on a bike path. Examining the location of accidents on a level, 3-mile bike path shows that they occur uniformly along the length of the path. Figure 16.5 displays the density curve that describes the distribution of accidents.

(a) Explain why this curve satisfies the two requirements for a density curve.

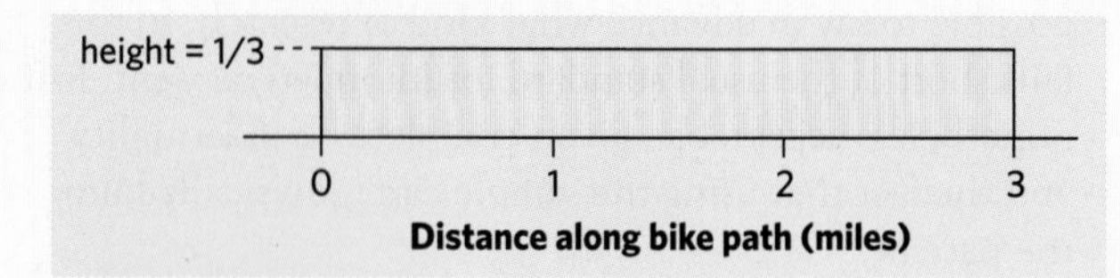

FIGURE 16.5 Density curve for the location of accidents along a 3-mile bike path, for Exercise 16.11.

(b) The proportion of accidents that occur in the first mile of the path is the area under the density curve between 0 and 1 mile. What is this area?

(c) Sue's property adjoins the bike path between the 0.8 mile mark and the 1.1 mile mark. What proportion of accidents happen in front of Sue's property?

16.12 Where are the quartiles? How many standard deviations above and below the mean do the quartiles of any Normal distribution lie? (Use the standard Normal distribution to answer this question.) Use the *Normal Curve* applet, software, or Table B to find the answers.

16.13 Weighing bean seeds. Many biological measurements on the same species follow a Normal distribution quite closely. The weights of seeds of a variety of winged bean are approximately Normal with mean 525 milligrams (mg) and standard deviation 110 mg.

(a) What percent of seeds weigh more than 500 mg?

(b) If we discard the lightest 10% of all seeds, what is the smallest weight among the remaining seeds?

Mohan Raj Edavattu Kanayamkott/iStockphoto

16.14 Are the data Normal? (optional) Table 2.1 (page 62) gives data on the penetrability of soil at each of three levels of compression. We might expect the penetrability of specimens of the same soil at the same level of compression to follow a Normal distribution. Make stemplots or dotplots of the data by soil type. Figure 16.6 shows the Normal quantile plot for the data. Do any of the three samples seem roughly Normal? Do any appear distinctly non-Normal? If so, what kind of departure from Normality do you notice in your plots?

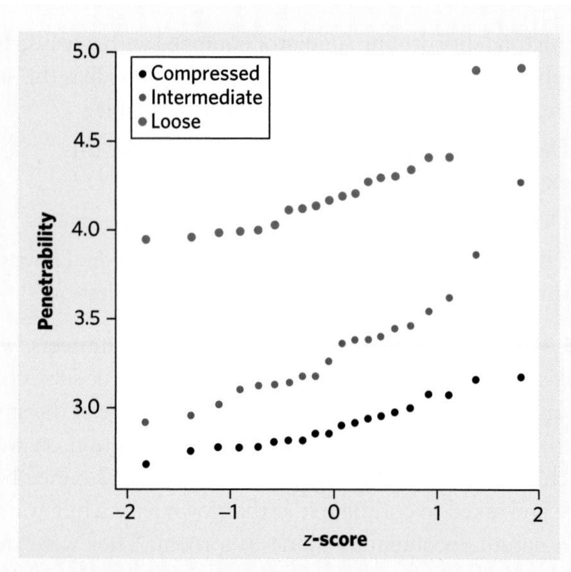

FIGURE 16.6 Normal quantile plot of three types of soil, for Exercise 16.14.

16.15 Distributions: means versus individuals. The z confidence interval and test are based on the sampling distribution of the sample mean $\bar{x}$. In Exercise 16.13 you examined the distribution of seed weights for a variety of winged bean. This distribution is approximately Normal with mean $\mu = 525$ mg and standard deviation $\sigma = 110$ mg.

(a) You take an SRS of 100 winged bean seeds. According to the 99.7 part of the 68–95–99.7 rule, what approximate *range of individual weights* x do you expect to see in your sample?

(b) You look at many SRSs of size 100. About what *range of sample mean weights* $\bar{x}$ do you expect to see?

16.16 Distributions: larger samples. In the setting of the previous exercise, how many winged-bean seeds must you sample to cut the range of values of $\bar{x}$ in half? This will also cut the margin of error of a confidence interval for μ in half. Do you expect the range of individual weights x in the new sample to also be much less than in a sample of size 100? Why?

16.17 Brains at work. When our brains store information, complicated chemical changes take place. In trying to understand these changes, researchers blocked some processes in brain cells taken from rats and compared these cells with a control group of normal cells. They report that "no differences were seen" between the two groups in four response variables. They give P-values of 0.45, 0.83, 0.26, and 0.84 for these four tests.[6]

(a) Say clearly what the P-value $P = 0.45$ says about the response that was observed.

(b) It isn't literally true that "no differences were seen." That is, the mean responses were not exactly alike in the two groups. Explain what the researchers mean when they give $P = 0.45$ and say "no difference was seen."

Use the following information for Exercises 16.18 to 16.21. The distribution of blood cholesterol level in the population of young men aged 20 to 34 years is close to Normal with standard deviation $\sigma = 41$ milligrams per deciliter (mg/dl).

16.18 Estimating blood cholesterol. You measure the blood cholesterol of 14 cross-country runners. The mean level is $\bar{x} = 172$ mg/dl. Assuming that σ is the same as in the general population, give a 90% confidence interval for the mean level μ among cross-country runners.

16.19 Testing blood cholesterol. The mean blood cholesterol level for all men aged 20 to 34 years is $\mu = 188$ mg/dl. We suspect that the mean for cross-country runners is lower.

(a) State hypotheses, find the test statistic, and give the P-value. Is the result significant at the $\alpha = 0.10$ level? At $\alpha = 0.05$? At $\alpha = 0.01$?

(b) Explain how you can use the confidence interval you computed in the previous exercise to test these hypotheses. What significance level does that 90% confidence interval yield when testing the hypotheses you stated in (a)? Remember that your test was one-sided.

16.20 Smaller margin of error (optional). How large a sample is needed to cut the margin of error in Exercise 16.18 in half? How large a sample is needed to cut the margin of error to ± 5 mg/dl?

16.21 More significant results. Suppose, hypothetically, that a larger sample of 56 cross-country runners yields the same mean level, $\bar{x} = 172$ mg/dl. Redo the test of hypotheses in Exercise 16.19. What is the P-value now? At which of the levels $\alpha = 0.10$, $\alpha = 0.05$, and $\alpha = 0.01$ is the result significant? What general fact about significance tests does comparing your results here and in Exercise 16.19 illustrate?

16.22 Support groups for breast cancer. Does participating in a support group extend the lives of women with breast cancer? Most of the evidence on this topic is either anecdotal or observational. The first randomized comparative experiment performed assigned 235 women with advanced breast cancer to two groups: 158 to "expressive group therapy" and 77 to a control group. Women in the treatment group reported less pain ($P = 0.04$), but there was no significant difference between the groups in median survival time ($P = 0.72$). Explain carefully why $P = 0.04$ is evidence that the treatment *does* make a difference in quality of life and why $P = 0.72$ means that there is no significant evidence that support groups prolong life.

SUPPLEMENTARY EXERCISES

Supplementary exercises apply the skills you have learned in ways that require more thought or more elaborate use of technology.

16.23 Do you believe in evolution? What proportion of all American adults believe in the Darwinian theory of evolution? In 2009, on Darwin's 200th birthday, Gallup published the result of a random sample survey of 1018 national adults contacted by telephone. When asked, "Do you, personally, believe in the theory of evolution?" 39% of the people in the sample answered "yes."

(a) Is it reasonable to state that these results are representative of the American adult population? Explain your answer.

(b) Gallup states in the report that "question wording and practical difficulties in conducting surveys can introduce error or bias into the findings of public opinion polls." Explain what "practical difficulties" were likely experienced during the conduction of this poll. What aspect of the "question wording" might bias responses? How might these affect the conclusions?

(c) A Harris Poll of 2158 adults surveyed online during the same period asked, "Do you believe Charles Darwin's theory which states that plants, animals and human beings have evolved over time?" The report tells us that 53% of respondents answered "yes." What aspects of the Gallup and Harris surveys might explain their rather substantially different results?

16.24 Sampling students. You want to investigate the attitudes of students at your school toward getting vaccinated against the flu. You have a grant that will pay the costs of contacting about 500 students.

(a) Specify the exact population for your study. For example, will you include part-time students?

(b) Describe your sample design. Will you use a stratified sample?

(c) Briefly discuss the practical difficulties that you anticipate. For example, how will you contact the students in your sample?

16.25 The placebo effect. A survey of physicians found that some doctors give a placebo to a patient who complains of pain for which the physician can find no cause. If the patient's pain improves, these doctors conclude that it had no physical basis. The medical school researchers who conducted the survey claimed that these doctors do not understand the placebo effect. Why?

16.26 Arsenic and lung cancer. Arsenic is frequently found both in the natural environment and in food. A study of the relationship between arsenic in drinking water and deaths from lung cancer measured arsenic levels in drinking water in 138 villages in Taiwan and examined death certificates to identify lung cancer deaths. The study summary says that "arsenic levels above 0.64 mg/l were associated with a significant increase in the mortality of lung cancer in both genders, but no significant effect was observed at lower levels."[7]

(a) Explain why this is an observational study rather than an experiment.

(b) The word "significant" in the conclusion has its statistical meaning, not its everyday meaning. Restate the study conclusion without using the word "significant" in a way that is clear to readers who know no statistics.

(c) Suggest some lurking variables that may be confounded with the arsenic content. Why can't we conclude that high arsenic levels in drinking water cause death from lung cancer?

16.27 Alcohol consumption and the shape of your glass. Go back to Exercise 16.6.

(a) What are the factors and their levels, the treatments, and the response variable?

(b) Describe and outline the design of this experiment.

(c) Can you conclude that the combination of drink type and glass shape *causes* the significant difference in the time it takes to finish a 12-ounce drink? Explain your reasoning.

Use the following information for Exercises 16.28 to 16.30. The level of pesticides found in the blubber of whales is a measure of pollution of the oceans by runoff from land and can also be used to identify different populations of whales. A sample of 8 male minke whales in the West Greenland area of the North Atlantic found the mean concentration of the insecticide dieldrin to be $\bar{x} = 357$ nanograms per gram of blubber (ng/g).[8] Suppose that the concentration in all such whales varies Normally with standard deviation $\sigma = 50$ ng/g.

16.28 Pesticides in whale blubber. 4 STEP Use a 95% confidence interval to estimate the mean dieldrin level in whale blubber. Follow the four-step process for confidence intervals (page 342) in your work.

Pete Atkinson/Getty Images

16.29 Testing pesticide levels. 4 STEP The Food and Drug Administration regulates the amount of dieldrin in raw food. For some foods, no more than 100 ng/g is allowed. Is there good evidence that the mean concentration in whale blubber is above this level? Follow the four-step process for significance tests (page 354) in your work.

16.30 Other confidence levels. Give an 80% confidence interval and a 90% confidence interval for the mean concentration of dieldrin in the whale population. What general fact about confidence intervals do the margins of error of your three intervals illustrate?

16.31 Birth weight and IQ: estimation. 4 STEP Infants weighing less than 1500 grams at birth are classed as "very low birth weight." Low birth weight carries many risks. One study followed 113 male infants with very low birth weight to adulthood. At age 20, the mean IQ score for these men was $\bar{x} = 87.6$.[9] IQ scores vary Normally with standard deviation $\sigma = 15$. Give a 95% confidence interval for the mean IQ score at age 20 for all very-low-birth-weight males. Use the four-step process for confidence intervals (page 342) as a guide.

16.32 Birth weight and IQ: testing. 4 STEP IQ tests are scaled so that the mean score in a large population should be $\mu = 100$. We suspect that the very-low-birth-weight population has mean score less than 100. Does the study described in the previous exercise give good evidence that this is true? Use the four-step process for significance tests (page 354) as a guide.

16.33 Birth weight and IQ: causation? Very-low-birth-weight babies are more likely to be born to unmarried mothers and to mothers who did not complete high school.

(a) Explain why the study of Exercise 16.31 was not an experiment.

(b) Explain clearly why confounding prevents us from concluding that very low birth weight in itself reduces adult IQ.

16.34 Can we trust this interval? Here are data on the percent change in the total mass (in tons) of wildlife in several West African game preserves in the years 1971 to 1999:[10]

1971	1972	1973	1974	1975	1976	1977	1978	1979	1980
2.9	3.1	−1.2	−1.1	−3.3	3.7	1.9	−0.3	−5.9	−7.9
1981	**1982**	**1983**	**1984**	**1985**	**1986**	**1987**	**1988**	**1989**	**1990**
−5.5	−7.2	−4.1	−8.6	−5.5	−0.7	−5.1	−7.1	−4.2	0.9
1991	**1992**	**1993**	**1994**	**1995**	**1996**	**1997**	**1998**	**1999**	
−6.1	−4.1	−4.8	−11.3	−9.3	−10.7	−1.8	−7.4	−22.9	

Software gives the 95% confidence interval for the mean annual percent change as −6.66% to −2.55%. There are several reasons why we might not trust this interval.

(a) Examine the distribution of the data. What feature of the distribution throws doubt on the validity of statistical inference?

(b) Plot the percents against year. What trend do you see in this time series? Explain why a trend over time casts doubt on the condition that the years 1971 to 1999 can be treated as an SRS from a larger population of years.

16.35 The first kid has higher IQ (optional). Does the birth order of a family's children influence their IQ scores? A careful study of 241,310 Norwegian 18- and 19-year-olds found that firstborn children scored 2.3 points higher on the average than second children in the same family. This difference was highly significant ($P < 0.001$). A commentator said, "One puzzle highlighted by these latest findings is why certain other within-family studies have failed to show equally consistent results. Some of these previous null findings, which have all been obtained in much smaller samples, may be explained by inadequate statistical power."[11]

(a) Explain in simple language why tests having low power often fail to give evidence against a null hypothesis even when the hypothesis is really false.

(b) A population mean IQ is typically 100. The Norwegian study found an effect size of 2.3 points. Is that a large effect size?

(c) The study examined 18- and 19-year-olds. By definition, firstborn children are older than second children in the same family. What confounding variable likely plays a role in the findings of this study?

16.36 Low power? (optional) It appears that eating oat bran lowers cholesterol slightly. At a time when oat bran was something of a fad, a paper in the *New England Journal of Medicine* found that it had no significant effect on cholesterol.[12] The paper reported a study with just 20 subjects. Letters to the journal denounced publication of a negative finding from a study with very low power. Explain why lack of significance in a study with low power gives no reason to accept the null hypothesis that oat bran has no effect.

16.37 Type I and Type II errors (optional). Exercise 16.32 asks for a significance test of the null hypothesis that the mean IQ of very-low-birth-weight male babies is 100 against the alternative hypothesis that the mean is less than 100. State in words what it means to make a Type I error and a Type II error in this setting.

OPTIONAL CHAPTERS EXERCISES

These exercises concern the optional material in Chapters 10 and 12.

16.38 Smoking and social class. As the dangers of smoking have become more widely known, clear class differences in smoking have emerged. British government statistics classify adult men by occupation as "managerial and professional" (30% of the population), "intermediate" (29%), or "routine and manual" (34%). A survey finds that 20% of men with managerial and professional occupations smoke, 29% of the intermediate group smoke, and 38% in routine and manual occupations smoke.[13]

(a) Make a tree diagram that displays the percent of all adult British men who smoke.

(b) Write each of the percents provided here, using probability notations such as P(event).

16.39 Exercising in high school. A large Canadian survey asked high school children, "How many days last week did you do an intense physical activity?" Call the response X for short. Based on the findings, here is a probability model for the answer among girls and among boys:[14]

Number of days (X)	0	1	2	3	4	5	6	7
Probability for girls	0.09	0.15	0.19	0.21	0.13	0.13	0.05	0.05
Probability for boys	0.08	0.09	0.14	0.17	0.15	0.16	0.08	0.13

(a) Verify that these are two legitimate discrete probability models.

(b) Describe the event $X < 3$ in words. What is $P(X < 3 \mid \text{girl})$? What is $P(X < 3 \mid \text{boy})$?

(c) Express the event "worked out at least one day" in terms of X. What is the probability of this event for a randomly selected boy?

(d) Without doing any calculations, determine whether X and gender are independent in this study.

16.40 Life tables. The National Center for Health Statistics produces a "life table" for the American population. For each year of age, the table gives the probability that a randomly chosen U.S. resident will die during that year of life. These are *conditional* probabilities, given that the person lived to the birthday that marks the beginning of the year. Here is an excerpt from the table:

Year of life	Probability of death
51	0.00439
52	0.00473
53	0.00512
54	0.00557
55	0.00610

What is the probability that a person who lives to age 50 (the beginning of the 51st year) will live to age 55?

16.41 Cystic fibrosis. Cystic fibrosis (CF) is a lung disorder that often results in death. It is inherited, but it can be inherited only if both parents are carriers of an abnormal gene. In 1989, the CF gene that is abnormal in carriers of cystic fibrosis was identified. The probability that a randomly chosen person of European ancestry carries an abnormal CF gene is 1/25. (The probability is less in other ethnic groups.) The CF20m test detects most but not all harmful mutations of the CF gene. The test is positive for 90% of people who are carriers. It is (ignoring human error) never positive for people who are not carriers. What is the probability that a randomly chosen person of European ancestry tests positive?

16.42 Teenage drivers. An insurance company has the following information about drivers aged 16 to 18 years: 20% are involved in accidents each year; 10% in this age group are A students; among those involved in an accident, 5% are A students.

(a) Let A be the event that a young driver is an A student and C the event that a young driver is involved in an accident this year. State the information given in terms of probabilities and conditional probabilities for the events A and C.

(b) What is the probability that a randomly chosen young driver is an A student and is involved in an accident?

16.43 Teenage drivers, continued. Use your work from the previous exercise to find the percent of A students who are involved in accidents. (Start by expressing this as a conditional probability.)

16.44 Freckles and sun exposure. Researchers in Germany examined the link between pigmentation (hair color, eye color, freckles) and reaction to sun exposure. After examining the reaction to midday summer sun exposure for thousands of Caucasian children, they found the following distribution in their sample:[15]

Reaction type to sun	No freckles	Freckles
(I) Always reddens, never tans	79	73
(II) Always reddens, slight tan	581	367
(III) Sometimes reddens, always tans	1025	324
(IV) Never reddens, always tans	1022	135

(a) Draw a tree diagram to show the possible outcomes for a Caucasian child in this study. Let's use the results of this study to describe a randomly chosen Caucasian child in Germany.

(b) What is the probability that a child has freckles? What is the probability that a child has a type I reaction to sun exposure (always reddens, never tans)?

(c) What is the conditional probability that a child has a type I reaction to sun exposure, knowing that this child has freckles? What is the conditional probability that a child has freckles, knowing that this child has a type I reaction to sun exposure?

16.45 Family planning. Many parents wish to have at least one boy and one girl. For simplicity, let's ignore twins and other multiple births and assume that successive births are independent events with a 50-50 chance that the baby is a boy or a girl.

(a) What is the distribution of the number of boys in a family with 5 children?

(b) What is the probability that a family with 5 children has 5 boys? What is the probability that a family with 5 children has 4 boys and 1 girl?

(c) What is the conditional probability that a couple will have a baby girl next if they already have four boys? Explain how and why this probability is different from the probabilities you calculated in (b).

16.46 Many tests. Exercise 15.40 (page 389) described a study that performed 77 separate significance tests and found that 2 were significant at the 5% level. Suppose that these tests are independent of each other. (In fact, they were not independent, because all involved the same subjects.) If all the null hypotheses are true, each test has probability 0.05 of being significant at the 5% level. Use the binomial distribution to find the probability that 2 or more of the tests are significant at that level.

16.47 Brain injuries. The state of Florida reports that 26% of individuals newly diagnosed with a brain injury are children. A sample survey is designed to interview an SRS of 500 patients newly diagnosed with a brain injury.

(a) What is the actual distribution of the number X in the sample who are children?

(b) What is the probability that 100 or fewer of the patients in the sample are children? 150 or fewer? (Use software or a suitable approximation.)

EESEE CASE STUDIES

The Electronic Encyclopedia of Statistical Examples and Exercises (EESEE) is available on the companion website. These more elaborate stories, with data, provide settings for longer case studies. Here are some suggestions for EESEE stories that apply the ideas you have learned in Chapters 7 through 15.

16.48 **Anecdotes of Bias.** Answer all the questions posed about these incidents. (Cautions about sample surveys.)

16.49 **Checkmating and Reading Skills.** Answer Questions 1(b), 2, and 3 for this story. (Normal distribution, sampling.)

16.50 **Surgery in a Blanket.** Write a response to Questions 1 and 2. (Design of experiments.)

16.51 **Visibility of Highway Signs.** Answer Questions 1, 2, and 3(a) for this study. (Design of experiments, data analysis.)

16.52 **Anecdotes of Significance Testing.** Answer all three questions. (Interpreting P-values.)

16.53 **Blinded Knee Doctors.** Answer Questions 1, 2, and 3 for this story. (Design of experiments, bias.)

16.54 **Is Caffeine Dependence Real?** Answer Questions 1, 2, and 3 for this story. (Design of experiments, bias.)

Statistical Inference

With the principles in hand, we proceed to practice, that is, to inference in fully realistic settings. In the remaining chapters of this book, we will meet many of the most commonly used statistical procedures. The first five chapters concern inference about the distribution of a single variable and inference for comparing the distributions of two variables. Chapters 17 and 18 deal with quantitative variables, whereas Chapters 19 through 21 deal with categorical variables. The last three chapters concern inference for relationships among variables. In Chapter 22, both variables are categorical, with data given as a two-way table of counts of outcomes. Chapter 23 considers inference in the setting of regressing a quantitative response variable on a quantitative explanatory variable. In Chapter 24 we meet methods for comparing the mean response in more than two groups. Here, the explanatory variable (group) is categorical and the response variable is quantitative. Chapter 25 reviews this part of the text and provides more comprehensive exercises.

With greater data complexity comes greater reliance on technology. In the last chapters of Part III you will more often be interpreting the output of statistical software or using software yourself. Fortunately, you can grasp the ideas without step-by-step arithmetic. These chapters introduce elaborate methods on the foundation we have laid without introducing fundamentally new concepts.

The four-step process for approaching a statistical problem can guide much of your work in all Part III chapters. You should review the outlines of the four-step process for a confidence interval (page 342) and for a test of significance (page 354). Many examples and exercises in these chapters involve both carrying out inference and thinking about inference in practice. Remember that any inference method is useful only under certain conditions, and that you must judge these conditions before rushing to inference.

PART III

INFERENCE ABOUT VARIABLES

INFERENCE ABOUT RELATIONSHIPS

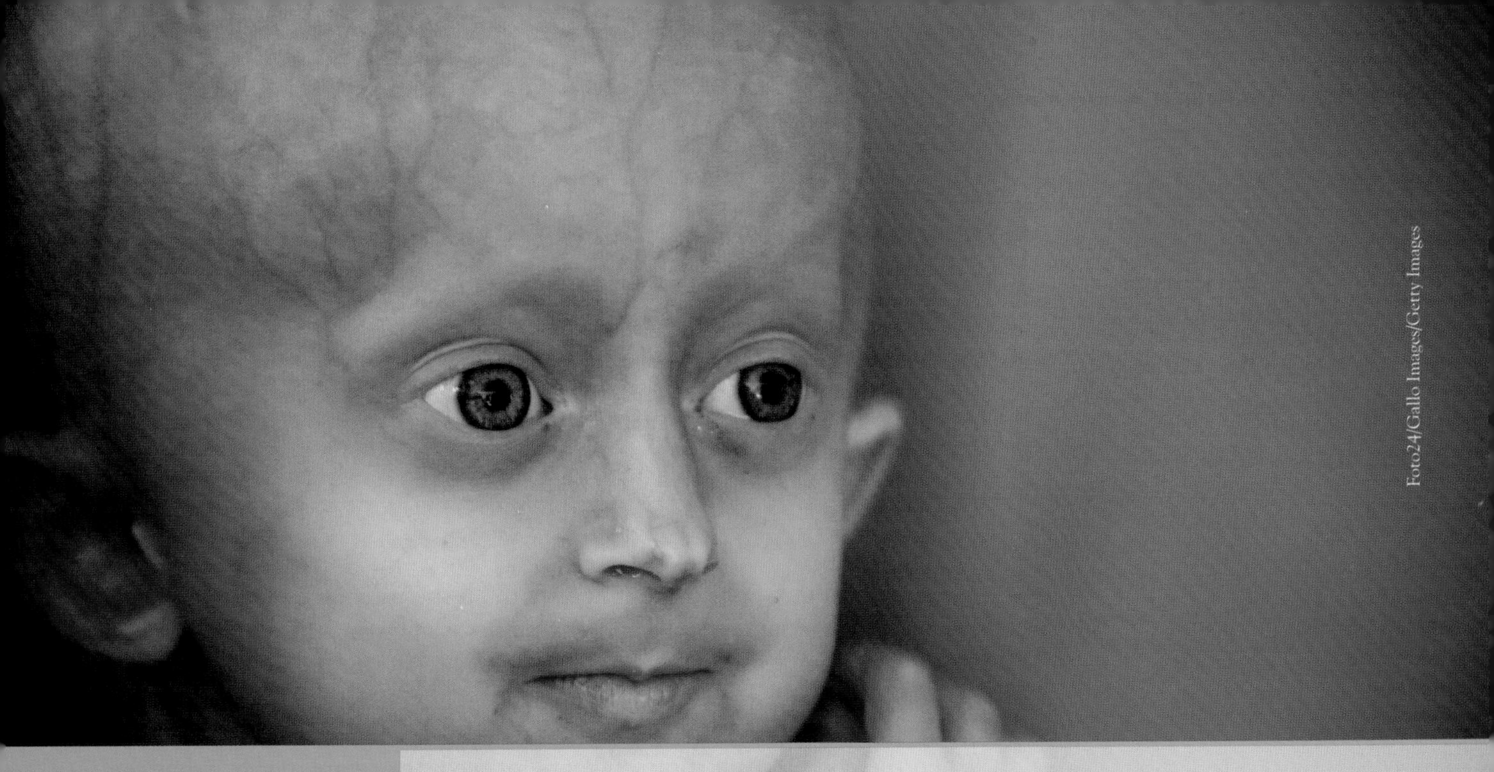

Foto24/Gallo Images/Getty Images

CHAPTER 17 Inference about a Population Mean

This chapter describes confidence intervals and significance tests for the mean μ of a population. We used the z procedures in this setting to introduce the ideas of confidence intervals and tests. Now we discard the unrealistic condition that we know the population standard deviation σ and present procedures for practical use. We also pay more attention to the real-data setting of our work. The details of confidence intervals and tests change only slightly when you don't know σ. More important, you can interpret your results just as before.

IN THIS CHAPTER WE COVER...

- Conditions for inference
- The t distributions
- The one-sample t confidence interval
- The one-sample t test
- Using technology
- Matched pairs t procedures
- Robustness of t procedures

Conditions for inference

Confidence intervals and tests of significance for the mean μ of a Normal population are based on the sample mean $\overline{x}$. Confidence levels and P-values are probabilities calculated from the sampling distribution of $\overline{x}$. Here are the conditions needed for realistic inference about a population mean.

CONDITIONS FOR INFERENCE ABOUT A MEAN

- We can regard our data as a **simple random sample** (SRS) from the population. This condition is very important.
- Observations from the population have a **Normal distribution** with mean μ and standard deviation σ. Both μ and σ are unknown parameters. In practice, inference procedures can accommodate some deviations from the Normality condition when the sample is large enough.

There is another condition that applies to all the inference methods in this book: *The population must be much larger than the sample,* say at least 20 times as large.[1] All our examples and exercises satisfy this condition. Practical settings in which the sample is a large part of the population are rather special, and we will not discuss them.

When the conditions for inference are satisfied, the sample mean $\bar{x}$ has the Normal distribution with mean μ and standard deviation $\sigma/\sqrt{n}$. Because we don't know σ, we estimate it by the sample standard deviation s. We then estimate the standard deviation of $\bar{x}$ by $s/\sqrt{n}$. This quantity is called the *standard error* of the sample mean $\bar{x}$.

STANDARD ERROR

When the standard deviation of a statistic is estimated from data, the result is called the **standard error** of the statistic. The standard error of the sample mean $\bar{x}$ is $s/\sqrt{n}$.

APPLY YOUR KNOWLEDGE

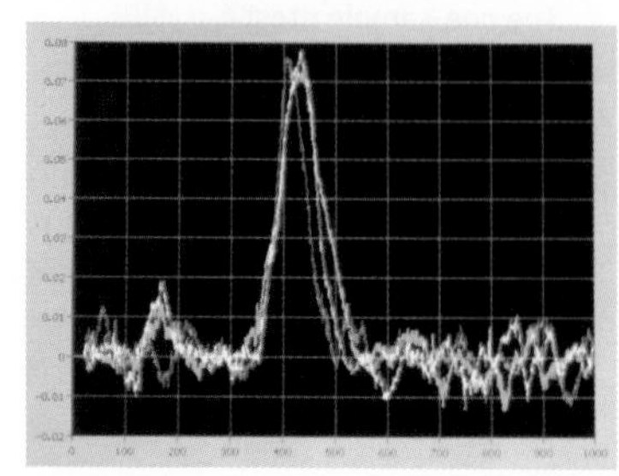

Donald Pye/Alamy

17.1 Neuronal activity. Neurons fire action potentials ("spikes") at different frequencies under different circumstances. A neuron is recorded in an awake-behaving monkey looking at a luminous dot 40 centimeters away to determine the neuron's baseline activity. Over the course of 72 trials randomly interspersed with other conditions, that neuron is found to have an average baseline activity of $\bar{x} = 27.8$ spikes per second with standard deviation $s = 32.2$ spikes per second. What is the standard error of the mean?

17.2 Spider webs. Researchers examined the mechanical properties of webs spun by the orb spider, *Nephelia edulis*.[2] A table in the publication reports summary statistics for a number of variables in the form of the mean plus or minus the standard error of the mean. This form is very common in scientific publications. For 15 adult female spiders, web areas (in centimeter^2) are 634 ± 30. What are $\bar{x}$ and s for these 15 spider webs?

The *t* distributions

If we knew the value of σ, we would base confidence intervals and tests for μ on the one-sample z statistic

$$z = \frac{\bar{x} - \mu}{\sigma/\sqrt{n}}$$

This z statistic has the standard Normal distribution $N(0, 1)$. In practice, we don't know σ, so we substitute the standard error $s/\sqrt{n}$ of $\bar{x}$ for its standard deviation $\sigma/\sqrt{n}$. The statistic that results does not have a Normal distribution. It has a distribution that is new to us, called a *t distribution*.

THE ONE-SAMPLE *t* STATISTIC AND THE *t* DISTRIBUTIONS

Draw an SRS of size n from a large population that has the Normal distribution with mean μ and standard deviation σ. The **one-sample *t* statistic**

$$t = \frac{\bar{x} - \mu}{s/\sqrt{n}}$$

has the ***t* distribution** with $n - 1$ degrees of freedom.

The t statistic has the same interpretation as any standardized statistic: It says how far $\bar{x}$ is from its mean (μ) in standard deviation units ($s/\sqrt{n}$).

Sample standard deviations obtained with larger samples are better estimates of the unknown population standard deviations σ. This is reflected in the fact that there is a different t distribution for each sample size. We specify a particular t distribution by giving its **degrees of freedom (df).** When doing inference about a single population mean μ, the t statistic follows a t distribution with $n - 1$ degrees of freedom.[3] (Other inference methods use different degrees of freedom.) We will write the t distribution with $n - 1$ degrees of freedom as $t(n - 1)$ for short.

degrees of freedom

Figure 17.1 compares the density curves of the standard Normal distribution and of the t distributions with 2 and 9 degrees of freedom. The figure illustrates these facts about the t distributions:

- The density curves of the t distributions are similar in shape to the standard Normal curve. They are symmetric about 0, single-peaked, and bell-shaped.
- The spread of the t distributions is a bit greater than that of the standard Normal distribution. The t distributions in Figure 17.1 have more probability in the tails and less in the center than does the standard Normal distribution. This is true because substituting the estimate s for the fixed parameter σ introduces more variation into the statistic. (Therefore, inference will be generally less precise.)
- As the degrees of freedom increase, the t density curve approaches the $N(0, 1)$ curve ever more closely. This happens because s estimates σ more accurately as the sample size increases. So using s in place of σ causes little extra variation when the sample is large.

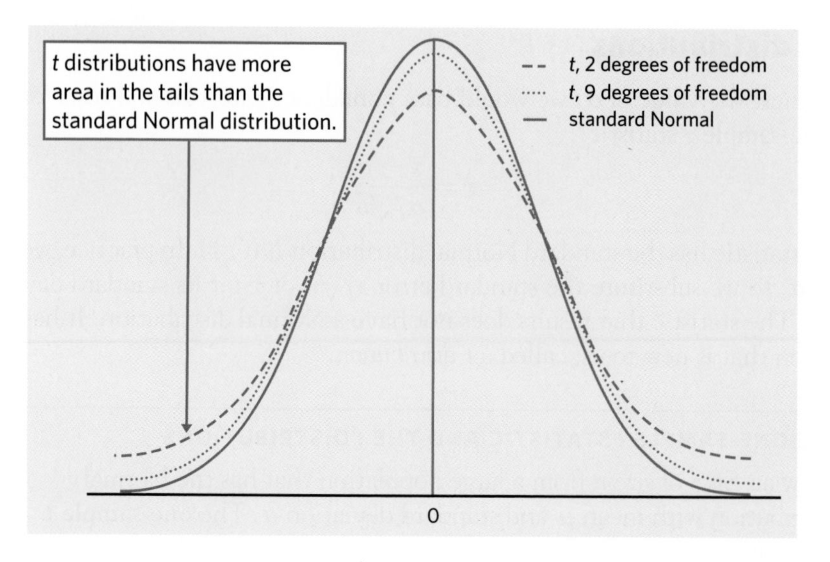

FIGURE 17.1 Density curves for the t distributions with 2 and 9 degrees of freedom and for the standard Normal distribution. All are symmetric with center 0. The t distributions are somewhat more spread out.

Students often wonder about the meaning of the degrees of freedom of the t distribution. The degrees of freedom help define the exact shape of a t distribution. For the purpose of performing inference procedures, all you need to know is that there are many t distributions and you must specify, through the degrees of freedom, which t distribution is relevant for your computations.

Table C in the back of the book gives critical values for the t distributions. Each row in the table contains critical values for the t distribution whose degrees of freedom appear at the left of the row. Because t distributions are symmetric about zero, only positive critical values are displayed. (You can easily figure out a negative critical value by symmetry.) For convenience, we label the table entries both by the confidence level C (in percent) required for confidence intervals and by the one-sided and two-sided P-values for each critical value. You have already used the standard Normal critical values z^* in the bottom row of Table C. By looking down any column, you can check that the t critical values approach the Normal values as the degrees of freedom increase, but that the t^* values are always larger than the corresponding z^*. It is the price we pay for replacing the unknown population standard deviation σ with the computed sample standard deviation s. As in the case of the Normal table, statistical software makes Table C unnecessary.

Better statistics, better beer The t distribution and the t inference procedures were invented by William S. Gosset (1876–1937). Gosset worked for the Guinness brewery to make better beer. He used his new t procedures to find the best varieties of barley and hops. Gosset's statistical work helped him become head brewer. Because of Guinness's strict nondisclosure clause, Gosset published his statistical work under the pen name "Student." You will often see the t distribution called "Student's t" in his honor.

EXAMPLE 17.1 *t* critical values

Figure 17.1 shows the density curve for the t distribution with 9 degrees of freedom. What point on this distribution has probability 0.05 to its right? In Table C, look in the df = 9 row above one-sided P-value .05 and you will find that this critical value is $t^* = 1.833$.

To use software, enter the degrees of freedom and the probability you want to the *left* (the cumulative area), 0.95 in this case. Excel gives $t^* = 1.833113$ for the command "=T.INV(0.95,9)." And here is Minitab's output:

```
Student's t distribution with 9 DF
P(X<=x)        x
   0.95  1.83311
```

■

APPLY YOUR KNOWLEDGE

17.3 Critical values. Use Table C or software to find

(a) the critical value for a one-sided test (in the greater-than direction) with level $\alpha = 0.05$ based on the $t(5)$ distribution.

(b) the critical value for a 98% confidence interval based on the $t(21)$ distribution.

17.4 More critical values. You have an SRS of size 25 and calculate the one-sample t statistic. Using Table C or software, what is the critical value t^* such that

(a) t has probability 0.025 to the right of t^*?

(b) t has probability 0.75 to the left of t^*?

The one-sample *t* confidence interval

To analyze samples from Normal populations with unknown σ, just replace the standard deviation $\sigma/\sqrt{n}$ of $\bar{x}$ by its standard error $s/\sqrt{n}$ in the z procedures of Chapter 14. The confidence interval and test that result are *one-sample t procedures*. Critical values and P-values come from the t distribution with $n - 1$ degrees of freedom. The one-sample t procedures are similar in both reasoning and computational detail to the z procedures.

THE ONE-SAMPLE *t* CONFIDENCE INTERVAL

Draw an SRS of size n from a large population having unknown mean μ. A level C **confidence interval for μ** is

$$\bar{x} \pm t^* \frac{s}{\sqrt{n}}$$

where t^* is the critical value for the $t(n - 1)$ density curve with area C between $-t^*$ and t^*. This interval is exact when the population distribution is Normal and is approximately correct for large n in other cases.

EXAMPLE 17.2 Testosterone and obesity in adolescent males

We follow the four-step process for a confidence interval, outlined on page 342.

STATE: Obesity in adult males is associated with lower levels of sex hormone. A study investigated a possible link between obesity and plasma testosterone concentrations in adolescent males between the ages of 14 and 20 years. Here are the data for 25 obese adolescent males, measured in nanomoles per liter of blood (nmol/l):[4]

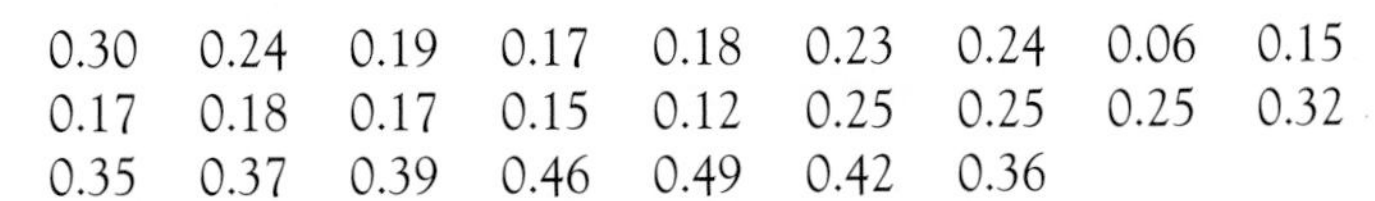

0.30	0.24	0.19	0.17	0.18	0.23	0.24	0.06	0.15
0.17	0.18	0.17	0.15	0.12	0.25	0.25	0.25	0.32
0.35	0.37	0.39	0.46	0.49	0.42	0.36		

We want to estimate the mean testosterone level in obese adolescent males.

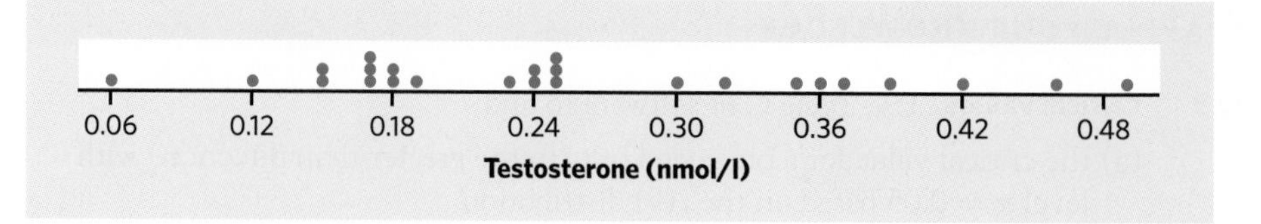

FIGURE 17.2 Dotplot of plasma testosterone levels in obese adolescent males, for Example 17.2.

PLAN: We will estimate the mean testosterone level μ for all obese adolescent males by giving a 95% confidence interval.

SOLVE: We must first check the conditions for inference.

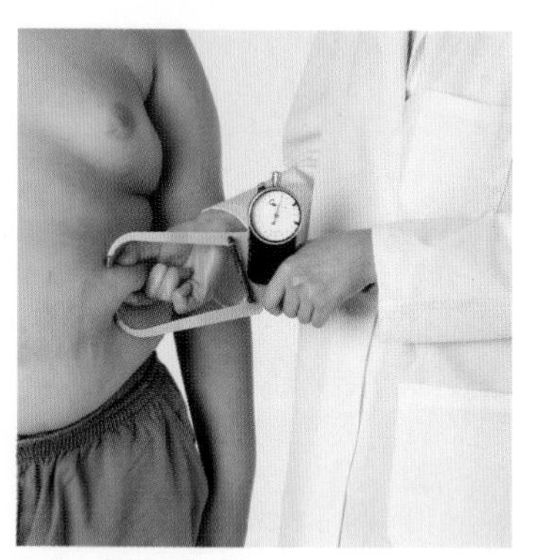

Dorling Kindersley/Getty Images

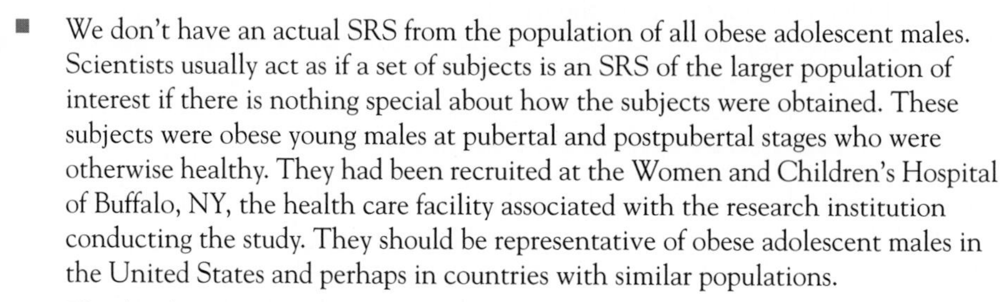

- We don't have an actual SRS from the population of all obese adolescent males. Scientists usually act as if a set of subjects is an SRS of the larger population of interest if there is nothing special about how the subjects were obtained. These subjects were obese young males at pubertal and postpubertal stages who were otherwise healthy. They had been recruited at the Women and Children's Hospital of Buffalo, NY, the health care facility associated with the research institution conducting the study. They should be representative of obese adolescent males in the United States and perhaps in countries with similar populations.

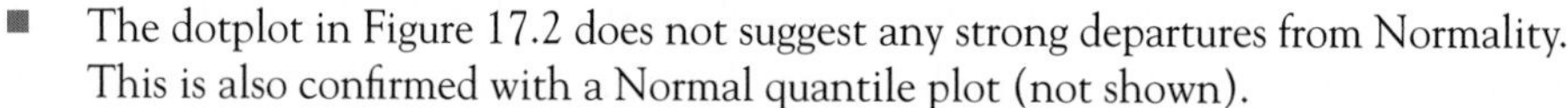

- The dotplot in Figure 17.2 does not suggest any strong departures from Normality. This is also confirmed with a Normal quantile plot (not shown).

We can proceed to calculation. For these data,

$$\bar{x} = 0.2584 \quad \text{and} \quad s = 0.1115$$

The degrees of freedom are $n - 1 = 24$. From Table C we find that for 95% confidence, $t^* = 2.064$. Similarly, Excel returns the t^* value 2.063898562 to the command "`=T.INV(0.975,24)`." The confidence interval is

$$\begin{aligned}\bar{x} \pm t^*\frac{s}{\sqrt{n}} &= 0.2584 \pm 2.064\frac{0.1115}{\sqrt{25}} \\ &= 0.2584 \pm 0.0460 \\ &= 0.2124 \text{ to } 0.3044 \text{ nmol/l}\end{aligned}$$

CONCLUDE: We are 95% confident that the mean plasma testosterone level for all obese adolescent males in the United States is between 0.212 and 0.304 nmol/l. ■

Our work in Example 17.2 is very similar to what we did in Chapter 14. But because we do not know the true value of the population standard deviation σ, we use the sample standard deviation $s = 0.1115$ calculated from the data. We also use the t critical value $t^* = 2.064$ for 24 degrees of freedom instead of the standard Normal critical value $z^* = 1.960$.

The one-sample t confidence interval has the form

$$\text{estimate} \pm t^*\text{SE}_{\text{estimate}}$$

where "SE" stands for "standard error." We will encounter a number of confidence intervals that have this common form. In Example 17.2 the estimate is the sample mean $\overline{x}$, and its standard error is

$$\mathrm{SE}_{\overline{x}} = \frac{s}{\sqrt{n}} = \frac{0.1115}{\sqrt{25}} = 0.0223$$

Software will find $\overline{x}$, s, $\mathrm{SE}_{\overline{x}}$, and the confidence interval from the data. Figure 17.4 displays typical software output for Example 17.2.

APPLY YOUR KNOWLEDGE

17.5 Critical values. Using Table C or technology, what critical value t^* would you use for a confidence interval for the mean of the population in each of the following situations?

(a) A 95% confidence interval based on $n = 10$ observations

(b) A 99% confidence interval from an SRS of 20 observations

(c) An 80% confidence interval from a sample of size 7

17.6 Spider webs. Exercise 17.2 gave the summary data (mean 634, standard error 30 cm^2) for the web area of 15 female orb spiders. Assuming an approximately Normal distribution of web areas, and assuming that we can regard these 15 female orb spiders as an SRS of the population of all female orb spiders, give the 95% confidence interval for the web area μ in this population.

17.7 Healing of skin wounds. Biologists studying the healing of skin wounds measured the rate at which new cells closed a razor cut made in the skin of an anesthetized newt. Here are data from 18 newts, measured in micrometers per hour (μm/h):[5]

29	27	34	40	22	28	14	35	26
35	12	30	23	18	11	22	23	33

Use a 95% confidence interval to estimate the mean healing rate of razor-cut skin wounds in anesthetized newts. Follow the four-step process as illustrated in Example 17.2.

17.8 Trout habitat. A student in wildlife management studied trout habitat in the upper Shavers Fork watershed in West Virginia. The springtime water pH of 29 randomly selected tributary sample sites was found to have the following values:[6]

6.2	6.3	5.0	5.8	4.6	4.7	4.7	5.4	6.2	6.0
5.4	5.9	6.2	6.1	6.0	6.3	6.2	5.8	6.2	6.3
6.3	6.3	6.4	6.5	6.6	6.1	6.3	4.4	6.7	

Use a 90% confidence interval to estimate the mean springtime water pH of the tributary water basin around the Shavers Fork watershed. Follow the four-step process as illustrated in Example 17.2.

The one-sample *t* test

Like the confidence interval, the *t* test is very similar to the *z* test we studied earlier.

THE ONE-SAMPLE *t* TEST

Draw an SRS of size n from a large population having unknown mean μ. To **test the hypothesis H_0: $\mu = \mu_0$**, compute the **one-sample *t* statistic**

$$t = \frac{\bar{x} - \mu_0}{s/\sqrt{n}}$$

In terms of a variable T having the $t(n-1)$ distribution, the P-value for a test of H_0 against

H_a: $\mu > \mu_0$ is $P(T \geq t)$

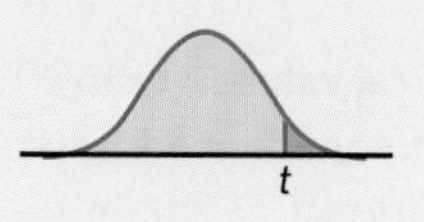

H_a: $\mu < \mu_0$ is $P(T \leq t)$

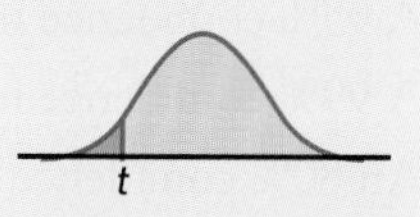

H_a: $\mu \neq \mu_0$ is $2P(T \geq |t|)$

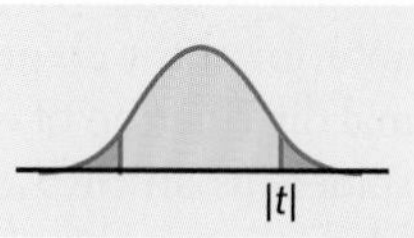

These P-values are exact if the population distribution is Normal; they are approximately correct for large n in other cases.

EXAMPLE 17.3 Children with progeria

Hutchinson-Gilford progeria syndrome is a rare genetic condition that produces rapid aging in children. As a result, cardiovascular disease is a common cause of death in the teenage years. A clinical study examined the effect of treatment with the drug lonafarnib on a number of physiological outcomes. To begin with, the researchers measured the pulse wave velocity (PWV) of 18 children diagnosed with progeria. PWV is the standard measure of vascular stiffness, an important factor in cardiovascular health. Here are the findings (in meters per second, m/s):[7]

18.8	17.6	17.5	16.0	14.8	14.1	13.7	13.1	12.9
12.9	12.4	10.1	9.3	9.1	8.3	8.3	7.9	7.2

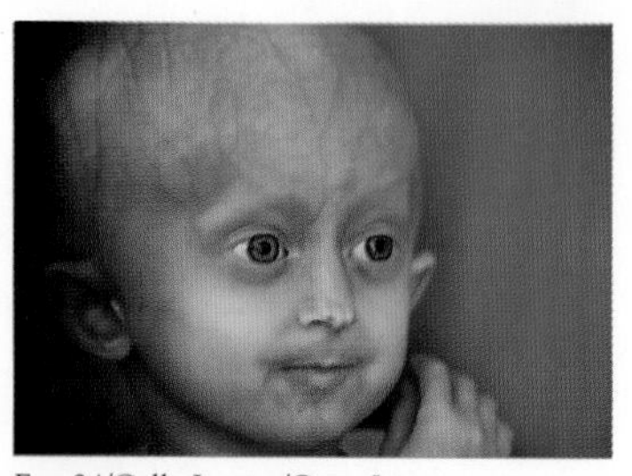
Foto24/Gallo Images/Getty Images

These data have mean $\bar{x} = 12.44$ m/s and standard deviation $s = 3.638$ m/s. In healthy children, PWV values above 6.6 m/s would be considered abnormal. We follow the four-step process for a significance test, outlined on page 354.

STATE: A study reports that the mean PWV of 18 children diagnosed with progeria is $\bar{x} = 12.44$ m/s with standard deviation $s = 3.638$ m/s. Is this significant evidence that the mean PWV of children with progeria is abnormally high, that is, above 6.6 m/s?

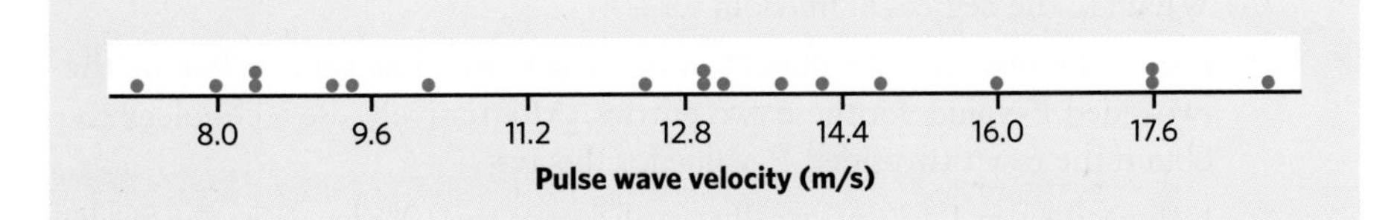

FIGURE 17.3 Dotplot of pulse wave velocity (PWV) in children with progeria, for Example 17.3.

PLAN: Individual PWV values vary, so we ask the question in terms of the mean PWV μ for the population of all children diagnosed with progeria. The null hypothesis is "no difference" from the mean $\mu_0 = 6.6$ still considered normal for healthy children. The alternative is one-sided, because we expect progeria to increase vascular stiffness and thus increase PWV on average.

$$H_0: \mu = 6.6$$
$$H_a: \mu > 6.6$$

SOLVE: First check the conditions for inference. As in Example 17.2, we are willing to regard these 18 children with progeria as an SRS from the population of all children with progeria. The dotplot in Figure 17.3 shows no obvious departure from Normality and no outlier. The t test should be accurate.

The basic statistics are

$$\bar{x} = 12.44 \quad \text{and} \quad s = 3.638$$

The one-sample t statistic is

$$t = \frac{\bar{x} - \mu_0}{s/\sqrt{n}} = \frac{12.44 - 6.6}{3.638/\sqrt{18}} = 6.81$$

Since the alternative is one-sided in the greater-than direction, the P-value for $t = 6.81$ is the area to the right of 6.81 under the t distribution curve with degrees of freedom $n - 1 = 17$. A TI-83 calculator gives this area as $P = 0.0000015$ using the command `tcdf(6.81,1E99,17)`. Excel gives the same answer for the command `=T.DIST.RT(6.81,17)`. You can also use technology to perform the whole set of computations directly from the raw data (see Figure 17.5).

Without software, we can obtain an approximate P-value by using Table C. In Table C, use the df = 17 row for entries closest to $t = 6.81$. The observed t is greater than the critical value for one-sided P-value 0.0005.

df = 17

t^*	3.965
One-sided P	.0005

CONCLUDE: There is very strong evidence ($P < 0.0005$) that the true mean PWV of children with progeria is greater than 6.6 m/s. Therefore, on average, children with progeria have abnormally high PWV, which is indicative of abnormally high vascular stiffness. ■

APPLY YOUR KNOWLEDGE

17.9 Is it significant? The one-sample t statistic from a sample of $n = 25$ observations for the two-sided test of

$$H_0: \mu = 64$$
$$H_a: \mu \neq 64$$

has the value $t = 1.12$.

(a) What are the degrees of freedom for t?

(b) Locate the two critical values t^* from Table C that bracket t. What are the two-sided P-values for these two entries? Alternatively, use technology to obtain the exact two-sided P-value for this test.

(c) Is the value $t = 1.12$ statistically significant at the 10% level? At the 5% level?

17.10 Is it significant? The one-sample t statistic for testing

$$H_0: \mu = 0$$
$$H_a: \mu > 0$$

from a sample of $n = 15$ observations has the value $t = 1.82$.

(a) What are the degrees of freedom for this statistic?

(b) Give the two critical values t^* from Table C that bracket t. What are the one-sided P-values for these two entries? Alternatively, use technology to obtain the exact one-sided P-value for this test.

(c) Is the value $t = 1.82$ significant at the 5% level? Is it significant at the 1% level?

17.11 Trout habitat. Do the data of Exercise 17.8 give good reason to think that the springtime water in the tributary water basin around the Shavers Fork watershed is not neutral (a neutral pH is the pH of pure water, pH 7)? Follow the four-step process as illustrated in Example 17.3.

Using technology

Any technology suitable for statistics will implement the one-sample t procedures. You can read and use almost any output now that you know what to look for. Figure 17.4 displays output for the 95% confidence interval of Example 17.2 from a graphing calculator, three statistical programs, and a spreadsheet program. The TI-83, Minitab, R, and SPSS outputs are straightforward. All three give the estimate $\bar{x}$ and the confidence interval plus a selection of other information. The confidence interval agrees with our hand calculation in Example 17.2. In general, software results are more accurate because of the rounding in hand calculations. Excel gives several descriptive measures but does not give the confidence interval. The entry labeled "Confidence Level (95.0%)" is the margin of error m, and the first entry, labeled "Mean," is the sample average $\bar{x}$. You can combine these values to compute the confidence interval $\bar{x} \pm m$.

Figure 17.5 displays output from four statistical programs and a graphing calculator for the t test in Example 17.3. They all give the sample mean $\bar{x}$, the t statistic, and its P-value. Note that SPSS always gives a two-sided P-value ("Sig. (2-tailed)"), whereas JMP gives a two-sided P-value called "Prob > |t|" as well as two one-sided P-values called "Prob > t" and "Prob < t." It is up to you to know what area your alternative hypothesis requires and to adjust the P-value provided if necessary. JMP also provides a graphical display of the sampling distribution under the null hypothesis and the position of the sample mean in this context. You can use it to make sure that you select the correct P-value from the information

SPSS

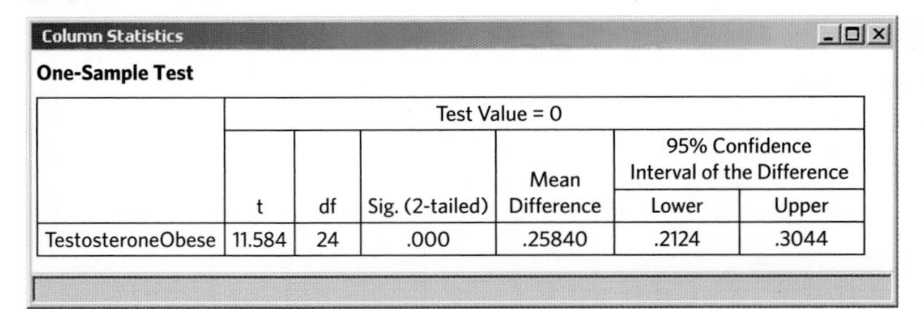

Column Statistics

One-Sample Test

	Test Value = 0					
					95% Confidence Interval of the Difference	
	t	df	Sig. (2-tailed)	Mean Difference	Lower	Upper
TestosteroneObese	11.584	24	.000	.25840	.2124	.3044

TI-83

```
TInterval
 (.21236,.30444)
 x̄=.2584
 Sx=.111530265
 n=25
```

Minitab

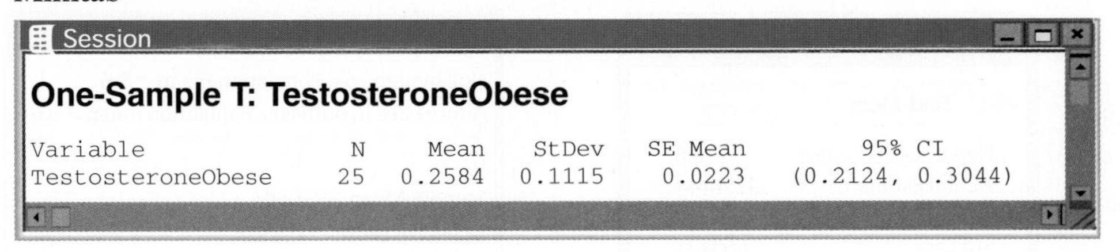

Session

One-Sample T: TestosteroneObese

```
Variable            N    Mean   StDev  SE Mean        95% CI
TestosteroneObese  25  0.2584  0.1115   0.0223  (0.2124, 0.3044)
```

R

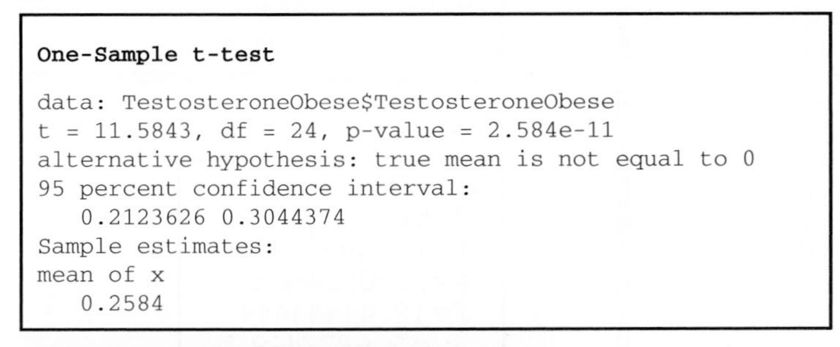

```
One-Sample t-test

data: TestosteroneObese$TestosteroneObese
t = 11.5843, df = 24, p-value = 2.584e-11
alternative hypothesis: true mean is not equal to 0
95 percent confidence interval:
   0.2123626 0.3044374
Sample estimates:
mean of x
   0.2584
```

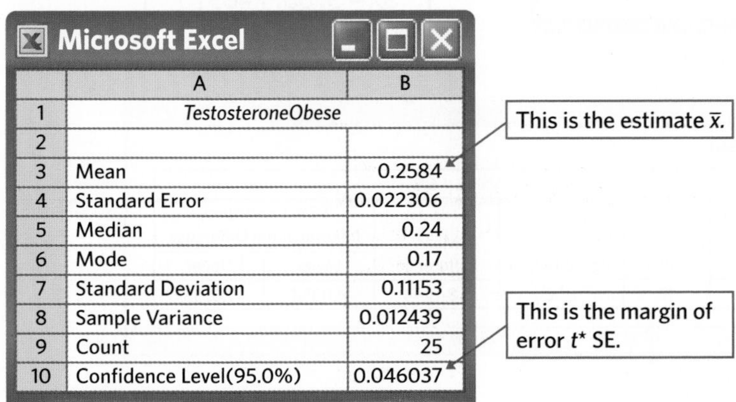

Microsoft Excel

	A	B
1	*TestosteroneObese*	
2		
3	Mean	0.2584
4	Standard Error	0.022306
5	Median	0.24
6	Mode	0.17
7	Standard Deviation	0.11153
8	Sample Variance	0.012439
9	Count	25
10	Confidence Level(95.0%)	0.046037

This is the estimate $\bar{x}$.
This is the margin of error t^* SE.

FIGURE 17.4 The t confidence interval for Example 17.2: output from a graphing calculator, three statistical programs, and a spreadsheet program.

provided. Some software programs, like Minitab and SPSS, round the P-value to three or four decimal places, which means that you sometimes see output like the "P 0.000" in Figure 17.5. This doesn't mean that the P-value is zero, but that it is closer to 0.0009 than to 0.001 (in other words, read as "$P < 0.0005$").

Some software programs also provide a confidence interval for the population mean μ in their output (for example, SPSS). Notice, however, that the Minitab

Minitab

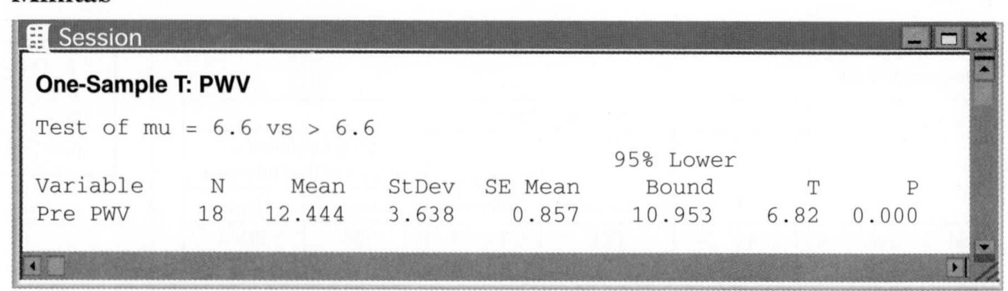

Session

One-Sample T: PWV

Test of mu = 6.6 vs > 6.6

Variable	N	Mean	StDev	SE Mean	95% Lower Bound	T	P
Pre PWV	18	12.444	3.638	0.857	10.953	6.82	0.000

JMP

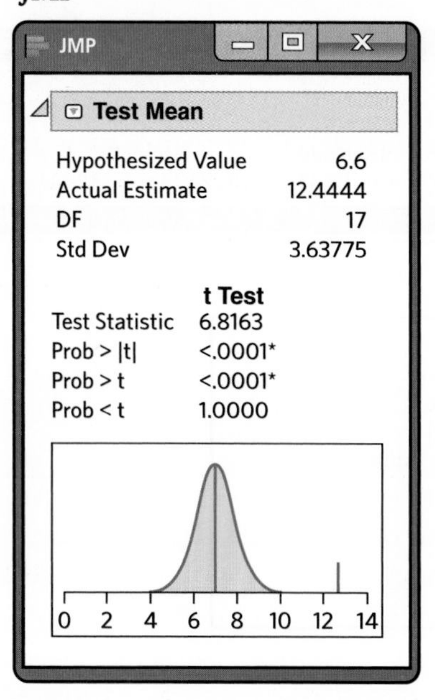

JMP

Test Mean

Hypothesized Value	6.6
Actual Estimate	12.4444
DF	17
Std Dev	3.63775

	t Test
Test Statistic	6.8163
Prob > |t|	<.0001*
Prob > t	<.0001*
Prob < t	1.0000

CrunchIt!

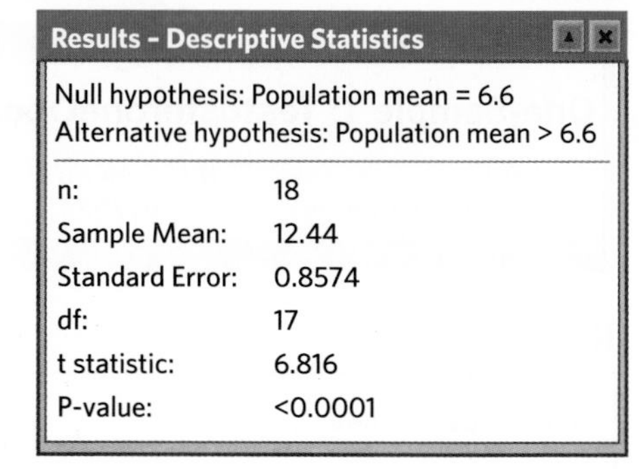

Results - Descriptive Statistics

Null hypothesis: Population mean = 6.6
Alternative hypothesis: Population mean > 6.6

n:	18
Sample Mean:	12.44
Standard Error:	0.8574
df:	17
t statistic:	6.816
P-value:	<0.0001

TI-83

T-Test
μ>6.6
t=6.816273473
p=1.501248E-6
x̄=12.44444444
Sx=3.637746915
n=18

SPSS

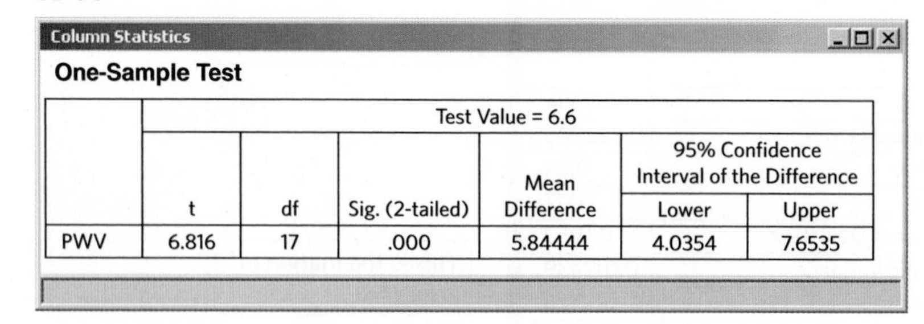

Column Statistics

One-Sample Test

	Test Value = 6.6					
	t	df	Sig. (2-tailed)	Mean Difference	95% Confidence Interval of the Difference Lower	Upper
PWV	6.816	17	.000	5.84444	4.0354	7.6535

FIGURE 17.5 The t test for Example 17.3: output from a graphing calculator and four statistical programs.

output in Figure 17.5 displays "95%C Lower Bound 10.953." This is not the lower bound of a 95% confidence interval but the lower bound of the upper 95% of the sampling distribution (because the test is one-sided). Don't hesitate to use the software's help menu if you are not entirely sure that you understand the output you get.

Lastly, some software programs let you use either the raw data or summary statistics ($\overline{x}$, s, and n) for t procedures. This can lead to very small differences in output values for t and P when the summary statistics are rounded. Accurate

P-values are the biggest advantage of software for the t procedures—although speedy computations are nice too.

Matched pairs t procedures

The study of healing in Exercise 17.7 estimated the mean healing rate for newts under natural conditions, but the researchers then compared results under several conditions (as illustrated in Exercises 17.15 and 17.45). Comparative studies are more convincing than single-sample investigations. For that reason, one-sample inference is less common than comparative inference. However, a common design to compare two conditions makes use of one-sample procedures. In a **matched pairs design,** subjects are matched in pairs, and each treatment is given to one subject in each pair. Another situation calling for matched pairs arises when we record two responses for each subject, as in before-and-after observations or when two treatments are given at different times. We first introduced the matched pairs design in Chapter 8.

matched pairs design

MATCHED PAIRS t PROCEDURES

To compare the responses to the two treatments in a matched pairs design, find the difference between the responses within each pair. Then apply the one-sample t procedures to these differences.

The parameter μ in a matched pairs t procedure is the mean difference in the responses to the two treatments within matched pairs of subjects in the entire population.

EXAMPLE 17.4 A treatment for progeria

STATE: In Example 17.3 a clinical study recorded PWV, a measure of vascular stiffness, in 18 children diagnosed with progeria. The objective of the study was to test the effectiveness of treatment with the drug lonafarnib, an enzyme inhibitor that interferes with the mutated protein progerin responsible for the syndrome. The 18 children in Example 17.3 all received a daily dose of lonafarnib for two years. At the end of the two-year treatment, PWV was assessed once more. Table 17.1 gives the PWV values (in m/s) at the beginning ("Untreated") and at the end ("Treated") of the study for each child. Is there evidence that a two-year treatment with lonafarnib lowers PWV in children with progeria?

PLAN: Take μ to be the mean difference (PWV treated minus PWV untreated) in the population of children with progeria. The null hypothesis says that the drug lonafarnib has no effect, and H_a says that the drug is effective and PWV values are smaller after treatment, on average. So we test the hypotheses

$$H_0: \mu = 0$$
$$H_a: \mu < 0$$

SOLVE: The subjects are not an actual SRS from the population of all children with progeria. But we are willing to regard them as an SRS. To analyze the data, subtract the

TABLE 17.1 PWV (m/s) of 18 children with progeria, before and after treatment

Child	Untreated	Treated	Differerence	Child	Untreated	Treated	Differerence
1	18.8	12.6	−6.2	10	12.9	7.5	−5.4
2	17.6	10.8	−6.8	11	12.4	9.4	−3.0
3	17.5	10.1	−7.4	12	10.1	6.4	−3.7
4	16.0	10.1	−5.9	13	9.3	9.0	−0.3
5	14.8	9.2	−5.6	14	9.1	6.2	−2.9
6	14.1	7.6	−6.5	15	8.3	7.2	−1.1
7	13.7	10.8	−2.9	16	7.9	7.3	−0.6
8	13.1	7.7	−5.4	17	8.3	5.7	−2.6
9	12.9	6.8	−6.1	18	7.2	9.1	1.9

untreated PWV values from the treated PWV values for each child. The 18 differences form a single sample from the population with unknown mean μ. They appear in the "Difference" column in Table 17.1. A negative difference shows that a subject's PWV has decreased after treatment. Figure 17.6 is a stemplot of the differences. The distribution is very mildly skewed, but with 18 values, the t procedures will still be valid.

The 18 differences have

$$\bar{x} = -3.917 \quad \text{and} \quad s = 2.644$$

The one-sample t statistic is therefore

$$t = \frac{\bar{x} - 0}{s/\sqrt{n}} = \frac{-3.917 - 0}{2.644/\sqrt{18}} = -6.29$$

−7	4
−6	8521
−5	9644
−4	
−3	70
−2	996
−1	1
−0	63
0	
1	9

FIGURE 17.6 Stemplot of PWV differences for 18 children with progeria, for Example 17.4.

df = 17

t^*	3.965
One-sided P	.0005

Find the P-value from the $t(17)$ distribution. (Remember that the degrees of freedom are 1 less than the sample size.) The alternative is one-sided. Table C shows that 6.29 is greater than the critical value 3.965 corresponding to the one-sided $P = 0.0005$. The P-value of our test is therefore less than 0.0005. Software gives the value $P = 0.000004$.

CONCLUDE: The data show very strong evidence that a two-year treatment with lonafarnib lowers PWV (and thus vascular stiffness) in children with progeria, on average. In this study, the 18 children benefited from an average reduction in PWV of 3.9 m/s from their earlier average PWV of 12.4 m/s. Using technology (as in Example 17.2), we obtain a 95% confidence interval for the population mean change in PWV of (−5.23, −2.60) m/s. That is, we are 95% confident that the true mean reduction in PWV after taking lonafarnib for two years is between 2.6 and 5.2 m/s. ■

Example 17.4 illustrates how to turn matched pairs data into single-sample data by calculating a difference within each pair.[8] We are making inferences about a single population, the population of all differences within matched pairs. *It is incorrect to ignore the matching and analyze the data as if we had two samples*, one from children treated with lonafarnib and a second from children left untreated. Inference procedures for comparing two samples assume that the samples are selected independently of each other, which is not true when the same subjects are measured twice. The proper analysis depends on the design used to produce the data. We will describe the procedures used for two independent samples in the next chapter.

APPLY YOUR KNOWLEDGE

Many exercises from this point on ask you to give the P-value of a t test. If you have suitable technology, give the exact P-value. Otherwise, use Table C to give an approximate P-value.

17.12 Water diet. Water consumption before a meal has been shown to help reduce the caloric intake from that meal, but does it translate into actual weight loss? An experiment randomly assigned overweight and obese individuals aged 55 to 75 years to one of two diet groups for 12 weeks: a hypocaloric diet supplemented with 16 ounces of bottled water prior to each of the three daily meals (water group) or a hypocaloric diet alone (nonwater group).[9]

(a) Each subject had his or her body mass index (BMI) computed before and after the 12-week diet. For the 23 subjects assigned to the water group, the mean BMIs before and after treatment were 32.16 and 29.5, respectively. Explain briefly why a matched pairs procedure must be used to assess whether the combination of a hypocaloric diet and pre-meal water intake significantly lowers BMI.

(b) The 25 patients assigned to the nonwater group had mean BMIs before and after the 12-week diet of 31.8 and 29.9, respectively. Explain briefly why we cannot use a matched pairs procedure to test the hypothesis that the combination of a hypocaloric diet and pre-meal water intake is significantly better than a hypocaloric diet alone at lowering BMI.

17.13 The brain responds to sound. The usual way to study the brain's response to sounds is to have subjects listen to "pure tones." The response to recognizable sounds may differ. To compare responses, researchers anesthetized macaque monkeys. They fed pure tones and also monkey calls directly to the monkeys' brains through electrodes. Response to the stimulus was measured by the firing rate (electrical spikes per second) of neurons in various areas of the brain. Table 17.2 contains the responses for 37 neurons.[10] Researchers suspected that the response to a monkey call would be stronger than the response to a pure tone. Do the data support this idea? Complete the Plan, Solve, and Conclude steps of the four-step process, following the model of Example 17.4.

Hiroyuki Matsumoto/Getty Images

TABLE 17.2 Neuron response (spikes per second) to tones and monkey calls

Tone	Call	Tone	Call	Tone	Call	Tone	Call
474	500	145	42	71	134	35	103
256	138	141	241	68	65	31	70
241	485	129	194	59	182	28	192
226	338	113	123	59	97	26	203
185	194	112	182	57	318	26	135
174	159	102	141	56	201	21	129
176	341	100	118	47	279	20	193
168	85	74	62	46	62	20	54
161	303	72	112	41	84	19	66
150	208						

17.14 The brain responds to sound, continued. How much more strongly do monkey brains respond to monkey calls than to pure tones? Give a 90% confidence interval to answer this question.

Robustness of *t* procedures

The *t* confidence interval and test are exactly correct when the distribution of the population is exactly Normal. No real data are exactly Normal. The usefulness of the *t* procedures in practice therefore depends on how strongly they are affected by lack of Normality.

> **ROBUST PROCEDURES**
>
> A confidence interval or significance test is called **robust** if the confidence level or *P*-value does not change very much when the conditions for use of the procedure are violated.

The condition that the population is Normal rules out outliers, so the presence of outliers shows that this condition is not fulfilled. *The t procedures are not robust against outliers unless the sample is very large, because $\overline{x}$ and s are not resistant to outliers.*

Fortunately, the *t* procedures are quite robust against non-Normality of the population except when outliers or strong skewness are present. (Skewness is more serious than other kinds of non-Normality.) As the size of the sample increases, the central limit theorem ensures that the distribution of the sample mean $\overline{x}$ becomes more nearly Normal and that the *t* distribution becomes more accurate for critical values and *P*-values of the *t* procedures.

Always make a plot to check for skewness and outliers before you use the *t* procedures for small samples. For most purposes, you can safely use the one-sample *t* procedures when $n \geq 15$ unless an outlier or quite strong skewness is present. Here are practical guidelines for inference on a single mean.[11]

> **USING THE *t* PROCEDURES**
>
> - Except in the case of small samples, the condition that the data are an SRS from the population of interest is more important than the condition that the population distribution is Normal.
> - *Sample size less than 15:* Use *t* procedures if the data appear close to Normal (roughly symmetric, single peak, no outliers). If the data are skewed or if outliers are present, do not use *t*.
> - *Sample size at least 15:* The *t* procedures can be used except in the presence of outliers or strong skewness.
> - *Large samples:* The *t* procedures can be used even for clearly skewed distributions when the sample is large, roughly $n \geq 40$.

EXAMPLE 17.5 Can we use *t*?

Figure 17.7 shows plots of several data sets. For which of these can we safely use the *t* procedures?

- Figure 17.7(a) is a histogram of the probability at birth of not surviving to age 40, in developing countries. The data come from the United Nations' *Human Development Report 2005* and include all 121 countries in the developing world. *We have data on the entire population of 121 developing countries, so inference is not needed.* We can calculate the exact mean for the population. There is no uncertainty due to having only a sample from the population, and no need for a confidence interval or test.
- Figure 17.7(b) is a stemplot of the percents of nitrogen found in the gas bubbles of 9 specimens of amber from the late Cretaceous era (75 to 95 million years ago).[12]

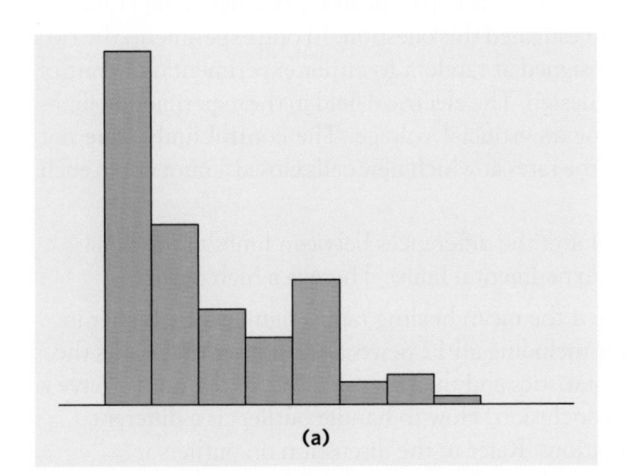

4	9
5	1
5	
5	4
5	
5	
6	0
6	3 3
6	4 4 5

(b)

37	4 8 9
38	0 0 1 1 2 2 8 9
39	2 6 8
40	6 7
41	5 7 9 9
42	0 2
43	1

(c)

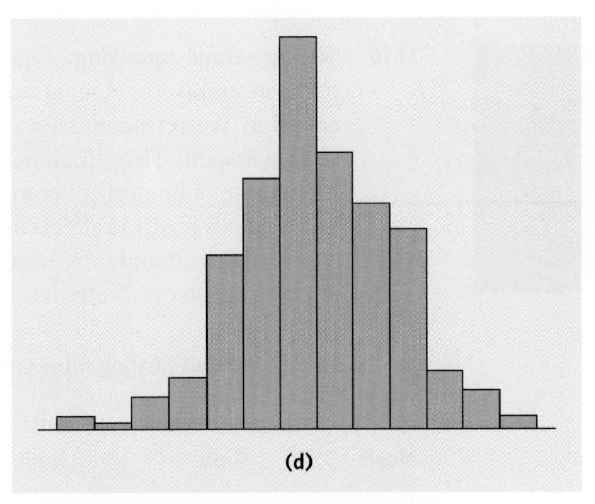

FIGURE 17.7 Can we use *t* procedures for these data? (a) Probability at birth of not surviving to age 40 in developing countries. *No*, this is an entire population (all 121 developing countries), not a sample. (b) Percent of nitrogen found in the gas bubbles of 9 specimens of amber from the late Cretaceous era. *No*, there are just 9 observations and strong skewness. (c) Lengths of 23 tropical flowers of the same variety. *Yes*, the sample is large enough to overcome the mild skewness. (d) Heights of college students. *Yes, for any size sample*, because the distribution is close to Normal.

The data are strongly skewed to the left with possible low outliers, so we cannot trust the t procedures for $n = 9$.

- Figure 17.7(c) is a stemplot of the lengths of 23 specimens of the red variety of the tropical flower *Heliconia*.[13] The data are mildly skewed to the right and there are no outliers. We can use the t distributions for such data.
- Figure 17.7(d) is a histogram of the heights of the students in a college class. This distribution is quite symmetric and appears close to Normal. We can use the t procedures for any sample size if the students can be considered an SRS of a larger population. ■

APPLY YOUR KNOWLEDGE

17.15 An outlier strikes. Our bodies have a natural electrical field that is known to help wounds heal. Might higher or lower levels impair healing? A series of experiments, introduced in Exercise 17.7, investigated this question. In one experiment, the two hind limbs of 12 newts were assigned at random to either experimental or control groups. This is a matched pairs design. The electrical field in the experimental limbs was reduced to zero by applying an artificial voltage. The control limbs were not manipulated. Table 17.3 gives the rates at which new cells closed a razor cut in each limb.

(a) Make a stemplot or a dotplot of the differences between limbs of the same newt (control limb minus experimental limb). There is a high outlier.

(b) Carry out two t tests to see if the mean healing rate is significantly higher in the control limbs, one test including all 12 newts and another that omits the outlier. What are the test statistics and their P-values? Does the outlier have a strong influence on your conclusion? How to handle outliers is a different challenge in different situations. Refer to the discussion on outliers in Chapter 2 (page 53) for an in-depth discussion.

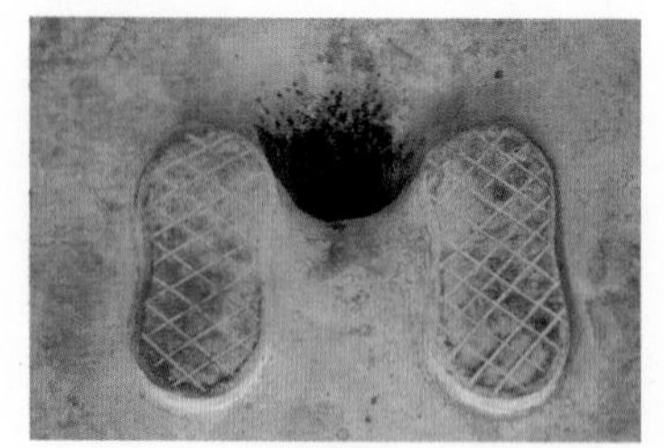

Oote Boe Photography/Alamy

17.16 Sitting versus squatting. Squatting for defecation continues to be the traditional position in most of Asia and Africa. In contrast, sitting on toilet seats has proliferated in Western countries since the 19th century with the development of the sewage system. This difference, along with differences in diet, might explain the higher rate of hemorrhoids and constipation in Western countries. A recent study asked if body position affects defecation in humans. Researchers recruited 6 healthy adult volunteers and took X-rays of their intestinal systems to measure the anorectal angle (in degrees). Note that, as with a water hose, an angle of 180 degrees should

TABLE 17.3 Newt healing rates (μm/h) for zero voltage condition

Newt	Experimental limb	Control limb	Newt	Experimental limb	Control limb
1	28	36	7	45	39
2	31	41	8	25	56
3	27	39	9	28	33
4	33	42	10	33	20
5	33	44	11	47	49
6	38	39	12	23	30

provide the least amount of flow disruption. For each subject, an X-ray was taken once in a sitting position and once in a squatting position. Here are the findings:[14]

Subject	Sitting	Squatting
1	109	122
2	73	127
3	117	141
4	77	103
5	98	121
6	125	142

(a) State the null and alternative hypotheses. Explain why the investigators used a two-sided alternative.

(b) Carry out a test and report your conclusion in simple language.

(c) The investigators used the test you just carried out. Any use of the t procedures with samples this size is risky. Why?

CHAPTER 17 SUMMARY

- Tests and confidence intervals for the mean μ of a Normal population are based on the sample mean $\overline{x}$ of an SRS. Because of the central limit theorem, the resulting procedures are approximately correct for other population distributions when the sample is large.
- The standardized sample mean is the **one-sample z statistic**

$$z = \frac{\overline{x} - \mu}{\sigma/\sqrt{n}}$$

  If we knew σ, we would use the z statistic and the standard Normal distribution.
- In practice, we do not know σ. Replace the standard deviation $\sigma/\sqrt{n}$ of $\overline{x}$ by the **standard error** $s/\sqrt{n}$ to get the **one-sample t statistic**

$$t = \frac{\overline{x} - \mu}{s/\sqrt{n}}$$

  The t statistic has the **t distribution** with $n - 1$ degrees of freedom.
- There is a t distribution for every positive **degrees of freedom.** All are symmetric distributions similar in shape to the standard Normal distribution. The t distribution approaches the $N(0, 1)$ distribution as the degrees of freedom increase.
- A level C **confidence interval for the mean μ** of a Normal population is

$$\overline{x} \pm t^* \frac{s}{\sqrt{n}}$$

  The **critical value** t^* is chosen so that the t density curve with $n - 1$ degrees of freedom has area C between $-t^*$ and t^*.

- **Significance tests** for H_0: $\mu = \mu_0$ are based on the t statistic. Use P-values or fixed significance levels from the $t(n-1)$ distribution.
- Use these one-sample procedures to analyze **matched pairs** data by first taking the difference within each matched pair to produce a single sample.
- The t procedures are quite **robust** when the population is non-Normal, especially for larger sample sizes. The t procedures are useful for non-Normal data when $n \geq 15$ unless the data show outliers or strong skewness (in which case, a larger n might be needed).

THIS CHAPTER IN CONTEXT

Chapters 14 and 15 introduced the idea of statistical inference for a population mean. We relied on the assumption that we knew the value of the population standard deviation σ. Knowing the value of σ makes the computations for a confidence interval or a P-value easier because we can use a Normal distribution for the sampling distribution of the sample mean (detailed in Chapter 13). In this chapter, we describe how the t distribution can be used instead when σ is unknown—a much more common situation by far. The reasoning behind statistical inference about the population mean is similar; only the details of the computations differ.

Optional sections in Chapter 15 showed how to determine the sample size for a desired margin of error or for a desired test power. This can be done for t procedures as well, although the actual computations are more complicated because the degrees of freedom for the t procedures also depend on the sample size (df $= n-1$). In practice, these computations are left to dedicated statistical software.

This chapter also introduces the concept of inference for a matched pairs design (introduced in Chapter 8), when two conditions are compared within pairs—either by using meaningful pairs of individuals (for example, siblings) or by obtaining a pair of measurements from each individual in the sample. In Chapter 18 we will describe a t inference procedure that lets us compare the means of two populations when the data have been collected using two independent samples. And in Chapters 24 and 26 we will introduce a new method (ANOVA) allowing the comparison of the means from more than two populations.

Before using a t procedure or an ANOVA on quantitative data, you must check that the procedure's assumptions are met. This requires plotting your data and identifying possible skew or outliers, as you first learned in Chapters 1 and 2. Optional Chapter 27 on nonparametric tests describes alternative inference procedures that can be used when the assumptions for a t procedure or an ANOVA are not met—most commonly when sample sizes are small.

CHECK YOUR SKILLS

17.17 We prefer the t procedures to the z procedures for inference about a population mean because

(a) z can be used only for large samples.

(b) z requires that you know the population standard deviation σ.

(c) z requires that you can regard your data as an SRS from the population.

17.18 You are testing H_0: $\mu = 10$ against H_a: $\mu < 10$ based on an SRS of 20 observations from a Normal population. The data give $\bar{x} = 8$ and $s = 4$. The value of the t statistic is
(a) −0.5. (b) −10. (c) −2.24.

17.19 You are testing H_0: $\mu = 10$ against H_a: $\mu < 10$ based on an SRS of 20 observations from a Normal population. The t statistic is $t = -2.25$. The degrees of freedom for this statistic are
(a) 19. (b) 20. (c) 21.

17.20 The P-value for the statistic in the previous exercise
(a) falls between 0.01 and 0.02.
(b) falls between 0.02 and 0.04.
(c) is greater than 0.25.

17.21 You are testing H_0: $\mu = 240$ against H_a: $\mu \neq 240$ based on an SRS of 16 observations from a Normal population. The t statistic is $t = -2.112$. The P-value for this test
(a) falls between 0.025 and 0.05.
(b) falls between 0.05 and 0.10.
(c) is greater than 0.50.

17.22 You have an SRS of 15 observations from a Normally distributed population. What critical value would you use to obtain a 98% confidence interval for the mean μ of the population?
(a) 2.326 (b) 2.602 (c) 2.624

17.23 Data on the blood cholesterol levels of 24 rats (in milligrams per deciliter) give $\bar{x} = 85$ and $s = 12$. A 95% confidence interval for the mean blood cholesterol of rats under this condition is
(a) 78.7 to 91.3. (b) 79.9 to 90.1. (c) 80.2 to 89.8.

17.24 Which of the following would be most worrisome for the validity of the confidence interval you calculated in the previous exercise?
(a) The presence of a clear outlier in the raw data
(b) A mild skew in the stemplot of the raw data
(c) Not knowing the population standard deviation σ

17.25 Which of these settings does *not* allow use of a matched pairs t procedure?
(a) You interview both the husband and the wife in 64 married couples and ask each about their ideal number of children.
(b) You interview a sample of 64 unmarried male students and another sample of 64 unmarried female students and ask each about their ideal number of children.
(c) You interview 64 female students in their freshman year and again in their senior year and ask each about their ideal number of children.

17.26 Because the t procedures are robust, the most important condition for their safe use is that
(a) the population standard deviation σ is known.
(b) the population distribution is exactly Normal.
(c) the data can be regarded as an SRS from the population.

CHAPTER 17 EXERCISES

17.27 Sharks. Great white sharks are big and hungry. Here are the lengths in feet of 44 great whites:[15]

18.7	12.3	18.6	16.4	15.7	18.3	14.6	15.8	14.9	17.6	12.1
16.4	16.7	17.8	16.2	12.6	17.8	13.8	12.2	15.2	14.7	12.4
13.2	15.8	14.3	16.6	9.4	18.2	13.2	13.6	15.3	16.1	13.5
19.1	16.2	22.8	16.8	13.6	13.2	15.7	19.7	18.7	13.2	16.8

(a) Examine these data for shape, center, spread, and outliers. The distribution is reasonably Normal except for one outlier in each direction. Because these are not extreme and preserve the symmetry of the distribution, use of the t procedures is safe with 44 observations.
(b) Give a 95% confidence interval for the mean length of great white sharks. Based on this interval, is there significant evidence at the 5% level to reject the claim "Great white sharks average 20 feet in length"?
(c) Before accepting the conclusions of (b), you need more information about the data. What would you like to know?

17.28 Alcohol in wine. The alcohol content of wine depends on the grape variety, the way in which the wine is produced from the grapes, the weather, and other influences. Here are data on the percent of alcohol in wine produced from the same grape variety in the same year by 48 winemakers in the same region of Italy:[16]

12.86	13.36	13.11	12.96	13.40	12.25	12.87	14.34
12.93	12.79	12.85	13.71	12.60	13.88	12.45	13.58
13.50	13.69	14.16	12.51	13.16	13.84	12.82	12.84
12.36	12.77	12.70	12.25	13.17	13.45	14.13	13.08
12.20	12.81	13.62	12.58	13.73	13.17	13.49	13.48
12.88	13.52	13.23	13.78	13.27	12.53	13.32	13.40

(a) Make a histogram of the data, using class width 0.25. The shape of the distribution is a bit irregular, but there are no outliers or strong skewness. There is no reason to avoid use of t procedures for $n = 48$.
(b) Give a 95% confidence interval for the mean alcohol content of wine of this type.
(c) Based on your confidence interval, is the mean alcohol content significantly different at the $\alpha = 0.05$ level from 12%? From 13%?

17.29 Wood lice. Students in a physiology lab collected 50 wood lice ("pill bugs") from the university grounds. Regard these as a random sample of wood lice on the university grounds. Here are the lengths in millimeters:[17]

11.0	11.1	10.1	10.8	9.2	9.0	11.0	9.5	8.0	14.6
9.1	9.7	12.0	12.5	13.1	11.7	8.1	8.7	11.9	10.3
8.1	8.3	8.2	12.1	12.3	10.9	8.9	7.5	9.0	8.2
8.3	9.0	7.9	10.5	10.0	9.8	8.9	11.0	11.0	10.9
11.0	11.5	11.0	13.9	9.3	8.5	10.2	8.3	9.1	7.0

Verify that there are no outliers in the data. What is a 95% confidence interval for the mean body length in this population?

17.30 A big-toe problem. 4 STEP Hallux abducto valgus (call it HAV) is a deformation of the big toe that often requires surgery. Doctors used X-rays to measure the angle (in degrees) of deformity in 38 consecutive patients under the age of 21 years who came to a medical center for surgery to correct HAV. The angle is a measure of the seriousness of the deformity. Here are the data:[18]

28	32	25	34	38	26	25	18	30	26	28	13	20
21	17	16	21	23	14	32	25	21	22	20	18	26
16	30	30	20	50	25	26	28	31	38	32	21	

It is reasonable to regard these patients as a random sample of young patients who require HAV surgery. Carry out the Solve and Conclude steps of a 95% confidence interval for the mean HAV angle in the population of all such patients.

17.31 Conditions for inference. Exercise 17.1 gave the summary data ($\bar{x} = 27.8$, $s = 32.2$ spikes per second) for the baseline activity of a neuron from 72 separate recordings. Neural activity is expressed in number of action potentials per second and cannot be less than 0. You notice that the standard deviation is larger than the mean. What does this suggest about the shape of the distribution? Is it nonetheless acceptable to calculate a t confidence interval for the population mean? Explain why or why not.

17.32 An outlier's effect. The data in Exercise 17.30 follow a Normal distribution quite closely except for one patient with HAV angle 50 degrees, a high outlier.
(a) Find the 95% confidence interval for the population mean based on the 37 patients who remain after you drop the outlier.
(b) Compare your interval in (a) with your interval from Exercise 17.30. What is the most important effect of removing the outlier? In general, you shouldn't remove an outlier unless you have reason to believe it is an error or unless you are interested only in "typical" cases. For a more in-depth discussion of how to handle outliers in various situations, read the discussion on outliers in Chapter 2 (page 53).

17.33 DNA on the ocean floor. We think of DNA as the stuff that stores the genetic code. It turns out that, outside living cells, DNA mainly occurs on the ocean floor. It is important in nourishing seafloor life. A random sample of ocean floor specimens from 116 locations around the world gives mean DNA concentration $\bar{x} = 0.2781$ and standard deviation $s = 0.1803$ (in grams per square meter).[19]
(a) Compute a 95% confidence interval for mean DNA concentration on the ocean floor all around the world.
(b) The sample distribution of 116 concentrations is strongly right-skewed. Is it appropriate to compute a t interval for these data? Briefly explain your answer.

17.34 Oligofructose and calcium absorption. Nondigestible oligosaccharides are known to stimulate calcium absorption in rats. A double-blind, randomized experiment investigated whether the consumption of oligofructose similarly stimulates calcium absorption in healthy male adolescents 14 to 16 years old. The subjects took a pill for nine days and had their calcium absorption tested on the last day. The experiment was repeated three weeks later. Some subjects received the oligofructose pill in the first round and then a pill containing sucrose (which served as a control). The order was switched for the remaining subjects. Here are the fractional calcium absorption data (in percent of intake) for 11 subjects:[20]

Subject	1	2	3	4	5	6
Control	78.4	76.6	57.4	51.5	49.0	46.6
Oligofructose	62.0	95.1	46.5	49.4	89.7	43.8

Subject	7	8	9	10	11
Control	44.2	42.9	37.2	34.1	24.6
Oligofructose	50.3	51.6	66.6	52.7	54.0

(a) Examine the data. Is it reasonable to use the t procedures?
(b) Do the data support the hypothesis that oligofructose facilitates calcium absorption? Use significance level $\alpha = 5\%$.

17.35 Genetic engineering for cancer treatment. Here's a new idea for treating advanced melanoma, the most serious kind of skin cancer. Genetically engineer white blood cells to better recognize and destroy cancer cells, then infuse these cells into patients. The subjects in a small initial study were 11 patients whose melanoma had not responded to existing treatments. One question was how rapidly the new cells would multiply after infusion, as measured by the doubling time in days. Here are the doubling times:[21]

1.4 1.0 1.3 1.0 1.3 2.0 0.6 0.8 0.7 0.9 1.9

(a) Examine the data. Is it reasonable to use the t procedures?
(b) Give a 90% confidence interval for the mean doubling time. Are you willing to use this interval to make an inference about the mean doubling time in a population of similar patients?

17.36 Genetic engineering for cancer treatment, continued. Another outcome in the cancer experiment described in the previous exercise is measured by a test for the presence of cells that trigger an immune response in the body and so may help fight cancer. The table below gives the counts of active cells per 100,000 cells before and after infusion for 11 subjects. The difference (after minus before) is the response variable.

Before	14	0	1	0	0	0	0	20	1	6	0
After	41	7	1	215	20	700	13	530	35	92	108
Difference	27	7	0	215	20	700	13	510	34	86	108

(a) Examine the data. Is it reasonable to use the t procedures?
(b) If your conclusion in part (a) is "yes," do the data give convincing evidence that the count of active cells is higher after treatment?

17.37 The placebo effect. The placebo effect is particularly strong in patients with Parkinson's disease. To understand the workings of the placebo effect, scientists measure activity at a key point in the brain when patients receive a placebo that they think is an active drug and also when no treatment is given.[22] The same 6 patients are measured both with and without the placebo, at different times.

(a) Explain why the proper procedure to compare the mean response to placebo with that to control (no treatment) is a matched pairs t test.
(b) The six differences (treatment minus control) had $\bar{x} = -0.326$ and $s = 0.181$. Is there significant evidence of a difference between treatment and control?

17.38 Blood pressure. At the beginning of a study on the effect of calcium in the diet on blood pressure, researchers measured the seated systolic blood pressure of 27 healthy white males. The paper reporting the study gives $\bar{x} = 114.9$ and $s = 9.3$.

(a) Give a 95% confidence interval for the mean seated systolic blood pressure in the population from which the subjects were recruited.
(b) What conditions for the population and the study design are required by the procedure you used in (a)? Which of these conditions are important for the validity of the procedure in this case?

17.39 Hand size. A clever way to determine hand size in three dimensions is to measure the volume (in millimeters, ml) of water displaced when the hand is dipped in a water container. A study used this method to gather the hand volumes of 12 male college students. Here are the measurements:[23]

400 360 420 520 460 350 500 420 450 430 395 400

(a) We can consider this an SRS of all male college students. Make a stemplot or a dotplot. Is there any sign of major deviation from Normality?
(b) Give a 95% confidence interval for the mean hand size.

17.40 Weeds among the corn. Velvetleaf is a particularly annoying weed in corn fields. It produces lots of seeds, and the seeds wait in the soil for years until conditions are right. How many seeds do velvetleaf plants produce? Here are counts from 28 plants that came up in a corn field when no herbicide was used:[24]

2450	2504	2114	1110	2137	8015	1623	1531	2008	1716
721	863	1136	2819	1911	2101	1051	218	1711	164
2228	363	5973	1050	1961	1809	130	880		

We would like to give a confidence interval for the mean number of seeds produced by velvetleaf plants. Alas, the t interval can't be safely used for these data. Why not?

17.41 Evolution and relative fitness. Can bacteria evolve a preference for the pH of their environment? An evolutionary biologist examined the relative fitness of *E. coli* bacteria grown for 2000 generations (about 300 days) at stressful acidic pH 5.5 and their parental generation grown and preserved at pH 7.2. Both types were later grown together in an acidic medium, and their relative fitness was computed. The experiment was replicated with 6 different lines of *E. coli*, giving the following relative fitness values:[25]

1.24 1.22 1.23 1.24 1.18 1.09

A relative fitness of 1 indicates that both bacteria types are equally fit. A relative fitness larger than 1 indicates that the acid-evolved line is more fit than the parental line kept at neutral pH when both are grown in acidic conditions (that is, the acid-evolved bacteria grew the most). Do the data provide evidence that bacteria evolved in acidic pH are better adapted to acidic conditions? Do a complete analysis, following the four-step process as illustrated in Example 17.3.

17.42 Radioactive contamination of tuna. Radioactive cesium (cesium-134 and cesium-137) is a waste product of nuclear reactors. A study examined the radioactive cesium tissue concentration of a random sample of 15 Pacific bluefin tuna, *Thunnus orientalis*, captured off the coast of California four months after the Fukushima (Japan) nuclear reactor meltdown of 2011. Here are the findings, in becquerels per kilogram of dry tissue (Bq/kg):[26]

4.6	6.9	7.9	8.0	8.0	9.2	10.4	10.6
11.1	11.1	11.3	12.3	12.7	14.2	15.6	

Historically, radioactive cesium tissue concentrations in bluefin tuna have been below 2 Bq/kg, on average. Do a complete analysis of the data, including a 95% confidence interval for the mean tissue concentration of radioactive cesium in bluefin tuna off the coast of California four months after the 2011 nuclear meltdown. Use your confidence interval to compare these radioactive cesium levels with the historical average. Follow the four-step process as illustrated in Example 17.2.

17.43 Do chimpanzees collaborate? Humans often collaborate to solve problems. Would a chimpanzee recruit another chimp when solving a problem that requires collaboration? Researchers presented 8 chimpanzee subjects with food outside their cage that they could bring within reach by pulling two ropes, one attached to each end of the food tray. If a chimp pulled only one rope, the rope came loose and the food was lost. Another chimp was available as a partner, but only if the subject unlocked a door joining two cages. (Chimpanzees learn these things quickly.) The same 8 chimpanzee subjects faced this problem in two versions: The two ropes were close enough together that one chimp could pull both (no collaboration needed), or the two ropes were too far apart for one chimp to pull both (collaboration needed). Table 17.4 shows how often in 24 trials each subject opened the door to recruit another chimp as partner.[27] Do the data provide evidence that chimpanzees recruit partners more often when a problem requires collaboration?

TABLE 17.4 Trials (out of 24) on which chimpanzees recruited a partner

	Collaboration Needed		
Chimpanzee	Yes	No	Difference
Namuiska	16	0	16
Kalema	16	1	15
Okech	23	5	18
Baluku	19	3	16
Umugenzi	15	4	11
Indi	20	9	11
Bili	24	16	8
Asega	24	20	4

17.44 Comparing two drugs. Makers of generic drugs must show that they do not differ significantly from the "reference" drugs that they imitate. One aspect in which drugs might differ is the extent of their absorption in the blood. Table 17.5 gives data taken from 20 healthy nonsmoking male subjects for one pair of drugs.[28] This is a matched pairs design. Numbers 1 to 20 were assigned at random to the subjects. Subjects 1 to 10 received the generic drug first, and Subjects 11 to 20 received the reference drug first. In all cases, a washout period separated the two drugs so that the first had disappeared from the blood before the subject took the second. Do the drugs differ significantly in absorption?

17.45 Does nature heal better? Exercise 17.7 introduced a series of experiments on the healing of skin wounds in newts. In one experiment, the two hind limbs of 14 newts were assigned at random to either experimental or control groups. The electrical field in the experimental limbs was reduced to half its natural value by applying a voltage. The

TABLE 17.5 Absorption extent for two versions of a drug

Subject	Reference	Generic	Subject	Reference	Generic
15	4108	1755	4	2344	2738
3	2526	1138	16	1864	2302
9	2779	1613	6	1022	1284
13	3852	2254	10	2256	3052
12	1833	1310	5	938	1287
8	2463	2120	7	1339	1930
18	2059	1851	14	1262	1964
20	1709	1878	11	1438	2549
17	1829	1682	1	1735	3340
2	2594	2613	19	1020	3050

TABLE 17.6 Newt healing rates (μm/h) for reduced-voltage condition

Newt	Experimental limb	Control limb	Newt	Experimental limb	Control limb
1	24	25	8	33	36
2	23	13	9	28	35
3	47	44	10	28	38
4	42	45	11	21	43
5	26	57	12	27	31
6	46	42	13	25	26
7	38	50	14	45	48

control limbs were not manipulated. Table 17.6 gives the rates at which new cells closed a razor cut in each limb. Is there good evidence that changing the electrical field from its natural level slows healing?

17.46 Lung capacity of children. LARGE DATA SET The forced expiratory volume (FEV, measured in liters) is a primary indicator of lung function and corresponds to the volume of air that can forcibly be blown out in the first second after full inspiration. The *Large.FEV* data file contains the FEV values of a large sample of children, along with some categorical descriptors of each individual. In Exercise 1.44 (page 37) you created a new variable based on age: *preschool* for ages 3 to 5, *elementary* for ages 6 to 10, *middle* for ages 11 to 13, and *highschool* for ages 14 and above. A school district is looking for reference FEV values for children in elementary, middle, and high school. Obtain a 95% confidence interval for the mean FEV of children in each age group.

17.47 Estimating body fat in men. LARGE DATA SET The data file *Large.Bodyfat* contains data on the percent of body fat, age, weight, height, and 10 body circumference measurements for 252 adult men. In Exercise 2.46 (page 64), you examined the distribution of weights and heights and identified a high outlier in each.

(a) The outlier in the distribution of men's heights was a typo. Obtain a 95% confidence interval for the mean height (in inches) of adult men, making sure to exclude the incorrect value.

(b) Estimate the mean body weight (in pounds) of adult men with 95% confidence, first using all 252 data points and then without the high outlier. Compare the two intervals. What is the impact of the outlier?

17.48 Elderly health. Exercise 2.45 (page 64) describes a study of calcium or inorganic phosphorus blood levels (both measured in millimoles per liter, mmol/l) in elderly patients. The data file *Large.Calcium* contains the data. Find a 90% confidence interval for the mean calcium level of all elderly men (in the data file, *sex* = 1 corresponds to men) and also compute a 90% confidence interval for the mean calcium level of all elderly women (*sex* = 2). Do the same for the inorganic phosphorus levels. Write a short paragraph describing your findings. (Describe each result separately. We will see how to determine if two sample means are significantly different in Chapter 18.)

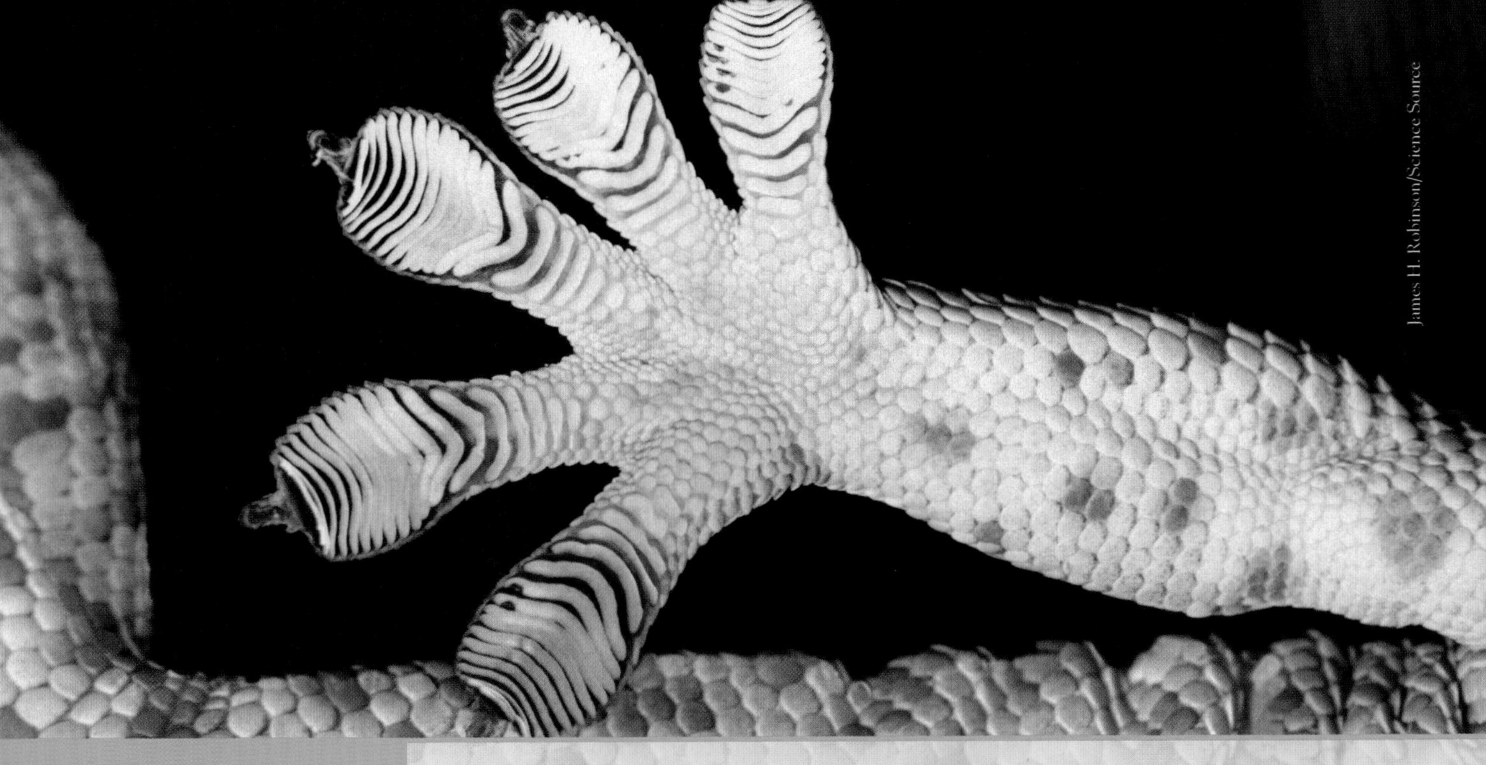

CHAPTER 18 Comparing Two Means

Comparing two populations or two treatments is one of the most common situations encountered in statistical practice. We call such situations *two-sample problems*.

> **TWO-SAMPLE PROBLEMS**
>
> - The goal of inference is to compare the responses to two treatments or to compare the characteristics of two populations.
> - We have a separate sample from each treatment or each population.

IN THIS CHAPTER WE COVER...

- Two-sample problems
- Comparing two population means
- Two-sample t procedures
- Using technology
- Robustness again
- Avoid the pooled two-sample t procedures*
- Avoid inference about standard deviations*

Two-sample problems

A two-sample problem can arise from a randomized comparative experiment that randomly divides subjects into two groups and exposes each group to a different treatment. Comparing random samples selected separately from two populations is also a two-sample problem. Unlike the matched pairs designs studied earlier, there is no matching of the individuals in the two samples, and the two samples can be of different sizes. Inference procedures for two-sample data differ from those for matched pairs. Here are some typical two-sample problems.

Can I trust this study? An anesthesiologist thought he might ease menopausal hot flashes by injecting a local anesthetic directly into a specific nerve in the neck, hoping to influence the brain's temperature regulation center. The women who elected to have the procedure done experienced some benefit, which lasted from weeks to months. The results were reported on NBC. But there were no controls and only 22 women, who were not selected at random. Medical standards require randomized comparative experiments and statistically significant results. Only then can we be confident that a new procedure really helps.

EXAMPLE 18.1 Two-sample problems

(a) Does regular physical therapy help with lower-back pain? A randomized experiment assigned patients with lower-back pain to two groups: 142 received an examination and advice from a physical therapist; another 144 received regular physical therapy for up to five weeks. After a year, the change in their level of disability (0% to 100%) was assessed by a doctor who did not know which treatment the patients had received.

(b) A field biologist observes gender-based behavior in wild chimpanzees. Twelve randomly chosen young chimpanzees are tagged remotely with a dart and their behavior is monitored. The amount of time each young chimpanzee spends in contact with its mother is recorded. The biologist then compares the amount of time spent in contact with the mother by young male and by young female chimpanzees.

(c) A physiologist compares the fermentation rates of yeast as it metabolizes glucose (a simple carbohydrate) or starch (a complex carbohydrate). Live yeast suspensions are placed in 20 fermentation flasks. Ten of the flasks contain a solution of water and glucose, and the other 10 flasks contain a solution of water and starch. The volume of CO_2 emitted per minute is measured for each flask. ■

We may wish to compare either the *centers* or the *spreads* of the two groups in a two-sample setting. This chapter emphasizes the most common inference procedures, those for comparing two population means. We comment briefly on the issue of comparing spreads (standard deviations), where simple inference is much less satisfactory.

APPLY YOUR KNOWLEDGE

Which data design? *Each situation described in Exercises 18.1 to 18.4 requires inference about a mean or means. Identify each as involving (1) a single sample, (2) matched pairs, or (3) two independent samples. The procedures of Chapter 17 apply to designs (1) and (2). We are about to learn procedures for (3).*

18.1 Acid rain. You contact a random sample of 24 hydrological stations in California and obtain data on the acidity of rainwater collected at each station in 2000 and in 2010. Acidity is measured by pH on a scale of 0 to 14 (the pH of distilled water is 7.0). You want to estimate the average change in acidity of rainwater in California between the year 2000 and the year 2010.

18.2 Agricultural pests. Agricultural pests can be controlled to some extent either by using pesticides to kill them or by introducing a large number of sterilized males to diminish the species' reproductive potential. Ten large corn fields are treated with pesticides, and 10 have sterilized males introduced. Corn yield at harvest time is then compared for the two treatments.

18.3 Chemical analysis. To check a new analytical method, a chemist obtains a reference specimen of known concentration from the National Institute of Standards and Technology. She then makes 20 measurements of the concentration of this specimen with the new method and checks for bias by comparing the mean result with the known concentration.

18.4 Chemical analysis again. Another chemist is checking the same new method. He has no reference specimen, but a familiar analytic method is available. He wants to know if the new and old methods agree. He takes a specimen of unknown concentration and measures the concentration 10 times with the new method and 10 times with the old method.

Comparing two population means

Comparing two populations or the responses to two treatments starts with data analysis: Make boxplots, stemplots (for small samples), or histograms (for larger samples) and compare the shapes, centers, and spreads of the two samples. The most common goal of inference is to compare the average or typical responses in the two populations. When data analysis suggests that both population distributions are symmetric, and especially when they are at least approximately Normal, we want to compare the population means. Here are the conditions for inference when comparing two means.

CONDITIONS FOR INFERENCE COMPARING TWO MEANS

- We have **two SRSs,** from two distinct populations. The samples are **independent.** That is, one sample has no influence on the other (matching violates independence, for example). We measure the same variable for both samples.
- Both populations are **Normally distributed.** The means and standard deviations of the populations are unknown. In practice, it is enough that the distributions have similar shapes and that the data have no strong outliers.

Call the variable we measure x_1 in the first population and x_2 in the second, because the variable may have different distributions in the two populations. Here is the notation we will use to describe the two populations:

Population	Variable	Population mean	Population standard deviation
1	x_1	μ_1	σ_1
2	x_2	μ_2	σ_2

There are four unknown parameters, the two means and the two standard deviations. The subscripts remind us which population a parameter describes. We want to compare the two population means, either by giving a confidence interval for their difference $\mu_1 - \mu_2$ or by testing the hypothesis of no difference, H_0: $\mu_1 = \mu_2$, which is the same as H_0: $\mu_1 - \mu_2 = 0$.

We use the sample means and standard deviations to estimate the unknown parameters. Again, subscripts remind us which sample a statistic comes from. Here is the notation that describes the samples:

Population	Sample size	Sample mean	Sample standard deviation
1	n_1	$\overline{x}_1$	s_1
2	n_2	$\overline{x}_2$	s_2

To do inference about the difference $\mu_1 - \mu_2$ between the means of the two populations, we start from the difference $\overline{x}_1 - \overline{x}_2$ between the means of the two samples.

EXAMPLE 18.2 Daily activity and obesity

STATE: People gain weight when they take in more energy from food than they expend. James Levine and his collaborators at the Mayo Clinic investigated the link between obesity and energy spent on daily activity.[1]

Choose 20 healthy volunteers who don't exercise. Deliberately choose 10 who are lean and 10 who are mildly obese but still healthy. Attach sensors that monitor the subjects' every move for 10 days. Table 18.1 presents data on the time (in minutes per

TABLE 18.1 Time (minutes per day) spent in three different postures by lean and obese subjects

Group	Subject	Stand/walk	Sit	Lie
Lean	1	511.100	370.300	555.500
Lean	2	607.925	374.512	450.650
Lean	3	319.212	582.138	537.362
Lean	4	584.644	357.144	489.269
Lean	5	578.869	348.994	514.081
Lean	6	543.388	585.312	506.500
Lean	7	677.188	268.188	467.700
Lean	8	555.656	322.219	567.006
Lean	9	374.831	537.031	531.431
Lean	10	504.700	528.838	396.962
Obese	11	260.244	646.281	521.044
Obese	12	464.756	456.644	514.931
Obese	13	367.138	578.662	563.300
Obese	14	413.667	463.333	532.208
Obese	15	347.375	567.556	504.931
Obese	16	416.531	567.556	448.856
Obese	17	358.650	621.262	460.550
Obese	18	267.344	646.181	509.981
Obese	19	410.631	572.769	448.706
Obese	20	426.356	591.369	412.919

day) that the subjects spent standing or walking, sitting, and lying down. Do lean and obese people differ in the average time they spend standing and walking?

Catchlight Visual Services/Alamy

PLAN: Examine the data and carry out a test of hypotheses. We suspect in advance that lean subjects (Group 1) are more active than obese subjects (Group 2), so we test the hypotheses

$$H_0: \mu_1 = \mu_2$$
$$H_a: \mu_1 > \mu_2$$

SOLVE (first steps): Are the conditions for inference met? The subjects are volunteers, so they are not SRSs from all lean and mildly obese adults. The study tried to recruit comparable groups: all worked in sedentary jobs, none smoked or were taking medication, and so on. Setting clear standards like these helps make up for the fact that we can't reasonably get SRSs for so invasive a study. The subjects were not told that they were chosen from a larger group of volunteers because they did not exercise and were either lean or mildly obese. Because their willingness to volunteer isn't related to the purpose of the experiment, we will treat them as two independent SRSs.

A dotplot of the two groups (Figure 18.1) displays the data in detail. The distributions are a bit irregular, as we expect with just 10 observations. There are no clear departures from Normality such as extreme outliers or skewness. This is confirmed in a Normal quantile plot (not shown). The lean subjects as a group spend much more time standing and walking than do the obese subjects. Calculating the group means confirms this:

Group	n	Mean $\bar{x}$	Std. dev. s
Group 1 (lean)	10	525.751	107.121
Group 2 (obese)	10	373.269	67.498

The observed difference in mean time per day spent standing or walking is

$$\bar{x}_1 - \bar{x}_2 = 525.751 - 373.269 = 152.482 \text{ minutes}$$

To complete the Solve step, we must learn the details of inference comparing two means. ■

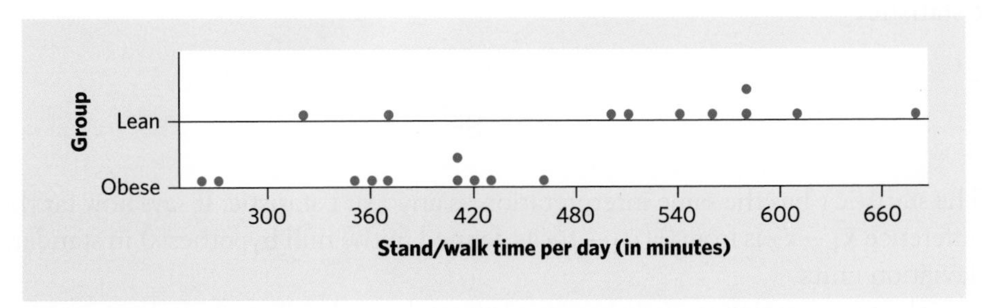

FIGURE 18.1 Dotplot of the time spent walking or standing, for Example 18.2.

Two-sample *t* procedures

Whether an observed difference in sample means is surprising depends on the spread of the individual observations as well as on the two means. Widely different means can arise just by chance if the individual observations vary a great deal. How much the difference $\bar{x}_1 - \bar{x}_2$ can vary from sample to sample is given by its sampling distribution.

When two random variables are Normally distributed, the new variable "difference in sample means" also follows a Normal distribution, centered on the difference in the two variables' means and with variance equal to the sum of the two variables' variances. We already know from Chapter 14 that the sampling distributions of $\bar{x}_1$ and $\bar{x}_2$ have standard deviations $\sigma_1/\sqrt{n_1}$ and $\sigma_2/\sqrt{n_2}$, respectively. Therefore, when we look at the difference $\bar{x}_1 - \bar{x}_2$, the standard deviation of its sampling distribution is

$$\sqrt{\frac{\sigma_1^2}{n_1} + \frac{\sigma_2^2}{n_2}}$$

This standard deviation gets larger as either population gets more variable, that is, as σ_1 or σ_2 increases. It gets smaller as the sample sizes n_1 and n_2 increase.

standard error

Because we don't know σ_1 and σ_2, we estimate them by the sample standard deviations s_1 and s_2. The result is the **standard error,** or estimated standard deviation, of the difference in sample means:

$$\text{SE} = \sqrt{\frac{s_1^2}{n_1} + \frac{s_2^2}{n_2}}$$

When we standardize the estimate, we get

$$t = \frac{(\bar{x}_1 - \bar{x}_2) - (\mu_1 - \mu_2)}{\text{SE}}$$

two-sample t statistic

but, because in a typical two-sample test $\mu_1 - \mu_2 = 0$, the result is the **two-sample *t* statistic:**

$$t = \frac{\bar{x}_1 - \bar{x}_2}{\text{SE}}$$

The statistic t has the same interpretation as any z or t statistic: It says how far the difference $\bar{x}_1 - \bar{x}_2$ is from 0 ($\mu_1 - \mu_2$ as defined in the null hypothesis) in standard deviation units.

The two-sample t statistic has approximately a t distribution. It does not have exactly a t distribution even if the populations are both exactly Normal. In practice, however, the approximation is very accurate.

THE TWO-SAMPLE t PROCEDURES

Draw an SRS of size n_1 from a large Normal population with unknown mean μ_1, and draw an independent SRS of size n_2 from another large Normal population with unknown mean μ_2. The distribution of the two-sample t statistic is very close to the t distribution with degrees of freedom df given by

$$\text{df} = \frac{\left(\dfrac{s_1^2}{n_1} + \dfrac{s_2^2}{n_2}\right)^2}{\dfrac{1}{n_1 - 1}\left(\dfrac{s_1^2}{n_1}\right)^2 + \dfrac{1}{n_2 - 1}\left(\dfrac{s_2^2}{n_2}\right)^2}$$

This approximation is accurate when both sample sizes n_1 and n_2 are 5 or larger.

A level C **confidence interval for $\mu_1 - \mu_2$** is given by

$$(\bar{x}_1 - \bar{x}_2) \pm t^* \sqrt{\frac{s_1^2}{n_1} + \frac{s_2^2}{n_2}}$$

Here t^* is the critical value with area C between $-t^*$ and t^* under the t density curve with df degrees of freedom.

To **test the hypothesis H_0: $\mu_1 = \mu_2$**, calculate the **two-sample t statistic**

$$t = \frac{\bar{x}_1 - \bar{x}_2}{\sqrt{\dfrac{s_1^2}{n_1} + \dfrac{s_2^2}{n_2}}}$$

Find P-values from the t distribution with df degrees of freedom.

Unlike the one-sample situation, the degrees of freedom of the two-sample t distribution are calculated from the sample data. The degrees of freedom, df, are generally not a whole number. The degrees of freedom are always at least as large as the smaller of $n_1 - 1$ and $n_2 - 1$ and at most equal to $n_1 + n_2 - 2$. There is a t distribution for any positive degrees of freedom, even though Table C contains entries only for whole-number degrees of freedom. When using Table C, round your calculation down to err on the safe side (that is, when the calculated degrees of freedom fall between two values in the table, the smaller value will be the more conservative choice and should be favored).

The formula for calculating the degrees of freedom from scratch may seem a bit lengthy. However, notice that you will have already computed s_1^2/n_1 and s_2^2/n_2 to obtain the t statistic. So here's a tip: Write down these two values and use them to calculate df faster.

EXAMPLE 18.3 Daily activity and obesity

We can now complete Example 18.2.

SOLVE (inference): The two-sample t statistic comparing the average minutes spent standing and walking in Group 1 (lean) and Group 2 (obese) is

$$t = \frac{\bar{x}_1 - \bar{x}_2}{\sqrt{\frac{s_1^2}{n_1} + \frac{s_2^2}{n_2}}}$$

$$= \frac{525.751 - 373.269}{\sqrt{\frac{107.121^2}{10} + \frac{67.498^2}{10}}} = \frac{152.482}{\sqrt{1147.491 + 455.598}}$$

$$= \frac{152.482}{40.039} = 3.808$$

Software gives one-sided P-value $P = 0.0008$ based on df $= 15.174$. The degrees of freedom df are given by

$$\text{df} = \frac{\left(\frac{107.121^2}{10} + \frac{67.498^2}{10}\right)^2}{\frac{1}{9}\left(\frac{107.121^2}{10}\right)^2 + \frac{1}{9}\left(\frac{67.498^2}{10}\right)^2} = \frac{(1147.491 + 455.598)^2}{\frac{1147.491^2}{9} + \frac{455.598^2}{9}}$$

$$= \frac{2{,}569{,}894}{169{,}367.2} = 15.1735$$

df = 15

t^*	3.733	4.073
One-sided P	.001	.0005

To use Table C, we round the degrees of freedom down to 15. Because H_a is one-sided, the P-value is the area to the right of $t = 3.808$ under the $t(15)$ curve. Figure 18.2 illustrates this P-value. In Table C, $t = 3.808$ lies between the t^* critical values 3.733 and 4.073. So $0.001 < P < 0.0005$.

CONCLUDE: There is very strong evidence ($P = 0.0008$) that lean people spend more time walking and standing than do moderately obese people. ■

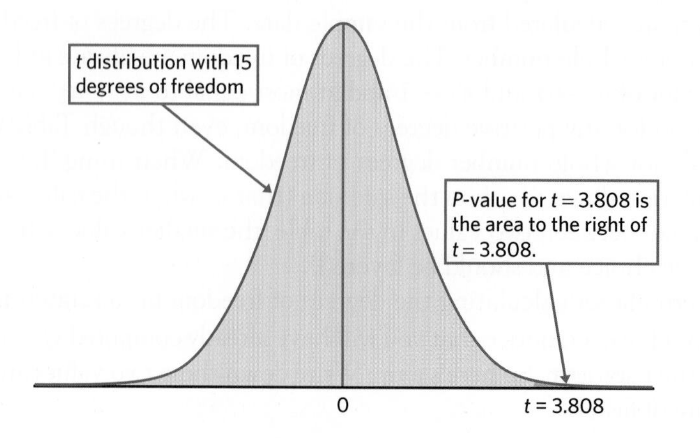

FIGURE 18.2 The P-value as the upper-tail area, for Example 18.3. We rounded down the degrees of freedom to 15.

Does lack of everyday activity *cause* obesity? It may be that some people are naturally more active and are therefore less likely to gain weight. Or it may be that people who gain weight reduce their activity level. The study went on to enroll most of the obese subjects in a weight-reduction program and most of the lean subjects in a supervised program of overeating. After eight weeks, the obese subjects had lost weight (mean 8 kg) and the lean subjects had gained weight (mean 4 kg). But both groups kept their original allocation of time to the different postures. This suggests that time allocation is biological and influences weight, rather than the other way around. The authors remark: "It should be emphasized that this was a pilot study and that the results need to be confirmed in larger studies."

EXAMPLE 18.4 How much more active are lean people?

PLAN: Give a 90% confidence interval for $\mu_1 - \mu_2$, the difference in average daily minutes spent standing and walking between lean and mildly obese adults.

SOLVE and CONCLUDE: As in Example 18.3, we use 15 degrees of freedom to find t^*. Table C shows that the $t(15)$ critical value is $t^* = 1.753$. We are 90% confident that $\mu_1 - \mu_2$ lies in the interval

$$(\overline{x}_1 - \overline{x}_2) \pm t^*\sqrt{\frac{s_1^2}{n_1} + \frac{s_2^2}{n_2}}$$

$$= (525.751 - 373.269) \pm 1.753\sqrt{\frac{107.121^2}{10} + \frac{67.498^2}{10}}$$

$$= 152.482 \pm 70.188$$

$$= 82.29 \text{ to } 222.67 \text{ minutes}$$

This interval is quite wide because the samples are small and the variation among individuals, as measured by the two sample standard deviations, is large. ■

EXAMPLE 18.5 Transgenic chickens

Tom Myers/agefotostock

STATE: Infection of chickens with the avian flu is a threat to both poultry production and human health. A research team created transgenic chickens resistant to avian flu infection. Could the modification affect chickens in other ways? The researchers compared the hatching weights (in grams, g) of 45 transgenic chickens and 54 independently selected commercial chickens of the same breed.[2] The data are displayed in Table 18.2.

PLAN: We had no specific direction for the difference in weight before looking at the data, so the alternative is two-sided. We will test the hypotheses

$$H_0: \mu_1 = \mu_2 (\text{that is, } \mu_1 - \mu_2 = 0)$$
$$H_a: \mu_1 \neq \mu_2 (\text{that is, } \mu_1 - \mu_2 \neq 0)$$

SOLVE: The two samples can be regarded as SRSs from two populations, although the population of transgenic chickens is a potential population not yet created. Figure 18.3 shows side-by-side boxplots of the two data sets. We find no outliers (which would appear as asterisks on the graph) or skew, and we can reasonably assume a Normal distribution

TABLE 18.2 Hatching weights (g) of transgenic and commercial chickens

Transgenic	38.8	39.0	39.7	40.0	40.8	40.9	41.0	41.0	41.0	42.5	42.6	43.0
	43.0	43.4	43.5	43.5	43.8	44.4	44.7	44.7	44.7	45.3	45.7	45.8
	46.4	46.5	46.6	46.7	46.7	46.8	46.9	47.1	47.1	47.1	47.3	47.6
	47.7	48.1	48.3	49.3	49.3	49.8	50.3	50.9	52.1			
Commercial	36.7	37.1	38.9	39.5	39.5	39.8	40.0	40.2	40.3	40.5	40.5	40.7
	41.1	41.2	41.5	41.5	41.6	41.6	41.7	42.4	43.1	43.3	43.3	43.4
	43.7	44.1	44.2	45.2	45.3	45.4	46.0	46.1	46.4	46.6	46.6	46.9
	47.3	47.5	48.1	48.2	48.4	48.6	49.0	49.1	49.3	49.6	50.1	50.2
	50.4	50.6	52.2	53.0	55.5	56.4						

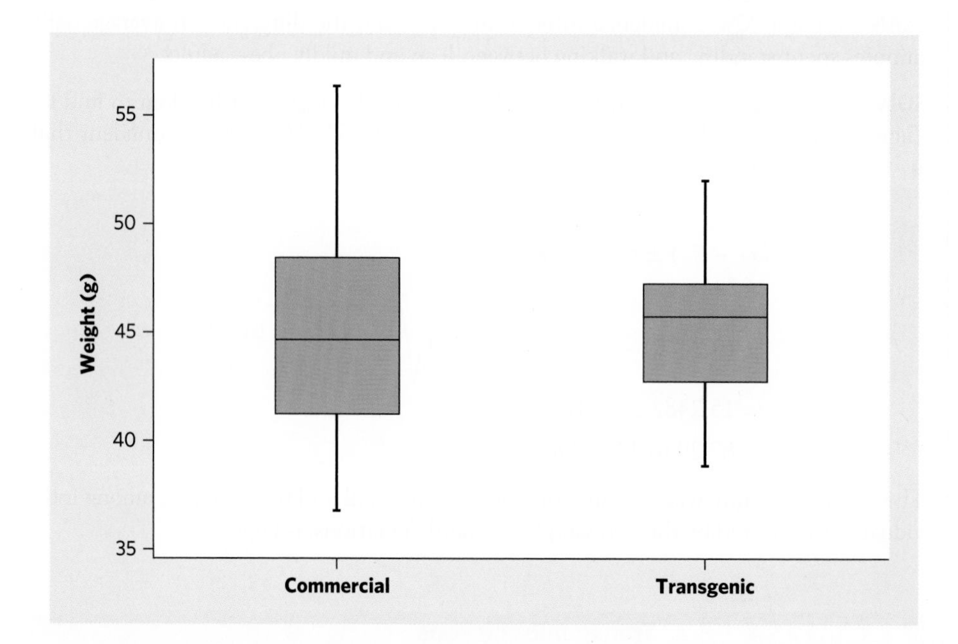

FIGURE 18.3 Boxplots of hatching weights of commercial and transgenic chickens, for Example 18.5.

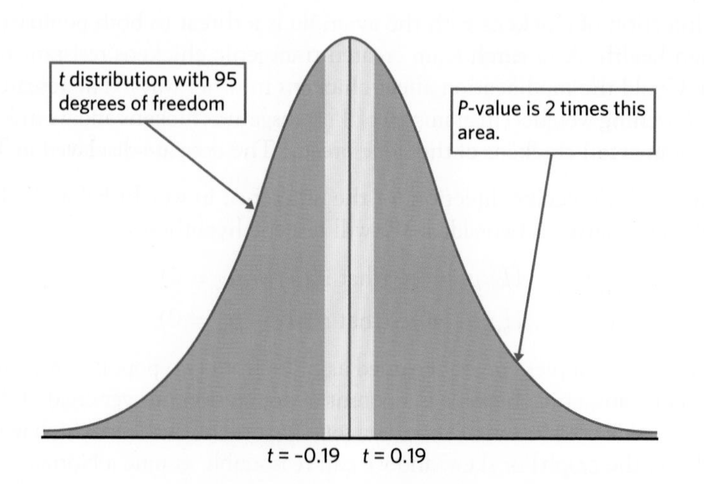

FIGURE 18.4 The P-value for Example 18.5. Because the alternative is two-sided, the P-value is double the area to the right of $t = 0.19$. We rounded the degrees of freedom to 95.

model for both groups. Thus, the t procedures should be valid. Here are the calculations leading to the two sample t-test:

Group	Condition	n	$\bar{x}$	s
1	Transgenic	45	45.14	3.32
2	Commercial	54	44.99	4.57

The two-sample t statistic is

$$t = \frac{\bar{x}_1 - \bar{x}_2}{\sqrt{\dfrac{s_1^2}{n_1} + \dfrac{s_2^2}{n_2}}} = \frac{45.14 - 44.99}{\sqrt{\dfrac{3.32^2}{45} + \dfrac{4.57^2}{54}}} = \frac{0.15}{0.795} = 0.19$$

Software says that the two-sided P-value is $P = 0.8507$ based on df $= 95.3$ (or rounded down to 95, depending on the software). Figure 18.4 illustrates the P-value. The degrees of freedom df are given by

$$\text{df} = \frac{\left(\dfrac{3.32^2}{45} + \dfrac{4.57^2}{54}\right)^2}{\dfrac{1}{44}\left(\dfrac{3.32^2}{45}\right)^2 + \dfrac{1}{53}\left(\dfrac{4.57^2}{54}\right)^2} = 95.3$$

Using Table C and df $= 80$ (it is more conservative to round df down to the nearest value available), compare $t = 0.19$ with the two-sided critical values for the $t(80)$ distribution. Using Table C and degrees of freedom 80, we conclude that the P-value for our test is greater than 0.50 (a two-sided P).

df $= 80$

t^*	0.678
Two-sided P	.50

CONCLUDE: The data give *no* significant evidence ($P > 0.10$) that the population of transgenic chickens differs in mean hatching weight from the population of commercial chickens of the same breed. ■

APPLY YOUR KNOWLEDGE

18.5 Whelks on the Pacific coast. Published reports of statistical analyses are often very terse. A knowledge of basic statistics helps you decode what you read. In a study of the presence of whelks (a type of sea snail) along the Pacific coast, investigators put down a frame that covered 0.25 square meter and counted the whelks on the sea bottom inside the frame. They did this at 7 locations off the coast of California and at 6 locations off the coast of Oregon. The report says that whelk densities "were twice as high in Oregon as in California (mean ± SEM, 26.9 ± 1.56 versus 11.9 ± 2.68 whelks per 0.25 m^2, Oregon versus California, respectively; Student's t test, $P < 0.001$)."[3]

(a) SEM stands for the standard error of the mean, $s/\sqrt{n}$. Fill in the values in this summary table:

Group	Location	n	$\bar{x}$	s
1	Oregon	?	?	?
2	California	?	?	?

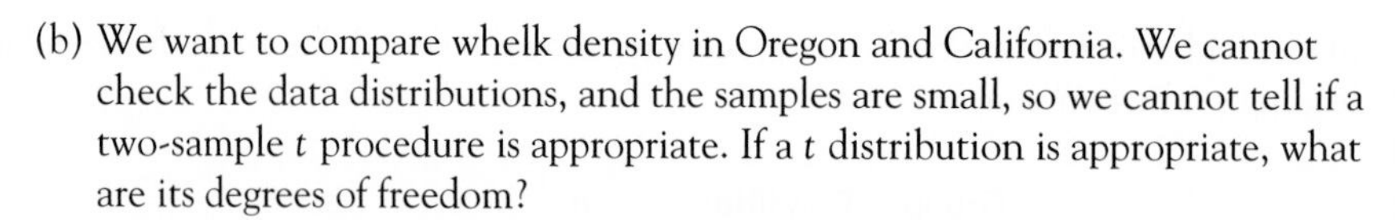

(b) We want to compare whelk density in Oregon and California. We cannot check the data distributions, and the samples are small, so we cannot tell if a two-sample t procedure is appropriate. If a t distribution is appropriate, what are its degrees of freedom?

D. Hurst/Alamy

18.6 Echinacea for the common cold? Echinacea is widely used as an herbal remedy for the common cold, but does it work? In a double-blind experiment, healthy volunteers agreed to be exposed to common-cold-causing rhinovirus type 39 and have their symptoms monitored. The volunteers were randomly assigned to take either a placebo or an echinacea supplement daily from seven days before until five days after viral exposure. A symptom score was recorded for each subject over the five days following exposure, with higher scores indicating more severe symptoms. The published results reported the mean ± SEM for both groups as 13.21 ± 1.91 (echinacea) and 15.05 ± 1.43 (placebo).[4]

(a) The two-sample t statistic for $\bar{x}_1 - \bar{x}_2$ was $t = -0.771$. You can draw a conclusion from this t without using a table and even without knowing the sizes of the samples (remember to specify your null and alternative hypotheses first). What is your conclusion? Why don't you need the sample sizes and a table?

(b) In fact, 52 subjects were assigned to the echinacea treatment and 103 to the placebo. Fill in the values in this summary table:

Group	Treatment	n	$\bar{x}$	s
1	Echinacea	?	?	?
2	Placebo	?	?	?

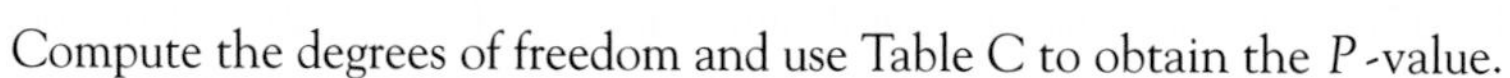

Compute the degrees of freedom and use Table C to obtain the P-value.

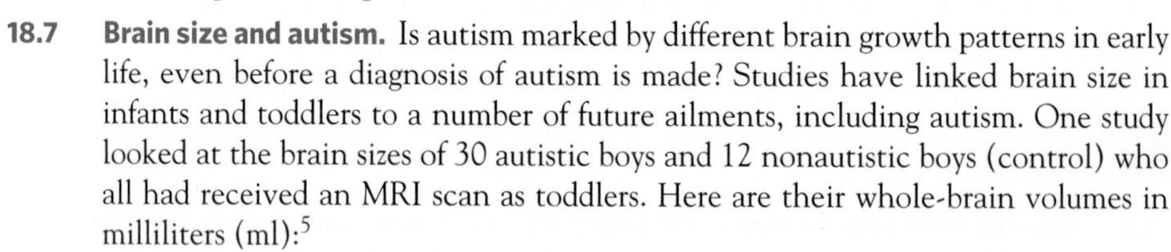

18.7 Brain size and autism. Is autism marked by different brain growth patterns in early life, even before a diagnosis of autism is made? Studies have linked brain size in infants and toddlers to a number of future ailments, including autism. One study looked at the brain sizes of 30 autistic boys and 12 nonautistic boys (control) who all had received an MRI scan as toddlers. Here are their whole-brain volumes in milliliters (ml):[5]

Autistic									
1311	1250	1292	1419	1401	1297	1202	1336	1308	1353
1515	1461	1365	1364	1362	1303	1278	1247	1333	1340
1319	1286	1223	1241	1229	1209	1171	1154	1128	1230
Control									
1040	1180	1207	1179	1115	1133	1298	1263	1194	1198
1230	1114								

Do the data give significant evidence that, on average, the brain volume of autistic boys is different from that of nonautistic boys during the toddler years? Use the four-step process to answer this question. Follow the model of Examples 18.2 and 18.3.

18.8 The autistic brain: how much larger? Go back to the previous exercise and give a 90% confidence interval for $\mu_1 - \mu_2$, the difference in mean brain size during the toddler years between all autistic and nonautistic boys.

Using technology

Software programs give accurate confidence intervals and P-values. However, there is some variation in software output for the two-sample t procedures due to how precisely the degrees of freedom are computed. Figure 18.5 displays output from a graphing calculator, three statistical programs, and a spreadsheet program for the test conducted in Example 18.3. The two-sample t statistic is exactly as in Example 18.3, $t = 3.808$. You can find this in all five outputs (Minitab rounds to 3.81). The different technologies use different methods to find the P-value for $t = 3.808$.

- The TI-83, R, and JMP results are the most accurate because they use the t distribution with 15.17 degrees of freedom. The P-value is $P = 0.00084$.
- Minitab and Excel both round the degrees of freedom to a whole number. Minitab truncates the exact degrees of freedom to the next-smallest whole number and therefore always errs a bit on the conservative side, whereas Excel rounds the exact degrees of freedom to the nearest whole number (in this case, both round df = 15.17 to df = 15). As a result, Excel offers $P = 0.00086$, and Minitab shows $P = 0.001$ after rounding the P-value to three decimal places.

JMP

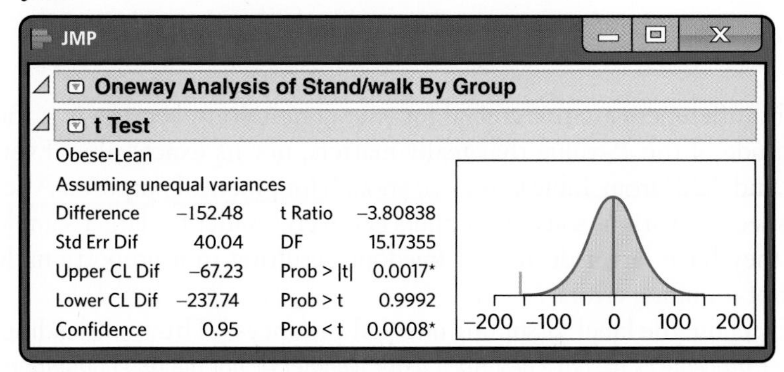

R

```
Welch Two Sample t-test

data: StandWalk by Group
t = 3.8084, df = 15.174, p-value = 0.000841
alternative hypothesis: true difference in means is greater than 0
sample estimates:
 mean in group Lean mean in group Obese
          525.7513             373.2692
```

FIGURE 18.5 The two-sample t procedures applied to the data on activity and obesity of Example 18.3: output from a graphing calculator, three statistical programs, and a spreadsheet program. (*Continued on next page*)

TI-83

2-Samp TTest
μ1>μ2
t=3.808375604
p=8.4101904E-4
df=15.17355038
$\bar{x}_1$=525.7513
↓$\bar{x}_2$=373.2692

Minitab

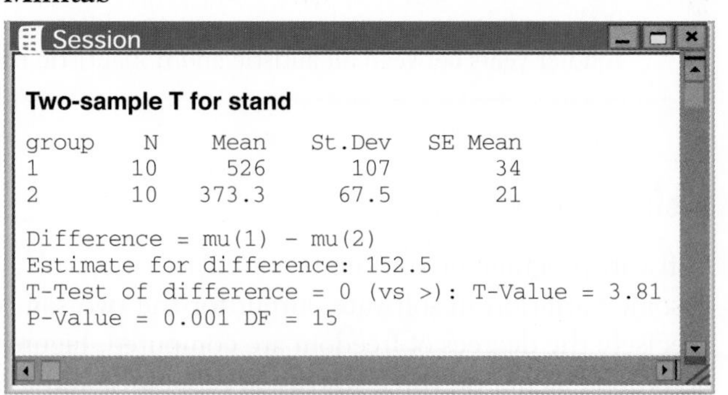

Session

Two-sample T for stand

group	N	Mean	St.Dev	SE Mean
1	10	526	107	34
2	10	373.3	67.5	21

Difference = mu(1) - mu(2)
Estimate for difference: 152.5
T-Test of difference = 0 (vs >): T-Value = 3.81
P-Value = 0.001 DF = 15

Microsoft Excel

	A	B	C
1	t-Test: Two-Sample Assuming Unequal Variances		
2			
3		*Lean*	*Obese*
4	Mean	525.7513	373.2692
5	Variance	11474.8903	4556.019849
6	Observations	10	10
7	Hypothesized Mean	0	
8	df	15	
9	t Stat	3.808375604	
10	P(T<=t)one-tail	0.000856818	
11	t Critical one-tail	1.753051038	
12	P(T<=t) two-tail	0.001713635	
13	P(T<t) two-tail	2.131450856	

FIGURE 18.5 *(Continued)*

These slight differences are not crucial for your conclusions, because it is the order of magnitude of the P-value that really matters, not its exact value. Even "between 0.15 and 0.20" from Table C is close enough for practical purposes. Also, the differences seen in various software outputs get even smaller for larger sample sizes, because they have larger degrees of freedom, resulting in a proportionally smaller impact of rounding errors.

JMP and Excel use the label "assuming unequal variances." This is misleading. *The two-sample t procedures we have described work whether or not the two populations have the same variance*. There is a variant of the two-sample t test we have described that works only when the two variances are equal. We discuss this method briefly on page 453, but you should favor the test that does not assume equal variances, because it will give correct results no matter what the population variances are.

Lastly, some software outputs show both the one-sided ("one-tail") and two-sided ("two-tail") P-values or just the two-sided P-value. Remember to specify your null and alternative hypotheses before examining the sample data, and then select or calculate the one- or two-sided P-value that matches your alternative.

APPLY YOUR KNOWLEDGE

18.9 Highly Superior Autobiographical Memory. Some individuals have the ability to recall accurately vast amounts of autobiographical information without mnemonic tricks or extra practice. This ability is called HSAM, for Highly Superior Autobiographical Memory. A study recruited adults with confirmed HSAM and control individuals of similar age without HSAM. All study participants were given a battery of cognitive and behavioral tests in the hope of finding out how this extraordinary ability works. Here are the participants' results for a visual memory test:[6]

HSAM	4	4	5	9	6	6	7	6	8	5
Control	5	1	2	4	2	2	3	3	3	9
	4	2	4	5	5	1	6	4		

(a) Below is the Minitab output for the two-sample t test. What are $\bar{x}$ and s for each of the two samples? Starting from these values, obtain the two standard errors of the means and the t test statistic. Your work should agree with the output.

```
        N    Mean   StDev     SE Mean
HSAM   10    6.00    1.63        0.52
CON    18    3.61    1.97        0.47
Difference = mu (HSAM) - mu (CON)
Estimate for difference:  2.389
T-Test of difference = 0 (vs >): T-Value = 3.44
                                 P-Value = 0.001  DF = 21
```

(b) Do individuals with HSAM have significantly greater abilities on visual memory tests than typical individuals? Write a summary in a sentence or two, including t, df, P, and a conclusion, as if you were preparing a report for publication.

18.10 Highly Superior Autobiographical Memory, continued. The study in Exercise 18.9 also gave a verbal association test to the study participants. Here are the findings:

HSAM	1	7	2	2	2	3	6	1	7	
Control	1	6	1	1	2	2	3	0	3	3
	4	4	1	6	6	8	0	3		

(a) Below is the CrunchIt! output for the two-sample t test. Starting from the sample means and standard deviations, find the value of the t test statistic. Check that your calculation agrees with the output.

```
Null hypothesis:          Difference of means = 0
Alternative hypothesis: Difference of means > 0
        n       Sample Mean    Std Dev
HSAM    9       3.444            2.506
CON     18      3                2.301
df:                    14.92
Difference of means:   0.4444
t statistic:           0.4463
P-value:               0.3309
```

(b) Do individuals with HSAM have significantly greater abilities on verbal association tests than typical individuals? Write a summary in a sentence or two, including t, df, P, and a conclusion, as if you were preparing a report for publication.

Robustness again

The two-sample t procedures are more robust than the one-sample t methods, particularly when the distributions are not symmetric. When the sizes of the two samples are equal and the two populations being compared have distributions with similar shapes, probability values from the t table are quite accurate for a broad range of distributions when the sample sizes are as small as $n_1 = n_2 = 5$.[7] When the two population distributions have different shapes, larger samples are needed.

As a guide to practice, adapt the guidelines given on page 426 for the use of one-sample t procedures to two-sample procedures by replacing "sample size" with "sum of the sample sizes," $n_1 + n_2$. These guidelines err on the side of safety, especially when the two samples are of equal size. *In planning a two-sample study, choose equal sample sizes whenever possible. The two-sample t procedures are most robust against non-Normality in this case.*

APPLY YOUR KNOWLEDGE

18.11 Bone loss in nursing mothers. Exercise 2.36 (page 62) gives the percent change in the mineral content of the spine for 47 mothers during three months of nursing a baby and for a control group of 22 women of similar age who were neither pregnant nor lactating.

(a) What two populations did the investigators want to compare? We must be willing to regard the women recruited for this observational study as SRSs from these populations.

(b) Do these data give good evidence that the average bone mineral loss is higher in the population of nursing mothers? Be sure to consider that a bone mineral loss is reflected here by a negative value. Complete the Plan, Solve, and Conclude steps of the four-step process as illustrated in Examples 18.2 and 18.3.

18.12 Geckos. Geckos are lizards with specialized toe pads that enable them to easily climb all sorts of surfaces. A research team examined the adhesive properties of 7 Tokay geckos, *Gekko gecko*. Below are the toe pad areas (in square centimeters, cm^2) for the 4 male and 3 female geckos:[8]

Males	5.57	4.94	5.98	7.45
Females	5.11	5.50	5.31	

Explain carefully why a two-sample t confidence interval for the difference in mean toe pad area between male and female geckos may not be accurate.

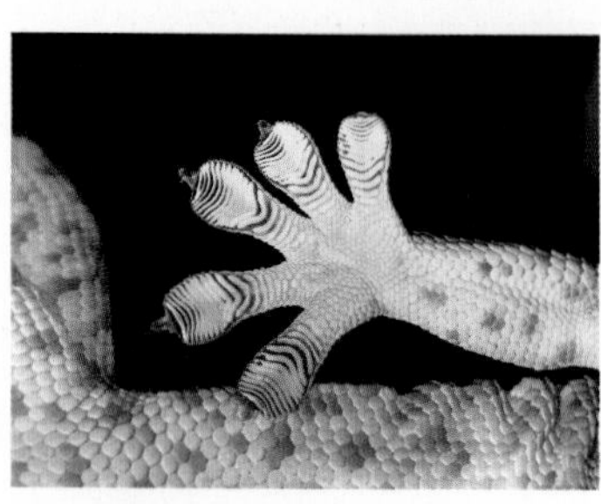
James H. Robinson/Science Source

Avoid the pooled two-sample *t* procedures*

Most software packages, including those illustrated in Figure 18.5, offer a choice of two-sample t statistics. One is often labeled for "unequal" variances, the other for "equal" variances. The "unequal" variance procedure is our two-sample t. *This test is valid whether or not the population variances are equal.* The other choice is a special version of the two-sample t statistic that assumes that the two populations have the same variance. It averages (the statistical term is "pools") the two sample variances to estimate the common population variance. The resulting statistic is called the *pooled two-sample t statistic*. It has the equation

$$t_{\text{pooled}} = \frac{\bar{x}_1 - \bar{x}_2}{s_{\text{pooled}}\sqrt{\dfrac{1}{n_1} + \dfrac{1}{n_2}}} = \frac{\bar{x}_1 - \bar{x}_2}{\sqrt{\dfrac{(n_1 - 1)s_1^2 + (n_2 - 1)s_2^2}{n_1 + n_2 - 2}}\sqrt{\dfrac{1}{n_1} + \dfrac{1}{n_2}}}$$

where s^2_{pooled} is the pooled sample variance. The pooled t statistic (which is directly related to the ANOVA procedure we will discuss in Chapter 24) is equal to our t statistic if the two sample sizes are the same but not otherwise.

We could choose to use the pooled t for tests and confidence intervals. The pooled t statistic has exactly the t distribution with $n_1 + n_2 - 2$ degrees of freedom *if* the two population variances really are equal and the population distributions are exactly Normal. Of course, in the real world, distributions are not exactly Normal, and population variances are not exactly equal. The requirement for equal population variances is one more test assumption to check, and simulations show that the "unequal" variances t procedures are almost always more accurate than the pooled procedures.[9] Our advice: *Always use the t procedures for "unequal" variances if you can.*

Meta-analysis Small samples have large margins of error. Large samples are expensive. Often we can find several studies of the same issue; if we could combine their results, we would have a large sample with a small margin of error. That is the idea of "meta-analysis." Of course, we can't just lump the studies together, because of differences in design and quality. Statisticians have more sophisticated ways of combining the results. Meta-analysis has been applied to issues ranging from the effect of secondhand smoke to whether echinacea can prevent the common cold.

Avoid inference about standard deviations*

Two basic features of a distribution are its center and its spread. In a Normal population, we measure the center by the mean and the spread by the standard deviation. We use the t procedures for inference about population means for Normal populations, and we know that t procedures are widely useful for non-Normal populations as well. It is natural to turn next to inference about the standard deviations of Normal populations. Our advice here is short and clear: *Don't do it without expert advice.*

There are methods for inference about the standard deviations of Normal populations. The most common such method is the F test for comparing the spread of two Normal populations. *Unlike the t procedures for means, the F test is extremely sensitive to non-Normal distributions.* This lack of robustness does not improve in large samples. It is difficult in practice to tell whether a significant test result is

*These two short sections concern optional special topics. They are not needed to read the rest of the book but can serve as background for Chapter 24 on one-way ANOVA.

evidence of unequal population spreads or simply a sign that the populations are not Normal.

The deeper difficulty underlying the very poor robustness of Normal population procedures for inference about spread has already been seen in our work on describing data. The standard deviation is a natural measure of spread for Normal distributions, but not for distributions in general. In fact, because skewed distributions have unequally spread tails, no single numerical measure does a good job of describing the spread of a skewed distribution. In summary, the standard deviation is not always a useful parameter, and even when it is (for symmetric distributions), the results of inference are not always trustworthy. Consequently, *we do not recommend trying to do inference about population standard deviations in basic statistical practice.*[10]

CHAPTER 18 SUMMARY

- The data in a **two-sample problem** are two independent SRSs, each drawn from a separate population.
- Tests and confidence intervals for the difference between the means μ_1 and μ_2 of two Normal populations start from the difference $\bar{x}_1 - \bar{x}_2$ between the two sample means. Because of the central limit theorem, the resulting procedures are approximately correct for other population distributions when the sample sizes are large.
- Draw independent SRSs of sizes n_1 and n_2 from two Normal populations with parameters μ_1, σ_1 and μ_2, σ_2. The **two-sample *t* statistic** is

$$t = \frac{(\bar{x}_1 - \bar{x}_2) - (\mu_1 - \mu_2)}{\sqrt{\dfrac{s_1^2}{n_1} + \dfrac{s_2^2}{n_2}}}$$

- The statistic t has approximately a t distribution with **degrees of freedom**

$$\text{df} = \frac{\left(\dfrac{s_1^2}{n_1} + \dfrac{s_2^2}{n_2}\right)^2}{\dfrac{1}{n_1 - 1}\left(\dfrac{s_1^2}{n_1}\right)^2 + \dfrac{1}{n_2 - 1}\left(\dfrac{s_2^2}{n_2}\right)^2}$$

  In practice, use technology to obtain the degrees of freedom and accurate probability values.

- The **confidence interval for $\mu_1 - \mu_2$** is

$$(\bar{x}_1 - \bar{x}_2) \pm t^* \sqrt{\frac{s_1^2}{n_1} + \frac{s_2^2}{n_2}}$$

The critical value t^* for the computed degrees of freedom gives a confidence level very close to the desired level C.

- **Significance tests for H_0: $\mu_1 = \mu_2$** are based on

$$t = \frac{\bar{x}_1 - \bar{x}_2}{\sqrt{\frac{s_1^2}{n_1} + \frac{s_2^2}{n_2}}}$$

P-values obtained with the computed degrees of freedom are very accurate.

- The guidelines for practical use of two-sample t procedures are similar to those for one-sample t procedures. Equal sample sizes are recommended.

THIS CHAPTER IN CONTEXT

In Chapter 17 we described the one-sample t procedures used to perform statistical inference for a population mean μ. We also saw that the method can be applied to matched pairs designs when we are interested in the population mean difference μ of paired observations. This chapter extends the use of the t procedures to compare the means of two distinct populations, μ_1 and μ_2. In Chapter 24 we will see how another method (ANOVA) can be used to compare the means from several populations. As the procedures described become more complex, our reliance on statistical software increases.

Which procedure to use depends on how the data were collected; we discussed this in Chapters 7 and 8. It is also worth remembering that conclusions from observational studies and experiments are substantially different because observational studies have confounding variables that prevent us from assigning causality. Therefore statistical significance should always be interpreted in the context of a study's design, strengths, and weaknesses.

While Chapters 17 and 18 have dealt with inference for a quantitative variable (with the population mean μ as the key parameter), the next two chapters will introduce inference procedures used for a categorical variable (with the population proportion p as the key parameter). You will notice that Chapters 19 and 20 closely mirror Chapters 17 and 18 in their approach, first addressing one-sample procedures and then two-sample procedures. Often what students find the most challenging is figuring out whether a variable is quantitative or categorical, a topic that we discussed in great detail in Chapter 1.

CHECK YOUR SKILLS

18.13 How many calories does a fruit juice popsicle labeled "no sugar added" pack? A random sample of 10 popsicles of a particular brand is selected, and the calories per serving are measured. To get a confidence interval for the mean calories per serving of all no-sugar-added popsicles from this brand, you would use

(a) the one-sample t interval.
(b) the matched pairs t interval.
(c) the two-sample t interval.

18.14 A study is designed to compare the amount of vitamin C (in milligrams per serving) in oranges that reach the stores either 1 day after being picked or 3 days after being picked. A random sample of 15 oranges is taken for each group. To test whether there is a difference in the mean vitamin C amount of all oranges reaching stores either 1 day or 3 days after being picked, you would use

(a) the one-sample t test.
(b) the matched pairs t test.
(c) the two-sample t test.

18.15 There are two common methods for measuring the concentration of a pollutant in fish tissue. Do the results of the two methods differ on the average? You apply both methods to one sample of 18 carp and use

(a) the one-sample t test.
(b) the matched pairs t test.
(c) the two-sample t test.

Use the following information for Exercises 18.16 to 18.20. A study of the effects of exercise used rats bred to have high or low capacity for exercise. The 8 high-capacity rats had mean blood pressure 89 and standard deviation 9; the 8 low-capacity rats had mean blood pressure 105 with standard deviation 13. (Blood pressure is measured in millimeters of mercury, mm Hg.)

18.16 To compare the mean blood pressure of the two types of rats with a t procedure, the correct degrees of freedom is

(a) 7. (b) 12.46. (c) 15.

18.17 The two-sample t statistic for comparing the two population means has value

(a) 0.5. (b) 2.86. (c) 9.65.

18.18 We suspect that rats with a low capacity for exercise tend to have a higher blood pressure than rats with a high capacity for exercise. To see if this is true, test these hypotheses for the mean blood pressure of all rats with high capacity (HC) and all rats with low capacity (LC) for exercise:

(a) H_0: $\mu_{HC} = 89$; $\mu_{LC} = 105$ versus
H_a: $\mu_{HC} \neq 89$; $\mu_{LC} \neq 105$.
(b) H_0: $\mu_{HC} = \mu_{LC}$ versus H_a: $\mu_{HC} \neq \mu_{LC}$.
(c) H_0: $\mu_{HC} = \mu_{LC}$ versus H_a: $\mu_{HC} < \mu_{LC}$.

18.19 The P-value for testing the hypotheses from the previous exercise satisfies

(a) $0.005 < P < 0.01$.
(b) $0.01 < P < 0.02$.
(c) $0.02 < P < 0.05$.

18.20 The margin of error of a 95% confidence interval for $\mu_{HC} - \mu_{LC}$ is approximately

(a) 1.96. (b) 10.96. (c) 12.18.

18.21 One major reason that the two-sample t procedures are widely used is that they are quite *robust*. This means that

(a) t procedures do not require that we know the standard deviations of the populations.
(b) confidence levels and P-values from the t procedures are quite accurate even if the population distribution is not exactly Normal.
(c) t procedures compare population means, a comparison that answers many practical questions.

18.22 A study of road rage asked samples of 596 men and 523 women about their behavior while driving. Based on their answers, each subject was assigned a road rage score on a scale of 0 to 20. The subjects were chosen by random digit dialing of telephone numbers. Are the conditions for two-sample t inference satisfied?

(a) Maybe: The SRS condition is OK but we need to look at the data to check Normality.
(b) No: Scores in a range between 0 and 20 can't be Normal.
(c) Yes: The SRS condition is OK and large sample sizes make the Normality condition unnecessary.

CHAPTER 18 EXERCISES

In exercises that call for two-sample t procedures, either do your calculations by hand and use Table C after rounding the degrees of freedom down, or use technology. (Remember that different software programs handle the degrees of freedom slightly differently, some rounding it and others not.) Many of the following exercises ask you to think about issues of statistical practice as well as to carry out procedures.

18.23 Sexual dimorphism in birds. The central tail feathers of the longtailed finch (*Poephila acuticauda*) are a sexually dimorphic trait hypothesized to play a role in sexual selection. The longer tail feathers in males cost energy to produce, and this is thought to serve as a token of the male's excellent health condition. Here are the lengths of the central tail feathers (average of the two central feathers, in millimeters) of 20 male and 21 female longtailed finches:[11]

Males									
87	77	95	73	74	85	56	86	95	108
75	87	73.5	82	89	64	74.5	87	85	86

Females									
60	59	72	54	65	58	59	60	65	60
68	70.5	80	87	65	59	65	67	62	66
70									

(a) Treat these data as SRSs from the population of adult longtailed finches in the wild. Make stemplots or dotplots of both data sets and determine whether the use of a two-sample t procedure is appropriate.

(b) How much longer, on average, are the central tail feathers of male longtail finches than those of the females? Give a 95% confidence interval for the difference in population mean length between the male and the female adult longtail finches.

18.24 Food counseling for obese children. According to the Centers for Disease Control and Prevention, obesity rates among children have increased dramatically over the past three decades, from a low of about 5% to 18% as of 2010. A study examined the effect of family-based food-counseling sessions provided by trained professionals. The study randomly assigned obese children aged 9 to 12 years to either the counseling intervention or a control group not receiving any food counseling. The children's weight changes (in pounds, lb) after 15 weeks are displayed in Table 18.3 for both conditions.[12]

(a) Plot both data sets using stemplots, dotplots, or histograms. Because the sample sizes are quite large, the t procedures will work well even in the presence of skew or mild outliers.

(b) Is there good evidence that obese children receiving food counseling gain less weight over a 15-week period? State the hypotheses, carry out the corresponding test, and conclude using a significance level of 0.05.

18.25 IQ scores for boys and girls. Here are the IQ test scores of 31 seventh-grade girls in a midwestern school district:[13]

114	100	104	89	102	91	114	114	103	105	
108	130	120	132	111	128	118	119	86	72	
111	103	74	112	107	103	98	96	112	112	93

The IQ test scores of 47 seventh-grade boys in the same district are

111	107	100	107	115	111	97	112	104	106	113
109	113	128	128	118	113	124	127	136	106	123
124	126	116	127	119	97	102	110	120	103	115
93	123	79	119	110	110	107	105	105	110	77
90	114	106								

(a) Make stemplots, dotplots, or histograms of both sets of data. Because the distributions are reasonably symmetric with no extreme outliers, the t procedures will work well.

(b) Treat these data as SRSs from all seventh-grade students in the district. Is there good evidence that girls and boys differ in their mean IQ scores?

18.26 Food counseling for obese children, continued.

(a) Use the data in Exercise 18.24 to give a 95% confidence interval for the difference in mean weight change after 15 weeks for obese children who receive professional food counseling and obese children who do not.

TABLE 18.3 Weight change (lb) of obese children, by group

Intervention									
−16.7	−14.8	−11.9	−9.7	−9.6	−8.8	−8.0	−7.1	−6.6	−6.0
−5.6	−5.6	−5.5	−5.5	−5.1	−5.0	−5.0	−4.8	−4.4	−4.4
−4.1	−4.0	−4.0	−3.6	−3.5	−3.2	−2.8	−2.0	−1.8	−1.8
−1.4	−1.2	−0.2	−0.1	0.0	0.2	0.6	1.0	1.2	1.2
1.4	1.8	2.0	2.2	2.5	2.8	3.3	4.2	5.4	5.8
6.0	6.4	8.4							
Control									
12.0	10.0	9.0	8.6	7.1	6.7	6.1	4.8	4.6	4.5
2.8	2.8	2.8	2.7	1.8	1.6	1.0	0.7	0.4	0.1
0.1	−5.1	−5.1							

(b) Give a 95% confidence interval for the mean weight change after 15 weeks for obese children who receive professional food counseling.

18.27 IQ scores for boys and girls, continued. Use the data in Exercise 18.25 to give a 95% confidence interval for the difference between the mean IQ scores of all boys and girls in the district.

18.28 Testosterone and obesity in adolescent males. Obesity in adult males is associated with lower levels of sex hormone. Example 17.2 described a study investigating a possible link between obesity and testosterone levels in adolescent males between the ages of 14 and 20 years. The study compared 25 obese adolescent males and 25 adolescent males with a healthy weight. Here are their plasma testosterone concentrations (in nanomoles per liter, nmol/l):

Obese adolescent males								
0.30	0.24	0.19	0.17	0.18	0.23	0.24	0.06	0.15
0.17	0.18	0.17	0.15	0.12	0.25	0.25	0.25	0.32
0.35	0.37	0.39	0.46	0.49	0.42	0.36		

Adolescent males with a healthy weight								
0.78	0.70	0.63	0.60	0.60	0.69	0.76	0.58	0.50
0.48	0.49	0.43	0.42	0.38	0.35	0.35	0.32	0.31
0.28	0.25	0.23	0.24	0.24	0.26	0.27		

(a) Plot each data set to investigate the shapes of the distributions. Are the data appropriate for a t procedure?
(b) We suspect that obese adolescent males have a lower plasma testosterone level, on average. Do the data support this suspicion? Carry out the appropriate test and state your conclusion.
(c) Give a 95% confidence interval for the difference in mean plasma testosterone level between adolescent males who are obese and those who have a healthy weight.

18.29 Bird songs. Bird songs have been hypothesized to be a secondary sexual character signaling an individual's health status. Researchers designed an experiment in which they randomly assigned male collared flycatchers (*Ficedula albicollis*) to two groups: One group received an immune challenge in the form of an injection of sheep red blood cells, and the other group received a placebo injection. Here are the changes in song rate (in strophes per minute) after the injection for 15 male collared flycatchers in the immune-challenge group and 12 males in the placebo group:[14]

Immune challenge							
−1.6	−3.1	−2.7	−3.7	−3.1	−3.6	−1.9	−1.5
−0.1	0.8	−0.1	−0.2	−1.2	−1.9	0.2	

Placebo							
−1.5	1.7	0.4	−1.8	0.0	0.4	0.8	2.0
0.0	−2.4	−1.5	−0.1				

(a) Make stemplots or dotplots to investigate the shape of the distributions. Is the use of a two-sample t procedure appropriate?
(b) Do the data provide significant evidence, at $\alpha = 5\%$, that an immune challenge reduces the male song rate, on average, more than a placebo injection does? Obtain the test statistic, degrees of freedom, and P-value, and state your conclusion. Does the study support the hypothesis that male bird songs advertise the male's health status?

18.30 Do subliminal messages work? A "subliminal" message is below our threshold of awareness but may nonetheless influence us. A study looked at the effect of subliminal messages on math skills. The messages were flashed on a screen too rapidly to be consciously read. Twenty-eight students who had failed the mathematics part of the City University of New York Skills Assessment Test were randomly assigned to receive daily either a positive subliminal message ("Each day I am getting better in math") or a neutral subliminal message ("People are walking on the street"). All students participated in a summer program designed to raise their math skills, and all took the assessment test again at the end of the program. Table 18.4 gives

TABLE 18.4 Mathematics skills scores before and after a subliminal message

Positive Message		Neutral Message	
Before	After	Before	After
18	24	18	29
18	25	24	29
21	33	20	24
18	29	18	26
18	33	24	38
20	36	22	27
23	34	15	22
23	36	19	31
21	34		
17	27		

data on the subjects' scores before and after the program.[15] Is there good evidence that the positive message brought about a greater improvement in math scores than the neutral message? How large is the mean difference in gains between the two groups? (Use 90% confidence.)

Exercises 18.31 to 18.37 are based on summary statistics rather than raw data. This information is typically all that is presented in published reports. Inference procedures can be calculated by hand from the summaries. You must trust that the authors understood the conditions for inference and verified that they apply. This isn't always true.

CAUTION

18.31 Pharmacokinetics of a drug. Avandia (rosiglitazone maleate) is an oral antidiabetic drug produced by the pharmaceutical company GlaxoSmithKline. Before a drug can be prescribed, we must know how the body absorbs and excretes it. Patients were given a single dose of either 1 milligram (mg) or 2 mg of rosiglitazone maleate, and the maximum plasma concentration of the drug (in nanograms per milliliter, ng/ml) was assessed.[16]

Treatment	n	$\bar{x}$	s
1 mg	32	76	13
2 mg	32	156	42

Is there significant evidence that maximum plasma concentration is dose dependent (higher doses resulting in higher concentrations)?

18.32 More on pharmacokinetics. The study in the previous exercise also looked at how fast the body gets rid of the drug by measuring the elimination half-life (in hours). Here are the results:

Treatment	n	$\bar{x}$	s
1 mg	32	3.16	0.72
2 mg	32	3.15	0.39

Is there significant evidence of a difference in elimination half-life depending on drug dose? Write your conclusions from this and the previous exercise in the context of the drug's pharmacokinetic properties.

18.33 Tongue piercing and cracked teeth. A study compared tooth health and periodontal damage in a group of 46 young adult males wearing a tongue piercing and a control group of 46 young adult males without tongue piercing. One question of interest was whether individuals with tongue piercing had more enamel cracks, on average. Here are the summary statistics:[17]

Group	n	$\bar{x}$	s
Tongue piercing	46	4.0	3.5
No tongue piercing	46	1.2	1.3

(a) Is there evidence at the 1% level that young adult males with tongue piercing have significantly more enamel cracks? State hypotheses in terms of the two population means, obtain the two-sample t statistic and P-value, and conclude.

(b) Can you conclude that tongue piercing causes more enamel cracks? Explain your reasoning.

18.34 Distraction and snacking. A study assessed the effects of playing a computer game (solitaire) during lunch on behavioral and physiological variables. A sample of 44 healthy adults was randomly assigned to eat lunch either with the computer game distraction or without distraction. The food served for lunch was exactly the same for everyone. All participants also had the same breakfast. Thirty minutes after lunch, participants were offered cookies as a snack and were instructed to eat as many or as few cookies as they liked. Here are the summary statistics for the findings:[18]

Group	n	$\bar{x}$	s
Distraction	22	52.1	45.1
No distraction	22	27.1	26.4

(a) Do the findings support the hypothesis of greater snack intake after meals taken with a distraction? State hypotheses in terms of the two population means, obtain the two-sample t statistic and P-value, and conclude using a significance level of 0.05.

(b) Can you conclude that distraction during meals causes the greater snack intake? Explain your reasoning.

18.35 Acupuncture for treating migraines. Acupuncture is widely used to prevent migraine attacks, but does it work? A study investigated the effectiveness of acupuncture compared with sham acupuncture and with no acupuncture (patients left on the wait list) in individuals with migraines. Patients were randomly assigned to 1 of the 3 "treatments" for a period of eight weeks, after which they monitored their migraine attacks over the next three weeks. Here are summary data about the number of days in which each patient used pain medication for migraines or headaches during the monitoring period:[19]

Acupuncture			Sham			Wait list		
n	$\overline{x}$	s	n	$\overline{x}$	s	n	$\overline{x}$	s
132	3.2	3.0	76	3.4	2.9	64	4.4	3.6

(a) Let's first ask if patients who received the acupuncture treatment had significantly fewer pain medication days than patients who stayed on the wait list. What do you conclude?

(b) Give a 95% confidence interval for the mean difference. Describe the effect size. Would you consider recommending acupuncture for the preventive treatment of headaches?

18.36 Acupuncture for treating migraines, continued. The placebo effect has been shown to be particularly strong in the treatment of pain.

(a) Is there significant evidence that patients who received the acupuncture treatment had significantly fewer pain medication days than patients who received the sham acupuncture (needles inserted at random locations)? What do you conclude?

(b) Explain what makes this study a randomized, controlled experiment. Finding statistical significance in the previous exercise does not imply that acupuncture is causing the improvement. Explain briefly why.

18.37 Acupuncture for treating migraines: critique. The initial pool of volunteers for the study of Exercise 18.35 was larger. A few individuals dropped out for various reasons after the study started. Here are the retention rates for all three groups:

Acupuncture 91%   Sham 94%   Wait list 84%

Comparatively more patients dropped out of the study in the wait list group. This could be just chance; or maybe individuals with fewer migraines lost patience with being left on the wait list for several weeks; or maybe those who least believed in the effectiveness of acupuncture decided to drop out. How could this fact affect the conclusions of the study?

The following exercises ask you to answer real questions from real data without having your work outlined in the exercise statement. Follow the ***Plan, Solve,*** *and* ***Conclude*** *steps of the four-step process. It may be helpful to restate in your own words the* ***State*** *information given in the exercise.*

4 STEP

18.38 Student drinking. A professor asked her sophomore students, "How many drinks do you typically have per session?" (A drink is defined as one 12-ounce beer, one 4-ounce glass of wine, or one 1-ounce shot of liquor.) Some of the students didn't drink. Table 18.5 gives the responses of the female and male students who did drink.[20] It is likely that some of the

TABLE 18.5 Drinks per session claimed by female and male students

Female students

2.5	9	1	3.5	2.5	3	1	3	3	3	3	2.5	2.5
5	3.5	5	1	2	1	7	3	7	4	4	6.5	4
3	6	5	3	8	6	6	3	6	8	3	4	7
4	5	3.5	4	2	1	5	5	3	3	6	4	2
7	7	7	5.5	3	2.5	10	5	4	9	8	1	6
2	5	2.5	3	4.5	9	5	4	4	3	4	6	7
4	5	1	5	3	4	10	7	3	4	4	4	4
2	1	2.5	2.5									

Male students

7	7.5	8	15	3	4	1	5	11	4.5	6	4	10
16	4	8	5	9	7	7	3	5	6.5	1	12	4
6	8	8	4.5	10.5	8	6	10	1	9	8	7	8
15	3	10	7	4	6	5	2	10	7	9	5	8
7	3	7	6	4	5	2	5	5.5	9	10	10	4
8	4	2	4	12.5	3	15	2	6	3	4	3	10
6	4.5	5										

students exaggerated a bit. The sample is all students in one large sophomore-level class. The class is popular, so we are tentatively willing to regard its members as an SRS of sophomore students at this college. Do a complete analysis that reports on

(a) the drinking behavior claimed by sophomore women.
(b) the drinking behavior claimed by sophomore men.
(c) a comparison of the behavior of women and men.

18.39 Ink toxicity. The National Toxicology Program evaluates the toxicity of chemicals found in manufacturing, in consumer products, or in the environment after disposal. Toxicity is assessed through a battery of tests. Here are some results from a study of the toxicity of black newsprint ink in 7-week-old female rats. The rats' fur was locally clipped twice a week for 13 weeks. One group of rats received a dermal application of ink right after each clipping, and a control group of rats was left untreated. Table 18.6 contains the body weights (in grams) of female rats at the beginning of the study and at the end of the 13 weeks.[21]

(a) Verify that the two experimental groups are not significantly different at the beginning of the study.
(b) Is there good evidence that ink application impairs growth in female rats between 7 and 20 weeks of age? Estimate the difference between the two-population mean weight gain. (Use 95% confidence.)

18.40 Magnets for pain relief. A randomized, double-blind experiment studied whether magnetic fields applied over a painful area can reduce pain intensity. The subjects were 50 volunteers with postpolio syndrome who reported muscular or arthritic pain. The pain level when pressing a painful area was graded subjectively on a scale from 0 to 10 (0 is no pain; 10 is maximum pain). Patients were randomly assigned to wear either a magnetic device or a placebo device over the painful area for 45 minutes. Here is a summary of the pain scores for this experiment, expressed as means ± standard deviations:[22]

	Magnetic device ($n = 29$)	Placebo device ($n = 21$)
Pretreatment	9.6 ± 0.7	9.5 ± 0.8
Posttreatment	4.4 ± 3.1	8.4 ± 1.8
Change	5.2 ± 3.2	1.1 ± 1.6

(a) Is there good evidence that the magnetic device is better than a placebo device at reducing pain?
(b) How much reduction in pain is achieved with the magnetic device? Give a 95% confidence interval for the mean difference in pain scores pre- and posttreatment among patients given the magnetic device. What procedure did you use and why?

TABLE 18.6 Body weights (g) at beginning and end of ink toxicity study

Control Group		Treatment Group	
Week 0	Week 13	Week 0	Week 13
111.2	191.6	107.3	187.0
105.4	191.2	116.7	189.5
110.8	210.7	112.2	179.2
105.6	185.2	103.4	172.2
106.1	195.0	113.2	178.7
104.4	188.3	110.6	180.9
114.0	188.4	110.6	188.3
115.1	195.6	100.5	188.9
109.2	204.6	106.3	183.1
111.3	195.7	112.5	184.5

18.41 Magnets for pain relief, continued. Conclusions from any study should be more comprehensive than a simple statement of significance or a confidence interval.

(a) The random assignment of subjects to treatments can sometimes lead, by chance, to an unbalanced split of subjects. Is there evidence of a significant difference in mean pain scores between the two groups at the beginning of the experiment?
(b) In all your calculations, you have been using summary data from the experiment. What would you like to know about the raw data to support the legitimacy of your statistical results?

Do birds learn to time their breeding? *Blue titmice eat caterpillars. The birds would like lots of caterpillars around when they have young to feed, but the birds breed earlier than peak caterpillar season. Do the birds learn from one year's experience when they time breeding the next year? Researchers randomly assigned 7 pairs of birds to have the natural caterpillar supply supplemented while the birds were feeding their young and another 6 pairs to serve as a control group relying on the natural food supply. The next year, they measured how many days after the caterpillar peak the birds produced their nestlings.*[23] *Exercises 18.42 to 18.44 are based on this experiment.*

18.42 Did the randomization produce similar groups? First,

compare the two groups in the first year. The only difference should be the chance effect of the random assignment. The study report says: "In the experimental year, the degree of synchronization did not differ between food-supplemented and control females." For this comparison, the report gives $t = -1.05$, df $= 11$. What type of t statistic (paired or two-sample) is this? Show that this t leads to the quoted conclusion.

18.43 Did the treatment have an effect? The investigators expected the control group to adjust their breeding date the next year, whereas the well-fed supplemented group had no reason to change. The report continues: "But in the following year food-supplemented females were more out of synchrony with the caterpillar peak than the controls." Here are the data (days behind the caterpillar peak):

Control	4.6	2.3	7.7	6.0	4.6	−1.2	
Supplemented	15.5	11.3	5.4	16.5	11.3	11.4	7.7

Carry out a t test and show that it leads to the quoted conclusion.

18.44 Year-to-year comparison. Rather than comparing the two groups in each year, we could compare the behavior of each group in the first and second years. The study report says: "Our main prediction was that females receiving additional food in the nestling period should not change laying date the next year, whereas controls, which (in our area) breed too late in their first year, were expected to advance their laying date in the second year."

Comparing days behind the caterpillar peak in Years 1 and 2 gave $t = 0.63$ for the control group and $t = -2.63$ for the supplemented group. Are these paired or two-sample t statistics? What are the degrees of freedom for each t? Show that these t-values do *not* agree with the prediction.

18.45 The cost of sex. Evolution theory predicts a physiological cost of increased reproduction that translates into a shorter lifespan. A study examined the cost of sexual activity in male fruit flies by comparing males allowed to be sexually active with those that were not. Table 18.7 gives the longevity in days for a sample of males randomly assigned to one or the other condition.[24] Do a complete analysis that reports on

(a) the longevity of sexually active male fruit flies.
(b) the longevity of sexually inactive male fruit flies.
(c) a comparison of the longevity of sexually active and sexually inactive male fruit flies.

18.46 Lung capacity of children. Go back to Exercise 17.46 (page 435), which uses the data file *Large.FEV*.

LARGE DATA SET

(a) How much do boys and girls differ in mean forced expiratory volume (FEV)? Verify the conditions for inference, run the appropriate analysis, and conclude. Use a 95% confidence level.
(b) Your previous work with this data set showed that FEV varies substantially over different age ranges. You could repeat your two-sample analysis from part (a) for each of the four age groups. Explain what problem this would create. Chapter 26 describes the appropriate method (two-way ANOVA) to handle situations like this one with two explanatory variables.

18.47 Children's growth. The data file *Large.FEV* from the previous exercise also includes data on children's height (in inches). For each of the four age groups, assess how much boys and girls differ in height by computing a 95% confidence interval for $\mu_B - \mu_G$. Verify the conditions for inference and write a short description of your findings.

LARGE DATA SET

18.48 Elderly health. Go back to Exercise 17.48 (page 436). Do elderly men and women differ, on average, in their blood calcium levels? Do they differ in their blood inorganic phosphorus levels? Do a complete analysis using significance level $\alpha = 5\%$.

LARGE DATA SET

TABLE 18.7 Longevity (days) of male *Drosophila*

Sexually inactive					Sexually active				
35	49	64	70	76	16	30	35	42	54
37	56	65	70	76	19	33	35	44	54
39	56	65	70	81	19	34	40	46	54
46	64	65	76	85	26	34	42	46	56
46	64	70	76		30	34	42	46	61

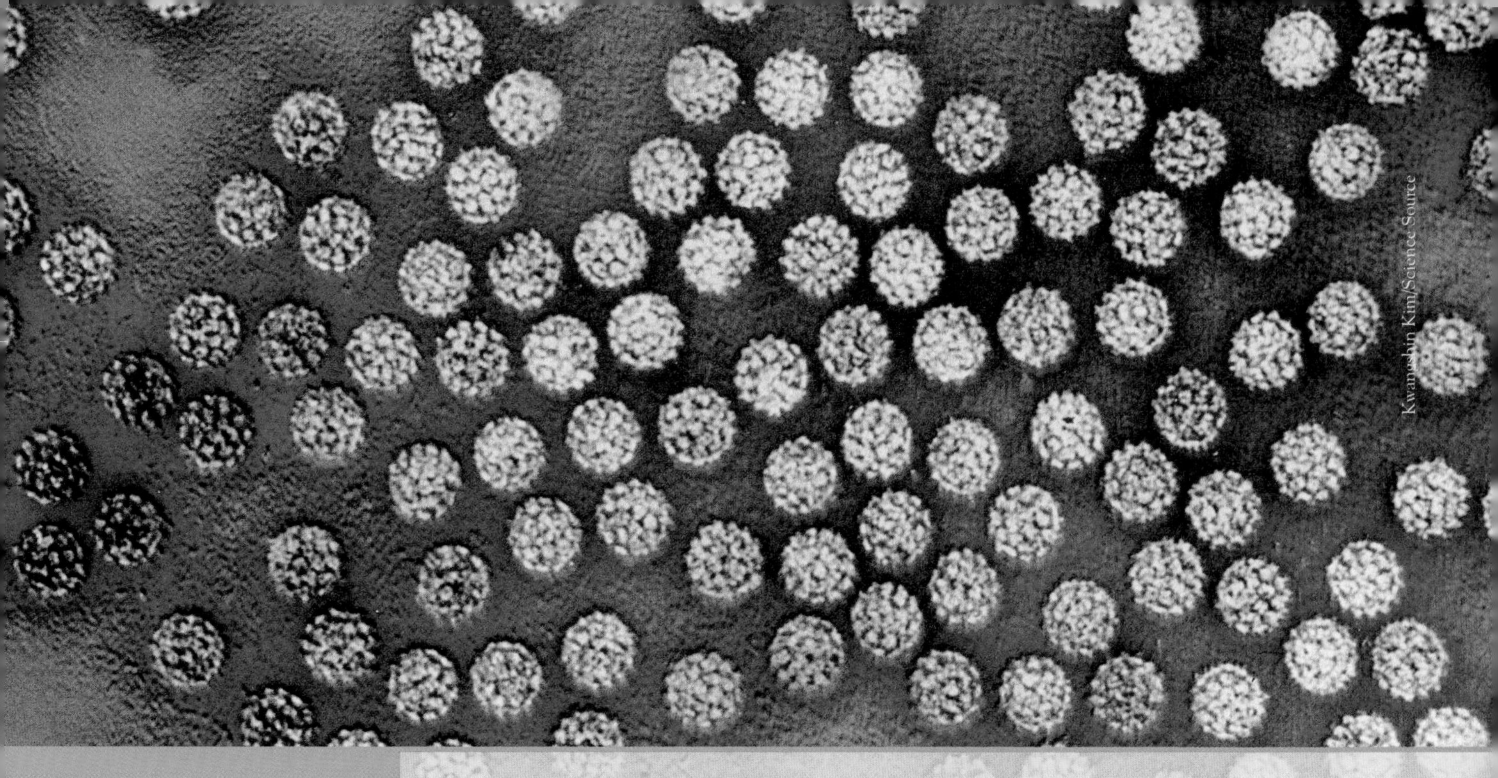

CHAPTER 19 Inference about a Population Proportion

Our discussion of statistical inference up to this point has concerned making inferences about population *means*. Now we turn to questions about the *proportion* of some outcome in a population when studying a categorical variable. Here are some examples that call for inference about population proportions.

IN THIS CHAPTER WE COVER...

- The sample proportion $\hat{p}$
- Large-sample confidence intervals for a proportion
- Accurate confidence intervals for a proportion
- Choosing the sample size*
- Significance tests for a proportion
- Using technology

EXAMPLE 19.1 Assessing the prevalence of HPV infections

Human papillomavirus (HPV) infection is the most common sexually transmitted infection. Certain types of HPV can cause genital warts in both men and women and cervical cancer in women. Whether or not to vaccinate all young women against HPV before they become sexually active is still the subject of heated debate. But just how common is HPV infection in women? The U.S. National Health and Nutrition Examination Survey (NHANES) contacted a representative sample of 1921 women between the ages of 14 and 59 years and asked them to provide a self-collected vaginal swab specimen. Of these, 515 tested positive for HPV, indicating a current HPV infection. That's 26.8% of the sample.[1] Based on these data, what can we say about the percent of all women aged 14 to 59 who test positive for HPV? We want to *estimate a single population proportion*. This chapter concerns inference about one proportion. ■

EXAMPLE 19.2 Treatment for type 1 diabetes

Chronic disease can lead to medical complications even more life impairing than the disease itself. The Diabetes Control and Complications Trial (DCCT) is the largest long-term study of type 1 diabetes (early-onset, or insulin-dependent, diabetes). Volunteers were randomly assigned to either a conventional treatment or a more intensive treatment aimed at maintaining the blood glucose level as close to normal as possible. One complication of diabetes can be the progressive damage to the retina (retinopathy), which can eventually lead to blindness. Among the patients without retinopathy at the beginning of the study, 24% in the conventional treatment group developed retinopathy, but only 7% in the intensive treatment group did.[2] Is this significant evidence that the proportions of diabetes patients who develop retinopathy differ in the population of all patients receiving the conventional treatment and the population all patients receiving the intensive treatment? We want to *compare two population proportions*. This is the topic of Chapter 20. ■

To do inference about a population mean μ, we use the mean $\overline{x}$ of a random sample from the population. The reasoning of inference starts with the sampling distribution of $\overline{x}$. Now we follow the same pattern, replacing means with proportions.

The sample proportion $\hat{p}$

We are interested in the unknown proportion p of a population that has some outcome. For convenience, call the outcome we are looking for a "success." In Example 19.1 the population is women aged 14 to 59, and the parameter p is the proportion who currently test positive for HPV. To estimate p, NHANES used a nationally representative sample of 1921 women. Of these, 515 tested positive for HPV. The statistic that estimates the parameter p is the **sample proportion**

sample proportion

$$\hat{p} = \frac{\text{number of successes in the sample}}{\text{total number of individuals in the sample}}$$
$$= \frac{515}{1921} = 0.2681$$

Read the sample proportion $\hat{p}$ as "p-hat."

How good is the statistic $\hat{p}$ as an estimate of the parameter p? To find out, we ask, "What would happen if we took many samples?" The sampling distribution of $\hat{p}$ answers this question.

We described the properties of the sampling distribution of $\hat{p}$ in Chapter 13 (page 328), and Figure 13.5 (page 328) summarizes them graphically. The behavior of sample proportions $\hat{p}$ is similar to the behavior of sample means $\overline{x}$. When the sample size n is large, the sampling distribution is approximately Normal. The larger the sample, the more nearly Normal the distribution is. *Don't use the Normal approximation to the distribution of $\hat{p}$ when the sample size n is small.*

The mean of the sampling distribution of $\hat{p}$ is the true value of the population proportion p. That is, $\hat{p}$ is an unbiased estimator of p. The standard deviation of $\hat{p}$ is $\sqrt{p(1-p)/n}$, and it gets smaller as the sample size n gets larger, so that estimation is likely to be more accurate when the sample is larger. As is the case for $\bar{x}$, the standard deviation gets smaller only at the rate $\sqrt{n}$. We need four times as many observations to cut the standard deviation in half, and 100 times as many observations to cut it 10 times.

The Normal approximation to the sampling distribution of $\hat{p}$ is least accurate when p is close to 0 or 1. If $p = 0$, successes are impossible; every sample has $\hat{p} = 0$, and no Normal distribution can model that. In the same way, the approximation works poorly when p is close to 1. In practice, this means that we need larger n for values of p near 0 or 1.

Inference about a population proportion p starts with using the sample proportion $\hat{p}$ to estimate p. Confidence levels and P-values are probabilities calculated from the sampling distribution of $\hat{p}$. We will consider only situations that allow us to use the Normal approximation to this sampling distribution. Here is a summary of the conditions we need.

CONDITIONS FOR INFERENCE ABOUT A PROPORTION

- We can regard our data as a **simple random sample** (SRS) from the population. This is, as usual, the most important condition.
- The **sample size n is large enough** to ensure that the distribution of $\hat{p}$ is close to Normal. We will see that different inference procedures require different answers to the question "How large is large enough?"

Remember also that *all our inference procedures require that the population be much larger than the sample.*[3] This condition is usually satisfied in practice and is satisfied in all our examples and exercises.

APPLY YOUR KNOWLEDGE

19.1 Antibiotic resistance. A sample of 1714 cultures from individuals in Florida diagnosed with a strep infection was tested for resistance to the antibiotic penicillin; 973 showed partial or complete resistance to the antibiotic.[4]

(a) Describe the population and explain in words what the parameter p is.

(b) Give the numerical value of the statistic $\hat{p}$ that estimates p.

19.2 Marijuana and medicinal purposes. A March 2013 SurveyUSA poll interviewed by phone a random sample of 500 adults in Florida; 420 said that marijuana should be legal when used for medicinal purposes.

(a) What is the population studied and what parameter are we interested in?

(b) Give the numerical value of the statistic $\hat{p}$ that estimates p.

Justin Sullivan/Getty Images

19.3 Obesity among high school students. The 2010 National Youth Physical Activity and Nutrition Study interviewed an SRS of 9701 high school students about a number of health issues. According to the Centers for Disease Control and Prevention (CDC), 18.4% of adolescents were obese in 2010. Let's assume that the proportion of obesity among all American high school students is $p = 0.184$.

Who is a smoker? When estimating a proportion p, be sure you know what counts as a "success." The news says that 20% of adolescents smoke. Shocking. It turns out that this is the percent who smoked at least once in the past month. If we say that a smoker is someone who smoked on at least 20 of the past 30 days and smoked at least half a pack on those days, fewer than 4% of adolescents qualify.

(a) What are the mean and standard deviation of the proportion $\hat{p}$ of the sample who are obese?

(b) Are the conditions for inference satisfied? Explain why or why not.

19.4 **No inference.** An article published on National Public Radio's website in the fall of 2012 discussed the recent advances in genome sequencing and offered an opt-in poll. The question asked was, "Would you have your genome sequenced if you could afford it?" The poll stayed open for two weeks, during which time 6627 clicks were received (5398 for yes, 656 for undecided, and 573 for no).[5] We can't use these data as the basis for inference about the proportion of all Americans who would have their genomes sequenced if they could afford it. Why not? Could we use these data as the basis for inference about the proportion of all NPR online readers who would have their genomes sequenced if they could afford it?

Large-sample confidence intervals for a proportion

To estimate a population proportion p, use the sample proportion $\hat{p}$. If our conditions for inference apply, the sampling distribution of $\hat{p}$ is close to Normal with mean p and standard deviation $\sqrt{p(1-p)/n}$. To obtain a level C confidence interval for p, we would like to use

$$\hat{p} \pm z^* \sqrt{\frac{p(1-p)}{n}}$$

with the critical value z^* chosen to cover the central area C under the standard Normal curve. Figure 19.1 shows why.

The major challenge of confidence intervals for a population proportion is that we don't know the value of p. One approach consists of replacing the standard deviation with the **standard error of $\hat{p}$**

standard error of $\hat{p}$

$$\text{SE} = \sqrt{\frac{\hat{p}(1-\hat{p})}{n}}$$

FIGURE 19.1 With probability C, $\hat{p}$ lies within $\pm z^* \sqrt{p(1-p)/n}$ of the unknown population proportion p. That is to say that in these samples p lies within $\pm z^* \sqrt{p(1-p)/n}$ of $\hat{p}$.

to get the confidence interval

$$\hat{p} \pm z^* \sqrt{\frac{\hat{p}(1-\hat{p})}{n}}$$

This interval has the form

$$\text{estimate} \pm z^* \text{SE}_{\text{estimate}}$$

We can trust this confidence interval only for large samples. Because the number of successes must be a whole number, using a continuous Normal distribution to describe the behavior of $\hat{p}$ may not be accurate unless n is large. Because the approximation is least accurate for populations that are almost all successes or almost all failures, we require that the sample have both enough successes and enough failures rather than that the overall sample size be large. *Pay attention to both conditions for inference in the box below that summarizes the confidence interval: We must as usual be willing to regard the sample as an SRS from the population, and the sample must have both enough successes and enough failures.*

LARGE-SAMPLE CONFIDENCE INTERVAL FOR A POPULATION PROPORTION

Draw an SRS of size n from a large population that contains an unknown proportion p of successes. An approximate level C **confidence interval for p** is

$$\hat{p} \pm z^* \sqrt{\frac{\hat{p}(1-\hat{p})}{n}}$$

where z^* is the critical value for the standard Normal density curve with area C between $-z^*$ and z^*.

Use this interval only when the number of successes and the number of failures in the sample are each at least 15.[6]

Why not t? Notice that we *don't* change z^* to t^* when we replace the standard deviation by the standard error. When the sample mean $\bar{x}$ estimates the population mean μ, a separate parameter σ describes the spread of the distribution of $\bar{x}$. We separately estimate σ, and this leads to a t distribution. When the sample proportion $\hat{p}$ estimates the population proportion p, the spread depends on p, not on a separate parameter. There is no t distribution—we just make the Normal approximation a bit less accurate when we replace p in the standard deviation by $\hat{p}$.

The next section introduces a more accurate method for computing a confidence interval for a population proportion, and we recommend that you always use the more accurate method. However, you may still come across publications that use the large-sample method, so, here is an example to help you understand how it works.

EXAMPLE 19.3 Assessing the prevalence of HPV infections

The four-step process for any confidence interval is outlined on page 342.

STATE: NHANES found that 515 of a sample of 1921 women aged 14 to 59 years currently tested positive for HPV. That is, $\hat{p} = 0.2681$. What can we say about the population of all women aged 14 to 59 in the United States?

PLAN: We will give a 99% confidence interval to estimate the proportion p of all women aged 14 to 59 in the United States who currently test positive for HPV.

SOLVE: First verify the conditions for inference:

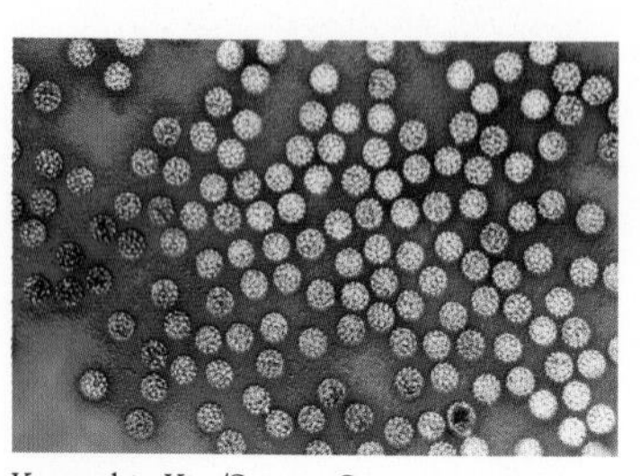
Kwangshin Kim/Science Source

- The sampling design was a complex stratified sample, and the survey used inference procedures for that design. The overall effect is close to an SRS, however.
- The sample is large enough: The numbers of successes (515) and failures (1406) in the sample are both much larger than 15.

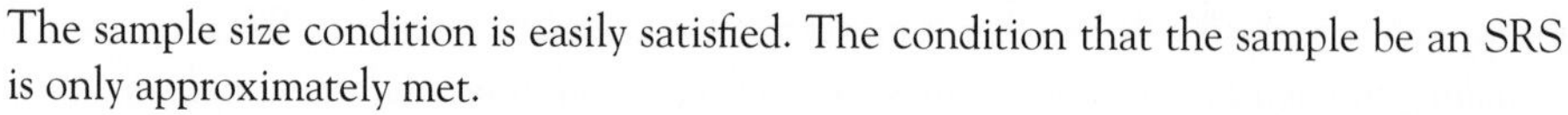

The sample size condition is easily satisfied. The condition that the sample be an SRS is only approximately met.

A 99% confidence interval for the proportion p of all women aged 14 to 59 who test positive for HPV uses the standard Normal critical value $z^* = 2.576$. The confidence interval is

$$\begin{aligned} \hat{p} \pm z^*\sqrt{\frac{\hat{p}(1-\hat{p})}{n}} &= 0.2681 \pm 2.576\sqrt{\frac{(0.2681)(0.7319)}{1921}} \\ &= 0.2681 \pm 0.0260 \\ &= 0.2421 \text{ to } 0.2941 \end{aligned}$$

CONCLUDE: We are 99% confident that the percent of women aged 14 to 59 in the United States who are currently experiencing an HPV infection (testing positive for HPV) lies between about 24.2% and 29.4%. ■

As usual, the practical problems of a large sample survey weaken our confidence in the NHANES conclusions. Only women who agreed to provide a self-collected vaginal swab and collected it properly were included in the final sample. According to the study authors, nonrespondents represented 23% of the original sample and tended to be younger and of ethnic minority. A nonresponse rate of 23% is not unusual in large sample surveys, but the fact that nonrespondents were somewhat different from respondents overall may cause some bias in the sample results. To address this issue, the survey used statistical methods that adjust for unequal response rates in different groups. Reading the report of a large study like NHANES reminds us that statistics in practice involves much more than simple formulas for inference.

APPLY YOUR KNOWLEDGE

19.5 Antibiotic resistance. Exercise 19.1 gave data about the resistance of strep cultures to the antibiotic penicillin. Give a 95% confidence interval for the proportion of strep cultures from Florida patients showing partial or complete resistance to penicillin. Follow the four-step process as illustrated in Example 19.3.

19.6 **Marijuana and medicinal purposes.** Exercise 19.2 gave the results of a 2013 SurveyUSA poll on opinions about the legality of marijuana use for medicinal purposes. Give a 95% confidence interval for the proportion of adults in Florida who think that marijuana should be legal when used for medicinal purposes. Follow the four-step process as illustrated in Example 19.3.

19.7 **No confidence interval.** Severe acute respiratory syndrome (SARS) is a viral respiratory illness that was first reported in Asia in 2003 and that spread worldwide to near pandemic proportions before eventually being contained. A study examined a random sample of 75 patients diagnosed with SARS to describe the disease's clinical progression. One finding was that, while fever and pneumonia initially improved after treatment in all 75 patients, 64 patients developed recurrent fever.[7] Explain why we can't use the large-sample confidence interval to estimate the proportion p in the population of patients treated for SARS who develop recurrent fever.

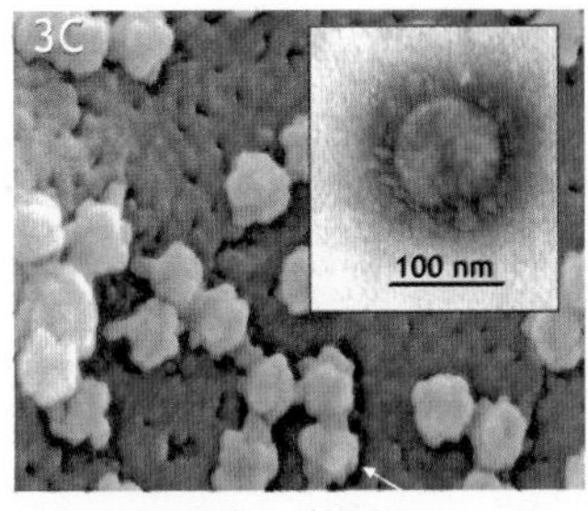

Dr. Mary Ng Mah Lee/CDC

Accurate confidence intervals for a proportion

The confidence interval $\hat{p} \pm z^* \sqrt{\hat{p}(1-\hat{p})/n}$ for a sample proportion p is easy to calculate. It is also easy to understand, because it rests directly on the approximately Normal distribution of $\hat{p}$. Unfortunately, confidence levels from this interval are often quite inaccurate unless the sample is very large. Simulations show that the actual confidence level is usually *less* than the confidence level you asked for in choosing the critical value z^*. That's bad. What is worse, accuracy does not consistently get better as the sample size n increases. There are "lucky" and "unlucky" combinations of the sample size n and the true population proportion p.

Fortunately, there is a simple modification that has been shown experimentally to successfully improve the accuracy of the confidence interval. We call it the "plus four" method, because all you need to do is *add four imaginary observations, two successes and two failures*. With the added observations, the **plus four estimate** of p is

plus four estimate

$$\tilde{p} = \frac{\text{number of successes in the sample} + 2}{n + 4}$$

The formula for the confidence interval is exactly as before, with the new sample size and number of successes.[8] You do not need software that offers the plus four interval—just enter the new sample size (actual size + 4) and number of successes (actual number + 2) into the large-sample procedure.

> **PLUS FOUR CONFIDENCE INTERVAL FOR A PROPORTION**
>
> Draw an SRS of size n from a large population that contains an unknown proportion p of successes. To get the **plus four confidence interval for p**, add four imaginary observations, two successes and two failures. Then use the large-sample confidence interval with the new sample size ($n + 4$) and count of successes (actual count + 2).
>
> Use this interval when the confidence level is at least 90% and the sample size n is at least 10, with any counts of success and failure.

EXAMPLE 19.4 Blinding in medical trials

STATE: Many medical trials randomly assign patients to either an active treatment or a placebo. These trials are always double-blind. Sometimes the patients can tell whether or not they are getting the active treatment. This defeats the purpose of blinding. Reports of medical research usually ignore this problem. Investigators looked at a random sample of 97 articles reporting on placebo-controlled randomized trials in the top five general medical journals. Only 7 of the 97 discussed the success of blinding—and in 5 of these the blinding was imperfect.[9] What proportion of all such studies discuss the success of blinding?

PLAN: Take p to be the proportion of articles that discuss the success of blinding. Give a 95% confidence interval for p.

SOLVE: The conditions for use of the large-sample interval are not met, because there are fewer than 15 successes in the sample. Add two successes and two failures to the original data. The plus four estimate of p is

$$\tilde{p} = \frac{7+2}{97+4} = \frac{9}{101} = 0.0891$$

The plus four confidence interval is the same as the large-sample interval based on 9 successes in 101 observations. Here it is:

$$\begin{aligned}\tilde{p} \pm z^* \sqrt{\frac{\tilde{p}(1-\tilde{p})}{n+4}} &= 0.0891 \pm 1.960\sqrt{\frac{(0.0891)(0.9109)}{101}} \\ &= 0.0891 \pm 0.0556 \\ &= 0.0335 \text{ to } 0.1447\end{aligned}$$

Software (see Figure 19.3) gives the same interval.

CONCLUDE: We estimate with 95% confidence that between about 3.4% and 14.5% of all such articles discuss whether the blinding succeeded. ■

For comparison, the ordinary sample proportion is

$$\hat{p} = \frac{7}{97} = 0.0722$$

The plus four estimate $\tilde{p} = 0.0891$ in Example 19.4 is farther away from zero than $\hat{p} = 0.0722$. The plus four estimate gains its added accuracy by always moving toward 0.5 and away from 0 or 1, whichever is closer. This is particularly helpful when the sample contains only a few successes or a few failures. The numerical difference between a large-sample interval and the corresponding plus four interval is often small. Remember that the confidence level is the probability that the interval will catch the true population proportion *in very many uses*. Small differences every time add up to accurate confidence levels from plus four intervals versus inaccurate levels from large-sample intervals.

How much more accurate is the plus four interval? Computer studies have asked how large n must be to guarantee an accurate confidence interval. If $p = 0.1$, for example, the answer is $n \geq 646$ for the large-sample interval and $n \geq 11$ for the plus four interval when we want a 95% confidence level.[10] The consensus of computational and theoretical studies is that plus four is very much better than the large-sample interval for many combinations of n and p. **We recommend that you always use the plus four interval for estimating a proportion.**

APPLY YOUR KNOWLEDGE

19.8 Antibiotic resistance. In Exercise 19.5 you gave a 95% confidence interval for the proportion of strep cultures from Florida patients showing partial or complete resistance to penicillin. This estimate was based on the finding of 973 cultures with partial or complete resistance out of a random sample of 1714 throat cultures.

(a) What is the plus four estimate $\tilde{p}$ for this population proportion? Give the plus four 95% confidence interval for p.

(b) How do the two confidence intervals compare? Explain why the two methods give similar results when the sample size is large. For large samples, both methods give similar and accurate estimates. For smaller samples, the plus four estimate is always more accurate.

19.9 Whelks and mussels. Sample surveys usually contact large samples, so we can use the large-sample confidence interval if the sample design is close to an SRS. Scientific studies often use small samples that require the plus four method. For example, the small round holes you often see in seashells were drilled by other sea creatures, who ate the former owners of the shells. Whelks often drill into mussels, but this behavior appears to be more or less common in different locations. Investigators collected whelk eggs from the coast of Oregon, raised the whelks in the laboratory, then put each whelk in a container with some delicious mussels. Only 9 of 98 whelks drilled into a mussel.[11]

(a) Why can't we use the large-sample confidence interval?

(b) Give the plus four 90% confidence interval for the proportion of Oregon whelks that will spontaneously drill into mussels.

19.10 Clinical progression of SARS. Go back to Exercise 19.7, which describes the progression of fever symptoms in individuals treated for SARS.

(a) The plus four method adds four observations, two successes and two failures. What are the sample size and the count of successes after you do this? What is the plus four estimate $\tilde{p}$ of the proportion p in the population of patients treated for SARS who develop recurrent fever?

(b) Give the plus four 95% confidence interval for p.

Choosing the sample size*

In planning a study, we may want to choose a sample size that will allow us to estimate the parameter within a given margin of error. We saw earlier (page 378) how to do this for a population mean. The method is similar for estimating a population proportion.

*This section presents more advanced material that is not needed to read the rest of the book. The idea of planning studies is, however, very important in practice.

Kids on bikes In the most recent year for which data are available, 77% of children killed in bicycle accidents were boys. You might take these data as a sample and start from $\hat{p} = 0.77$ to do inference about bicycle deaths in the near future. What you should not do is conclude from these data that boys on bikes are in greater danger than girls. We don't know how many boys and how many girls ride bikes—it may be that most fatalities are boys because most riders are boys.

The margin of error in the large-sample confidence interval for p is

$$m = z^*\sqrt{\frac{\hat{p}(1-\hat{p})}{n}}$$

Here z^* is the standard Normal critical value for the level of confidence we want. Because the margin of error involves the sample proportion of successes $\hat{p}$, we need to guess this value when choosing n. Call our guess p^*. Here are two ways to get p^*:

1. Use a guess p^* based on a pilot study or on past experience with similar studies. You can do several calculations to cover the range of values of $\hat{p}$ you might get.
2. Use $p^* = 0.5$ as the guess. The margin of error m is largest when $\hat{p} = 0.5$, so this guess is conservative in the sense that if we get any other $\hat{p}$ when we do our study, we will get a margin of error smaller than planned.

Once you have a guess p^*, the recipe for the margin of error can be solved to give the sample size n needed. Here is the result for the large-sample confidence interval. For simplicity, use this result even if you plan to use the plus four interval.

SAMPLE SIZE FOR DESIRED MARGIN OF ERROR

The level C confidence interval for a population proportion p will have margin of error approximately equal to a specified value m when the sample size is

$$n = \left(\frac{z^*}{m}\right)^2 p^*(1-p^*)$$

where p^* is a guessed value for the sample proportion. The margin of error will be less than or equal to m if you take the guess p^* to be 0.5.

Which method for finding the guess p^* should you use? The n you get doesn't change much when you change p^* as long as p^* is not too far from 0.5. You can use the conservative guess $p^* = 0.5$ if you expect the true $\hat{p}$ to be roughly between 0.3 and 0.7. If the true $\hat{p}$ is close to 0 or 1, using $p^* = 0.5$ as your guess will yield a sample much larger than you need. Try to use a better guess from a pilot study when you suspect that $\hat{p}$ will be less than 0.3 or greater than 0.7.

EXAMPLE 19.5 Planning a survey

STATE: Flossing helps prevent tooth decay and gum disease. Yet reports suggest that millions of Americans never floss. You are planning a sample survey to determine what percent of students at your university never floss. You obtain permission to use the university roster to contact an SRS of students. You want to estimate the proportion p of students who never floss with 95% confidence and a margin of error no greater than 4% (0.04). How large a sample do you need?

PLAN: Find the sample size n needed for margin of error $m = 0.04$ and 95% confidence. Based on your readings on the topic, you guess that the proportion of individuals who never floss is between 30% and 70% for any given population. You can use the guess $p^* = 0.5$.

SOLVE: The sample size you need is

$$n = \left(\frac{1.96}{0.04}\right)^2 (0.5)(1 - 0.5) = 600.25$$

Round the result up to $n = 601$. (Rounding down would yield a margin of error slightly greater than 0.04.)

CONCLUDE: An SRS of 601 students from your university is adequate for a margin of error of ±4% with a confidence level of 95%. ■

What sample size would you have needed for a margin of error of 3% rather than 4%? The answer is, after rounding up,

$$n = \left(\frac{1.96}{0.03}\right)^2 (0.5)(1 - 0.5) = 1068$$

For a 2% margin of error, the sample size you would have needed is

$$n = \left(\frac{1.96}{0.02}\right)^2 (0.5)(1 - 0.5) = 2401$$

As usual, smaller margins of error call for larger samples. Remember, though, that the population of interest must be much larger than the sample in order to conduct inference for proportions legitimately. Are there more than 24,000 students enrolled at your university? If not, you cannot legitimately calculate a plus four or large-sample confidence interval with only a 2% margin of error. Inference requires taking everything into consideration.

Notice also how large a sample size is required for a 2% or 3% margin of error. When dealing with proportions, the margin of error is always a value well below 1. Because it goes in the denominator of the sample size equation and gets squared, it drives the sample size up. You will find that, typically, studies of categorical variables (for example, clinical trials looking for improvement versus no improvement) tend to have rather large sample sizes compared with studies of quantitative variables.

APPLY YOUR KNOWLEDGE

19.11 Planning a study of influenza. The CDC estimates that, each year in the United States, on average, 5% to 20% of the population gets the flu. Using 20% as our best guess (p^*), find how large a sample would be needed to get a ±2 percentage point margin of error on a 95% confidence interval for the population proportion of Americans who get the flu in a given year.

19.12 Can you taste PTC? PTC is a substance that has a strong bitter taste for some people and is tasteless for others. The ability to taste PTC is inherited. About 75% of Italians can taste PTC, for example. You want to estimate the proportion of Americans who have at least one Italian grandparent and can taste PTC. Starting with the 75% estimate for Italians, how large a sample must you collect in order to estimate the proportion of PTC tasters within ±0.04 with 90% confidence?

Significance tests for a proportion

We now turn to tests of significance and will assume that the conditions for inference we described earlier are met. When the null hypothesis $H_0: p = p_0$ is true, the sampling distribution of the sample proportion $\hat{p}$ is approximately Normal, centered on p_0 with standard deviation $\sqrt{p_0(1 - p_0)/n}$, as illustrated in Figure 19.2.

The test statistic for the null hypothesis $H_0: p = p_0$ is the sample proportion $\hat{p}$ standardized using the value p_0 specified by H_0,

$$z = \frac{\hat{p} - p_0}{\sqrt{\dfrac{p_0(1 - p_0)}{n}}}$$

This z statistic has approximately the standard Normal distribution when H_0 is true. P-values therefore come from the standard Normal distribution. Because H_0 gives us a value for p, the inaccuracy that plagues the large-sample confidence interval does not affect tests (therefore, no adjustment is required). Here is the procedure for tests.

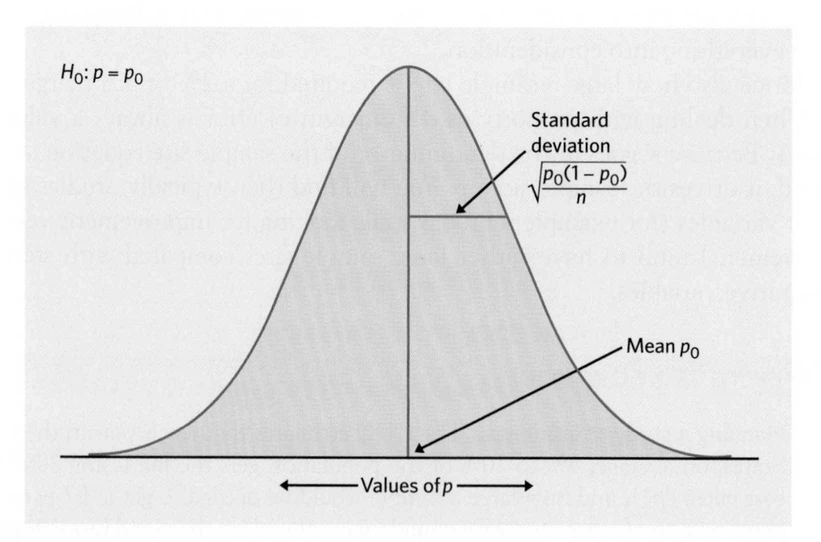

FIGURE 19.2 When H_0: $p = p_0$ is true, the sampling distribution of $\hat{p}$ is approximately Normal with mean p_0 and standard deviation $\sqrt{p_0(1 - p_0)/n}$.

SIGNIFICANCE TESTS FOR A PROPORTION

Draw an SRS of size n from a large population that contains an unknown proportion p of successes. To **test the hypothesis H_0: $p = p_0$**, compute the z statistic

$$z = \frac{\hat{p} - p_0}{\sqrt{\frac{p_0(1 - p_0)}{n}}}$$

In terms of a variable Z having the standard Normal distribution, the approximate P-value for a test of H_0 against

H_a: $p > p_0$ is $P(Z \geq z)$

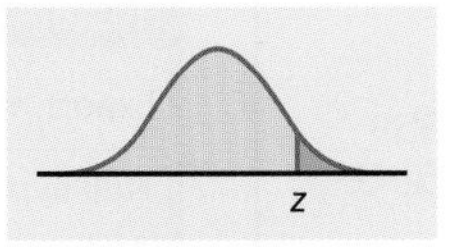

H_a: $p < p_0$ is $P(Z \leq z)$

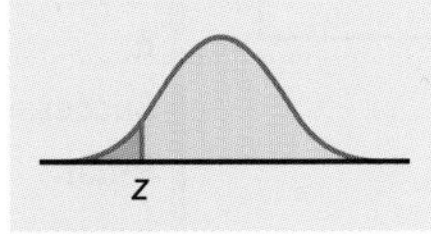

H_a: $p \neq p_0$ is $2P(Z \geq |z|)$

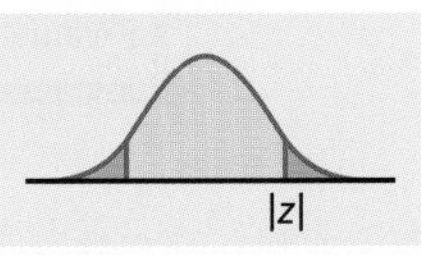

Use this test when the sample size n is so large that both np_0 and $n(1 - p_0)$ are 10 or more.[12]

EXAMPLE 19.6 Do aphids right themselves when dropped?

The four-step process for any significance test is outlined on page 354.

STATE: Pea aphids, *Acyrthosiphon pisum*, are small, wingless, sap-sucking insects that live on plants. They evade predators (such as ladybugs) by dropping off. A study examined the mechanism of aphid drops. Researchers hung live aphids upside down from delicate tweezers and then released them. The videotaped drops show that 19 of the 20 aphids landed on their legs.[13] Is this evidence that live aphids land right side up (on their legs) more often than chance alone would predict?

PLAN: Take p to be the proportion of all live aphids that land on their legs when dropped. Aphids can land either on their legs or on their backs. If chance alone was at work, we would expect aphids to land on their legs half the time. Therefore, we test the hypotheses

$$H_0: p = 0.5$$
$$H_a: p > 0.5$$

SOLVE: The aphids in the study were reared in the lab and selected without bias. We can consider them to be the equivalent of an SRS. The expected counts of successes and failures under the null hypothesis are both 10 ($np_0 = n(1 - p_0) = 20 \times 0.5 = 10$).

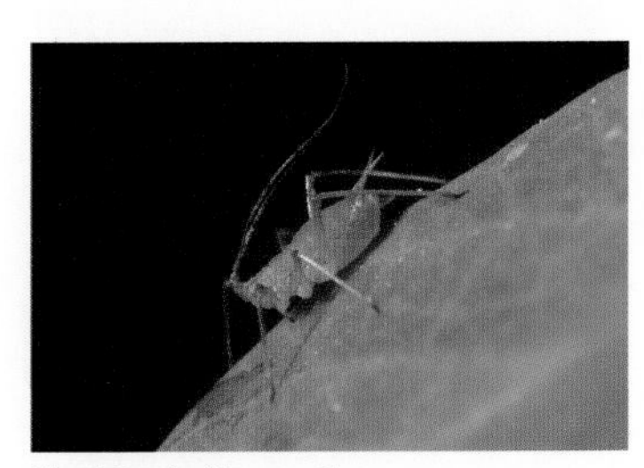
Nigel Cattlin/Science Source

TI-83

```
1-PropZInt
 (.03355, .14467)
p̂=.0891089109
n=101
```

Minitab

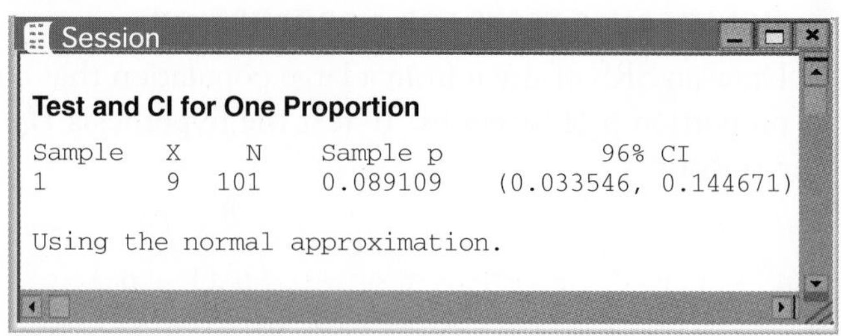

Session
Test and CI for One Proportion
Sample X N Sample p 96% CI
1 9 101 0.089109 (0.033546, 0.144671)
Using the normal approximation.

TI-83

```
1-PropZTest
 prop >.5
 z=4.024922359
 p=2.8510885E-5
 p̂=.95
 n=20
```

CrunchIt!

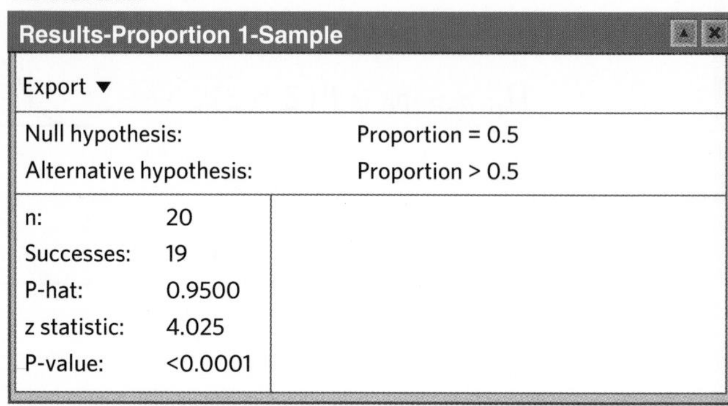

Results-Proportion 1-Sample
Export ▼
Null hypothesis: Proportion = 0.5
Alternative hypothesis: Proportion > 0.5
n: 20
Successes: 19
P-hat: 0.9500
z statistic: 4.025
P-value: <0.0001

FIGURE 19.3 Output from a graphing calculator and statistical software for the plus four 95% confidence interval in Example 19.4 and for the P-value in Example 19.6.

Therefore, the conditions for inference are met. The sample proportion of aphids landing on their legs was

$$\hat{p} = \frac{19}{20} = 0.95$$

Next we obtain the z test statistic:

$$z = \frac{\hat{p} - p_0}{\sqrt{\frac{p_0(1 - p_0)}{n}}} = \frac{0.95 - 0.5}{\sqrt{\frac{(0.5)(0.5)}{20}}} = 4.02$$

The P-value is the area under the standard Normal curve to the right of $z = 4.02$. We know that this is a very small area. Table C shows that $P < 0.0005$. The TI-83 output in Figure 19.3 indicates that, in fact, $P = 0.00003$.

CONCLUDE: There is very strong evidence that more than half of live aphids dropped upside down from tweezers land on their legs ($P < 0.0005$). This supports the idea that live aphids possess a mechanism for righting themselves when dropped (as cats do). ■

APPLY YOUR KNOWLEDGE

19.13 More on dropped aphids. The study in Example 19.6 also asked whether aphid aerial-righting behavior is an active or a passive mechanism (body shape and simple aerodynamics). So they used dead aphids in the same setting. Of the 23 dead aphids hung upside down from tweezers and then released, 12 landed on their legs.

(a) Is this evidence that dead aphids land right side up (on their legs) more often than chance alone would predict? Follow the four-step process as illustrated in Example 19.6.

(b) Write a short paragraph discussing aphid aerial righting behavior based on these results and those described in Example 19.6. Be sure to include sample statistics, test statistics, and P-values in your report.

19.14 Formulating the theory of inheritance. Gregor Mendel has been called the "father of genetics," having formulated laws of inheritance even before the function of DNA was discovered. In one famous experiment with peas, he crossed pure breeds of plants that produced either smooth peas or wrinkled peas. The first-generation peas were all smooth. In the second generation (F2), he obtained 5474 smooth peas and 1850 wrinkled peas. Do these data agree with the conclusion of 75% dominant-trait (in this case "smooth") occurrence in F2? Follow the four-step process as illustrated in Example 19.6.

19.15 No test. Explain why we can't use the z test for a proportion in these situations:

(a) You want to know if people have predictive psychic abilities when answering their phones. You ask 10 subjects to guess which of 5 possible friends is calling them before they pick up the phone in order to test the hypotheses H_0: $p = 0.2$ (no psychic ability) versus H_a: $p > 0.2$.

(b) You bet a friend that 90% or more of the students taking the Bio 101 class know their blood types. You contact an SRS of 80 of the 1211 students taking Bio 101 this term to test the hypotheses H_0: $p = 0.9$ versus H_a: $p > 0.9$.

Using technology

Not all software packages provide tests and confidence intervals for a sample proportion. Excel and SPSS, for instance, do not have an easy way to perform these procedures, while JMP and R use a slightly different methodology. Figure 19.3 displays output from a graphing calculator and two statistical programs for the plus four 95% confidence interval of Example 19.4 and for the test of Example 19.6. All three use the Normal approximation for the sampling distribution of $\hat{p}$, although this is something you must specify with Minitab. The programs give the same numerical values except for rounding differences.

For tests, remember to always specify your null and alternative hypotheses before examining the sample data and then obtain the one- or two-sided P-value that matches your alternative.

The only information required to obtain a confidence interval for p with technology is the count of successes in the sample, the sample size, and your desired confidence level. Therefore, you do not need special technology to use the plus four method: simply type in the count of successes plus 2 and the sample size plus 4 in the software input window. This is noticeable in the TI-83 and Minitab outputs, which match all our calculations for Example 19.4.

CHAPTER 19 SUMMARY

- When the data are an SRS of size n, tests and confidence intervals for a population proportion p are based on the **sample proportion $\hat{p}$.**
- When n is large, $\hat{p}$ has approximately the Normal distribution with mean p and standard deviation $\sqrt{p(1-p)/n}$.
- The level C **large-sample confidence interval for p** is

$$\hat{p} \pm z^* \sqrt{\frac{\hat{p}(1-\hat{p})}{n}}$$

  where z^* is the critical value for the standard Normal curve with area C between $-z^*$ and z^*.
- The true confidence level of the large-sample interval can be substantially less than the planned level C unless the sample is very large. *We recommend using the plus four interval instead.*
- To get a more accurate confidence interval, add four imaginary observations, two successes and two failures, to your sample. Then use the same formula for the confidence interval. This is the **plus four confidence interval.** Use this interval in practice for confidence level 90% or higher and sample size n at least 10.
- The **sample size** needed to obtain a confidence interval with approximate margin of error m for a population proportion is

$$n = \left(\frac{z^*}{m}\right)^2 p^*(1-p^*)$$

  where p^* is a guessed value for the sample proportion $\hat{p}$ and z^* is the standard Normal critical point for the level of confidence you want. If you use $p^* = 0.5$ in this formula, the margin of error of the interval will be less than or equal to m no matter what the value of $\hat{p}$ is.
- **Significance tests for H_0: $p = p_0$** are based on the z statistic

$$z = \frac{\hat{p} - p_0}{\sqrt{\dfrac{p_0(1-p_0)}{n}}}$$

  with P-values calculated from the standard Normal distribution. Use this test in practice when $np_0 \geq 10$ and $n(1-p_0) \geq 10$.

THIS CHAPTER IN CONTEXT

Over the course of Chapters 19 through 22, we describe statistical inference procedures used when the response variable is categorical. That is, for each individual or observation in the sample(s) we record an attribute (such as gender or opinion) rather than a numerical value, as explained in Chapter 1. Ultimately, we are interested in the proportion p of successes in one or more populations.

In Chapter 19 we compute confidence intervals and test hypotheses about one population proportion p, just as we computed confidence intervals and tested hypotheses about a population mean μ in Chapter 17 when the response variable was quantitative. In the next chapter we will describe inference procedures that let us compare the proportions of successes in two populations when the data have been collected using two independent samples.

The computations described in this chapter build on the Normal calculations you learned in Chapter 11 and on the properties of the sampling distribution of $\hat{p}$ introduced in Chapter 13. These Normal calculations are approximately correct only when the counts of successes and failures in the sample (observed or expected from a null hypothesis) are large enough. So you must be careful that these count assumptions are met before performing the inference procedure you want.

CHECK YOUR SKILLS

19.16 Suppose that a population has 30% of individuals displaying a particular phenotype. In repeated random samples of 400 individuals, the sample proportion $\hat{p}$ would follow a Normal distribution with mean

(a) 20. (b) 0.3. (c) 0.023.

19.17 The standard deviation of the distribution of $\hat{p}$ in the previous exercise is about

(a) 0.0011. (b) 0.3. (c) 0.023.

The March 2013 SurveyUSA poll from Exercise 19.2 also asked its random sample of adults in Florida their opinion on marijuana use for recreational purposes. Of the 500 interviewed, 259 said that marijuana should be legal when used for recreational purposes. Use this information to answer Exercises 19.18 to 19.20.

19.18 The sample proportion $\hat{p}$ who said that marijuana use for recreational purposes should be legal is

(a) 259. (b) 52. (c) 0.518.

19.19 Based on the SurveyUSA sample, the margin of error of the 95% confidence interval for the proportion of all adults in the state of Florida who think that marijuana use for recreational purposes should be legal is

(a) 0.088. (b) 0.044. (c) 0.002.

19.20 (Optional) How many adults must be interviewed to estimate that population proportion to within ± 0.02 with 95% confidence? Use 0.5 as the conservative guess for p.

(a) $n = 25$ (b) $n = 1225$ (c) $n = 2401$

A farmer wants to test the effectiveness of a pest control method in allowing strawberry blooms to yield marketable strawberries. Out of a random sample of 100 blooms, 77 yield marketable strawberries. Is this evidence that more than two-thirds of all blooms grown with this pest control method end up as marketable strawberries? Use this information to answer Exercises 19.21 to 19.23.

19.21 The hypotheses for a test to answer this question are

(a) H_0: $p = 0.667$, H_a: $p > 0.667$.
(b) H_0: $p = 0.667$, H_a: $p \neq 0.667$.
(c) H_0: $p = 0.77$, H_a: $p \neq 0.77$.

19.22 The value of the z statistic for this test is about

(a) 1.03. (b) 2.19. (c) 2.46.

19.23 The P-value of this test is about

(a) 0.014. (b) 0.029. (c) 0.047.

19.24 A March 2013 Gallup poll reports that 67% of American adults say that, if given the opportunity, they would vote for a law that limits food sold in public schools to food that meets standards for high nutritional value. The poll's stated margin of error was 4 percentage points with 95% confidence. This means that

(a) the poll used a method that gets an answer within 4 percentage points of the truth about the population parameter 95% of the time.
(b) we can be sure that the percent of all adults who say they would vote this way is between 63% and 71%.
(c) if Gallup took another poll using the same method, the results of the second poll would lie between 63% and 71%.

19.25 The Gallup poll in the previous exercise probably had at least 80% nonresponse. The nonresponse could cause the survey result to be in error. The error due to nonresponse

(a) is in addition to the margin of error given in Exercise 19.24.
(b) is included in the margin of error given in Exercise 19.24.
(c) can be ignored because it isn't random.

CHAPTER 19 EXERCISES

We recommend using the plus four method for all confidence intervals for a proportion. However, the large-sample method is acceptable when the guidelines for its use are met.

19.26 Detecting genetically modified soybeans. Most soybeans grown in the United States are genetically modified to, for example, resist pests and so reduce the use of pesticides. Because some nations do not accept genetically modified (GM) foods, grain-handling facilities routinely test soybean shipments for the presence of GM beans. In a study of the accuracy of these tests, researchers submitted lots of soybeans containing 1% of GM beans to 23 randomly selected facilities. Eighteen detected the GM beans.[14]

(a) Show that the conditions for the large-sample confidence interval are not met. Show that the conditions for the plus four interval are met.

(b) Use the plus four method to give a 90% confidence interval for the percent of all grain-handling facilities that will correctly detect 1% of GM beans in a shipment.

19.27 White-eyed fruit flies. Wild-type fruit flies have red eyes, but a recessive mutation produces white-eyed individuals. A researcher wants to assess the frequency of heterozygous individuals among red-eyed fruit flies. A heterozygous red-eyed fly crossed with a white-eyed mutant will have mixed progeny. Of the 100 red-eyed fruit flies crossed with white-eyed mutants, 11 produced mixed progeny.

(a) Show that the conditions for the large-sample confidence interval are not met. Show that the conditions for the plus four interval are met.

(b) Give the plus four 95% confidence interval for the proportion of heterozygotes in the population of red-eyed fruit flies studied.

19.28 Looking for health information online. A Pew Research Center survey found that 58% of the 1931 adult American internet users interviewed said that the internet has helped them in some way with obtaining health information.[15]

(a) Verify that the conditions for inference are met. Give a 99% confidence interval for the proportion of all adult American internet users who feel this way.

(b) Can you generalize this finding to all adult Americans? Explain why or why not. Do you think that the proportion who would say that the internet helped them obtain health information is higher among all adult American internet users or among all adult Americans? Explain your answer.

19.29 White-eyed fruit flies, continued (optional). Go back to the experiment of Exercise 19.27. How many red-eyed fruit flies should be used in this experiment to obtain a 95% confidence interval with a margin of error no more than 3%? Use 20% as a conservative (worst-case) estimate of the percent of heterozygotes in the population.

19.30 Color blindness (optional). Color blindness, or dyschromatopsia, is a form of genetic deficiency in color perception. The condition is much more prevalent among males than females, pointing to a genetic connection with the X chromosome. The frequency of dyschromatopsia in the Caucasian American male population is about 8%. However, it is thought that this proportion might be smaller among males of other ethnicities. We want to estimate the proportion of Asian American males who are color-blind. How large a sample size do we need to obtain a 95% confidence interval with a margin of error no greater than 2%, or 0.02? Use 0.1 for p^*.

19.31 Fast food for baby. The Gerber Products Company sponsored a large survey of the eating habits of American infants and toddlers. Among the many questions parents were asked was whether their child had eaten fried potatoes on the day before the interview. Among the 679 infants 9 to 11 months old, 9% had eaten fried potatoes that day.[16] Verify that the conditions for inference are met and give the plus four 95% confidence interval for the population proportion of 9- to 11-month-old infants who eat fried potatoes on a given day.

19.32 Side effects. An experiment on the side effects of pain relievers assigned arthritis patients to take one of several over-the-counter pain medications. Of the 440 patients who took one brand of pain reliever, 23 suffered some "adverse symptom."

(a) If 10% of all patients suffer adverse symptoms, what would be the sampling distribution of the proportion with adverse symptoms in a sample of 440 patients?

(b) Does the experiment provide strong evidence that fewer than 10% of patients who take this medication have adverse symptoms? Verify that the conditions for inference are met. State the hypotheses, calculate the test statistic, then obtain and interpret the P-value.

19.33 Cancer-detecting dogs. A study was designed to determine whether dogs can be trained to identify urine specimens from individuals with bladder cancer. Dogs were first trained to discriminate between urine specimens from patients with bladder cancer and urine specimens from patients with other conditions. After the training was completed, the dogs had to pick one of seven new urine specimens. Each time, only one of the seven urine specimens came from a patient with bladder cancer. Out of 54 trials, the dogs identified the correct urine specimen 22 times.[17]

(a) If the dogs were simply picking a urine specimen at random, we would expect them to be correct, on average, 1 out of 7 times. The experiment was designed to test whether dogs can perform better than chance. State the null and alternative hypotheses for this test.
(b) Obtain the test statistic and the P-value. What do you conclude?

19.34 Matched pairs. One-sample procedures for proportions, like those for means, are used to analyze data from matched pairs designs. Here is an example. Each of 50 subjects tastes two unmarked cups of coffee and says which he or she prefers. One cup in each pair contains instant coffee; the other, fresh-brewed coffee. Thirty-one of the subjects prefer the fresh-brewed coffee. Take p to be the proportion of the population who would prefer fresh-brewed coffee in a blind tasting.

(a) Test the claim that a majority of people prefer the taste of fresh-brewed coffee. State hypotheses and report the z statistic and its P-value. Is your result significant at the 5% level? What is your practical conclusion?
(b) Find a 90% confidence interval for p.
(c) When you do an experiment like this, in what order should you present the two cups of coffee to the subjects?

In responding to Exercises 19.35 through 19.41, follow the ***Plan, Solve,*** *and* ***Conclude*** *steps of the four-step process. It may be helpful to restate in your own words the* ***State*** *information given in the exercise.*

19.35 Lead in tap water. 4 STEP The U.S. Environmental Protection Agency (EPA) defines the Action Level for lead in tap water as 15 parts per billion (ppb; 15 ppb = 0.015 mg/l). This is a level above which further analysis and public education campaigns must be done, but not yet a level causing a real health risk. A random sample of 10,341 water specimens from the Washington, DC, area found that 7496 water specimens had lead contents up to 15 ppb.[18] What is the proportion of all possible water specimens in DC that would exceed the Action Level? Use a 95% confidence level.

19.36 Vote for the best face? 4 STEP We often judge other people by their faces. It appears that some people judge candidates for elected office by their faces. Researchers showed head-and-shoulders photos of the two main candidates in 32 races for the U.S. Senate to many subjects (dropping subjects who recognized one of the candidates) to see which candidate was rated "more competent" based on nothing but the photos. On Election Day, the candidates whose faces looked more competent won 22 of the 32 contests.[19] If faces don't influence voting, half of all races in the long run should be won by the candidate with the better face. Is there evidence that the candidate with the better face wins more than half the time? Follow the four-step process as illustrated in Example 19.6.

19.37 Depression. 4 STEP The Gallup-Healthways Well-Being Index is a comprehensive survey of the health status of Americans. In 2009, the survey asked a random sample of 258,141 adults, "Have you ever been told by a physician or a nurse that you have depression?" Of these, 43,884 answered "yes." Estimate with 95% confidence the proportion of adults in America who have ever been told by a health professional that they have depression.

19.38 The polarizing taste of cilantro. 4 STEP The culinary herb cilantro, *Coriandrum sativum*, is very polarizing: Some people love it and others hate it. A genetic component is suspected to be at play. A survey of 12,087 American adults of European ancestry asked whether they like or dislike the taste of cilantro. A total of 3181 said that they dislike the taste of cilantro.[20] Estimate with 95% confidence the proportion of all American adults of European ancestry who dislike the taste of cilantro.

19.39 Children's food choices. 4 STEP Do popular cartoon characters on food packages influence children's food choices? A study asked 40 young children (ages four to six) to taste graham crackers presented in a package either with or without a popular cartoon character. When asked to indicate which of the two options they would prefer to eat for a snack, 35 chose the version with a cartoon on the package.[21] Is this evidence that young children prefer snacks packaged with popular cartoon images more than chance alone would explain? Estimate with 95% confidence the proportion of young children who prefer graham crackers packaged with popular cartoon images. Answer both questions with appropriate inference methods.

19.40 The polarizing taste of cilantro, continued. 4 STEP One reason sometimes given for the dislike of cilantro is that it tastes like soap. The survey in Exercise 19.38 asked 14,604 American adults of European ancestry whether cilantro tastes soapy to them and found that 1994 answered that it does. Estimate with 95% confidence the proportion of all American adults of European ancestry who find that cilantro tastes soapy.

19.41 Staph infections. 4 STEP A study investigated ways to prevent staph infections in surgery patients. In a first step, the researchers examined the nasal secretions of a random sample of 6771 patients admitted to various hospitals for surgery. They found that 1251 of these patients tested positive for *Staphylococcus aureus*, a bacterium responsible for most staph infections.[22] Calculate a 95% confidence interval for the proportion of all patients admitted for surgery who are carriers of *S. aureus*.

19.42 Estimating body fat in men. LARGE DATA SET In Exercises 1.45 (page 37) and 2.46 (page 64) you examined the data from the data file *Large.Bodyfat* containing the percent of body fat, age, weight, height, and 10 body circumference measurements for 252 adult men.

(a) Use the columns for weight (in pounds) and height (in inches) to compute a new variable representing an individual's body mass index (BMI). The formula for BMI is

$$\text{BMI} = \frac{703 \times \text{weight}}{\text{height}^2}$$

(b) Obesity is defined as a BMI of 30.0 or above. Find the proportion of individuals in this sample who are considered obese based on their BMI. Make sure to exclude the individual with an incorrect height value that you identified in Exercise 2.46.

(c) The data are a random sample of 252 adult males taken in 1985. Give a 95% confidence interval for the proportion of obese adult males in 1985.

(d) The most recent data from the CDC indicate that, currently, 28.4% of adult men are obese.[23] How do the 1985 data compare with the current estimate? Use your results from (c) to answer this question.

CHAPTER 20 Comparing Two Proportions

IN THIS CHAPTER WE COVER...

- Two-sample problems: proportions
- The sampling distribution of a difference between proportions
- Large-sample confidence intervals for comparing proportions
- Accurate confidence intervals for comparing proportions
- Significance tests for comparing proportions
- Relative risk and odds ratio*
- *Discussion: Assessing and understanding health risks*

In a **two-sample problem,** we want to compare two populations or the responses to two treatments based on two independent samples. When the comparison involves the *means* of two populations, we use the two-sample t methods of Chapter 18. Now we turn to methods to compare the *proportions* of successes in two populations.

two-sample problem

Two-sample problems: proportions

We will use notation similar to that used in our study of two-sample t statistics. The groups we want to compare are Population 1 and Population 2. We have a separate SRS from each population or responses from two treatments in a randomized comparative experiment. A subscript shows which group a parameter or statistic describes. Here is our notation:

Population	Population proportion	Sample size	Sample proportion
1	p_1	n_1	$\hat{p}_1$
2	p_2	n_2	$\hat{p}_2$

We compare the populations by doing inference about the difference $p_1 - p_2$ between the population proportions. The statistic that estimates this difference is the difference between the two sample proportions, $\hat{p}_1 - \hat{p}_2$.

EXAMPLE 20.1 How to treat type 1 diabetes

STATE: Chronic diseases, such as type 1 diabetes (early onset, or insulin-dependent, diabetes), can lead to severe medical complications. For a long time, the medical community was divided on whether or not to aggressively control patients' blood glucose levels. The Diabetes Control and Complications Trial (DCCT) randomly assigned volunteers with type 1 diabetes without retinopathy (damage to the retina that can lead to blindness) either to a conventional treatment or to a more intensive treatment aimed at maintaining the blood glucose level as close to normal as possible. The health of 378 patients in the conventional care group and 348 in the intensive care group was closely monitored for about six years. By the end of the study, 91 patients in the conventional treatment group had developed retinopathy, compared with only 23 in the intensive treatment group.[1] Is this good evidence that the proportion of diabetes patients developing retinopathy differs in the two populations? How large is the difference?

Voisin/Phanie/Superstock

PLAN: Take patients receiving the conventional treatment to be Population 1 and patients receiving the intensive treatment to be Population 2. The population proportions who develop retinopathy within six years are p_1 for the conventional treatment and p_2 for the intensive treatment. We want to test the hypotheses

$$H_0: p_1 = p_2 \text{ (the same as } H_0: p_1 - p_2 = 0)$$

$$H_a: p_1 \neq p_2 \text{ (the same as } H_a: p_1 - p_2 \neq 0)$$

Given the medical knowledge at the time, there was no reason to anticipate better results with one treatment or the other; therefore, the alternative hypothesis should be two-sided. We also want to give a confidence interval for the difference $p_1 - p_2$.

SOLVE: Inference about population proportions is based on the sample proportions

$$\hat{p}_1 = \frac{91}{378} = 0.2407 \quad \text{(conventional)}$$

$$\hat{p}_2 = \frac{23}{348} = 0.0661 \quad \text{(intensive)}$$

We see that about 24% of patients receiving the conventional treatment but only about 7% of patients receiving the intensive treatment developed retinopathy during the six years of the study. Because the samples are large and the sample proportions are quite different, we expect that a test will be highly significant. So we will concentrate on the confidence interval (which will also reveal whether or not the test is significant). To estimate $p_1 - p_2$, start from the difference between the sample proportions

$$\hat{p}_1 - \hat{p}_2 = 0.2407 - 0.0661 = 0.1746$$

To complete the Solve step, we must know how this difference behaves. ■

The sampling distribution of a difference between proportions

To use $\hat{p}_1 - \hat{p}_2$ for inference, we must know its sampling distribution. We discussed in Chapter 18 that when two random variables are Normally distributed, the new variable "difference" also follows a Normal distribution, centered on the difference between the two variables' means and with variance equal to the sum of the two variables' variances. In Chapter 19 we learned that the sampling distribution of $\hat{p}$ is approximately Normal when the sample size n is large, with mean and standard deviation p and $\sqrt{p(1-p)/n}$, respectively. So here are the facts on which we rely when conducting inference for the difference between two population proportions:

- When the samples are large, the distribution of $\hat{p}_1 - \hat{p}_2$ is **approximately Normal.**
- The **mean** of the sampling distribution is $p_1 - p_2$. That is, the difference between sample proportions is an unbiased estimator of the difference between population proportions.
- The **standard deviation** of the distribution is

$$\sqrt{\frac{p_1(1-p_1)}{n_1} + \frac{p_2(1-p_2)}{n_2}}$$

Figure 20.1 displays the distribution of $\hat{p}_1 - \hat{p}_2$. The standard deviation of $\hat{p}_1 - \hat{p}_2$ involves the unknown parameters p_1 and p_2. Just as in the previous chapter, we must replace these by estimates in order to do inference. And just as in the previous chapter, we do this a bit differently for confidence intervals and for tests.

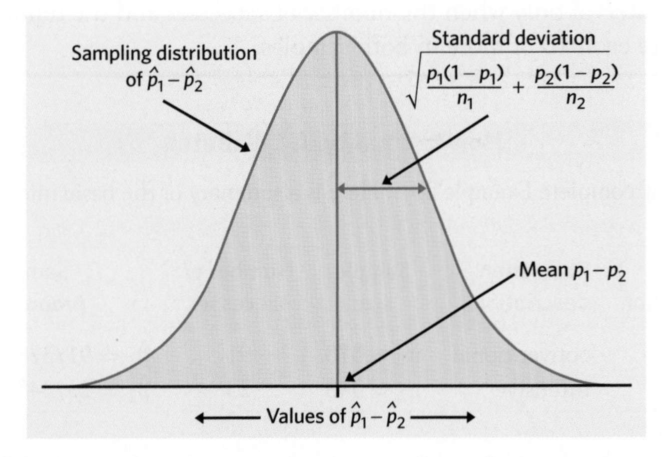

FIGURE 20.1 Select independent SRSs from two populations having proportions of successes p_1 and p_2. The proportions of successes in the two samples are $\hat{p}_1$ and $\hat{p}_2$. When the samples are large, the sampling distribution of the difference $\hat{p}_1 - \hat{p}_2$ is approximately Normal.

Large-sample confidence intervals for comparing proportions

To obtain a confidence interval, one option is to replace the population proportions p_1 and p_2 in the standard deviation with the sample proportions. The result is the **standard error** of the statistic $\hat{p}_1 - \hat{p}_2$:

standard error

$$SE = \sqrt{\frac{\hat{p}_1(1-\hat{p}_1)}{n_1} + \frac{\hat{p}_2(1-\hat{p}_2)}{n_2}}$$

The confidence interval has the same form we saw in the previous chapter,

$$\text{estimate} \pm z^* SE_{\text{estimate}}$$

LARGE-SAMPLE CONFIDENCE INTERVAL FOR COMPARING TWO PROPORTIONS

Draw an SRS of size n_1 from a large population having proportion p_1 of successes and draw an independent SRS of size n_2 from another large population having proportion p_2 of successes. When n_1 and n_2 are large, an approximate level C **confidence interval for $p_1 - p_2$** is

$$(\hat{p}_1 - \hat{p}_2) \pm z^* SE$$

In this formula the standard error SE of $\hat{p}_1 - \hat{p}_2$ is

$$SE = \sqrt{\frac{\hat{p}_1(1-\hat{p}_1)}{n_1} + \frac{\hat{p}_2(1-\hat{p}_2)}{n_2}}$$

and z^* is the critical value for the standard Normal density curve with area C between $-z^*$ and z^*.

Use this interval only when the number of successes and the number of failures are each 10 or more in both samples.

EXAMPLE 20.2 How to treat type 1 diabetes

We can now complete Example 20.1. Here is a summary of the basic information:

Population	Population description	Sample size	Number of successes	Sample proportion
1	conventional	$n_1 = 378$	91	$\hat{p}_1 = 91/378 = 0.2407$
2	intensive	$n_2 = 348$	23	$\hat{p}_2 = 23/348 = 0.0661$

SOLVE: We will give a 95% confidence interval for $p_1 - p_2$, the difference between the proportions developing retinopathy for the two treatment alternatives. To check that the large-sample confidence interval is safe, look at the counts of successes and failures in the two samples. All of these four counts are much larger than 10, so the large-sample method will be accurate. The standard error is

$$\begin{aligned} SE &= \sqrt{\frac{\hat{p}_1(1-\hat{p}_1)}{n_1} + \frac{\hat{p}_2(1-\hat{p}_2)}{n_2}} \\ &= \sqrt{\frac{(0.2407)(0.7593)}{378} + \frac{(0.0661)(0.9339)}{348}} \\ &= \sqrt{0.000661} = 0.02571 \end{aligned}$$

The 95% confidence interval is

$$\begin{aligned} (\hat{p}_1 - \hat{p}_2) \pm z^*SE &= (0.2407 - 0.0661) \pm (1.960)(0.02571) \\ &= 0.1746 \pm 0.0504 \\ &= 0.1242 \text{ to } 0.2250 \end{aligned}$$

CONCLUDE: We are 95% confident that the percent of diabetes patients who would develop retinopathy over a six-year period is between 12.5 and 22.5 percentage points higher when receiving a conventional treatment than when receiving an intensive treatment. ■

Figure 20.2 displays software output for Example 20.2 from a graphing calculator and two statistical software programs (not all software packages provide a simple way to obtain confidence intervals for proportions). As usual, you can understand the output even without knowledge of the program that produced it.

TI-83

```
2-PropZInt
 (.12426,.22504)
p̂1=.2407407407
p̂2=.066091954
n1=378
n2=348
```

CrunchIt!

Results-Proportion 2-Sample

Export ▼

	n	Successes	P-hat
Sample 1	378	91	0.2407
Sample 2	348	23	0.06609

Difference: 0.1746
StdErr: 0.02571
95% ConfInt: (0.1243, 0.2250)

Minitab

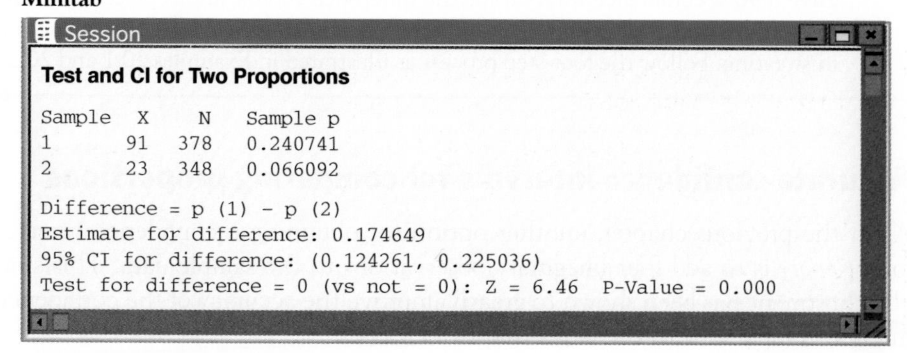
Session

Test and CI for Two Proportions

```
Sample  X    N    Sample p
1       91   378  0.240741
2       23   348  0.066092

Difference = p (1) - p (2)
Estimate for difference: 0.174649
95% CI for difference: (0.124261, 0.225036)
Test for difference = 0 (vs not = 0): Z = 6.46  P-Value = 0.000
```

FIGURE 20.2 Output from the TI-83 graphing calculator, CrunchIt!, and Minitab for the 95% confidence interval of Example 20.2.

Minitab gives the test as well as the confidence interval, confirming that the difference between the two treatment types is highly significant.

Like the large-sample confidence interval for a single proportion p, the large-sample interval for $p_1 - p_2$ generally has a true confidence level less than the level you asked for. The inaccuracy is not as serious as in the one-sample case, however, at least if our guidelines for use are followed.

APPLY YOUR KNOWLEDGE

20.1 More about the DCCT. The Diabetes Control and Complications Trial described in Example 20.1 also followed diabetes patients diagnosed with retinopathy *before* joining the study. They too were randomly assigned to one of the two treatments and monitored for six years. The study found that 143 of the 352 patients assigned to the conventional treatment showed a sustained progression of their original retinopathy. In contrast, only 77 of the 363 patients assigned to the intensive treatment had sustained retinopathy progression. Give a 95% confidence interval for the difference between the proportions of patients with retinopathy progression when diabetes patients with preexisting retinopathy receive either conventional or intensive treatment. Follow the four-step process as illustrated in Examples 20.1 and 20.2.

20.2 Botox for excessive sweating. Hyperhidrosis is a stressful medical condition characterized by chronic excessive sweating. Primary hyperhidrosis is inherited and typically starts during adolescence. Botox was approved for treatment of hyperhidrosis in adults in 2004. A clinical trial examined the effectiveness of botox for excessive armpit sweating in teenagers aged 12 to 17 years. Participants received either a Botox injection or a placebo injection (a simple saline solution) and were examined four weeks later to see if their sweating had been reduced by 50% or more. Here are the study's findings:[2]

	At Least 50% Reduction		
Treatment	Yes	No	Total
Botox	84	20	104
Placebo	44	64	108
Total	128	84	212

© PYMCA/SuperStock

Give a 90% confidence interval for the difference (botox minus placebo) in the proportions of teenagers with hyperhidrosis who experience at least 50% reduction in sweating. Follow the four-step process as illustrated in Examples 20.1 and 20.2.

Accurate confidence intervals for comparing proportions

As in the previous chapter, another option for obtaining a confidence interval for $p_1 - p_2$ is to add four imaginary observations to the sample data. This simple adjustment has been shown to greatly improve the accuracy of the confidence interval compared with the large-sample method.[3]

PLUS FOUR CONFIDENCE INTERVAL FOR COMPARING TWO PROPORTIONS

Draw independent SRSs from two populations with population proportions of successes p_1 and p_2. To get the **plus four confidence interval for the difference $p_1 - p_2$**, add four imaginary observations, one success and one failure in each of the two samples. Then use the large-sample confidence interval with the new sample sizes (actual sample sizes + 2) and counts of successes (actual counts + 1).

Use this interval when the sample size is at least 5 in each group, with any counts of successes and failures.

If your software does not offer the plus four method, just enter the new plus four sample sizes and success counts into the large-sample procedure.

EXAMPLE 20.3 Shrubs that withstand fire

STATE: Some shrubs can resprout from their roots after their tops are destroyed. Fire is a serious threat to shrubs in dry climates, as it can injure the roots as well as destroy the tops. One study of resprouting took place in a dry area of Mexico.[4] The investigators randomly assigned shrubs to treatment and control groups. They clipped the tops of all the shrubs. They then applied a propane torch to the stumps of the treatment group to simulate a fire. A shrub is a success if it resprouts. Here are the data for the shrub *Xerospirea hartwegiana:*

Population	Population description	Sample size	Number of successes	Sample proportion
1	control	$n_1 = 12$	12	$\hat{p}_1 = 12/12 = 1.000$
2	treatment	$n_2 = 12$	8	$\hat{p}_2 = 8/12 = 0.667$

How much does burning reduce the proportion of shrubs of this species that resprout?

PLAN: Give a 90% confidence interval for the difference in population proportions, $p_1 - p_2$.

SOLVE: The conditions for the large-sample interval are not met. In fact, there are *no* failures in the control group. We will use the plus four method. Add four imaginary observations. The new data summary is

Population	Population description	Sample size	Number of successes	Plus four sample proportion
1	control	$n_1 + 2 = 14$	$12 + 1 = 13$	$\tilde{p}_1 = 13/14 = 0.9286$
2	treatment	$n_2 + 2 = 14$	$8 + 1 = 9$	$\tilde{p}_2 = 9/14 = 0.6429$

The standard error based on the new facts is

$$\begin{aligned} SE &= \sqrt{\frac{\tilde{p}_1(1-\tilde{p}_1)}{n_1+2} + \frac{\tilde{p}_2(1-\tilde{p}_2)}{n_2+2}} \\ &= \sqrt{\frac{(0.9286)(0.0714)}{14} + \frac{(0.6429)(0.3571)}{14}} \\ &= \sqrt{0.02113} = 0.1454 \end{aligned}$$

The plus four 90% confidence interval, using $z^* = 1.645$ from Table C, is

$$\begin{aligned} (\tilde{p}_1 - \tilde{p}_2) \pm z^*SE &= (0.9286 - 0.6429) \pm (1.645)(0.1454) \\ &= 0.2857 \pm 0.2392 \\ &= 0.047 \text{ to } 0.525 \end{aligned}$$

CONCLUDE: We are 90% confident that burning reduces the percent of these shrubs that resprout by between 4.7% and 52.5%. ■

The plus four interval may be conservative (that is, the true confidence level may be *higher* than you asked for) for very small samples and population p's close to 0 or 1, as in this example. It is generally much more accurate than the large-sample interval when the samples are small. Nevertheless, the plus four interval in Example 20.3 cannot save us from the fact that small samples produce wide confidence intervals.

APPLY YOUR KNOWLEDGE

imagebroker/Alamy

20.3 Echinacea for the common cold? Echinacea is widely used as an herbal remedy for the common cold, but does it work? In a double-blind experiment, healthy volunteers agreed to be exposed to common-cold-causing rhinovirus type 39 and have their symptoms monitored. The volunteers were randomly assigned to take either a placebo or an echinacea supplement daily for 5 days following viral exposure. Among the 103 subjects taking a placebo, 88 developed a cold, whereas 44 of the 48 subjects taking echinacea developed a cold.[5]

(a) Explain why the large-sample confidence interval is not appropriate for these data.

(b) Give the 95% plus four confidence interval for the difference in proportions of individuals developing a cold after viral exposure between the echinacea treatment and the placebo.

20.4 Vaccine protection against cervical cancer. Most cases of cervical cancer are linked to a few strains of the human papillomavirus (HPV). The pharmaceutical company Merck developed a vaccine (Gardasil) against these HPV strains. Worldwide clinical trials followed young women 16 to 26 years of age after vaccination or administration of a placebo for two to four years for signs of HPV-caused cervical cancer. Of 8460 women given Gardasil, none developed HPV-caused cervical cancer. In contrast, 53 of the 8487 women given a placebo did.[6]

(a) Why should we not use the large-sample confidence interval for these data?

(b) Give the plus four 95% confidence interval for the difference between the two population proportions of women developing HPV-caused cervical cancer.

Significance tests for comparing proportions

An observed difference between two sample proportions can reflect an actual difference between the populations, or it may just be due to chance variation in random sampling. Significance tests help us decide if the effect we see in the samples is really there in the populations. The null hypothesis says that there is no difference between the two populations:

$$H_0: p_1 = p_2$$

The alternative hypothesis says what kind of difference we expect.

EXAMPLE 20.4 Toward an HIV vaccine

STATE: After decades of research and clinical trials, a recent study brings hope that we will one day have effective vaccination against HIV. The study, conducted in Thailand, enrolled HIV-negative men and women between the ages of 18 and 30 years old who were at an average risk of infection. The design was randomized and double-blind, assigning half of the subjects to a placebo and the other half to a combination of two vaccines, ALVAC and AIDSVAX, which on their own had previously not worked. All participants received counseling on how to avoid becoming infected with HIV and were regularly tested for HIV infection over a three-year period. Of 8198 subjects given a placebo, 74 became infected with HIV, compared with 51 of 8197 vaccinated subjects.[7] Do the data support the hypothesis that the vaccination combo is better than a placebo in preventing HIV infections?

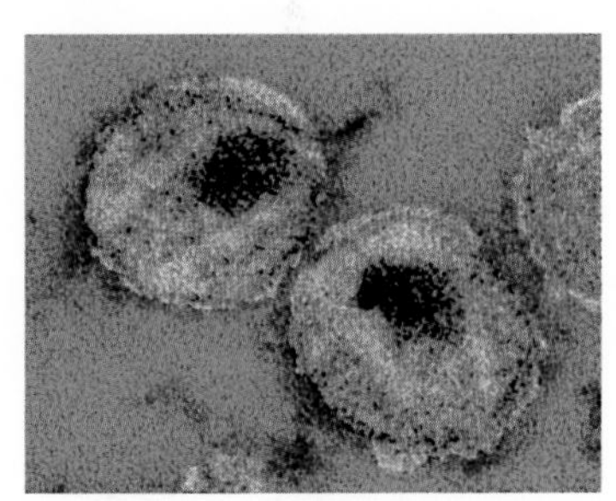
CDC/A. Harrison; Dr. P. Feorino

PLAN: Take individuals given a placebo to be Population 1 and vaccinated individuals to be Population 2. Our alternative is clearly one-sided, so we test

$$H_0: p_1 = p_2$$
$$H_a: p_1 > p_2$$

SOLVE: The study design is randomized so we can treat the two samples as if they were two independent SRSs of the two populations. The sample proportions who become infected with HIV are

$$\hat{p}_1 = \frac{74}{8198} = 0.0090 \quad \text{(placebo)}$$
$$\hat{p}_2 = \frac{51}{8197} = 0.0062 \quad \text{(vaccine)}$$

That is, about 0.9% of the subjects given a placebo but only about 0.6% of the vaccinated subjects became infected with HIV over a three-year period. Is this apparent difference statistically significant? To continue the solution, we must learn the proper test. ■

To do a test, standardize $\hat{p}_1 - \hat{p}_2$ to get a z statistic. If H_0 is true, all the observations in both samples come from a single population of Thai adults of whom a single unknown proportion p would become infected with HIV over a three-year period. So instead of estimating p_1 and p_2 separately, we pool the two samples and use the overall sample proportion to estimate the single population parameter p. Call this the **pooled sample proportion.** It is

pooled sample proportion

$$\hat{p} = \frac{\text{number of successes in both samples combined}}{\text{number of individuals in both samples combined}}$$

Use $\hat{p}$ in place of both $\hat{p}_1$ and $\hat{p}_2$ in the expression for the standard error SE of $\hat{p}_1 - \hat{p}_2$ to get a z statistic that has the standard Normal distribution when H_0 is true. Here is the test.

SIGNIFICANCE TEST FOR COMPARING TWO PROPORTIONS

Draw an SRS of size n_1 from a large population having proportion p_1 of successes, and draw an independent SRS of size n_2 from another large population having proportion p_2 of successes. To **test the hypothesis H_0: $p_1 = p_2$**, first find the pooled proportion $\hat{p}$ of successes in both samples combined. Then compute the z statistic

$$z = \frac{\hat{p}_1 - \hat{p}_2}{\sqrt{\hat{p}(1-\hat{p})\left(\frac{1}{n_1} + \frac{1}{n_2}\right)}}$$

In terms of a variable Z having the standard Normal distribution, the P-value for a test of H_0 against

H_a: $p_1 > p_2$ is $P(Z \geq z)$

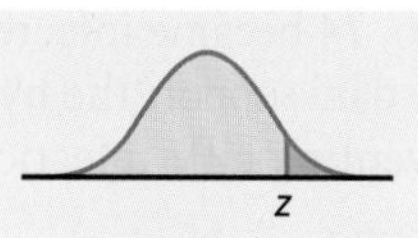

H_a: $p_1 < p_2$ is $P(Z \leq z)$

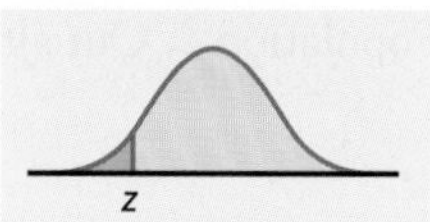

H_a: $p_1 \neq p_2$ is $2P(Z \geq |z|)$

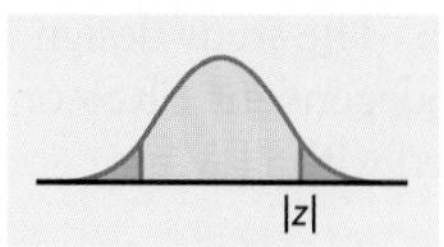

Use this test when the counts of successes and failures are each 5 or more in both samples.[8]

EXAMPLE 20.5 Toward an HIV vaccine, continued

SOLVE: The data come from a randomized experiment, and the counts of successes and failures are all larger than 5. The pooled proportion of individuals getting infected with HIV is

$$\hat{p} = \frac{\text{number of individuals infected with HIV in both samples}}{\text{number of individuals in both samples combined}}$$

$$= \frac{74 + 51}{8198 + 8197} = \frac{125}{16{,}395} = 0.0076$$

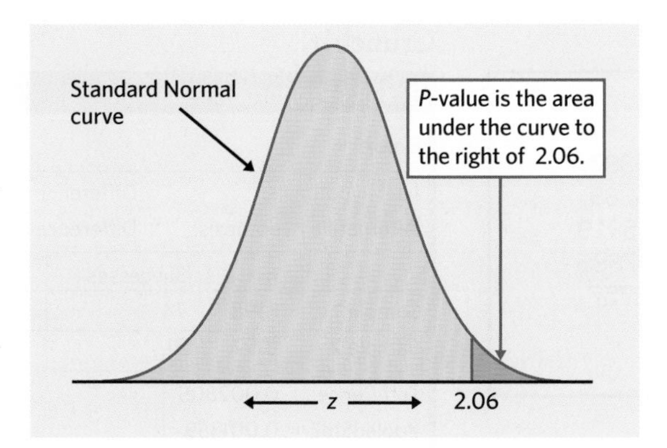

FIGURE 20.3 The P-value for the one-sided test of Example 20.5.

The z test statistic is

$$z = \frac{\hat{p}_1 - \hat{p}_2}{\sqrt{\hat{p}(1-\hat{p})\left(\frac{1}{n_1}+\frac{1}{n_2}\right)}}$$

$$= \frac{0.0090 - 0.0062}{\sqrt{(0.0076)(0.9924)\left(\frac{1}{8198}+\frac{1}{8197}\right)}}$$

$$= \frac{0.0028}{0.00136} = 2.06$$

The one-sided, upper-tail P-value is the area under the standard Normal curve more than 2.06 distant from 0. Figure 20.3 shows this area. Figure 20.4 shows the output from the TI-83 graphing calculator and from the statistical program CrunchIt!, indicating that $P = 0.0195$.

Without software, you can use the bottom of Table C (standard Normal critical values) to approximate P with no calculations: $z = 2.06$ is between the critical z-values 2.054 and 2.326. Therefore, the one-sided P-value is between 0.01 and 0.02.

CONCLUDE: There is strong evidence ($P < 0.02$) that the HIV infection rate is significantly lower among vaccinated individuals than among individuals given a placebo. The HIV vaccine combo offers better protection against HIV infection than a placebo. ■

APPLY YOUR KNOWLEDGE

20.5 Do free samples influence prescription decisions? To answer this question, a study randomly assigned resident physicians to either accept free drug samples or refuse them. The researchers then examined prescriptions for drugs that are available both by prescription and in an over-the-counter (OTC) version. Of the 202 relevant prescriptions written by physicians accepting free samples, 51 stipulated the OTC version. Of the 188 relevant prescriptions written by physicians refusing free samples, 73 stipulated the OTC version.[9] Consider these data as two random samples from the populations of prescriptions written by resident physicians with or without access to free samples. Is there good evidence that physicians are influenced by

TI-83

```
2-PropZTest
 p1>p2
 z=2.064385387
 P=.0194905289
 p̂1=.0090265919
 p̂2=.0062217885
 p̂=.0076242757
 n1=8198
 n2=8197
```

CrunchIt!

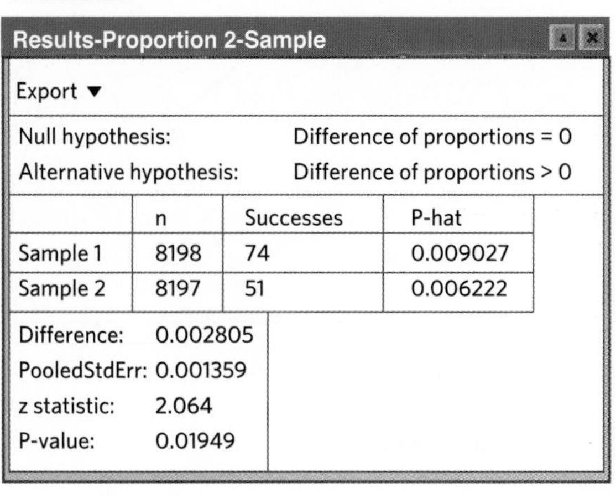

Results-Proportion 2-Sample

Export ▼

Null hypothesis: Difference of proportions = 0
Alternative hypothesis: Difference of proportions > 0

	n	Successes	P-hat
Sample 1	8198	74	0.009027
Sample 2	8197	51	0.006222

Difference: 0.002805
PooledStdErr: 0.001359
z statistic: 2.064
P-value: 0.01949

FIGURE 20.4 Output from the TI-83 graphing calculator and CrunchIt! for Example 20.5

free samples and prescribe proportionally fewer OTC versions when they receive free samples? Follow the four-step process as illustrated in Examples 20.4 and 20.5.

20.6 FDA warning on insomnia prescriptions. In a January 2013 safety announcement, the Food and Drug Administration warned drug manufacturers that the recommended dose of the widely prescribed insomnia drug zolpidem should be halved for women. The warning came after pharmacokinetic trials found that 15% of 250 women and 3% of 250 men who had taken the recommended dose of zolpidem still had enough drug in their blood 8 hours later to cause impairment. Do the data support the claim that women are more likely than men to have impairing zolpidem levels 8 hours after dosing? Follow the four-step process as illustrated in Examples 20.4 and 20.5.

Relative risk and odds ratio*

Categorical variables are particularly widespread in the health sciences. Many studies describe the presence or absence of a disease in a particular population or an improvement or a lack of improvement after treatment. In fact, the objective of most scientific reports in this field is to study either risk factors or treatment efficacy.

Studies evaluating the potential impact of a risk factor look for the association between exposure to the risk factor and a negative health outcome in sample individuals. As discussed in Chapter 7, these epidemiological studies can have different designs, but they are typically observational.

Studies evaluating the efficacy of a treatment compare the outcome of a sample given the treatment with the outcome of a sample given a placebo or some other control. These are typically experimental studies. We covered their design in detail in Chapter 8. The Diabetes Control and Complications Trial described in Example 20.1 is one such experiment.

*The remainder of this chapter presents more advanced material that is not needed to read the rest of the book. Relative risk and odds ratio are, however, important concepts in the health sciences.

Both types of study examine a categorical variable in two populations, but they rarely report the difference in sample proportions as we have studied in this chapter so far. Instead, you typically read about a "relative risk" or an "odds ratio" computed from the sample data. The following section gives a brief introduction to these topics.

Sounds good—but no comparison Most women have mammograms to check for breast cancer once they reach middle age. Could a fancier test do a better job of finding cancers early? PET scans are a fancier (and more expensive) test. Doctors used PET scans on 14 women with tumors and got the detailed diagnosis right in 12 cases. That's promising. But there were no controls, and 14 cases are not statistically significant. Medical standards require randomized comparative experiments and statistically significant results. Only then can we be confident that the fancy test really is better.

Relative risk and odds ratio Both observational and experimental studies in the health sciences typically compare a sample from a population of interest with a sample from a control population. The comparison is then represented by either a relative risk or an odds ratio.

Chapter 9 introduced the concepts of risk and odds in probability. When comparing two groups, the relative risk, RR, is simply the ratio of the two risks with the control group in the denominator. Likewise, the odds ratio, OR, is the ratio of both odds with the control group in the denominator. So, if a study finds that 50% of individuals in the treatment group had some symptoms compared with only 25% in the control group, the relative risk would be $RR = 50\%/25\% = 2$ and the odds ratio would be $OR = (1/1)/(1/3) = 3$. Relative risk is easy to interpret—individuals in the treatment group are twice as likely to have symptoms as similar control individuals. Interpreting the odds ratio is more challenging, because it compares unfamiliar odds rather than probabilities. Odds ratios can also give an exaggerated impression of the relative difference between the groups, particularly when the events studied are not rare. However, odds ratios offer mathematical advantages for advanced statistical analysis. Odds ratios are found most often in observational, case-control (or historical control) epidemiological studies, while relative risks are typically used in prospective cohort studies and randomized controlled experiments.

RELATIVE RISK AND ODDS RATIO IN THE HEALTH SCIENCES

Draw an SRS of size n_1 from a large population having proportion p_1 of a medical outcome, and draw an independent SRS of size n_2 from another large population having proportion p_2 of that outcome. Let the first sample represent the group of interest and the second sample represent the reference, or control, group.

The relative risk for the medical outcome in the group of interest compared with the reference group is

$$RR = \frac{\hat{p}_1}{\hat{p}_2}$$

The relative risk is sometimes also called the hazard ratio, or HR.

The odds ratio for the medical outcome in the group of interest compared with the reference group is

$$OR = \frac{\text{odds}_1}{\text{odds}_2} = \frac{\hat{p}_1(1 - \hat{p}_2)}{\hat{p}_2(1 - \hat{p}_1)}$$

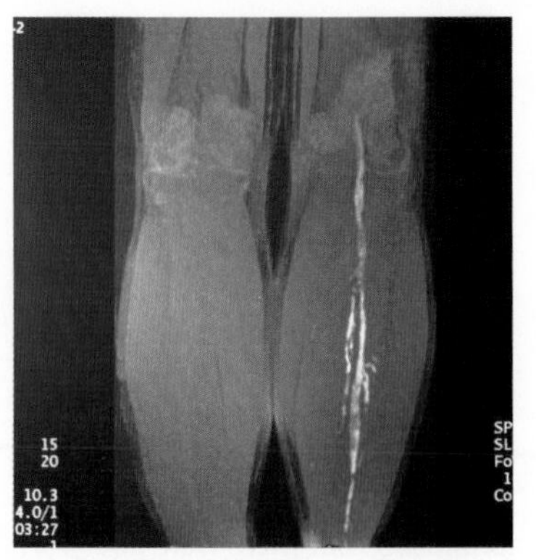

James King-Holmes/Photo Researchers, Inc.

EXAMPLE 20.6 Preventing blood clots in immobilized patients

Patients immobilized for a substantial amount of time can develop deep vein thrombosis (DVT), a blood clot in a leg or pelvis vein. DVT can have serious adverse health effects and can be difficult to diagnose. On its website, drug manufacturer Pfizer reports the outcome of a study looking at the effectiveness of the drug Fragmin (dalteparin) compared with that of a placebo in preventing DVT in immobilized patients.

In a double-blind, multinational study, severely immobilized patients were randomly assigned to receive daily subcutaneous injections of either Fragmin or a placebo for 12 to 14 days and were monitored for 90 days. The results, in number of patients experiencing a complication from DVT (including death), are summarized in the table below:

	Outcome		
Treatment	Complication	No complication	Sample size
Fragmin	42	1476	1518
Placebo	73	1400	1473

The proportion of subjects experiencing DVT complications in the two samples are

$$\hat{p}_{\text{Fragmin}} = 42/1518 = 0.0277$$
$$\hat{p}_{\text{placebo}} = 73/1473 = 0.0496$$

The relative risk of DVT complications is the ratio of the two sample proportions, with the placebo group in the denominator:

$$RR = \frac{\hat{p}_{\text{Fragmin}}}{\hat{p}_{\text{placebo}}} = \frac{0.0277}{0.0496} = 0.558$$

That is, patients receiving the Fragmin injections are less likely to develop complications from DVT than patients in the control group.

The odds of a subject experiencing DVT complications in the two samples are

$$\text{odds}_{\text{Fragmin}} = 42/1476, \text{ or about 1 to 35 against}$$
$$\text{odds}_{\text{placebo}} = 73/1400, \text{ or about 1 to 19 against}$$

The odds ratio of DVT complications is the ratio of these two odds, with the placebo group in the denominator, but it can also be computed from the two sample proportions:

$$OR = \frac{\text{odds}_{\text{Fragmin}}}{\text{odds}_{\text{placebo}}} = \frac{42/1476}{73/1400} = 0.546$$
$$= \frac{\hat{p}_{\text{Fragmin}}\,(1 - \hat{p}_{\text{placebo}})}{\hat{p}_{\text{placebo}}\,(1 - \hat{p}_{\text{Fragmin}})} = \frac{0.0277 \times (1 - 0.0496)}{0.0496 \times (1 - 0.0277)} = 0.546$$

This value is quite similar to the RR value. When samples are large and the outcome studied is rare (here 3% to 5%), RR and OR are very close. ■

What do the values for an RR or an OR mean? A relative risk of 1 means that there is no difference in risk between the two groups. However, relative risk is computed from sample data and is only an estimate of the true population risks. Therefore, an RR different from 1 does not necessarily imply a difference between the two populations. How large or how small a relative risk—or an odds ratio—needs to be in order to be statistically significant depends on the sampling distributions of these statistics.

Because relative risk and odds ratio are expressed as ratios of random variables, their sampling distributions are not symmetric. Therefore, confidence intervals for the population parameter are not symmetric and cannot be expressed as "estimate ± margin of error." For instance, in the Fragmin study of Example 20.6, special software intended for the analysis of medical data gives the following confidence intervals:

Statistic	Value	Low 95% CI	High 95% CI
Odds ratio	0.546	0.364	0.816
Relative risk	0.558	0.378	0.823

The calculations required to obtain such confidence intervals are beyond the scope of this introductory statistics textbook. The interpretation of a confidence interval or a P-value is the same, however, whether we study an average, a proportion, an odds ratio, or a relative risk. As always, *statistical significance does not imply relevance or importance. When sample sizes are very large, statistical significance can be reached for even very small effects. And observational studies always carry potential confounding factors that make a clear-cut interpretation of cause and effect between a studied risk factor and an observed outcome impossible.*

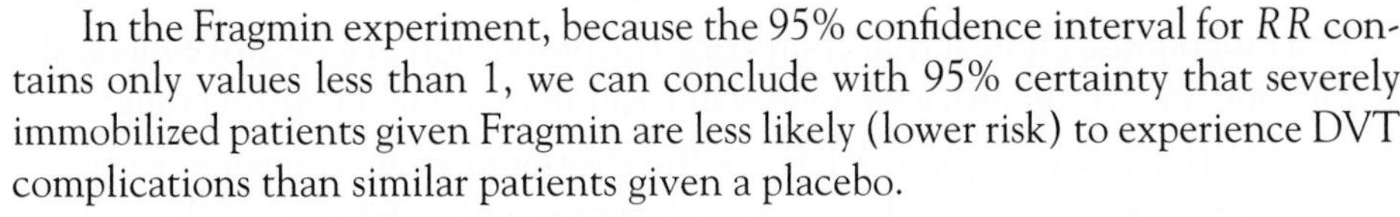

In the Fragmin experiment, because the 95% confidence interval for RR contains only values less than 1, we can conclude with 95% certainty that severely immobilized patients given Fragmin are less likely (lower risk) to experience DVT complications than similar patients given a placebo.

APPLY YOUR KNOWLEDGE

20.7 Aspirin and heart attacks. The landmark Physicians' Health Study randomly assigned 22,071 healthy male physicians at least 40 years old to take either an aspirin every other day or a placebo pill every other day. Of the 11,037 physicians who took aspirin regularly for 5 years, 10 had a fatal heart attack. Of the 11,034 who took the placebo, 26 suffered a fatal heart attack.[10]

(a) What are the proportions of physicians who suffered a fatal heart attack in the two groups?

(b) Calculate the relative risk of a fatal heart attack for physicians taking aspirin compared with those taking a placebo. Explain what this value means concretely.

(c) Calculate the odds ratio for a fatal heart attack comparing physicians taking aspirin to those taking a placebo. How does this value compare with the RR you calculated in (b)?

20.8 Antibiotics in the ICU. Patients admitted to the intensive care unit (ICU) requiring mechanical ventilation can develop fatal bacterial infections. A clinical trial randomly assigned 934 patients admitted to an ICU and requiring mechanical ventilation to two groups. The control group received a standard medical treatment, whereas the antibiotics group received antibiotics prophylactically in addition to standard treatment. The study found that 107 of the 468 patients in the control group died, compared with 69 of the 466 patients in the antibiotics group.[11]

(a) What are the proportions of ICU patients who died in the two groups?

(b) Calculate the relative risk of death for ICU patients given the antibiotics prophylactically compared with those given only a standard treatment. Explain what this value means concretely.

(c) Calculate the odds ratio for death, comparing ICU patients given the antibiotics prophylactically compared with those given only a standard treatment. How does this value compare with the RR you calculated in (b)?

Effect size: relative versus absolute measures The relative risk gives a measure of how effective a treatment is or how influential a risk factor is in a group compared with a control group without the treatment or the risk factor. One topic of debate in the biomedical community centers on how to interpret it. We will compare the relative risk to other measures of treatment effectiveness in comparative, randomized experiments.

RELATIVE RISK REDUCTION, ABSOLUTE RISK REDUCTION, AND NUMBER NEEDED TO TREAT

In a randomized clinical trial comparing the proportion of a negative medical outcome in a group provided a medical treatment and in a control group:

The relative risk reduction, RRR, for the treatment group compared with the control group is the reduction in risk expressed as a fraction of the control group's risk:

$$RRR = \frac{\hat{p}_{\text{control}} - \hat{p}_{\text{treatment}}}{\hat{p}_{\text{control}}}$$

The absolute risk reduction, ARR, for the treatment group compared with the control group is simply the difference in risks:

$$ARR = \hat{p}_{\text{control}} - \hat{p}_{\text{treatment}}$$

The number needed to treat, NNT, is the expected number of subjects who must be treated before 1 subject in the treatment group can be spared the negative medical outcome compared with what would have been expected in the control group. It is given by

$$NNT = \frac{1}{ARR} = \frac{1}{\hat{p}_{\text{control}} - \hat{p}_{\text{treatment}}}$$

when ARR is a strictly positive value. (When studying treatment side effects, ARR is typically negative and the number needed to harm, NNH, is reported instead.)

EXAMPLE 20.7 Preventing blood clots: relative risk reduction

Example 20.6 described the efficacy of Fragmin for preventing complications from DVT in immobilized patients compared with those receiving a placebo. Here are the proportions of adverse events in the two groups:

$$\hat{p}_{\text{Fragmin}} = 42/1518 = 0.0277$$
$$\hat{p}_{\text{placebo}} = 73/1473 = 0.0496$$

The Fragmin website reports a relative risk reduction of 44%. How was this number computed?

$$RRR = \frac{\hat{p}_{\text{placebo}} - \hat{p}_{\text{Fragmin}}}{\hat{p}_{\text{placebo}}} = \frac{0.0496 - 0.0277}{0.0496} = 0.442$$

That is, immobilized patients treated with Fragmin have a 44% lower probability of suffering complications from DVT than similar patients given a placebo instead.

How does this value compare with the absolute risk reduction for this study?

$$ARR = \hat{p}_{\text{placebo}} - \hat{p}_{\text{Fragmin}} = 0.0496 - 0.0277 = 0.0219$$

Compared with receiving a placebo, receiving Fragmin injections helps prevent DVT complications in about 2% of all immobilized patients.

The number needed to treat, NNT, is gradually becoming a standard in communicating clinical results. In the Fragmin study it is

$$NNT = \frac{1}{ARR} = \frac{1}{0.0219} = 45.7$$

That is, we expect that, on average, for every 46 immobilized patients treated with Fragmin instead of a placebo, DVT complications can be prevented in 1 patient. ■

The 2% ARR is radically different from the 44% RRR, and both give quite a different impression of treatment success. RRR tends to inflate the impression of success, but ARR can be difficult to comprehend. There is a growing trend toward always providing the NNT, as it gives a clear, practical medical interpretation of the relative benefit of treatment. Of course, the decision to provide a preventive or curative treatment depends not only on the relative benefit of one particular treatment but also on such things as the availability, costs, and benefits of alternative treatments, the number and seriousness of potential adverse events, and the seriousness and consequences of the condition (for example, death versus minor discomfort).

APPLY YOUR KNOWLEDGE

20.9 Aspirin and heart attacks, continued. Go back to Exercise 20.7 about the Physicians' Health Study.

(a) Calculate the absolute and relative risk reduction in fatal heart attacks for physicians taking aspirin compared with those taking a placebo. Explain briefly what these two values represent.

(b) Calculate the number needed to treat with aspirin compared with a placebo to prevent 1 fatal heart attack. What does this value tell you?

20.10 Antibiotics in the ICU, continued. Go back to Exercise 20.8 about the prophylactic use of antibiotics in the ICU.

(a) Calculate the absolute and relative risk reduction in death for ICU patients given the antibiotics prophylactically compared with those in the control group. Explain briefly what these two values represent.

(b) Calculate the number needed to treat with a prophylactic antibiotic regimen compared with doing nothing beyond a standard treatment to prevent 1 death. What does this value tell you?

DISCUSSION Assessing and understanding health risks

We have all heard news stories about factors that may increase or decrease the risk of getting a particular disease. At times, press releases even seem conflicting. For instance, we hear that coffee may reduce the risk of heart disease but also that it may increase the risk of a heart attack. So where does this information come from? And what does it really mean? Epidemiological studies and clinical trials are two major tools used to study health risk. Here we discuss how health risks are assessed in these two different contexts and what information they really convey.

Studying risk factors in epidemiological studies

Epidemiology is the science of understanding disease patterns through observations. Epidemiologists examine data to see what behaviors, locations, or exposures are associated with various diseases and causes of death. Because these studies are observational, confounding is a common problem and causality cannot be directly established. As discussed in Chapter 7, there are several types of epidemiological studies. Prospective cohort studies are the least prone to confounding errors, because subject selection does not depend on the factors studied. Despite their weaknesses, epidemiological studies can be very effective when many different studies converge to the same conclusion, as with tobacco smoking and the risk of developing lung and several other cancers.

Because confounding can never be completely ruled out in observational studies, finding a significant difference between some groups does not necessarily imply that a risk factor has been uncovered. This is especially true if the effect found is small. For example, a Harvard study found a two-and-a-half times higher risk of pancreatic cancer among coffee drinkers than among non–coffee drinkers. However, the association was not substantiated by later research from this group and countless others. By comparison, smoking multiplies the risk of cancer 20 to 30 times. On the other hand, even a small increase that shows up consistently in many studies with different designs can be credible and may have a large societal impact if the risk is widespread. Meta-analysis is an advanced statistical technique developed to find significant trends in the results of many different studies of the same variable. It has become a fairly common and important tool of epidemiology.

The information about new health risks that reaches the general public depends on scientific publications and how they are reported in the media. Unfortunately, studies that find no evidence of an association are often unreported or altogether abandoned. Significant results can be published more easily and in more prestigious papers and are much more likely to be funded and reported in the news. This is an important bias that cannot be compensated for by running a meta-analysis. A related and potentially devastating bias is due to financial conflicts of interest. Studies show that results may go unreported or conclusions may be unreasonably speculative as a consequence. For example, a 1998 observational study of 12 patients speculated about a possible link between the measles, mumps, rubella (MMR) vaccine and autism, creating quite a stir. As a result, many parents refused to have their children given the MMR vaccine. It was later revealed that the lead author had received funding from lawyers attempting class action suits against the vaccine manufacturer. Many well-designed studies since have found no evidence of a link between the MMR vaccine and autism, yet public skepticism persists. The public can be scared off easily, and it may take a whole generation to undo the damage.

Health risks and treatment options in clinical trials

Clinical trials are randomized controlled experiments involving human subjects. For obvious ethical reasons, clinical trials are limited to studying ways to reduce risk and not risk factors themselves. Because subjects are randomly assigned to various treatments and controls, differences in response can be directly attributed to differences in treatment. However, study participants are typically volunteers, and they may not always represent the target population adequately. The undercoverage of women and minority groups in clinical trials, for instance, is a serious issue, because trial results are often generalized to the whole target population despite their poor representation.

Patients are increasingly involved in treatment decisions, especially since the appearance of internet health sites and pharmaceutical advertising. So how do patients make sense of claims about risk reduction for a given treatment? Studies show that our understanding of risk is strongly influenced by the way it is presented and framed. Like physicians and funding institutions, patients are much more likely to favor treatment when the information about risk is presented in a relative way—for instance, by using relative risk reduction RRR. Patients also tend to choose treatment more often when risk is described in terms of negative outcomes (death or disease) rather than as a positive outcome (survival or remaining event free).

Let's examine a clinical trial for pravastatin, a cholesterol-lowering drug and one of the highest-selling prescription drugs in the United States (under its brand names). The study can be summarized with the statement that (a) pravastatin leads to a 22% reduction of the risk of coronary death (RRR), (b) the reduction is 0.9 percentage points (absolute risk reduction, ARR), or (c) 111 patients need to take pravastatin daily for five years to prevent

1 coronary death (number needed to treat, *NNT*). These three statements focus on the negative outcome, coronary death. The information can also be reframed to say that taking the drug daily for five years increases the probability of *avoiding* coronary death from 95.9% to 96.8%. All four summaries reflect the same data, but they give very different impressions. Historically, the relative risk reduction has been predominant, largely pushed by the pharmaceutical industry. However, there is a growing recognition of the need to provide a more comprehensive and balanced description of health risks.

One reason why the way risk is presented can be so influential is that patients have difficulty grasping the difference between average risk and personal outcomes. They may perceive a stated percent risk reduction as a guaranteed benefit. But there is no such guarantee. In fact, only a fraction of all patients treated will avoid developing the disease thanks to the treatment; all other patients would have the same outcome (disease or not) no matter whether they took the treatment or not. There is no way to tell which it will be for any given patient. At the same time, some proportion of all patients treated will experience negative side effects because of the treatment.

Patients should also understand that clinical trials are conducted over a defined time period and that this impacts risk calculations. For instance, a woman's risk of breast cancer over the next 10 years is substantially lower than her risk of breast cancer over a lifetime, yet this distinction is rarely emphasized. Long-term adverse events due to treatment are often monitored more extensively in phase IV clinical trials, after a treatment has been made publicly available. This is the dilemma between bringing the benefits of a new treatment to the target population as soon as possible and remaining cautious about side effects that may be too rare to see in limited trials or may take many years to develop. The withdrawal of the pain reliever VIOXX and warnings about suicide attempts among teenagers taking antidepressants are unfortunate examples of side effects discovered too late.

A blanket presumption for dietary supplements The Food and Drug Administration (FDA) regulates dietary supplements (vitamin, mineral, herb) as foods rather than drugs. Under the stated presumption that these are good for you, the Dietary Supplement Health and Education Act of 1994 exempted dietary supplements from existing premarket safety evaluations (which still apply to other foods) and shifted the burden of proof to the FDA for issues of safety and false or misleading claims. Manufacturers are only required to follow health claims with this warning: "This statement has not been evaluated by the Food and Drug Administration. This product is not intended to diagnose, treat, cure, or prevent any disease."

Health claims and you

The take-home message is that being properly informed is critical to understanding health risks. Clearing misconceptions about risk and risk reduction should become an important part of the patient-physician dialogue, but this will also require educating physicians about the meaning of risk and how to clearly communicate it. If the information matters to you, you should find the original publication and assess the study design and its possible flaws. No single study is entirely conclusive by itself, so you may also want to examine the findings from related studies.

Last, you should know that all over-the-counter herbal treatments are available for sale without ever having to undergo the rigors of a clinical trial to establish their effectiveness or safety. Their claims to reduce certain health risks do not have to be substantiated scientifically. And, being sold without prescription, they also come without medical or pharmaceutical counsel. You can find some objective information on the National Center for

Complementary and Alternative Medicine website at `nccam.nih.gov`. A few brands do carry a quality seal, but this typically represents only a voluntary evaluation of manufacturing standards (for example, to attest lack of dangerous contaminants). Buyers beware!

CHAPTER 20 SUMMARY

- The data in a **two-sample problem** are two independent SRSs, each drawn from a separate population.
- Tests and confidence intervals to compare the proportions p_1 and p_2 of successes in the two populations are based on the difference $\hat{p}_1 - \hat{p}_2$ between the sample proportions of successes in the two SRSs.
- When the sample sizes n_1 and n_2 are large, the sampling distribution of $\hat{p}_1 - \hat{p}_2$ is close to Normal with mean $p_1 - p_2$.
- The level C **large-sample confidence interval for $p_1 - p_2$** is

$$(\hat{p}_1 - \hat{p}_2) \pm z^*\text{SE}$$

where the standard error of $\hat{p}_1 - \hat{p}_2$ is

$$\text{SE} = \sqrt{\frac{\hat{p}_1(1-\hat{p}_1)}{n_1} + \frac{\hat{p}_2(1-\hat{p}_2)}{n_2}}$$

and z^* is a standard Normal critical value.
- The true confidence level of the large-sample interval can be substantially less than the planned level C. Use this interval only if the counts of successes and failures in both samples are each 10 or greater.
- To get a more accurate confidence interval, add four imaginary observations, one success and one failure in each sample. Then use the same formula for the confidence interval. This is the **plus four confidence interval.** You can use it whenever the samples have 5 or more observations each.
- **Significance tests for H_0: $p_1 = p_2$** use the **pooled sample proportion**

$$\hat{p} = \frac{\text{number of successes in both samples combined}}{\text{number of individuals in both samples combined}}$$

and the z statistic

$$z = \frac{\hat{p}_1 - \hat{p}_2}{\sqrt{\hat{p}(1-\hat{p})\left(\frac{1}{n_1} + \frac{1}{n_2}\right)}}$$

P-values come from the standard Normal distribution. Use this test when there are 5 or more successes and 5 or more failures in each sample.

- In the health sciences, the value $\hat{p}_2 - \hat{p}_1$ is called the **absolute risk reduction** (ARR) for a medical outcome in a target population (1) compared with a reference population (2). Other measures for comparing proportions are also used in this context:
  - The **relative risk** is

$$RR = \frac{\hat{p}_1}{\hat{p}_2}$$

  - The **odds ratio** is

$$OR = \frac{\hat{p}_1(1 - \hat{p}_2)}{\hat{p}_2(1 - \hat{p}_1)}$$

  - The **relative risk reduction** is

$$RRR = \frac{\hat{p}_2 - \hat{p}_1}{\hat{p}_2}$$

  - The **number needed to treat** is

$$NNT = \frac{1}{ARR} = \frac{1}{\hat{p}_2 - \hat{p}_1}$$

- Odds ratios are found most often in observational, retrospective studies, while RR, RRR, and NNT are typically used in prospective designs such as randomized controlled experiments. The number needed to treat is the number of subjects that must be treated before 1 subject in the treatment group can be spared a negative medical outcome compared with what would have been expected in the control group.

THIS CHAPTER IN CONTEXT

In Chapter 19 we computed confidence intervals and tested hypotheses about the proportion p of successes in one population. This chapter expands the concept and z procedures to situations where we want to compare the proportions of successes p_1 and p_2 in two distinct populations. The computations described in this chapter build on Normal calculations you learned in Chapter 11, provided that the counts of successes and failures in the samples are large enough.

When comparing two populations using two independent random samples, it is important to keep in mind that what you can conclude depends on how the data were collected. No matter how statistically significant a P-value is, you cannot directly conclude about causation if the data came from an observational study, because of confounding variables (as emphasized in Chapters 7 and 8). Statistical significance also doesn't imply practical significance, and it is important to note the effect size as well as its statistical significance. This chapter briefly discusses approaches that help evaluate the relative and absolute differences between two population proportions (RR, OR, RRR, NNT), an important first step in assessing practical relevance.

In Chapters 21 and 22 we will describe another method (chi-square) to test hypotheses for categorical variables beyond the simple case of binary success/failure. That is, we will look at categorical variables that can have more than two levels, such as a person's blood type (A, B, AB, O) or weight status (underweight, normal weight, overweight, obese). You will notice the parallels between the one-sample z test of Chapter 19 and the chi-square test for goodness of fit of Chapter 21, and between the two-sample z test described in this chapter and the chi-square test for two-way tables of Chapter 22.

CHECK YOUR SKILLS

The culinary herb cilantro, Coriandrum sativum, *is very polarizing: Some people love it and others hate it. A genetic component is suspected. A survey of American adults of European ancestry found that 1129 of 7295 women and 865 of 7309 men said that cilantro tastes soapy.*[12] *Exercises 20.11 to 20.15 are based on this survey.*

20.11 Take p_W and p_M to be, respectively, the proportions of all women and men among American adults of European ancestry who think that cilantro tastes soapy. We want to know if women and men in this population differ in how they taste cilantro. The hypotheses to be tested are

(a) H_0: $p_W = p_M$ versus H_a: $p_W \neq p_M$.
(b) H_0: $p_W = p_M$ versus H_a: $p_W > p_M$.
(c) H_0: $p_W = p_M$ versus H_a: $p_W < p_M$.

20.12 The pooled sample proportion of women and men who say that cilantro tastes soapy is about

(a) $\hat{p} = 0.50$. (b) $\hat{p} = 0.27$. (c) $\hat{p} = 0.14$.

20.13 The z statistic for a test comparing the proportions of women and men in this population who say that cilantro tastes soapy is about (in absolute value)

(a) $z = 6.41$. (b) $z = 3.20$. (c) $z = 1.48$.

20.14 The P-value for this test is

(a) > 0.05.
(b) between 0.005 and 0.05.
(c) < 0.005.

20.15 The margin of error of the 95% large-sample confidence interval for the difference $p_W - p_M$ in the proportions of women and men in this population who say that cilantro tastes soapy is about

(a) 0.048. (b) 0.022. (c) 0.011.

In an experiment to learn if Substance M can help restore memory, the brains of 20 rats were treated to damage their memories. The rats were then trained to run a maze. After a day, 10 rats were given Substance M and 7 of them succeeded in the maze; only 2 of the 10 control rats were successful. Exercises 20.16 to 20.18 are based on this study.

20.16 The z test for "no difference" in this case

(a) may be inaccurate because the populations are too small.
(b) may be inaccurate because some counts of successes and failures are too small.
(c) is reasonably accurate because the conditions for inference are met.

20.17 The plus four 90% confidence interval for the difference between the proportion of rats that succeed when given Substance M and the proportion that succeed without it has for its center the value

(a) 0.455. (b) 0.417. (c) 0.5.

20.18 The plus four 90% confidence interval described above has for its margin of error the value

(a) 0.312. (b) 0.304. (c) 0.185.

Glycoprotein IIb/IIIa inhibitors are platelet aggregation inhibitors used during angioplasty for the treatment of heart attacks. You read that eptifibatide is an effective and less expensive GPIIb/IIIa inhibitor. A randomized, double-blind clinical trial found that 5.4% of subjects in the angioplasty-with-eptifibatide group and 9.2% in the angioplasty-with-placebo group had either died or experienced another heart attack following treatment.[13] *Exercises 20.19 and 20.20 are based on this study.*

20.19 (Optional) The relative risk reduction RRR is

(a) 0.587. (b) 0.413. (c) 0.704.

20.20 (Optional) The number needed to treat NNT is approximately

(a) 4. (b) 9. (c) 26.

CHAPTER 20 EXERCISES

We recommend using the plus four method for all confidence intervals for proportions. However, the large-sample method is acceptable when the guidelines for its use are met.

20.21 Genetically altered mice. Genetic influences on cancer can be studied by manipulating the genetic makeup of mice. One of the processes that turn genes on or off (so to speak) in particular locations is called "DNA methylation." Do low levels of this process help cause tumors? Compare mice altered to have low levels with normal mice. Of 33 mice with lowered levels of DNA methylation, 23 developed tumors. None of the control group of 18 normal mice developed tumors in the same time period.[14]

(a) Explain why we cannot safely use either the large-sample confidence interval or the test for comparing the proportions of normal and altered mice that develop tumors.

(b) The plus four method adds two observations, a success and a failure, to each sample. What are the sample sizes and the numbers of mice with tumors after you do this? Give a plus four 99% confidence interval for the difference in the proportions of the two populations that develop tumors.

(c) Based on your confidence interval, is the difference between normal and altered mice significant at the 1% level?

20.22 Drug testing in schools. In 2002, the Supreme Court ruled that schools could require random drug tests of students participating in competitive after-school activities such as athletics. Does drug testing reduce use of illegal drugs? A study compared two similar high schools in Oregon. Wahtonka High School tested athletes at random, and Warrenton High School did not. In a confidential survey, 7 of 135 athletes at Wahtonka and 27 of 141 athletes at Warrenton said they were using drugs.[15] Regard these athletes as SRSs from the populations of athletes at similar schools with and without drug testing.

(a) You should not use the large-sample confidence interval. Why not?

(b) The plus four method adds two observations, a success and a failure, to each sample. What are the sample sizes and the numbers of drug users after you do this?

(c) Give the plus four 95% confidence interval for the difference between the proportion of athletes using drugs at schools with and without testing.

20.23 Treating AIDS. The drug AZT was the first drug that seemed effective in delaying the onset of AIDS in HIV-positive patients. Evidence for AZT's effectiveness came from a large randomized comparative experiment. The subjects were 1300 HIV-positive volunteers who had not yet developed AIDS. The study assigned 435 of the subjects at random to take 500 milligrams of AZT each day, and another 435 to take a placebo. (The others were assigned to a higher dose of AZT, but we will compare only the first two groups.) At the end of the study, 38 of the placebo subjects and 17 of the AZT subjects had developed AIDS. We want to test the claim that taking AZT lowers the proportion of infected people who will develop AIDS in a given period of time.

(a) State the hypotheses and check that you can safely use the z procedures.

(b) How significant is the evidence that AZT is effective?

(c) The experiment was double-blind. Explain what this means.

(*Comment:* Medical experiments on treatments for AIDS and other fatal diseases raise hard ethical questions. Some people argue that because AIDS is always fatal, infected people should get any drug that has any hope of helping them. The counter-argument is that then we will never find out which drugs really work. The placebo patients in this study were given AZT as soon as the results indicated that AZT was clearly more effective.)

20.24 Drug testing in schools, continued. Exercise 20.22 describes a study that compared the proportions of athletes who use illegal drugs in two similar high schools, one that tests for drugs and one that does not. Drug testing is intended to reduce the use of drugs. Do the data give good reason to think that drug use among athletes is lower in schools that test for drugs? State hypotheses, find the test statistic, and use either software or Table C for the P-value. Be sure to state your conclusion. (Because the study is not an experiment, the conclusion depends on the condition that athletes in these two schools can be considered SRSs from all similar schools.)

Does involving a statistician to help with statistical methods improve the chance that a medical research paper will be published? A study of a random sample of papers submitted to two medical journals found that 135 of the 190 papers that lacked statistical assistance were rejected without even being reviewed in detail. In contrast, 293 of the 514 papers with statistical help were sent back without review.[16] Exercises 20.25 to 20.27 are based on this study.

20.25 Does statistical help make a difference? Is there a significant difference in the proportions of papers with and without statistical help that are rejected without review? Use software or the bottom row of Table C to get a P-value. (This observational study does not establish causation: Studies that include statistical help may also be better in other ways than those that do not.)

20.26 How often are statisticians involved? Give a 95% confidence interval for the proportion of papers submitted to these journals that include help from a statistician.

20.27 How big a difference? Give a 95% confidence interval for the difference between the proportions of papers rejected without review when a statistician is and is not involved in the research.

20.28 Staph infections. Exercise 19.41 (page 482) described a study of staph infections in surgery patients. The researchers recruited 917 patients who had tested positive for *Staphylococcus aureus* and randomly assigned them to receive a nasal ointment that contained either a staph-killing solution or a placebo. Postsurgery infections with *S. aureus* were then recorded. In the treatment group, 17 of 504 patients developed a staph infection, compared with 32 of 413 patients in the placebo group.[17]

(a) Is there good evidence that treatment of staph-positive patients before surgery reduces the rate of postsurgical staph infection? State the null and alternative hypotheses, find the test statistic, and use either software or Table C for the P-value.

(b) Calculate a 95% confidence interval for the difference between the proportions of postsurgical staph infections among patients receiving a staph-killing ointment or a placebo.

20.29 Children's food choices. Exercise 19.39 (page 481) describes a study in which 40 young children were asked to taste graham crackers presented in a package either with or without a popular cartoon character. When asked to indicate which of the two options they would prefer to eat for a snack, 35 chose the version with a cartoon on the package. When asked to choose between packages of baby carrots with or without a popular cartoon character, 29 said they would prefer to eat the version with a cartoon on the package. Explain why we *cannot* use the methods of this chapter to compare the proportions of young children selecting the packaging with a popular cartoon when the food item is either graham crackers or baby carrots.

In responding to Exercises 20.30 to 20.40, follow the ***Plan, Solve,*** *and* ***Conclude*** *steps of the four-step process. It may be helpful to restate in your own words the* ***State*** *information given in the exercise.*

20.30 Knuckle cracking. 4 STEP Urban legend says that knuckle cracking is bad for you and will eventually lead to osteoarthritis of the hand joints. A retrospective case-control study of individuals who had received an X-ray of the right hand in the past five years enrolled 135 subjects with a confirmed diagnosis of hand osteoarthritis and 80 control subjects in the same age group. The study found that 24 subjects with hand osteoarthritis and 19 control subjects were habitual knuckle crackers.[18] Is there good evidence that the proportion of knuckle crackers is greater among individuals with hand osteoarthritis? Does this support the urban legend about knuckle cracking?

20.31 Altruism in prairie dogs. 4 STEP Is altruistic behavior influenced by its potential cost? Prairie dogs are social rodents who warn each other of a predator's presence with barking calls. A researcher examined whether close proximity to a predator results in less frequent alarm calls because the callers bear more personal risk. The "predator," a stuffed badger controlled remotely, was placed multiple times randomly either near or far from a prairie dog colony. The number of warning calls made by isolated foraging colony members is displayed in the following table:[19]

	Distance to Predator	
	Near	**Far**
Alert call	29	55
No call	80	78

Do the data support the researcher's hypothesis?

20.32 More on altruism in prairie dogs. 4 STEP The researcher in the previous exercise also hypothesized that if prairie dogs make fewer calls when they feel individually more at risk, it would be reflected in their behavior. That is, individuals remaining aboveground would make a higher proportion of alert calls than individuals seeking cover upon appearance of a predator. Here are the study's findings:

	Behavioral Response	
	Seek cover	**Stay above ground**
Alert call	11	95
No call	63	121

Do the data support the researcher's second hypothesis?

20.33 Did the random assignment work? A large clinical trial of the effect of diet on breast cancer assigned women at random to either a normal diet or a low-fat diet. To check that the random assignment did produce comparable groups, we can compare the two groups at the start of the study. Ask if there is a family history of breast cancer: 3396 of the 19,541 women in the low-fat group and 4929 of the 29,294 women in the control group said "yes."[20] If the random assignment worked well, there should *not* be a significant difference in the proportions with a family history of breast cancer. How significant is the observed difference?

20.34 Aflatoxicosis in Kenya. Aflatoxins are toxic compounds secreted by fungus found most often in damaged crops. Kenya experienced an outbreak of aflatoxicosis in 2004 that resulted in several hundred cases of liver failure, including 125 deaths. The Kenyan Ministry of Health suspected that improper maize (corn) storage conditions were at least in part responsible for the outbreak. A random sample of 27 case-patients with aflatoxicosis and 43 healthy controls were asked whether they stored their maize in the house or in a dedicated granary. Prolonged storage in the house exposes crops to levels of humidity that foster fungal growth. The study found that 22 of the case-patients and 23 of the controls had stored maize in the house.[21] Choose carefully a valid statistical method to compare the effectiveness of both storage methods against aflatoxicosis. Do your results support the ministry's hypothesis?

20.35 Surgical treatment of migraines. Migraine headaches are a neurological condition with a number of symptoms, including severe pain on one half of the head that can last from 4 to 72 hours. A double-blind, sham-surgery-controlled study examined the effectiveness of facial surgery that removed small portions of migraine-triggering muscle or nerve tissue in the hope of providing permanent relief from migraines. Seventy-five patients with moderate to severe migraines were randomly assigned to receive either actual or sham surgery in their predominant trigger site.[22]

(a) A year later, 15 of the 26 subjects in the sham surgery group and 41 of the 49 subjects in the actual surgery group experienced at least a 50% reduction in migraine headaches. How significant is the observed difference between the proportions of subjects with an improved condition?

(b) The study also reported that 28 of 49 patients in the actual surgery group reported complete elimination of migraines a year after treatment, compared with only 1 of 26 patients in the sham surgery group. How much better is the surgical treatment than the sham procedure in completely eliminating migraines? Give a 95% confidence interval using the plus four method.

20.36 Smoking cessation. Chantix is a medication prescribed for smoking cessation that works by targeting nicotine receptors in the brain. Chantix's website describes a study of the drug efficacy.[23]

(a) In a first stage, the study enrolled 1927 smokers who wanted to quit smoking and provided them with Chantix for twelve weeks. At the end of this first period, 1236 subjects had successfully quit smoking. Give a 95% confidence interval for the proportion of smokers using Chantix for twelve weeks who successfully quit smoking by the end of the twelve-week period.

(b) In a second stage, the 1236 subjects who had successfully quit smoking after twelve weeks of Chantix were randomly assigned to remain on Chantix or to receive a placebo for another twelve weeks. By the end, 425 of the 602 subjects given Chantix and 301 of the 604 subjects given a placebo remained tobacco free. Is there a significant difference in the proportions of ex-smokers who remain tobacco free in the two groups?

20.37 Duct tape for wart removal. A study compared the effectiveness of duct tape with cryotherapy with liquid nitrogen in the treatment of common warts in children and young adults. A total of 61 patients aged 3 to 22 years were randomly assigned to either treatment (duct tape applied directly on warts continuously for up to two months or up to six applications of liquid nitrogen every two or three weeks). Of the 26 patients treated with duct tape, 22 showed complete wart remission. In comparison, 15 of the 25 patients treated with cryotherapy reached complete wart remission.[24]

(a) Explain why the study cannot be double-blind.

(b) Choose carefully a valid statistical method to compare the effectiveness of both methods. What do you conclude?

20.38 Head injuries. Most alpine skiers and snowboarders do not use helmets. Do helmets reduce the risk of head injuries? A study in Norway compared skiers and snowboarders who suffered head injuries with a control group who were not injured. Of 578 injured subjects, 96 had worn a helmet. Of the 2992 in the control group, 656 had worn a helmet.[25] Is helmet use less common among skiers and snowboarders who have head injuries? (Because this is an observational study, the conclusion depends on how comparable the injured and uninjured groups are.)

20.39 Carcinogenicity of electromagnetic fields. The U.S. National Toxicology Program studies the toxicity and carcinogenicity of agents potentially causing a risk to human health. Electromagnetic fields around electrical installations are harmless in theory, but they might have a long-term effect on health—for instance, in triggering tumors. Observational studies in human populations have suggested a potential association. One study looked at the occurrence of cancers in otherwise healthy rats after two years of daily exposure to 60-hertz electromagnetic fields. Several groups were studied, each exposed to a different electromagnetic intensity, as well as a control group kept in the same conditions but not exposed to any electromagnetic field. Here are the number of rats with a tumor after the two-year study period for the control group and for the group exposed to 2 gauss, which is approximately 1000-fold the intensity level considered high exposure for humans:[26]

Treatment	Male rats	Female rats
Control: no exposure	16 (n = 99)	19 (n = 100)
2-gauss exposure	30 (n = 100)	22 (n = 100)

In rats not exposed to an electromagnetic field, is there significant evidence of a difference in the tumor rates between males and females?

20.40 Carcinogenicity study, continued. Use the data from the previous exercise to answer the following questions.

(a) Is there significant evidence of a difference in the tumor rates in male rats either exposed to 2 gauss or not exposed?
(b) Is there significant evidence of a difference in the tumor rates in female rats either exposed to 2 gauss or not exposed?
(c) Write a brief summary of your findings from this and the previous exercise.

20.41 Melatonin and menopause. Melatonin is a naturally occurring hormone involved in the regulation of the body's internal clock and is available as a dietary supplement in the United States. An experiment was designed to assess the effect of nighttime melatonin intake on the levels of luteinizing hormone (LH) in perimenopausal and menopausal women. Women were randomly assigned to take either melatonin or a placebo daily for six months. Hormone levels were determined from blood samples taken at the beginning and at the end of the study. The response variable was whether or not a woman had an LH increase of at least 10%. Here are the results, by age group:[27]

Treatment	Women in their 40s	Women in their 50s and 60s
Melatonin	10 (n = 38)	22 (n = 30)
Placebo	20 (n = 34)	18 (n = 38)

(a) For women in their 40s, is there significant evidence of a difference in the proportions of women with increased LH levels between women taking melatonin and women taking a placebo?
(b) For women in their 50s and 60s, is there significant evidence of a difference in the proportions of women with increased LH levels between women taking melatonin and women taking a placebo?
(c) Write a short description of your findings, contrasting your results for parts (a) and (b).
(d) If we had ignored age, the results would be 32 out of 68 women with increased LH in the melatonin condition compared with 38 out of 72 in the placebo condition. Show that these results are not statistically significant. This is an example of Simpson's paradox. The paradox was first described in Chapter 5.

The following exercises concern the optional material on relative risk and odds ratio in the health sciences.

20.42 Cardiovascular disease prevention. The Heart Outcomes Prevention Evaluation trial studied the effectiveness of angiotensin-converting enzyme (ACE) inhibitors in cardiovascular disease prevention. Part of the study randomly assigned diabetic patients over age 55 to receive either the ACE inhibitor ramipril daily for five years or a placebo. Of the 1808 subjects taking ramipril daily, 112 died of a cardiovascular accident during the five-year study period. In contrast, 172 of the 1769 subjects receiving a placebo died of a cardiovascular accident.[28]

(a) Calculate the relative risk reduction and write a short statement using this measure to summarize the study findings.
(b) Calculate the number needed to treat to save 1 life from cardiovascular death, and write a short statement using this measure to summarize the study findings.
(c) Some studies also report the number of pills that must be taken to save 1 life. Calculate this value considering, for simplicity, that each subject in the treatment group took 1 pill every day for five years.

20.43 The DCCT and diabetes treatment. Go back to Example 20.1 comparing the effectiveness of a conventional treatment with that of an intensive treatment in preventing complications from type 1 diabetes.

(a) Give the absolute and relative risk reduction for the development of retinopathy in this study. Also compute the number needed to treat.

(b) Choose one of the values you calculated in (a) and write a brief statement explaining the study findings to a patient.

20.44 The DCCT and diabetes treatment, continued. Exercise 20.1 describes the results of the DCCT study for individuals who already suffered from retinopathy at the beginning of the study, comparing the effectiveness of the conventional treatment with that of the intensive treatment in preventing retinopathy progression.

(a) Give the absolute and relative risk reduction for the progression of retinopathy in this study. Also compute the number needed to treat.

(b) Choose one of the values you calculated in (a) and write a brief statement explaining the study findings to a patient.

20.45 Treating AIDS. Go back to Exercise 20.23, which describes the first study of AZT effectiveness in delaying the onset of AIDS.

(a) Calculate and interpret the odds ratio and the relative risk for the AZT group compared with the placebo group. How do they compare?

(b) Calculate and interpret the absolute and the relative risk reduction for AIDS onset.

(c) Calculate the number needed to treat with AZT compared with a placebo to prevent 1 onset of AIDS.

CHAPTER 21 The Chi-Square Test for Goodness of Fit

In Chapter 19 we used the one-sample z procedures to obtain confidence intervals and test hypotheses about the proportion of successes in a population. The definition of "success" was arbitrary and simply referred to the outcome of interest, with any other outcome constituting a "failure." Such dichotomous labeling is sometimes an obvious choice, as in Example 19.6, in which we examined whether live aphids land on their legs after a drop. In other situations, a more detailed description of possible outcomes would be helpful. For instance, we might want to study the response to a new treatment by counting the number of subjects whose health status improved, remained stable, or deteriorated. In this chapter we introduce a new statistical method, the chi-square test, that allows us to test hypotheses about a categorical variable with two *or more* levels. We will see in the next chapter that this versatile method can also be adapted to study the relationship between two independent categorical variables.

IN THIS CHAPTER WE COVER...

- Hypotheses for goodness of fit
- The chi-square test for goodness of fit
- Using technology
- Interpreting chi-square results
- Conditions for the chi-square test
- The chi-square distributions
- The chi-square test and the one-sample z test*

Hypotheses for goodness of fit

We often have a choice in defining a response variable. For instance, if we suspect that births are less common on the weekend, we could take a random sample of births and record whether they occurred on a weekday or on the weekend. However, this distinction might not be fine enough. Saturday and Sunday could

Weekend outing? You work all week. Then it rains on the weekend. Can there really be a statistical truth behind our perception that the weather is against us? At least on the East Coast of the United States, the answer is "Yes." Going back to 1946, it seems that 22% more precipitation falls on Sundays than on Mondays. The likely explanation is that the pollution from all those workday cars and trucks forms the seeds for raindrops—with just enough delay to cause rain on the weekend.

have very different birthrates, but this would be masked by pooling them into a single outcome. Likewise, Monday and Friday might differ substantially from midweek days. Therefore, a better way to look for any nonrandom pattern in the distribution of births would be to consider all seven days of the week.

EXAMPLE 21.1 Never on Sunday?

A random sample of 700 births from local records shows this distribution across the days of the week:

Day	Sun.	Mon.	Tue.	Wed.	Thu.	Fri.	Sat.
Births	84	110	124	104	94	112	72

As expected, the two smallest counts of births are on Saturday and Sunday. But do these data give significant evidence that local births are not equally likely on all days of the week?

The null hypothesis says that births *are* evenly distributed. To state the hypotheses carefully, write the discrete probability distribution for days of birth:

Day	Sun.	Mon.	Tue.	Wed.	Thu.	Fri.	Sat.
Probability	p_1	p_2	p_3	p_4	p_5	p_6	p_7

The null hypothesis says that the probabilities are the same on all days. In that case, each day would get one-seventh of all births. That is, all 7 probabilities must be 1/7. So the null hypothesis is

$$H_0\text{: } p_1 = p_2 = p_3 = p_4 = p_5 = p_6 = p_7 = \frac{1}{7} \; \blacksquare$$

Beware of wanting to state the null hypothesis in terms of the sample proportions. This is a mistake often made by students. The null hypothesis must reflect your assumptions, not the data used to test it. The alternative hypothesis says that days are *not* all equally probable:

$$H_a\text{: not all } p_i = \frac{1}{7}$$

The alternative hypothesis is nonspecific and simply says that H_0 is not true.

In this example, we hypothesized under H_0 that births would be equally distributed across all 7 days of the week. However, we could have specified any distribution for our choice for the null hypothesis, as long as there was a sound biological argument for it. Here is an example in which it would not make sense to choose a null hypothesis of equal proportions.

EXAMPLE 21.2 Genetics of seed color

Epistasis is the control of a phenotype by two or more interacting genes. In a dominant epistatic model, one gene can mask the effect of the second gene, leading to the expression of one main phenotype and two rarer phenotype variants.

Geneticists examined the distribution of seed coat color in cultivated amaranth grains, *Amaranthus caudatus*. Crossing black-seeded and pale-seeded *A. caudatus* populations gave the following counts of black, brown, and pale seeds in a second generation (F2):[1]

Seed coat color	black	brown	pale
Seed count	321	77	31

iStockphoto/Thinkstock

According to genetics laws, dominant epistasis should lead to a 12:3:1 distribution in F2. That is, out of 16 individuals, 12 would be expected to express the dominant phenotype (black), 3 the intermediate phenotype (brown), and only 1 the recessive phenotype (pale). We want to know if seed color could follow a dominant epistatic model. The null hypothesis is therefore

$$H_0\colon p_{\text{black}} = \frac{12}{16} = \frac{3}{4} \text{ and } p_{\text{brown}} = \frac{3}{16} \text{ and } p_{\text{pale}} = \frac{1}{16}$$

Again, the alternative hypothesis is simply that H_0 is not true. ■

In Example 21.1, H_0 postulates equal proportions of births for the seven days (a uniform distribution), whereas in Example 21.2, the three seed colors are assumed to have different proportions under H_0. In both cases we want to test whether a categorical variable (day of birth or seed color) has a particular distribution outlined by H_0. The statistical test we will use for that is the chi-square test for *goodness of fit*. The idea is that the test assesses whether the observed counts "fit" the distribution outlined by H_0.

APPLY YOUR KNOWLEDGE

21.1 **Saving birds from windows.** Many birds are injured or killed by flying into windows. It appears that birds don't see windows. Can tilting windows down so that they reflect earth rather than sky reduce bird strikes? Researchers placed six windows at the edge of a woods: two vertical, two tilted 20 degrees, and two tilted 40 degrees. During the next four months, there were 53 bird strikes, 31 on the vertical windows, 14 on the 20-degree windows, and 8 on the 40-degree windows.[2] Does the tilt have an effect? State the null and alternative hypotheses.

21.2 **Heat resistance in rice.** Rice is one of the most widely consumed grain worldwide. As increasing global temperatures hamper rice production, scientists are searching for varieties exhibiting some degree of heat resistance. The HT54 variety of the indica rice, *Oryza sativa*, was found to tolerate several days of high temperatures up to 48 degrees Celsius (118 degrees Fahrenheit) during its growth. Researchers performed a genetic crossing experiment between the heat-resistant HT54 variety and a control variety, HT13, without heat resistance. All seedlings in the first generation survived a heat challenge of 48 degrees Celsius. In the second generation (F2), only 548 of the 744 plants survived the heat challenge.[3] This would suggest that heat resistance is carried by a single gene with a dominant allele and a recessive allele. Are the findings consistent with a simple dominant-recessive Mendelian genetic model in which F2 is made up of 75% dominant-trait (heat resistance) and 25% recessive-trait rice plants? State the null and alternative hypotheses.

The chi-square test for goodness of fit

We have stated the null and alternative hypotheses, and now we need to test whether the observed results differ significantly from expectations under H_0. To test H_0, we compare the observed counts with the *expected counts*, the counts we would expect—except for random variation—if H_0 was true. When the observed counts are far from the expected counts, that is evidence against H_0.

The expected count in n independent trials for a particular outcome with probability p is simply np. (We covered this in great detail in Chapters 9, 10, and 12.) This doesn't mean that we expect to see exactly np counts of that outcome in n trials. Rather, we expect np counts on average over the long run if we repeated the n trials many, many times.

EXPECTED COUNTS

A categorical variable has k possible outcomes, with probabilities p_1, p_2, $p_3, \ldots, p_k$. That is, p_i is the probability of the ith outcome. We have n independent observations from this categorical variable.

To test the null hypothesis that the probabilities have specified values

$$H_0: p_1 = p_{10},\ p_2 = p_{20},\ \ldots,\ p_k = p_{k0}$$

first compute an **expected count** for each of the k outcomes as follows:

$$\text{expected count of outcome } i = np_{i0}$$

EXAMPLE 21.3 Never on Sunday? Expected counts

In Example 21.1 the observations are counts of births for each of the 7 days of the week. That is, the outcomes are days of the week, with $k = 7$.

The null hypothesis says that the probability of a birth on the ith day is $p_{i0} = 1/7$ for all days. There were 700 births in our sample. If H_0 was true, we would expect to see an even distribution of these 700 births across all 7 days of the week, that is,

$$\text{expected count}_i = np_{i0} = 700 \times \frac{1}{7} = 100$$

Each of the seven expected counts is equal to 100. ■

Expected counts do not have to be round numbers. In fact, if there had been only 699 observations instead of 700, the expected counts would each have been equal to $699/7 = 99.86$.

The statistical test that tells us whether the observed differences between the 7 days of the week are statistically significant compares the observed and expected counts. The test statistic that is used for this comparison is the *chi-square statistic*.

CHI-SQUARE STATISTIC

The **chi-square statistic** is a measure of how far observed counts are from expected counts under the null hypothesis. The formula for the statistic is

$$X^2 = \sum \frac{(\text{observed count} - \text{expected count})^2}{\text{expected count}}$$

$$= \frac{(\text{observed}_1 - \text{expected}_1)^2}{\text{expected}_1} + \cdots + \frac{(\text{observed}_k - \text{expected}_k)^2}{\text{expected}_k}$$

where k is the number of different outcomes the categorical variable can take. Each of the k terms in the sum is called a **chi-square component.**

The chi-square statistic is a sum of terms, one for each possible outcome. In the birth example, 84 babies were born on Sunday. The expected count for each of the seven days is 100. So the term of the chi-square statistic for the Sunday outcome is

$$\frac{(\text{observed count} - \text{expected count})^2}{\text{expected count}} = \frac{(84 - 100)^2}{100} = \frac{256}{100} = 2.56$$

To compute the chi-square statistic, we need to calculate each of the seven chi-square components and then sum them. That is,

$$X^2 = \sum \frac{(\text{observed count} - \text{expected count})^2}{\text{expected count}}$$

$$= \frac{(84 - 100)^2}{100} + \frac{(110 - 100)^2}{100} + \frac{(124 - 100)^2}{100} + \frac{(104 - 100)^2}{100} + \frac{(94 - 100)^2}{100} + \frac{(112 - 100)^2}{100} + \frac{(72 - 100)^2}{100}$$

$$= 19.12$$

In this particular example, each outcome has the same expected count because the null hypothesis assumed equal probability for all seven days. When H_0 does not assume a uniform distribution, the expected counts vary across outcomes and this is reflected in the chi-square calculations. Here is an example.

EXAMPLE 21.4 Genetics of seed color: calculating X^2

Example 21.2 described the color of 429 second-generation *A. caudatus* seeds after crossing black-seeded and pale-seeded populations. The null hypothesis, based on a dominant epistatic model of genetic inheritance, assumed that

$$H_0\text{: } p_{\text{black}0} = \frac{12}{16} = \frac{3}{4} \text{ and } p_{\text{brown}0} = \frac{3}{16} \text{ and } p_{\text{pale}0} = \frac{1}{16}$$

If H_0 was true, we would expect the distribution of seed color to be

$$\text{black seeds: } np_{\text{black}0} = 429 \times 3/4 = 321.75$$
$$\text{brown seeds: } np_{\text{brown}0} = 429 \times 3/16 = 80.4375$$
$$\text{pale seeds: } np_{\text{pale}0} = 429 \times 1/16 = 26.8125$$

We can now calculate the chi-square statistic to test H_0.

$$X^2 = \sum \frac{(\text{observed count} - \text{expected count})^2}{\text{expected count}}$$
$$= \frac{(321 - 321.75)^2}{321.75} + \frac{(77 - 80.4375)^2}{80.4375} + \frac{(31 - 26.8125)^2}{26.8125}$$
$$= 0.8026$$ ■

Think of the chi-square statistic X^2 as a measure of the distance of the observed counts from the expected counts. Like any distance, it is always zero or positive, and it is zero only when the observed counts are exactly equal to the expected counts. Small values of X^2 represent small deviations from H_0 that do not provide sufficient evidence to reject H_0. Inversely, large values of X^2 are evidence against H_0, because they say that the observed counts are far from what we would expect if H_0 was true. *The alternative hypothesis H_a for the chi-square test is nonspecific, or nondirectional, because any violation of H_0 tends to produce a large value of X^2.*

APPLY YOUR KNOWLEDGE

Carol Bloomfield/Alamy

21.3 Saving birds from windows, continued. Exercise 21.1 described an experiment designed to figure out whether tilting windows down so that they reflect earth rather than sky can help reduce accidental bird strikes. Calculate the expected counts in each of the three conditions and compute the chi-square statistic.

21.4 Heat resistance in rice, continued. Go back to Exercise 21.2 investigating the genetic model of heat resistance in rice plants. What are the expected counts for each of the two phenotypes? Use these expected counts to compute the chi-square statistic.

Using technology

As usual, after calculating a statistic we want the probability of finding a statistic at least as extreme as the one obtained if H_0 was true. We first turn to technology for this computation. Figure 21.1 shows output for the chi-square goodness-of-fit test for the birth data in Example 21.1 from two statistical programs and a spreadsheet program.

EXAMPLE 21.5 Never on Sunday? Chi-square from software

The outputs differ in the type and amount of information they give. Minitab and SPSS tell us that the chi-square statistic is $X^2 = 19.12$. All three give a P-value of about 0.004. That is, there is very strong evidence that the distribution of births is *not* evenly ("uniformly") distributed across all 7 days of the week. The data observed are not consistent with a null hypothesis of equal proportions. ■

Minitab

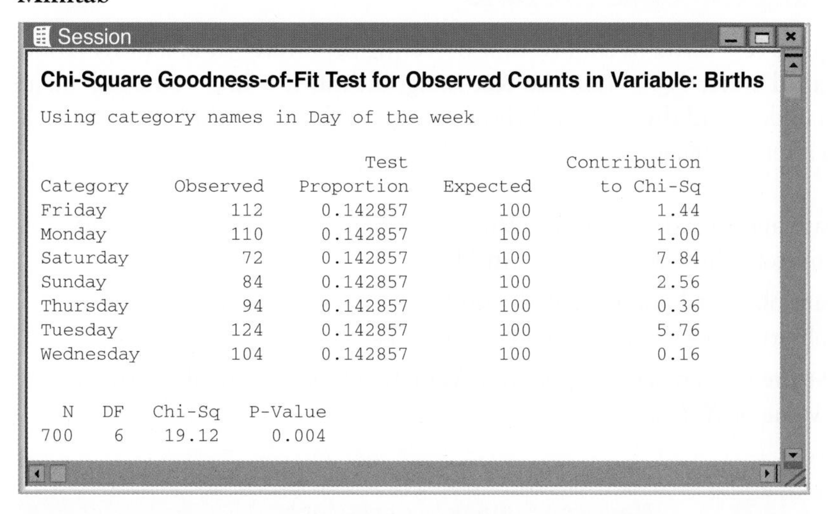

Session

Chi-Square Goodness-of-Fit Test for Observed Counts in Variable: Births

Using category names in Day of the week

Category	Observed	Test Proportion	Expected	Contribution to Chi-Sq
Friday	112	0.142857	100	1.44
Monday	110	0.142857	100	1.00
Saturday	72	0.142857	100	7.84
Sunday	84	0.142857	100	2.56
Thursday	94	0.142857	100	0.36
Tuesday	124	0.142857	100	5.76
Wednesday	104	0.142857	100	0.16

N	DF	Chi-Sq	P-Value
700	6	19.12	0.004

SPSS

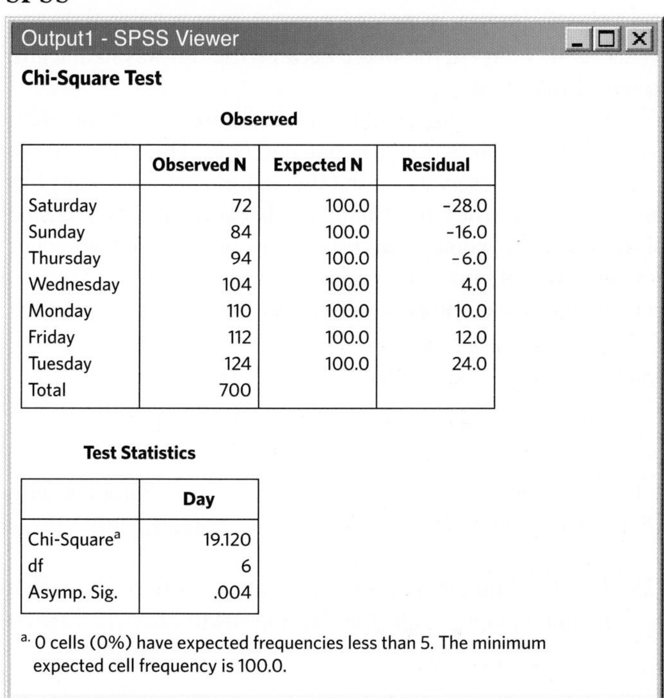

Output1 - SPSS Viewer

Chi-Square Test

Observed

	Observed N	Expected N	Residual
Saturday	72	100.0	-28.0
Sunday	84	100.0	-16.0
Thursday	94	100.0	-6.0
Wednesday	104	100.0	4.0
Monday	110	100.0	10.0
Friday	112	100.0	12.0
Tuesday	124	100.0	24.0
Total	700		

Test Statistics

	Day
Chi-Square[a]	19.120
df	6
Asymp. Sig.	.004

[a] 0 cells (0%) have expected frequencies less than 5. The minimum expected cell frequency is 100.0.

Microsoft Excel

	A	B	C
1	Day	Observed Births	Expected Births
2	Sunday	84	100
3	Monday	110	100
4	Tuesday	124	100
5	Wednesday	104	100
6	Thursday	94	100
7	Friday	112	100
8	Saturday	72	100
9			
10			
11	0.003965699	=CHISQ.TEST(B2 : B8, C2 : C8)	

FIGURE 21.1 Output from Minitab, SPSS, and Excel for the birth data, for Example 21.5.

Minitab and SPSS provide additional information. SPSS offers some remarks about expected counts that are related to the validity of this chi-square test, while Minitab gives the details of the chi-square calculations by providing the chi-square components. We will discuss their respective relevance in the following sections.

Interpreting chi-square results

The chi-square test is the overall test for detecting departures from a distribution model assumed under H_0. **When the test is significant,** it is important to look at the data to understand the nature of the distribution. Here are three ways to look at the data after significance has been established:

- **Compare appropriate percents:** Which outcomes occur in percents quite different from those hypothesized in H_0?
- **Compare observed and expected counts:** Which outcomes have more or fewer observations than we would expect if H_0 was true?
- **Look at the chi-square components:** Which outcomes contribute the most to the value of X^2?

EXAMPLE 21.6 Never on Sunday? Conclusions

We found in Example 21.5 a significant departure from the uniform model assuming an equal probability of births across all 7 days of the week. Because the test was significant, we can look deeper to explain this finding.

Look at the seven X^2 components provided by Minitab in Figure 21.1. About 40% of the value of X^2 (7.84 out of 19.12) comes from just one outcome. This points to the most important difference between the observations and the uniform model: a really low count of births on Saturday, representing the largest difference between any observed and expected counts. Most of the rest of X^2 comes from two other outcomes: a high count of births on Tuesday and a low count of births on Sunday.

Computing the percent of births occurring on each day we get

Day	Sun.	Mon.	Tue.	Wed.	Thu.	Fri.	Sat.
Percent	12%	16%	18%	15%	13%	16%	10%

With 18% of the 700 births, Tuesday has nearly twice as many births as Saturday, with only 10% of births, and one and a half times as many births as Sunday, with 12% of births.

Thus, we can conclude that the main difference from a uniform birth model is that weekend births are particularly uncommon, while Tuesday appears to have the highest frequency of births. ■

Looking at the details of the X^2 test is appropriate only when the test has been shown to be statistically significant. If not, any deviation from the model assumed in H_0 would represent only the kind of random variations we would expect to see when H_0 is true. That is, differences between chi-square components or differences between sample proportions are meaningless when the overall X^2 test is not significant.

We have discussed how to interpret a significant X^2 test. Now we turn to interpreting lack of significance. *Lack of significance in any statistical test means that there is not enough evidence to reject the null hypothesis. It does not, however, imply*

that the null hypothesis is true. This is particularly important to remember when the goodness-of-fit test is used in an attempt to confirm the suspicion that a variable has a particular distribution stated under H_0.

EXAMPLE 21.7 Genetics of seed color, continued

Example 21.2 described the seed color of 429 second-generation *A. caudatus* plants after crossing black-seeded and pale-seeded populations. The null hypothesis, based on a dominant epistatic model of genetic inheritance, assumed that

$$H_0\text{: } p_{\text{black}0} = \frac{12}{16} = \frac{3}{4} \quad \text{and} \quad p_{\text{brown}0} = \frac{3}{16} \quad \text{and} \quad p_{\text{pale}0} = \frac{1}{16}$$

We found in Example 21.4 that $X^2 = 0.8026$. With df $= 2$, software gives a P-value of 0.669. This P-value is not significant, and therefore, we cannot reject the null hypothesis of a distribution of seed color based on a dominant epistatic model. The genetic determination of seed color in *A. caudatus* **could** indeed represent a case of dominant epistasis. ■

The fact that we cannot reject H_0 does not imply that it is true. We cannot claim that seed color does follow a dominant epistatic model of genetic inheritance, only that such a model is possible. That is, the data gathered are *consistent with* dominant epistasis.

Always remember that a large P-value can arise under any of the following conditions:

- H_0 is indeed true. Only further scientific investigation can confirm this.
- H_0 is not actually true but it is too close to the real population distribution for us to tell them apart statistically.
- H_0 is definitely not true but the sample size is too small or the variability too great to reach significance. Example 21.8 illustrates this case.

EXAMPLE 21.8 More on birth days

Based on a sample of 700 births in Example 21.6, we clearly *rejected* the null hypothesis that births are uniformly distributed over the 7 days of the week. Figure 21.2 shows the SPSS output for the same test using only preliminary data based on a smaller sample of 140 births. With a P-value of 0.269, the test is not significant, and we cannot reject H_0. It would be a mistake to conclude that births are uniformly distributed over the week. Instead, we should conclude that the preliminary data did not give convincing evidence that births are not equally likely on all days of the week. ■

Bird fatalities According to the U.S. Fish and Wildlife Service (FWS), more birds die because of window collisions than because of any other factor associated with human activity. The FWS estimates that anywhere from 100 million to 1 billion birds are killed by window strikes each year, representing up to 5% of the North American autumn bird population. Actual deaths are likely underestimated by the general public, because some injured birds fly away before dying from injuries sustained and natural predators feed on injured and dying birds.

APPLY YOUR KNOWLEDGE

21.5 Saving birds from windows, continued. Exercise 21.1 described an experiment designed to figure out whether tilting windows down so that they reflect earth rather than sky could help reduce accidental bird strikes. Software gives a P-value less than 0.001. Interpret this result and draw a conclusion about the efficacy of tilting windows. Refer to your earlier calculations in Exercise 21.3.

SPSS

Output1 - SPSS Viewer

Chi-Square Test

Day

	Observed N	Expected N	Residual
Sunday	13	20.0	-7.0
Monday	23	20.0	3.0
Tuesday	24	20.0	4.0
Wednesday	20	20.0	.0
Thursday	27	20.0	7.0
Friday	18	20.0	-2.0
Saturday	15	20.0	-5.0
Total	140		

Test Statistics

	Day
Chi-Square[a]	7.600
df	6
Asymp. Sig.	.269

[a] 0 cells (0%) have expected frequencies less than 5. The minimum expected cell frequency is 20.0.

FIGURE 21.2 SPSS output for the smaller birth data set, for Example 21.8.

21.6 Heat resistance in rice, continued. Go back to Exercise 21.2 investigating the genetic model of heat resistance in rice plants. Software gives a P-value of 0.397. Interpret this result. Can you conclude that the distribution of heat-resistant phenotypes in F2 follows a dominant-recessive model (75%-25%)?

Conditions for the chi-square test

The chi-square test for goodness of fit is a one-sample test for categorical data. As with all the statistical tests we have studied, this test is valid only when certain conditions are met. First of all, the data must satisfy a *multinomial setting*.

THE MULTINOMIAL SETTING

1. There is a fixed number n of observations.
2. The n observations are all **independent.** That is, knowing the result of one observation does not change the probabilities we assign to other observations.
3. Each observation falls into just one of a finite number k of complementary and mutually exclusive outcomes.
4. The probability of a given outcome is the same for each observation.

If you covered optional Chapter 12, you will notice the similarity to the binomial setting. In fact, the binomial setting is a particular case of the multinomial setting. In Example 21.1, each birth falls on one of the 7 days of the week, the 7 possible outcomes for this variable. Births are independent events: Knowing which day one particular birth occurred says nothing about the day of another birth.

The second set of requirements for a valid test is related to issues of sample size. The chi-square test, like the z procedures, is an approximate method that becomes more accurate as the counts for each outcome get larger. We must therefore check that the counts are large enough to trust the P-value. Fortunately, the chi-square approximation is accurate for quite modest counts. Here is a practical guideline.[4]

COUNTS REQUIRED FOR THE CHI-SQUARE TEST

You can safely use the chi-square test with critical values from the chi-square distribution when no more than 20% of the *expected counts* computed are smaller than 5.0 and all individual *expected counts* are 1.0 or greater.

Note that the guideline uses *expected* counts. The expected counts for the day-of-birth study of Example 21.1 appear in the Minitab, SPSS, and Excel outputs in Figure 21.1. The smallest expected count is 100, so the data easily meet the guideline for safe use of chi-square.

Now that we have covered the requirements for the chi-square goodness-of-fit test, we have all the information needed to conduct such a test from beginning to end. Here is an example treated using the four-step approach.

EXAMPLE 21.9 More on the genetics of seed color

In Example 21.2 we tested the hypothesis of a dominant epistatic model for the seed color of amaranth, expressed as black, brown, and pale phenotypes. The chi-square test was not significant, implying that the data were consistent with such a genetic model—*but not proving* that this was indeed the genetic basis of amaranth seed color inheritance. Research papers often test two possible, competing genetic models for the phenotype studied. Here we test another model for the same amaranth color data.

STATE: A simpler, single-gene mode of inheritance can also produce three phenotypes in F2. When one gene locus has two codominant alleles, crossing pure breeds of the dominant and recessive traits gives rise to the two original traits as well as an intermediate phenotype in F2. The expected ratios for such a model are 1:2:1. That is, we would expect to see 1/4 (25%) of the dominant phenotype, 2/4 (50%) of the intermediate phenotype, and 1/4 (25%) of the recessive phenotype in F2. Example 21.2 showed the results in F2 of crossing black- and pale-seeded amaranth plants.

PLAN: Carry out a chi-square test for

$$H_0\text{: } p_{\text{black}} = 1/4 \text{ and } p_{\text{brown}} = 2/4 = 1/2 \text{ and } p_{\text{pale}} = 1/4$$

versus H_a: H_0 is not true. If the test is significant, compare observed outcome percents, observed versus expected counts, or chi-square components to describe the nature of the distribution.

SOLVE: First, check the guideline for use of chi-square. The data represent a single sample with 3 complementary and mutually exclusive outcomes. The expected counts appear in the Minitab output in Figure 21.3. All the expected counts are quite large;

Minitab

```
Session

                            Test                      Contribution
Category   Observed   Proportion   Expected          to Chi-Sq
black           321         0.25     107.25            426.005
brown            77         0.50     214.50             88.141
pale             31         0.25     107.25             54.210

  N    DF     Chi-Sq    P-Value
429     2    568.357      0.000
```

FIGURE 21.3 Minitab output for Example 21.9.

therefore, we can safely use chi-square. The output shows that the test is highly significant ($X^2 = 568.357$, $P = 0.000$). Note that Minitab rounds the P-value to three decimal places, which means that $P < 0.0005$ (a P-value is never exactly zero). The chi-square components indicate that black-seeded plants, in particular, are much too frequent for such a model.

CONCLUDE: We find that the F2 generation resulting from crossing black- and pale-seeded amaranth plants is composed of about 75% black-, 18% brown-, and 7% pale-seeded plants. A single-gene theory with codominant alleles is clearly not a good model of genetic inheritance of seed color in amaranth (1:2:1 model, $P < 0.0005$). Previous analysis of the data with a dominant epistatic model of genetic inheritance provided a good fit for the data (12:3:1 model, $P = 0.669$), suggesting that seed color in amaranth is governed by two separate gene loci. ■

APPLY YOUR KNOWLEDGE

21.7 **Saving birds from windows, continued.** Go back to Exercise 21.1, describing an experiment with tilted windows to help reduce accidental bird strikes. Explain why a chi-square procedure is appropriate.

21.8 **Heat resistance in rice, continued.** Go back to Exercise 21.2 investigating the genetic model of heat resistance in rice plants. Explain why a chi-square procedure is appropriate.

21.9 **Are the conditions met?** You want to run a chi-square goodness-of-fit test on multinomial data from one random sample. Here are the observed and expected counts:

Observed	3	8	9	4
Expected	6	6	6	6

Are the conditions met for this test? Explain your answer.

21.10 **Are the conditions met?** You want to run a chi-square goodness-of-fit test on multinomial data from one random sample. Here are the observed and expected counts:

Observed	7	9	7	13
Expected	4	8	8	16

Are the conditions met for this test? Explain your answer.

The chi-square distributions

Software usually finds P-values for us. The P-value for a chi-square test comes from comparing the value of the chi-square statistic with critical values for a *chi-square distribution*.

THE CHI-SQUARE DISTRIBUTIONS

The **chi-square distributions** are a family of distributions that take only positive values and are skewed to the right. A specific chi-square distribution is specified by giving its **degrees of freedom.**

The chi-square goodness-of-fit test involving k outcomes uses critical values from the chi-square distribution with $k - 1$ degrees of freedom. The P-value is the area to the right of X^2 under the density curve of this chi-square distribution.

Figure 21.4 shows the density curves for three members of the chi-square family of distributions. As the degrees of freedom increase, the density curves become less skewed and larger values become more probable. Table D in the back of the book gives critical values for chi-square distributions. You can use Table D if you do not have software that gives you P-values for a chi-square test.

EXAMPLE 21.10 Never on Sunday? Using the chi-square table

Example 21.1 described a test for the equal distribution of births over the 7 days of the week. That is, $k = 7$. The chi-square statistic therefore has $k - 1 = 6$ degrees of freedom. Two of the outputs in Figure 21.1 provide the degrees of freedom.

The observed value of the chi-square statistic is $X^2 = 19.12$. Look in the df = 6 row of Table D. The value $X^2 = 19.12$ falls between the 0.005 and 0.0025 critical values of the chi-square distribution with 6 degrees of freedom. Remember that the chi-square test is always nondirectional. So the P-value of $X^2 = 19.12$ is between 0.005 and 0.0025. The outputs in Figure 21.1 show that the P-value is about 0.004. ■

df = 6

p	.005	.0025
X^*	18.55	20.25

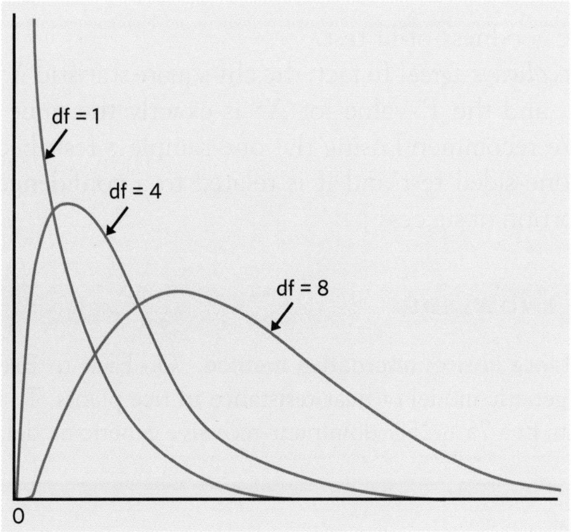

df = 1
df = 4
df = 8
0

FIGURE 21.4 Density curves for the chi-square distributions with 1, 4, and 8 degrees of freedom. Chi-square distributions take only positive values and are right-skewed.

The chi-square test for goodness of fit can also be used to test whether some data might follow a known probability distribution, such as the binomial or Poisson distributions we described in Chapter 12. Doing so first requires estimating the distribution parameters from the sample data. For each parameter that must be estimated this way (like the probability of success p for binomial, or the mean number of occurrences μ for Poisson), the chi-square test for goodness of fit loses an additional degree of freedom. Exercise 21.38 walks you through step-by-step.

APPLY YOUR KNOWLEDGE

21.11 Saving birds from windows, continued. Go back to Exercise 21.1, describing an experiment with tilted windows to help reduce accidental bird strikes. What are the degrees of freedom for this test? Use Table D to find the critical values closest to the chi-square statistic you calculated in Exercise 21.3.

21.12 Heat resistance in rice, continued. Go back to Exercise 21.2 investigating the genetic model of heat resistance in rice plants. What are the degrees of freedom for this test? Use Table D to find the critical values closest to the chi-square statistic you calculated in Exercise 21.4.

The chi-square test and the one-sample z test*

The chi-square goodness-of-fit test can be used to evaluate the distribution of a categorical random variable in a one-sample setting. When the random variable can take only one of two values—arbitrarily labeled success or failure—we can run either a chi-square goodness-of-fit test with 1 degree of freedom or a one-sample z test as described in Chapter 19. The null hypotheses tested are

$$H_0\colon p_{\text{success}} = p_0$$

for the one-sample z test, and

$$H_0\colon p_{\text{success}} = p_0,\ \ p_{\text{failure}} = 1 - p_0$$

for the chi-square goodness-of-fit test.

These two tests always agree. In fact, the chi-square statistic X^2 is just the square of the z statistic, and the P-value for X^2 is exactly the same as the two-sided P-value for z. We recommend using the one-sample z test, because it gives you the choice of a one-sided test and it is related to a confidence interval for the population proportion of success p.

P&R Fotos/age fotostock/SuperStock

APPLY YOUR KNOWLEDGE

21.13 Heat resistance in rice: alternative method. Go back to Exercise 21.2 investigating the genetic model of heat resistance in rice plants. To test the hypothesis that the data fit a 75%-25% dominant-recessive genetic model, you computed the

*This section is optional.

chi-square statistic in Exercise 21.4 and were told in Exercise 21.6 that $P = 0.397$. Test the same null hypothesis using a one-sample z test.

(a) How do the X^2 and the z statistics compare for these two tests?

(b) How do the P-value for the chi-square test and the two-sided P-value for the z test compare?

CHAPTER 21 SUMMARY

- The **chi-square test for goodness of fit** tests the null hypothesis H_0 that the data come from a population with a given distribution. It specifies a proportion for each of the k possible outcomes in the distribution, and the sum of all these proportions equals 1. The alternative hypothesis H_a simply says that H_0 is not true.
- The test compares the observed counts of observations with the counts that would be expected if H_0 was true. With n total observations, the **expected count** for any given outcome i is

$$\text{expected count}_i = n \times p_{i0}$$

The **chi-square statistic** is a sum of components computed separately for each of the k possible outcomes in the distribution:

$$X^2 = \sum \frac{(\text{observed count} - \text{expected count})^2}{\text{expected count}}$$

- The chi-square test compares the value of the statistic X^2 with critical values from the **chi-square distribution** with $k - 1$ **degrees of freedom.** Large values of X^2 are evidence against H_0, so the P-value is the area under the chi-square density curve to the right of X^2.
- The chi-square distribution is an approximation to the distribution of the statistic X^2. You can safely use this approximation when all *expected counts* are at least 1 and no more than 20% are less than 5.
- When the chi-square test finds a statistically significant departure from the hypothesized distribution, do data analysis to describe the nature of the distribution. You can do this by comparing the percents of each outcome, comparing the observed counts with the expected counts, and looking for the largest **components of the chi-square statistic.**

THIS CHAPTER IN CONTEXT

In Chapter 19 we computed confidence intervals and tested hypotheses about the proportion p of successes in one population based on data from one SRS from this population. That is, for each individual in the sample, we had a binary categorical variable of a success/failure type. This could be, for example, a patient's survival (yes/no) or a person's gender (female/male). For more complex categorical variables, such as a survey respondent's opinion (support/oppose/not sure), we selected one outcome as success (support) and treated all other outcomes as failure (not support).

The chi-square method introduced in this chapter lets us deal with categorical variables in a more nuanced way, by considering all levels of the variable. This is particularly relevant for genetics testing when we examine a trait expressed in more than just two phenotypes (for example, black, brown, and pale seeds). The approach compares the counts of each outcome observed in the sample with hypothetical counts we would expect for these outcomes if our null hypothesis was true.

The null hypothesis of the chi-square test for goodness of fit represents how you believe (or wonder if) the population of interest breaks down into k levels of the categorical variable. This is similar to the concept of a pie chart for a categorical variable, which we discussed in Chapter 1. Under the null hypothesis, the pie depicts the entire population, and the sizes of the slices correspond to the specified proportions for each level ($p_1, p_2, \ldots, p_k$). This explains why you must be sure that these hypothesized proportions all add up to 1 (or 100%, the entirety of the population).

In Chapter 22 we will see how the chi-square statistic can be used to test hypotheses about the relationship between two categorical variables, a comprehensive expansion of the z test for comparing two population proportions described in Chapter 20.

CHECK YOUR SKILLS

Example 21.2 showed a case of dominant epistasis with a 12:3:1 phenotypic ratio. A cross of white and green summer squash plants gives the following numbers of squash in the second generation (F2): 131 white squash, 34 yellow squash, and 10 green squash. Are these data consistent with a 12:3:1 dominant epistatic model of genetic inheritance (white being dominant)? Use this information for Exercises 21.14 through 21.21.

21.14 The null hypothesis for the chi-square goodness-of-fit test is

(a) H_0: the distribution is not uniform.
(b) H_0: the distribution is uniform: $p_{\text{white}} = p_{\text{yellow}} = p_{\text{green}}$.
(c) H_0: $p_{\text{white}} = 12/16$ and $p_{\text{yellow}} = 3/16$ and $p_{\text{green}} = 1/16$.

21.15 The alternative hypothesis is

(a) H_a: the distribution is not uniform.
(b) H_a: $p_{\text{white}} \neq p_{\text{yellow}} \neq p_{\text{green}}$.
(c) H_a: H_0 is not true.

21.16 The expected count of green squash is about

(a) 10.9. (b) 1. (c) 0.06.

21.17 The chi-square component for the green squash is about

(a) 10. (b) 1. (c) 0.08.

21.18 The degrees of freedom for this chi-square goodness-of-fit test are

(a) 15. (b) 3. (c) 2.

21.19 Software gives a chi-square statistic $X^2 = 0.124$ for this test. From the table of critical values, we can say that the P-value is

(a) less than 0.05.
(b) between 0.05 and 0.1.
(c) greater than 0.1.

21.20 Your conclusion based on this test is that

(a) the test results prove that squash color is determined by dominant epistasis.
(b) the test results are consistent with the idea that squash color is determined by dominant epistasis.
(c) the test results suggest that squash color is not determined by dominant epistasis.

21.21 You can trust the validity of this test because

(a) all expected counts are larger than 5.

(b) all observed counts are larger than 5.
(c) the sample size is larger than 40.

A very large pediatrics office is open for business Monday through Friday. Management wants to know if antibiotics are prescribed uniformly over the workweek (Monday through Friday). A random sample of 200 antibiotic prescriptions written over the past 12 months is taken. Use this information for Exercises 21.22 and 21.23.

21.22 The null hypothesis for the chi-square goodness-of-fit test is

(a) H_0: the distribution is not uniform.
(b) H_0: the distribution is uniform: $p_{\text{Mon}} = p_{\text{Tue}} = p_{\text{Wed}} = p_{\text{Thu}} = p_{\text{Fri}} = 1/5$.
(c) H_0: $p = 0.5$.

21.23 Software gives $P < 0.01$. Your conclusion based on this test is that

(a) the data are consistent with a uniform distribution.
(b) the data provide strong evidence that all five proportions are different.
(c) the data provide strong evidence that the distribution is not uniform.

CHAPTER 21 EXERCISES

If you have access to software or a graphing calculator, use it to speed your analysis of the data in these exercises. Exercises 21.24 to 21.28 are suitable for hand calculation if necessary. Exercises 21.29 to 21.33 provide software output for their respective tests.

21.24 Mendel's first law. Gregor Mendel has been called the "father of genetics," having formulated laws of inheritance even before the function of DNA was discovered. In one famous experiment with peas, he crossed pure breeds of tall and dwarf pea plants. The first-generation pea plants were all tall. In the second generation (F2), he obtained 787 tall plants and 277 dwarf plants. Mendel's first law of inheritance is a model with a dominant and a recessive phenotype. The first law says that we can expect a 3:1 ratio, that is, 3/4 with the dominant and 1/4 with the recessive phenotype, in F2.

(a) Are the conditions for a chi-square goodness-of-fit test satisfied?
(b) What are the null and alternative hypotheses for this test?
(c) What are the expected counts under the null hypothesis?
(d) Compute the chi-square statistic and use software or Table D to find the P-value. What can you conclude? Do these data support Mendel's first law?

21.25 Mendel's second law. Mendel formulated the law of independent assortment as his second law of inheritance. It involves two genes independently segregated during reproduction, each independently determining one aspect of the phenotype. In one experiment, Mendel crossed pea plants producing yellow, round seeds with pea plants producing green, wrinkled seeds. The first generation resulted in only plants producing yellow, round seeds. Self-crossing of the F1 yielded the following phenotypes in F2:

Phenotype	Yellow, round	Yellow, wrinkled	Green, round	Green, wrinkled
Counts	315	101	108	32

Assuming two independent genes, each with a dominant allele and a recessive allele, we would expect to find a 9:3:3:1 phenotypic ratio.

(a) Are the conditions for a chi-square goodness-of-fit test satisfied?
(b) What are the null and alternative hypotheses for this test?
(c) What are the expected counts under the null hypothesis?
(d) The chi-square statistic for this test is $X^2 = 0.47$. Use software or Table D to find the P-value. What can you conclude? Do the data agree with Mendel's second law of genetic inheritance?

21.26 Nematode resistance in soybeans. The soybean cyst nematode is a roundworm pest affecting the yield of soybean crops worldwide. Nematode-resistant soybean lines have been used for agriculture, but nematode populations have adapted. One recent soybean line we will call A is resistant to most nematodes except the supervirulent nematode LY1. A new soybean line we will call B has been shown to be resistant to LY1 and to be genetically different from soybean line A. Understanding the genetics of soybean resistance to LY1 is an important step in creating a genetically engineered soybean that will resist most nematodes, including the virulent LY1. Researchers crossed soybean lines A and B. They obtained 105 second-generation plants, of which 5 were resistant to LY1 and 100 were not. The researchers state that the data are "a good fit to an expected three-gene model, Rhg, rhg, rhg (one dominant and two recessive genes) with a segregation ratio of 3 (Resistant):61 (Susceptible)."[5]

(a) The researchers describe a 3:61 distribution. What are the null and alternative hypotheses for the corresponding chi-square goodness-of-fit test?
(b) What are the expected counts under the null hypothesis? Are the conditions for the test satisfied? Let's assume that they are very nearly satisfied.
(c) Use software or Table D to find the chi-square statistic and P-value for this test. Do the data support the researchers proposed model?
(d) Can you conclude that the genetic basis of soybean resistance to the virulent LY1 nematode is carried by three genes, one dominant and two recessive? Explain your answer.

21.27 Plan B emergency contraception. Plan B is the common name in the United States and Canada for the emergency contraception drug levonorgestrel. The high dose of progestogen it delivers has been shown in the lab to inhibit ovulation, thus rendering fertilization impossible. Some groups, however, are concerned that if the drug interfered with the fertilized egg, it could be considered abortive rather than strictly contraceptive. Researchers enrolled a cohort of fertile women attending a family-planning clinic for emergency contraception after unprotected intercourse during their fertile period. From interviews, blood samples, and ultrasounds, the researchers were able to establish the time of intercourse relative to the women's menstrual cycles. In this cohort, 87 women had unprotected intercourse before ovulation (pre-ovulation group) and 35 women had unprotected intercourse on or after ovulation (post-ovulation group). Based on standard clinical assessments of fertility rates by day of the menstrual cycle, the researchers expected 13.2 pregnancies in the pre-ovulation group and 7.1 pregnancies in the post-ovulation group. Instead, they observed 0 and 6 pregnancies, respectively.[6]

(a) Is there evidence that Plan B impacts the chance of pregnancy when taken before ovulation? What are the observed and expected counts of pregnancy and no pregnancy in the pre-ovulation group? Use these counts to compute the chi-square statistic. Are the test assumptions met? What are the test degrees of freedom? Use technology or Table D to obtain the test P-value and conclude.
(b) Is there evidence that Plan B impacts the chance of pregnancy when taken after ovulation? What are the observed and expected counts of pregnancy and no pregnancy in the post-ovulation group? Use these counts to compute the chi-square statistic. Are the test assumptions met? What are the test degrees of freedom? Use technology or Table D to obtain the test P-value and conclude.
(c) Write a short paragraph describing the study findings and what they suggest about the mechanism of action of the Plan B drug. Include the test statistics and P-values in your report.

21.28 Characteristics of a forest. 4 STEP Forests are complex, evolving ecosystems. For instance, pioneer tree species can be displaced by successional species better adapted to the changing environment. Ecologists mapped a large Canadian forest plot dominated by Douglas fir with an understory of western hemlock and western red cedar. Sapling trees (young trees shorter than 1.3 meters) are indicative of the future of a forest. The 246 sapling trees recorded in this sample forest plot were of the following types:[7]

Tree species	Count of sapling trees
Western red cedar (RC)	164
Douglas fir (DF)	0
Western hemlock (WH)	82

Are there equal proportions of RC, DF, and WH among sapling trees in this forest? What does your analysis suggest about this forest's successional stage? Follow the four-step process as illustrated in Example 21.9.

21.29 Melanoma. 4 STEP Melanoma is a rare form of skin cancer that accounts for the great majority of skin cancer fatalities. UV exposure is a major risk factor for melanoma. Some body parts are regularly more exposed to the sun than others. Is this reflected in the population distribution of melanoma location on the body? A random sample of 310 women diagnosed with melanoma were classified according to the known location of the melanoma on their bodies. Here are the results:[8]

Location	Head/neck	Trunk	Upper limbs	Lower limbs
Count	45	80	34	151

Assuming that these four body parts represent roughly equal skin areas, do the data support the hypothesis that melanoma occurs evenly on the body? Figure 21.5(a) shows the Minitab output for this test. Follow the four-step process as illustrated in Example 21.9.

Minitab

```
Session

                                  Test                    Contribution
Category        Observed    Proportion     Expected        to Chi-Sq
Head/neck             45          0.25         77.5          13.6290
Lower limbs          151          0.25         77.5          69.7065
Trunk                 80          0.25         77.5           0.0806
Upper limbs           34          0.25         77.5          24.4161

  N    DF     Chi-Sq    P-Value
310     3    107.832      0.000
```

(a)

```
Session

                                  Test                    Contribution
Category        Observed    Proportion     Expected        to Chi-Sq
Head/neck             36          0.25           56            7.143
Lower limbs           32          0.25           56           10.286
Trunk                139          0.25           56          123.018
Upper limbs           17          0.25           56           27.161

  N    DF     Chi-Sq    P-Value
224     3    167.607      0.000
```

(b)

FIGURE 21.5 Minitab outputs (a) for the melanoma data for women for Exercise 21.29 and (b) for the melanoma data for men for Exercise 21.30.

21.30 More on melanoma. [4 STEP] The previous exercise described data about melanoma body locations in a random sample of women. The study also reported data for a random sample of 224 men diagnosed with melanoma on a known body location. The results are as follows:

Location	Head/neck	Trunk	Upper limbs	Lower limbs
Count	36	139	17	32

Assuming that these four body parts represent roughly equal skin areas, do the data support the hypothesis that melanoma occurs evenly on the body? Figure 21.5(b) shows the Minitab output for this test. Follow the four-step process.

21.31 Are clinical trials representative of the American population? [4 STEP] Individuals of different ethnicities are somewhat more genetically variable than those of the same ethnicity and may respond to treatments differently. The FDA has recently requested that every effort be made to include individuals of all major ethnic backgrounds when enrolling subjects in clinical trials. A randomly chosen article in the *New England Journal of Medicine* describes the efficacy of calcium and vitamin D supplementation in preventing hip and other fractures in healthy postmenopausal women. Here is the breakdown by major ethnic group of the subjects whose ethnicity is known, along with the percent of individuals from each ethnicity in the general American population:[9]

Ethnicity	Count in study	Percent in U.S.
White	30,153	75.6
Black	3,317	10.8
Hispanic	1,507	9.1
Asian or Pacific Islander	722	3.8
Native American	149	0.7

Did this study successfully match the ethnic diversity of the American population? Figure 21.6 shows the Minitab output for this test. What does your analysis suggest? Follow the four-step process.

21.32 Weight perception. [4 STEP] A Gallup survey asked 1014 adults aged 18 and over about their ideal weight. The survey found that 116 interviewees thought that they were currently under their ideal weight, 180 thought that they were at about their ideal weight, and 718 thought that they were over their ideal weight. Do these self-perceptions match

Minitab

Session

Category	Observed	Test Proportion	Expected	Contribution to Chi-Sq
Asian-Pacific Isl.	722	0.038	1362.2	300.895
Black	3317	0.108	3871.6	79.441
Hispanic	1507	0.091	3262.2	944.346
Native American	149	0.007	250.9	41.409
White	30153	0.756	27101.1	343.682

N	DF	Chi-Sq	P-Value
35848	4	1709.77	0.000

FIGURE 21.6 Minitab output for Exercise 21.31.

the reality of the weight distribution in the United States? The National Health Interview Survey estimates, based on people's actual body mass index (computed from both weight and height), that 1.8% of the U.S. adult population is underweight, 36.7% has a healthy weight, and 61.5% is either overweight or obese. Use these official percents to compute the expected counts. Follow the four-step process. (Software gives $X^2 = 636.9$.)

21.33 Inbreeding in prairie dogs. 4 STEP A field biologist recorded events of inbreeding in a colony of prairie dogs. Based on a sample of 44 estrous females and 17 sexually mature males, the researcher computed the coefficient of relatedness of all possible male-female pairs in this sample. If individuals avoided inbreeding, we would expect to observe breeding mostly between individuals with little genetic relatedness. On the other hand, if breeding occurred randomly, without regard for the genetic relatedness of the mating couple, we would expect breeding observations among individuals of various relatedness to be proportional to the number of possible pairs with a given coefficient of relatedness. Here is the table of observed and expected breeding pairs in this study:[10]

Coefficient of relatedness	Observed pairs	Expected pairs
$r \geq 1/4$ (siblings)	6	6
$1/4 > r \geq 1/8$	4	5
$1/8 > r \geq 1/16$	18	10
$1/16 > r \geq 1/32$	12	12
$1/32 > r \geq 1/64$	10	11
$1/64 > r \geq 1/128$	3	7
$1/128 > r \geq 0$	8	8
No known kinship	8	10

Software computes $X^2 = 9.377$. Do the data support the hypothesis that breeding among prairie dogs occurs randomly, without regard for genetic relatedness? Follow the four-step process.

21.34 Genetics of plant variegation. 4 STEP *Arabidopsis thaliana* is a green-leafed plant, but a variegated type has green-and-yellow leaves. To investigate the mechanism of plant variegation, researchers created, in the DNA of the variegated *Arabidopsis*, a recessive mutation that suppresses variegation.[11] When the green mutants were crossed with the variegated types, the first-generation plants were all variegated, thus confirming the recessive aspect of the mutation. In the second generation (F2), the researchers obtained 84 green plants and 216 variegated plants. Are these data consistent with a single recessive mutation for which we would expect a 3:1 ratio in F2? Follow the four-step process.

21.35 Genetics of plant variegation, continued. 4 STEP The researchers in the previous exercise also crossed mutant green plants with wild-type green plants. Not surprisingly, the first generation was 100% green. In the second generation (F2), however, a new phenotype was observed in some of the plants: a transitional variegation occurring only on the first few leaves. (We will call them "transitional.") Of the 85 plants obtained in F2, 65 were wild-type green, 15 variegated, and 5 transitional. The researchers hypothesized that these phenotypes are the result of two genes following a dominant epistatic model. Are these data consistent with a dominant epistatic model for which we would expect a 12:3:1 distribution in F2? Follow the four-step process.

21.36 Transgenic chickens. 4 STEP Infection of chickens with the avian flu is a threat to both poultry production and human health. A research team created transgenic chicken resistant to avian flu infection. If the transgene is inherited normally without lethal effects, then transgenic and wild-type crosses should provide, in the second generation (F2), 25% homozygote transgenic offspring, 50% heterozygote offspring, and 25% homozygote wild-type offspring. The researchers obtained 102 chickens in F2, 25 of which were homozygote transgenic, 45 heterozygote, and 32 homozygote wild-type.[12] Are these data consistent with a 1:2:1 ratio in F2 as expected from a transgene inherited normally without lethal effects? Follow the four-step process.

21.37 Frizzled feathers. The Frizzle fowl is a striking variety of chicken with curled feathers. In a 1930 experiment, Frizzle fowls were crossed with a Leghorn variety exhibiting straight feathers. The first generation (F1) produced all slightly frizzled chicks. When the F1 was interbred, the following characteristics were observed in F2:[13]

4 STEP

Phenotype (feather type)	Observed counts
Frizzled	23
Slightly frizzled	50
Straight	20

The most likely genetic model for these results is that of a single gene locus with two codominant alleles. Under such a model, we would expect a 1:2:1 ratio in F2. Do the data support a codominance model of inheritance for this feather phenotype? Follow the four-step process as illustrated in Example 21.9.

21.38 Poisson distribution? (optional) As part of a class project, students dropped split peas onto a surface marked with a 20 cm by 20 cm grid made up of 100 equal-sized quadrats (each 4 cm^2 in surface area). The peas were dropped blindly from a height of one foot above the grid. Then students counted how many peas had landed on each of the 100 quadrats. Here are the results in summarized form:[14]

Number of peas per quadrat	0	1	2	3	≥ 4
Number of quadrats	45	36	16	3	0

(a) Verify that the number of quadrats adds up to 100 and that the number of peas that landed on the grid is 77. What is the average number of peas per quadrat?

(b) We described the Poisson probability distribution in Chapter 12. You might want to refresh your memory. If the throwing of the split peas was truly random, we would expect the number of peas per quadrat to follow a Poisson distribution with mean equal to the mean number of peas per quadrat. The value you computed in (a) based on the sample data is your best guess for the true value of the Poisson parameter. Obtain $P(X = k)$ for X equals 0, 1, 2, 3, and 4 or more using this Poisson distribution.

(c) We now want to test whether there is significant evidence that the distribution of split peas on the grid is not Poisson. For that, we can use the chi-square test for goodness of fit. Use the values you computed in (b) to obtain the expected counts. Are the conditions for inference satisfied?

(d) We need to group the data so that the expected counts are not too low. We can pool the results for 2, 3, and 4 or more peas per quadrat into a single category "2 or more peas per quadrat." What is $P(X \geq 2)$? Verify that the conditions for inference are now satisfied.

(e) Figure 21.7 shows the details of the chi-square test using Excel. Verify that you obtain the same value for the chi-square statistic on your own. Notice that we used only 1 degree of freedom to find the P-value. That's because we lost 1 degree of freedom when we used sample data to estimate the true population parameter μ of the Poisson distribution.

(f) Does the test provide significant evidence that the distribution of split peas on the grid is not Poisson? That is, is there significant evidence that the throwing was not completely random?

Microsoft Excel

	A	B	C	D
1	Category (peas/quadrat)	Observed counts	Expected counts	Chi-square component
2	0	45	46.30	0.0365
3	1	36	35.65	0.0034
4	2 or more	19	18.05	0.0500
5				
6			Chi-square	0.0899
7			P-value	0.7643
8				
9			=CHISQ.DIST.RT(0.0899,1)	

FIGURE 21.7 Excel calculations for Exercise 21.38.

CHAPTER 22

The Chi-Square Test for Two-Way Tables

In Chapter 21 we introduced a new statistic, X^2, and used it to test whether a categorical variable follows (or "fits") a particular distribution. This is a situation often found in genetic analysis. However, the most common use of the chi-square statistic is to test the hypothesis that there is *no relationship between two categorical variables*.

Some variables—such as sex, race, and blood type—are categorical by nature. Other categorical variables are created by grouping values of a quantitative variable into classes—such as age groups (for example, 20–29, 30–39, and so on) and weight categories (for example, underweight, healthy, overweight, obese). As usual, we are interested in both data analysis (describe the relationship) and inference (is the relationship statistically significant?).

IN THIS CHAPTER WE COVER...

- Two-way tables
- The problem of multiple comparisons
- Expected counts in two-way tables
- The chi-square test
- Using technology
- Conditions for the chi-square test
- Uses of the chi-square test
- Using a table of critical values*
- The chi-square test and the two-sample z test*

Two-way tables

We saw in the optional Chapter 5 that we can present data on two categorical variables in a **two-way table** of counts. That's our starting point. Here is an example.

two-way table

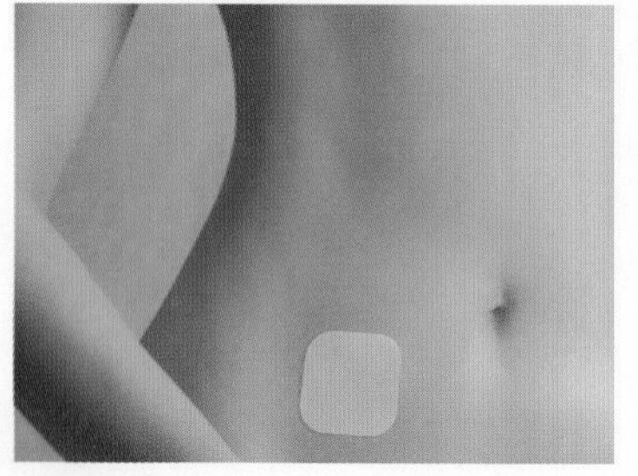
Getty Images

EXAMPLE 22.1 Hormone therapy and thromboembolism

Oral estrogen-replacement therapy (ERT) activates blood coagulation and increases the risk of venous thromboembolism (a traveling blood clot, sometimes fatal) in postmenopausal women. Transdermal ERT might be less problematic, at least in theory. A retrospective, case-control study compared a random sample of 155 postmenopausal women with a first documented thromboembolism and a control random sample of 381 postmenopausal women with no history of thromboembolism. The women were asked about their ERT history. Here are the results:[1]

	Thromboembolism Status	
ERT History	**Cases**	**Controls**
Never use	71	208
Past use	22	53
Current oral ERT use	32	27
Current transdermal ERT use	30	93
Total	155	381

■

The two-way table in Example 22.1 shows the relationship between two categorical variables. The explanatory variable is the patient's ERT history, with 4 levels. The response variable is the patient's thromboembolism status: first thromboembolism (cases) or no thromboembolism (controls). The two-way table gives the counts for all 8 combinations of values of these variables. Each of the 8 counts occupies a **cell** of the table.

cell

It is hard to compare the counts because the control sample is much larger than the case sample. Here are the percents of each outcome within each sample:

	Thromboembolism Status	
ERT History	**Cases**	**Controls**
Never use	46%	55%
Past use	14%	14%
Current oral ERT use	21%	7%
Current transdermal ERT use	19%	24%
Total	100%	100%

In the language of Chapter 5 (page 125), these are the *conditional distributions* of outcomes, given the patients' thromboembolism status. The differences are not large, but higher percents of women in the case group are current oral ERT users. Figure 22.1 compares the two distributions. We want to know if there is a significant difference between the two distributions of ERT history.

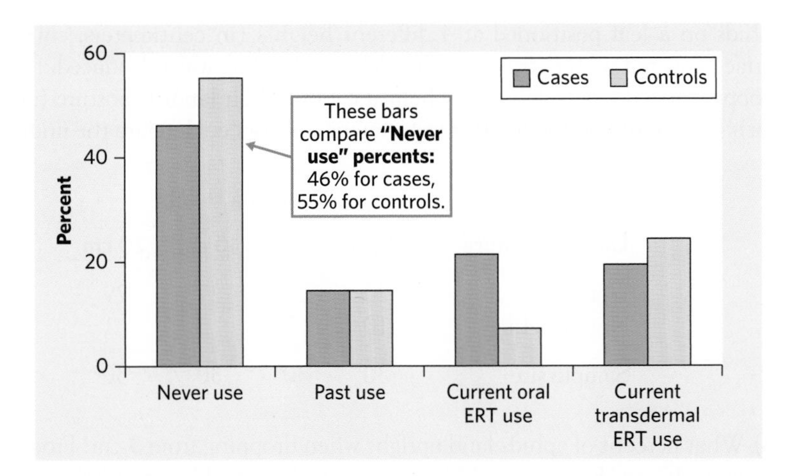

FIGURE 22.1 Bar graph comparing hormone therapy (ERT) history for women with (cases) or without (controls) a first diagnosis of thromboembolism, for Example 22.1.

APPLY YOUR KNOWLEDGE

22.1 Age and HPV infections. Human papillomavirus (HPV) infection is the most common sexually transmitted infection. Most people who become infected with HPV do not even know they have it. Most often, the body's immune system clears HPV naturally within two years. However, certain types of HPV can cause genital warts in both men and women and cervical cancer in women. Example 19.1 (page 463) examined the prevalence of HPV infections among women between the ages of 14 and 59 years based on the U.S. National Health and Nutrition Examination Survey (NHANES). Here are the HPV test results broken down by age group:

	Age Group					
HPV Test	14-19	20-24	25-29	30-39	40-49	50-59
Positive	160	85	48	90	82	50
Negative	492	104	126	238	242	204
Total	652	189	174	328	324	254

(a) What percent of women who test positive for HPV are in the 14–19 years age group? 20–24? 25–29? 30–39? 40–49? 50–59? These percents should add to 100% (aside from roundoff error). They form the conditional distribution of age, given a positive HPV test.

(b) In a similar way, find the conditional distributions of age, given a negative HPV test. Make a table that presents the two conditional distributions. Be sure to include a "Total" column showing that each row adds to 100%.

(c) Compare the two conditional distributions. The study enrolled a different number of women in each age group. If there was no relationship between HPV test and age group, we would expect the two conditional distributions to be very similar. Interpret your results.

22.2 Aphid righting behavior. Pea aphids, *Acyrthosiphon pisum*, are small, wingless, sap-sucking insects that live on plants. They evade predators (such as ladybugs) by dropping off. A study examined the mechanism of aphid drops. Researchers placed

Perennou Nuridsany/Science Source

aphids on a leaf positioned at 4 different heights (in centimeters, cm) above a surface covered with petroleum jelly. When a ladybug was introduced, the aphids dropped, and the petroleum jelly helped capture their landing posture (upright or not). Each aphid performed this experiment only once. Here are the findings:[2]

	Dropping Height			
Landing Posture	**3 cm**	**5 cm**	**10 cm**	**20 cm**
Upright	20	23	27	29
Not upright	10	7	3	1
Sample size	30	30	30	30

(a) What percent of aphids land upright when dropping from 3 cm? From 5 cm? From 10 cm? From 20 cm? These percents do not add up to 100% because these groups represent separate samples.

(b) Make a bar graph that compares your percents for all four samples. Is there any clear relationship between dropping height and landing posture? Explain your answer.

The problem of multiple comparisons

The null hypothesis in Example 22.1 is that there is *no difference* between the distributions of ERT history among women with or without a first thromboembolism diagnosis. Put more generally, the null hypothesis is that there is *no relationship* between two categorical variables,

H_0: there is no relationship between thromboembolism status and ERT history

The alternative hypothesis says that there *is* a relationship but does not specify any particular kind of relationship,

H_a: there is some relationship between thromboembolism status and ERT history

Any significant difference between the distributions among women with or without a first thromboembolism diagnosis means that the null hypothesis is false and the alternative hypothesis is true. The alternative hypothesis is not one-sided or two-sided. It is nonspecific and allows any kind of difference.

Using the z procedures of Chapter 20, we might start by comparing the proportions of patients in the two groups who have never used ERT. We could similarly compare the proportions for each of the other ERT histories: four tests in all, with four P-values. This is a bad idea. *When we do many individual tests or confidence intervals, the individual P-values and confidence levels don't tell us how confident we can be in all of the inferences taken together.*

Because of this, it's cheating to pick out the largest of the four differences and then test its significance as if it were the only comparison we had in mind. For example, the "current oral ERT use" proportions in Example 22.1 are significantly different ($P < 0.0001$) if we compare just this one outcome. But is it surprising that the *most different* proportions among four ERT history levels differ by this much? That's a different question.

The problem of how to do many comparisons at once with an overall measure of confidence in all our conclusions is common in statistics. This is the problem of **multiple comparisons.** Statistical methods for dealing with multiple comparisons usually have two steps:

multiple comparisons

1. An *overall test* to see if there is good evidence of *any* differences among the parameters that we want to compare.
2. A detailed *follow-up analysis* to decide which of the parameters differ and to estimate how large the differences are.

The overall test, though more complex than the tests we saw earlier, is often reasonably straightforward. The follow-up analysis can be quite elaborate. In our basic introduction to statistical practice, we will concentrate on the overall test, along with descriptive data analysis that points to the nature of the differences.

APPLY YOUR KNOWLEDGE

22.3 Aphid righting behavior, continued. Exercise 22.2 describes a study of aphids dropping from various heights to escape predators.

(a) Using the plus four method from Chapter 19, give four 95% confidence intervals for the percents of aphids that land upright when dropping from 3, 5, 10, and 20 cm.

(b) Explain clearly why we are *not* 95% confident that *all four* of these intervals capture their respective population proportions.

22.4 Mating behavior of female prairie dogs. Gunnison's prairie dogs (*Cynomys gunnisoni*) are small social rodents found in the American Midwest. Careful observations in the wild have shown that both males and females can have multiple sex partners during the mating season. Male reproductive success is clearly related to the number of inseminations completed, but is there such a relationship for the females? A field biologist observed the mating behavior and reproductive success of females in one colony of Gunnison's prairie dogs:[3]

Wendy Shattil and Bob Rozinski/Getty Images

	Number of Mother's Sexual Partners				
	1	2	3	4	5
Number of copulating females	87	93	61	17	5
Number that gave birth	81	85	61	17	5
Proportion that gave birth	0.93	0.91	1.0	1.0	1.0

(a) Ten different pairwise comparisons of proportions are possible here. Write down the corresponding 10 null hypotheses.

(b) The females that copulated with 2 males or with 3 males represent two extremes in these observations. Give the P-value for a two-sided test comparing these two proportions. Why is it cheating to select the most extreme results and test only these for significance?

(c) If you calculated all 10 P-values, they would not tell you how often the five proportions would be spread this far apart just by chance. Explain why.

Expected counts in two-way tables

Our general null hypothesis H_0 is that there is *no relationship* between the two categorical variables that label the rows and columns of a two-way table. That is, the two variables are independent. To test H_0, we compare the observed counts in the table with the *expected counts*, the counts we would expect—except for random variation—if H_0 was true. If the observed counts are far from the expected counts, that is evidence against H_0. It is easy to find the expected counts.

EXPECTED COUNTS

The **expected count** in any cell of a two-way table when H_0 is true is

$$\text{expected count} = \frac{\text{row total} \times \text{column total}}{\text{table total}}$$

EXAMPLE 22.2 Observed versus expected counts

Let's find the expected counts for the thromboembolism study. Here is the two-way table with row and column totals:

	Thromboembolism Status		
ERT History	**Cases**	**Controls**	**Total**
Never use	71	208	279
Past use	22	53	75
Current oral ERT use	32	27	59
Current transdermal ERT use	30	93	123
Total	155	381	536

The expected count of women with thrombosis (cases) who have never used ERT is

$$\frac{\text{row 1 total} \times \text{column 1 total}}{\text{table total}} = \frac{(279)(155)}{536} = 80.68$$

Here is the table of all 8 expected counts:

	Thromboembolism Status		
ERT History	**Cases**	**Controls**	**Total**
Never use	80.68	198.32	279
Past use	21.69	53.31	75
Current oral ERT use	17.06	41.94	59
Current transdermal ERT use	35.57	87.43	123
Total	155	381	536

As this table shows, *the expected counts have exactly the same row and column totals (aside from roundoff error) as the observed counts*. That's a good way to check your work.

To see how the data diverge from the null hypothesis, compare the observed counts with these expected counts. You see, for example, that 208 control women with no thromboembolism diagnosis reported never having used ERT, whereas we would expect only 198.32 if the null hypothesis was true. ■

Why the formula works Where does the formula for an expected cell count come from? As stated in Chapter 21, if we have n independent tries and the probability of a given outcome on each try is p, we expect np such outcomes. (We also covered this in Chapters 9, 10, and 12.) That is, we expect np counts on average over the long run. In the thromboembolism study, the proportion of all 536 women who reported never having used ERT is

$$\frac{\text{count of never use}}{\text{table total}} = \frac{\text{row 1 total}}{\text{table total}} = \frac{279}{536}$$

Think of this as p, the overall proportion of women who never used ERT. If H_0 is true, we expect (aside from random variation) the same proportion of successes in both groups. So the expected count of never use among the 155 case women with a thromboembolism diagnosis is

$$np = (155)\left(\frac{279}{536}\right) = 80.68$$

That's the formula in the Expected Counts box.

Here is another way of thinking about the expected counts. If H_0 is true and the two variables are independent, then we expect the joint probability of two outcomes to be equal to the product of the corresponding marginal probabilities for each outcome. This is the multiplication rule for independent events described in Chapter 10. In a two-way table, this translates to

$$\frac{\text{expected count}}{\text{table total}} = \frac{\text{row total}}{\text{table total}} \times \frac{\text{column total}}{\text{table total}}$$

Simplifying by "table total" gives the formula in the Expected Counts box.

APPLY YOUR KNOWLEDGE

22.5 Age and HPV infections, continued. The two-way table in Exercise 22.1 displays data on HPV infection among women of different age groups. The null hypothesis says that there is no relationship between these variables. That is, the proportion of women who test postive for HPV is the same for all six age groups.

(a) Find the two expected counts for women in the 14–19 age group who test either positive or negative for HPV. This is one column of the two-way table of expected counts. Find the column total and verify that it agrees with the column total for the observed counts.

(b) Compare the observed and expected counts in this column. Are young women in the 14–19 age group more likely or less likely to test positive for HPV than would be predicted by the null hypothesis?

22.6 Aphid righting behavior, continued. Exercise 22.2 describes a study of aphids dropping from various heights to escape predators. The null hypothesis "no relationship" says that in the population of aphids, the proportions that land upright are the same when the dropping height is 3, 5, 10, and 20 cm.

(a) Find the expected cell counts if this hypothesis is true and display them in a two-way table. Add the row and column totals to your table and check that they agree with the totals for the observed counts.

(b) Are there any large deviations between the observed counts and the expected counts? What kind of relationship between the two variables do these deviations point to?

The chi-square test

The statistical test that tells us whether the observed differences between women in the case and control groups described in Example 22.1 are statistically significant compares the observed and expected counts. The test statistic that makes the comparison is the *chi-square statistic*.

CHI-SQUARE STATISTIC

The **chi-square statistic** is a measure of how far the observed counts in a two-way table are from the expected counts. The formula for the statistic is

$$X^2 = \sum \frac{(\text{observed count} - \text{expected count})^2}{\text{expected count}}$$

The sum is over all cells in the table. That is, there are as many terms in the sum as there are cells in the table. Each term in the sum is called a $\boldsymbol{X^2}$ **component.**

The chi-square statistic is a sum of terms, one for each cell in the table. There are 8 cells in the thromboembolism table of Example 22.1 (2 groups with 4 ERT histories each), and therefore, we first need to compute 8 X^2 components separately. Here is how to compute the first one: 71 case women with a thromboembolism diagnosis reported never having used ERT. The expected count for this cell is 80.68. So the X^2 component from this cell is

$$\frac{(\text{observed count} - \text{expected count})^2}{\text{expected count}} = \frac{(71 - 80.68)^2}{80.68} = \frac{93.7024}{80.68} = 1.16$$

To obtain the chi-square statistic for this test we must add all 8 components, so that X^2 is

$$\begin{aligned} X^2 &= \frac{(71 - 80.68)^2}{80.68} + \frac{(208 - 198.32)^2}{198.32} \\ &+ \frac{(22 - 21.69)^2}{21.69} + \frac{(53 - 53.31)^2}{53.31} \\ &+ \frac{(32 - 17.06)^2}{17.06} + \frac{(27 - 41.94)^2}{41.94} \\ &+ \frac{(30 - 35.57)^2}{35.57} + \frac{(93 - 87.43)^2}{87.43} \\ &= 21.268 \end{aligned}$$

Think of the chi-square statistic X^2 as a measure of the distance between the observed counts and the expected counts. Like any distance, it is always zero or positive, and it is zero only when the observed counts are exactly equal to the expected counts. Small values of X^2 represent small deviations from H_0 that do not provide sufficient evidence to reject H_0. Inversely, large values of X^2 are evidence against H_0, because they say that the observed counts are far from what we would expect if H_0 was true.

Using technology

Calculating the expected counts and then the chi-square statistic by hand is a bit time-consuming. As usual, software saves time and always gets the arithmetic right. Figure 22.2 shows output for the chi-square test for the thromboembolism data from a graphing calculator, three statistical programs, and a spreadsheet program.

EXAMPLE 22.3 Chi-square from software

The outputs differ in the information they give. All except the Excel spreadsheet tell us that the chi-square statistic is $X^2 = 21.268$; all show a P-value very close to zero, about 0.00009. (Notice that the three statistical programs refer to this test as **Pearson chi-square.**) There is very strong evidence that the distributions of ERT history are different in women with or without a first episode of thromboembolism.

Pearson chi-square

The three statistical programs repeat the two-way table of observed counts and add the row and column totals. They offer additional information on request, such as

Minitab

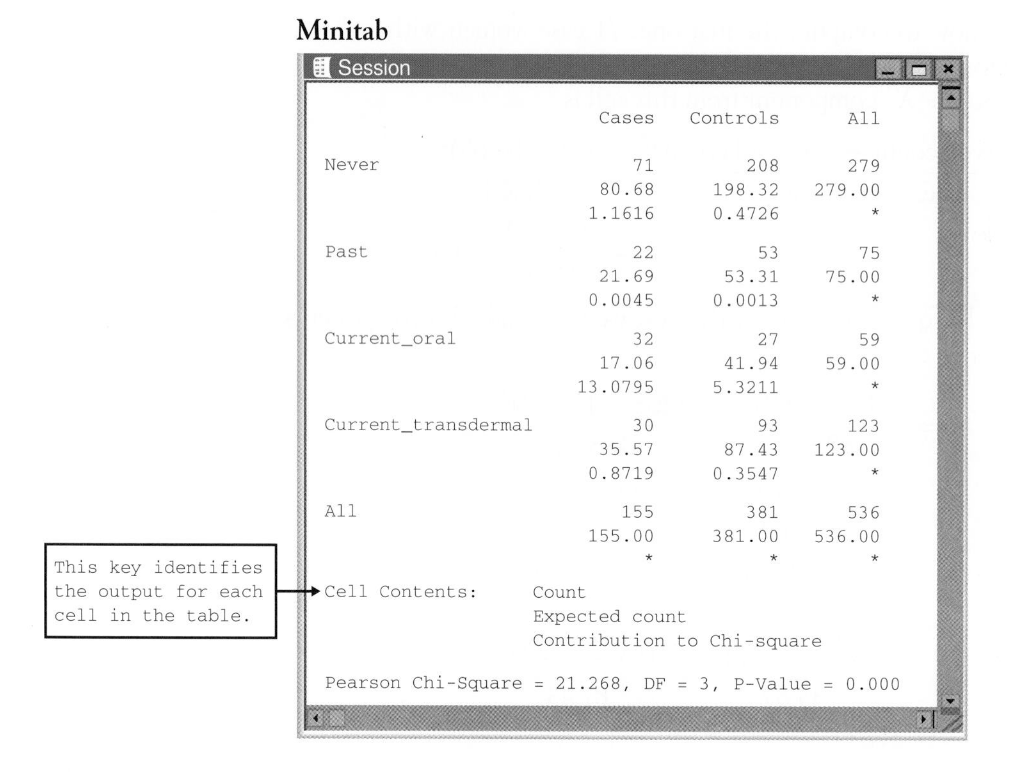

	Cases	Controls	All
Never	71	208	279
	80.68	198.32	279.00
	1.1616	0.4726	*
Past	22	53	75
	21.69	53.31	75.00
	0.0045	0.0013	*
Current_oral	32	27	59
	17.06	41.94	59.00
	13.0795	5.3211	*
Current_transdermal	30	93	123
	35.57	87.43	123.00
	0.8719	0.3547	*
All	155	381	536
	155.00	381.00	536.00
	*	*	*

Cell Contents: Count
Expected count
Contribution to Chi-square

Pearson Chi-Square = 21.268, DF = 3, P-Value = 0.000

JMP

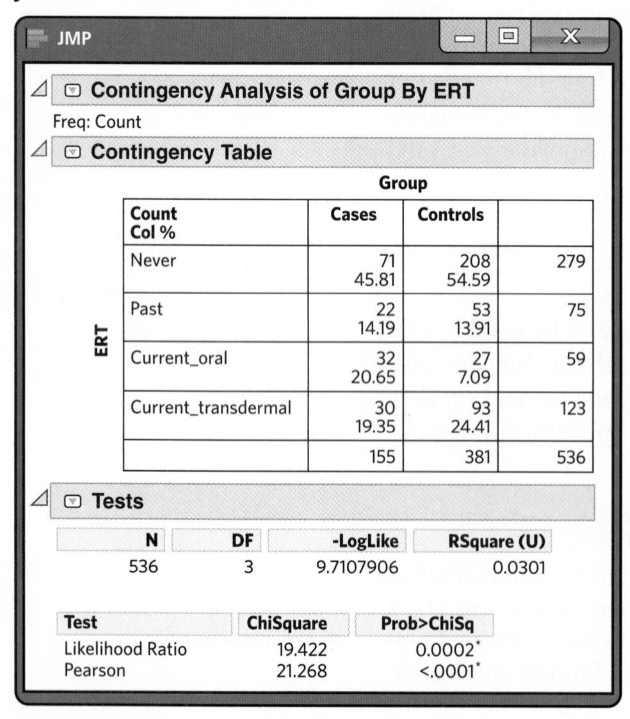

Count Col %	Cases	Controls	
Never	71 45.81	208 54.59	279
Past	22 14.19	53 13.91	75
Current_oral	32 20.65	27 7.09	59
Current_transdermal	30 19.35	93 24.41	123
	155	381	536

N	DF	-LogLike	RSquare (U)
536	3	9.7107906	0.0301

Test	ChiSquare	Prob>ChiSq
Likelihood Ratio	19.422	0.0002*
Pearson	21.268	<.0001*

Microsoft Excel

	A	B	C
1		Observed counts	
2	ERT	Cases	Controls
3	Never	71	208
4	Past	22	53
5	Current_oral	32	27
6	Current_transdermal	30	93
7			
8		Expected counts	
9	ERT	Cases	Controls
10	Never	80.68	198.32
11	Past	21.69	53.31
12	Current_oral	17.06	41.94
13	Current_transdermal	35.57	87.43
14			
15	=CHISQ.TEST(B3:C6,B10:C13)		9.24E-05
16			

FIGURE 22.2 Output from the TI-83 graphing calculator, Minitab, JMP, SPSS, and Excel for the two-way table in the thromboembolism study, for Example 22.3.

SPSS

Column Statistics

ERT * Groups Crosstabulation

			Groups		
			Cases	Controls	Total
ERT	1Never	Count	71	208	279
		Expected Count	80.7	198.3	279.0
	2Past	Count	22	53	75
		Expected Count	21.7	53.3	75.0
	3Current_oral	Count	32	27	59
		Expected Count	17.1	41.9	59.0
	4Current_transdermal	Count	30	93	123
		Expected Count	35.6	87.4	123.0
Total		Count	155	381	536
		Expected Count	155.0	381.0	536.0

Chi-Square Tests

	Value	df	Asymp. Sig. (2-sided)
Pearson Chi-Square	21.268[a]	3	.000
Likelihood Ratio	19.422	3	.000
N of valid Cases	536		

[a] 0 cells (0%) have expected count less than 5. The minimum expected count is 17.06.

TI-83

```
X²-Test                [B]
 X²=21.26772612        [[80.68...  198.31...
 p=.92622322E-5         [21.69...  53.31...
 df=3                   [17.06...  41.94...
                        [35.57...  87.43...
```

FIGURE 22.2 *(Continued)*

the expected counts. We asked JMP to add the column percents ("Col %") that enable us to compare the cases and the controls. The chi-square statistic is a sum of 8 terms, one for each cell in the table. We asked Minitab to give the expected counts and contribution to chi-square for each cell. The top-left cell has expected count 80.68 and X^2 component 1.1616, as we calculated. Look at the 8 terms. More than half the value of X^2 (13.0795 out of 21.268) comes from just one cell. This points to the most important difference between the two groups: A higher proportion of women with a diagnosed thromboembolism currently take ERT orally. Most of the rest of X^2 comes from one other cell: Fewer women with no thromboembolism (control) currently take ERT orally.

Excel lacks a menu selection for the chi-square test. You must program the spreadsheet to calculate the expected cell counts and then use the CHISQ.TEST worksheet formula. This gives the P-value but not the test statistic itself. You can, of course, program the spreadsheet to find the value of X^2. The Excel output shows the observed and expected cell counts and the P-value. ■

The chi-square test is the overall test for detecting relationships between two categorical variables. If the test is significant, it is important to look at the data to learn the nature of the relationship. We have three ways to look at the thromboembolism data from a descriptive standpoint:

- **Compare appropriate percents:** Which outcomes occur in quite different percents of case- and control-subjects? This is the method we learned in Chapter 5.
- **Compare observed and expected cell counts:** Which cells have more or fewer observations than we would expect if H_0 was true?
- **Look at the components of the chi-square statistic:** Which cells contribute the most to the value of X^2?

EXAMPLE 22.4 ERT and thromboembolism: conclusions

There is a significant difference between the distributions of ERT history of women with or without a first diagnosis of thromboembolism. All three ways of comparing the distributions show that the main difference is that a higher proportion of women diagnosed with a thromboembolism currently take ERT orally. On the other hand, the contributions of the two "current transdermal" cells to the chi-square statistic are very small, suggesting that taking ERT transdermally may be safer with regard to thromboembolism.

The broader conclusion, however, is more challenging. The study is observational. That is, the association revealed in the analysis does not show that oral ERT use "causes" a higher rate of thromboembolism. There could be lurking variables responsible for both a higher rate of oral ERT prescription and a higher rate of thromboembolism. In addition, the study has a retrospective, case-control design. As we discussed in Chapter 7, a major challenge of such studies is to find cases and controls that are as similar as possible in every respect except for the variable studied. The researchers did take great care in their sample selection. The 381 controls were enrolled at the same medical centers and at the same time as the cases, and they were selected to be similar in age (mean 62.1 years for the cases and 62.0 for the controls) and overall health (body mass index, hypertension, diabetes). A thorough analysis and discussion of possible confounding variables were also included in the research paper.

There is an important message here: The comprehensive examination of the potential weaknesses of a study is often the most challenging aspect of statistics—and maybe the most critical. ■

APPLY YOUR KNOWLEDGE

22.7 Age and HPV infections, continued. In Exercises 22.1 and 22.5 you began to analyze data on the HPV status of American women of different ages. Figure 22.3 displays the Minitab output for the chi-square test applied to these data.

(a) Starting from the observed and expected counts in the output, calculate the two components of the chi-square statistic for the first column (14–19 year old). Verify that your work agrees with Minitab's "Chi-square contributions" aside from roundoff error.

(b) According to Minitab, what are the value of the chi-square statistic X^2 and the P-value of the chi-square test?

Minitab

```
Session

Expected counts are printed below observed counts
Chi-Square contributions are printed below expected counts

            14-19   20-25   25-29   30-39   40-49   50-59   Total
Positive      160      85      48      90      82      50     515
           174.79   50.67   46.65   87.93   86.86   68.09
            1.252  23.261   0.039   0.049   0.272   4.808

Negative      492     104     126     238     242     204    1406
           477.21  138.33  127.35  240.07  237.14  185.92
            0.459   8.520   0.014   0.018   0.100   1.761

Total         652     189     174     328     324     254    1921

Chi-Sq = 40.554, DF = 5, P-Value = 0.000
```

FIGURE 22.3 Minitab output for the two-way table of HPV status by age group among American women, for Exercise 22.7.

R

```
> .Table # count
              3 cm   5 cm  10 cm  20 cm
Upright         20     23     27     29
Not upright     10      7      3      1

> .Test <- chisq.test(.Table, correct=FALSE)
> .Test
         Pearson's Chi-squared test
data: .Table
X-squared = 11.2554, df = 3, p-value = 0.01042

> .Test$expected # Expected Count
              3 cm   5 cm  10 cm  20 cm
Upright      24.75  24.75  24.75  24.75
Not upright   5.25   5.25   5.25   5.25
```

FIGURE 22.4 R output for the two-way table in the study of aphid righting behavior, for Exercise 22.8.

(c) Look at the "Chi-square contributions" entries in Minitab's display. Which components contribute the most to X^2? Write a brief summary of the nature and significance of the relationship between age and HPV.

22.8 Aphid righting behavior, continued. In Exercises 22.2 and 22.6 you began to analyze data on the landing posture of aphids dropping from various heights to escape predators. Figure 22.4 gives the R output for these data.

(a) Starting from the observed and expected counts, find the eight components of the chi-square statistic and then the statistic X^2 itself. Check your work against the computer output.

(b) What is the P-value for the test? Explain in simple language what it means to reject H_0 in this setting.

(c) Which cells contribute the most to X^2? What kind of relationship do these components in combination with the column percents in the table point to?

Conditions for the chi-square test

The chi-square test, like the z procedures for comparing two proportions, is an approximate method that becomes more accurate as the counts in the cells of the table get larger. We must therefore check that the counts are large enough to trust the P-value. Fortunately, the chi-square approximation is accurate for quite modest counts. Here is a practical guideline.[4]

CELL COUNTS REQUIRED FOR THE CHI-SQUARE TEST

You can safely use the chi-square test with critical values from the chi-square distribution when no more than 20% of the *expected counts* computed are smaller than 5.0 and all individual *expected counts* are 1.0 or greater. In particular, all four expected counts in a 2×2 table should be values equal to 5.0 or greater.

Note that the guideline uses *expected* cell counts. The expected counts for the thromboembolism study of Example 22.1 appear in the Minitab output in Figure 22.2. The smallest expected count is 17.06, so the data easily meet the guideline for safe use of chi-square.

APPLY YOUR KNOWLEDGE

22.9 Age and HPV infections, continued. Figure 22.3 displays Minitab output for data on HPV and age among American women. Using the information in the output, verify that the data meet the cell count requirement for use of chi-square.

22.10 Aphid righting behavior, continued. Figure 22.4 displays the R output for data on the landing posture of aphids dropping from various heights to escape predators. Using the information in the output, check whether the data meet the cell count requirement for use of chi-square.

22.11 Mating behavior of female prairie dogs, continued. Exercise 22.4 described an observational study of the mating behavior and reproductive success of females in one colony of Gunnison's prairie dogs. Here is a table providing the observed counts (top) and expected counts (bottom) for each cell in the two-way table under a null hypothesis of no relationship between mating behavior and reproductive success:

	Number of Mother's Sexual Partners				
	1	2	3	4	5
Gave birth	81 82.37	85 88.05	61 57.75	17 16.1	5 4.73
Did not give birth	6 4.63	8 4.95	0 3.25	0 0.9	0 0.27

With a total of 263 observations, this is a rather large data set. Nonetheless, you should *not* perform a chi-square test on these data. Explain why not.

Uses of the chi-square test

Two-way tables can arise in several ways. The thromboembolism study compared two independent random samples, one of women who had experienced a first thromboembolism (cases) and the other of women who had not (controls). The design of the study fixed the sizes of the two samples. The next example illustrates a different setting, in which all the observations come from just one sample.

EXAMPLE 22.5 Characteristics of a forest

STATE: Forests are complex, evolving ecosystems. For instance, pioneer tree species can be displaced by successional species better adapted to the changing environment. Ecologists mapped a large Canadian forest plot dominated by Douglas fir with an understory of western hemlock and western red cedar. The two-way table below records all 2050 trees in the plot by species and by life stage. Sapling trees (young trees shorter than 1.3 meters) are indicative of the future of a forest.[5]

4 STEP

John Sylvester/Alamy

Tree Species	Dead	Live	Sapling
Western red cedar (RC)	48	214	154
Douglas fir (DF)	326	324	2
Western hemlock (WH)	474	420	88

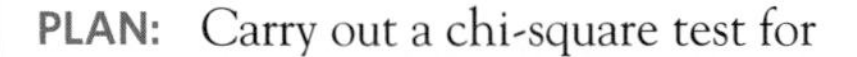

PLAN: Carry out a chi-square test for

H_0: there is no relationship between species and life stage
H_a: there is some relationship between these two variables

Compare column percents or observed versus expected cell counts or X^2 components to see the nature of the relationship.

SOLVE: First check the guideline for use of chi-square. The expected cell counts appear in the Minitab output in Figure 22.5. The smallest expected count is 49.51, which definitely meets the requirement. So we can safely use chi-square. The output shows that

Minitab

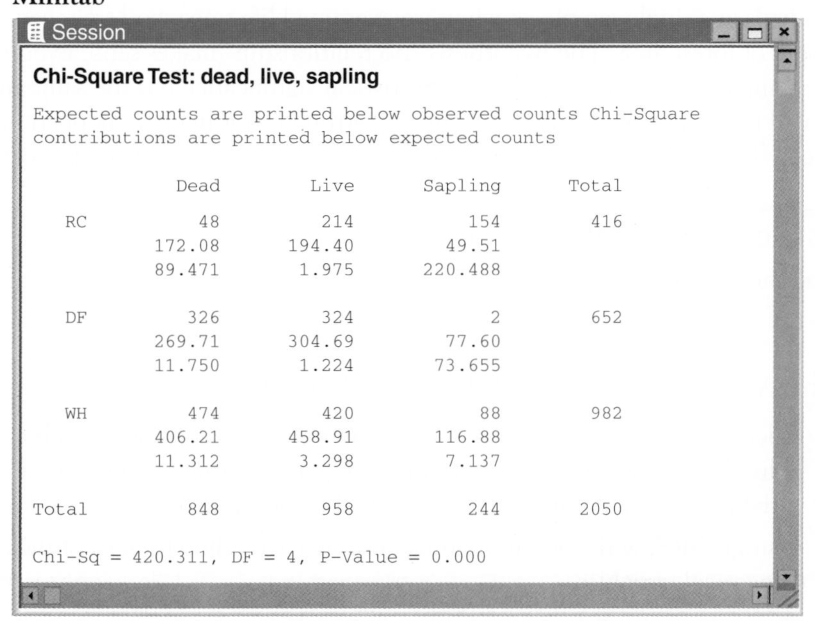

Session

Chi-Square Test: dead, live, sapling

```
Expected counts are printed below observed counts Chi-Square
contributions are printed below expected counts

          Dead      Live   Sapling    Total
  RC        48       214       154      416
        172.08    194.40     49.51
        89.471     1.975   220.488

  DF       326       324         2      652
        269.71    304.69     77.60
        11.750     1.224    73.655

  WH       474       420        88      982
        406.21    458.91    116.88
        11.312     3.298     7.137

Total      848       958       244     2050

Chi-Sq = 420.311, DF = 4, P-Value = 0.000
```

FIGURE 22.5 Minitab output for the two-way table in the forest study, for Example 22.5.

there is a significant relationship ($X^2 = 420.311$, $P < 0.001$). The largest contributor to X^2 comes from the cell for sapling western red cedar (RC) (220 out of 420). There are many more sapling western red cedar trees than would be expected if H_0 was true. The next largest chi-square component is for the RC and dead cell (89 out of 420). Western red cedars appear to have an interesting pattern, with many more sapling trees and far fewer dead ones than expected.

CONCLUDE: We find that 63% of sapling trees are western red cedar trees, 36% are western hemlock, and only 1% are Douglas fir. The species western red cedar is thus taking over this forest. In addition, western red cedars have the lowest percent of dead trees among all three species, with only about 12% of western red cedar trees in this forest found dead. Together, these data suggest that the western red cedar species is a successional species displacing the other established species in this forest. ■

Pay attention to the nature of the data in Example 22.5:

- We do not have three separate samples of trees, one for each species. We have a single plot of forest with 2050 trees in it, each classified in two ways (species and life stage).
- The data cover *all* the trees found in this forest plot. We might regard this as one sample from a population of similar forest areas in the same successional stage. But we might also regard these 2050 trees as the entire population of interest rather than a sample from a larger population.

One of the most useful properties of chi-square is that it tests the null hypothesis "the row and column variables are not related to each other" whenever this hypothesis makes sense for a two-way table. It makes sense when we are comparing a categorical response in two or more samples, as when we compared the history of ERT use for women with or without a diagnosis of thromboembolism. The hypothesis also makes sense when we have data on two categorical variables for the individuals in a single sample, as when we examined life stages and tree species for a sample of forest area. The hypothesis "no relationship" makes sense even if the single sample is an entire population. Statistical significance has the same meaning in all these settings: "A relationship this strong is not likely to happen just by chance." This makes sense whether the data are a sample or an entire population.

USES OF THE CHI-SQUARE TEST

Use the chi-square test to test the null hypothesis

H_0: there is no relationship between two categorical variables

when you have a two-way table from one of these situations:

- Independent SRSs from each of two or more populations, with each individual classified according to one categorical variable. (The other variable says which sample the individual comes from.)
- A single SRS, with each individual classified according to both of two categorical variables.

APPLY YOUR KNOWLEDGE

22.12 Aphid righting behavior, continued. Exercise 22.2 describes a study of aphids dropping from various heights to escape predators. Did the study use a single SRS or multiple SRSs? Would you say that the study represents its own population or that it hopes to represent a larger population?

22.13 Age and HPV infections, continued. Exercise 22.1 describes data on the HPV status of American women in different age groups. Did the study use a single random sample or multiple random samples? Would you say that the study represents its own population or that it hopes to represent a larger population?

22.14 Hearing impairment in dalmatians. Congenital sensorineural deafness is the most common form of deafness in dogs and is often associated with congenital pigmentation deficiencies. A study of hearing impairment in dogs examined over 5000 dalmatians for both hearing impairment and iris color. Figure 22.6 shows the SPSS output for a chi-square test on these data.[6] Dogs with either one or both of their irises blue (a trait due to low iris pigmentation) were labeled "blue-eyed." This is an example of a single sample classified according to two categorical variables (hearing impairment and iris color).

Petra Wegner/Alamy

SPSS

Column Statistics

Hearing * EyeColor Crosstabulation

			EyeColor		
			Brown eyed	**Blue eyed**	**Total**
Hearing	Deaf	Count	324	102	426
		Expected Count	381.7	44.3	426.0
		% within EyeColor	6.8%	18.4%	8.0%
	Unilaterally impaired	Count	988	179	1167
		Expected Count	1045.8	121.2	1167.0
		% within EyeColor	20.7%	32.3%	21.9%
	Bilateral hearing	Count	3467	273	3740
		Expected Count	3351.5	388.5	3740.0
		% within EyeColor	72.5%	49.3%	70.1%
Total		Count	4779	554	5333
		Expected Count	4779.0	554.0	5333.0
		% within EyeColor	100.0%	100.0%	100.0%

Chi-Square Tests

	Value	**df**	**Asymp. Sig. (2-sided)**
Pearson Chi-Square	153.138[a]	2	.000
Likelihood Ratio	133.597	2	.000
Linear-by-Linear Association	153.052	1	.000
N of Valid Cases	5333		

[a] 0 cells (.0%) have expected count less than 5. The minimum expected count is 44.25.

FIGURE 22.6 SPSS output for the two-way table in the study of hearing impairment among dalmatians, for Exercise 22.14.

(a) Describe the differences between the distributions of hearing impairments among brown-eyed and blue-eyed dogs with percents, with a bar graph, and in words.

(b) Verify that the expected cell counts satisfy the requirement for use of chi-square.

(c) Test the null hypothesis that there is no relationship between hearing impairment and iris color. Give a P-value.

(d) How do the observed and expected counts differ? Relate this to your conclusions in (a).

Using a table of critical values*

Software usually finds P-values for us. The P-value for a chi-square test comes from comparing the value of the chi-square statistic with critical values for a *chi-square distribution*.

We saw in Chapter 21 that the chi-square distributions are a family of distributions that take only positive values and are skewed to the right. A specific chi-square distribution is specified by giving its *degrees of freedom*.

CHI-SQUARE DISTRIBUTIONS FOR TWO-WAY TABLES

The chi-square test for a two-way table with r rows and c columns uses critical values from the chi-square distribution with $(r - 1)(c - 1)$ **degrees of freedom.**

The P-value is the area to the right of X^2 under the density curve of this chi-square distribution.

Figure 21.4 (page 523) shows the density curves for three members of the chi-square family of distributions. As the degrees of freedom increase, the density curves become less skewed and larger values become more probable. Table D in the back of the book gives critical values for chi-square distributions. You can use Table D if you do not have software that gives you P-values for a chi-square test.

EXAMPLE 22.6 Using the chi-square table

The two-way table of 4 ERT histories by 2 groups for the thromboembolism study has 4 rows and 2 columns. That is, $r = 4$ and $c = 2$. The chi-square statistic therefore has degrees of freedom

$$(r - 1)(c - 1) = (4 - 1)(2 - 1) = (3)(1) = 3$$

Four of the outputs in Figure 22.2 give 3 as the degrees of freedom.

*This section is optional.

The observed value of the chi-square statistic is $X^2 = 21.268$. Look in the df = 3 row of Table D. The value $X^2 = 21.268$ falls off the table, to the right of the 0.0005 critical value of the chi-square distribution with 3 degrees of freedom. Remember that the chi-square test is always one-sided. So the P-value of $X^2 = 21.268$ is less than 0.0005 and, therefore, highly significant. The outputs in Figure 22.2 show that the P-value is 0.00009, less than 0.0005. ■

df = 3

p	.0005
X^*	17.73

APPLY YOUR KNOWLEDGE

22.15 Age and HPV infections, continued. The Minitab output in Figure 22.3 gives the degrees of freedom for the table of age group and HPV status as `DF = 5`.

(a) Show that this is correct for a table with 2 rows and 6 columns.

(b) Minitab gives the chi-square statistic as `Chi-Square = 40.554`. Use Table D to approximate the P-value of the test. Does this value agree with Minitab's result?

22.16 Aphid righting behavior, continued. The R output in Figure 22.4 gives 3 degrees of freedom for the table in Exercise 22.2.

(a) Verify that this is correct.

(b) The computer gives the value of the chi-square statistic as $X^2 = 11.2554$. Between what two entries in Table D does this value lie? What does the table tell you about the P-value?

The chi-square test and the two-sample *z* test*

One use of the chi-square test is to compare the proportions of successes in any number of groups. When comparing the proportions of successes in only two groups, the data can be summarized in a 2 × 2 table like this one:

	Success	Failure
Group 1		
Group 2		

We then have two inference procedures available: the two-sample z test from Chapter 20 and the chi-square test with 1 degree of freedom for a 2 × 2 table. The null hypotheses tested are

$$H_0: p_1 = p_2$$

for the two-sample z test, where p_1 and p_2 are the proportion of successes in each group, and

$$H_0 : \text{there is no relationship between the row and column variables}$$

for the chi-square test for two-way tables.

He started it! A study of deaths in bar fights showed that in 90% of the cases, the person who died started the fight. You shouldn't believe this. If you killed someone in a fight, what would you say when the police ask you who started the fight? After all, dead men tell no tales.

*This section is optional.

These two tests always agree. In fact, the chi-square statistic X^2 is just the square of the z statistic, and the P-value for X^2 is exactly the same as the two-sided P-value for z. We recommend using the z test to compare two proportions, because it gives you the choice of a one-sided test and is related to a confidence interval for the difference $p_1 - p_2$.

APPLY YOUR KNOWLEDGE

22.17 Effectiveness of endovascular therapy. New surgical procedures are sometimes performed without evidence from randomized controlled experiments. Endovascular therapy has been increasingly used to supplement treatment with a clot-dissolving drug (t-PA) in patients with an acute ischemic stroke. It is not clear, however, if the addition of this surgical intervention truly benefits patients. A recent study randomly assigned patients with acute ischemic stroke to receive either a t-PA injection alone or a t-PA injection supplemented with endovascular therapy. Patients were examined 90 days later for evidence of functional independence. Of the 434 patients receiving t-PA and endovascular therapy, 177 were functionally independent at 90 days, compared with 86 of the 222 patients receiving t-PA alone.[7] We can test the hypothesis of "no difference" between the two groups in either of two ways: using the two-sample z statistic or using the chi-square statistic.

(a) Check the conditions required for both tests, given in the boxes on pages 492 and 546. The conditions are very similar, as they ought to be.

(b) State the null hypothesis with a two-sided alternative and carry out the z test. What is the P-value, exactly from software or approximately from the bottom of Table C?

(c) Present the data in a 2×2 table. Use the chi-square test to test the hypothesis from (b). Verify that the X^2 statistic is the square of the z statistic. Use software or Table D to verify that the chi-square P-value agrees with the z result (up to the accuracy of the tables if you do not use software).

(d) What do you conclude about the effectiveness of supplementing t-PA with endovascular therapy in patients with an acute ischemic stroke?

CHAPTER 22 SUMMARY

- The **chi-square test** for a two-way table tests the null hypothesis H_0 that there is no relationship between the row variable and the column variable. The alternative hypothesis H_a says that there is some relationship but does not say what kind.
- The test compares the observed counts of observations in the cells of the table with the counts that would be expected if H_0 was true. The **expected count** in any cell is

$$\text{expected count} = \frac{\text{row total} \times \text{column total}}{\text{table total}}$$

The **chi-square statistic** is

$$X^2 = \sum \frac{(\text{observed count} - \text{expected count})^2}{\text{expected count}}$$

- The chi-square test compares the value of the statistic X^2 with critical values from the **chi-square distribution** with $(r - 1)(c - 1)$ **degrees of freedom.** Large values of X^2 are evidence against H_0, so the P-value is the area under the chi-square density curve to the right of X^2.
- The chi-square distribution is an approximation of the distribution of the statistic X^2. You can safely use this approximation when all expected cell counts are at least 1 and no more than 20% are less than 5.
- If the chi-square test finds a statistically significant relationship between the row and column variables in a two-way table, use descriptive data analysis to explain the nature of the relationship. You can do this by comparing well-chosen percents, comparing the observed counts with the expected counts, and looking for the largest **components of the chi-square statistic.**

THIS CHAPTER IN CONTEXT

In Chapter 19 we computed confidence intervals and tested hypotheses about the proportion p of successes in one population. In Chapter 20, we expanded the use of z procedures to situations where we want to compare the proportions of successes p_1 and p_2 in two distinct populations. In this chapter we describe how the chi-square procedure can be used to test the hypothesis that several populations have the same proportion of successes.

The chi-square procedures are very versatile and can also be used for categorical variables beyond the simple case of binary success/failure. In Chapter 21, we saw how the chi-square test for goodness of fit can be used to test the hypothesis that a population breaks down into set proportions of defined outcomes (the way a pie chart breaks down into its slices). In this chapter, we examine the use of the chi-square method for categorical data that can be represented in two-way tables. That is, we can break down the data according to two categorical variables, each with any number of levels (the r rows and c columns in the table).

We used two-way tables to describe relationships between two *categorical* variables in the optional Chapter 5. We examined the marginal distribution of each variable and the conditional distribution of one variable among only individuals who have a given value of the other variable. In the optional Chapter 10 we addressed similar situations from a probability point of view and described the concept of independence. These ideas are directly relevant to the chi-square test for two-way tables, which is sometimes also called the chi-square test for independence.

As in Chapter 5, we must be careful not to assume that an observed association implies a cause-and-effect relation. A lurking variable may influence both variables studied and create the appearance of an association or change the nature of the association. Simpson's paradox is an extreme example in which an association that holds for all of several groups reverses direction when these groups are combined into a single group.

CHECK YOUR SKILLS

Use the following information for Exercises 22.18 through 22.27. S-Adenosyl-l-methionine (SAMe) is a dietary supplement used to treat depression. A study randomly assigned smokers who wanted to quit smoking to one of three groups: a low dose of SAMe, a high dose of SAMe, or a placebo. After 8 weeks of treatment and behavioral smoking cessation intervention, smoking abstinence was assessed biochemically. Here are the findings:[8]

	Abstinent	Still smoking
Placebo	7	33
Low-dose SAMe	7	33
High-dose SAMe	5	35

22.18 The percent of subjects abstaining from smoking in the placebo group is about

(a) 36.8%. (b) 17.5%. (c) 5.8%.

22.19 The percent of subjects abstaining from smoking in the placebo group is

(a) higher than the percent of subjects abstaining from smoking in the high-dose SAMe group.
(b) about the same as the percent of subjects abstaining from smoking in the high-dose SAMe group.
(c) lower than the percent of subjects abstaining from smoking in the high-dose SAMe group.

22.20 The expected count of subjects abstaining from smoking in the placebo group is about

(a) 6.3. (b) 17.5. (c) 19.

22.21 The component of the chi-square statistic for the cell of subjects abstaining from smoking in the placebo group is about

(a) 6.33 (b) 0.77. (c) 0.07.

22.22 The degrees of freedom for the chi-square test for this two-way table are

(a) 2. (b) 3. (c) 6.

22.23 The null hypothesis for the chi-square test for this two-way table is

(a) There are equal proportions of subjects abstaining and still smoking in the placebo group.
(b) There is no difference in the proportions of subjects abstaining from smoking among the three treatments.
(c) There are equal numbers of subjects abstaining and still smoking.

22.24 The alternative hypothesis for the chi-square test for this two-way table is

(a) The proportions of subjects abstaining from smoking are not the same for the three treatments.
(b) The placebo group has the lowest proportion of subjects abstaining from smoking.
(c) The high-dose SAMe group has the lowest proportion of subjects abstaining from smoking.

22.25 Software gives the chi-square statistic $X^2 = 0.5003$ for this table. Based on software or Table D, we can say that the P-value is

(a) greater than 0.05.
(b) between 0.01 and 0.05.
(c) less than 0.01.

22.26 The most important fact that allows us to trust the results of the chi-square test is that

(a) the study is large, 120 individuals in all.
(b) all the cell counts are greater than 5.
(c) all the expected cell counts are greater than 5.

22.27 What can we reasonably conclude from the study findings?

(a) Smokers who take the high dose of SAMe for 8 weeks are significantly more likely to abstain from smoking than smokers who take a lower dose or a placebo.
(b) The study fails to find significant evidence that SAMe in either a low or a high dose performs differently from a placebo.
(c) Smokers who take the high dose of SAMe for 8 weeks are significantly less likely to abstain from smoking than smokers who take a lower dose or a placebo.

CHAPTER 22 EXERCISES

If you have access to software or a graphing calculator, use it to speed your analysis of the data in these exercises. Exercises 22.28 to 22.32 are suitable for hand calculation if necessary. About half of the remaining exercises provide software output for their respective tests.

22.28 Foot health. Foot pain is a very common musculoskeletal complaint in the United States, especially among older adults. Foot pain is more frequent among women than among men, possibly because men and women wear different types of shoes over their lifetimes. Example 5.1 (page 121) describes the findings of a survey based on the Framingham cohort, a group now made up of mainly older individuals (65 years of age, on average). The participants were asked to select from a list which type of footwear they had worn most regularly in the past. The answers were then categorized as shoes providing good, average, or poor structural foot support. Table 5.1 (page 122) presents the results for the men and the women in the study.

(a) What percent of men wore shoes offering poor structural support? What percent of women did?
(b) We want to see if there is a relationship between gender and shoe type. Can we safely use the chi-square test?
(c) What null and alternative hypotheses does X^2 test? What are the degrees of freedom for this test?
(d) Software gives $X^2 = 1375.8$. Is this evidence of a significant difference in shoe support type between men and women? Use Table D or software to obtain the P-value of the test.

22.29 Tell us if you use cocaine. Sample surveys on sensitive issues can give different results depending on how the question is asked. A University of Wisconsin study divided 2400 respondents into 3 groups at random. All were asked if they had ever used cocaine. One group of 800 was interviewed by phone; 21% said they had used cocaine. Another 800 people were asked the question in a one-on-one personal interview; 25% said "yes." The remaining 800 were allowed to make an anonymous written response; 28% said "yes."[9]

(a) Make a two-way table of answers (yes or no) by survey method.
(b) Can we safely use the chi-square test? What null and alternative hypotheses does X^2 test? What are the degrees of freedom for this test?
(c) Compute the chi-square component for each cell and give X^2. Use Table D or software to obtain the P-value of the test.
(d) What do you conclude from these data?

22.30 Cranberry juice for preventing urinary tract infections. American cranberry (*Vaccinium macrocarpon*) was used medicinally by Native Americans for the treatment of bladder and kidney ailments. A randomized controlled study was designed to test whether regular drinking of cranberry juice can prevent the recurrence of urinary tract infections (UTIs) in women. One hundred and fifty women with a urinary tract infection were treated with antibiotic and then randomly assigned to one of three groups. One group drank cranberry juice concentrate daily for six months; another group took a lactobacillus drink daily for six months (a lactose-fermenting bacterium thought to help inhibit the growth of UTI-causing bacteria); the last group served as the control group and drank neither cranberry juice nor lactobacillus drinks for six months. After six months, the number of women in each group with recurring symptomatic UTI (defined as one or more new infections) was recorded. Here are the results:[10]

	Outcome		
Treatment	**Recurring UTI**	**No new UTI**	**Total**
Cranberry juice	8	42	50
Lactobacillus drink	19	30	49
Control	18	32	50

(a) Find the percent of women experiencing no new UTI in each of the three treatment groups. Make a graph to compare these percents. Describe the association between treatment and outcome.
(b) Explain in words what the null hypothesis for the chi-square test says about treatment outcome. Verify that the conditions are met for a chi-square test.
(c) Compute the chi-square statistic and find the P-value for this test. Examine the X^2 components to confirm the pattern you saw in (a). What is your overall conclusion? Interestingly, recent pharmacological studies have shown that some cranberry compounds prevent bacteria from binding to the membrane of host cells.

22.31 Stress and heart attacks. You read a newspaper article that describes a study of whether stress management can help reduce heart attacks. The 107 subjects all had reduced blood flow to the heart and so were at risk of a heart attack. They were assigned at random to one of three groups. The article goes on to say:

One group took a four-month stress management program, another underwent a four-month exercise program and the third received usual heart care from their personal physicians.

In the next three years, only three of the 33 people in the stress management group suffered "cardiac events," defined as a fatal or non-fatal heart attack or a surgical procedure such as a bypass or angioplasty. In the same period, seven of the 34 people in the exercise group and 12 out of the 40 patients in usual care suffered such events.[11]

(a) Use the information in the news article to make a two-way table that describes the study results. What are the success rates of the three treatments in preventing cardiac events?
(b) Find the expected cell counts under the null hypothesis that there is no difference among the treatments. Verify

that the expected counts meet our guideline for use of the chi-square test.

(c) Is there a significant difference among the success rates for the three treatments? Compute the chi-square statistic, use Table D to approximate the P-value of the test, and conclude.

22.32 HIV vaccine boost. Creating an effective HIV vaccine has been very challenging. A new approach combines a DNA vaccine with a delivery system using electrical impulses (electroporation) to boost the immune response. A preliminary study of the method's efficacy was conducted on adult volunteers. Each volunteer received one of five possible treatment options and a follow-up clinical examination to assess his or her immune response to the vaccine. Here is a two-way table that shows the results of the clinical exam after two immunization sessions:[12]

	Immune response	No immune response
0.2 mg vaccine via electroporation	3	5
1 mg vaccine via electroporation	7	1
4 mg vaccine via electroporation	6	2
4 mg vaccine intramuscularly	0	8
Saline solution via electroporation	0	8

We would like to know if the treatment type significantly influences the proportion of individuals showing an immune response. Can we use the chi-square test with these data? Check the conditions for inference and explain your answer.

22.33 Sex ratios and birth order in geese. 4 STEP Is sex determination truly a random process? The laws of genetics would suggest so because of the random distribution of X and Y chromosomes among sperm cells. However, physiological or environmental factors could favor one gender over the other under specific circumstances. A field biologist surveyed a random sample of lesser snow geese nests containing exactly four eggs. For each egg resulting in a live gosling, the laying order and the gender of the gosling were recorded. Here are the data:[13]

	Order in Which Egg is Laid				
Sex	1	2	3	4	Total
Male	17	16	7	5	45
Female	10	9	17	14	50
Total	27	25	24	19	95

Is there evidence of a relationship between sex and laying order in lesser snow geese?

22.34 Sickle-cell anemia and malaria. 4 STEP Sickle-cell anemia is a hereditary condition that is more common among blacks and can cause medical problems. A landmark study examined the relationship between sickle-cell anemia and malaria, a parasitic disease common in Africa. The study tested 543 African children for the sickle-cell gene and also for malaria. Here are the results:[14]

	Heavy Malaria Infection	
	Yes	No
Sickle-cell gene	36	100
No sickle-cell gene	152	255

Is there good evidence of a relationship between the sickle-cell gene and heavy malaria infection? It is now known that individuals with the sickle-cell gene have abnormally rigid red blood cells, making it more difficult for the parasite responsible for malaria to multiply and complete its life cycle.

22.35 Tongue piercing and cracked teeth. 4 STEP Researchers compared tooth health and periodontal damage in a group of 46 young adult males wearing a tongue piercing and a control group of 46 young adult males without tongue piercing. They found that 38 individuals in the tongue-piercing group and 26 in the control group had enamel cracks.[15] Does the study provide significant evidence of an association between enamel cracks and tongue piercing? Can you conclude that there is a causal relationship between tongue piercing and enamel cracks?

22.36 Comparing acne treatments. 4 STEP Acne is a common skin disease that affects most adolescents and can continue into adulthood. A study compared the effectiveness of three acne treatments and a placebo, all in gel form, applied twice daily for twelve weeks. The study's 517 teenage volunteers were randomly assigned to one of the four treatments. Success was assessed as clear or almost clear skin at the end of the twelve-week period. Figure 22.7 gives counts and chi-square output for this study.[16] Is there significant evidence that the four treatments perform differently? If so, how do they compare?

22.37 Treating cocaine addiction. 4 STEP Cocaine produces short-term feelings of physical and mental well-being followed by feelings of tiredness and depression once the

Minitab

Session

Chi-Square Test: epiduo, adapalene, benzoyl peroxide, placebo

```
Expected counts are printed below observed counts
Chi-Square contributions are printed below expected counts

                                      benzoyl
              epiduo    adapalene    peroxide    placebo    Total
Success           32           18          18          4       72
               20.75        20.61       20.75       9.89
               6.099        0.331       0.365      3.506

Failure          117          130         131         67      445
              128.25       127.39      128.25      61.11
               0.987        0.054       0.059      0.567

Total            149          148         149         71      517

Chi-Sq = 11.967, DF = 3, P-Value = 0.007
```

FIGURE 22.7 Minitab output for the acne data of Exercise 22.36.

drug wears off. This cycle of positive and negative feelings is the root problem of addiction and can be very hard to break. Perhaps giving cocaine addicts a medication that fights depression can help them break the addiction.

A three-year study compared an antidepressant called desipramine with lithium (a standard treatment for cocaine addiction) and a placebo. The subjects were 72 chronic users of cocaine who wanted to break their drug habit. Twenty-four of the subjects were randomly assigned to each treatment. Here are the counts and proportions of the subjects who avoided relapse into cocaine use during the study:[17]

Group	Treatment	Subjects	No relapse	Proportion
1	Desipramine	24	14	0.583
2	Lithium	24	6	0.250
3	Placebo	24	4	0.167

Does data analysis suggest that desipramine is more successful than the other two treatments? Are there significant differences among the outcomes for the treatments?

22.38 Preventing strokes. Individuals who have experienced a stroke are at an increased risk of having another stroke. Prevention is particularly important for high-risk populations. The Second European Stroke Prevention Study asked whether daily doses of an anti-blood-clotting agent, such as aspirin or dipyridamole, would help prevent strokes among patients who had just experienced a first stroke. Volunteer patients were randomly assigned to one of 4 treatment options, including a placebo. Researchers recorded how many patients experienced another stroke and how many died of a stroke during the two years of the study. Figure 22.8 gives the observed and expected counts for this study, along with the chi-square calculations. Is there a relationship between treatment and strokes? Write a careful summary of your overall findings.

22.39 More on preventing strokes. The study described in the previous exercise also recorded the number of patients in each group who died following a stroke during the two years of the study. Is there a relationship between treatment and stroke-related deaths? Write a careful summary of your overall findings.

22.40 Do angry people have more heart disease? People who get angry easily tend to have more heart disease. That's the conclusion of a study that followed a random sample of 12,986 people from three locations for about four years. All subjects were free of heart disease at the beginning of the study. The subjects took the Spielberger Trait Anger Scale test, which measures how prone a person is to sudden anger. Here are data for the 8474 people in the sample who had normal blood pressure.[18] CHD stands for "coronary heart disease." This includes people who had heart attacks and those who needed medical treatment for heart disease.

Minitab

```
Session

            Stroke  NoStroke  Total  |            Death  NoDeath  Total
Placebo        250      1399   1649  | Placebo      202     1447   1649
            205.81   1443.19         |           189.08  1459.92

Aspirin        206      1443   1649  | Aspirin      182     1467   1649
            205.81   1443.19         |           189.08  1459.92

Dipyr.         211      1443   1654  | Dipyr.       188     1466   1654
            206.44   1447.56         |           189.65  1464.35

Both           157      1493   1650  | Both         185     1465   1650
            205.94   1444.06         |           189.19  1460.81

Total          824      5778   6602  | Total        757     5845   6602

ChiSq = 9.487 + 1.353 +              | ChiSq = 0.883 + 0.114 +
        0.000 + 0.000 +              |         0.265 + 0.034 +
        0.101 + 0.014 +              |         0.014 + 0.002 +
       11.629 + 1.658 = 24.243       |         0.093 + 0.012 = 1.418
df = 3, p = 0.000                    | df = 3, p = 0.701
```

FIGURE 22.8 Minitab output for the experimental results of Exercise 22.38.

	Low anger	Moderate anger	High anger	Total
CHD	53	110	27	190
No CHD	3057	4621	606	8284
Total	3110	4731	633	8474

Do these data support the study's conclusion about the relationship between anger and heart disease?

22.41 Echinacea for the common cold? The National Center for Complementary and Alternative Medicine (NCCAM, a branch of the NIH) offers on its website comprehensive information about various complementary and alternative healing practices, including a summary of research findings for each practice. Echinacea is a flowering plant related to daisies and has a long history of medicinal use starting with Native Americans. Today it is the top-selling herbal remedy in the United States, mainly used against the common cold. In a double-blind experiment, healthy volunteers agreed to be exposed to common-cold-causing rhinovirus type 39. Three types of echinacea extracts were produced (labeled E1, E2, and E3 in the table below). Subjects received either an echinacea extract or a placebo ("P") over two different periods: for seven consecutive days before viral exposure (infection and symptom prevention) and for seven consecutive days following exposure (symptom treatment). Among the variables studied were whether or not a subject became infected following viral exposure and whether or not infected subjects developed cold symptoms. Here are the results:[19]

Treatment	Infected	Not infected	Cold symptoms	No symptoms
E1 then E1	40	5	25	15
E2 then E2	42	10	24	18
E3 then E3	48	4	24	24
P then E1	43	5	27	16
P then E2	44	4	33	11
P then E3	44	7	28	16
P then P	88	15	58	30

Running a chi-square test for each variable separately gave $X^2 = 4.745$ and $X^2 = 7.120$ for the first and the second variable, respectively. Write a careful summary of the overall findings from this study.

22.42 Melanoma in men and women. Intense or repetitive sun exposure can lead to melanoma, a rare but severe form of skin cancer. Patterns of sun exposure, however, differ

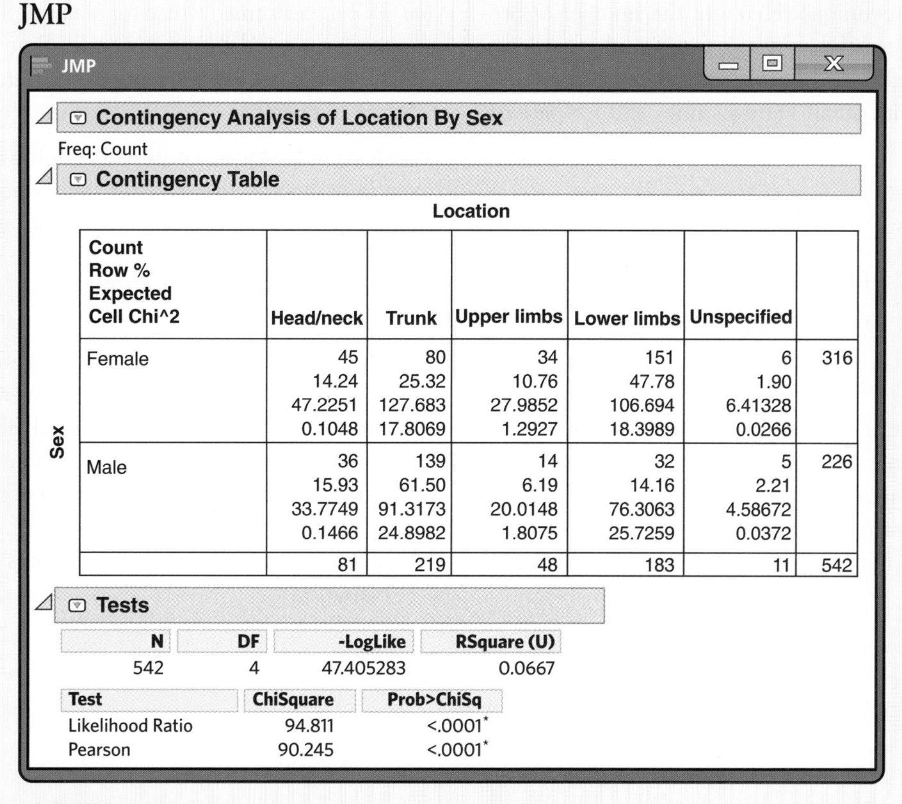

JMP

JMP

Contingency Analysis of Location By Sex

Freq: Count

Contingency Table

Sex	Count Row % Expected Cell Chi^2	Head/neck	Trunk	Upper limbs	Lower limbs	Unspecified	
		Location					
	Female	45 14.24 47.2251 0.1048	80 25.32 127.683 17.8069	34 10.76 27.9852 1.2927	151 47.78 106.694 18.3989	6 1.90 6.41328 0.0266	316
	Male	36 15.93 33.7749 0.1466	139 61.50 91.3173 24.8982	14 6.19 20.0148 1.8075	32 14.16 76.3063 25.7259	5 2.21 4.58672 0.0372	226
		81	219	48	183	11	542

Tests

N	DF	-LogLike	RSquare (U)
542	4	47.405283	0.0667

Test	ChiSquare	Prob>ChiSq
Likelihood Ratio	94.811	<.0001*
Pearson	90.245	<.0001*

FIGURE 22.9 JMP output for the epidemiological data of Exercise 22.42.

between men and women. A study of cutaneous malignant melanoma in the Italian population recorded the distribution of melanoma by body location for men and for women.[20] Figure 22.9 displays the observed and expected counts for this study and the chi-square test from JMP. Write a careful summary of your overall findings.

22.43 Carcinogenicity of electromagnetic fields. The U.S. National Toxicology Program (NTP) studies the toxicity and carcinogenicity of agents potentially causing a risk to human health. Electromagnetic fields around electrical installations are harmless in theory, but they might have a long-term effect on health (for instance, in triggering tumors). Observational studies in human populations have suggested a potential association.

An NTP study looked at the occurrence of cancers in otherwise healthy rats after two years of daily exposure to 60-hertz electromagnetic fields of various intensities. Here are the numbers of male and female rats with a tumor after the two-year study period.[21] Note that 2 gauss is approximately 1000-fold greater than what is considered high exposure for humans.

	Male rats	Female rats
Control: no exposure	16 ($n = 99$)	19 ($n = 100$)
0.02-gauss exposure	31 ($n = 100$)	22 ($n = 100$)
2-gauss exposure	30 ($n = 100$)	22 ($n = 100$)

In rats not exposed to an electromagnetic field (sample size in parentheses), is there significant evidence of a difference in the tumor rates between males and females?

22.44 Carcinogenicity study, continued. Use the data from the previous exercise to answer the following questions.

(a) Is there significant evidence of a relationship between tumor rate and electromagnetic exposure in male rats?

(b) Is there significant evidence of a relationship between tumor rate and electromagnetic exposure in female rats?

(c) Write a brief summary of your findings from this and the previous exercise.

22.45 Kidney stones. A study compared the success rates of two different procedures for removing kidney stones: open surgery and percutaneous nephrolithotomy (PCNL), a minimally invasive technique. Here are the number of procedures that were successful or not at getting rid of patients' kidney stones for each type of procedure. A separate table is given for patients with small kidney stones and for patients with large stones.[22]

Small stones

	Open surgery	PCNL
Success	81	234
Failure	6	36

Large stones

	Open surgery	PCNL
Success	192	55
Failure	71	25

(a) Give the percent of successful procedures of each type for small kidney stones. Do the same for large kidney stones. What trend emerges?
(b) Is there a significant difference between the two methods for small stones? For large stones? Use either a chi-square or a z test.
(c) PCNL performed worse for *both* small and large kidney stones, yet it did better overall. That sounds impossible. Explain carefully, referring to the data, how this paradox can happen.

22.46 Kidney stones, continued. Going back to the previous exercise, combine the data for small and large stones into a single 2 × 2 table. Now answer the following questions using the pooled data.

(a) Which procedure had the higher overall success rate? Is the difference significant?
(b) How do your conclusions in this exercise compare with your conclusions in the previous exercise? This is an example of **Simpson's paradox**, when pooling categorical data that are not homogeneous can create a confounding variable and reverse the direction of an association (Chapter 5 describes this paradox in more detail). It is also a reminder that association does not imply causation.

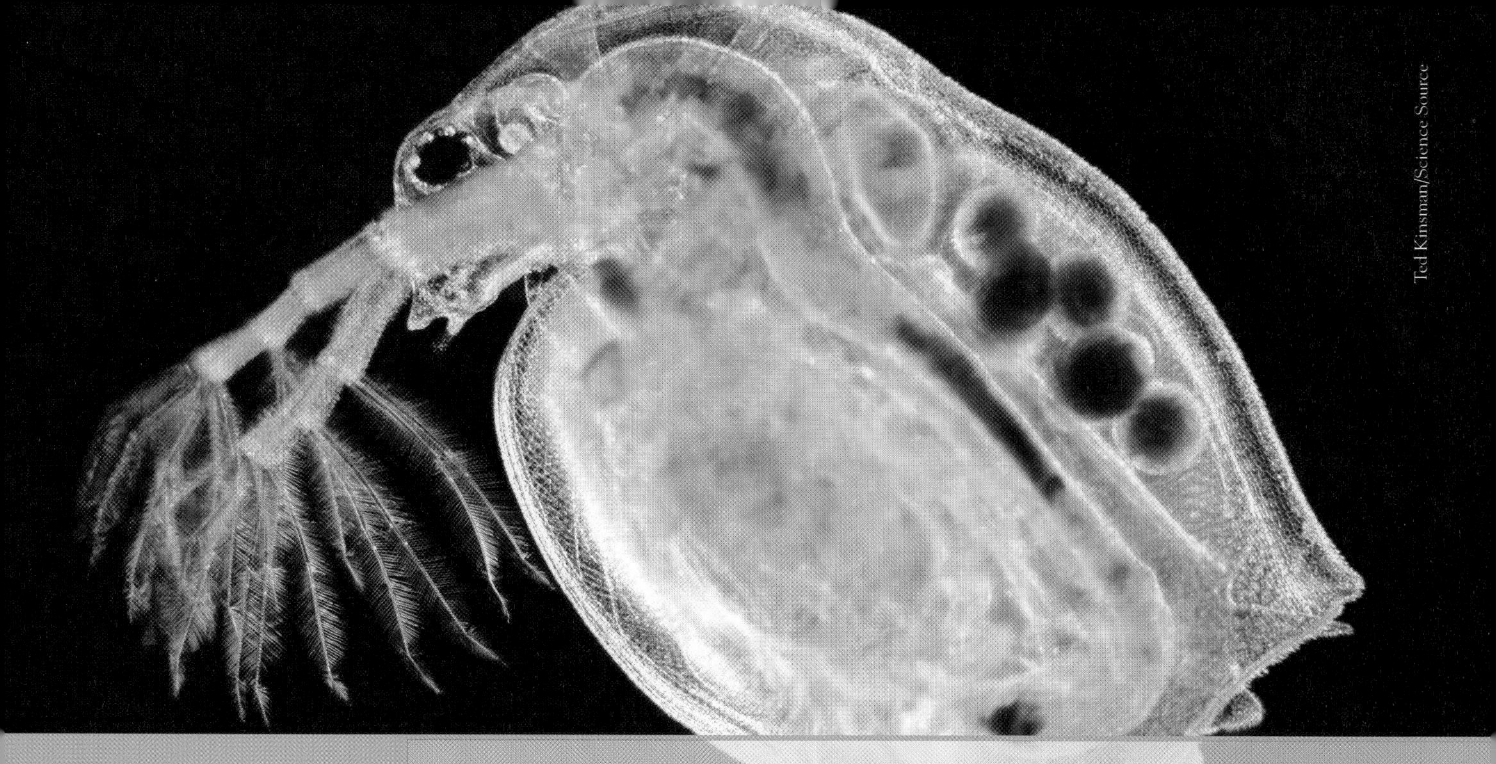

CHAPTER 23 Inference for Regression

When a scatterplot shows a linear relationship between a quantitative explanatory variable x and a quantitative response variable y, we can use the least-squares regression line fitted to the data to predict y for a given value of x. We discussed the approach in great detail in Chapters 3 and 4. When the data are a sample from a larger population, we need statistical inference to answer the following questions about the population:

- Is there really a linear relationship between x and y in the population, or might the pattern we see in the scatterplot plausibly arise just by chance?
- What is the slope (rate of change) that relates y to x in the population, including a margin of error for our estimate of the slope?
- If we use the least-squares line to predict y for a given value of x, how accurate is our prediction (again, with a margin of error)?

This chapter shows you how to answer these questions. Here is an example we will explore.

IN THIS CHAPTER WE COVER...

EXAMPLE 23.1 Grading oysters

STATE: Oysters are categorized for retail as small, medium, or large based on their volume. The grading process is slow and expensive when done by hand. A computer reconstruction of oyster volume is designed based on two-dimensional (2D) image processing

TABLE 23.1 Actual volume and computer reconstruction of 30 oysters

Actual	2D	3D	Actual	2D	3D	Actual	2D	3D
13.04	47.907	5.136699	10.53	31.216	3.942783	10.95	37.156	4.707532
11.71	41.458	4.795151	10.84	41.852	4.052638	7.97	29.070	3.019077
17.42	60.891	6.453115	13.12	44.608	5.334558	7.34	24.590	2.768160
7.23	29.949	2.895239	8.48	35.343	3.527926	13.21	48.082	4.945743
10.03	41.616	3.672746	14.24	47.481	5.679636	7.83	32.118	3.138463
15.59	48.070	5.728880	11.11	40.976	4.013992	11.38	45.112	4.410797
9.94	34.717	3.987582	15.35	65.361	5.565995	11.22	37.020	4.558251
7.53	27.230	2.678423	15.44	50.910	6.303198	9.25	39.333	3.449867
12.73	52.712	5.481545	5.67	22.895	1.928109	13.75	51.351	5.609681
12.66	41.500	5.016762	8.26	34.804	3.450164	14.37	53.281	5.292105

Norbert Bieberstein/istockphoto.com

of the oysters. We want to know if this method is a good predictor of actual oyster volume. The results for 30 oysters are displayed in Table 23.1. Actual volumes are expressed in cubic centimeters (cm^3) and 2D reconstructions in thousands of pixels.[1]

PLAN: Make a scatterplot. If the relationship appears linear, use correlation and regression to describe it. Finally, ask whether there is a *statistically significant* relationship between 2D reconstruction and actual volume.

scatterplot

SOLVE (first part): Chapters 3 and 4 introduced the data analysis that must come before inference. Let's first review it. Figure 23.1 is a **scatterplot** of the oyster data. Plot the explanatory variable (2D reconstruction) horizontally and the response variable (actual volume) vertically. Look for the form, direction, and strength of the relationship, as well as for outliers or other deviations. There is a strong positive linear relationship, with no extreme outliers or potentially influential observations.

correlation

Because the scatterplot shows a roughly linear (straight-line) pattern, the **correlation** describes the direction and strength of the relationship. The correlation between

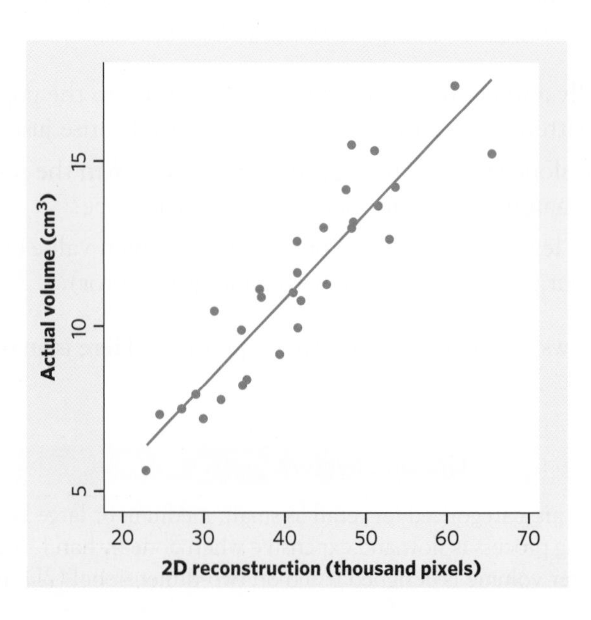

FIGURE 23.1 Scatterplot of the actual volumes of 30 oysters against their 2D reconstructions using a computer-imaging program, with the least-squares regression line, for Example 23.1.

actual volume and 2D reconstruction is $r = 0.9198$. We are interested in predicting the response from information about the explanatory variable. So we find the **least-squares regression line** for predicting actual volume from 2D reconstruction. The equation of the regression line is

least-squares line

$$\hat{y} = a + bx$$
$$= 0.3367 + 0.2649x$$

CONCLUDE (first part): The 2D reconstruction method does provide a good way of assessing actual oyster volumes. Because $r^2 = 0.846$, about 85% of the variation in actual oyster volumes is explained by 2D reconstruction. Prediction of actual oyster volumes will be reasonably accurate. We next ask whether this relationship, observed on only 30 oysters, is statistically significant. We must now develop tools for inference in the regression setting. ■

Conditions for regression inference

We can fit a regression line to *any* data relating two quantitative variables, though the results are useful *only* if the scatterplot shows a linear pattern. Statistical inference requires more detailed conditions. Because the conclusions of inference always concern some *population*, the conditions describe the population and how the data are produced from it. The slope b and intercept a of the least-squares line are *statistics*. That is, we calculated them from the sample data. These statistics would take somewhat different values if we repeated the analysis with different oysters. To do inference, think of a and b as estimates of unknown *parameters* that describe the population of all oysters.

CONDITIONS FOR REGRESSION INFERENCE

We have n pairs of observations on an explanatory variable x and a response variable y. Our goal is to study or predict the behavior of y for given values of x.

- For any fixed value of x, the response y varies according to a **Normal distribution.** Repeated responses y are **independent** of each other.
- The mean response μ_y has a **straight-line relationship** with x given by a **population regression line**

$$\mu_y = \alpha + \beta x$$

  The slope β and intercept α are unknown parameters.
- The **standard deviation** of y (call it σ) is the same for all values of x. The value of σ is unknown.

There are thus three population parameters that we must estimate from the data: α, β, and σ.

First of all, notice the important distinction between x and y. Unlike correlation, regression treats the two quantitative variables very differently. The

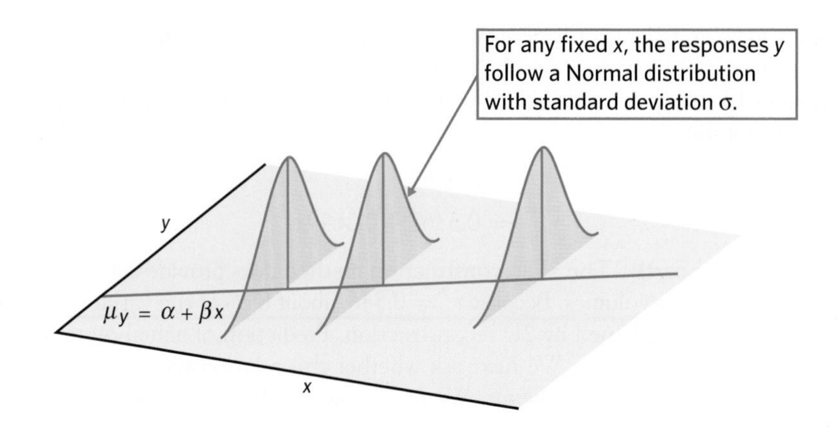
For any fixed x, the responses y follow a Normal distribution with standard deviation σ.
y
μy = α + βx
x

FIGURE 23.2 The nature of regression data when the conditions for inference are met. The line is the population regression line, which shows how the mean response μ_y changes as the explanatory variable x changes. For any fixed value of x, the observed response y varies according to a Normal distribution having mean μ_y and standard deviation σ.

distinction is particularly important for regression inference, because predictions are valid only for y, and the conditions for inference all concern the response variable y.

These conditions say that in the population there is "on average" a straight-line relationship between y and x. The population regression line $\mu_y = \alpha + \beta x$ says that the *mean* response μ_y moves along a straight line as the explanatory variable x changes. We can't observe the population regression line. The values of y that we do observe vary about their means according to a Normal distribution. If we hold x fixed and take many observations on y, the Normal pattern will eventually appear in a stemplot or histogram. In practice, we observe y for many different values of x, so that we see an overall linear pattern formed by points scattered about the population regression line. The standard deviation σ determines whether the points fall close to the population regression line (small σ) or are widely scattered (large σ).

Figure 23.2 shows the nature of regression data in picture form. The line in the figure is the population regression line. The mean of the response y moves along this line as the explanatory variable x takes different values. The Normal curves are the distribution of y at different fixed values of x. All the curves have the same σ, so the variability of y is the same for all values of x. You should check the conditions for inference when you do inference about regression. We will see later how to do that.

Estimating the parameters

The first step in inference is to estimate the unknown parameters α, β, and σ.

ESTIMATING THE POPULATION REGRESSION LINE

When the conditions for regression are met and we calculate the least-squares line $\hat{y} = a + bx$, the slope b of the least-squares line is an unbiased estimator of the population slope β, and the intercept a of the least-squares line is an unbiased estimator of the population intercept α.

EXAMPLE 23.2 Grading oysters: slope and intercept

The data in Figure 23.1 satisfy the condition of scatter about an invisible population regression line. The least-squares line is $\hat{y} = 0.3367 + 0.2649x$. The slope is particularly important. A *slope is a rate of change*. The population slope β says how much larger an average actual volume is for oysters with one thousand more pixels in their 2D reconstruction. Because $b = 0.2649$ estimates the unknown β, we estimate that, on average, actual volume is about 0.26 cm^3 larger for each additional thousand pixels in 2D reconstruction.

We need the intercept $a = 0.3367$ to draw the line, but it has no statistical meaning in this example. No oyster had a 2D reconstruction less than 22 thousand pixels, so we have no data near $x = 0$. Newly formed oysters are not sold for retail, so that we will never observe x close to 0. ■

The remaining parameter is the standard deviation σ, which describes the variability of the response y about the population regression line. The least-squares line estimates the population regression line. So the **residuals** estimate how much y varies about the population line. Recall that the residuals are the vertical deviations of the data points from the least-squares line:

residuals

$$\begin{aligned} \text{residual} &= \text{observed } y - \text{ predicted } y \\ &= y - \hat{y} \end{aligned}$$

There are n residuals, one for each data point. Because σ is the standard deviation of responses about the population regression line, we estimate it by a sample standard deviation of the residuals. We call this sample standard deviation the *regression standard error* to emphasize that it is estimated from data. The residuals from a least-squares line always have a mean of zero. That simplifies their standard error.

REGRESSION STANDARD ERROR

The **regression standard error** is

$$\begin{aligned} s &= \sqrt{\frac{1}{n-2}\sum \text{residual}^2} \\ &= \sqrt{\frac{1}{n-2}\sum (y - \hat{y})^2} \end{aligned}$$

Use s to estimate the standard deviation σ of responses about the mean given by the population regression line.

Because we use the regression standard error so often, we just call it s. Notice that s^2 is an average of the squared deviations of the data points from the line, so it qualifies as a variance. We average the squared deviations by dividing by $n - 2$, 2 less than the number of data points. It turns out that if we know s and $n - 2$ of the n residuals, the other two residuals are determined. That is, $n - 2$ are the **degrees of freedom** of s. We first encountered the idea of degrees of freedom in the case of

degrees of freedom

the ordinary sample standard deviation of n observations, which has $n - 1$ degrees of freedom. Now we observe two variables rather than one, and the proper degrees of freedom are $n - 2$ rather than $n - 1$.

Calculating s is unpleasant. You must find the predicted response for each x in your data set, then the residuals, and then s. In practice, you will use software that does this arithmetic instantly. Nonetheless, here is an example to help you understand the standard error s.

EXAMPLE 23.3 Grading oysters: residuals and standard error

Table 23.1 shows that the first oyster studied had 47.907 thousand pixels in 2D reconstruction and an actual volume of 13.04 cm^3. The predicted actual volume for $x =$ 47.907 is

$$\begin{aligned}\hat{y} &= 0.3367 + 0.2649x \\ &= 0.3367 + 0.2649(47.907) = 13.03\end{aligned}$$

The residual for this observation is

$$\begin{aligned}\text{residual} &= y - \hat{y} \\ &= 13.04 - 13.03 = 0.01\end{aligned}$$

That is, the observed actual volume for this oyster lies 0.01 cm^3 above the least-squares line on the scatterplot.

Repeat this calculation 29 more times, once for each oyster. The 30 residuals are

0.01	0.39	0.95	−1.04	−1.33	2.52	0.41	−0.02	−1.57	1.33
1.92	−0.58	0.97	−1.22	1.33	−0.08	−2.30	1.62	−0.73	−1.30
0.77	−0.07	0.49	0.14	−1.01	−0.91	1.08	−1.51	−0.19	−0.08

Negative residuals indicate observed points that lie below the least-squares regression line on the scatterplot. Check the calculations by verifying that the sum of the residuals is zero. It is −0.1, not quite zero, because of roundoff error. Another reason to use software in regression is that roundoff errors in hand calculations can accumulate to make the results inaccurate.

The variance about the regression line is

$$\begin{aligned}s^2 &= \frac{1}{n-2}\sum \text{residual}^2 \\ &= \frac{1}{30-2}[(0.01)^2 + (0.39)^2 + \cdots + (-0.08)^2] \\ &= \frac{1}{28}(39.22) = 1.401\end{aligned}$$

Finally, the regression standard error is

$$s = \sqrt{1.401} = 1.18$$ ■

We will study several kinds of inference in the regression setting. The regression standard error s is the key measure of the variability of the responses in regression. It is part of the standard error of all the statistics we will use for inference.

APPLY YOUR KNOWLEDGE

23.1 Algal bloom control. Algal blooms can have negative effects on an ecosystem by dominating its phytoplankton communities. *Gonyostomum semen* is a nuisance alga infesting many parts of northern Europe. Could the overall biomass of *G. semen* be controlled by grazing zooplankton species? A research team examined the relationship between the net growth rate of *G. semen* and the number of *Daphnia magna* grazers introduced in test tubes. Net growth rate was computed by comparing the initial and final abundance of *G. semen* in the experiment, with a negative value indicative of a decrease in abundance. Here are the findings:[2]

Justin Kase z12z/Alamy

Number of *D. magna* grazers	1	2	3	4	5	6
Net growth rate of *G. semen*	−1.9	−2.5	−2.2	−3.9	−4.1	−4.3

(a) Examine the data. Make a scatterplot with number of grazers as the explanatory variable and find the correlation. There is a reasonably strong linear relationship.

(b) Explain in words what the slope β of the population regression line would tell us if we knew it. Based on the data, what are the estimates of β and the intercept α of the population regression line?

(c) Calculate by hand the residuals for the six data points. Check that their sum is 0 (aside from roundoff error). Use the residuals to estimate the standard deviation σ that measures variation in the responses (net growth rate) about the means given by the population regression line. You have now estimated all three parameters.

Using technology

Basic "two-variable statistics" calculators will find the slope b and intercept a of the least-squares line from keyed-in data. Inference about regression requires in addition the regression standard error s. At this point, software or a graphing calculator that includes procedures for regression inference becomes almost essential for practical work.

Figure 23.3 shows regression output for the data of Table 23.1 from a graphing calculator, three statistical programs, and a spreadsheet program. When we entered the data into the programs, we called the explanatory variable "Reconstruction2D." The programs use that label for the information relating to the slope. The TI-83 just uses "x" and "y" to label the explanatory and response variables. You can locate the basic information in all the outputs. The regression slope is $b = 0.2649$, and the regression intercept is $a = 0.3367$. The equation of the least-squares line is therefore (after rounding) just as given in Example 23.1. The regression standard error is $s = 1.183$ (JMP calls it "Root Mean Square Error," a term we will discuss in Chapter 24), and the squared correlation is $r^2 = 0.846$. Both of these results reflect the reasonably small scatter of the points in Figure 23.1 about the least-squares line.

Microsoft Excel

	A	B	C	D	E	F	G
1	SUMMARY OUTPUT						
2							
3	*Regression Statistics*						
4	Multiple R	0.919822041					
5	R Square	0.846072587					
6	Adjusted R Square	0.84057518					
7	Standard Error	1.183113597					
8	Observations	30					
9							
10							
11		*Coefficients*	*Standard Error*	*t Stat*	*P-value*	*Lower 95%*	*Upper 95%*
12	Intercept	0.336699371	0.907625559	0.3709673	0.71345396	−1.5224894	2.19588813
13	Reconstruction2D	0.26488506	0.021351708	12.405802	6.7703E-13	0.22114802	0.3086221

SPSS

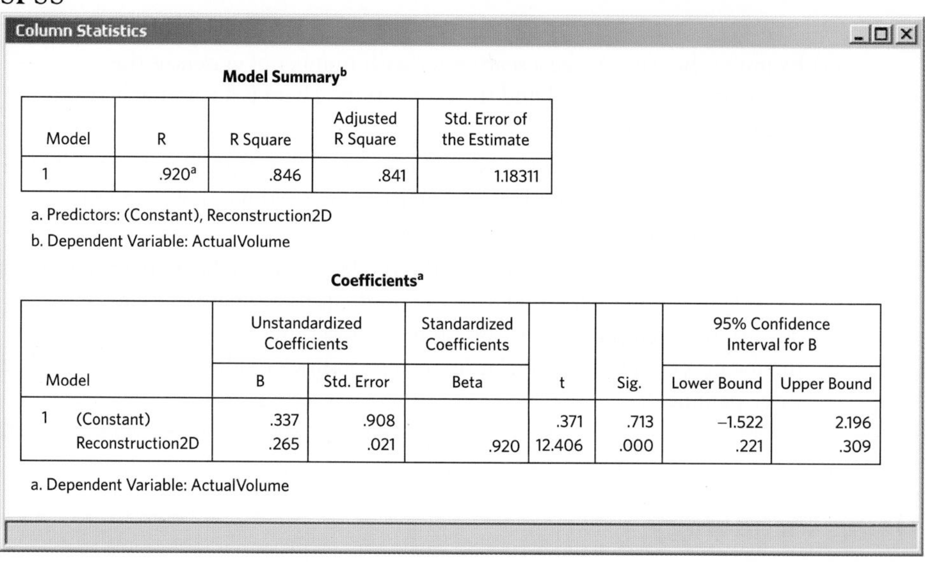

Column Statistics

Model Summary[b]

Model	R	R Square	Adjusted R Square	Std. Error of the Estimate
1	.920[a]	.846	.841	1.18311

a. Predictors: (Constant), Reconstruction2D
b. Dependent Variable: ActualVolume

Coefficients[a]

Model		Unstandardized Coefficients		Standardized Coefficients			95% Confidence Interval for B	
		B	Std. Error	Beta	t	Sig.	Lower Bound	Upper Bound
1	(Constant)	.337	.908		.371	.713	−1.522	2.196
	Reconstruction2D	.265	.021	.920	12.406	.000	.221	.309

a. Dependent Variable: ActualVolume

R

```
Coefficients:
                    Estimate   Std.Error   t value   Pr(>|t|)
(Intercept)          0.33670     0.90763     0.371      0.713
Reconstruction2D     0.26489     0.02135    12.406   6.77e-13

Residual standard error: 1.183 on 28 degrees of freedom
Multiple R-squared: 0.8461, Adjusted R-squared: 0.8406
```

TI-83

```
LinRegTTest              LinRegTTest
 y=a+bx                   y = a+bx
 β>0 and ρ>0              β>0 and ρ>0
 t=12.40580193           ↑b=.2648850597
 p=3.385137E-13           s=1.183113597
 df=28                    r²=.8460725872
↓a=.3366993706            r=.9198220411
```

FIGURE 23.3 Regression of actual oyster volume on 2D reconstruction: output from a graphing calculator, three statistical programs, and a spreadsheet program.

JMP

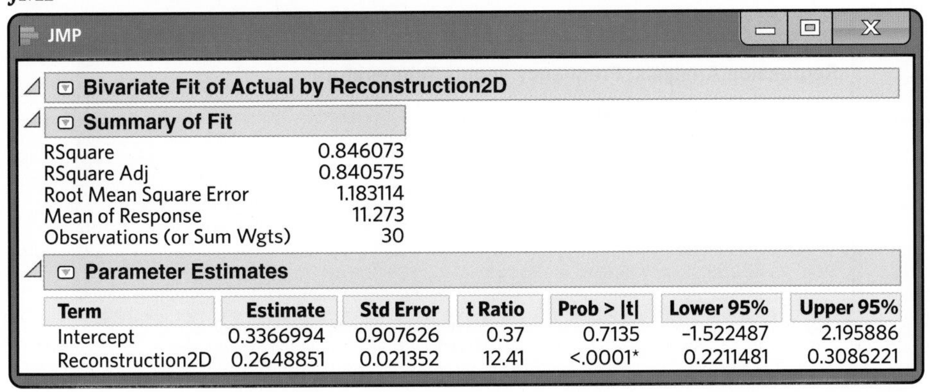
JMP

Bivariate Fit of Actual by Reconstruction2D

Summary of Fit

RSquare	0.846073
RSquare Adj	0.840575
Root Mean Square Error	1.183114
Mean of Response	11.273
Observations (or Sum Wgts)	30

Parameter Estimates

Term	Estimate	Std Error	t Ratio	Prob > |t|	Lower 95%	Upper 95%
Intercept	0.3366994	0.907626	0.37	0.7135	-1.522487	2.195886
Reconstruction2D	0.2648851	0.021352	12.41	<.0001*	0.2211481	0.3086221

FIGURE 23.3 *(Continued)*

Each output contains other information, some of which we will need shortly and some of which we don't need. In fact, we left out some output to save space. Once you know what to look for, you can find what you want in almost any output.

APPLY YOUR KNOWLEDGE

23.2 Frog mating calls. *Hyla chrysoscelis* is a type of gray tree frog with a distinct mating call. Frogs are cold-blooded animals, and their physiology is affected by variations in temperature. Field biologists wanted to know if the mating call of *H. chrysoscelis* is affected by the temperature of its natural habitat. Here is the note-repetition frequency (in notes per second) of the mating call as a function of habitat temperature (in degrees Celsius, °C) for 20 *H. chrysoscelis* in the wild:[3]

Ryan M. Bolton/Alamy

Temperature (°C)	19	21	22	22	23	23	23	23	23	24
Frequency (notes/sec)	38	42	45	45	41	45	48	50	53	51
Temperature (°C)	24	24	24	25	25	25	25	26	26	27
Frequency (notes/sec)	48	53	47	53	49	56	53	55	55	54

We want to predict call frequency from habitat temperature. Figure 23.4 shows Minitab regression output for these data.

(a) Make a scatterplot suitable for predicting frequency from temperature. The pattern is linear. What is the squared correlation r^2? Temperature explains a lot of the variation in call frequency.

(b) For regression inference, we must estimate the three parameters α, β, and σ. From the output, what are the estimates of these parameters?

(c) What is the equation of the least-squares regression line of frequency on temperature? Add this line to your plot. We will continue the analysis of these data in later exercises.

23.3 Great Arctic rivers. One effect of global warming is to increase the flow of water into the Arctic Ocean from rivers. Such an increase may have major effects on the

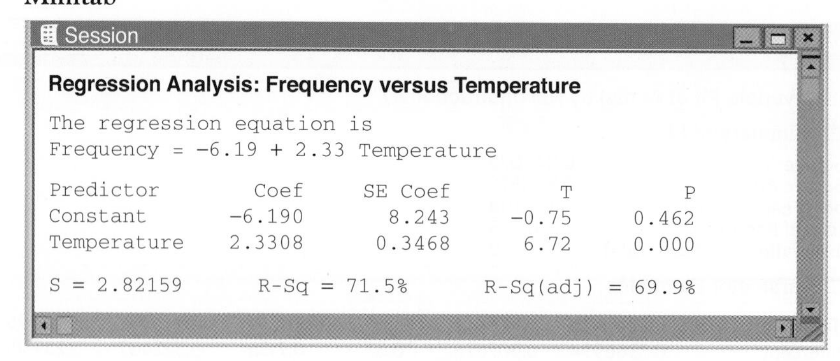

FIGURE 23.4 Minitab output for the frog mating-call data, for Exercise 23.2.

world's climate. Six rivers (Yenisey, Lena, Ob, Pechora, Kolyma, and Severnaya Dvina) drain two-thirds of the Arctic in Europe and Asia. Several of these are among the largest rivers on earth. Table 23.2 presents the total discharge from these rivers each year from 1936 to 2010.[4] Discharge is measured in cubic kilometers of water. Use software to analyze these data.

(a) Make a scatterplot of river discharge against time. Is there a clear increasing trend? Using technology, find r^2 and briefly interpret its value. There is considerable year-to-year variation, so we wonder if the trend is statistically significant.

(b) As a first step, find the least-squares line and add it to your plot. Then find the regression standard error s, which measures scatter about this line. We will continue the analysis in later exercises.

TABLE 23.2 Arctic river discharge (cubic kilometers), 1936 to 2010

Year	Discharge	Year	Discharge	Year	Discharge	Year	Discharge	Year	Discharge
1936	1721	1951	1864	1966	1883	1981	1774	1996	1849
1937	1713	1952	1829	1967	1642	1982	1728	1997	2007
1938	1860	1953	1652	1968	1713	1983	1920	1998	1903
1939	1739	1954	1589	1969	1742	1984	1823	1999	1970
1940	1615	1955	1656	1970	1751	1985	1822	2000	1905
1941	1838	1956	1721	1971	1879	1986	1860	2001	1894
1942	1762	1957	1762	1972	1736	1987	1732	2002	2077
1943	1709	1958	1936	1973	1861	1988	1906	2003	1781
1944	1921	1959	1906	1974	2000	1989	1932	2004	1900
1945	1581	1960	1736	1975	1928	1990	1861	2005	1937
1946	1834	1961	1970	1976	1653	1991	1801	2006	1914
1947	1890	1962	1849	1977	1698	1992	1793	2007	2254
1948	1898	1963	1774	1978	2008	1993	1845	2008	2074
1949	1958	1964	1606	1979	1970	1994	1902	2009	1917
1950	1830	1965	1735	1980	1758	1995	1842	2010	1819

Testing the hypothesis of no linear relationship

Example 23.1 asked, "Is 2D computer reconstruction a good assessment of actual oyster volume?" That is, do oysters with larger 2D reconstruction values tend to be larger oysters? Data analysis supports this conjecture. But is the positive association statistically significant? That is, is it too strong to have occurred just by chance? To answer this question, test hypotheses about the slope β of the population regression line:

$$H_0: \beta = 0$$
$$H_a: \beta > 0$$

A regression line with slope 0 is horizontal. That is, the mean of y does not change at all when x changes. So H_0 says that there is *no linear relationship* between x and y in the population. Put another way, H_0 says that *linear regression of y on x is of no value for predicting y*. The alternative hypothesis is one-sided because we are looking for evidence of a positive relationship (2D reconstruction values should be larger for larger oysters).

The test statistic is just the standardized version of the least-squares slope b, using the hypothesized value $\beta = 0$ for the mean of b. It is another t statistic. Here are the details.

SIGNIFICANCE TEST FOR REGRESSION SLOPE

To **test the hypothesis H_0: $\beta = 0$,** compute the t statistic

$$t = \frac{b}{\text{SE}_b}$$

In this formula, the standard error of the least-squares slope b is

$$\text{SE}_b = \frac{s}{\sqrt{\sum(x - \bar{x})^2}}$$

The sum runs over all observations on the explanatory variable x. In terms of a random variable T having the $t(n-2)$ distribution, the P-value for a test of H_0 against

H_a: $\beta > 0$ is $P(T \geq t)$

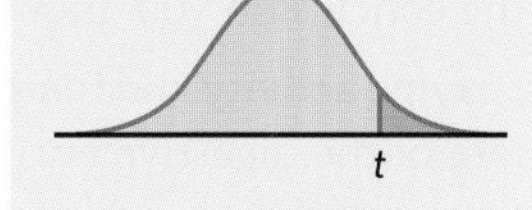

H_a: $\beta < 0$ is $P(T \leq t)$

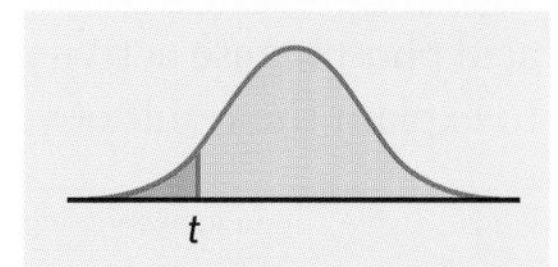

H_a: $\beta \neq 0$ is $2P(T \geq |t|)$

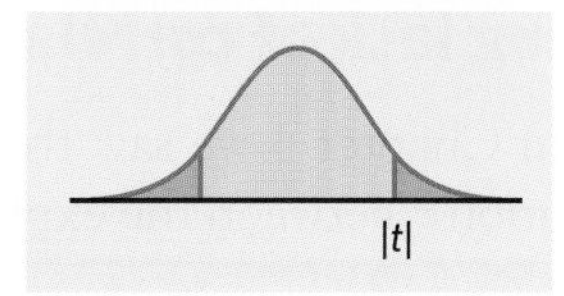

The standard error of b is a multiple of the regression standard error s. The degrees of freedom $n - 2$ are the degrees of freedom of s. Although we give the formula for this standard error, you should not try to calculate it by hand. Regression software gives the standard error SE_b along with b itself.

EXAMPLE 23.4 Oyster grading: Is the relationship significant?

The hypothesis H_0: $\beta = 0$ says that 2D reconstruction has no straight-line relationship with actual volume. In all logic, we anticipate a positive relationship, so we use the one-sided alternative H_a: $\beta > 0$.

Figure 23.1 shows that there is a positive relationship, so it is not surprising that all the outputs in Figure 23.3 give $t = 12.41$ with a very small P-value. Software programs give the two-sided P-value, whereas the TI-83 makes us choose a one-sided or two-sided alternative. In our case, the alternative is one-sided, and we need the one-sided P-value (which is half of the two-sided P-value). The P-value for the one-sided test is $P = 0.0000000000003$, which can be simplified to $P < 0.001$, highly significant. There is very strong evidence that actual volume increases as the 2D reconstruction increases. Remember, however, that strong statistical significance does not imply a strong effect, only that the observed effect is highly unlikely to have arisen just by chance because of the random sampling process. ■

APPLY YOUR KNOWLEDGE

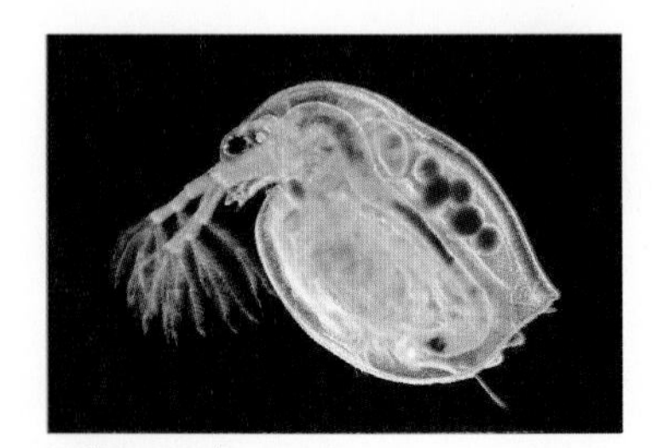

Ted Kinsman/Science Source

23.4 Algal bloom control: testing. Exercise 23.1 gives data on the number of *D. magna* grazers introduced in test tubes and the net growth rate of *G. semen* algae. Software tells us that the least-squares slope is $b = -0.5286$ with standard error $SE_b = 0.1059$.

(a) What is the t statistic for testing H_0: $\beta = 0$?

(b) How many degrees of freedom does t have? Use Table C to approximate the P-value of t against the one-sided alternative H_a: $\beta < 0$ (as the grazers would be expected to reduce algal growth). What do you conclude?

23.5 Great Arctic rivers: testing. The most important question we ask of the data in Table 23.2 is this: Is the increasing trend visible in your plot (Exercise 23.3) statistically significant? If so, changes in the Arctic may already be affecting the earth's climate. Use software to answer this question. Give a test statistic, its P-value, and the conclusion you draw from the test.

23.6 Enzyme activity and temperature. Exercise 3.6 (page 71) gives data on the activity rate of the digestive enzyme acid phosphatase in vitro at varying temperatures. Is there significant evidence of straight-line dependence between temperature and acid phosphatase activity rate? If you haven't already, make a scatterplot and use it to explain the result of your test.

Testing lack of correlation*

Back in Chapter 4 we saw that the least-squares slope b is closely related to the correlation r between the explanatory and response variables x and y. In the same

*This section is optional.

way, the slope β of the population regression line is closely related to the correlation between x and y in the population. In particular, the slope is 0 exactly when the correlation is 0.

Testing the null hypothesis H_0: $\beta = 0$ is therefore exactly the same as testing that there is *no correlation* between x and y in the population from which we drew our data. You can use the test for zero slope to test the hypothesis of zero correlation between any two quantitative variables. That's a useful trick.

Because correlation also makes sense when there is no explanatory-response distinction, it is handy to be able to test correlation without doing regression. The statistic r follows a particular distribution with $n - 2$ degrees of freedom. We don't cover the specifics of this distribution here, but Table E in the back of the book gives critical values of the sample correlation r under the null hypothesis that the correlation is 0 in the population. Use this table when *both variables* have at least approximately Normal distributions or when the sample size is large.

EXAMPLE 23.5 IQ and dyslexia in young children

STATE: IQ tests are sometimes used to assess children's cognitive abilities. However, typical IQ tests draw on reading skills, which may translate into lower scores for dyslexic children. Nonverbal IQ tests should circumvent this issue, allowing us to ask whether there is a relationship between reading skills and cognitive abilities in dyslexic children. Figure 23.5 displays the reading-skill scores (on a standardized scale) and the nonverbal IQ scores (also on a standardized scale) of 22 dyslexic children aged 7 to 10 years.[5] The relationship, if any, appears rather weak.

Historically, reading difficulties in children with low IQ test scores have often been attributed to the low IQ. However, recent research suggests that dyslexia has a neuronal foundation and tends to occur regardless of other cognitive abilities. Therefore, we want to test whether reading-skill score and nonverbal IQ score are significantly *correlated* in young dyslexic children.

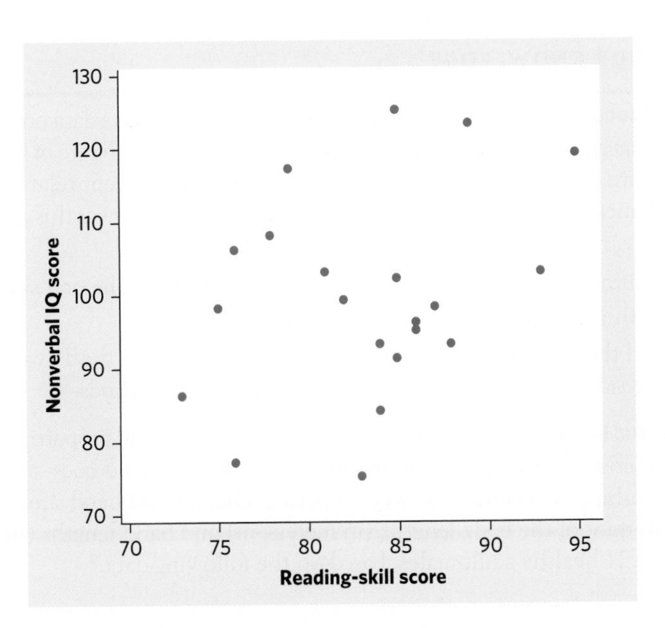

FIGURE 23.5 Scatterplot of the reading-skill scores and nonverbal IQ scores of 22 dyslexic children, for Example 23.5.

PLAN: We test the hypotheses

$$H_0\text{: population correlation} = 0$$
$$H_a\text{: population correlation} \neq 0$$

SOLVE: The data are displayed in the table below. Dotplots (we don't display them) show that the distributions of both variables are roughly symmetric, without outliers, and Normal quantile plots indicate that the deviations from Normality are minor. With 22 observations, a hypotheses test is safe.

Reading-skill score	84	81	85	82	87	85	89	86	85	83	76
Nonverbal IQ score	93	103	102	99	98	91	123	95	125	75	106
Reading-skill score	76	75	78	86	73	79	95	85	93	84	88
Nonverbal IQ score	77	98	108	96	86	117	119	91	103	84	93

The correlation coefficient for these data is $r = 0.308$. Compare this value with the critical values in the $n = 20$ row of Table E. The closest critical values in that row are 0.2992 and 0.3783, for tail areas 0.10 and 0.05, respectively. H_a is two-sided, because we have no a priori expectation. Therefore, the P-value is $0.10 < P < 0.20$.

We could also have tested H_0: $\beta = 0$. The t statistic for this test is $t = 1.45$ with two-sided P-value 0.163 given by software. **This is also exactly the P-value for testing population correlation = 0.**

CONCLUDE: The study failed to find evidence of a correlation between reading-skill scores and nonverbal IQ scores in young dyslexic children. Absence of evidence is not evidence of absence, and we can never prove the null hypothesis. However, the data already suggest that, even if there was a relationship in the population, nonverbal IQ score would predict only about 9.5% of variations in reading-skill score, and vice versa (a rather weak predictor). ■

APPLY YOUR KNOWLEDGE

23.7 Algal bloom control: testing correlation. Exercise 23.1 gives data on the number of *D. magna* grazers introduced in test tubes and the net growth rate of *G. semen* algae. There are only 6 observations, so we worry that the apparent relationship may be just chance. Is the correlation significantly less than 0? Answer this question in two ways.

(a) Return to your t statistic from Exercise 23.4. What is the one-sided P-value for this t? Apply your result to test the correlation.

(b) Find the correlation r and use Table E to approximate the P-value of the one-sided test. Compare your answers using both methods.

23.8 Hand and body lengths. While the human body has overall proportions undeniably characteristic of the species, individuals do vary in size and body shape. Is there a relationship, for instance, between a person's height and hand size? Figure 23.6 is a scatterplot of the body lengths (in meters, m) and hand lengths (in millimeters, mm) of 21 healthy adult males, based on the following data.[6]

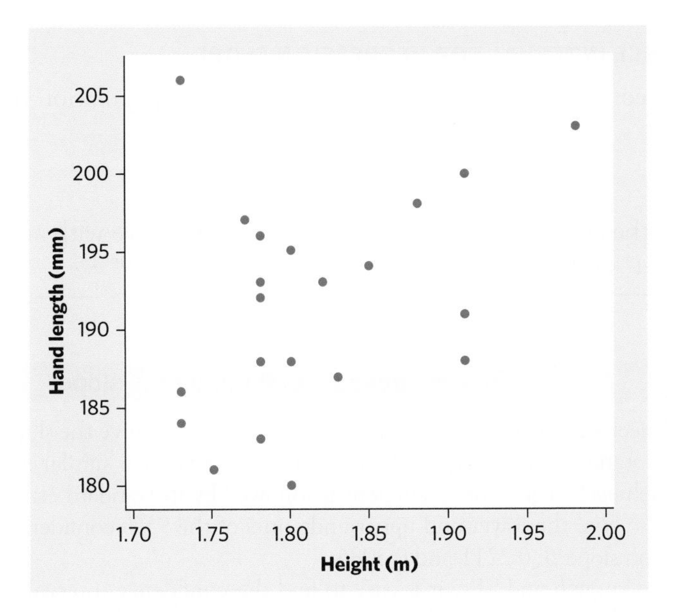

FIGURE 23.6 Scatterplot of hand lengths and body heights for 21 men, for Exercise 23.8.

Body length (m)	Hand length (mm)	Body length (m)	Hand length (mm)	Body length (m)	Hand length (mm)
1.75	181	1.80	195	1.82	193
1.91	188	1.78	188	1.73	186
1.73	184	1.80	188	1.85	194
1.80	180	1.83	187	1.77	197
1.78	183	1.91	191	1.98	203
1.88	198	1.78	196	1.91	200
1.78	192	1.78	193	1.73	206

Neither body length nor hand length can be considered a possible explanatory variable. Rather, genes and environmental factors influence both variables and are responsible for any relationship that may exist between body and hand lengths. Therefore, we want to test whether body length and hand length are significantly *correlated* in men. Follow the four-step process in your analysis, as shown in Example 23.5.

Confidence intervals for the regression slope

The slope β of the population regression line is usually the most important parameter in a regression problem. The slope is the rate of change of the mean response as the explanatory variable increases. We often want to estimate β. The slope b of the least-squares line is an unbiased estimator of β. A confidence interval is more useful because it shows how accurate the estimate b is likely to be. The confidence interval for β has the familiar form

$$\text{estimate } \pm t^* \text{SE}_{\text{estimate}}$$

Because b is our estimate, the confidence interval is $b \pm t^* \text{SE}_b$. Here are the details.

CONFIDENCE INTERVAL FOR REGRESSION SLOPE

A level C **confidence interval for the slope β** of the population regression line is

$$b \pm t^* \text{SE}_b$$

Here t^* is the critical value for the $t(n-2)$ density curve with area C between $-t^*$ and t^*.

EXAMPLE 23.6 Grading oysters: estimating the slope

The spreadsheet and the three software outputs in Figure 23.3 give the slope $b = 0.2649$ and also the standard error $\text{SE}_b = 0.02135$. The outputs use a similar arrangement, a table in which each regression coefficient is followed by its standard error. Excel, JMP, and SPSS also give the lower and upper endpoints of the 95% confidence interval for the population slope β, 0.2211 and 0.3086.

Once we know b and SE_b, it is easy to find the confidence interval. There are 30 data points, so the degrees of freedom are $n - 2 = 28$. Using Table C, we get $t^* = 2.048$. Using Excel, for 95% confidence enter the corresponding two-tailed area 0.05 and 28 degrees of freedom in the command `=TINV(0.05,28)`, which returns a value of 2.0484 for t^*. The 95% confidence interval for the population slope β is

$$\begin{aligned} b \pm t^* \text{SE}_b &= 0.2649 \pm (2.0484)(0.02135) \\ &= 0.2649 \pm 0.0437 \\ &= 0.2212 \text{ to } 0.3086 \end{aligned}$$

This agrees, up to rounding error, with the ouputs from Excel, JMP, and SPSS in Figure 23.3. We are 95% confident that mean actual volume increases by between about 0.22 and 0.31 cm^3 for each additional thousand pixels in 2D reconstruction. ■

You can find a confidence interval for the intercept α of the population regression line in the same way, using a and SE_a from the "Intercept" line in R. JMP, SPSS, and Excel already include this interval in their outputs. However, we rarely need to estimate α.

APPLY YOUR KNOWLEDGE

23.9 Algal bloom control: testing. Exercise 23.1 gives data on the number of *D. magna* grazers introduced in test tubes and the net growth rate of *G. semen* algae. Software tells us that the least-squares slope is $b = -0.5286$ with standard error $\text{SE}_b = 0.1059$. Because there are only 6 observations, the observed slope b may not be an accurate estimate of the population slope β. Give a 95% confidence interval for β.

23.10 Frog mating calls: estimating slope. Exercise 23.2 gives data on the mating-call frequency of *H. chrysoscelis* as a function of habitat temperature. We want a 95% confidence interval for the slope of the population regression line. Starting from the information in the Minitab output in Figure 23.4, find this interval. Say in words what the slope of the population regression line tells us about the frequency of this species' mating call in the wild under various temperatures.

23.11 Great Arctic rivers: estimating slope. Use the data in Table 23.2 to give a 90% confidence interval for the slope of the population regression of Arctic river discharge on year. Does this interval convince you that discharge is actually increasing over time? Explain your answer.

Inference about prediction

One of the most common reasons to fit a line to data is to predict the response to a particular value of the explanatory variable. This is another setting for regression inference: We want not simply a prediction but a prediction with a margin of error that describes how accurate the prediction is likely to be.

EXAMPLE 23.7 Beer and blood alcohol

STATE: The EESEE story "Blood Alcohol Content" on the companion website describes a study in which 16 student volunteers at the Ohio State University drank a randomly assigned number of cans of beer. Thirty minutes later, a police officer measured their blood alcohol content (BAC) in grams of alcohol per deciliter of blood. Here are the data:

Student	1	2	3	4	5	6	7	8
Beers	5	2	9	8	3	7	3	5
BAC	0.10	0.03	0.19	0.12	0.04	0.095	0.07	0.06
Student	**9**	**10**	**11**	**12**	**13**	**14**	**15**	**16**
Beers	3	5	4	6	5	7	1	4
BAC	0.02	0.05	0.07	0.10	0.085	0.09	0.01	0.05

The students were equally divided between men and women and differed in weight and usual drinking habits. Some students may not believe that number of drinks predicts BAC well. Steve did not participate in the study, but he thinks he can drive legally 30 minutes after he finishes drinking 5 beers. The legal limit for driving is a BAC of 0.08 in all states. We want to predict Steve's BAC, using no information except that he drinks 5 beers.

PLAN: Regress BAC on number of beers. Use the regression line to predict Steve's BAC. Give a margin of error that allows us to have 95% confidence in our prediction.

SOLVE: The scatterplot in Figure 23.7 and the regression output in Figure 23.8 show that student opinion is wrong: Number of beers predicts blood alcohol content quite well. In fact, $r^2 = 0.80$, so number of beers explains 80% of the observed variation in BAC. To predict Steve's BAC after 5 beers, use the equation of the regression line:

$$\begin{aligned}\hat{y} &= -0.0127 + 0.0180x \\ &= -0.0127 + 0.0180(5) = 0.077\end{aligned}$$

That's dangerously close to the legal limit 0.08. What about 95% confidence? The "Predicted Values for New Observations" part of the output in Figure 23.8 shows *two* 95% intervals. Which should we use? ■

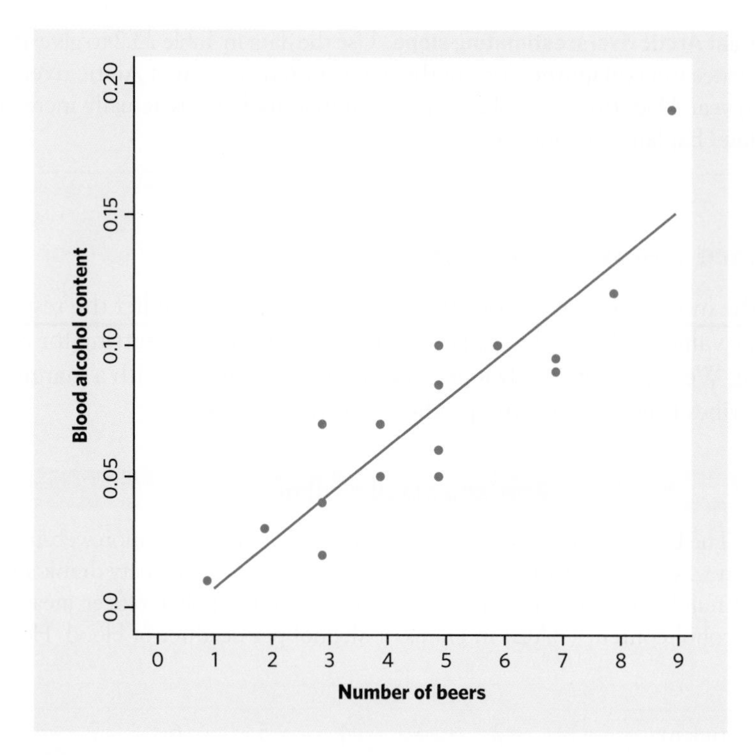

FIGURE 23.7 Scatterplot of students' blood alcohol content against the number of cans of beer consumed, with the least-squares regression line, for Example 23.7.

Minitab

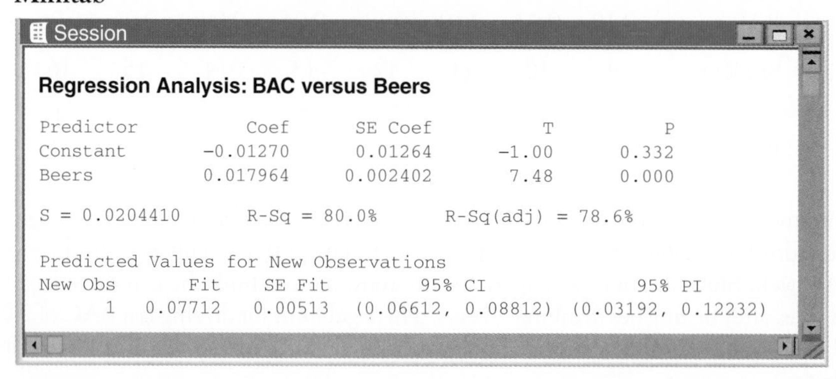

Session

Regression Analysis: BAC versus Beers

Predictor	Coef	SE Coef	T	P
Constant	-0.01270	0.01264	-1.00	0.332
Beers	0.017964	0.002402	7.48	0.000

S = 0.0204410 R-Sq = 80.0% R-Sq(adj) = 78.6%

Predicted Values for New Observations

New Obs	Fit	SE Fit	95% CI	95% PI
1	0.07712	0.00513	(0.06612, 0.08812)	(0.03192, 0.12232)

FIGURE 23.8 Partial Minitab output for the blood alcohol content data in Example 23.7.

To decide which interval to use, you must answer this question: Do you want to predict the *mean* BAC for *all students* who drink 5 beers, or do you want to predict the BAC of *one individual student* who drinks 5 beers? *Both of these predictions may be interesting, but they are two different problems.* The actual, sample-based prediction is the same, $\hat{y} = 0.077$. But the margin of error is different for the two kinds of prediction. Individual students who drink 5 beers don't all have the same BAC. So we need a larger margin of error to pin down Steve's result than we would to estimate the mean BAC for all students who would drink 5 beers.

Write the given value of the explanatory variable x as x^*. In Example 23.7 $x^* = 5$. The distinction between predicting a single outcome and predicting the mean of all outcomes when $x = x^*$ determines what margin of error is correct. To emphasize the distinction, we use different terms for the two intervals.

- To estimate the *mean* response, we use a *confidence interval*. It is an ordinary confidence interval for the mean response when x has the value x^*, which is $\mu_y = \alpha + \beta x^*$. This is a parameter, a fixed number whose value we don't know.
- To estimate an *individual* response y, we use a **prediction interval.** A prediction interval estimates a single random response y rather than a parameter like μ_y. The response y is not a fixed number. If we took more observations with $x = x^*$, we would get different responses.

prediction interval

EXAMPLE 23.8 Beer and blood alcohol: conclusion

Steve is one individual, so we must use the prediction interval. The output in Figure 23.8 helpfully labels the confidence interval as "CI" and the prediction interval as "PI." We are 95% confident that Steve's BAC after 5 beers will lie between 0.032 and 0.122. The upper part of that range will get him arrested if he drives. The 95% confidence interval for the mean BAC of all students who drink 5 beers is much narrower, 0.066 to 0.088. ■

The meaning of a prediction interval is very much like the meaning of a confidence interval. A 95% prediction interval, like a 95% confidence interval, is right 95% of the time in repeated use. "Repeated use" now means that we take an observation on y for each of the n values of x in the original data, and then take one more observation y with $x = x^*$. Form the prediction interval from the n observations, then see if it covers the one more y. It will in 95% of all repetitions.

The interpretation of prediction intervals is a minor point. The main point is that it is harder to predict one response than to predict a mean response. Both intervals have the usual form

$$\hat{y} \pm t^*\text{SE}$$

but the prediction interval is wider than the confidence interval because individuals are more variable than averages. You will rarely need to know the details, because software automates the calculation, but here they are.

Regression from garbage. The Census Bureau once asked if weighing a neighborhood's garbage would help count its people. So 63 households had their garbage sorted and weighed. It turned out that pounds of plastic in the trash gave the best garbage prediction of the number of people in a neighborhood. The margin of error for a 95% prediction interval in a neighborhood of about 100 households, based on five weeks' worth of garbage, was about ± 2.5 people. Alas, that is not accurate enough to help the Census Bureau.

CONFIDENCE AND PREDICTION INTERVALS FOR REGRESSION RESPONSE

A level C **confidence interval for the mean response μ_y** when x takes the value x^* is

$$\hat{y} \pm t^*\text{SE}_{\hat{\mu}}$$

The standard error $\text{SE}_{\hat{\mu}}$ is

$$\text{SE}_{\hat{\mu}} = s\sqrt{\frac{1}{n} + \frac{(x^* - \bar{x})^2}{\sum(x - \bar{x})^2}}$$

A level C **prediction interval for a single observation y** when x takes the value x^* is

$$\hat{y} \pm t^*\text{SE}_{\hat{y}}$$

The standard error for prediction $SE_{\hat{y}}$ is

$$SE_{\hat{y}} = s\sqrt{1 + \frac{1}{n} + \frac{(x^* - \overline{x})^2}{\sum(x - \overline{x})^2}}$$

In both intervals, t^* is the critical value for the $t(n-2)$ density curve with area C between $-t^*$ and t^*.

There are two standard errors: $SE_{\hat{\mu}}$ for estimating the mean response μ_y, and $SE_{\hat{y}}$ for predicting an individual response y. The only difference between the two standard errors is the extra 1 under the square root sign in the standard error for prediction. The extra 1 makes the prediction interval wider. Both standard errors are multiples of the regression standard error s. The degrees of freedom are again $n - 2$, the degrees of freedom of s.

APPLY YOUR KNOWLEDGE

23.12 Algal bloom control: prediction. Exercise 23.1 gives data on the number of *D. magna* grazers introduced in test tubes and the net growth rate of *G. semen* algae.

(a) If researchers were to set up another experiment with 3 grazers in one test tube, predict the net algal growth rate in this test tube using the partial Minitab output in Figure 23.9 for prediction when $x^* = 3$. Which interval in the output is the proper 95% interval for predicting the net algal growth rate in this test tube?

(b) Minitab gives only one of the two standard errors used in prediction. It is $SE_{\hat{\mu}}$, the standard error for estimating the mean response. Use this fact along with the output to give a 95% confidence interval for the mean net algal growth rate for all tests tubes containing 3 grazers.

23.13 Frog mating calls: prediction. Analysis of the data in Exercise 23.2 shows that mating-call frequency in *H. chrysoscelis* varies linearly with habitat temperature. We might want to predict the mean call frequency (in notes per second) at

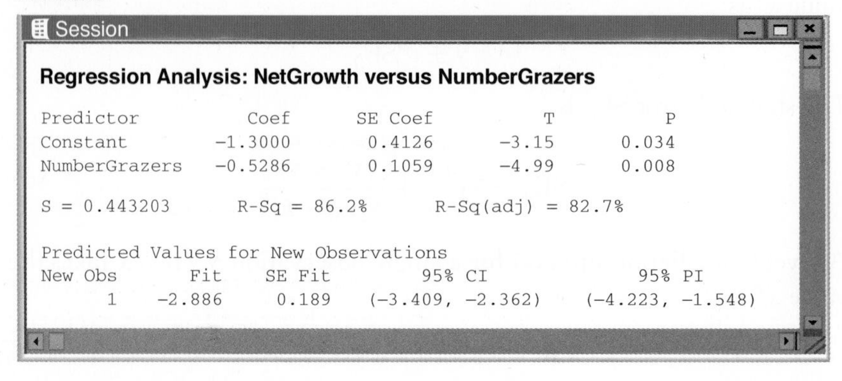
Minitab

Session

Regression Analysis: NetGrowth versus NumberGrazers

Predictor	Coef	SE Coef	T	P
Constant	-1.3000	0.4126	-3.15	0.034
NumberGrazers	-0.5286	0.1059	-4.99	0.008

S = 0.443203 R-Sq = 86.2% R-Sq(adj) = 82.7%

Predicted Values for New Observations

New Obs	Fit	SE Fit	95% CI	95% PI
1	-2.886	0.189	(-3.409, -2.362)	(-4.223, -1.548)

FIGURE 23.9 Partial Minitab output for regressing net algal growth rate on number of grazers in test tubes, for Exercise 23.12.

24 degrees Celsius in the wild. Here is the Minitab output for prediction when $x^* = 24$ degrees Celsius (labeled "New Obs 1" in the output):

Minitab

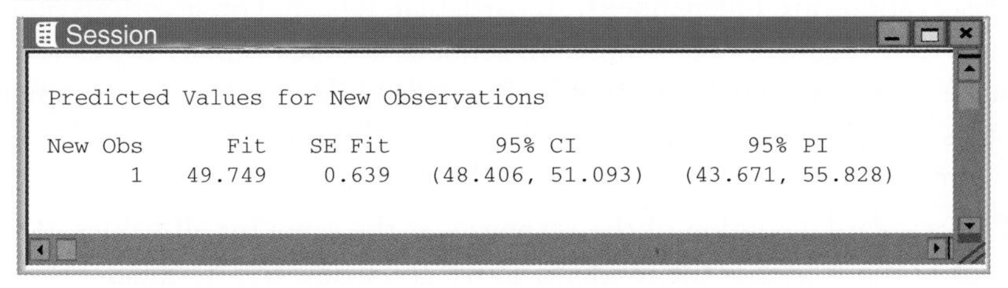

```
Session

Predicted Values for New Observations

New Obs     Fit   SE Fit        95% CI              95% PI
      1  49.749    0.639  (48.406, 51.093)  (43.671, 55.828)
```

(a) Use the regression line from Figure 23.4 to verify that "Fit" is the predicted value for $x^* = 24$. (Start with the results in the "Coef" column of Figure 23.4 to reduce roundoff error.)

(b) What is the 95% interval we want?

Checking the conditions for inference

You can fit a least-squares line to any set of explanatory-response data when both variables are quantitative and at least the response variable is continuous. If the scatterplot doesn't show a roughly linear pattern, the fitted line may be almost useless. But it is still the line that fits the data best in the least-squares sense. To use regression inference, however, the data must satisfy additional conditions. *Before you can trust the results of inference, you must check the conditions for inference one by one.* There are ways to deal with violations of any of the conditions. If you see a clear violation, get expert advice.

Although the conditions for regression inference are a bit elaborate, it is not hard to check for major violations. The conditions involve the population regression line and the deviations of responses from this line. We can't observe the population line, but the least-squares line estimates it, and the residuals estimate the deviations from the population line. *You can check all the conditions for regression inference by looking at graphs of the residuals.* Most regression software will calculate and save the residuals for you, and even plot them. Start by examining a dotplot or histogram of the residuals and also a **residual plot,** a plot of the residuals against the explanatory variable x, with a horizontal line at the "residual = 0" position. The "residual = 0" line represents the position of the least-squares line in the scatterplot of y against x. Let's look at each condition in turn.

residual plot

- **The relationship is linear in the population.** Look for curved patterns or other departures from a straight-line overall pattern in the residual plot. You can also use the original scatterplot, but the residual plot magnifies any effects.
- **The response varies Normally about the population regression line.** Because different y-values usually come from different x-values, the responses themselves need not be Normal. It is the deviations from the

population line—estimated by the residuals—that must be Normal. Check for clear skewness or other major departures from Normality in a histogram of the residuals or by making a Normal quantile plot of the residuals.

- **Observations are independent.** In particular, repeated observations on the same individual are not allowed. You should not use ordinary regression to make inferences about the growth of a single child over time, for example. Signs of dependence in the residual plot are a bit subtle, so we usually rely on common sense and information from the data collection process.
- **The standard deviation of the responses is the same for all values of x.** Look at the scatter of the residuals above and below the "residual = 0" line in the residual plot. The scatter should be roughly the same from one end to the other. You will sometimes find that as the response y gets larger, so does the scatter of the residuals. Rather than remaining fixed, the standard deviation σ about the line changes with x as the mean response changes with x. There is no fixed σ for s to estimate. You cannot trust the results of inference when this happens.

You will always see some irregularity when you look for Normality and fixed standard deviation in the residuals, especially when you have few observations. Don't overreact to minor violations of the conditions. Like other t procedures, inference for regression is (with one exception) not very sensitive to lack of Normality, especially when we have many observations. Do beware of influential observations, which can greatly affect the results of inference.

The exception is the prediction interval for a single response y. This interval relies on Normality of individual observations, not just on the approximate Normality of statistics like the slope a and intercept b of the least-squares line. The statistics a and b become more Normal as we take more observations. This contributes to the robustness of regression inference, but it isn't enough for the prediction interval. We will not study methods that carefully check Normality of the residuals, so *you should regard prediction intervals as rough approximations*.

EXAMPLE 23.9 Does climate change chase fish north?

STATE: As the climate grows warmer, we expect many animal species to move toward the poles in an attempt to maintain their preferred temperature range. Do data on fish in the North Sea confirm this expectation? Here are data for 25 years, 1977 through 2001, on mean winter temperatures at the bottom of the North Sea (degrees Celsius) and on the center of the distribution of anglerfish in degrees of north latitude:[7]

Temperature	6.26	6.26	6.27	6.31	6.34	6.32	6.37	6.39	6.42
Latitude	57.20	57.96	57.65	57.59	58.01	59.06	56.85	56.87	57.43
Temperature	6.52	6.68	6.76	6.78	6.89	6.90	6.93	6.98	7.02
Latitude	57.72	57.83	57.87	57.48	58.13	58.52	58.48	57.89	58.71
Temperature	7.09	7.13	7.15	7.29	7.34	7.57	7.65		
Latitude	58.07	58.49	58.28	58.49	58.01	58.57	58.90		

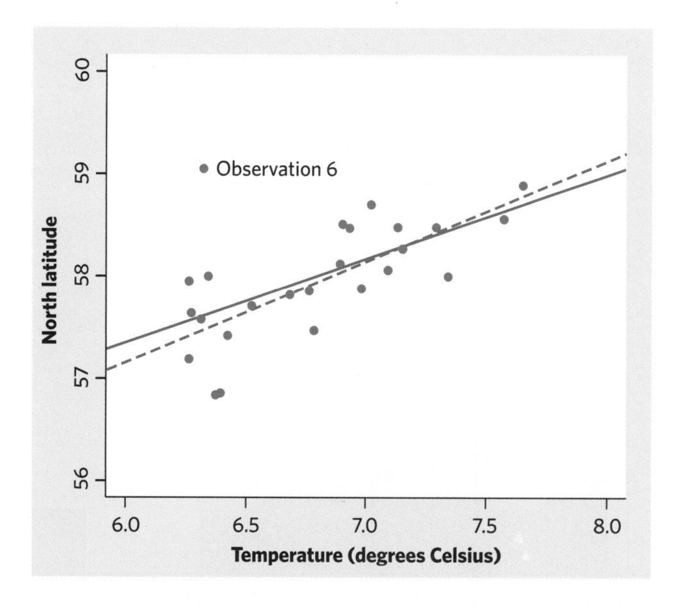

FIGURE 23.10 Scatterplot of the latitude of the center of the distribution of anglerfish in the North Sea against mean winter temperature at the bottom of the sea, for Example 23.9. The two regression lines are for the data with (solid) and without (dashed) Observation 6.

PLAN: Regress latitude on temperature. Look for a positive linear relationship and assess its significance. Be sure to check the conditions for regression inference.

SOLVE: The scatterplot in Figure 23.10 shows a clear positive linear relationship. The solid line in the plot is the least-squares regression line of the center of the fish distribution (north latitude) on winter ocean temperature. Software shows that the slope is $b = 0.818$. That is, each degree of ocean warming moves the fish about 0.8 degree of latitude farther north. The t statistic for testing H_0: $\beta = 0$ is $t = 3.6287$ with one-sided P-value $P = 0.0007$ and $r^2 = 0.364$. There is very strong evidence that the population slope is positive, $\beta > 0$.

CONCLUDE: The data give highly significant evidence that anglerfish have moved north as the ocean has grown warmer. Before relying on this conclusion, we must check the conditions for inference. ■

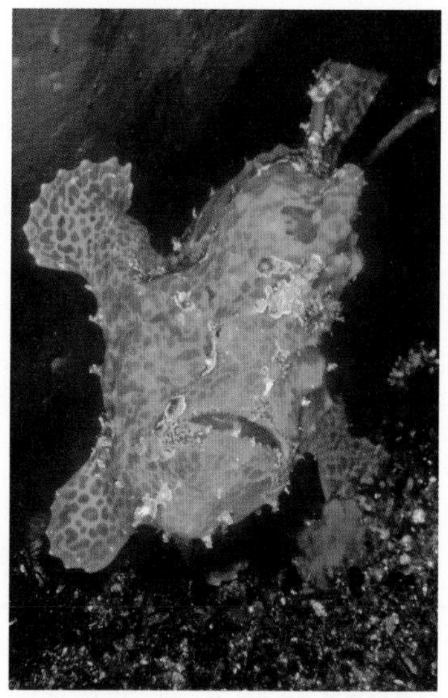

Fionaayers/Dreamstime.com

The software that did the regression calculations also finds the 25 residuals. In the same order as the observations in Example 23.9, they are

```
-0.3731  0.3869  0.0687 -0.0240  0.3714  1.4378 -0.8131
-0.8095 -0.2740 -0.0658 -0.0867 -0.1121 -0.5185  0.0415
 0.4234  0.3588 -0.2721  0.5152 -0.1821  0.2052 -0.0211
 0.0743 -0.4466 -0.0747  0.1899
```

Graphs play a central role in checking the conditions for inference. We begin by making two graphs of the residuals. Figure 23.11 is a histogram of the residuals. Figure 23.12 is the residual plot, a plot of the residuals against the explanatory variable, sea-bottom temperature. The "residual = 0" line marks the position of the regression line. Notice that the vertical scale in Figure 23.12 is expanded beyond what is necessary to simply show the points. Patterns in residual plots are often easier to see if you use a wider vertical scale than your software's default plot. Both

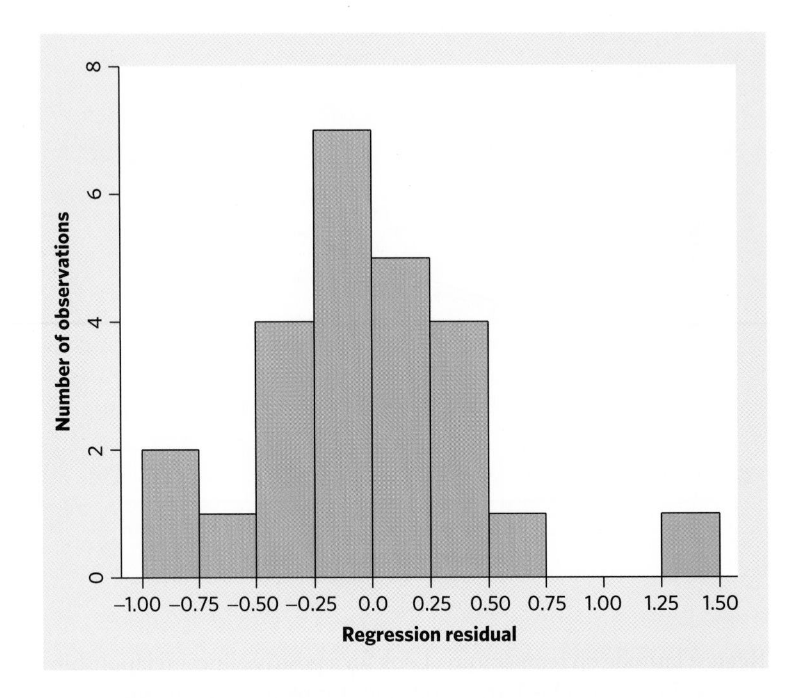

FIGURE 23.11 Histogram of the residuals from the regression of latitude on temperature in Example 23.9.

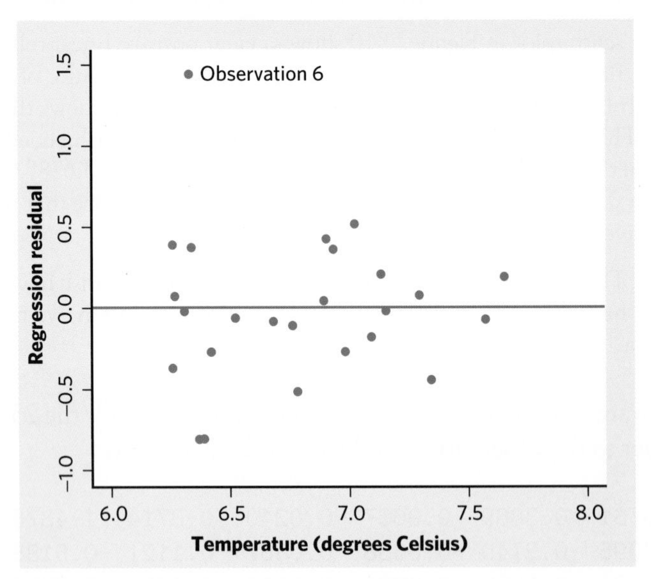

FIGURE 23.12 Residual plot for the regression of latitude on temperature in Example 23.9.

graphs show that Observation 6 is a high outlier. Let's check the conditions for regression inference.

- **Linear relationship.** The scatterplot in Figure 23.10 and the residual plot in Figure 23.12 both show a linear relationship except for the outlier.
- **Normal residuals.** The histogram in Figure 23.11 is roughly symmetric and single-peaked. There are no important departures from Normality except for the outlier.

- **Independent observations.** The observations were taken a year apart, so we are willing to regard them as close to independent. The residual plot shows no obvious pattern of dependence, such as runs of points all above or all below the line.
- **Constant standard deviation.** Again excepting the outlier, the residual plot shows no unusual variation in the scatter of the residuals above and below the line as x varies.

The outlier is the only serious violation of the conditions for inference. How influential is the outlier? The dashed line in Figure 23.10 is the regression line without Observation 6. Because there are several other observations with similar values of temperature, dropping Observation 6 does not move the regression line very much. *Even though the outlier is not very influential for the regression line, it influences regression inference because of its effect on the regression standard error.* The standard error is $s = 0.4734$ with Observation 6 and $s = 0.3622$ without it. When we omit the outlier, the t statistic changes from $t = 3.6287$ to $t = 5.5599$, and the one-sided P-value changes from $P = 0.0007$ to $P < 0.00001$. Fortunately, the outlier does not affect the conclusion we drew from the data. Dropping Observation 6 makes the test for the population slope *more* significant and *increases* the percent of variation in fish location explained by ocean temperature.

One more caution about inference in this example: As usual in an observational study, the possibility of lurking variables makes us hesitant to conclude that rising temperature is *causing* anglerfish to move north. Ocean temperature was steadily rising during these years. The effect on fish latitude of any lurking variable that increased over time—perhaps increased commercial fishing—is confounded with the effect of temperature.

APPLY YOUR KNOWLEDGE

23.14 Oyster grading: residuals. The residuals for the study of oyster actual volume and 2D reconstruction appear in Example 23.3.

(a) Make a dotplot to display the distribution of the residuals. Are there outliers or signs of strong departures from Normality?

(b) Make a plot of the residuals against the explanatory variable. Draw a horizontal line at height 0 on your plot. Does the plot show a nonrandom pattern?

23.15 Frog mating calls: residuals. Figure 23.4 gives part of the Minitab output for the data on mating-call frequency and temperature in wild *H. chrysoscelis* from Exercise 23.2. Table 23.3 comes from another part of the output. It gives the predicted response $\hat{y}$ and the residual $y - \hat{y}$ for each of the 20 observations. Most statistical software packages provide similar output. Examine the conditions for regression inference one by one.

(a) **Linear relationship.** Your plot and r^2 from Exercise 23.2 show that the relationship is indeed linear. Residual plots magnify effects. Plot the residuals against temperature. Are there clear deviations from a linear relationship?

TABLE 23.3 Frog mating calls: predictions and residuals

Obs.	Temp. x	Freq. y	Prediction $\hat{y}$	Residual $y - \hat{y}$
1	19	38	38.10	−0.10
2	21	42	42.76	−0.76
3	22	45	45.09	−0.09
4	22	45	45.09	−0.09
5	23	41	47.42	−6.42
6	23	45	47.42	−2.42
7	23	48	47.42	0.58
8	23	50	47.42	2.58
9	23	53	47.42	5.58
10	24	51	49.75	1.25
11	24	48	49.75	−1.75
12	24	53	49.75	3.25
13	24	47	49.75	−2.75
14	25	53	52.08	0.92
15	25	49	52.08	−3.08
16	25	56	52.08	3.92
17	25	53	52.08	0.92
18	26	55	54.41	0.59
19	26	55	54.41	0.59
20	27	54	56.74	−2.74

(b) **Normal variation about the line.** Make a histogram of the residuals. Do strong skewness or outliers suggest lack of Normality?

(c) **Independent observations.** The data come from observations of 20 different male *H. chrysoscelis* in the wild. They are therefore independent observations.

(d) **Spread about the line stays the same.** Your plot in (b) shows that it does not. Variation is greater for warmer temperatures. However, the reason for less variation at lower temperatures is that there are very few observations. Rather than a problem with constant spread, this suggests that the low-temperature observations might be more influential.

CHAPTER 23 SUMMARY

- **Least-squares regression** fits a straight line to data in order to predict a response variable y from an explanatory variable x. Inference about regression requires more conditions.
- The **conditions for regression inference** say that there is a **population regression line** $\mu_y = \alpha + \beta x$ that describes how the mean response varies as x changes. The observed response y for any x has a Normal distribution with

mean given by the population regression line and with the same standard deviation σ for any value of x. Observations on y are independent.

- The **parameters to be estimated** are the intercept α and the slope β of the population regression line and also the standard deviation σ. The slope a and intercept b of the least-squares line estimate α and β. Use the **regression standard error s** to estimate σ.
- The regression standard error s has $n - 2$ **degrees of freedom.** All t procedures in regression inference have $n - 2$ degrees of freedom.
- To test **the hypothesis that the slope is zero in the population,** use the t statistic $t = b/\text{SE}_b$. This null hypothesis says that straight-line dependence on x has no value for predicting y. In practice, use software to find the slope b of the least-squares line, its standard error SE_b, and the t statistic.
- The t test for regression slope is also a test for **the hypothesis that the population correlation between x and y is zero.** To do this test without software, use the sample correlation r and Table E.
- **Confidence intervals for the slope** of the population regression line have the form $b \pm t^*\text{SE}_b$.
- **Confidence intervals for the mean response** when x has value x^* have the form $\hat{y} \pm t^*\text{SE}_{\hat{\mu}}$. **Prediction intervals** for an individual future response y have a similar form with a larger standard error, $\hat{y} \pm t^*\text{SE}_{\hat{y}}$. Software often gives these intervals.

THIS CHAPTER IN CONTEXT

In Chapters 3 and 4 we discussed how to describe the relationship between *two quantitative variables*. We used scatterplots to visualize patterns and identify linear relationships in particular. We then quantified the strength and direction of a linear relationship with the correlation coefficient r and obtained the equation of the least-squares regression line to model that relationship. In this chapter we expand these concepts beyond the domain of descriptive statistics into that of statistical inference to ask whether the linear pattern in the sample data would hold for the entire population.

The statistical inference methods described in this chapter draw on t procedures similar to those seen in Chapter 17. We can use them to estimate the parameters of the regression line or to test the hypothesis of no association (a slope of zero) between two variables. Even if the association is statistically significant, we cannot necessarily conclude that there is an underlying cause-and-effect relation between the response and explanatory variables. As we saw in Chapters 7 and 8, the way the data were collected can substantially limit the extent of our conclusions.

In companion Chapter 28 we will expand the concept of simple linear regression to include multiple explanatory variables used to describe and predict one response variable (multiple linear regression). We will use inference procedures to determine which of these explanatory variables, alone or in sets, has a statistically significant effect on the response variable. We will also see how a categorical variable can be coded with zeros and ones to allow it into the regression model (in multiple linear regression and in logistic regression).

CHECK YOUR SKILLS

Our first example of regression (Example 4.1, page 90) presented data showing that people who increased their nonexercise activity (NEA) when they were deliberately overfed gained less fat than other people. The scatterplot and regression line based on 16 overfed subjects are displayed in Figure 4.1, page 90. Here is part of the Minitab output for regressing fat gain on NEA change in this study, along with prediction for a person adding 400 NEA calories:

```
Predictor          Coef     SE Coef       T      P
Constant         3.5051      0.3036   11.54  0.000
NEA change   -0.0034415   0.0007414   -4.64  0.000

S = 0.739853    R-Sq = 60.6%    R-Sq(adj) = 57.8%

Predicted Values for New Observations
New Obs    Fit SE Fit          95% CI             95% PI
      1  2.129  0.193  (1.714, 2.543)  (0.488, 3.769)
```

Exercises 23.16 to 23.24 are based on this information.

23.16 The equation of the least-squares regression line for predicting fat gain from NEA change is

(a) fat = 11.54 − 4.64 × NEA change.
(b) fat = −0.0034 + 3.5051 × NEA change.
(c) fat = 3.5051 − 0.0034 × NEA change.

23.17 What is the correlation between fat gain and NEA change?

(a) 0.606 (b) 0.778 (c) −0.778

23.18 Is there significant evidence that fat gain decreases as NEA change increases? To answer this question, we test the hypotheses

(a) H_0: $\beta = 0$ versus H_a: $\beta < 0$.
(b) H_0: $\beta = 0$ versus H_a: $\beta \neq 0$.
(c) H_0: $\alpha = 0$ versus H_a: $\alpha < 0$.

23.19 Minitab shows that the P-value for this test is

(a) 0.7398. (b) 0.3036. (c) less than 0.001.

23.20 The regression standard error for these data is

(a) 0.0007. (b) 0.3036. (c) 0.7399.

23.21 Confidence intervals and tests for these data use the t distribution with degrees of freedom

(a) 16. (b) 15. (c) 14.

23.22 A 95% confidence interval for the population slope β is

(a) -0.0034 ± 0.00145.
(b) -0.0034 ± 0.00159.
(c) -0.0034 ± 0.00344.

23.23 The Minitab output includes a prediction for y when $x^* = 400$. If an overfed adult burned an additional 400 NEA calories, we can be 95% confident that the person's fat gain would be between

(a) 1.71 and 2.54 kg.
(b) 1.75 and 2.51 kg.
(c) 0.49 and 3.77 kg.

23.24 If a whole population of overfed adults burned an additional 400 NEA calories, we can be 95% confident that the population mean fat gain would be between

(a) 1.71 and 2.54 kg.
(b) 1.75 and 2.51 kg.
(c) 0.49 and 3.77 kg.

CHAPTER 23 EXERCISES

23.25 Too much nitrogen? Intensive agriculture and burning of fossil fuels increase the amount of nitrogen deposited on the land. Too much nitrogen can reduce the variety of plants by favoring rapid growth of some species—think of putting fertilizer on your lawn to help grass choke out weeds. A study of 68 grassland sites in Britain measured nitrogen deposited (kilograms of nitrogen per hectare of land area per year) and also the "richness" of plant species (based on number of species and how abundant each species is). The authors reported a regression analysis as follows:[8]

$$\text{plant species richness} = 23.3 - 0.408 \times \text{nitrogen deposited}$$
$$r^2 = 0.55 \quad P < 0.0001$$

(a) What does the slope $b = -0.408$ say about the effect of increased nitrogen deposits on species richness?
(b) What does $r^2 = 0.55$ add to the information given by the equation of the least-squares line?
(c) What null and alternative hypotheses do you think the P-value refers to? What does this P-value tell you?

23.26 Beavers and beetles. Ecologists sometimes find rather strange relationships in our environment. One study seems to show that beavers benefit beetles. The researchers laid out 23 circular plots, each 4 meters in diameter, in an area where beavers were cutting down cottonwood trees. In each plot, they measured the number of stumps from trees cut by beavers and the number of clusters of beetle larvae. Here are the data:[9]

Stumps	2	2	1	3	3	4	3	1	2	5	1	3
Beetle larvae	10	30	12	24	36	40	43	11	27	56	18	40

Stumps	2	1	2	2	1	1	4	1	2	1	4
Beetle larvae	25	8	21	14	16	6	54	9	13	14	50

(a) Make a scatterplot that shows how the number of beaver-caused stumps influences the number of beetle larvae clusters. What does your plot show?
(b) Here is part of the Minitab regression output for these data:

Minitab

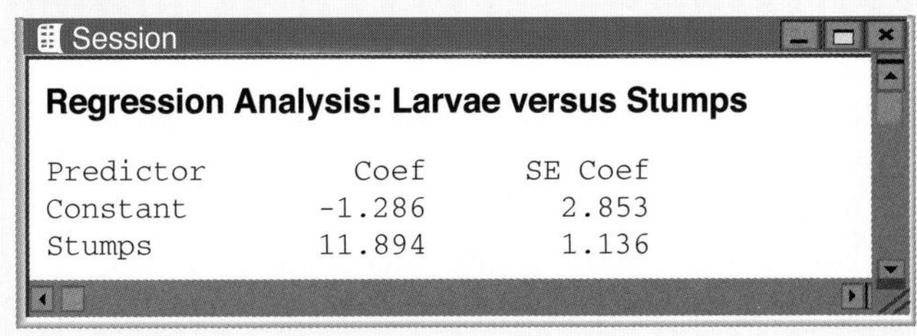

Session
Regression Analysis: Larvae versus Stumps
Predictor Coef SE Coef
Constant -1.286 2.853
Stumps 11.894 1.136

Find the least-squares regression line and draw it on your plot.
(c) Is there strong evidence that beaver stumps help explain beetle larvae counts? State hypotheses, give a test statistic and its P-value, and state your conclusion.

23.27 The amygdala and memory. The amygdala is a brain structure involved in the processing and memory of emotional reactions. In a research project, 10 subjects were shown emotional video clips. Then they had their brains scanned with positron emission tomography (PET), and their memory of the clips was assessed quantitatively. Here are the relative amygdala activity and the memory score for each subject:[10]

Relative activity	−0.417	−0.258	−0.234	−0.249	−0.156
Memory score	31	29	29	30	33

Relative activity	−0.031	0.120	0.240	0.342	0.654
Memory score	32	31	35	34	33

(a) Make a scatterplot that shows the relationship between relative amygdala activity (explanatory) and memory score (response).
(b) The scatterplot shows a positive linear relationship. How strong is it? Compute r.
(c) With only 10 data points, we wonder if the relationship is statistically significant. Run the appropriate test, find the P-value, and conclude. The standard error of the least-squares slope is $SE_b = 1.607$.

23.28 Heritability of phenotype. Figure 23.13 shows the relationship between parent nest size and offspring nest size for 100 parent-offspring observations of barn swallows, *Hirundo rustica*.[11] The line is the least-squares-regression line. Note that the graph actually plots the logarithm of nest size, because nest sizes, like many measures of volume, are otherwise strongly skewed to the right. Here is a partial Excel regression output for these data:

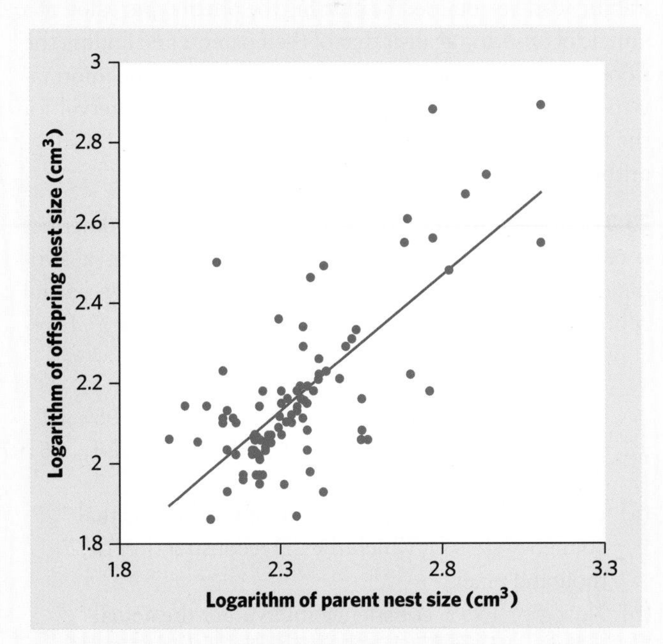

FIGURE 23.13 Scatterplot of the log of offspring nest size against the log of parent nest size, for Exercise 23.28.

```
Regression Statistics
Multiple R        0.7229
R Square          0.5226
Standard Error    0.1396
Observations      100

                          Standard
            Coefficients     Error   t Stat   P-value
Intercept         0.5764    0.1542
Parents           0.6762    0.0653
```

(a) Describe the relationship from the graph and assess the validity of inference for these data.

(b) Give the value of r^2 and the equation of the least-squares line.

(c) Based on the information given, test the hypothesis that there is no straight-line relationship between parent and offspring nest size. Give a test statistic, its approximate P-value, and your conclusion.

23.29 The amygdala and memory: estimating the slope. Refer to Exercise 23.27. Find the 95% confidence interval for the slope of the population regression line using either software or Table C.

23.30 Heritability of phenotype: estimating the slope. Regression is used in genetics to assess the heritability of a given phenotype from parents to their offspring. An estimate of heritability can be obtained by plotting the phenotypic value of a sample of offspring against that of their parents and finding the slope of the corresponding regression line. Using the information from Exercise 23.28, give a 95% confidence interval for the true population heritability (slope β) of nest size on a logarithmic scale.

23.31 Grading oysters: prediction. Example 23.1 describes a computerized assessment of oyster volume using a two-dimensional (2D) image-processing system. Here is a partial Minitab output for this example with prediction for a 2D reconstruction of 50 thousand pixels:

```
   Fit   SE Fit          95% CI                95% PI
13.581    0.285   (12.997, 14.165)   (11.088, 16.074)
```

(a) Give a 95% confidence interval for the mean actual volume oysters for which the 2D reconstruction is 50 thousand pixels.

(b) Now give a 95% confidence interval for the actual volume of 1 randomly selected oyster if that oyster has a 2D reconstruction of 50 thousand pixels.

(c) The purpose of this computer-imaging system is to avoid grading oysters by hand. Oysters are labeled as small, medium, or large if their volume is between 0 and 9.9, between 10 and 12.9, or above 13 cm^3, respectively. Which of the two confidence intervals would you use to assign a grade label to an oyster? In your judgment, is the 2D imaging system practically useful?

23.32 Grading oysters in 3D. Example 23.1 describes a 2D imaging system for grading oysters. Engineers were given the task of improving on the 2D reconstruction system. They designed a new program that estimates oyster volume using three-dimensional (3D) digital image processing. The results are displayed in Table 23.1.

(a) Make a scatterplot of 3D volume reconstruction (in millions of volume pixels) and actual volume (in cm^3), using 3D reconstruction for the explanatory variable. Find the equation for the least-squares regression line.

(b) Verify the conditions for inference.

(c) Following Examples 23.4 and 23.6, test whether the linear relationship is statistically significant and give a 95% confidence interval for the slope of the population regression line.

23.33 Grading oysters in 3D: prediction. Continue your work from the previous exercise. Here is a partial Minitab output with prediction for a 3D reconstruction of 5 million volume pixels:

```
   Fit   SE Fit          95% CI                95% PI
12.796    0.134   (12.520, 13.071)   (11.439, 14.153)
```

(a) Give a 95% confidence interval for the mean actual volume of oysters for which the 3D reconstruction is 5 million volume pixels.

(b) Now give a 95% confidence interval for the actual volume of 1 randomly selected oyster if that oyster has a 3D reconstruction of 5 million volume pixels.

(c) The purpose of this computer-imaging system is to avoid grading oysters by hand. Oysters are labeled as small, medium, or large if their volume is between 0 and 9.9, between 10 and 12.9, or above 13 cm^3, respectively. Which of the two confidence intervals would you use to assign a grade label to an oyster? In your judgment, is the 3D imaging system practically useful? How does it compare with the 2D system from Exercise 23.31?

23.34 Predicting tree height. Measuring tree height is not an easy task. How well might trunk diameter predict tree height? A survey of 958 live trees in an old-growth forest in Canada answered this question.[12] Here is part of the Minitab output based on these data for regressing height on diameter, both in centimeters (cm), along with prediction for a tree having a diameter of 50 cm:

```
Predictor      Coef    SE Coef       T      P
Constant     2.6696     0.1677   15.92  0.000
Diameter   0.550940   0.005058  108.93  0.000

S = 3.67427   R-Sq = 92.5%   R-Sq(adj) = 92.5%

New
Obs     Fit SE Fit          95% CI             95% PI
  1  30.217  0.179 (29.865,30.569) (22.997,37.436)
```

(a) A scatterplot of the data shows a reasonably linear relationship between tree height and diameter. Is this relationship statistically significant? How strong is the relationship?

(b) Give a 95% confidence interval for the height of 1 tree randomly selected from this forest if the tree has a diameter of 50 cm.

(c) Now give a 95% confidence interval for the mean height of all the trees in this forest that have a diameter of 50 cm. How does this interval compare with the one you calculated in (b)?

23.35 DNA on the ocean floor. 4 STEP We think of DNA as the stuff that stores the genetic code. It turns out that DNA occurs, mainly outside living cells, on the ocean floor. It is important in nourishing seafloor life. Scientists think that this DNA comes from organic matter that settles to the bottom from the top layers of the ocean. "Phytopigments," which come mainly from algae, are a measure of the amount of organic matter that has settled to the bottom. Table 23.4 contains data on concentrations of DNA and phytopigments (both in grams per square meter) in 116 ocean locations around the world.[13] Examine the relationship between DNA and phytopigments. Do the data in Table 23.4 give good reason to think that phytopigment concentration helps explain DNA concentration? Describe the data and follow the four-step process in answering this question.

TABLE 23.4 DNA and phytopigment concentrations (g/m^2) on the ocean floor

DNA	Phyto	DNA	Phyto	DNA	Phyto	DNA	Phyto	DNA	Phyto	DNA	Phyto
0.148	0.010	0.276	0.056	0.156	0.032	0.300	0.022	0.344	0.009	0.090	0.003
0.108	0.009	0.214	0.023	0.112	0.016	0.116	0.008	0.464	0.030	0.206	0.001
0.180	0.008	0.330	0.016	0.280	0.005	0.120	0.004	0.538	0.055	0.307	0.001
0.218	0.006	0.240	0.007	0.308	0.005	0.064	0.006	0.153	0.001	0.552	0.040
0.152	0.006	0.100	0.006	0.238	0.011	0.228	0.010	0.095	0	0.168	0.005
0.589	0.050	0.463	0.038	0.461	0.034	0.333	0.020	0.901	0.075	0.112	0.003
0.357	0.023	0.382	0.032	0.414	0.034	0.241	0.012	0.116	0.003	0.179	0.002
0.458	0.036	0.396	0.033	0.307	0.018	0.236	0.002	0.132	0.005	0.264	0.026
0.076	0.001	0.001	0.002	0.009	0	0.099	0	0.302	0.014	1.056	0.082
0.187	0.001	0.104	0.004	0.088	0.009	0.072	0.002	0.360	0.010	0.207	0.001
0.192	0.005	0.152	0.028	0.152	0.006	0.272	0.004	0.296	0.010	0.131	0.001
0.288	0.046	0.232	0.006	0.368	0.003	0.216	0.011	0.130	0.001	0.822	0.058
0.248	0.006	0.280	0.002	0.336	0.062	0.320	0.006	0.172	0.001	0.172	0.006
0.896	0.055	0.200	0.017	0.408	0.018	0.472	0.017	0.171	0.001	0.100	0.001
0.648	0.034	0.384	0.008	0.440	0.042	0.592	0.032	0.391	0.026	0.162	0.001
0.392	0.036	0.312	0.002	0.312	0.003	0.208	0.001	0.074	0	0.213	0.007
0.128	0.001	0.264	0.001	0.264	0.008	0.328	0.010	0.121	0.004		
0.264	0.002	0.376	0.003	0.288	0.001	0.208	0.024	0.369	0.023		
0.224	0.017	0.376	0.010	0.600	0.024	0.168	0.014	0.184	0.010		
0.264	0.018	0.152	0.010	0.184	0.016	0.312	0.017	0.328	0.028		

23.36 Sparrowhawk colonies. One of nature's patterns connects the percent of adult birds in a colony that return from the previous year and the number of new adults that join the colony. Here are data for 13 colonies of sparrowhawks:[14]

Percent return x	74	66	81	52	73	62	52	45	62	46	60	46	38
New adults y	5	6	8	11	12	15	16	17	18	18	19	20	20

In Exercise 4.9 (page 103) you found a moderately strong linear relationship between x and y. Figure 23.14 shows part of the Minitab regression output, including a prediction for y when 60% of the previous year's adult birds return.

(a) Write the equation of the least-squares line and use it to check that the "Fit" in the output is the predicted response for $x^* = 60\%$.

(b) Which 95% interval in the output gives us a margin of error for predicting the average number of new birds in colonies to which 60% of the past year's adults return?

23.37 DNA on the ocean floor: residuals. Save the residuals from the regression of DNA concentration on phytopigment concentration (Exercise 23.35). Examine the residuals to see how well the conditions for regression inference are met.

(a) Plot the residuals against phytopigment concentration (the explanatory variable), using vertical limits −1 to 1 to make the pattern clearer. Add a horizontal line at height 0 to represent the regression line. What do you conclude about the conditions of linear relationship and constant standard deviation?

(b) Make a histogram of the residuals. What do you conclude about Normality?

23.38 Sparrowhawk colonies: residuals. The regression of number of new birds that join a sparrowhawk colony on the percent of adult birds in the colony that return from the previous year is an example of data that satisfy the conditions for regression inference well. Here are the residuals for the 13 colonies in Exercise 23.36:

Percent return	74	66	81	52	73	62	52
Residual	−4.44	−5.87	0.69	−5.13	2.26	1.92	−0.13

Percent return	45	62	46	60	46	38
Residual	−1.25	4.92	0.05	5.31	2.05	−0.38

(a) **Independent observations.** Why are the 13 observations independent?

(b) **Linear relationship.** A plot of the residuals against the explanatory variable x magnifies the deviations from the least-squares line. Does the plot show any systematic deviation from a roughly linear pattern?

(c) **Constant spread about the line.** Does your plot in (b) show any systematic change in spread as x changes?

(d) **Normal variation about the line.** Make a histogram of the residuals. With only 13 observations, no clear shape emerges. Do strong skewness or outliers suggest lack of Normality?

Minitab

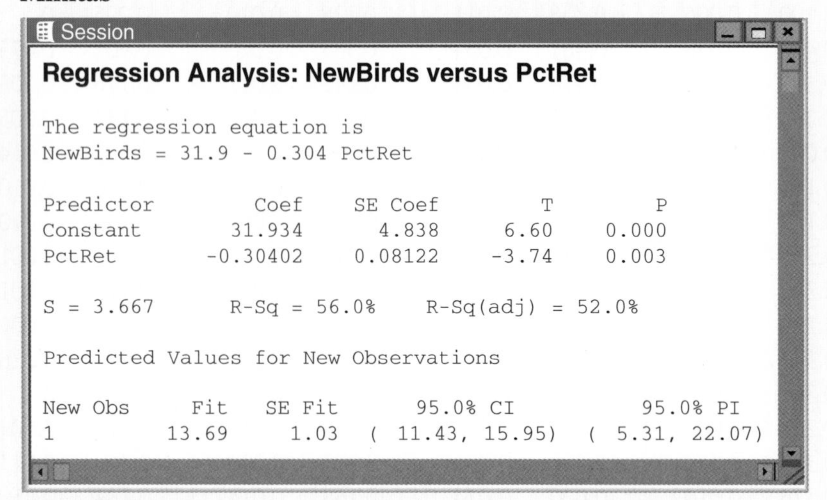

FIGURE 23.14 Partial Minitab output for predicting number of new birds in a sparrowhawk colony from percent of birds returning, for Exercise 23.36.

TABLE 23.5 **Body mass index (BMI, in kg/m^2) and testosterone concentration (in nmol/l) of 50 adolescent males**

BMI	Testosterone	BMI	Testosterone	BMI	Testosterone	BMI	Testosterone	BMI	Testosterone
21.4	0.78	20.4	0.49	19.5	0.23	33.2	0.23	41.5	0.25
19.0	0.70	16.2	0.43	20.2	0.24	34.7	0.24	38.0	0.25
18.3	0.63	17.8	0.42	21.2	0.24	35.8	0.06	38.1	0.32
19.5	0.60	21.0	0.38	21.3	0.26	37.0	0.15	37.8	0.35
20.9	0.60	18.6	0.35	22.2	0.27	37.0	0.17	34.9	0.37
23.4	0.69	20.9	0.35	28.3	0.30	39.0	0.18	34.8	0.39
25.0	0.76	22.4	0.32	27.7	0.24	41.6	0.17	34.7	0.46
24.1	0.58	23.5	0.31	28.1	0.19	42.4	0.15	32.0	0.49
24.2	0.50	18.8	0.28	29.2	0.17	47.7	0.12	31.9	0.42
22.6	0.48	19.3	0.25	33.3	0.18	45.7	0.25	30.5	0.36

23.39 Testosterone and BMI in adolescent males. 4 STEP Obesity in adult males is associated with lower levels of sex hormone. A study investigated a possible link between body mass index (BMI, in kilograms per square meter, kg/m^2) and plasma testosterone concentrations (in nanomoles per liter of blood, nmol/l) in a sample of 50 adolescent males between the ages of 14 and 20 years. The data are displayed in Table 23.5.[15] Follow the four-step process in the following analysis.

(a) Describe the relationship in a graph and with a regression line, using BMI as the explanatory variable. Be sure to save the regression residuals.

(b) Check the conditions for inference. Parts (a) to (d) of Exercise 23.38 provide a handy outline.

(c) Give a 95% confidence interval to estimate how plasma testosterone concentration changes as BMI increases.

23.40 Foot problems. 4 STEP Hallux abducto valgus (call it HAV) is a deformation of the big toe that is not common in youth and often requires surgery. Metatarsus adductus (call it MA) is a turning in of the front part of the foot that is common in adolescents and usually corrects itself. Doctors used X-rays to measure the angle of deformity (HAV angle and MA angle, both measured in degrees) in 38 consecutive patients under the age of 21 who came to a medical center for surgery to correct HAV. The data appear in Table 23.6.[16]

Metatarsus adductus may help predict the severity of hallux abducto valgus. The paper that reports this study says, "Linear regression analysis, using the hallux abductus angle as the response variable, demonstrated a significant correlation between the metatarsus adductus and hallux abductus angles." Do a suitable analysis to verify this finding, following the four-step process. (Be sure to check the conditions for inference as

TABLE 23.6 **Angle of deformity (degrees) for two types of foot deformity**

HAV angle	MA angle	HAV angle	MA angle	HAV angle	MA angle	HAV angle	MA angle	HAV angle	MA angle
28	18	30	19	21	7	18	15	26	30
32	16	26	10	23	11	26	16	28	22
25	22	28	17	14	15	16	10	31	24
34	17	13	14	32	12	30	12	38	20
38	33	20	20	25	16	30	10	32	37
26	10	21	15	21	16	20	10	21	23
25	18	17	16	22	18	50	12		
18	13	16	10	20	10	25	25		

part of the Solve step. Parts (a) through (d) of Exercise 23.38 provide a handy outline. The authors note that the scatterplot suggests that the variation in y may change as x changes, so they offer a more elaborate analysis as well.)

23.41 Testosterone and BMI in adolescent males: prediction. An adolescent male with a BMI of 30 gets a physical checkup. Using the information provided in Exercise 23.39, obtain a 95% interval to predict this adolescent's plasma testosterone concentration.

23.42 Chickadee alarm calls. The black-capped chickadee (*Poecile atricapilla*) is a small songbird commonly found in the northern United States and Canada. Chickadees often live in cooperative flocks, using a complex language to communicate about food sources and predator threats. In an experiment, researchers recorded chickadee vocalizations in an aviary when presented with predators of various sizes. The data below represent the average number of D notes per chickadee warning call for each type of predator, along with the predator wingspan (in centimeters):[17]

Predator	Predator wingspan (cm)	Number of D notes per call
Pygmy owl	31.2	3.96
Saw-whet owl	38.8	4.09
Kestrel	57.6	2.76
Merlin	60.6	3.04
Cooper's hawk	80.6	3.18
Short-eared owl	89.2	2.28
Prairie falcon	109.9	2.20
Gyrfalcon	115.1	2.25
Peregrine falcon	120.0	2.80
Red-tailed hawk	120.0	2.56
Great horned owl	120.4	2.46
Great gray owl	132.2	2.06
Rough-leg hawk	138.0	1.36

Use software to analyze these data.

(a) Make a scatterplot and find the least-squares line. Is the linear relationship obtained in this experiment statistically significant? Run the appropriate test and check the assumptions for inference.

(b) Give a 95% confidence interval for the population slope.

23.43 Revenge! Does revenge feel good? Or do people take revenge just because they are mad about being harmed? Different areas in the brain are active in the two cases, so brain scans can help decide which explanation is correct. Here's a game that supports the first explanation.

Player A is given $10. If he gives it to Player B, it turns into $40. B can keep all the money or give half to A, who naturally feels that B owes him half. If B keeps all $40, A can take revenge by removing up to $20 of B's ill-gotten gains, at no cost or gain to himself. Scan A's brain at that point, recording activity in the caudate nucleus, a region involved in "making decisions or taking actions that are motivated by anticipated rewards." Only the A players who took $20 from B play again, with different partners B. So all the A players have shown the same level of revenge when cheated by B.

The new B also keeps all the money—but punishing B now costs A $1 for every $2 he takes from B. The researchers predicted that A players who get more kicks from revenge, as measured by caudate activity, will punish B more severely even when it costs them money to do it. Here are data for 11 players:[18]

Caudate activity	−0.057	−0.011	−0.032	−0.025	−0.012	0.028
A takes from B	$0	$0	$5	$5	$10	$10

Caudate activity	−0.002	0.008	0.029	0.037	0.043
A takes from B	$10	$20	$20	$20	$20

(a) Make a scatterplot with caudate activity as the explanatory variable. Add the least-squares regression line to your plot to show the overall pattern.

(b) The research report mentions positive correlation and its significance. What is the correlation r? Is it significantly greater than zero?

(c) The nature of the data gives some reason to doubt the accuracy of the significance level. Why?

23.44 Standardized residuals (optional). Software often calculates **standardized residuals** as well as the actual residuals from regression. Because the standardized residuals have the standard z-score scale, it is easier to judge whether any are extreme. Here are the standardized residuals from Exercise 23.26 (beavers and beetles), rounded to two decimal places:

−1.99	1.20	0.23	−1.67	0.26	−1.06	1.38	0.06
0.72	−0.40	1.21	0.90	0.40	−0.43	−0.24	−1.36
0.88	−0.75	1.30	−0.26	−1.51	0.55	0.62	

(a) Find the mean and standard deviation of the standardized residuals. Why do you expect values close to those you obtain?

(b) Make a stemplot of the standardized residuals. Are there any striking deviations from Normality? The most extreme residual is $z = -1.99$. Would this be surprisingly large if the 23 observations had a Normal distribution? Explain your answer.
(c) Plot the standardized residuals against the explanatory variable. Are there any suspicious patterns?

23.45 Tests for the intercept (optional). Figure 23.8 gives the Minitab output for the regression of blood alcohol content (BAC) on number of beers consumed. The t test for the hypothesis that the population regression line has *slope* $\beta = 0$ has $P < 0.0001$. The data show a positive linear relationship between BAC and beers. We might expect the *intercept* α of the population regression line to be 0, because no beers ($x = 0$) should produce no alcohol in the blood ($y = 0$). To test

$$H_0: \alpha = 0$$
$$H_a: \alpha \neq 0$$

we use a t statistic formed by dividing the least-squares intercept a by its standard error SE_a. Locate this statistic in the output of Figure 23.8 and verify that it is in fact a divided by its standard error. What is the P-value? Do the data suggest that the intercept is not 0?

23.46 Confidence intervals for the intercept (optional). The output in Figure 23.8 allows you to calculate confidence intervals for both the slope β and the intercept α of the population regression line of BAC on beers in the population of all students. Confidence intervals for the intercept α have the familiar form $a \pm t^*SE_a$ with degrees of freedom $n - 2$. What is the 95% confidence interval for the intercept? Does it contain 0, the value we might guess for α?

23.47 Estimating body fat in men (optional). LARGE DATA SET The data file *Large.Bodyfat* contains data on the percent of body fat, age, weight, height, and ten body circumference measurements for 252 adult men. In Exercise 3.42 (page 88) you examined the two assessments of a person's percent body fat (percent fat1 and percent fat2) based on two computation methods using underwater weighing. If you haven't done so already, make a scatterplot of the two variables. Neither variable is explanatory here, so you should test the hypothesis of no correlation.

23.48 More on estimating body fat in men. LARGE DATA SET Continue your work with the data set from the previous exercise. In Exercise 4.51 (page 120) you found the equation for the least-squares regression predicting percent fat1 from abdomen circumference.
(a) Is the relationship between percent fat1 and abdomen circumference statistically significant? Run the appropriate test and check the conditions for inference.
(b) A physician needs to know the average percent body fat of adult men with abdomen circumferences of 100 cm. Obtain a 95% confidence interval for your answer.
(c) One adult man with abdomen circumference of 100 cm wants to predict his own percent body fat from his abdomen circumference value. Obtain a 95% prediction interval for your answer.

23.49 Lung capacity of children. LARGE DATA SET Continue your work from Exercise 4.49 (page 119) with the data file *Large.FEV*. Based on your earlier analysis of the data, let's study adolescents (middle school and high school: ages 11 and older) separately from the younger children (preschool and elementary school: ages 10 and younger); among adolescents, let's study boys and girls separately.
(a) Among the younger children, boys and girls are similar enough to be studied together. Find a 95% confidence interval for the slope of the population regression line between FEV and height for these younger children. Verify that the conditions for inference are satisfied. Obtain a 95% confidence interval for the mean FEV of children who are 5 feet tall (60 inches).
(b) Study the relationship between FEV and height for adolescent girls. Verify that the conditions for inference are satisfied and obtain a 95% confidence interval for the slope of the population regression line. Find a 95% confidence interval for the mean FEV of adolescent girls who are 68 inches tall.
(c) Study the relationship between FEV and height for adolescent boys. Verify that the conditions for inference are satisfied and find a 95% confidence interval for the slope of the population regression line. Obtain a 95% confidence interval for the mean FEV of adolescent boys who are 68 inches tall. Compare this interval with the one you obtained in (b).

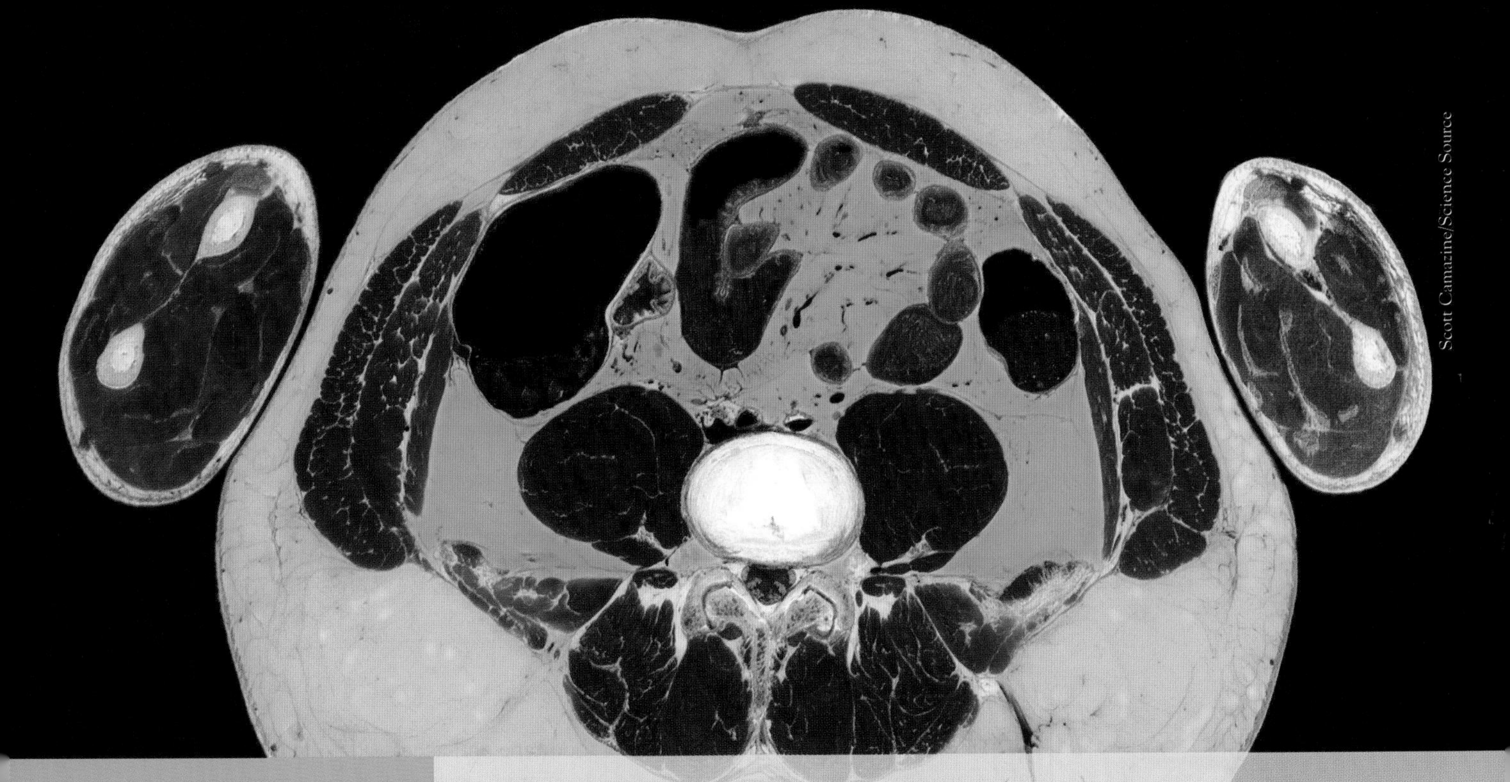

CHAPTER 24 One-Way Analysis of Variance: Comparing Several Means

The two-sample t procedures of Chapter 18 compare the means of two populations or the mean responses to two treatments in an experiment. Of course, studies don't always compare just two groups. We need a method for comparing any number of means.

IN THIS CHAPTER WE COVER...

- Comparing several means
- The analysis of variance F test
- Using technology
- The idea of analysis of variance
- Conditions for ANOVA
- F distributions and degrees of freedom
- The one-way ANOVA and the pooled two-sample t test*
- Details of ANOVA calculations*

EXAMPLE 24.1 Rats on a cafeteria-style diet

Recent studies have suggested that palatable, high-fat food can be addictive. A research team provided rats access to a cafeteria-style diet consisting of bacon, sausage, cheesecake, pound cake, frosting, and chocolate. To study the neural and behavioral impact of extended access to such food, the researchers first randomly assigned 50 adult male Wistar rats to 3 diet groups: unlimited access to rat chow only (chow group), unlimited access to rat chow with access to cafeteria food for 1 hour each day (restricted group), or unlimited access to rat chow with extended access to cafeteria food for 18 hours each day (extended group). Table 24.1 gives the weights in grams (g) of the rats in each group several weeks later.[1]

TABLE 24.1 Body weights (in grams) of adult male rats fed one of three diets

Chow only									
516	547	546	564	577	570	582	594	597	599
606	606	624	623	641	655	667	690	703	
Chow plus restricted access to cafeteria food									
546	599	612	627	629	638	645	635	660	676
695	687	693	713	726	736				
Chow plus extended access to cafeteria food									
564	611	625	644	660	679	687	688	706	714
738	733	744	780	794					

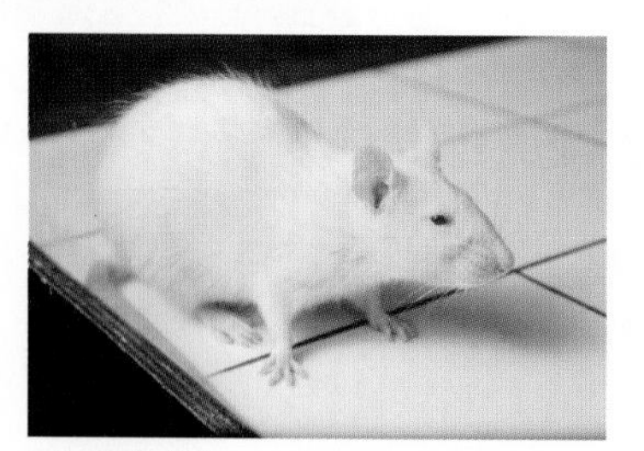

Luka Culig/Alamy

STATE: We are interested in the effect of food type and access on body weight in adult male rats. Are there differences in average weight among the 3 experimental groups?

PLAN: Use graphs and numerical summaries to describe and compare the 3 distributions of body weight. Finally, ask whether the differences among the mean weights of the 3 experimental groups are *statistically significant*.

SOLVE (first steps): Figure 24.1 displays the data in side-by-side dotplots. Here are the summary measures we will use in further analysis:

Sample	Group	Sample size	Mean weight	Standard deviation
1	Chow	19	605.6	49.64
2	Restricted	16	657.3	50.68
3	Extended	15	691.1	63.41

CONCLUDE (first steps): The body weights in the three experimental groups overlap substantially. However, the weights in the extended-access group seem higher overall than the weights in the restricted-access group, which themselves seem higher overall than the weights in the chow-only group. This is reflected in mean weights of 691.1, 657.3, and 605.6 g, respectively. Are these differences in sample means statistically significant? We must develop a test for comparing more than two population means. ■

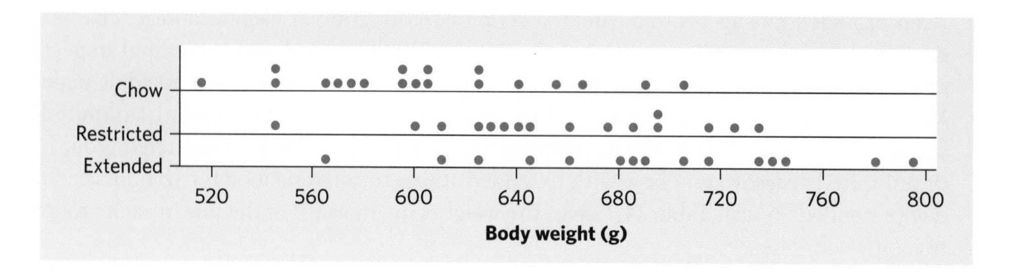

FIGURE 24.1 Dotplots comparing the body weights in grams of rats randomly assigned to 3 diet groups, for Example 24.1.

Comparing several means

Call the mean weights for the three experimental populations of diet types μ_1 for chow, μ_2 for restricted, and μ_3 for extended. The subscript reminds us which group a parameter or statistic describes. To compare these three population means, we might use the two-sample t test several times:

- Test H_0: $\mu_1 = \mu_2$ to see if the mean weight for chow differs from the mean weight for restricted.
- Test H_0: $\mu_1 = \mu_3$ to see if the mean weight for chow differs from the mean weight for extended.
- Test H_0: $\mu_2 = \mu_3$ to see if the mean weight for restricted differs from the mean weight for extended.

The weakness of doing three tests is that we get three P-values, one for each test performed. That doesn't tell us how likely it is that *three* sample means are spread apart as far as these are. It may be that $\bar{x}_1 = 605.6$ and $\bar{x}_3 = 691.1$ are significantly different if we look at just two groups but not significantly different if we know that they are the largest and the smallest means in three groups. (Think of comparing the tallest and shortest person in a large group of people. They may have quite different heights, yet they simply represent two extremes of a continuum of heights.) The problem gets worse as we compare more groups, because we expect the gap between the largest and smallest sample mean to get larger just by chance even if they are all samples from the same population. That is, *we can't safely compare many groups by doing tests or confidence intervals for two parameters at a time*.

The problem of how to do many comparisons at once with an overall measure of confidence in all our conclusions is common in statistics. This is the problem of **multiple comparisons.** Statistical methods for dealing with multiple comparisons usually have two steps:

multiple comparisons

1. An *overall test* to see if there is good evidence of *any* differences among the parameters that we want to compare.
2. A detailed *follow-up analysis* to decide which of the parameters differ and to estimate how large the differences are.

The overall test, though more complex than the tests we learned about earlier, is often reasonably straightforward. The follow-up analysis can be quite elaborate. This chapter concentrates on the overall test, along with data analysis that points to the nature of the differences. In companion Chapter 26 (available on the text's website) we cover the details of some common follow-up analyses used when comparing several means.

The analysis of variance F test

We want to test the null hypothesis that there are *no differences* among the mean weights for the three experimental populations of diet types:

$$H_0: \mu_1 = \mu_2 = \mu_3$$

The alternative hypothesis is that there is *some difference*. That is, not all three population means are equal:

$$H_a: \text{not all of } \mu_1,\ \mu_2, \text{ and } \mu_3 \text{ are equal}$$

analysis of variance F test

H_a simply says that H_0 is not true. The alternative hypothesis is no longer one-sided or two-sided. It is nondirectional, because it allows any relationship other than "all three equal." For example, H_a includes the case in which $\mu_2 = \mu_3$ but μ_1 has a different value. The test of H_0 against H_a is called the **analysis of variance *F* test.** Analysis of variance is usually abbreviated as ANOVA. The ANOVA F test is almost always carried out with software that reports the test statistic and its P-value.

EXAMPLE 24.2 Rats on a cafeteria-style diet: ANOVA

4 STEP

SOLVE (inference): Software tells us that for the weight data in Table 24.1, the test statistic is $F = 10.71$ with P-value $P = 0.0001$. There is very strong evidence that the three experimental groups do not all have the same mean weight.

The F test does not say *which* of the three means are significantly different. Our preliminary data analysis indicates that, overall, rats in the extended-access group have the largest weights and that rats in the chow-only group have the lowest weights.

CONCLUDE: There is strong evidence ($P = 0.0001$) that the population means are not all equal. The most important difference among the means is that rats given extended access to cafeteria food are heavier than rats given access only to rat chow. ■

Example 24.2 illustrates our approach to comparing means. The ANOVA F test (done with software) assesses the evidence for *some* difference among the population means. Even if you strongly suspect from looking at the data that you will find some effect, a formal test is important to guard against being misled by chance variation. We will not do the formal follow-up analysis that is often the most useful part of an ANOVA study. Follow-up analysis would allow us to say which means differ and by how much, with (say) 95% confidence that *all* our conclusions are correct. Companion Chapter 26 shows you how to do and interpret such an analysis. In this chapter we rely instead on a descriptive examination of the data to further interpret the findings in context.

Dr. Larry Jernigan/Getty Images

APPLY YOUR KNOWLEDGE

24.1 Do fruit flies sleep? Mammals and birds sleep. Insects such as fruit flies rest, but is this rest sleep? Biologists now think that insects do sleep. One experiment gave caffeine to fruit flies to see if it affected their rest. We know that caffeine reduces sleep in mammals, so if it reduces rest in fruit flies, that's another hint that the rest is really sleep. The paper reporting the study contains a graph similar to Figure 24.2

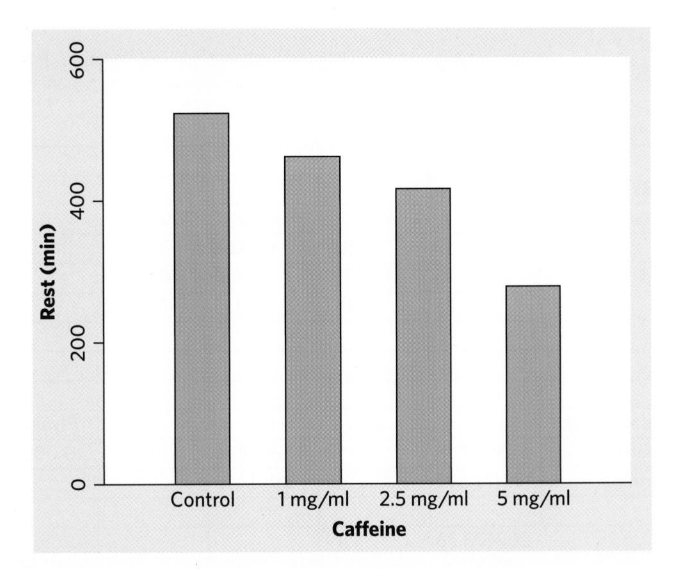

FIGURE 24.2 Bar graph comparing the mean rest time of fruit flies given different amounts of caffeine, for Exercise 24.1.

and states, "Flies given caffeine obtained less rest during the dark period in a dose-dependent fashion ($n = 36$ per group, $P < 0.0001$)."[2]

(a) The explanatory variable is amount of caffeine, in milligrams per milliliter (mg/ml) of blood. The response variable is minutes of rest (measured by an infrared motion sensor) during a 12-hour dark period. Outline the design of this experiment.

(b) The P-value in the report comes from the ANOVA F test. What means does this test compare? State in words the null and alternative hypotheses for the test in this setting. What do the graph and the statistical test together lead you to conclude?

24.2 Brain plasticity in songbirds. A study examined the plasticity in the adult brains of songbirds related to seasonal behavioral changes. The researchers captured adult male song sparrows at four different times in the year representing milestones in the birds' reproductive cycle. A brain nucleus, the higher vocal center, that controls song in songbirds was measured right after capture. Here are the mean nucleus volumes in cubic millimeters (mm^3):[3]

Early spring	Late spring	Early fall	Late fall
1.73	1.60	1.11	1.06

The publication states that $F = 7.7$ and $P < 0.001$.

(a) What are the null and alternative hypotheses for the ANOVA F test? Be sure to explain what means the test compares.

(b) Based on the sample means and the F test, what do you conclude?

Using technology

Any technology used for statistics should perform analysis of variance. Figure 24.3 displays ANOVA output for the data of Table 24.1 from a graphing calculator, three statistical programs, and a spreadsheet program.

Microsoft Excel

	A	B	C	D	E	F	G
1	Anova: Single Factor						
2							
3	SUMMARY						
4	*Groups*	*Count*	*Sum*	*Average*	*Variance*		
5	Chow	19	11507	605.63	2464.13		
6	Restricted	16	10517	657.31	2568.63		
7	Extended	15	10367	691.13	4020.70		
8							
9							
10	ANOVA						
11	*Source of variation*	*SS*	*df*	*MS*	*F*	*P-value*	*F crit*
12	Between Groups	63399.788	2	31699.894	10.705	0.00015	3.195
13	Within Groups	139173.592	47	2961.140			
14							
15	Total	202573.38	49				

Minitab

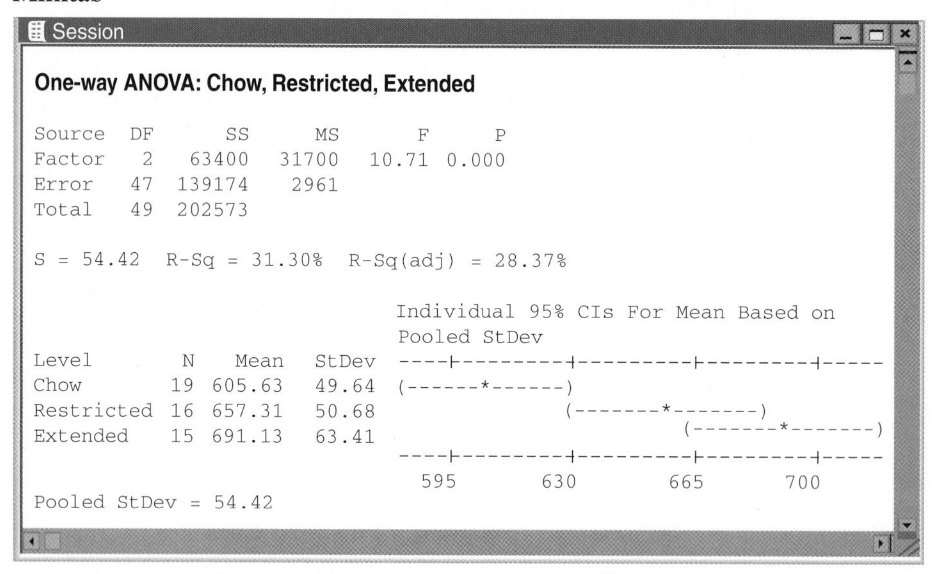

Session

One-way ANOVA: Chow, Restricted, Extended

```
Source  DF      SS     MS      F      P
Factor   2   63400  31700  10.71  0.000
Error   47  139174   2961
Total   49  202573

S = 54.42   R-Sq = 31.30%   R-Sq(adj) = 28.37%

                                Individual 95% CIs For Mean Based on
                                Pooled StDev
Level        N    Mean   StDev  ----+---------+---------+---------+-----
Chow        19  605.63   49.64  (------*------)
Restricted  16  657.31   50.68                (-------*-------)
Extended    15  691.13   63.41                          (-------*-------)
                                ----+---------+---------+---------+-----
                                  595       630       665       700

Pooled StDev = 54.42
```

FIGURE 24.3 ANOVA for the rat weight data: output from a graphing calculator, three statistical programs, and a spreadsheet program.

The software outputs give the sizes of the three samples and their means. They agree with those in Example 24.1. Minitab and SPSS also give the standard deviations. You should be able to recover the standard deviations from the variances (Excel) or the standard errors (JMP). All outputs report the most important part, the F test statistic, $F = 10.705$, and its P-value. Minitab and SPSS report the P-value as 0 to three decimal places, which is all we need to know in practice. JMP, Excel, and the TI-83 provide more decimal places. There is very strong evidence that rats in the three diet treatments do not all have the same mean weight.

All outputs also report degrees of freedom (df), sums of squares (SS), and mean squares (MS). We don't need this information now.

TI-83

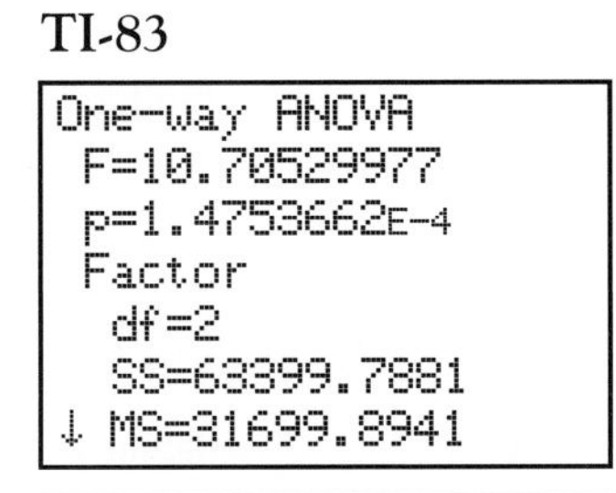

```
One-way ANOVA
 F=10.70529977
 p=1.4753662E-4
 Factor
  df=2
  SS=63399.7881
↓ MS=31699.8941
```

```
One-way ANOVA
↑ MS=31699.8941
 Error
  df=47
  SS=139173.592
  MS=2961.14025
 Sxp=54.4163602
```

JMP

JMP

Oneway Analysis of Weight by Diet

Oneway Anova

Summary of Fit

Rsquare	0.312972
Adj Rsquare	0.283737
Root Mean Square Error	54.41636
Mean of Response	647.82
Observations (or Sum Wgts)	50

Analysis of Variance

Source	DF	Sum of Squares	Mean Square	F Ratio	Prob > F
Diet	2	63399.79	31699.9	10.7053	0.0001*
Error	47	139173.59	2961.1		
C. Total	49	202573.38			

Means for Oneway Anova

Level	Number	Mean	Std Error	Lower 95%	Upper 95%
Chow	19	605.632	12.484	580.52	630.75
Extended	15	691.133	14.050	662.87	719.40
Restricted	16	657.313	13.604	629.94	684.68

Std Error uses a pooled estimate of error variance

SPSS

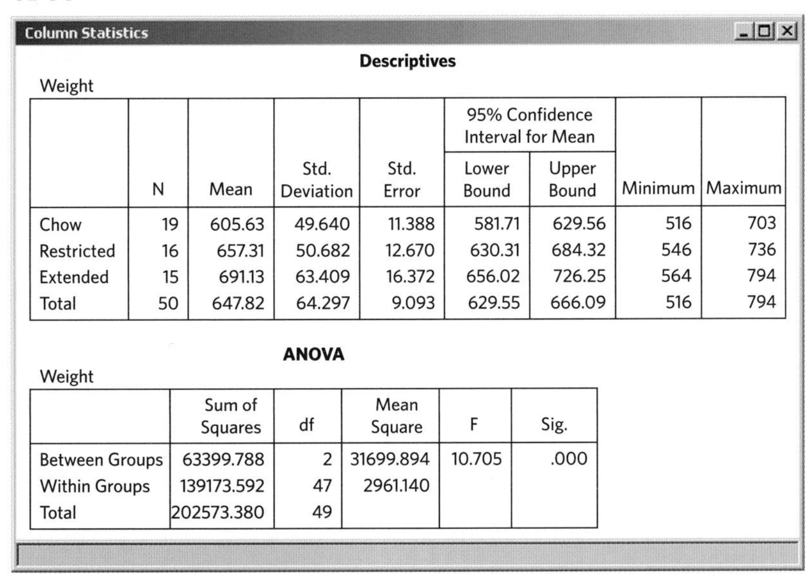

Column Statistics

Descriptives

Weight

	N	Mean	Std. Deviation	Std. Error	95% Confidence Interval for Mean Lower Bound	95% Confidence Interval for Mean Upper Bound	Minimum	Maximum
Chow	19	605.63	49.640	11.388	581.71	629.56	516	703
Restricted	16	657.31	50.682	12.670	630.31	684.32	546	736
Extended	15	691.13	63.409	16.372	656.02	726.25	564	794
Total	50	647.82	64.297	9.093	629.55	666.09	516	794

ANOVA

Weight

	Sum of Squares	df	Mean Square	F	Sig.
Between Groups	63399.788	2	31699.894	10.705	.000
Within Groups	139173.592	47	2961.140		
Total	202573.380	49			

FIGURE 24.3 *(Continued)*

Minitab, JMP, and SPSS also give confidence intervals for all three means that help us see which means differ and by how much. (We will see on page 618 how these are derived.) The chow and restricted intervals overlap only slightly, and the interval for extended is well above that for chow. These are 95% confidence intervals for each mean separately. We are *not* 95% confident that *all three* intervals cover the three population means. This is another example of the perils of multiple comparisons.

APPLY YOUR KNOWLEDGE

24.3 Logging in the rain forest. How does logging in a tropical rain forest affect the forest several years later? Researchers compared forest plots in Borneo that had never been logged (Group 1) with similar plots nearby that had been logged 1 year earlier (Group 2) and 8 years earlier (Group 3). Although the study was not an experiment, the authors explain why we can consider the plots to be randomly selected. The data appear in Table 24.2. The variable Trees is the count of trees in

TABLE 24.2 Data from a study of logging in Borneo

Observation	Group	Trees	Species	Richness
1	1	27	22	0.81481
2	1	22	18	0.81818
3	1	29	22	0.75862
4	1	21	20	0.95238
5	1	19	15	0.78947
6	1	33	21	0.63636
7	1	16	13	0.81250
8	1	20	13	0.65000
9	1	24	19	0.79167
10	1	27	13	0.48148
11	1	28	19	0.67857
12	1	19	15	0.78947
13	2	12	11	0.91667
14	2	12	11	0.91667
15	2	15	14	0.93333
16	2	9	7	0.77778
17	2	20	18	0.90000
18	2	18	15	0.83333
19	2	17	15	0.88235
20	2	14	12	0.85714
21	2	14	13	0.92857
22	2	2	2	1.00000
23	2	17	15	0.88235
24	2	19	8	0.42105
25	3	18	17	0.94444
26	3	4	4	1.00000
27	3	22	18	0.81818
28	3	15	14	0.93333
29	3	18	18	1.00000
30	3	19	15	0.78947
31	3	22	15	0.68182
32	3	12	10	0.83333
33	3	12	12	1.00000

Microsoft Excel

	A	B	C	D	E	F	G
1	Anova: Single Factor						
2							
3	SUMMARY						
4	*Groups*	*Count*	*Sum*	*Average*	*Variance*		
5	Group 1	12	285	23.75	25.6591		
6	Group 2	12	169	14.0833	24.8106		
7	Group 3	9	142	15.7778	33.1944		
8							
9							
10	ANOVA						
11	*Source of variation*	*SS*	*df*	*MS*	*F*	*P-value*	*F crit*
12	Between Groups	625.1566	2	312.57828	11.4257	0.000205	3.31583
13	Within Groups	820.7222	30	27.3574			
14							
15	Total	1445.879	32				

FIGURE 24.4 Excel output for analysis of variance on the number of trees in forest plots, for Exercise 24.3.

a plot; Species is the count of tree species in a plot. The variable Richness is the number of species divided by the number of individual trees, Species/Trees.[4]

(a) Make side-by-side stemplots or dotplots of Trees for the three groups. Use stems 0, 1, 2, and 3 and split the stems if you choose to make a stemplot (see page 21 for an example of a split-stem stemplot). What effects of logging are visible?

(b) Figure 24.4 shows Excel ANOVA output for Trees. What do the group means show about the effects of logging?

(c) What are the values of the ANOVA F statistic and its P-value? What hypotheses does F test? What conclusions about the effects of logging on number of trees do the data lead to?

24.4 Dogs, friends, and stress. If you are a dog lover, perhaps having your dog along reduces the effect of stress. To examine the effect of pets in stressful situations, researchers recruited 45 women who said they were dog lovers. The EESEE story "Stress among Pets and Friends" describes the results. Fifteen of the subjects were randomly assigned to each of three groups to do a stressful task alone (the control group), with a good friend present, or with their dog present. The subject's mean heart rate (in beats per minute) during the task is one measure of the effect of stress. Table 24.3 contains the data.

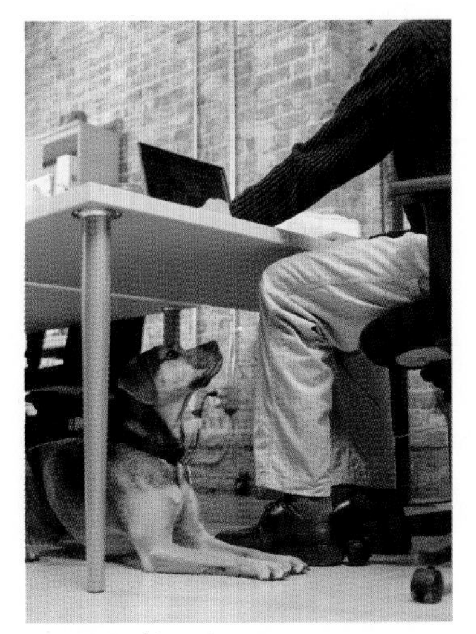

Ryan McVay/Getty Images

(a) Make stemplots or dotplots of the heart rates for the three groups (round to the nearest whole number of beats). Do any of the groups show outliers or extreme skewness?

(b) Figure 24.5 gives the Minitab ANOVA output for these data. Do the mean heart rates for the groups appear to show that the presence of a pet or a friend reduces heart rate during a stressful task?

(c) What are the values of the ANOVA F statistic and its P-value? What hypotheses does F test? Briefly describe the conclusions that you draw from these data. Did you find anything surprising?

TABLE 24.3 Mean heart rates during stress with a pet (P), with a friend (F), and for the control group (C)

Group	Rate	Group	Rate	Group	Rate
P	69.169	P	68.862	C	84.738
F	99.692	C	87.231	C	84.877
P	70.169	P	64.169	P	58.692
C	80.369	C	91.754	P	79.662
C	87.446	C	87.785	P	69.231
P	75.985	F	91.354	C	73.277
F	83.400	F	100.877	C	84.523
F	102.154	C	77.800	C	70.877
P	86.446	P	97.538	F	89.815
F	80.277	P	85.000	F	98.200
C	90.015	F	101.062	F	76.908
C	99.046	F	97.046	P	69.538
C	75.477	C	62.646	P	70.077
F	88.015	F	81.600	F	86.985
F	92.492	P	72.262	P	65.446

Minitab

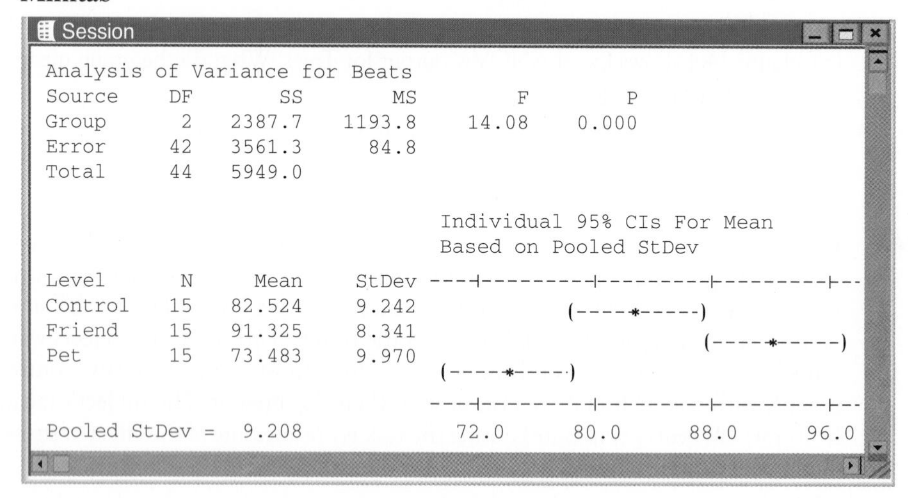

```
Session
Analysis of Variance for Beats
Source     DF       SS        MS        F        P
Group       2   2387.7    1193.8    14.08    0.000
Error      42   3561.3      84.8
Total      44   5949.0

                                  Individual 95% CIs For Mean
                                  Based on Pooled StDev
Level       N     Mean     StDev  ---+---------+---------+---------+--
Control    15   82.524     9.242            (----*-----)
Friend     15   91.325     8.341                       (----*-----)
Pet        15   73.483     9.970  (----*----)
                                  ---+---------+---------+---------+--
Pooled StDev =   9.208               72.0      80.0      88.0      96.0
```

FIGURE 24.5 Minitab output for the data in Table 24.3 on heart rates during stress, for Exercise 24.4. The "Control" group worked alone, the "Friend" group had a friend present, and the "Pet" group had a pet dog present.

The idea of analysis of variance

The details of ANOVA are a bit heavier computationally than those of other tests we have studied so far (they appear in an optional section at the end of this chapter). The main idea of ANOVA is both more accessible and much more important. Here it is: When we ask if a set of sample means gives evidence for differences among the population means, what matters is not how far apart the sample means are but how far apart they are *relative to the variability of individual observations*.

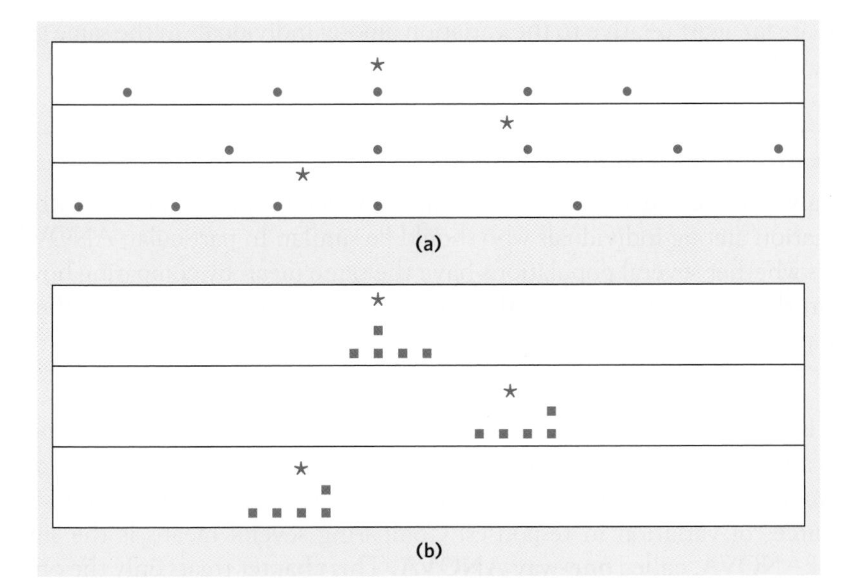

FIGURE 24.6 Dotplots for two sets of three samples each, using the same scale. The sample means, marked with green asterisks, are the same in (a) and (b). Analysis of variance will find a more significant difference among the means in (b) because there is less variation among the individuals within those samples.

Look at the two sets of dotplots in Figure 24.6. The sample means are marked with green asterisks for ease of comparison. Both sets of dotplots use the same scale and compare three samples with the same three means. Could differences this large easily arise just due to chance, or are they statistically significant?

- The dotplots in Figure 24.6(a) show lots of variation among the individuals in each group. With this much variation among individuals, we would not be surprised if another set of samples gave quite different sample means. The observed differences among the sample means could easily happen just by chance.
- The dotplots in Figure 24.6(b) have the same means as those in Figure 24.6(a), but there is much less variation among the individuals in each group. It is unlikely that any sample from the second (b) group would have a mean as small as the mean of the third (b) group. Because means as far apart as those observed would rarely arise just by chance in repeated sampling, they are good evidence of real differences among the means of the three populations from which we are sampling.

You can use the *One-Way ANOVA* applet to demonstrate the analysis of variance idea for yourself. The applet allows you to change both the group means and the spread within groups. You can watch the ANOVA F statistic and its P-value change as you work.

The comparison of the two parts of Figure 24.6 is, of course, rudimentary. Our discussion ignored the effect of the sample sizes. *Small differences among sample means can be significant if the samples are large. Large differences among sample means can fail to be significant if the samples are small.* All we can be sure of is that, because the sample sizes in (a) and (b) are the same, Figure 24.6(b) will give a much smaller P-value than Figure 24.6(a). Despite this caveat, the big idea remains: If sample

means are far apart relative to the variation among individuals in the same groups, that's evidence that something other than chance is at work.

> **THE ANALYSIS OF VARIANCE IDEA**
>
> **Analysis of variance** compares the variation due to specific sources with the variation among individuals who should be similar. In particular, ANOVA tests whether several populations have the same mean by comparing how far apart the sample means are with how much variation there is within the samples.

It is one of the oddities of statistical language that methods for comparing means are named after the variance. The reason is that the test works by comparing two kinds of variation. Analysis of variance is a general method for studying sources of variation in responses. Comparing several means is the simplest form of ANOVA, called **one-way ANOVA.** This chapter treats only the one-way ANOVA. Companion Chapter 26 introduces the two-way ANOVA, used when data can be organized in two-way layouts representing two factors.

one-way ANOVA

> **THE ANOVA *F* STATISTIC**
>
> The **analysis of variance *F* statistic** for testing the equality of several means has this form:
>
> $$F = \frac{\text{variation among the sample means}}{\text{variation among individuals in the same sample}}$$

The numerator is fundamentally a variance calculated using the sample means, while the denominator is fundamentally an average of the sample variances. Both are weighted to take into account different sample sizes. If you want more detail, read the optional section at the end of this chapter.

The F statistic can take only values that are zero or positive. It is zero only when all the sample means are identical and gets larger as they move farther apart. When H_0 is true, the numerator and denominator should have similar values and we would expect F to be small. F is large when the sample means are much more variable than individuals in the same sample. Large values of F are evidence against the null hypothesis H_0 that all population means are the same. Like the alternative hypothesis H_a, the ANOVA F test is nondirectional, because any violation of H_0 tends to produce a large value of F.

APPLY YOUR KNOWLEDGE

24.5 ANOVA compares several means. The *One-Way ANOVA* applet displays the observations in three groups, with the group means highlighted by black dots. When you open or reset the applet, the scale at the bottom of the display shows the

ANOVA F statistic and P-value for these groups. (The P-value is marked by a red dot that moves along the scale.)

(a) One group has a larger mean than the other two. Grab its mean point with the mouse. How small can you make F? What did you do to the mean to make F small? Roughly how significant is your small F?

(b) Starting with the three means aligned from your configuration at the end of (a), drag any one of the group means either up or down. What happens to F? What happens to the P-value? Convince yourself that the same thing happens if you move any one of the means, or if you move one slightly and then another slightly in the opposite direction.

24.6 ANOVA uses within-group variation. Reset the *One-Way ANOVA* applet to its original state.

(a) Use the mouse to slide the Standard Deviation at the top of the display to the right (to increase it). You see that the group means do not change, but the spread of the observations in each group increases. What happens to F and P as the spread among the observations in each group increases? What are the values of F and P when the slider is all the way to the right? This is similar to Figure 24.6(a): Variation within groups hides the differences among the group means.

(b) Leave the Standard Deviation slider at the extreme right of its scale, so that spread within groups stays fixed. Use the mouse to move the group means apart. What happens to F and P as you do this?

Conditions for ANOVA

Like all inference procedures, ANOVA is valid only in some circumstances. Here are the conditions under which we can use ANOVA to compare population means.

CONDITIONS FOR APPLYING ANOVA

- We have **k independent SRSs,** one from each of k populations.
- Each of the k populations has a **Normal distribution** with an unknown mean. The parameter μ_i is the unknown mean of the ith population. The means may be different in the different populations.
- All the populations have the **same standard deviation σ**, whose value is unknown.

The first two requirements are familiar from our study of the two-sample t procedures for comparing two means. As usual, the design of the data production is the most important condition for inference. Biased sampling or confounding can make any inference meaningless. *If we do not actually draw separate SRSs from each population or carry out a randomized comparative experiment, it may be unclear to what population the conclusions of inference apply.* In the life sciences, samples are often not true SRSs but, rather, individuals randomly chosen from a pool of available individuals. You must judge each use on its merits, a judgment that usually

requires some knowledge of the subject of the study in addition to some knowledge of statistics.

Because no real population has an exactly Normal distribution, the usefulness of inference procedures that assume Normality depends on how sensitive they are to departures from Normality. Fortunately, procedures for comparing means are not very sensitive to lack of Normality. The ANOVA F test, like the t procedures, is **robust.** What matters is Normality of the sample means, so ANOVA becomes safer as the sample sizes get larger, because of the central limit theorem effect. Remember to check for outliers that change the value of sample means and for extreme skewness. When there are no outliers and the distributions are roughly symmetric, you can safely use ANOVA for sample sizes as small as 4 or 5.

robustness

The third condition is more of a challenge: ANOVA assumes that the variability of observations, measured by the standard deviation, is the same in all populations. You may recall from Chapter 18 (page 453) that there is a special version of the two-sample t test that assumes equal standard deviations in both populations. The ANOVA F for comparing two means is exactly the square of this special t statistic. We prefer the t test that does not assume equal standard deviations, but for comparing more than two means there is no general alternative to the ANOVA F. It is not easy to check the condition that the populations have equal standard deviations. Statistical tests for equality of standard deviations are very sensitive to lack of Normality, so much so that they are of little practical value. You must either seek expert advice or rely on the robustness of ANOVA.

How serious are unequal standard deviations? ANOVA is not too sensitive to violations of the condition, especially when all samples have the same or similar sizes and no sample is very small. When designing a study, try to take samples of about the same size from all the groups you want to compare. The sample standard deviations estimate the population standard deviations, so check before doing ANOVA that the sample standard deviations are similar to each other. We expect some variation among them due to chance. Here is a conservative rule of thumb that is safe in almost all situations.

CHECKING STANDARD DEVIATIONS IN ANOVA

The results of the ANOVA F test are approximately correct when the largest sample standard deviation is no more than twice as large as the smallest sample standard deviation.

EXAMPLE 24.3 Rats on a cafeteria-style diet: conditions for ANOVA

The rat diet study is based on the random assignment of adult male Wistar lab rats to three treatment groups, thus creating independent samples that the researchers consider to be random samples from all adult male Wistar rats. The dotplots in Figure 24.1 show no obvious departure from Normality for the weights in each group, so the sample means of samples of sizes at least 15 will have distributions that are close to Normal. The sample standard deviations for the three experimental groups are

$$s_1 = 49.64 \quad s_2 = 50.68 \quad s_3 = 63.41$$

These standard deviations satisfy our rule of thumb:

$$\frac{\text{largest } s}{\text{smallest } s} = \frac{63.41}{49.64} = 1.28$$

We can safely use ANOVA to compare the mean weights for the three experimental populations. ■

EXAMPLE 24.4 Which color attracts beetles best?

STATE: To detect the presence of harmful insects in farm fields, we can put up boards covered with a sticky material and examine the insects trapped on the boards. Which colors attract insects best? Experimenters placed six boards of each of four colors at random locations in a field of oats and measured the number of cereal leaf beetles trapped. Here are the data:[5]

Board color	Beetles trapped					
Blue	16	11	20	21	14	7
Green	37	32	20	29	37	32
White	21	12	14	17	13	20
Yellow	45	59	48	46	38	47

Holt Studios International/Alamy

PLAN: Examine the data to determine the effect of board color on beetles trapped and check that we can safely use ANOVA. If the data allow ANOVA, assess the significance of the observed differences in mean counts of beetles trapped.

SOLVE: Because the samples are small, we plot the data in side-by-side stemplots in Figure 24.7. We can see that the yellow boards attract by far the most beetles ($\bar{x}_4 = 47.2$), with green next ($\bar{x}_2 = 31.2$) and blue and white far behind.

Check that we can safely use ANOVA to test equality of the four means. The largest of the four standard deviations is 6.795 and the smallest is 3.764. The ratio

$$\frac{\text{largest } s}{\text{smallest } s} = \frac{6.795}{3.764} = 1.8$$

is less than 2, so these data satisfy our rule of thumb. The shapes of the four distributions are irregular, as we expect with only 6 observations in each group, but there are no outliers. The ANOVA results will be approximately correct. The F statistic is $F = 42.84$, a large F with $P < 0.0001$. SPSS output for ANOVA appears in Figure 24.8.

CONCLUDE: Despite the small samples, the experiment gives very strong evidence of differences among the colors. A descriptive evaluation of the data suggests that yellow boards may be best at attracting leaf beetles. ■

Blue		Green		White		Yellow	
0	7	0		0		0	
1	1 4 6	1		1	2 3 4 7	1	
2	0 1	2	0 9	2	0 1	2	
3		3	2 2 7 7	3		3	8
4		4		4		4	5 6 7 8
5		5		5		5	9

FIGURE 24.7 Side-by-side stemplots comparing the counts of insects attracted by six boards of each of four colors, for Example 24.4.

SPSS

Column Statistics

Descriptives

	N	Mean	Std. Deviation	Std. Error	95% Confidence Interval for Mean	
					Lower Bound	Upper Bound
Blue	6	14.83	5.345	2.182	9.22	20.44
Green	6	31.17	6.306	2.574	24.55	37.78
White	6	16.17	3.764	1.537	12.22	20.12
Yellow	6	47.17	6.795	2.774	40.04	54.30
Total	24	27.33	14.412	2.942	21.25	33.42

ANOVA

	Sum of Squares	df	Mean Square	F	Sig.
Between Groups	4134.000	3	1378.000	42.839	.000
Within Groups	643.333	20	32.167		
Total	4777.333	23			

FIGURE 24.8 SPSS ANOVA output for comparing the four board colors in Example 24.4.

APPLY YOUR KNOWLEDGE

24.7 Checking standard deviations. Verify that the sample standard deviations for the following data sets do allow use of ANOVA to compare the population means.

(a) The counts of trees in Exercise 24.3 and Figure 24.4.

(b) The heart rates of Exercise 24.4 and Figure 24.5.

24.8 Species richness after logging. Table 24.2 gives data on the species richness in rain forest plots, defined as the number of tree species in a plot divided by the number of trees in the plot. ANOVA may not be trustworthy for the richness data. Do data analysis: Make side-by-side stemplots or dotplots to examine the distributions of the response variable in the three groups, and also compare the standard deviations. What characteristic of the data makes ANOVA risky?

24.9 Smoking during pregnancy. Cigarette labels warn pregnant women against smoking. Does nicotine actually reach the fetus, crossing the protective placental barrier? Researchers selected consecutive pregnant women delivering at an Egyptian hospital and categorized them as active smokers, passive smokers, or nonsmokers. They then analyzed the newborns' meconium for cotinine content, the metabolized form of nicotine. Meconium is a newborn's first stool right after birth and is a good biological marker for fetal exposure to drugs or other chemical agents. Here are the meconium cotinine levels (in nanograms per milliliter, ng/ml) for the three groups:[6]

Active smokers	490	418	405	328	700	292	295	272	240	232
Passive smokers	254	219	287	257	271	282	148	273	350	293
Nonsmokers	158	163	153	207	211	159	199	187	200	213

(a) In their published report the researchers wrote that they ran an ANOVA and that $F = 10.45$ and $P < 0.01$. Use technology to confirm that ANOVA does indeed produce these values.

(b) Examine the three samples. Do the standard deviations satisfy our rule of thumb? What are the overall shapes of the distributions? Are there outliers? Can ANOVA be used safely on these data?

(c) Suppose that the high outlier you found in part (b) was absent. Could we safely use ANOVA on the remaining data? Explain your answer. (It is not hopeless though. When samples have very different standard deviations, transforming the original scale into a logarithmic scale can help obtain more similar standard deviations. We discussed logarithmic transformations previously, in the context of correlation and regression in Chapters 3 and 4.)

F distributions and degrees of freedom

To find the P-value for the ANOVA F statistic, we must know the sampling distribution of F when the null hypothesis (all population means are equal) is true. This sampling distribution is an **F distribution.**

***F* distribution**

The F distributions are a family of right-skewed distributions with two parameters. The parameters are the degrees of freedom of the numerator and of the denominator of the F statistic. The numerator's degrees of freedom are always mentioned first. *Interchanging the degrees of freedom changes the distribution, so the order is important.* Our brief notation will be $F(\text{df1}, \text{df2})$ for the F distribution with df1 degrees of freedom in the numerator and df2 in the denominator.

EXAMPLE 24.5 Rats on a cafeteria-style diet: the *F* distribution

Look again at the software output for the body weight data in Figure 24.3. All outputs give the degrees of freedom for the F test, labeled "df" or "DF." There are 2 degrees of freedom in the numerator and 47 in the denominator of the F statistic. P-values for the F test therefore come from the F distribution with 2 and 47 degrees of freedom. Figure 24.9 shows the density curve of this distribution. The 5% critical value is 3.195, and the 1% critical value is 5.087. The observed value $F = 10.705$ of the ANOVA F statistic lies far to the right of these values, so the P-value is very small. ■

The degrees of freedom of the ANOVA F statistic depend on the number of means we are comparing and the number of observations in each sample. That is, the F test takes into account the number of observations. Here are the details.

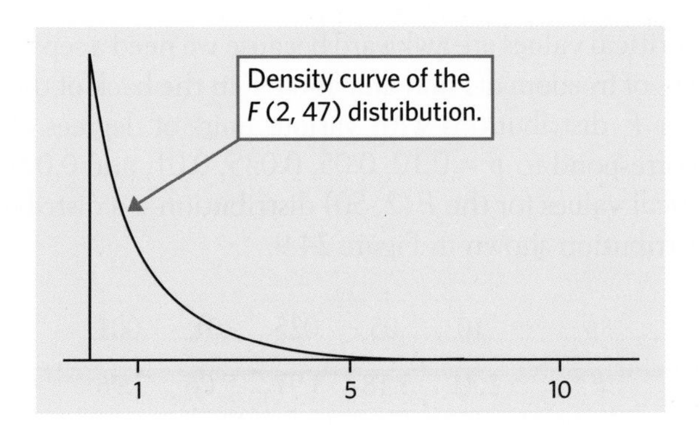

FIGURE 24.9 The density curve of the F distribution with 2 degrees of freedom in the numerator and 47 degrees of freedom in the denominator, for Example 24.5.

DEGREES OF FREEDOM FOR THE *F* TEST

We want to compare the means of k populations. We have an SRS of size n_i from the ith population, so that the total number N of observations in all samples combined is

$$N = n_1 + n_2 + \cdots + n_k$$

If the null hypothesis that all population means are equal is true, the ANOVA F statistic has the F distribution with $k - 1$ degrees of freedom in the numerator and $N - k$ degrees of freedom in the denominator.

This makes sense if you think back to the definitions for the sample standard deviation and sample variance in Chapter 2. For a sample of n data points, the variance has $n - 1$ degrees of freedom. The F statistic is a ratio of two measures of variance. The numerator assesses the variance of the k sample means and so has $k - 1$ degrees of freedom. The denominator assesses the variance of the data within each sample by combining information from the k sample variances (each with $n_i - 1$ degrees of freedom) and so has $n_1 - 1 + n_2 - 1 + \cdots + n_k - 1 = N - k$ degrees of freedom.

EXAMPLE 24.6 Degrees of freedom for *F*

In Examples 24.1 and 24.2 we compared the mean weights for rats randomly assigned to three diets, so $k = 3$. The three sample sizes are

$$n_1 = 19 \quad n_2 = 16 \quad n_3 = 15$$

The total number of observations is therefore

$$N = 19 + 16 + 15 = 50$$

The ANOVA F statistic has numerator degrees of freedom

$$k - 1 = 3 - 1 = 2$$

and denominator degrees of freedom

$$N - k = 50 - 3 = 47$$

These are the degrees of freedom given in the outputs in Figure 24.3. ■

Tables of F critical values are awkward because we need a separate table for every pair of degrees of freedom df1 and df2. Table F in the back of the book contains critical values for F distributions with various pairs of degrees of freedom. The critical values correspond to $p = 0.10, 0.05, 0.025, 0.01$, and 0.001. For example, here are the critical values for the $F(2, 50)$ distribution—a distribution similar to the $F(2, 47)$ distribution shown in Figure 24.9:

p	.10	.05	.025	.01	.001
F^*	2.41	3.18	3.97	5.06	7.96

With $F = 10.705$ we can conclude that the test P-value for Example 24.5 is smaller than 0.001.

You will rarely need Table F, however, because software gives P-values directly.

APPLY YOUR KNOWLEDGE

24.10 Logging in the rain forest, continued. Exercise 24.3 compares the number of trees in rain forest plots that had never been logged (Group 1) with similar plots nearby that had been logged 1 year earlier (Group 2) and 8 years earlier (Group 3).

(a) What are k, the n_i, and N for these data? Identify these quantities in words and give their numerical values.

(b) Find the degrees of freedom for the ANOVA F statistic. Check your work against the Excel output in Figure 24.4.

(c) For these data, $F = 11.43$. What does Table F tell you about the P-value of this statistic?

24.11 Do fruit flies sleep? Exercise 24.1 describes a study of the effect of caffeine on sleep in fruit flies.

(a) What are the degrees of freedom for this experiment?

(b) Find in Table F the F distribution with the most similar degrees of freedom. Using this distribution, how large would the test F statistic need to be to reach significance at $\alpha = 5\%$?

(c) The report gives $P < 0.0001$ for this ANOVA test. Based on Table F, we know that the test F statistic must have been larger than what value?

The one-way ANOVA and the pooled two-sample t test*

If the conditions for inference are met, the one-way ANOVA tests the hypothesis that all of k populations have the same mean:

$$H_0: \mu_1 = \mu_2 = \cdots = \mu_k$$
$$H_a: \text{not all of the } \mu_i \text{ are equal}$$

When there are just two populations, the hypotheses become

$$H_0: \mu_1 = \mu_2$$
$$H_a: \mu_1 \neq \mu_2$$

which can also be tested with a two-sample t procedure.

Because ANOVA requires equal population standard deviations, it is most closely related to the pooled two-sample t test. Recall from Chapter 18 (page 453)

*This more advanced section is optional.

that the pooled two-sample t statistic is

$$t = \frac{\bar{x}_1 - \bar{x}_2}{s_p\sqrt{\dfrac{1}{n_1} + \dfrac{1}{n_2}}}$$

where s_p, the pooled estimate of the population standard deviation σ, is

$$s_p = \sqrt{\frac{(n_1 - 1)s_1^2 + (n_2 - 1)s_2^2}{n_1 + n_2 - 2}}$$

In fact, in the two-sample case the F statistic is exactly the square of the pooled t statistic,

$$F = \frac{\text{variation among the sample means}}{\text{variation among individuals within samples}} = \frac{(\bar{x}_1 - \bar{x}_2)^2}{s_p^2\left(\dfrac{1}{n_1} + \dfrac{1}{n_2}\right)} = t^2$$

and both tests give exactly the same two-sided P-value.

We discussed in Chapter 18 the advantages of not having to assume equal population variances. When comparing two means, we recommend avoiding the pooled t procedure, because extensive studies show that t with its accurate degrees of freedom performs essentially as well as pooled t when the standard deviations really are equal and provides notably more accurate P-values when they are not. When we want to compare more than two means, however, there is no simple way to avoid the equal standard deviations condition.

The ANOVA F for more than two samples extends the idea of the two-sample procedure in a straightforward way. In fact, the denominator of F is exactly the weighted average of all the sample variances with degrees of freedom as weights (see for yourself that the sum of all the $n_i - 1$ is $N - k$). The numerator must compare k sample means rather than just 2, so it becomes a bit more complicated.

Details of ANOVA calculations*

Don't touch the plants We know that confounding can distort inference. We don't always recognize how easy it is to confound data. Consider the innocent scientist who visits plants in the field once a week to measure their size. A study of six plant species found that one touch a week significantly increased leaf damage by insects in two species and significantly decreased damage in another species.

Now we will give the general formula for the ANOVA F statistic. We have SRSs from each of k populations. Subscripts from 1 to k tell us which sample a statistic refers to:

Population	Sample size	Sample mean	Sample std. dev.
1	n_1	$\bar{x}_1$	s_1
2	n_2	$\bar{x}_2$	s_2
⋮	⋮	⋮	⋮
k	n_k	$\bar{x}_k$	s_k

*This more advanced section is optional if you are using software to find the F statistic.

You can find the F statistic from just the sample sizes n_i, the sample means $\overline{x}_i$, and the sample standard deviations s_i. You don't need to go back to the individual observations.

The ANOVA F statistic has the form

$$F = \frac{\text{variation among the sample means}}{\text{variation among individuals in the same sample}}$$

The measures of variation in the numerator and denominator of F are called **mean squares.** A mean square is a more general form of a sample variance. An ordinary sample variance s^2 is an average (or mean) of the squared deviations of observations from their mean, so it qualifies as a "mean square."

mean squares

The numerator of F is a mean square that measures variation among the k sample means $\overline{x}_1, \overline{x}_2, \ldots, \overline{x}_k$. Call the overall mean response (the mean of all N observations together) $\overline{x}$. You can find $\overline{x}$ from the k sample means by taking

$$\overline{x} = \frac{n_1\overline{x}_1 + n_2\overline{x}_2 + \cdots + n_k\overline{x}_k}{N}$$

The sum of each mean multiplied by the number of observations it represents is the sum of all the individual observations. Dividing this sum by N, the total number of observations, gives the overall mean $\overline{x}$. The numerator mean square in F is an average of the k squared deviations of the means of the samples from $\overline{x}$. We call it the **mean square for groups,** abbreviated as **MSG:**

MSG

$$\text{MSG} = \frac{n_1(\overline{x}_1 - \overline{x})^2 + n_2(\overline{x}_2 - \overline{x})^2 + \cdots + n_k(\overline{x}_k - \overline{x})^2}{k - 1}$$

Each squared deviation is weighted by n_i, the number of observations it represents.

The mean square in the denominator of F measures variation among individual observations in the same sample. For any one sample, the sample variance s_i^2 does this job. For all k samples together, we use an average of the individual sample variances. It is again a weighted average in which each s_i^2 is weighted by one fewer than the number of observations it represents, $n_i - 1$. Another way to put this is that each s_i^2 is weighted by its degrees of freedom $n_i - 1$. The resulting mean square is called the **mean square for error, MSE:**

MSE

$$\text{MSE} = \frac{(n_1 - 1)s_1^2 + (n_2 - 1)s_2^2 + \cdots + (n_k - 1)s_k^2}{N - k}$$

"Error" doesn't mean that a mistake has been made. It's a traditional term for chance variation. Here is a summary of the ANOVA test.

THE ANOVA F TEST

Draw an independent SRS from each of k populations. The ith population has the $N(\mu_i, \sigma)$ distribution, where σ is the common standard deviation in all the populations. The ith sample has size n_i, sample mean $\overline{x}_i$, and sample standard deviation s_i.

The **ANOVA F statistic** tests the null hypothesis that all k populations have the same mean:

$$H_0: \mu_1 = \mu_2 = \cdots = \mu_k$$

$$H_a: \text{not all of the } \mu_i \text{ are equal } (H_0 \text{ is not true})$$

The statistic is

$$F = \frac{\text{MSG}}{\text{MSE}}$$

The numerator of F is the **mean square for groups**

$$\text{MSG} = \frac{n_1(\overline{x}_1 - \overline{x})^2 + n_2(\overline{x}_2 - \overline{x})^2 + \cdots + n_k(\overline{x}_k - \overline{x})^2}{k - 1}$$

The denominator of F is the **mean square for error**

$$\text{MSE} = \frac{(n_1 - 1)s_1^2 + (n_2 - 1)s_2^2 + \cdots + (n_k - 1)s_k^2}{N - k}$$

When H_0 is true, F has the **F distribution** with $k - 1$ and $N - k$ degrees of freedom.

sums of squares

ANOVA table

The denominators in the formulas for MSG and MSE are the two degrees of freedom, $k - 1$ and $N - k$, of the F test. The numerators are called **sums of squares,** from their algebraic form. It is usual to present the results of ANOVA in an **ANOVA table.** Output from software usually includes an ANOVA table.

EXAMPLE 24.7 ANOVA calculations: software

Look again at the five outputs in Figure 24.3. The four software outputs give the ANOVA table. The TI-83, with its small screen, gives the same information as a list rather than a table. Each output uses slightly different language to identify the two sources of variation. The basic ANOVA table is

Source of variation	df	SS	MS	F statistic
Among samples (groups)	2	63,399.79	MSG = 31,699.89	10.705
Within samples (error)	47	139,173.59	MSE = 2,961.14	

You can check that each mean square MS is the corresponding sum of squares SS divided by its degrees of freedom df. The F statistic is MSG divided by MSE. ■

Because MSE is an average of the individual sample variances, it is also called the *pooled sample variance*, written as s_p^2. When all k populations have the same population variance σ^2, as ANOVA assumes that they do, s_p^2 estimates the common variance σ^2. The square root of MSE (root mean square error) is the **pooled standard deviation** s_p. It estimates the common standard deviation σ of observations in each group. The Minitab, JMP, and TI-83 outputs in Figure 24.3 give the value $s_p = 54.416$.

pooled standard deviation

The pooled standard deviation s_p is a better estimator of the common σ than any individual sample standard deviation s_i because it combines (pools) the information in all k samples. We can get a confidence interval for any of the means μ_i from the usual form

$$\text{estimate} \pm t^*\text{SE}_{\text{estimate}}$$

using s_p to estimate σ. The confidence interval for μ_i is

$$\overline{x}_i \pm t^* \frac{s_p}{\sqrt{n_i}}$$

Use the critical value t^* from the t distribution with $N - k$ degrees of freedom, because s_p has $N - k$ degrees of freedom. These are the confidence intervals that appear in the Minitab output in Figure 24.3.

EXAMPLE 24.8 ANOVA calculations: without software

We can do the ANOVA test comparing the mean weights of adult male Wistar rats eating the chow, restricted, and extended diets using only the sample sizes, sample means, and sample standard deviations. These appear in Example 24.1, but it is easy to find them with a calculator. There are $k = 3$ groups with a total of $N = 50$ rats.

The overall mean of the 50 weights in Table 24.1 is

$$\begin{aligned} \overline{x} &= \frac{n_1\overline{x}_1 + n_2\overline{x}_2 + n_3\overline{x}_3}{N} \\ &= \frac{(19)(605.63) + (16)(657.31) + (15)(691.13)}{50} \\ &= \frac{32{,}390.88}{50} = 647.82 \end{aligned}$$

The mean square for groups is

$$\begin{aligned} \text{MSG} &= \frac{n_1(\overline{x}_1 - \overline{x})^2 + n_2(\overline{x}_2 - \overline{x})^2 + n_3(\overline{x}_3 - \overline{x})^2}{k - 1} \\ &= \frac{1}{3-1}[(19)(605.63 - 647.82)^2 + (16)(657.31 - 647.82)^2 \\ &\quad + (15)(691.13 - 647.82)^2] \\ &= \frac{63{,}397.23}{2} = 31{,}698.62 \end{aligned}$$

The mean square for error is

$$\begin{aligned}\text{MSE} &= \frac{(n_1 - 1)s_1^2 + (n_2 - 1)s_2^2 + (n_3 - 1)s_3^2}{N - k} \\ &= \frac{(18)(49.64^2) + (15)(50.68^2) + (14)(63.41^2)}{47} \\ &= \frac{139{,}172.86}{47} = 2{,}961.12\end{aligned}$$

Finally, the ANOVA test statistic is

$$F = \frac{\text{MSG}}{\text{MSE}} = \frac{31{,}698.62}{2{,}961.12} = 10.705$$

Our work differs slightly in places from the output in Figure 24.3 because of roundoff error. We don't recommend doing these calculations, because tedium and roundoff errors cause frequent mistakes. ■

APPLY YOUR KNOWLEDGE

The calculations of ANOVA use only the sample sizes n_i, the sample means $\overline{x}_i$, and the sample standard deviations s_i. You can therefore re-create the ANOVA calculations when a report gives these summaries but does not give the actual data. These optional exercises ask you to do the ANOVA calculations starting with the summary statistics.

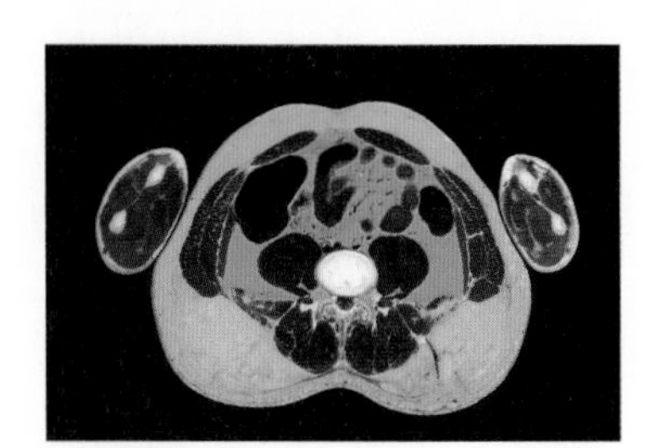

Scott Camazine/Science Source

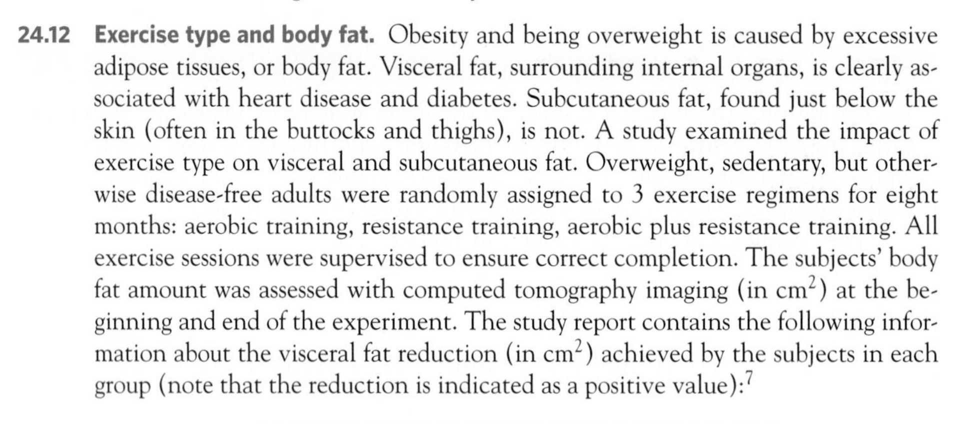

24.12 Exercise type and body fat. Obesity and being overweight is caused by excessive adipose tissues, or body fat. Visceral fat, surrounding internal organs, is clearly associated with heart disease and diabetes. Subcutaneous fat, found just below the skin (often in the buttocks and thighs), is not. A study examined the impact of exercise type on visceral and subcutaneous fat. Overweight, sedentary, but otherwise disease-free adults were randomly assigned to 3 exercise regimens for eight months: aerobic training, resistance training, aerobic plus resistance training. All exercise sessions were supervised to ensure correct completion. The subjects' body fat amount was assessed with computed tomography imaging (in cm^2) at the beginning and end of the experiment. The study report contains the following information about the visceral fat reduction (in cm^2) achieved by the subjects in each group (note that the reduction is indicated as a positive value):[7]

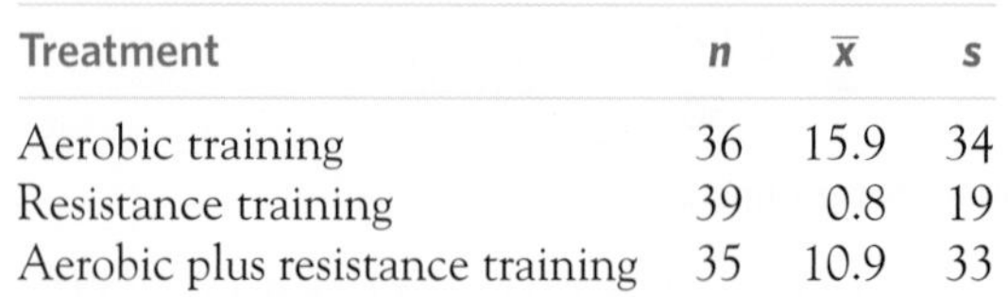

Treatment	n	$\overline{x}$	s
Aerobic training	36	15.9	34
Resistance training	39	0.8	19
Aerobic plus resistance training	35	10.9	33

(a) Do the standard deviations satisfy the rule of thumb for safe use of ANOVA?

(b) The report does not provide the distributions of visceral fat reduction. ANOVA is nonetheless safe for these data. Why?

(c) Calculate the overall mean response $\overline{x}$, the mean square for groups MSG, and the mean square for error MSE.

(d) Find the ANOVA F statistic and its approximate P-value. Is there evidence that the mean visceral fat reductions of overweight adults who follow the 3 exercise programs differ?

24.13 Exercise type and body fat, continued. The study in the previous exercise also reports the subcutaneous fat reduction (in cm^2) achieved by the subjects in each group (note that the reduction is indicated as a positive value):

Treatment	n	$\bar{x}$	s
Aerobic training	36	25.1	30
Resistance training	39	8.2	54
Aerobic plus resistance training	35	28.7	37

(a) Do the standard deviations satisfy the rule of thumb for safe use of ANOVA?

(b) The report does not provide the distributions of subcutaneous fat reduction. ANOVA is nonetheless safe for these data. Why?

(c) Calculate the overall mean response $\bar{x}$, the mean square for groups MSG, and the mean square for error MSE.

(d) Find the ANOVA F statistic and its approximate P-value. Is there evidence that the mean subcutaneous fat reductions of overweight adults who follow the 3 exercise programs differ?

24.14 Did the randomization work? When assigning patients with existing symptoms to various treatments in a clinical trial, it is important to verify that randomization did not, by chance, create uneven groups, with different overall levels of symptom severity. A study of low bone mineral density (BMD) among postmenopausal women randomly assigned subjects to 5 treatment groups. BMD at the lumbar spine was measured at the beginning of the study to assess symptom severity. BMD is expressed as a unitless, standardized value; a negative score indicates a BMD less than the normal healthy average of a population of young women. Here are the summarized BMD scores for the 5 groups at the beginning of the study:[8]

Treatment group	n	$\bar{x}$	s
Placebo	46	−2.2	0.7
Denosumab, 6 mg	44	−2.0	0.9
Denosumab, 14 mg	44	−2.0	0.8
Denosumab, 30 mg	41	−2.2	0.7
Alendronate	47	−2.0	0.9

Calculate the ANOVA table and the F statistic. Are the differences in mean BMD scores among the 5 groups significant? What does that say about the randomization process for this study?

CHAPTER 24 SUMMARY

- **One-way analysis of variance (ANOVA)** compares the means of several populations. The **ANOVA F test** tests the overall H_0 that all the populations have the same mean. If the F test shows significant differences, examine the data to see where the differences lie and whether they are large enough to be important.

- The **conditions for ANOVA** state that we have an **independent SRS** from each population; that each population has a **Normal distribution;** and that all populations have the **same standard deviation.**
- In practice, ANOVA inference is relatively **robust** when the populations are non-Normal, especially when the samples are large. Before doing the F test, check the observations in each sample for outliers or strong skewness. Also verify that the largest sample standard deviation is no more than twice as large as the smallest standard deviation.
- When the null hypothesis is true, the **ANOVA F statistic** for comparing k means from a total of N observations in all samples combined has the **F distribution** with $k - 1$ and $N - k$ degrees of freedom.
- ANOVA calculations are reported in an **ANOVA table** that gives sums of squares, mean squares, and degrees of freedom for variation among groups and for variation within groups. In practice, we use software to do the calculations.

THIS CHAPTER IN CONTEXT

In Chapter 17 we described the one-sample t procedures used to perform statistical inference for one population mean μ. This was followed in Chapter 18 by the two-sample t procedures used to perform statistical inference for the difference $\mu_1 - \mu_2$ between the means of two distinct populations. In this chapter, we address how analysis of variance, specifically one-way ANOVA, can be used to compare the means from several populations.

ANOVA is a first step in comparing the means of several populations. To avoid the problems associated with multiple comparisons, we must first ask if there is evidence that these population means are not all equal. This is what the one-way ANOVA tests. When the test is statistically significant, we have evidence that at least one population mean in the set is significantly different from at least one other population mean. But ANOVA does not tell us which ones. Companion Chapter 26 describes how we can perform follow-up tests that compare the population means two at a time while compensating for the resulting multiple comparisons.

Analysis of variance is a versatile statistical procedure that can be adapted to more complex settings. Companion Chapter 26 introduces the two-way ANOVA procedure, which allows us to compare population means obtained from more complex designs involving two different factors, or explanatory variables.

The t procedures and ANOVA are used when the response variable is quantitative. Which procedure to use depends on the study design, something that we discussed in Chapters 7 and 8. Concerns about how the data were collected should lead you to question the extent or even the validity of the conclusions from any statistical inference based on these data. And the conditions for inference also typically require that we examine the data for evidence of skew or outliers, as seen in Chapters 1 and 2.

CHECK YOUR SKILLS

24.15 The purpose of analysis of variance is to compare

(a) the variances of several populations.
(b) the proportions of successes in several populations.
(c) the means of several populations.

24.16 A study of the effects of smoking classifies subjects as nonsmokers, moderate smokers, or heavy smokers. The investigators interview a sample of 200 people in each group. Among the questions is "How many hours do you sleep on a typical night?" The degrees of freedom for the ANOVA F statistic comparing mean hours of sleep are

(a) 2 and 197. (b) 2 and 597. (c) 3 and 597.

24.17 The alternative hypothesis for the ANOVA F test in the previous exercise is

(a) the population mean hours of sleep in the groups are all the same.
(b) the population mean hours of sleep in the groups are all different.
(c) the population mean hours of sleep in the groups are not all the same.

A dental study evaluated the effect of tooth etch time on resin bonding strength. A total of 78 undamaged, recently extracted first molars (baby teeth) were randomly assigned to be etched with phosphoric acid gel for either 15, 30, or 60 seconds. Composite resin cylinders of identical size were then bonded to the tooth enamel. The researchers examined the bond strength after 24 hours by finding the failure load (in megapascals) for each bond. Here are the summary data and ANOVA table for this experiment:[9]

Etch time	n	$\bar{x}$	s
15 seconds	26	4.49	2.28
30 seconds	26	6.98	3.15
60 seconds	26	8.48	4.17

Source	DF	SS	MS	F	P
Etch time	2	211.208	105.604	9.745	0.0002
Error	75	812.745	10.837		
Total	77	1023.953			

Exercises 24.18 to 24.24 are based on this study.

24.18 The most striking conclusion from the numerical summaries for the three etch times is that

(a) there appears to be little difference among the etch times.
(b) on average, failure load increases with etch time.
(c) on average, failure load decreases with etch time.

24.19 The null hypothesis for this ANOVA F test is

(a) the population mean load failures for the three etch times are all the same.
(b) the population mean load failures for the three etch times are all different.
(c) the population mean load failure is lowest for the 15-second condition and highest for the 60-second condition.

24.20 If we used a series of two-sample t procedures to compare the three conditions, we would have to give three 95% confidence intervals to compare all three pairs of etch times. The weakness of doing this is that

(a) we won't be 95% confident that all 3 intervals cover the true differences in population means.
(b) the conclusions from the three intervals might not agree.
(c) the conditions for two-sample t inference are not met for all 3 pairs of etch times.

24.21 The conclusion of the ANOVA test is that

(a) there is quite strong evidence ($P = 0.0002$) that the mean failure loads are not the same in all three etch conditions.
(b) there is quite strong evidence ($P = 0.0002$) that the mean failure load is much lower for 15 seconds of etch time than for any other two etch times.
(c) the data give no evidence ($P = 0.0002$) to suggest that mean failure loads differ among the three etch times.

24.22 Without software, we would compare $F = 9.745$ with critical values from Table F. This comparison shows that

(a) the P-value is greater than 0.05.
(b) the P-value is between 0.05 and 0.001.
(c) the P-value is less than 0.001.

24.23 After checking the condition that all populations compared have the same standard deviation, you decide that this condition

(a) is not met, because the sample standard deviations are too different.
(b) is not met, because we don't know the population standard deviations.
(c) is met, because the sample standard deviations are similar enough. (The largest one is less than twice the smallest.)

24.24 The researchers stated that the distributions of failure loads were not Normally distributed but right-skewed (without extreme outliers). The test P-value of 0.0002 is

(a) reasonably accurate anyway because the sample sizes are large (26 each).

(b) inaccurate because the populations must be Normally distributed for ANOVA.

(c) unreliable no matter how the data are distributed because these data can't be regarded as random samples or a randomized experiment.

CHAPTER 24 EXERCISES

Exercises 24.25 to 24.27 describe situations in which we want to compare the mean responses in several populations. For each setting, identify the populations and the response variable. Then give k, the n_i, and N. Finally, give the degrees of freedom of the ANOVA F test.

24.25 Biological rhythms. Are you a morning person, an evening person, or neither? Does this personality trait affect how well you perform? A sample of 100 students took a psychological test that found 16 morning people, 30 evening people, and 54 who were neither. All the students then took a test of their ability to memorize at 8 A.M. and again at 9 P.M. You analyze the score at 8 A.M. minus the score at 9 P.M.

24.26 Caffeine and sugar. A double-blind randomized experiment assigned healthy undergraduate students to drink one of four beverages after fasting overnight: water, water with 75 mg of caffeine, water with 75 g of glucose, and water with 75 mg of caffeine and 75 g of glucose. Subjects performed a number of cognitive tasks, including the California Computerized Assessment Package, a computerized reaction time program measuring sustained attention, reaction time, and visual scanning speed. Here are the resulting reaction times (SEM is the standard error of the mean):[10]

Beverage	n	$\bar{x}$	SEM
Water	18	389.35	18.50
Water and caffeine	18	320.16	17.98
Water and glucose	18	318.16	17.04
Water, caffeine, and glucose	18	336.44	14.02

24.27 A medical study. The Québec (Canada) Cardiovascular Study recruited men aged 34 to 64 years at random from towns in the Québec City metropolitan area. Of these, 1824 met the criteria (no diabetes, free of heart disease, and so on) for a study of the relationship between being overweight and medical risks. The 719 normal-weight men had a mean triglyceride level of 1.5 millimoles per liter (mmol/l); the 885 overweight men had a mean of 1.7 mmol/l; and the 220 obese men had a mean of 1.9 mmol/l.[11]

24.28 Whose little bird is that? The males of many animal species signal their fitness to females by displays of various kinds. For male barn swallows, the signal is the color of their plumage. Does better color really lead to more offspring? A randomized comparative experiment assigned barn swallow pairs who had already produced a clutch of eggs to three groups: 13 males had their color "enhanced" by the investigators; 9 were handled but their color was left alone; and 8 were not touched. DNA testing showed whether the eggs were sired by the male of the pair or by another male. The investigators then destroyed the eggs. Barn swallows usually produce a second brood in such circumstances. Sure enough, the males with enhanced color fathered more young in the second brood, and males in the other groups fathered fewer. The investigators used ANOVA to compare the 3 groups.[12]

(a) What are the degrees of freedom for the F statistic that compares the number of young sired by each male among the eggs in the first brood? The statistic was $F = 0.07$. What can you say about the P-value? What do you conclude?

(b) Of the 30 original pairs, 27 produced a second brood. What are the degrees of freedom for F to compare differences in the number of young sired by each male between the first and second broods produced by these 27 pairs? The statistic was $F = 5.45$. How significant is this F?

24.29 Plants defend themselves. When some plants are attacked by leaf-eating insects, they release chemical compounds that both repel the leaf-eaters and attract other insects that prey on the leaf-eaters. An experiment examined the release of these protective compounds by a plant that grows in the Utah desert.[13] The investigators compared plants attacked by one of three types of leaf-eaters and other plants that were undamaged—32 plants of the same species in all, 8 per group. They then measured emissions of several compounds during seven hours. Here are data (mean ± standard error of the mean) for one compound. The emission rate is measured in nanograms per hour (ng/hr).

Group	Emission rate (ng/hr)
Control	9.22 ± 5.93
Hornworm	31.03 ± 8.75
Leaf bug	18.97 ± 6.64
Flea beetle	27.12 ± 8.62

(a) Make a graph that compares the mean emission rates for the four groups. Does it appear that emissions increase when the plant is attacked?
(b) What hypotheses does ANOVA test in this setting?
(c) We do not have the full data set. What would you look for in deciding whether you can safely use ANOVA?
(d) What is the relationship between the standard error of the mean and the standard deviation for a sample? Do the standard deviations satisfy our rule of thumb for safe use of ANOVA?

24.30 Can you hear these words? To test whether a hearing aid is right for a patient, audiologists play a tape on which words are pronounced at low volume. The patient tries to repeat the words. There are several different lists of words that are supposed to be equally difficult. Are the lists equally difficult when there is background noise? To find out, an experimenter had subjects with normal hearing listen to four lists with a noisy background. The response variable was the percent of the 50 words in a list that the subject repeated correctly. The data set contains 96 responses.[14] Here are two study designs that could produce these data:

Design A. The experimenter assigns 96 subjects to 4 groups at random. Each group of 24 subjects listens to one of the lists. All individuals listen and respond separately.
Design B. The experimenter has 24 subjects. Each subject listens to all four lists in random order. All individuals listen and respond separately.

Does Design A allow use of one-way ANOVA to compare the lists? Does Design B allow use of one-way ANOVA to compare the lists? Briefly explain your answers.

24.31 More rain for California? The changing climate will probably bring more rain to California, but we don't know whether the additional rain will come during the winter wet season or extend into the long dry season in spring and summer. Kenwyn Suttle of the University of California at Berkeley and his coworkers randomly assigned plots of open grassland to three treatments: added water equal to 20% of annual rainfall either during January to March (winter) or during April to June (spring), and no added water (control). Here are some of the data, for plant biomass (in grams per square meter) produced by each plot in a single year.[15]

Winter	Spring	Control
264.1514	318.4182	129.0538
187.7312	281.6830	144.6578
291.1431	288.8433	172.7772
176.2879	382.6673	113.2813
141.7525	326.8877	142.1562
169.9737	293.8502	117.9808

Figure 24.10 shows Minitab ANOVA output for these data.
(a) Make side-by-side stemplots or dotplots of plant biomass for the three treatments, as well as a table of the sample means and standard deviations. What do the data appear to show about the effect of extra water in winter and in spring on biomass? Do these data satisfy the conditions for ANOVA?
(b) State H_0 and H_a for the ANOVA F test, and explain in words what ANOVA tests in this setting.
(c) Report your overall conclusions about the effect of added water on plant growth in California. Be sure to report the test statistic and P-value.

24.32 Can you hear these words? continued. Figure 24.11 displays the Minitab output for one-way ANOVA applied to the hearing data described in Exercise 24.30. The response variable is "Percent," and "List" identifies the four lists of words. Based on this analysis, is there good reason to think that the four lists are not all equally difficult? Write a brief summary of the study findings.

24.33 Evolution in bacteria. 4 STEP Can bacteria evolve a preference for the pH of their environment? An evolutionary biologist took lines of *E. coli* bacteria grown and kept at a neutral pH of 7.2 and grew them for 2000 generations (about 300 days) at a stressful acidic pH of 5.5. The original bacteria (kept frozen all that time) and the "acid-evolved" bacteria were then grown together in various test environments, and a relative fitness score was computed. A score of 1 would indicate that both bacteria types were equally fit. A score higher than 1 would indicate that the acid-evolved line was more fit than the ancestral line in that environment (that is, that the acid-evolved bacteria grew the most). Here are the relative fitness scores obtained for test environments of pH 5.5 (acid), 7.2 (neutral), or 8.0 (basic). There are six replicates using different ancestor bacterial lines for each test environment:[16]

Minitab

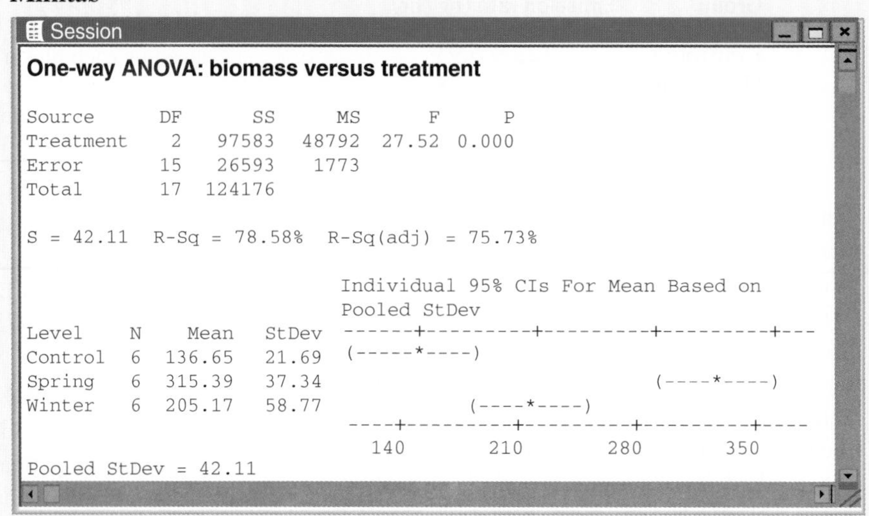

FIGURE 24.10 Minitab ANOVA output for comparing the total plant biomass of grassland plots under different water conditions, for Exercise 24.31.

Minitab

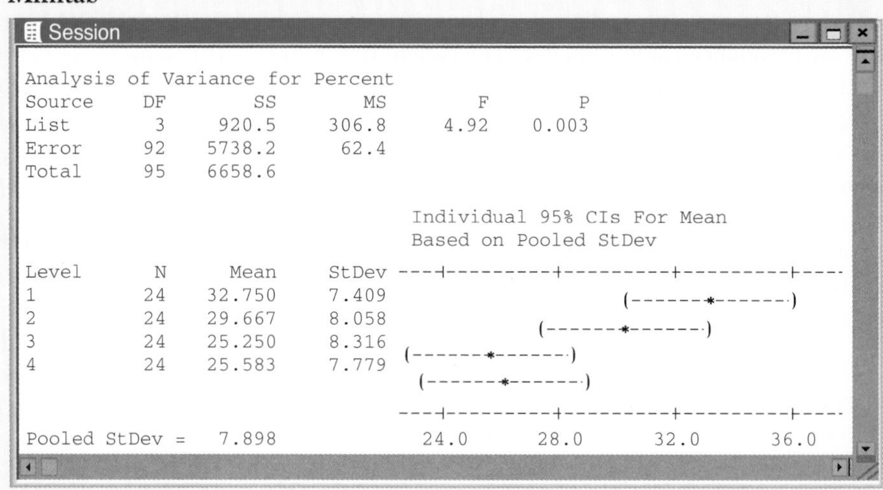

FIGURE 24.11 Minitab ANOVA output for comparing the percents heard correctly in four lists of words, for Exercise 24.32.

Test in acid pH	1.24	1.22	1.23	1.24	1.18	1.09
Test in neutral pH	0.99	0.99	0.98	0.94	0.95	0.95
Test in basic pH	0.56	0.83	0.82	0.72	0.86	0.84

Do the data provide evidence that *E. coli* can evolve adaptations to the pH of its environment over the course of 2000 generations? Follow the four-step process in answering this question. (Although the standard deviations do not quite satisfy our rule of thumb, that rule is conservative and small sample sizes can more easily lead to widely different variances. The analysis might be performed again without the low outlier to confirm the initial interpretation.)

24.34 Does nature heal best? 4 STEP Our bodies have a natural electrical field that helps wounds heal. Might higher or lower levels speed healing? An experiment with newts investigated this question. Newts were randomly assigned to five groups. In four of the groups, an electrode applied to one hind limb (chosen at random) changed the natural field, while the other hind limb was not manipulated. Both limbs in the fifth (control) group remained in their natural state.[17]

Table 24.4 gives data from this experiment. The "Group" variable shows the field applied as a multiple of the natural field for each newt. For example, "0.5" is half the natural field, "1" is the natural level (the control group), and "1.5" indicates

TABLE 24.4 Effect of electrical field on healing rate in newts

Group	Diff	Group	Diff	Group	Diff	Group	Diff	Group	Diff
0	−10	0.5	−1	1	−7	1.25	1	1.5	−13
0	−12	0.5	10	1	15	1.25	8	1.5	−49
0	−9	0.5	3	1	−4	1.25	−15	1.5	−16
0	−11	0.5	−3	1	−16	1.25	14	1.5	−8
0	−1	0.5	−31	1	−2	1.25	−7	1.5	−2
0	6	0.5	4	1	−13	1.25	−1	1.5	−35
0	−31	0.5	−12	1	5	1.25	11	1.5	−11
0	−5	0.5	−3	1	−4	1.25	8	1.5	−46
0	13	0.5	−7	1	−2	1.25	11	1.5	−22
0	−2	0.5	−10	1	−14	1.25	−4	1.5	2
0	−7	0.5	−22	1	5	1.25	7	1.5	10
0	−8	0.5	−4	1	11	1.25	−14	1.5	−4
		0.5	−1	1	10	1.25	0	1.5	−10
		0.5	−3	1	3	1.25	5	1.5	2
				1	6	1.25	−2	1.5	−5
				1	−1				
				1	13				
				1	−8				

a field 1.5 times the natural level. "Diff" is the response variable, the difference in the healing rate (in micrometers per hour) of cuts made in the experimental and control limbs of that newt. Negative values mean that the experimental limb healed more slowly. The investigators conjectured that nature heals best, so that changing the field from the natural state (the "1" group) will slow healing.

Do a complete analysis to see whether the groups differ in the effect of the electrical field level on healing. Follow the four-step process in your work.

24.35 Fertilizing bromeliads. Bromeliads are tropical flowering plants. Many are epiphytes that attach to trees and obtain moisture and nutrients from air and rain. Their leaf bases form cups that collect water and are home to the larvae of many insects. As a preliminary to a study of changes in the nutrient cycle, Jacqueline Ngai and Diane Srivastava examined the effects of adding nitrogen, phosphorus, or both to the cups. They randomly assigned 8 bromeliads growing in Costa Rica to each of four treatment groups, including an unfertilized control group. A monkey destroyed one of the plants in the control group, leaving 7 bromeliads in that group. Here are the numbers of new leaves on each plant over the seven months following fertilization:[18]

Nitrogen	Phosphorus	Both	Neither
15	14	14	11
14	14	16	13
15	14	15	16
16	11	14	15
17	13	14	15
18	12	13	11
17	15	17	12
13	15	14	

Does nitrogen or phosphorus have a greater effect on the growth of bromeliads? Analyze these data and discuss the results.

24.36 Neural basis of 3D vision. One way the brain assesses depth in a visual scene is by using the disparity of information received by both retinas. Stereograms, like 3D movies, contain a visual pattern duplicated and shifted a bit horizontally. Each pattern is seen by just one eye through the use of special 3D glasses, introducing retinal disparity (measured in degrees). The result is an image that appears to float either in front of or behind the screen, depending on the horizontal disparity introduced.

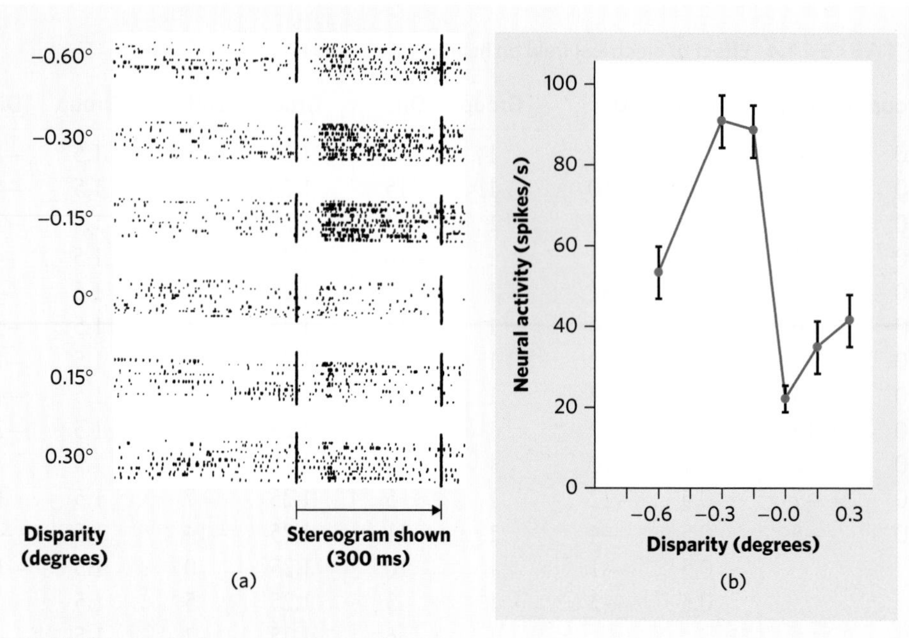

FIGURE 24.12 (a) Action potentials (dots) over time in response to 3D visual stimuli of different disparities and (b) graph of the mean response plus or minus standard error of the mean for each disparity value, for Exercise 24.36.

Some neurons in the primary visual cortex specialize in the perception of depth from binocular disparity. One such neuron was recorded in a monkey shown stereograms of various disparities in random order. Figure 24.12(a) shows the action potentials ("spikes") generated over time (horizontal axis) by that neuron for different disparities. Figure 24.12(b) shows the "tuning curve" representing the mean neural response (± SEM) for each disparity. Here are the raw data from 14 replications:[19]

Disparity	Neural response to stereogram (spikes/s)						
−0.6	10	46.7	83.3	80	40	36.7	46.7
−0.3	30	110	100	90	70	56.7	90
−0.15	70	93.3	86.7	76.7	103.3	53.3	106.7
0	10	13.3	36.7	16.7	36.7	16.7	20
0.15	13.3	63.3	10	63.3	93.3	16.7	10
0.3	6.7	23.3	50	43.3	76.7	10	46.7
−0.6	46.7	70	76.7	20	53.3	66.7	83.3
−0.3	90	130	103.3	60	86.7	126.7	123.3
−0.15	120	100	126.7	30	100	113.3	96.7
0	23.3	23.3	16.7	26.7	23.3	13.3	36.7
0.15	26.7	23.3	70	3.3	23.3	20	40
0.3	83.3	60	26.7	16.7	30	46.7	63.3

A negative disparity corresponds to the impression of a near object. This neuron clearly responds more strongly to objects that appear in the near space. Does your analysis support this interpretation? Use the four-step process to guide your discussion.

24.37 Compressing soil. Farmers know that driving heavy equipment on wet soil compresses the soil and injures future crops. Table 2.1 (page 62) gives data on the "penetrability" of the same soil at three levels of compression.[20] Penetrability is a measure of how much resistance plant roots will meet when they try to grow through the soil. Do the data suggest that penetrability decreases as soil is more compressed? Follow the four-step process in data analysis and ANOVA. Be sure to check the conditions for ANOVA, paying special attention to outliers.

24.38 Logging in the rain forest: species counts. Table 24.2 gives data on the number of trees per forest plot, the number of species per plot, and species richness. Exercise 24.3 analyzed the effect of logging on number of trees. Exercise 24.8 concluded that it would be risky to use ANOVA to analyze richness. Use software to analyze the effect of logging on the number of species.

(a) Make a table of the group means and standard deviations. Do the standard deviations satisfy our rule of thumb for safe use of ANOVA? What do the means suggest about the effect of logging on the number of species?

(b) Carry out the ANOVA. Report the F statistic and its P-value and state your conclusion.

24.39 Homeopathic treatment of wounds. Homeopathy seeks to stimulate the body's ability to heal itself by giving very small doses of highly diluted and generally harmful substances. The dilution can be so high that no molecules of the healing substance remain in any traceable amount. (Homeopathy proposes that the substance leaves an imprint called the "memory of water.") The National Center for Complementary and Alternative Medicine (NCCAM) warns that, except for a few studies, there is little evidence to support homeopathy as an effective treatment for any specific condition.

4 STEP

In one such study, researchers sutured a deep muscular incision that they had made in anaesthetized rats and then randomly assigned the rats to 5 treatments: a placebo (sugar water) and two homeopathic remedies (*Arnica montana* and *Staphisagria*) administered at two levels of dilution. The treatments were injected daily into the rats' stomachs via a gastric tube, and the wounds were examined daily to determine the time (in days) until complete healing of the wound. Table 24.5 gives the raw data.[21] Is there significant evidence that healing time depends on the treatment received? Use the four-step process to guide your discussion.

24.40 F versus t (optional). Acclimation is a process of adaptation to a persistent change in environment taking place typically over the course of weeks. When goldfish are relocated to a colder aquarium, for instance, their ventilation rate immediately slows down but, after weeks, moves back toward its previous value. This acclimation process can be complete (the rates before and after acclimation are exactly the same) or only partial. Here are ventilation rates (in beats per minute) for goldfish that have been acclimated to either a cold environment (12 degrees Celsius) or a warm environment (22 degrees Celsius):[22]

Group	n	$\bar{x}$	s
Cold-acclimated	18	55.8	18.4
Warm-acclimated	18	78.3	25.8

A significant difference in ventilation rate between warm- and cold-acclimated goldfish would indicate that acclimation was not complete.

(a) Calculate the two-sample t statistic for testing H_0: $\mu_1 = \mu_2$ against the two-sided alternative. Use the conservative method to find the P-value.

(b) Calculate MSG, MSE, and the ANOVA F statistic for the same hypotheses. What is the P-value of F?

(c) How close are these two P-values?

TABLE 24.5 Healing times (in days) for a deep surgical wound

Placebo	Arnica low dose	Arnica higher dose	Staphisagria low dose	Staphisagria higher dose
21	16	16	15	16
22	15	16	15	16
22	16	16	16	16
21	15	16	16	15
22	15	15	15	15
21	16	15	16	16
20	16	16	15	16
22	15	16	15	16
21	15	16	16	15
21	16	15	16	16
22	15	16	16	15
23	16	16	15	16
21	16	15	15	15
22	15	16	16	15
21	15	15	15	16

24.41 Plant defenses (optional). ANOVA calculations use only the sample sizes n_i, the sample means $\overline{x}_i$, and the sample standard deviations s_i. You can therefore re-create the ANOVA calculations when a report gives these summaries but does not give the actual data. Use the information in Exercise 24.29 to calculate the ANOVA table (sums of squares, degrees of freedom, mean squares, and the F statistic). Note that the report gives the standard error of the mean rather than the standard deviation. Are there significant differences among the mean emission rates for the four populations of plants?

24.42 Lung capacity of children. LARGE DATA SET The forced expiratory volume (FEV, measured in liters) is a primary indicator of lung function and corresponds to the volume of air that can forcibly be blown out in the first second after full inspiration. The *Large.FEV* data file contains the FEV values of a large sample of children, along with some categorical descriptors of each individual. In Exercise 1.44, you created a new variable based on age: *preschool* for ages 3 to 5, *elementary* for ages 6 to 10, *middle* for ages 11 to 13, and *highschool* for ages 14 and above. Is there evidence that FEV is not the same for all four age groups? Verify that the conditions for ANOVA are met, run the test, and conclude.

24.43 Elderly health. LARGE DATA SET A study examined the medical records of elderly patients for calcium or inorganic phosphorus blood levels (both in mmol/l). The data file *Large.Calcium*, which you examined in Exercise 18.48, contains the data. One purpose of the study was to determine if different laboratories yield substantially different results. Do the data support this hypothesis? Verify that the conditions for ANOVA are met for the calcium variable, run the test, and conclude. Then do the same with the inorganic phosphorus data. Write a short paragraph describing your findings.

24.44 The cost of sex. LARGE DATA SET Evolution theory predicts a physiological cost of increased reproduction that translates into shorter life span. The data file *Large.Fruitfly* describes a study of the cost of sexual activity in male *Drosophila*. The authors randomly assigned male fruit flies to 5 sex conditions. In the first condition, the males were kept alone. In the second condition, each male was placed in the company of one virgin female (replaced every other day with a new virgin female to provide reproductive opportunities). The third condition had each male placed with eight virgin females. In the fourth condition, each male was placed with a newly inseminated female that was replaced every other day; these females do not usually mate for at least two days. The fifth condition had each male placed in the company of eight newly inseminated females. All other living conditions were the same for all five groups. The researchers recorded the longevity (in days) of each male fruit fly and its thorax length.[23]

(a) The researchers confirmed by examining the females that the males kept with virgin females were indeed reproducing and that those kept with newly inseminated females were not. Is there evidence that the different sex conditions result in significantly different longevities in laboratory male fruit flies? Check the conditions for ANOVA, run the test, and conclude.

(b) Is there evidence that the different sex conditions result in significantly different thorax lengths in laboratory male fruit flies? Check the conditions for ANOVA, run the test, and conclude.

(c) Write a short paragraph describing your findings.

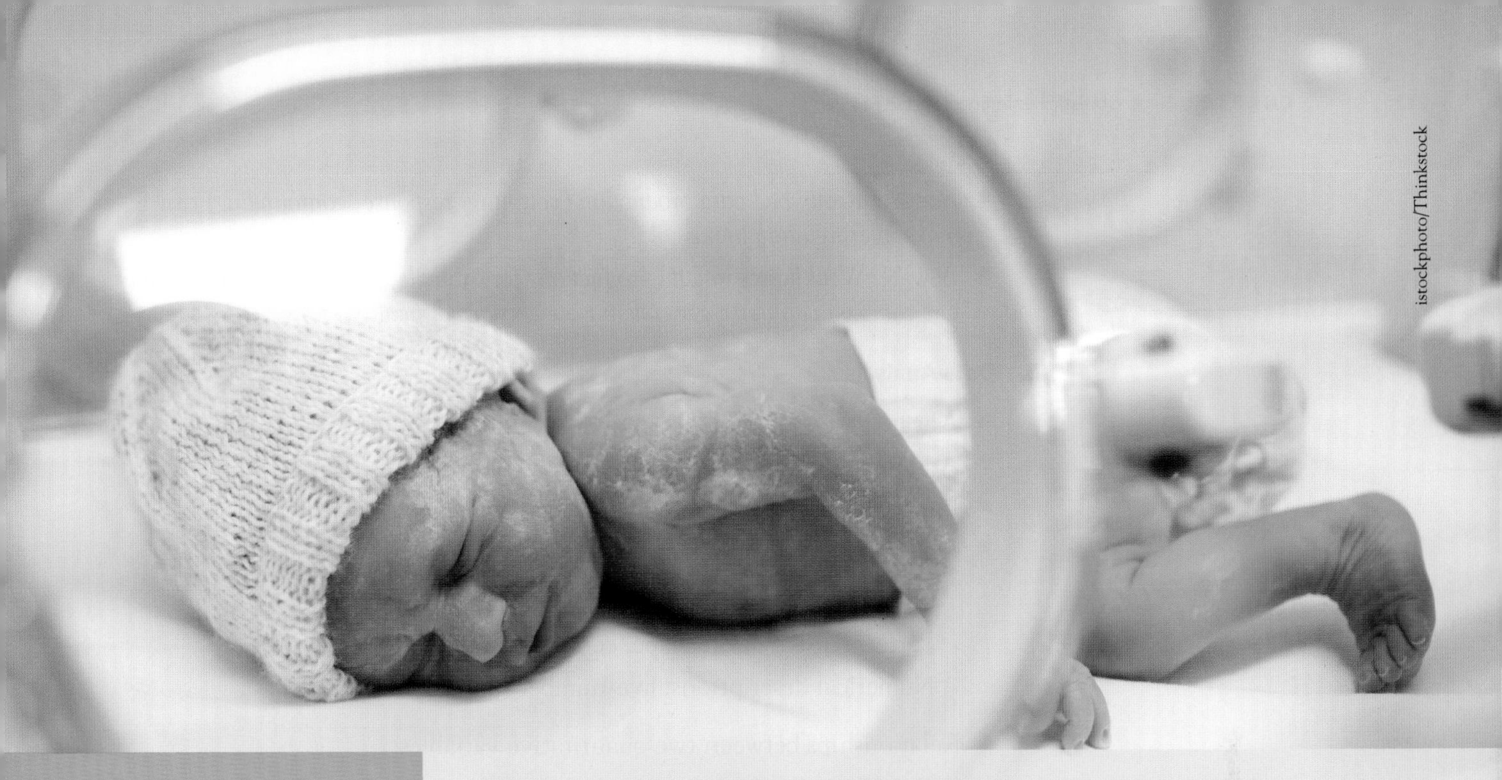
istockphoto/Thinkstock

CHAPTER 25 Statistical Inference: Part III Review

IN THIS CHAPTER WE COVER...

The procedures of Chapters 17 to 24 are among the most common of all statistical inference methods. Now that you have mastered important ideas and practical methods for inference, it's time to review the big ideas of statistics in outline form. Here is a summary of Parts I and II of this book, leading up to Part III. The outline contains important warnings: Look for the Caution icon.

A. Data Production

- Data basics:
  - Individuals (subjects).
  - Variables: categorical versus quantitative, units of measurement, explanatory versus response.
  - Purpose of study.
- Data production basics:
  - Observation versus experiment.
  - Simple random samples.
  - Comparative randomized experiments.

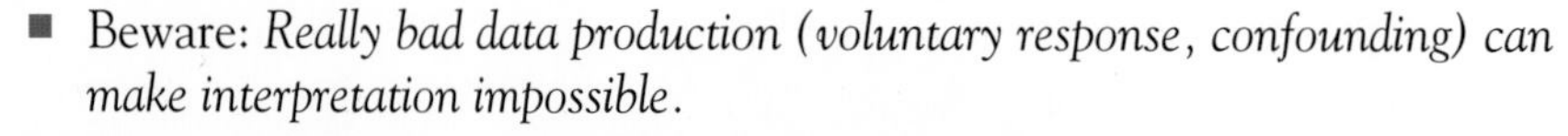

- Beware: *Really bad data production (voluntary response, confounding) can make interpretation impossible.*

- Beware: *Weaknesses in data production (for example, sampling students at only one campus) can make generalizing conclusions difficult.*

B. Data Analysis

- Always plot your data. Look for an overall pattern and striking deviations.
- Add numerical descriptions based on what you see.

- Beware: *Averages and other simple descriptions can miss the real story.*
- One quantitative variable:

  Graphs: stemplot, dotplot, time plot, histogram, boxplot.

  Pattern: distribution shape, center, spread. Outliers?

  Numerical descriptions: five-number summary or $\bar{x}$ and s.

- Relationships between two quantitative variables:

  Graph: scatterplot.

  Pattern: relationship form, direction, strength. Outliers? Influential observations?

  Numerical description for linear relationships: correlation, regression line.

  Beware the lurking variable: Correlation does not imply causation.

- *Beware the effects of outliers and influential observations.*

C. The Reasoning of Inference

- Inference uses data to infer conclusions about a wider population.

- When you do inference, you are acting as if your data come from random samples or randomized comparative experiments. Beware: *If they don't, you may have "garbage in, garbage out."*
- Always examine your data before doing inference. Inference often requires a regular pattern, such as roughly Normal with no strong outliers.
- Key idea: "What would happen if we did this many times?"
- Confidence intervals: Estimate a population parameter.

  95% confidence: I used a method that captures the true parameter 95% of the time in repeated use.

  Beware: *The margin of error of a confidence interval does not include the effects of practical errors such as undercoverage and nonresponse.*

- Significance tests: Assess evidence against H_0 in favor of H_a.

  P-value: If H_0 were true, how often would I get an outcome favoring the alternative this strongly? Smaller P is stronger evidence against H_0.

Statistical significance at the 5% level, $P < 0.05$, means that an outcome this extreme would occur less than 5% of the time if H_0 were true.

Beware: $P < 0.05$ *is not sacred.*

Beware: *Statistical significance is not the same as practical importance. Large samples can make small effects significant. Small samples can fail to declare large effects significant.*

Always try to estimate the size of an effect (for example, with a confidence interval), not just its significance.

D. Inference Methods

- Choose inference procedures by asking, "What parameter?" and "What study design?"

  It may be helpful to ask "How many variables?" and "How many groups compared?"

  Also ask, "What kind of variables?" Categorical and quantitative variables need very different inference procedures.

- Carry out the details by hand or using technology.
- State your conclusions in context.

Part III of this book expands on the last part of this outline (D. Inference Methods). To actually do inference, you must choose the right procedure. The Statistics in Summary flowcharts in Figures 25.1 and 25.2 help you decide. The flowchart for means and proportions is based on two key questions:

- **What population parameter** does your problem concern? You should be familiar in detail with inference for *means* (Chapters 17, 18, and 24) and for *proportions* (Chapters 19, 20, 21, and 22). This is the first branch point in the flowchart.
- **What type of design** produced the data? You should be familiar with data from a *single sample*, from *matched pairs*, and from *two or more independent samples*. This is the second branch point in the flowchart. Your choice here is the same for experiments and for observational studies—inference methods are the same for both. But don't forget that experiments give much better evidence that an effect uncovered by inference can be explained by direct causation.

The flowchart for relationships is based on this simple question: **What type of variables** make up the relationship? You should be familiar in detail with inference for relationships between *two categorical variables* (Chapter 22), between two quantitative variables (Chapter 23), and between *a quantitative and a categorical variable* (Chapter 24).

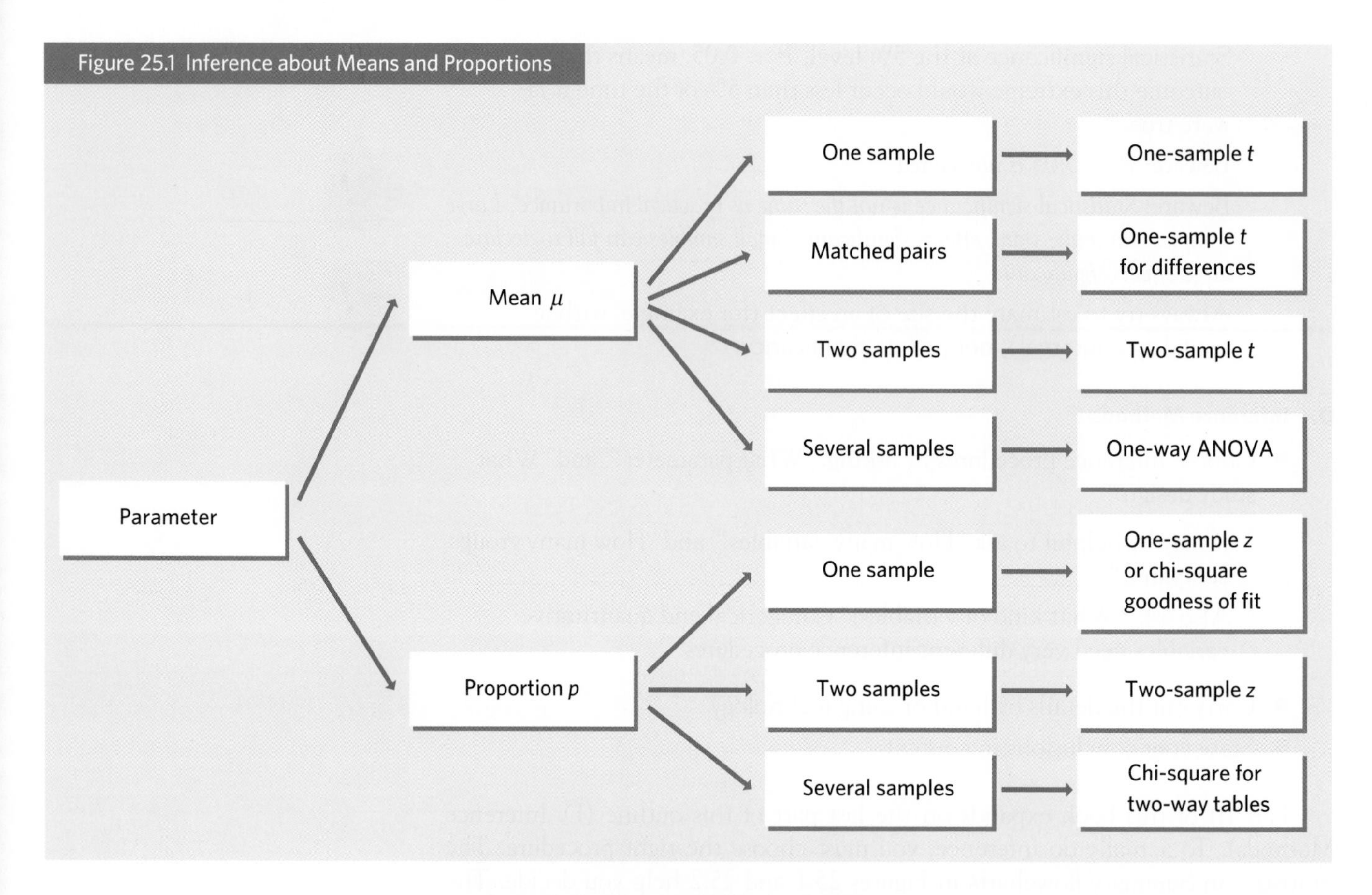

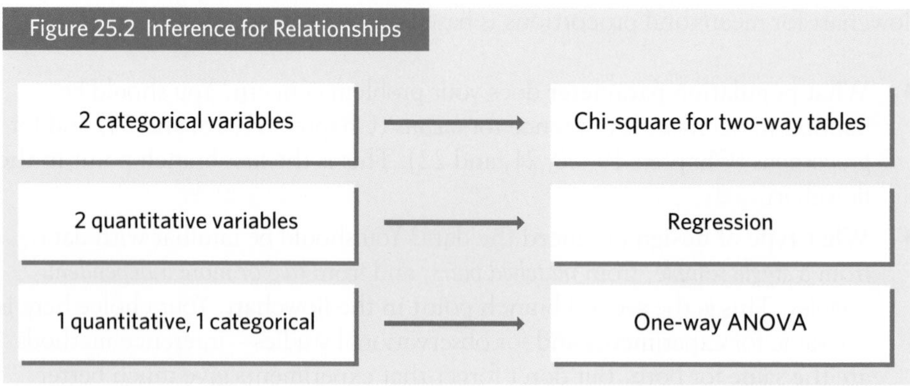

Answering these questions is part of the Plan step in the four-step process. To begin the Solve step, follow the flowcharts to the correct procedure. Then ask another question:

- **Are the conditions for this procedure met?** Can you act as if the data come from a random sample or randomized comparative experiment? Do the data show extreme outliers or strong skewness that forbid use of inference based on Normality? Do you have enough observations for your intended procedure?

Mastery of statistics comes with practicing. It is important to do some of the review exercises in this chapter because now, for the first time, you must decide which of several inference procedures to use. Learning to recognize problem settings in order to choose the right type of inference is a key step in advancing your mastery of statistics.

PART III SUMMARY

Here are the most important skills you should have acquired from reading Chapters 17 through 24.

A. Recognition

1. Recognize when a problem requires inference about population means (quantitative response variable), population proportions (usually categorical response variable), or relationships.
2. Recognize from the design of a study whether one-sample, matched pairs, or two-sample procedures for one variable are needed, or whether the design assesses the relationship between two variables.
3. For one-variable problems, choose among the one- and two-sample t procedures for means, the one- and two-sample z procedures for proportions, and the chi-square goodness-of-fit test for proportions. For relationships, choose among the chi-square procedures for two categorical variables, regression procedures for two quantitative variables, and the ANOVA F procedures for a quantitative and a categorical variable.

B. Inference About One Mean

1. Verify that the t procedures are appropriate in a particular setting. Check the study design and the distribution of the data, and take advantage of robustness against lack of Normality.
2. Recognize when poor study design, outliers, or a small sample from a skewed distribution make the t procedures risky.
3. Use the one-sample t procedure to obtain a confidence interval at a stated level of confidence for the mean μ of a population.
4. Carry out a one-sample t test for the hypothesis that a population mean μ has a specified value against either a one-sided or a two-sided alternative. Use software to find the P-value or Table C to get an approximate value.
5. Recognize matched pairs data and use the t procedures to obtain a confidence interval and to perform a test of significance for such data.

C. Comparing Two Means

1. Verify that the two-sample t procedures are appropriate in a particular setting. Check the study design and the distribution of the data, and take advantage of robustness against lack of Normality.
2. Give a confidence interval for the difference between two means. Use software if you have it. Use the two-sample t statistic with conservative degrees of freedom and Table C if you do not have statistical software.

3. Test the hypothesis that two populations have equal means against either a one-sided or a two-sided alternative. Use software if you have it. Use the two-sample t test with conservative degrees of freedom and Table C if you do not have statistical software.
4. Know that procedures for comparing the standard deviations of two Normal populations are available, but that these procedures are risky because they are not at all robust against non-Normal distributions.

D. Inference About One Proportion

1. Verify that you can safely use either the large-sample or the plus four z procedures in a particular setting. Check the study design and the guidelines for sample size.
2. Use the large-sample z procedure to give a confidence interval for a population proportion p. Understand that the true confidence level may be substantially less than you ask for unless the sample is very large and the true p is not close to 0 or 1.
3. Use the plus four modification of the z procedure to give a confidence interval for p that is accurate even for small samples and for any value of p.
4. Use the z statistic to carry out a test of significance for the hypothesis H_0: $p = p_0$ about a population proportion p against either a one-sided or a two-sided alternative. Use software or Table B to find the P-value, or Table C to get an approximate value.

E. Comparing Two Proportions

1. Verify that you can safely use either the large-sample or the plus four z procedures in a particular setting. Check the study design and the guidelines for sample sizes.
2. Use the large-sample z procedure to give a confidence interval for the difference $p_1 - p_2$ between proportions in two populations based on independent samples from the populations. Understand that the true confidence level may be less than you ask for unless the samples are quite large.
3. Use the plus four modification of the z procedure to give a confidence interval for $p_1 - p_2$ that is accurate even for very small samples and for any values of p_1 and p_2.
4. Use a z statistic to test the hypothesis H_0: $p_1 = p_2$ that proportions in two distinct populations are equal. Use software or Table B to find the P-value, or Table C to get an approximate value.

F. Chi-Square Test for Goodness of Fit

1. State the null hypothesis of your specific chi-square for goodness of fit test.
2. Locate the chi-square statistic, its P-value, and other useful values (percents, expected counts, chi-square components) in output from your software or calculator.

3. Use the expected counts to check whether you can safely use the chi-square test.
4. If the test is significant, compare percents, compare observed with expected counts, or look for the largest components of the chi-square statistic to see what deviations from the null hypothesis are most important.

G. Chi-Square Test for Two-Way Tables

1. Know how to organize data representing two categorical variables as a two-way table of counts of outcomes.
2. Use percents to describe the relationship between any two categorical variables, starting from the counts in a two-way table.
3. Locate the chi-square statistic, its P-value, and other useful values (row or column percents, expected counts, chi-square components) in output from your software or calculator.
4. Use the expected counts to check whether you can safely use the chi-square test.
5. Explain what null hypothesis the chi-square statistic tests in a specific two-way table.
6. If the test is significant, compare percents, compare observed with expected cell counts, or look for the largest components of the chi-square statistic to see what deviations from the null hypothesis are most important.

H. Chi-Square Computations by Hand

1. When doing a chi-square test by hand, first calculate all expected counts. Check whether you can safely use the chi-square test.
2. Know that expected counts for a goodness-of-fit test are based on the null hypothesis and sample sizes, whereas for a two-way test they are based on the observed counts displayed in the two-way table.
3. Calculate the components of the chi-square statistic, as well as the overall statistic.
4. Give the degrees of freedom of the chi-square statistic. Use the chi-square critical values in Table D to approximate the P-value of the chi-square test.

I. Preliminaries of Inference for Regression

1. Make a scatterplot to show the relationship between an explanatory and a response variable.
2. Use a calculator or software to find r^2 and the equation of the least-squares regression line.
3. Recognize the regression setting: a straight-line relationship between an explanatory variable x and a response variable y.

4. Recognize which type of inference you need in a particular regression setting.
5. Inspect the data to recognize situations in which inference isn't safe: a nonlinear relationship, influential observations, strongly skewed residuals in a small sample, or nonconstant variation of the data points about the regression line.

J. Inference for Regression Using Software Output

1. Explain in any specific regression setting the meaning of the slope β of the population regression line.
2. Understand software output for regression. Find in the output the slope and intercept of the least-squares line, their standard errors, and the regression standard error.
3. Use that information to carry out tests of H_0: $\beta = 0$ and calculate confidence intervals for β.
4. Explain the distinction between a confidence interval for the mean response and a prediction interval for an individual response.
5. If software gives output for prediction, use that output to give either confidence or prediction intervals.

K. Preliminaries of ANOVA

1. Recognize when testing the equality of several means is helpful in understanding data.
2. Recognize when you can safely use ANOVA to compare means. Check the data production, the presence of outliers, and the sample standard deviations for the groups you want to compare.
3. Know that the statistical significance of differences among sample means depends on the sizes of the samples and on how much variation there is within the samples.

L. Interpreting ANOVA Software Output

1. State the null hypothesis of the F test in a specific setting.
2. Locate the F statistic and its P-value on the output of analysis of variance software.
3. Find the degrees of freedom for the F statistic from the number and sizes of the samples. Use Table F of the F distributions to approximate the P-value when software does not give it.
4. If the test is significant, use graphs and descriptive statistics to see what differences among the means are most important (optional Chapter 26 on the companion website covers the details of some common follow-up tests used when the F test is significant).

REVIEW EXERCISES

Review exercises are short and straightforward exercises that help you solidify the basic ideas and skills from Chapters 17 through 24.

For exercises that call for inference, your answers should include the ***Plan, Solve,*** *and* ***Conclude*** *steps of the four-step process. It is helpful to also summarize in your own words the* ***State*** *information given in the exercise. For tests of significance, use a table of critical values to approximate P-values unless you use software that reports the P-value. For confidence intervals for proportions, use the plus four procedures.*

25.1 Comparing universities. You have data from two random samples of students, one SRS from each of two large universities. Which procedure from the Statistics in Summary flowchart would you use to compare

(a) the percents of students enrolled in a university-sponsored physical activity?

(b) the average amount of time spent in a university-sponsored physical activity in a typical week?

25.2 Acid rain? You have data on rainwater collected at 16 locations in the Adirondack Mountains of New York State. One measurement is the acidity of the water, measured by pH on a scale of 0 to 14 (the pH of distilled water is 7.0). Which procedure from the Statistics in Summary flowchart would you use to estimate the average acidity of rainwater in the Adirondacks?

25.3 Looking back on love. How do young adults look back on adolescent romance? Investigators interviewed 40 couples in their midtwenties. The female and male partners were interviewed separately. Each was asked about their current relationship and also about a romantic relationship that lasted at least two months when they were aged 15 or 16. One response variable was a measure on a numerical scale of how much the attractiveness of the adolescent partner mattered. Which of the tests in the Statistics in Summary flowchart should you use to compare the men and women? Use the selections in (a) and (b) to help you find the correct procedure.

(a) t for means or z for proportions

(b) one-sample, matched pairs, or two-sample

25.4 Preventing AIDS through education. The Multisite HIV Prevention Trial was a randomized comparative experiment to compare the effects of twice-weekly small-group AIDS discussion sessions (the treatment) with a single one-hour session (the control). Which of the procedures in the Statistics in Summary flowchart should you use to compare the effects of treatment and control on each of the following response variables?

(a) A subject does or does not use condoms six months after the education sessions.

(b) The number of unprotected intercourse acts by a subject between four and eight months after the sessions.

(c) A subject is or is not infected with a sexually transmitted disease six months after the sessions.

25.5 Exercise and weight loss. A randomized, comparative experiment assigned subjects to three types of exercise programs intended to help them lose weight. Some of the results of this study were analyzed using the chi-square test for two-way tables, and some others were analyzed using one-way ANOVA. For each of the following excerpts from the study report, say which analysis is appropriate and explain how you made your choice.

(a) "Overall, 115 subjects (78% of 148 subjects randomized) completed 18 months of treatment, with no significant difference in attrition rates between the groups ($P = .12$)."

(b) "In analyses using only the 115 subjects who completed 18 months of treatment, there were no significant differences in weight loss at 6 months among the groups."

(c) "The duration of exercise for weeks 1 through 4 was significantly greater in the SB compared with both LB and SBEQ groups ($P < .05$). . . . However, exercise duration was greater in SBEQ compared with both LB and SB groups for months 13 through 18 ($P < .05$)."

25.6 Monkeys and music. Humans generally prefer music to silence. What about monkeys? Allow a tamarin monkey to enter a V-shaped cage with food in both arms of the V. After the monkey eats the food, which arm will it prefer? The monkey's location determines what it hears, a lullaby played by a flute in one arm and silence in the other. Each of 4 monkeys was tested 6 times, on different days and with the music arm alternating between left and right (in case a monkey prefers one direction). In each trial, the researchers measured the amount of time the monkey spent in each arm, and they reported it as a percent of the total time spent in the cage (rather than in minutes). That is, each trial results in a meaningful numerical value, making the response variable quantitative. On average, the monkeys chose silence for about 65% of their time in the cage. The researchers reported a one-sample t test for the mean percent of time spent in the music arm, H_0: $\mu = 50\%$ against the two-sided alternative, $t = -5.26$, df $= 23$, $P < 0.0001$.[1]

Although the result is interesting, the statistical analysis is not correct. The degrees of freedom df = 23 show that the researchers assumed that they had 24 independent observations. Explain why the results of the 24 trials are not independent.

25.7 Pain from a rubber hand. People who have had limbs amputated sometimes feel sensations from the limb that is no longer there. To study this effect, psychologists asked subjects to place their right arm on a table. They then put a rubber arm and hand next to the real arm, with a high partition arranged so that the subject could see only the rubber arm. After a few minutes during which the real and the fake hand were both tapped by an experimenter, the subjects felt the taps coming from the location of the rubber hand they could see, not from the real hand they couldn't see. Now the experiment begins: Bend back a finger of the fake hand in a way that would cause pain, while at the same time merely lifting slightly a real finger. Do electrical measurements show a response to pain? Because there would be some response from the surprise of being touched, a control condition delayed the touch to the real hand to separate surprise from "pain." Here are summary data for 16 undergraduate students who were subjected to both stimuli:[2]

Stimulus	$\bar{x}$	s
Simultaneous	0.39	0.28
Control (delayed)	0.18	0.20

(a) Which t procedures are correct for comparing the mean response to treatment and control: one-sample, matched pairs, or two-sample?

(b) The data summary given is not enough information to carry out the correct t procedures. Explain why not.

25.8 Mouse endurance. A study of the inheritance of speed and endurance in mice found a trade-off between these two characteristics, both of which help mice survive. To test endurance, mice were made to swim in a bucket with a weight attached to their tails. (The mice were rescued when exhausted.) Here are data on endurance in minutes for female and male mice:[3]

Group	n	Mean	Std. dev.
Female	162	11.4	26.09
Male	135	6.7	6.69

(a) Both sets of endurance data are skewed to the right. Why are t procedures nonetheless reasonably accurate for these data?

(b) Give a 95% confidence interval for the mean endurance of female mice swimming.

(c) Give a 95% confidence interval for the mean difference (female minus male) in endurance times.

25.9 Mouse endurance, continued. Do the data in the previous exercise show that female mice have significantly higher endurance on the average than male mice?

25.10 Prey attract predators. Stable ecosystems should have enough predators to keep prey densities in check. An experiment tested this theory by studying kelp perch and their common predator, the kelp bass, in four large circular pens of identical size. The explanatory variable is the number of perch (the prey) in each enclosure. The response variable is the proportion of perch killed by bass (the predator) in two hours when the bass are allowed access to the perch.[4]

Perch	Proportion killed			
10	0.0	0.1	0.3	0.3
20	0.2	0.3	0.3	0.6
40	0.075	0.3	0.6	0.725
60	0.517	0.55	0.7	0.817

(a) Make a scatterplot and describe the relationship.

(b) Use software to obtain the equation of the least-squares line for predicting proportion killed from count of perch. What part of this equation shows that more perch do result in a higher proportion being killed by bass? What is the regression standard error s?

(c) Use software to obtain a 95% confidence interval for the slope of the population regression line. Does this interval give evidence of a significant relationship between proportion of perch killed by bass and initial count of perch?

(d) The correlation between proportion of perch killed by bass and initial count of perch is $r = 0.6821$. Use Table E to say how significant this correlation is for testing zero correlation against positive correlation in the population. Verify that your result is consistent with your answer in (c).

25.11 Prey attract predators: residuals. Here are the residuals (rounded to three decimal places) for the regression of proportion of perch killed by bass on initial count of perch:

Perch	10	10	10	10	20	20
Residual	−0.206	−0.106	0.094	0.094	−0.092	0.008
Perch	20	20	40	40	40	40
Residual	0.008	0.308	−0.388	−0.163	0.137	0.262
Perch	60	60	60	60		
Residual	−0.118	−0.085	0.065	0.182		

(a) Check the calculation of residuals by finding their sum. What should the sum be? Does the sum have that value (up to roundoff error)?

(b) Plot the residuals against initial count of perch (the explanatory variable). Does your plot show a systematically nonlinear relationship? Does it show systematic change in the spread about the regression line?

(c) Make a stemplot or dotplot of the residuals and describe the distribution. A Normal quantile plot shows that the distribution of residuals is reasonably Normal with one mild outlier.

25.12 Preventing drowning. Drowning in bathtubs is a major cause of death in children less than five years old. A random sample of parents was asked many questions related to bathtub safety. Overall, 85% of the sample said they used baby bathtubs for infants. What procedure would you use to estimate the percent of all parents of young children who use baby bathtubs? (You do not need to carry out the actual procedure.)

25.13 Ginkgo extract and the post-lunch dip. The post-lunch dip is the drop in mental alertness after a midday meal. Does an extract of the leaves of the ginkgo tree reduce the post-lunch dip? Assign healthy people aged 18 to 40 years to take either ginkgo extract or a placebo pill. After lunch, ask them to read seven pages of random letters and place an X over every e. Count the number of misses per line read.[5]

(a) What is a placebo and why was one group given a placebo?

(b) What is the double-blind method and why should it be used in this experiment?

(c) Here are summaries of performance after 13 weeks of either ginkgo extract or placebo:

Group	Group size	Mean	Std. dev.
Ginkgo	21	0.06383	0.01462
Placebo	18	0.05342	0.01549

Is there a significant difference between the two groups? What do these data show about the effect of ginkgo extract?

25.14 Nematodes and tomato plants. How do nematodes (microscopic worms) affect plant growth? A botanist prepares 16 identical planting pots and then introduces different numbers of nematodes into the pots. He transplants a tomato seedling into each pot. Here are data on the increase in height of the seedlings (in centimeters) 16 days after planting:[6]

Nematodes	Seedling growth			
0	10.8	9.1	13.5	9.2
1,000	11.1	11.1	8.2	11.3
5,000	5.4	4.6	7.4	5.0
10,000	5.8	5.3	3.2	7.5

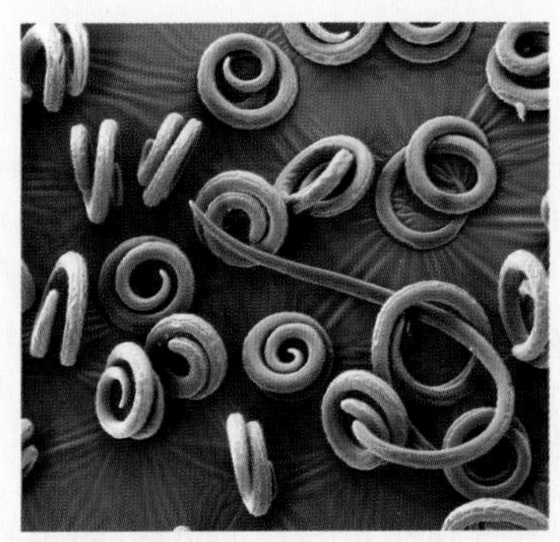

Eye of Science/Science Source

(a) Make a table of the means and standard deviations in the groups. Make side-by-side stemplots or dotplots to compare the treatments. What do the data appear to show about the effect of nematodes on growth? Is use of ANOVA justified?

(b) State H_0 and H_a for the ANOVA test for these data, and explain in words what ANOVA tests in this setting.

(c) Use software for the test and report your overall conclusions about the effect of nematodes on plant growth.

25.15 Butterflies mating. During mating in butterflies, a male passes to a female a packet of sperm called a spermatophore. Females may mate several times. Will they remate sooner if the first spermatophore they receive is small? Among 20 females who received a large spermatophore (greater than 25 milligrams), the mean time to the next mating was 5.15 days, with standard deviation 0.18 day. For 21 females who received a small spermatophore (about 7 milligrams), the mean was 4.33 days and the standard deviation was 0.31 day.[7] Is the observed difference in means statistically significant?

25.16 Very-low-birth-weight babies. Starting in the 1970s, medical technology allowed babies with very low birth weight (VLBW, less than 1500 grams, about 3.3 pounds) to survive without major handicaps. It was noticed that these children nonetheless had difficulties in school and as adults. A long-term study has followed 242 VLBW babies to age 20 years, along with a control group of 233 babies from the same population who had normal birth weight.[8]

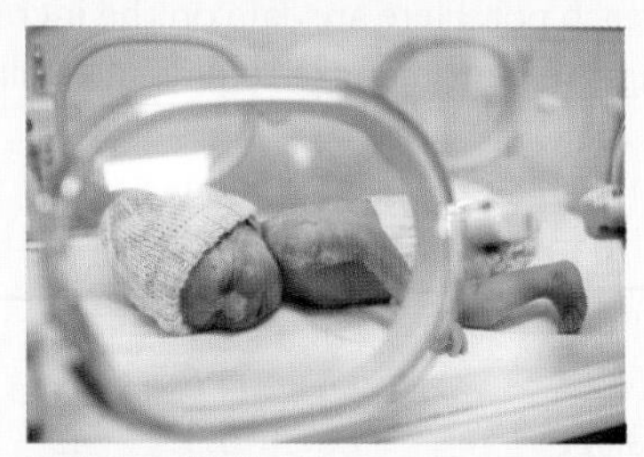

istockphoto/Thinkstock

(a) Is this an experiment or an observational study? Why?

(b) At age 20, 179 of the VLBW group and 193 of the control group had graduated from high school. Is the graduation rate among the VLBW group significantly lower than for the normal-birth-weight controls?

25.17 Very-low-birth-weight babies, continued. IQ scores were available for 113 men in the VLBW group. The mean IQ score was 87.6, and the standard deviation was 15.1. The 106 men in the control group had a mean IQ score of 94.7, with standard deviation 14.9. Is there good evidence that mean IQ is lower among VLBW men than among controls from similar backgrounds?

25.18 Very-low-birth-weight babies, continued. The 123 women in the VLBW group had a mean IQ score of 86.2 (standard deviation 13.4), whereas the 125 women in the control group had a mean IQ score of 89.8 (standard deviation 14.0). Compute a 95% confidence interval for the difference in mean IQ between VLBW women and normal-birth-weight women from similar backgrounds. What can you conclude from this interval?

25.19 Do fruit flies sleep? Mammals and birds sleep. Fruit flies show a daily cycle of rest and activity, but does the rest qualify as sleep? Researchers looking at brain activity and behavior finally concluded that fruit flies do sleep. A small part of the study used an infrared motion sensor to see if flies moved in response to vibrations. Here are results for low levels of vibration:[9]

	Response to Vibration?	
	No	**Yes**
Fly was walking	10	54
Fly was resting	28	4

Analyze these results. Is there good reason to think that resting flies respond differently than flies that are walking? (That would be a sign that the resting flies may actually be sleeping.)

25.20 California brushfires. We often see televised reports of brushfires threatening homes in California. Some people argue that the modern practice of quickly putting out small fires allows fuel to accumulate and so increases the damage done by large fires. A detailed study of historical data suggests that this is wrong—the damage has risen simply because there are more houses in risky areas.[10] As usual, the study report gives statistical information tersely. Here is the summary of a regression of number of fires x on decade y (9 data points, for the 1910s to the 1990s): "Collectively, since 1910, there has been a highly significant increase ($r^2 = 0.61$, $P < 0.01$) in the number of fires per decade." How would you explain this statement to someone who knows no statistics? Include an explanation of both the description given by r^2 and its statistical significance.

25.21 Soybean genetics. 4 STEP A soybean mutant exhibiting a tan-colored seed coat was recently identified. Crossing the mutant with the wild-type yellow-seeded soybean plant gave all yellow-seeded soybeans in the first generation, F1. Self-pollination of F1 resulted in a second generation, F2, consisting of 814 yellow-seeded and 251 tan-seeded plants. This pattern suggests a simple dominant-recessive model involving a single genetic mutation (3:1 expected ratio; 3/4 and 1/4 expected proportions).[11] Do the data support this interpretation? Run the appropriate chi-square test and conclude. Follow the four-step process in your answer.

25.22 Drug interactions. 4 STEP A drug treatment can sometimes be affected by the use of another drug taken concomitantly. This is why doctors and pharmacists routinely ask patients about all the drugs they use, even the ones available over the counter. One study examined the impact of nonsteroidal anti-inflammatory drugs (NSAIDs) on the efficacy of the antidepressant citalopram in clinically depressed individuals. Data from a large clinical trial were reviewed, and clinical remission from depressive symptoms was assessed for individuals who had taken NSAIDs concomitantly or not. Here are the findings:[12]

	Remission	**No remission**
NSAID use	182	227
No NSAID use	628	509

How strong is the evidence that concomitant NSAID use lowers the effectiveness of the antidepressant citalopram? Follow the four-step process as illustrated in Examples 20.4 and 20.5.

25.23 Opinions about evolution. 4 STEP A sample survey funded by the National Science Foundation asked a random sample of American adults about biological evolution.[13] One question asked subjects to answer "true," "false," or "not sure" to the statement "Human beings, as we know them today, developed from earlier species of animals." Of the 1484 respondents, 594 said "true." What can you say with 95% confidence about the percent of all American adults who think that humans developed from earlier species of animals?

25.24 Fertilizing a tropical plant. 4 STEP Bromeliads are tropical flowering plants. Many are epiphytes that attach to trees and obtain moisture and nutrients from air and rain. Their leaf bases form cups that collect water and are home to the larvae of many insects. In an experiment in Costa Rica, Jacqueline Ngai and Diane Srivastava studied whether added nitrogen increases the productivity of bromeliad plants. Bromeliads were randomly assigned to nitrogen or control groups. Here are data on the number of new leaves produced over a seven-month period:[14]

Control	11	13	16	15	15	11	12	
Nitrogen	15	14	15	16	17	18	17	13

Is there evidence that adding nitrogen increases the mean number of new leaves formed?

25.25 Fish sizes. Table 25.1 contains data on the size of perch caught in a lake in Finland.[15] Statistical software will help you analyze these data.

(a) How well can we predict the width of a perch from its length?

(b) The length of a typical perch is about $x^* = 27$ centimeters. Predict the mean width of such fish and give a 95% confidence interval.

(c) Perch number 143 had six newly eaten fish in its stomach. Examine the residuals. Does fish number 143 have an unusually large residual? How might this impact inference?

25.26 Transforming a variable (optional). We can also use the data in Table 25.1 to study the prediction of the weight of a perch from its length.

(a) Describe the pattern of the relationship between weight versus length, with length as the explanatory variable.

(b) It is reasonable to expect the one-third power of the weight to have a straight-line relationship with length since animals are three-dimensional. Use your software to create a new variable that is the one-third power of weight. Describe the pattern of this new response variable against length. Is the straight-line pattern now stronger or weaker than that in (a)?

(c) Find the least-squares regression line to predict the new weight variable from length. Predict the mean of the new variable for perch 27 centimeters long, and give a 95% confidence interval.

SUPPLEMENTARY EXERCISES

Supplementary exercises apply the skills you have learned in ways that require more thought or more use of technology. Many of these exercises, for example, start from the raw data rather than from data summaries. Remember that the **Solve** *step for a statistical problem includes checking the conditions for the inference you plan.*

25.27 Fishery management in Lake Malawi. Lake Malawi is a large African lake with unique fish species essential to the local ecosystem and economy. The endemic catfish species *Bagrus meridionalis*, or kampango, is an important part of the local fishing trade. Small-scale fisheries mainly draw from the shallower water, while large-scale commercial fisheries operate mainly in deeper water. To better manage fishing from an ecological perspective, a study examined the length of kampango found in shallow versus deep water. The study reports the following information about the body lengths of a random sample of male kampango extracted from shallow water: $n = 2131$, $\bar{x} = 39.9$ cm, and standard error 0.25 cm.[16]

(a) We don't have the 2131 individual fish lengths, but use of the t procedures is surely safe. Why?

(b) Give a 99% confidence interval for the mean body length of male kampango swimming in the shallow waters of Lake Malawi (Be careful: The report gives the standard error of $\bar{x}$, not the standard deviation s.)

25.28 Spider silk. Spider silk is the strongest known material, natural or man-made, on a weight basis. A study examined the mechanical properties of spider silk using 21 female golden

TABLE 25.1 Measurements on 56 perch

Observation number	Weight (g)	Length (cm)	Width (cm)	Observation number	Weight (g)	Length (cm)	Width (cm)
104	5.9	8.8	1.4	132	197.0	27.0	4.2
105	32.0	14.7	2.0	133	218.0	28.0	4.1
106	40.0	16.0	2.4	134	300.0	28.7	5.1
107	51.5	17.0	2.6	135	260.0	28.9	4.3
108	70.0	18.5	2.9	136	265.0	28.9	4.3
109	100.0	19.2	3.3	137	250.0	28.9	4.6
110	78.0	19.4	3.1	138	250.0	29.4	4.2
111	80.0	20.2	3.1	139	300.0	30.1	4.6
112	85.0	20.8	3.0	140	320.0	31.6	4.8
113	85.0	21.0	2.8	141	514.0	34.0	6.0
114	110.0	22.5	3.6	142	556.0	36.5	6.4
115	115.0	22.5	3.3	143	840.0	37.3	7.8
116	125.0	22.5	3.7	144	685.0	39.0	6.9
117	130.0	22.8	3.5	145	700.0	38.3	6.7
118	120.0	23.5	3.4	146	700.0	39.4	6.3
119	120.0	23.5	3.5	147	690.0	39.3	6.4
120	130.0	23.5	3.5	148	900.0	41.4	7.5
121	135.0	23.5	3.5	149	650.0	41.4	6.0
122	110.0	23.5	4.0	150	820.0	41.3	7.4
123	130.0	24.0	3.6	151	850.0	42.3	7.1
124	150.0	24.0	3.6	152	900.0	42.5	7.2
125	145.0	24.2	3.6	153	1015.0	42.4	7.5
126	150.0	24.5	3.6	154	820.0	42.5	6.6
127	170.0	25.0	3.7	155	1100.0	44.6	6.9
128	225.0	25.5	3.7	156	1000.0	45.2	7.3
129	145.0	25.5	3.8	157	1100.0	45.5	7.4
130	188.0	26.2	4.2	158	1000.0	46.0	8.1
131	180.0	26.5	3.7	159	1000.0	46.6	7.6

orb weavers, *Nephila clavipes*. Here are data on silk yield stress, which represents the amount of force per unit area needed to reach permanent deformation of the silk strand. The data are expressed in megapascals (MPa).[17]

164.0	478.7	251.3	351.7	173.0	448.9	300.6
362.0	272.4	740.2	329.0	327.2	270.5	332.1
288.8	176.1	282.2	236.1	358.2	270.5	290.7

Give a 95% confidence interval for the mean yield stress of golden orb weaver silk.

25.29 Highly Superior Autobiographical Memory. Some individuals have the ability to recall accurately vast amounts of autobiographical information without mnemonic tricks or extra practice. This ability is called HSAM, for Highly Superior Autobiographical Memory. A study recruited adults with confirmed HSAM and control individuals of similar age without HSAM. All study participants were given a battery of cognitive and behavioral tests in hope of finding out how this extraordinary ability works. Here are the participants' scores on a visual reproduction task in which participants were given ten seconds to look at an abstract design before drawing the design from memory on a blank sheet of paper:[18]

HSAM	15.5	18.0	18.3	18.6	21.0	22.1	23.6	24.0	24.5	26.0
Control	8.1	11.2	17.7	18.2	19.7	21.2	21.3	21.7	22.2	22.2
	22.7	23.2	24.2	24.7	25.2	25.2	26.1			

Do individuals with HSAM have higher scores, on average, at abstract visual reproduction tasks than typical individuals? Write a summary in a sentence or two, including the test statistic, P-value, and a conclusion, as if you were preparing a report for publication.

25.30 Smoking on campus. In February 2013, SurveyUSA interviewed a random sample of 800 adults in the Tampa, Florida, area. Two of the questions asked were "Should smoking be allowed? Or banned on college campuses?" and "What is your smoking status?" The two-way table below summarizes the answers to these two questions.[19]

	Smoking Status		
Opinion	**Current smoker**	**Former smoker**	**Never a smoker**
Allowed	115	93	79
Banned	20	213	188
Not sure	23	39	30

Is there significant evidence that opinion about smoking on college campuses depends on individual smoking habits?

25.31 Diabetic mice. The body's natural electrical field helps wounds heal. If diabetes changes this field, that might explain why people with diabetes heal slowly. A study of this idea compared normal mice and mice bred to spontaneously develop diabetes. The investigators attached sensors to the right hip and front feet of the mice and measured the difference in electrical potential (millivolts) between these locations. Here are the data:[20]

Diabetic mice					
14.70	13.60	7.40	1.05	10.55	16.40
10.00	22.60	15.20	19.60	17.25	18.40
9.80	11.70	14.85	14.45	18.25	10.15
10.85	10.30	10.45	8.55	8.85	19.20

Normal mice				
13.80	9.10	4.95	7.70	9.40
7.20	10.00	14.55	13.30	6.65
9.50	10.40	7.75	8.70	8.85
8.40	8.55	12.60		

(a) Make a stemplot of each sample of potentials. There is a low outlier in the diabetic group. Does it appear that potentials in the two groups differ in a systematic way?

(b) Is there significant evidence of a difference in mean potentials between the two groups?

(c) Repeat your inference without the outlier. Does the outlier affect your conclusion?

25.32 Comparing tropical flowers. Ethan Temeles of Amherst College, with his colleague W. John Kress, studied the relationship between varieties of the tropical flower *Heliconia* on the island of Dominica and the different species of hummingbirds that fertilize the flowers.[21] Over time, the researchers believe, the lengths of the flowers and the form of the hummingbirds' beaks have evolved to match each other. If that is true, flower varieties fertilized by different hummingbird species should have distinct distributions of length. Table 25.2 gives length measurements (in millimeters, mm) for samples of three varieties of *Heliconia*, each fertilized by a

TABLE 25.2 Flower lengths (millimeters) for three *Heliconia* varieties

H. bihai	47.12	46.75	46.81	47.12	46.67	47.43	46.44	46.64
	48.07	48.34	48.15	50.26	50.12	46.34	46.94	48.36
***H. caribaea* red**	41.90	42.01	41.93	43.09	41.47	41.69	39.78	40.57
	39.63	42.18	40.66	37.87	39.16	37.40	38.20	38.07
	38.10	37.97	38.79	38.23	38.87	37.78	38.01	
***H. caribaea* yellow**	36.78	37.02	36.52	36.11	36.03	35.45	38.13	37.10
	35.17	36.82	36.66	35.68	36.03	34.57	34.63	

different species of hummingbird. Do the three varieties have different flower lengths, on average?

Kevin Schafer/Alamy

25.33 Side effects of medication. A study of "adverse symptoms" in users of over-the-counter pain relief medications assigned subjects at random to one of two common pain relievers: acetaminophen and ibuprofen. (Both of these pain relievers are sold under various brand names, sometimes combined with other ingredients.) In all, 650 subjects took acetaminophen, and 44 experienced some adverse symptom. Of the 347 subjects who took ibuprofen, 49 had an adverse symptom. How strong is the evidence that the two pain relievers differ in the proportion of people who experience an adverse symptom?

25.34 Mouse genes. A study of genetic influences on diabetes compared normal mice with similar mice genetically altered to remove a gene called *aP2*. Mice of both types were allowed to become obese by eating a high-fat diet. The researchers then measured the levels of insulin and glucose in their blood plasma. Here are some excerpts from their findings.[22] The normal mice are called "wild-type" and the altered mice are called "$aP2^{-/-}$."

Each value is the mean ± SEM of measurements on at least 10 mice. Mean values of each plasma component are compared between $aP2^{-/-}$ mice and wild-type controls by Student's t test ($P < 0.05$ and **$P < 0.005$).*

Parameter	Wild-type	$aP2^{-/-}$
Insulin (ng/ml)	5.9 ± 0.9	0.75 ± 0.2**
Glucose (mg/dl)	230 ± 25	150 ± 17*

Despite much greater circulating amounts of insulin, the wild-type mice had higher blood glucose than the $aP2^{-/-}$ animals. These results indicate that the absence of aP2 interferes with the development of dietary obesity-induced insulin resistance.

Other biologists are supposed to understand the statistics reported so tersely.

(a) What does "SEM" mean? What is the expression for SEM based on n, $\bar{x}$, and s from a sample?

(b) Which of the tests we have studied did the researchers apply?

(c) Explain to a biologist who knows no statistics what $P < 0.05$ and $P < 0.005$ mean. Which is stronger evidence of a difference between the two types of mice?

25.35 Mouse genes, continued. The report quoted in the previous exercise says only that the sample sizes were "at least 10." Suppose that the results are based on exactly 10 mice of each type. Use the values in the table to find $\bar{x}$ and s for the insulin concentrations in the two types of mice. Carry out a test to assess the significance of the difference in mean insulin concentration. Does your P-value confirm the claim in the report that $P < 0.005$?

25.36 Squirrels and their food supply. 4 STEP Exercise 6.5 (page 143) gives data on the abundance of the pine cones that red squirrels feed on and the mean number of offspring per female squirrel over 16 years. The strength of the relationship is remarkable because females produce young before the food is available. How significant is the evidence that a greater number of cones leads to more offspring? (Use a vertical scale from −2 to 2 in your residual plot to show the pattern more clearly.)

25.37 Python eggs. How is the hatching of water python eggs influenced by the temperature of the snake's nest? Researchers assigned newly laid eggs to one of three temperatures: hot, neutral, or cold. Hot duplicates the extra warmth provided by the mother python, and cold duplicates the absence of the mother. Here are the data on the number of eggs and the number that hatched:[23]

Temperature	Eggs	Hatched
Cold	27	16
Neutral	56	38
Hot	104	75

The researchers anticipated that eggs would not hatch at cold temperatures. Do the data support that anticipation? Are there significant differences among the proportions of eggs that hatched in the three groups?

25.38 Weights of newly hatched pythons. The previous study also examined the little pythons that had hatched. The report summarized the data in the common form "mean ± standard error" as follows:

Temperature	n	Weight (grams) at hatching	Propensity to strike
Cold	16	28.89 ± 8.08	6.40 ± 5.67
Neutral	38	32.93 ± 5.61	5.82 ± 4.24
Hot	75	32.27 ± 4.10	4.30 ± 2.70

Is there evidence that nest temperature affects the mean weight of newly hatched pythons?

25.39 Python strikes. The data in the previous exercise also describe the "propensity to strike" of the hatched pythons at 30 days of age. This is the number of taps on the head with a small brush until the python launches a strike. (Don't try this with adult pythons.) The data are again summarized in the form "sample mean ± standard error of the mean." Does nest temperature appear to influence propensity to strike?

25.40 Cell-only versus landline users. Random digit dialing telephone surveys do not call cell phone numbers. If the opinions of people who have only cell phones differ from those of people who still have landline service, the poll results may not represent the entire adult population. The Pew Research Center interviewed separate random samples of cell-only and landline telephone users. Here is a two-way table that breaks down these samples by age group:

	Landline sample	Cell-only sample
Age 18–29	108	96
Age 30–49	264	70
Age 50–64	202	26
Age 65 or older	178	8
Total	752	200

Is there significant evidence that age distribution differs between landline users and cell phone–only users?

TABLE 25.3 Goldfish ventilation rates (number of opercular beats per minute)

Water temperature (°C)				
10	12	15	22	25
43	25	64	114	104
39	30	69	86	158
8	68	76	120	160
30	54	82	75	120
44	63	78	40	141
64	35	49	96	79
12	80	96	90	122
15	80	43	65	172
47	71	62	54	110
14	12	48	86	153
28	27	46	87	96
40	8	54	20	110
70	15	30	78	128
21	40	70	80	158
47	30	48	87	117
20	4	66	112	85
34	45	48	67	189
35	30	46	52	120

25.41 Temperature response in goldfish. Goldfish are cold-blooded and have strong physiological responses to sudden temperature changes ("acute response"). An experiment compared the acute response to a given test temperature (10, 12, 15, 22, or 25 degrees Celsius) in goldfish who had been kept for months in tanks with water that was 22 degrees Celsius. Table 25.3 gives the ventilation rate in number of opercular beats per minute for each fish in this experiment.[24] Is there evidence that goldfish ventilation rate varies significantly when the fish are exposed suddenly to different temperatures?

EESEE CASE STUDIES

The Electronic Encyclopedia of Statistical Examples and Exercises (EESEE) is available on the companion website. These more elaborate stories, with data, provide settings for longer case studies. Here are some suggestions for EESEE stories that apply the ideas you have learned in Part III.

25.42 Is Caffeine Dependence Real? Answer Questions 2, 3, 4, and 6 for this case study. (Matched pairs study.)

25.43 Seasonal Weevil Migration. Respond to Question 1. (Proportions.)

25.44 Radar Detectors and Speeding. Read this case study and answer Questions 1, 3, and 5. (Study design, proportions.)

25.45 Leave Survey after the Beep. Carefully answer Question 3; in part (a), use the preferred two-sample

t procedure rather than the pooled t. (Two-sample problems, choice of procedure.)

25.46 Passive Smoking and Respiratory Health. Write careful answers to Questions 1, 3, and 4. (Conditions for inference, choice of procedure.)

25.47 Emissions from an Oil Refinery. Answer both questions. (Conditions for inference.)

25.48 Surgery in a Blanket. Read this case study and answer all the questions. (Study design, choice of procedure, interpretation of results.)

25.49 Blood Alcohol Content. Read this case study and answer all the questions. (Regression.)

NOTES AND DATA SOURCES

"About This Book"

1. The GAISE College Report can be found at `www.amstat.org/education/gaise`.
2. G. Gigerenzer and A. Edwards, "Simple tools for understanding risks: from innumeracy to insight," *British Medical Journal*, 327 (2003), pp. 741–744.
3. L. D. Brown, T. Cai, and A. DasGupta, "Interval estimation for a binomial proportion," *Statistical Science*, 16 (2001), pp. 101–133.

"Statistical Thinking"

1. E. W. Campion, "Editorial: power lines, cancer, and fear," *New England Journal of Medicine*, 337, No. 1 (1997). The study report is M. S. Linet et al., "Residential exposure to magnetic fields and acute lymphoblastic leukemia in children," in the same issue. See also G. Taubes, "Magnetic field–cancer link: will it rest in peace?," *Science*, 277 (1997), p. 29.
2. K. Hartmann et al., "Outcomes of routine episiotomy: a systematic review," *Journal of the American Medical Association*, 293 (2005), pp. 2141–2148.
3. "Energy drink consumption and its association with sleep problems among U.S. service members on a combat deployment—Afghanistan, 2010," *Morbidity and Mortality Weekly Report*, 61 (2012), pp. 895–898.
4. See, for example, M. Enserink, "The vanishing promises of hormone replacement," *Science*, 297 (2002), pp. 325–326; and B. Vastag, "Hormone replacement therapy falls out of favor with expert committee," *Journal of the American Medical Association*, 287 (2002), pp. 1923–1924. A National Institutes of Health panel's comprehensive report is *International Position Paper on Women's Health and Menopause*, NIH Publication 02-3284, 2002.
5. The data are part of a larger data set in the *Journal of Statistics Education* archive, accessible online. The original source is P. Brofeldt, "Bidrag till kaennedom on fiskbestondet i vaara sjoear: Laengelmaevesi," in T. H. Jaervi, *Finlands fiskeriet*, Vol. 4, *Meddelanden utgivna av fiskerifoereningen i Finland*, Helsinki, 1917. The data were contributed to the archive (with information in English) by Juha Puranen of the University of Helsinki.
6. We thank Rudi Berkelhamer of the University of California at Irvine for the data. The data are part of a larger set collected for an undergraduate lab exercise in scientific methods.
7. A. C. Nielsen, Jr., "Statistics in marketing," in *Making Statistics More Effective in Schools of Business*, Graduate School of Business, University of Chicago, 1986.
8. J. S. Mandelblatt et al., "Effects of mammography screening under different screening schedules: model estimates of potential benefits and harms," *Annals of Internal Medicine*, 151 (2009), pp. 738–747.

Chapter 1

1. Discrete variables can also be countably infinite, as in the enumeration of all the whole numbers from zero to infinity. We will see examples of such variables in optional Chapter 12, in the context of Poisson distributions.

2. M. Joly et al., "Posture does not matter! Paw usage and grasping paw preference in a small-bodied rooting quadrupedal mammal," *PLoS ONE*, 7 (2012): e38228, doi:10.1371/journal.pone.0038228.
3. S. B. Willingham et al., "The CD47-signal regulatory protein alpha (SIRPa) interaction is a therapeutic target for human solid tumors," *PNAS*, 109 (2012), pp. 6662–6667.
4. The Electronic Encyclopedia of Statistical Examples and Exercises (EESEE) is a free resource for statistical education. Many of the examples in the EESEE database are available on the text's companion website, `www.whfreeman.com/psls3e`.
5. *National Vital Statistics Reports*, Vol. 61, No. 6 (October 10, 2012), Table B.
6. Substance Abuse and Mental Health Services Administration, "Results from the 2009 National Household Survey on Drug Use and Health," September 2010.
7. C. A. Roberto et al., "Influence of licensed characters on children's taste and snack preferences," *Pediatrics*, 126 (2010), pp. 88–93, doi:10.1542/peds.2009-3433.
8. Data provided by Chris Olsen, who found the information in scuba-diving magazines.
9. Our eyes do respond to area, but not quite linearly. It appears that we perceive the ratio of two bars to be about the 0.7 power of the ratio of their actual areas. See W. S. Cleveland, *The Elements of Graphing Data*, Wadsworth, 1985, pp. 278–284.
10. Data provided by Drina Iglesia, Purdue University. The data are part of a larger study reported in D. D. S. Iglesia, E. J. Cragoe, Jr., and J. W. Vanable, "Electric field strength and epithelization in the newt (*Notophthalmus viridescens*)," *Journal of Experimental Zoology*, 274 (1996), pp. 56–62.
11. T. Bjerkedal, "Acquisition of resistance in guinea pigs infected with different doses of virulent tubercle bacilli," *American Journal of Hygiene*, 72 (1960), pp. 130–148.
12. Centers for Disease Control and Prevention website, `www.cdc.gov`, lime disease statistics.
13. Statistical Society of Canada, 2004 case study II. Data contributed by Caroline Davis, Elizabeth Blackmore, Deborah Katzman, and John Fox.
14. D. L. Arsenau, "Comparison of diet management instruction for patients with non–insulin dependent diabetes mellitus: learning activity package vs. group instruction," MS thesis, Purdue University, 1993.
15. U.S. Geological Survey, National Water Information System, Shark River, Everglades National Park, Florida, `waterdata.usgs.gov`.
16. NASA Goddard Institute for Space Studies (GISS) Surface Temperature Analysis, `http://data.giss.nasa.gov/gistemp`.
17. Virginia Health Information, Vaginal Delivery, `www.vhi.org`.
18. National Oceanic and Atmospheric Administration, Earth System Research Laboratory, Trends in Atmospheric Carbon Dioxide, `www.esrl.noaa.gov`.
19. From the UCI Machine Learning Archive at `archive.ics.uci.edu/ml`.
20. G. P. Sutton and M. Burrows, "Biomechanics of jumping in the flea," *Journal of Experimental Biology*, 214 (2011), pp. 836–847, doi:10.1242/jeb.052399.
21. Surveillance, Epidemiology, and End Results (SEER) Program (`www.seer.cancer.gov`) SEER*Stat Databases.

22. See Note 4.
23. National Center for Health Statistics, "Deaths and death rates for the 10 leading causes of death in specified age groups: United States, preliminary 2010," *National Vital Statistics Reports*, 60, (2012), Table 7.
24. Florida Fish and Wildlife Conservation Commission, Marine Mammal Pathobiology Laboratory, Final Manatee Mortality Table, 2012.
25. A. K. R. LePort et al., "Behavioral and neuroanatomical investigation of Highly Superior Autobiographical Memory (HSAM)," *Neurobiology of Learning and Memory*, 98 (2012), pp. 78–92.
26. This data set was provided by Nicolas Fisher.
27. National Institutes of Health, Essential Fatty Acids Education site, `efaeducation.nih.gov`.
28. International Energy Agency, *Key World Energy Statistics 2007*, `www.iea.org`.
29. National Oceanic and Atmospheric Administration, `www.beringclimate.noaa.gov`.
30. Y.-H. Huang et al., "Nanoparticle-delivered suicide gene therapy effectively reduces ovarian tumor burden in mice," *Cancer Research*, 69 (2009), pp. 6184–6191.
31. I. Heap, International Survey of Herbicide Resistant Weeds, `www.weedscience.org`.
32. J. A. Martin et al., "Births: final data for 2010," *National Vital Statistics Reports*, 61 (2012).
33. From the website of the Bureau of Labor Statistics, `www.bls.gov/cpi`.
34. See Note 15.
35. This data set comes from the *Journal of Statistics Education* online data archive. It was originally submitted by M. J. Kahn of Wheaton College.
36. This data set comes from the *Journal of Statistics Education* online data archive. It was originally submitted by A. Garth Fisher of Brigham Young University.

Chapter 2

1. Data collected by Brigitte Baldi.
2. J. Lanett Beltran, "The effect of spider size and fibroin composition on the material properties of dragline silk from *Nephila clavipes*," MS thesis, University of California at Irvine, 2007.
3. See Note 1.
4. This isn't a mathematical theorem. The mean can be less than the median in right-skewed distributions that take only a few values, many of which lie exactly at the median. The rule almost never fails for distributions taking many values, and counterexamples don't appear clearly skewed in graphs even though they may be slightly skewed according to technical measures of skewness. See P. T. von Hippel, "Mean, median, and skew: correcting a textbook rule," *Journal of Statistics Education*, 13, No. 2 (2005), online journal, `www.amstat.org/publications/jse`.
5. A. Dell'Anno and R. Danovaro, "Extracellular DNA plays a key role in deep-sea ecosystem functioning," *Science*, 309 (2005), p. 2179.
6. EESEE story "Acorn Size and Oak Tree Range."

7. Although this is not mathematically correct, you might find it helpful to think of the standard deviation s as representing the average space (or absolute distance) between the n data points and their mean.
8. See Note 25 from Chapter 1.
9. D. J. Madigana et al., "Pacific bluefin tuna transport Fukushima-derived radionuclides from Japan to California," *PNAS*, 109 (2012), pp. 9483–9486, doi:10.1073/pnas.1204859109.
10. EESEE story "Mercury in Florida's Bass."
11. See Note 32 from Chapter 1 (Table 23).
12. M. A. Laskey et al., "Bone changes after 3 mo of lactation: influence of calcium intake, breast-milk output, and vitamin D–receptor genotype," *American Journal of Clinical Nutrition*, 67 (1998), pp. 685–692.
13. P. S. Gupta, "Reaction of plants to the density of soil," *Journal of Ecology*, 21 (1933), pp. 452–474.
14. J. A. Levine et al., "Inter-individual variation in posture allocation: possible role in human obesity," *Science*, 307 (2005), pp. 584–586. We thank James Levine for providing the data.
15. We thank Charles Cannon of Duke University for providing the data. The study report is C. H. Cannon, D. R. Peart, and M. Leighton, "Tree species diversity in commercially logged Bornean rainforest," *Science*, 281 (1998), pp. 1366–1367.
16. J. Tunga et al., "Social environment is associated with gene regulatory variation in the rhesus macaque immune system," *PNAS*, 109 (2012), pp. 6490–6495, doi:10.1073/pnas.1202734109.
17. This data set comes from the *Journal of Statistics Education* online data archive. It was originally submitted by John P. Holcomb, Jr. of Cleveland State University.

Chapter 3

1. From a graph in N. E. Cantin et al., "Ocean warming slows coral growth in the Central Red Sea," *Science*, 329 (2010), pp. 322–325, doi:10.1126/science.1190182.
2. Florida Fish and Wildlife Conservation Commission, "Boating accident statistics" and "Manatee mortality," `www.myfwc.com`.
3. C. N. Templeton, E. Greene, and K. Davis, "Allometry of alarm calls: black-capped chickadees encode information about predator size," *Science*, 308 (2005), pp. 1934–1937.
4. From a graph in J. A. Levine, N. L. Eberhardt, and M. D. Jensen, "Role of nonexercise activity thermogenesis in resistance to fat gain in humans," *Science*, 283 (1999), pp. 212–214.
5. M. E. Peterson et al., "The dependence of enzyme activity on temperature: determination and validation of parameters," *Biochemical Journal*, 402 (2007), pp. 331–337, doi:10.1042/BJ20061143.
6. From a graph in L. Partridge and M. Farquhar, "Sexual activity reduces lifespan of male fruitflies," *Nature*, 294 (1981), pp. 580–582.
7. A careful study of this phenomenon is W. S. Cleveland, P. Diaconis, and R. McGill, "Variables on scatterplots look more highly correlated when the scales are increased," *Science*, 216 (1982), pp. 1138–1141.

8. See Note 1.
9. D. M. Etheridge et al., "Atmospheric methane between 1000 A.D. and present: evidence of anthropogenic emissions and climatic variability," *Journal of Geophysical Research*, 103 (1998), p. 15,979.
10. These data were originally collected by L. M. Linde of UCLA but were first published by M. R. Mickey, O. J. Dunn, and V. Clark, "Note on the use of stepwise regression in detecting outliers," *Computers and Biomedical Research*, 1 (1967), pp. 105–111. The data have been used by several authors. These are taken from N. R. Draper and J. A. John, "Influential observations and outliers in regression," *Technometrics*, 23 (1981), pp. 21–26.
11. From graphs in K. Lebret et al., "Grazing resistance allows bloom formation and may explain invasion success of *Gonyostomum semen*," *Limnology and Oceanography*, 57 (2012), pp. 727–734, doi:10.4319/lo.2012.57.3.0727.
12. C. J. Lawrence, "Post-learning mechanisms related to memory for emotional stimuli in humans," MS thesis, University of California at Irvine, 2007.
13. M. Mozny et al., "The impact of climate change on the yield and quality of Saaz hops in the Czech Republic," *Agricultural and Forest Meteorology*, 149 (2009), pp. 913–919.
14. This data set comes from the *Journal of Statistics Education* online data archive. It was originally submitted by G. Andy Chang of Youngstown State University.
15. R. Margaria et al., "Energy cost of running," *Journal of Applied Physiology*, 18 (1963), pp. 367–370.
16. S. J. Husson et al., "Optogenetic analysis of a nociceptor neuron and network reveals ion channels acting downstream of primary sensors," *Current Biology* 22 (2012), pp. 743–752, doi:10.1016/j.cub.2012.02.066.
17. Data from many studies compiled in D. F. Greene and E. A. Johnson, "Estimating the mean annual seed production of trees," *Ecology*, 75 (1994), pp. 642–647.
18. From a graph in G. J. Tattersall et al., "Heat exchange from the toucan bill reveals a controllable vascular thermal radiator," *Science*, 325 (2009), pp. 468–470.
19. From a graph in C. G. Wiklund, "Food as a mechanism of density-dependent regulation of breeding numbers in the merlin *Falco columbarius*," *Ecology*, 82 (2001), pp. 860–867.
20. J. S. Brashares et al., "Bushmeat hunting, wildlife declines, and fish supply in West Africa," *Science*, 306 (2004), pp. 1180–1183. The data used here are found in the online supplementary material. The published analysis omits data for 1999, an extreme low outlier, without explanation.
21. From a graph in N. I. Eisenberger, M. D. Lieberman, and K. D. Williams, "Does rejection hurt? An fMRI study of social exclusion," *Science*, 302 (2003), pp. 290–292.

Chapter 4

1. Data from G. A. Sacher and E. F. Staffelt, "Relation of gestation time to brain weight for placental mammals: implications for the theory of vertebrate growth," *American Naturalist*, 108 (1974), pp. 593–613. These data were found in F. L. Ramsey and D. W. Schafer, *The Statistical Sleuth: A Course in Methods of Data Analysis*, Duxbury, 1997.

2. University of Oklahoma Police Department, "The Police Notebook," based on material from the National Highway Traffic Safety Administration.
3. Department of Motor Vehicles, "California Driver Handbook." Exact definition varies depending on alcohol content of the particular brand consumed.
4. K. McCrickerd et al., "Subtle changes in the flavour and texture of a drink enhance expectations of satiety," *Flavour*, 2012, pp. 1–20, doi:10.1186/2044-7248-1-20.
5. Data courtesy of David LeBauer, University of California, Irvine.
6. From a graph in T. Singer et al., "Empathy for pain involves the affective but not sensory components of pain," *Science*, 303 (2004), pp. 1157–1162. Data for other brain regions showed a stronger correlation and no outliers.
7. From a graph in B.-E. Saether, S. Engen, and E. Mattysen, "Demographic characteristics and population dynamical patterns of solitary birds," *Science*, 295 (2002), pp. 2070–2073.
8. Contributed by Marigene Arnold, Kalamazoo College.
9. D. P. Casey et al., "Relationship between muscle sympathetic nerve activity and aortic wave reflection characteristics in young men and women," *Hypertension*, 57 (2011), pp. 421–427.
10. F. H. Messerli, "Chocolate consumption, cognitive function, and Nobel laureates," *New England Journal of Medicine*, 367 (2012), pp. 1562–1564, doi:10.1056/NEJMon1211064.
11. L. L. Calderon et al., "Risk factors for obesity in Mexican-American girls: dietary factors, anthropometric factors, physical activity, and hours of television viewing," *Journal of the American Dietetic Association*, 96 (1996), pp. 1177–1179.
12. *The Health Consequences of Smoking: 1983*, Public Health Service, Washington, DC, 1983.
13. World Bank, "World Development Indicators," `data.worldbank.org`.
14. P. K. Mills et al., "Dietary habits and breast cancer incidence among Seventh-Day Adventists," *Cancer*, 64 (1989), pp. 582–590.
15. J. D. McLister, "The energetics of male reproductive behavior in the treefrog, *Hyla versicolor*," MS thesis, University of California at Irvine, 2000.
16. G. L. Kooyman et al., "Diving behavior and energetics during foraging cycles in king penguins," *Ecological Monographs*, 62 (1992), pp. 143–163.
17. From a presentation by Charles Knauf, Monroe County (New York) Environmental Health Laboratory.
18. C. Carbone and J. L. Gittleman, "A common rule for the scaling of carnivore density," *Science*, 295 (2002), pp. 2273–2276.
19. F. Sultan and V. Braitenberg, "Shapes and sizes of different mammalian cerebella: a study in quantitative comparative neuroanatomy," *Journal für Hirnforschung*, 34 (1993), pp. 79–92.
20. Statistical Society of Canada, 2001 Case Study II. Original publication: F. He and R. Duncan, "Density-dependent effects on tree survival in an old-growth Douglas fir forest," *Journal of Ecology*, 88 (2000), pp. 676–688.
21. Frank J. Anscombe, "Graphs in statistical analysis," *American Statistician*, 27 (1973), pp. 17–21.

22. National Toxicology Program, CAS No. 22839-47-0 (2005).
23. From a graph in Feng Sheng Hu et al., "Cyclic variation and solar forcing of Holocene climate in the Alaskan subarctic," *Science*, 301 (2003), pp. 1890–1893.
24. See Note 14 from Chapter 1.
25. From a graph in A. L. Perry et al., "Climate change and distribution shifts in marine fishes," *Science*, 308 (2005), pp. 1912–1915. The explanatory variable is the five-year running mean of winter (December to March) sea-bottom temperature.
26. From a graph in G. D. Martinsen, E. M. Driebe, and T. G. Whitham, "Indirect interactions mediated by changing plant chemistry: beaver browsing benefits beetles," *Ecology*, 79 (1998), pp. 192–200.
27. F. H. Simmons, "Physiology of the trade-off between fecundity and survival in *Drosophila melanogaster*, as revealed through dietary manipulation," MS thesis, University of California at Irvine, 1996.

Chapter 5

1. A. B. Dufour et al., "Foot pain: Is current or past shoewear a factor? The Framingham Foot Study," *Arthritis Care and Research*, 61 (2009), pp. 1352–1358, doi:10.1002/art.24733.
2. Results presented by S. Vasan and D. Ho at the AIDS Vaccine 2009 conference in Paris, France, `www.vaccineenterprise.org/conference_archive/2009`.
3. J. L. Hoogland, "Why do Gunnison's prairie dogs give anti-predator calls?," *Animal Behavior*, 51 (1996), pp. 871–880.
4. S. Oppe and F. De Charro, "The effect of medical care by a helicopter trauma team on the probability of survival and the quality of life of hospitalized victims," *Accident Analysis and Prevention*, 33 (2001), pp. 129–138. The authors give the data in this Example as a "theoretical example" to illustrate the need for their more elaborate analysis of actual data using severity scores for each victim.
5. C. R. Charig et al., "Comparison of treatment of renal calculi by operative surgery, percutaneous nephrolithotomy, and extracorporeal shock wave lithotripsy," *British Medical Journal*, 292 (1986), pp. 879–882.
6. Condensed from D. R. Appleton, J. M. French, and M. P. J. Vanderpump, "Ignoring a covariate: an example of Simpson's paradox," *American Statistician*, 50 (1996), pp. 340–341.
7. R. B. Turner et al., "An evaluation of *Echinacea angustifolia* in experimental rhinovirus infections," *New England Journal of Medicine*, 353 (2005), pp. 341–348.
8. C. Davison Ankney, "Sex-ratio varies with egg sequence in lesser snow geese," *Auk*, 99 (1982), pp. 662–666.
9. Information for physicians at `www.botoxseveresweating.com`.
10. D. M. Barnes, "Breaking the cycle of addiction," *Science*, 241 (1988), pp. 1029–1030.
11. Feeding Infants and Toddlers Study, a Gerber initiative (2002), `http://medical.gerber.com/nestlescience/fits.aspx`.
12. G. D. Myer et al., "Youth versus adult weightlifting injuries presenting to United States emergency rooms: accidental versus nonaccidental injury mechanisms," *Journal of Strength and Conditioning Research*, 23 (2009), pp. 2054–2060.

13. A. E. Czeizel and I. Dudas, "Prevention of the first occurrence of neural-tube defects by periconceptional vitamin supplementation," *New England Journal of Medicine*, 327 (1992), pp. 1832–1835.

14. W. Gatling, M. A. Mullee, and R. D. Hill, "The general characteristics of a community based population," *Practical Diabetes*, 5 (1989), pp. 104–107.

15. R. Shine, T. R. L. Madsen, M. J. Elphick, and P. S. Harlow, "The influence of nest temperatures and maternal brooding on hatchling phenotypes in water pythons," *Ecology*, 78 (1997), pp. 1713–1721.

16. J. E. Williams et al., "Anger proneness predicts coronary heart disease risk," *Circulation*, 101 (2000), pp. 2034–2039.

17. See Note 20 for Chapter 4.

18. N. J. O. Birkmeyer, "Hospital complication rates with bariatric surgery in Michigan," *Journal of the American Medical Association*, 304 (2010), pp. 435–442.

19. D. Gonzales et al., "Varenicline, an $\alpha 4 \beta 2$ nicotinic acetylcholine receptor partial agonist, vs sustained-release bupropion and placebo for smoking cessation: a randomized controlled trial," *Journal of the American Medical Association*, 296 (2006), pp. 47–55, doi:10.1001/jama.296.1.56

Chapter 6

1. Data for Cohort 2 in R. A. Morgan et al., "Cancer regression in patients after transfer of genetically engineered lymphocytes," *Science*, 314 (2006), pp. 126–129. The doubling time data are given in the paper and the immune response data appear in the supplementary online material.

2. From a graph in S. Boutin et al., "Anticipatory reproduction and population growth in seed predators," *Science*, 314 (2006), pp. 1928–1930.

3. L. H. Yang, "Periodical cicadas as resource pulses in North American forests," *Science*, 306 (2004), pp. 1565–1567. The data are simulated Normal values that match the means and standard deviations reported in this article.

4. International Food Information Council (IFIC) Foundation, 2010 Food and Health Survey: Consumer Attitudes toward Food Safety, Nutrition, and Health.

5. We thank Ethan J. Temeles of Amherst College for providing the data. His work is described in E. J. Temeles and W. J. Kress, "Adaptation in a plant-hummingbird association," *Science*, 300 (2003), pp. 630–633.

6. See Note 2 for Chapter 1.

7. A. S. Banks et al., "Juvenile hallux abducto valgus association with metatarsus adductus," *Journal of the American Podiatric Medical Association*, 84 (1994), pp. 219–224.

8. From a graph in S. M. Vallina and R. Simó, "Strong relationship between DMS and the solar radiation dose over the global surface ocean," *Science*, 315 (2007), pp. 506–508.

9. Data from a plot in J. P. Rauschecker, B. Tian, and M. Hauser, "Processing of complex sounds in the macaque nonprimary auditory cortex," *Science*, 268 (1995), pp. 111–114. The paper states that there are $n = 41$ observations, but only $n = 37$ can be read accurately from the plot.

10. From a graph in C. Packer et al., "Ecological change, group territoriality, and population dynamics in Serengeti lions," *Science*, 307 (2005), pp. 390–393.
11. T. W. Anderson, "Predator responses, prey refuges, and density-dependent mortality of a marine fish," *Ecology*, 81 (2001), pp. 245–257.
12. From the Nenana Ice Classic website, `www.nenanaakiceclassic.com`. See R. Sagarin and F. Micheli, "Climate change in nontraditional data sets," *Science*, 294 (2001), p. 811, for a careful discussion.
13. J. W. Grier, "Ban of DDT and subsequent recovery of reproduction in bald eagles," *Science*, 218 (1982), pp. 1232–1234.
14. From a plot in Jon J. Ramsey et al., "Energy expenditure, body composition, and glucose metabolism in lean and obese rhesus monkeys treated with ephedrine and caffeine," *American Journal of Clinical Nutrition*, 68 (1998), pp. 42–51.
15. Data provided by Samuel Phillips, Purdue University.
16. T. Kontiokari et al., "Randomised trial of cranberry-lingonberry juice and Lactobacillus GG drink for the prevention of urinary tract infections in women," *British Medical Journal*, 322 (2001), pp. 1–5.
17. SurveyUSA News Poll 20292, Tampa–Saint Petersburg (Sarasota) DMA, February 13, 2013.

Chapter 7

1. See, for example, M. Enserink, "The vanishing promises of hormone replacement," *Science*, 297 (2002), pp. 325–326; B. Vastag, "Hormone replacement therapy falls out of favor with expert committee," *Journal of the American Medical Association*, 287 (2002), pp. 1923–1924; and J. E. Manson et al., "Estrogen therapy and coronary-artery calcification, *New England Journal of Medicine*, 356 (2007), pp. 2591–2602.
2. J. C. Barefoot et al., "Alcoholic beverage preference, diet, and health habits in the UNC Alumni Heart Study," *American Journal of Clinical Nutrition*, 76 (2002), pp. 466–472.
3. J. E. Muscat et al., "Handheld cellular telephone use and risk of brain cancer," *Journal of the American Medical Association*, 284 (2000), pp. 3001–3007.
4. Committee on the Science of Climate Change, National Research Council, "Climate change science: an analysis of some key questions," *National Academy Press*, 2001, p. 17.
5. See Note 2 for Chapter 1.
6. See Note 20 for Chapter 1.
7. "Birth Control Methods," `WomensHealth.gov`.
8. California Department of Public Health, statewide data, Table 2.1.
9. The American Association for Public Opinon Research states that "surveys based on self-selected volunteers do not have [a] known relationship to the target population and are subject to unknown, non-measurable biases." Yet opt-in polls are quite common, from the dedicated `Twiigs.com` polling website to the occasional features of broadcast and web-based news organizations such as CNN, Fox, and NPR, and they have even been seen on the websites of scientific, peer-reviewed journals (as shown in Exercise 7.14).

10. R. Stein, "Scientists see upside and downside of sequencing their own genes, *NPR*, September 19, 2012.
11. From the website of the Gallup Organization, `www.gallup.com`. Press releases remain on this site for only a limited time.
12. R. C. Parker and P. A. Glass, "Preliminary results of double-sample forest inventory of pine and mixed stands with high- and low-density LiDAR," in Kristina F. Connoe (ed.), *Proceedings of the 12th Biennial Southern Silvicultural Research Conference*, U.S. Department of Agriculture, Forest Service, Southern Research Station, 2004.
13. "Legal and ethical issues in RDD cell phone surveys," `www.aapor.org`.
14. S. J. Blumberg and J. V. Luke, "Wireless substitution: early release of estimates from the National Health Interview Survey, July–December 2012," National Center for Health Statistics, June 2013, `www.cdc.gov/nchs/nhis.htm`.
15. For information on the American Community Survey of households (there is a separate sample of group quarters), go to `www.census.gov/acs`.
16. For information on the University of Chicago's General Social Survey, go to `http://www3.norc.org/gss+website`.
17. From the SurveyUSA website at `www.surveyusa.com/respondentcoop.html`.
18. "Bad samples," `www.aapor.org`.
19. For more detail on the limits of memory in surveys, see N. M. Bradburn, L. J. Rips, and S. K. Shevell, "Answering autobiographical questions: the impact of memory and inference on surveys," *Science*, 236 (1987), pp. 157–161.
20. The responses on welfare are from a *New York Times*/CBS News Poll reported in the *New York Times*, July 5, 1992. Many other examples appear in T. W. Smith, "That which we call welfare by any other name would smell sweeter," *Public Opinion Quarterly*, 51 (1987), pp. 75–83.
21. "Health reform and the decline of physician private practice," Physician Foundation, 2010. The cover page states "Includes results of Physicians and Health Reform, a survey of 100,000 physicians." The full report can be found at `www.physiciansfoundation.org`.
22. E. Azziz-Baumgartner et al., "Case-control study of an acute aflatoxicosis outbreak, Kenya, 2004," *Environmental Health Perspectives*, 113 (2005), pp. 1779–1783.
23. Most cohort studies are prospective. However, some cohort studies select a group of individuals currently enrolled in a health care plan and then search the subjects' past medical records for factors that may explain their current health conditions: Such cohorts are retrospective rather than prospective.
24. L. Trasande et al., "Infant antibiotic exposures and early-life body mass," *International Journal of Obesity*, 2012, pp. 1–8, doi:10.1038/ijo.2012.132.
25. P. Krakowiak et al., "Maternal metabolic conditions and risk for autism and other neurodevelopmental disorders," *Pediatrics*, 129 (2012), pp. 1–8, doi:10.1542/peds.2011-2583.
26. C. G. Bacon et al., "A prospective study of risk factors for erectile dysfunction," *Journal of Urology*, 176 (2006), pp. 217–221.
27. L. E. Moses and F. Mosteller, "Safety of anesthetics," in J. M. Tanur et al. (eds.), *Statistics: A Guide to the Unknown*, 3rd ed., Wadsworth, 1989, pp. 15–24.
28. Jane Goodall Institute, `www.janegoodall.org/`.

29. M. A. Parada et al., "The validity of self-reported seatbelt use: Hispanic and non-Hispanic drivers in El Paso," *Accident Analysis and Prevention*, 33 (2001), pp. 139–143.

30. See Note 11.

31. S. Sulheim et al., "Helmet use and risk of head injuries in alpine skiers and snowboarders," *Journal of the American Medical Association*, 295 (2006), pp. 919–924.

32. See Note 12 for Chapter 2.

33. M. Lucas et al., "Coffee, caffeine, and risk of depression among women," *Archives of Internal Medicine*, 171 (2011), pp. 1571–1578.

34. E. Courchesne et al., "Unusual brain growth patterns in early life in patients with autistic disorder: an MRI study," *Neurology*, 57 (2001), pp. 245–254.

35. S. Sutcliffe et al., "A prospective cohort study of red wine consumption and risk of prostate cancer," *International Journal of Cancer*, 120 (2007), pp. 1529–1535.

36. P.-Y. Scarabin, E. Oger, and G. Plu-Bureau, "Differential association of oral and transdermal oestrogen replacement therapy with venous thromboembolism risk," *Lancet*, 362 (2003), pp. 428–432.

Chapter 8

1. M. de Lorgeril et al., "Mediterranean dietary pattern in a randomized trial–Prolonged survival and possible reduced cancer rate," *Archives of Internal Medicine*, 158 (1998), pp. 1181–1187.

2. Y. Higuchia et al., "Day light quality affects the night-break response in the short-day plant chrysanthemum, suggesting differential phytochrome-mediated regulation of flowering," *Journal of Plant Physiology*, 169 (2012), pp. 1789–1796, doi:10.1016/j.jplph.2012.07.003.

3. L. L. Miao, "Gastric freezing: an example of the evaluation of medical therapy by randomized clinical trials," in J. P. Bunker, B. A. Barnes, and F. Mosteller (eds.), *Costs, Risks, and Benefits of Surgery*, Oxford University Press, 1977, pp. 198–211.

4. H.-C. Yeh et al.,"Smoking, smoking cessation, and risk for type 2 diabetes mellitus: a cohort study," *Annals of Internal Medicine*, 152 (2010), pp. 10–17.

5. F. Eippert et al., "Direct evidence for spinal cord involvement in placebo analgesia," *Science*, 326 (2009), p. 404.

6. K. Linde et al., "Acupuncture for patients with migraine: a randomized controlled trial," *Journal of the American Medical Association*, 293 (2005), pp. 2118–2125.

7. Paul E. O'Brien et al., "Laparascopic adjustable gastric banding in severely obese adolescents," *Journal of the American Medical Association*, 303 (2010), pp. 519–526.

8. K. B. Suttle, M. A. Thomsen, and M. E. Power, "Species interactions reverse grassland responses to changing climate," *Science*, 315 (2007), pp. 640–642. We thank Kenwyn Blakeslee Suttle for providing these data.

9. D. T. Kirkendall, "Creatinine, carbs, and fluids: how important in soccer nutrition?," *Sports Science Exchange*, 94 (2004), pp. 1–6.

10. J. S. Halterman et al., "Randomized controlled rrial to improve care for urban children with asthma," *Archives of Pediatrics and Adolescent Medicine*, 165 (2011), pp. 262–268.

11. See Note 10 for Chapter 1.

12. K. Thomas et al., "Improved endurance capacity following chocolate milk consumption compared with 2 commercially available sport drinks," *Applied Physiology, Nutrition, and Metabolism*, 34 (2009), pp. 78–82.

13. L.-P. Wang et al., "Efficacy of acupuncture for migraine prophylaxis: a single-blinded, double-dummy, randomized controlled trial," *Pain*, 152 (2011), pp. 1864–1871.

14. M. Enserink, "Fickle mice highlight test problems," *Science*, 284 (1999), pp. 1599–1600. There is a full report of the study in the same issue.

15. U.S. Department of Health and Human Services, Public Health Service, National Toxicology Program, *Report on Carcinogens*, 11th ed., Table B: Agents, substances, mixtures, or exposure circumstances delisted from the *Report on Carcinogens*, `ntp.niehs.nih.gov/ntp/roc/toc11.html`.

16. M. H. Emmelot-Vonk et al., "Effect of testosterone supplementation on functional mobility, cognition, and other parameters in older men," *Journal of the American Medical Association*, 299 (2008), pp. 39–52.

17. A. S. Attwood et al., "Glass shape influences consumption rate for alcoholic beverages," *PLoS ONE*, 7 (2012), e43007, doi:10.1371/journal.pone.0043007.

18. The difficulties of interpreting guidelines for informed consent and for the work of institutional review boards in medical research are a main theme of Beverly Woodward, "Challenges to human subject protections in U.S. medical research," *Journal of the American Medical Association*, 282 (1999), pp. 1947–1952. The references in this paper point to other discussions. Updated regulations and guidelines appear on the website for the Office for Human Research Protections of the Department of Health and Human Services, `www.hhs.gov/ohrp`.

19. NPR News Service article from *All Things Considered*, November 16, 2005, "'My Lobotomy': Howard Dully's Journey," `www.npr.org`.

20. K. Hartmann et al., "Outcomes of routine episiotomy a systematic review," *Journal of the American Medical Association*, 293 (2005), pp. 2141–2148.

21. "President discusses stem cell research," August 9, 2001, `www.whitehouse.gov`.

22. Proposition 71, "Stem Cell Research; Funding; Bonds; Initiative Constitutional Amendment and Statute." Proposition 71 was passed by California voters in November 2004 and resulted in the creation of the California Institute for Regenerative Medicine.

23. "Removing barriers to responsible scientific research involving human stem cells," March 9, 2009, `www.whitehouse.gov`.

24. P. C. Tan et al., "Dengue infection and miscarriage: a prospective case control study," *PLoS Neglected Tropical Diseases*, 6 (2012), e1637, doi:10.1371/journal.pntd.0001637.

25. N. D. Volkow et al., "Effects of cell phone radiofrequency signal exposure on brain glucose metabolism," *Journal of the American Medical Association*, 305 (2011), pp. 808-814.

26. NBC news report aired April 25, 2007. The study had been published in E. Lipov, S. Lipov, and J. T. Stark, "Stellate ganglion blockade provides relief from menopausal hot flashes: a case report series," *Journal of Women's Health*, 14 (2005), pp. 737–741.

27. Rita F. Redburg, "Vitamin E and cardiovascular health," *Journal of the American Medical Association*, 294 (2005), pp. 107–109.

28. R. C. Shelton et al., "Effectiveness of St. John's wort in major depression," *Journal of the American Medical Association*, 285 (2001), pp. 1978–1986.
29. J. N. Kheir et al., "Oxygen gas-filled microparticles provide intravenous oxygen delivery," *Science Translational Medicine*, 4 (2012), pp. 140ra88, doi:10.1126/scitranslmed.3003679.

Chapter 9

1. 2012 *Statistical Abstract of the United States*, Table 80.
2. From the American Red Cross website, `www.givelife2.org/aboutblood/bloodtypes.asp`. There are some other blood types besides the eight main types but they are extremely rare.
3. J. S. Schiller et al., "Summary health statistics for U.S. adults: National Health Interview Survey, 2011," *Vital Health Statistics*, 10 (2012), Table 31.
4. Florida Department of Health, Bureau of Epidemiology: "Animal rabies confirmed cases by county and animal type, January–December 2012," `www.myfloridaeh.com`.
5. Discrete sample spaces are most often, but not necessarily, a finite list of integers or individual outcomes. We will see in Chapter 12 the example of the Poisson distribution, a discrete distribution for $X = \{0, 1, 2, \ldots\}$.
6. G. M. Strain, "Deafness prevalence and pigmentation and gender associations in dog breeds at risk," *The Veterinary Journal*, 167 (2004), pp. 23–32.
7. "Nearly half of Americans drink soda daily," Gallup, July 23, 2012.
8. Data from the National Youth Risk Behavior Survey can be found on the CDC website, `www.cdc.gov/HealthyYouth/yrbs/data`.
9. 2004 *Statistical Abstract of the United States*, Table 193.
10. National Center for Health Statistics, "Health, United States, 2005," p. 38.
11. "Kids and Media," Kaiser Family Foundation Report, 1999.
12. A. Leizorovicz et al., "Randomized, placebo-controlled trial of dalteparin for the prevention of venous thromboembolism in acutely ill medical patients," *Circulation*, 110 (2004), pp. 874–879.
13. E. F. Dunne et al., "Prevalence of HPV infection among females in the United States," *Journal of the American Medical Association*, 297 (2007), pp. 813–819.
14. G. Dezecache and R. I. M. Dunbar, "Sharing the joke: the size of natural laughter groups," *Evolution and Human Behavior*, 33 (2012), pp. 775–779, doi:10.1016/j.evolhumbehav.2012.07.002.
15. See Note 24 for Chapter 1.
16. "Racial and ethnic distribution of ABO blood types," `www.bloodbook.com`.
17. "Phenylketonuria," from the Free Health Encyclopedia at `faqs.org/health`.
18. E. S. Siris et al., "Identification and fracture outcomes of undiagnosed low bone mineral density in postmenopausal women—results from the National Osteoporosis Risk Assessment," *Journal of the American Medical Association*, 286 (2001), pp. 2815–2822.
19. Health Resources and Services Administration, "Providing HIV/AIDS care in a changing environment: hepatitis C and HIV co-infection," September 2003.

20. A. N. Dey and B. Bloom, "Summary health statistics for U.S. children: National Health Interview Survey, 2003," *Vital Health Statistics*, 10 (2005), Table 17.
21. R. N. Rosenfiel et al., "Comparative relationships among eye color, age, and sex in three North American populations of Cooper's hawks," *Wilson Bulletin*, 115 (2003), pp. 225–230.
22. National Cancer Institute, "Fact sheet: probability of breast cancer in American women," `www.cancer.gov`.
23. N. Eriksson et al., "A genetic variant near olfactory receptor genes influences cilantro preference," arXiv:1209.2096v1 [q-bio.GN] (2012).

Chapter 10

1. See Note 2 for Chapter 9.
2. This is one of several tests discussed in Bernard M. Branson, "Rapid HIV testing: 2005 update," a presentation by the Centers for Disease Control and Prevention, at `www.cdc.gov`. The Malawi clinic result is reported by Bernard M. Branson, "Point-of-care rapid tests for HIV antibody," *Journal of Laboratory Medicine*, 27 (2003), pp. 288–295.
3. N. Ranjit et al., "Contraceptive failure in the first two years of use: differences across socioeconomic subgroups," *Family Planning Perspectives*, 33 (2001), pp. 19–27.
4. See Note 3.
5. From the Florida Department of Health website (`http://www.doh.state.fl.us`), Brain and Spinal Cord Injury Program.
6. See Note 20 for Chapter 4.
7. Chronic Kidney Disease Surveillance System, available from the CDC website at `www.cdc.gov/ckd`.
8. Epidemiologic Catchment Area Survey, 1990, diagnosing mental disorders using a random sample of 18,571 Americans.
9. See Note 6 for Chapter 9.
10. Royal Statistical Society news release, "Royal Statistical Society concerned by issues raised in Sally Clark case," October 23, 2001, `www.rss.org.uk`. For background, see an editorial and article in the *Economist*, January 22, 2004. The editorial is titled "The probability of injustice."
11. A. C. Allison and D. F. Clyde, "Malaria in African children with deficient erythrocyte dehydrogenase," *British Medical Journal*, 1 (1961), pp. 1346–1349.
12. L. Naldi et al., "Pigmentary traits, modalities of sun reaction, history of sunburns, and melanocytic nevi as risk factors for cutaneous malignant melanoma in the Italian population: results of a collaborative case-control study," *Cancer*, 88 (2000), pp. 2703–2710.
13. W. Uter et al., "Inter-relation between variables determining constitutional UV sensitivity in Caucasian children," *Photodermatology, Photoimmunology and Photomedicine*, 20 (2004), pp. 9–13.
14. Probabilities from trials with 2897 people known to be free of HIV antibodies and 673 people known to be infected, reported in J. Richard George, "Alternative

specimen sources: methods for confirming positives," 1998 Conference on the Laboratory Science of HIV, found online at the Centers for Disease Control and Prevention, www.cdc.gov.

15. G. Gigerenzer and A. Edwards, "Simple tools for understanding risks: from innumeracy to insight," *British Medical Journal*, 327 (2003), pp. 741–744.

16. See Note 22 for Chapter 9.

17. Different countries use different cutoff points for deciding whether to call a result positive or negative, and some countries have each mammogram reviewed by two different experts. Thus, sensitivity and specificity vary from country to country. See U.S. Preventive Services Task Force, "Screening for breast cancer: recommendations and rationale," February 2002, Agency for Healthcare Research and Quality, www.ahrq.gov.

18. OraQuick In-Home HIV Test customer package insert information, downloaded from www.orasure.com.

19. See Note 15.

20. A. A. Wright, and Ingrid T. Katz, "Home testing for HIV," *New England Journal of Medicine*, 354 (2006), pp. 437–440, doi:10.1056/NEJMp058302; "FDA approves first over-the-counter home use HIV test kit," FDA News Release, July 3, 2012.

21. "Breast guidelines test American tolerance for risk," Reuters, November 20, 2009, www.reuters.com.

22. "Silicosis: from public menace to litigation target," NPR News Service article, March 6, 2006, www.npr.org. The website also links to the judge's filed opinion, "Silica products liability litigation," MDL Docket No. 1553.

23. See Note 23 for Chapter 9.

24. www.cdc.gov/cancer/lung/basic_info/risk_factors.htm.

25. S. W. Guo and D. R. Reed, "The genetics of phenylthiocarbamide perception," *Annals of Human Biology*, 28 (2001), pp. 111–142.

26. Polydactyly is not reported systematically, so estimated rates vary. National Center for Health Statistics, Series 20, No. 31 (1996); "Supernumerary Digit," "Polydactyly of the Foot," Medscape, emedicine.medscape.com.

27. See Note 13.

28. M. McCulloch et al., "Diagnostic accuracy of canine scent detection in early- and late-stage lung and breast cancers," *Integrative Cancer Therapies*, 5 (2006), pp. 30–39.

29. National Institutes of Health's National Digestive Diseases Information Clearinghouse, wrongdiagnosis.com.

30. Data provided by Patricia Heithaus and the Department of Biology at Kenyon College.

31. See Note 18 for Chapter 9.

32. See Note 19 for Chapter 9.

33. S. D. Grosse et al., "Newborn screening for cystic fibrosis," *CDC Morbidity and Mortality Weekly Report*, October 15, 2004.

34. "Should you get a PSA test?," 2006 Kaiser Permanente pamphlet based on "Should you get a PSA test? A patient-doctor decision," from the American Academy of Family Physicians, www.aafp.org.

35. S. H. Sicherer, "Prevalence of peanut and tree nut allergy in the U.S. determined by random digit dial telephone survey," *Journal of Allergy and Clinical Immunology*, 103 (1999), pp. 559–562.

36. National Tay-Sachs and Allied Diseases Association, `www.ntsad.org`.

Chapter 11

1. M. Jakob, *Normal Values pocket*, Borm Bruckmeier Publishing, 2002.
2. J. Lyall et al., "Suppression of avian influenza transmission in genetically modified chickens," *Science*, 331 (2011), pp. 223–226, doi:10.1126/science.1198020.
3. R. C. Nelson, C. M. Brooks, and N. L. Pike, "Biomechanical comparison of male and female distance runners," in P. Milvy (ed.), *The Marathon: Physiological, Medical, Epidemiological, and Psychological Studies*, New York Academy of Sciences, 1977, pp. 793–807.
4. See Note 1.
5. See Note 1.
6. See Note 32 for Chapter 1.
7. See Note 18 for Chapter 9.
8. School-wide data for 2011.
9. M. A. McDowell et al., "Anthropometric reference data for children and adults: U.S. population, 1999–2002," National Center for Health Statistics, Advance Data from Vital and Health Statistics, No. 361, 2005, at `www.cdc.gov/nchs`.
10. See Note 1.
11. IPCC Third Assessment Report "Climate change 2001," Chapter 2, "Observed climate variability and change," page 155 for graph, doi:10.1038/NCLIMATE1452.
12. Data from the EESEE story "Acorn Size and Oak Tree Range."
13. See Note 6 for Chapter 3.

Chapter 12

1. B. Bekan Homawoo et al., "Chlamydia mapping by New York State school district boundaries, excluding New York City (2006–2007)," New York Bureau of STD Prevention and Epidemiology, `www.health.ny.gov`.
2. S. Kidd et al., "Cephalosporin resistant *Neisseria gonorrhoeae* public health response plan," CDC publication (2012).
3. See Note 3 for Chapter 10.
4. CDC, *Morbidity and Mortality Weekly Report*, March 30, 2006, `www.cdc.gov`.
5. See Note 4.
6. CDC, Division of Bacterial and Mycotic Diseases, "Typhoid Fever," `www.cdc.gov`.
7. Office of Research and Statistics, South Carolina State Budget and Control Board, "Top 50 reasons for emergency room visits fror residents of South Carolina."
8. United Nations, "Human development report 2005," Table 9: Leading global health crises and risks.
9. Associated Press news item dated December 9, 2007, found at `www.msnbc.msn.com`.

10. Florida Department of Health, Bureau of Epidemiology: "20-Year animal rabies summary by species," `www.myfloridaeh.com`.
11. CDC, *Morbidity and Mortality Weekly Report*, October 14, 1994, `www.cdc.gov`.
12. CDC, *Morbidity and Mortality Weekly Report*, February 12, 2010, `www.cdc.gov`.
13. See Note 18 for Chapter 9.
14. See Note 3 for Chapter 9.

Chapter 13

1. Strictly speaking, the formula $\sigma/\sqrt{n}$ for the standard deviation of $\overline{x}$ assumes that we draw an SRS of size n from an *infinite* population. If the population has finite size N, this standard deviation is multiplied by $\sqrt{1-(n-1)/(N-1)}$. This "finite population correction" approaches 1 as N increases. When the population is at least 20 times as large as the sample, the correction factor is between about 0.97 and 1. It is reasonable to use the simpler form $\sigma/\sqrt{n}$ in these settings.
2. Strictly speaking, the formula $\sqrt{p(1-p)/n}$ for the standard deviation of $\hat{p}$ assumes that we draw an SRS of size n from an *infinite* population. If the population has finite size N, this standard deviation is multiplied by $\sqrt{1-(n-1)/(N-1)}$. This "finite population correction" approaches 1 as N increases. When the population is at least 20 times as large as the sample, the correction factor is between about 0.97 and 1. It is reasonable to use the simpler form $\sqrt{p(1-p)/n}$ in these settings.

3. See Note 2 for Chapter 12.
4. W. H. Burt and R. P. Grossenheider, *A Field Guide to the Mammals: Field Marks of All North American Species Found North of Mexico*, 3rd ed., Houghton Mifflin, 1976.
5. See Note 3 for Chapter 10.

Chapter 14

1. This value of σ is based on the growth curves developed by the National Center for Health Statistics in collaboration with the National Center for Chronic Disease Prevention and Health Promotion (2000).

2. Data provided by Darlene Gordon, Purdue University.
3. See Note 17 for Chapter 2.
4. See Note 1 for Chapter 11.
5. T. Hesselberg and F. Vollrath, "The mechanical properties of the non-sticky spiral in *Nephila* orb webs (Araneae, Nephilidae)," *Journal of Experimental Biology*, 215 (2012), pp. 3362–3369, doi:10.1242/jeb.068890.
6. This data set comes from the *Journal of Statistics Education* online data archive. It was originally submitted by A. L. Shoemaker of Calvin College. The original data were published in P. A. Mackowiak et al., "A critical appraisal of 98.6 degrees F, the upper limit of the normal body temperature, and other legacies of Carl Reinhold August Wunderlich," *Journal of the American Medical Association*, 268 (1992), pp. 1578–1580.
7. J.-C. Lin et al., "Association between chronic hepatitis C virus infection and bone mineral density," *Calcified Tissue International*, 91 (2012), pp. 423–429, doi:10.1007/s00223-012-9653-y.

8. C. D. Gardner et al., "Effect of raw garlic vs. commercial garlic supplements on plasma lipid concentrations in adults with moderate hypercholesterolemia: a randomized clinical trial," *Archives of Internal Medicine*, 167 (2007), pp. 346–353.
9. See Note 4 for Chapter 13.
10. Avandia (rosiglitazone maleate) prescribing information can be found at `www.avandia.com`.
11. Self-reported heights (in inches); course project.
12. Information found on the USDA website at `www.ams.usda.gov/howtobuy/eggs.htm`.
13. See Note 3 for Chapter 6.
14. S. L. Webb and S. E. Scanga, "Windstorm disturbance without patch dynamics: twelve years of change in a Minnesota forest," *Ecology*, 82 (2001), pp. 893–897.
15. See Note 29 for Chapter 7.

Chapter 15

1. You can find detailed information on the CDC website, `www.cdc.gov`, about the sampling design and analytic guidelines. Here is an excerpt: "Because NHANES is a complex probability sample, analytic approaches based on data from simple random samples are usually not appropriate. Ignoring the complex design can lead to biased estimates and overstated significance levels. Sample weights and the stratification and clustering of the design must be incorporated into an analysis to get proper estimates and standard errors of estimates."
2. "Regulation of sugar-sweetened beverages," *New England Journal of Medicine*, doi:10.1056/NEJMclde1210278.
3. E. M. Barsamian, "The rise and fall of internal mammary artery ligation," in J. P. Bunker, B. A. Barnes, and F. Mosteller (eds.), *Costs, Risks, and Benefits of Surgery*, Oxford University Press, 1977, pp. 212–220.
4. G. I. Pinkhasov et al., "Complications following prostate needle biopsy requiring hospital admission or emergency department visits–experience from 1000 consecutive cases," *BJU International*, 110 (2012), pp. 369–374, doi:10.1111/j.1464-410X.2011.10926.x.
5. See Note 2 for Chapter 1.
6. Harvard School of Public Health College Alcohol Study, April 12, 2001, press release posted on the study website at `www.hsph.harvard.edu/cas`.
7. W. E. Leary, "Cell phones: questions but no answers," *New York Times*, October 26, 1999.
8. R. A. Laven and C. T. Livesey, "The effect of housing and methionine intake on hoof horn hemorrhages in primiparous lactating Holstein cows," *Journal of Dairy Science*, 87 (2004), pp. 1015–1023.
9. P. Weiner et al. "The cumulative effect of long-acting bronchodilators, exercise, and inspiratory muscle training on the perception of dyspnea in patients with advanced COPD," *Chest*, 118 (2000), pp. 672–678.
10. INTERPHONE Study Group, "Brain tumour risk in relation to mobile telephone use: results of the INTERPHONE international case-control study," *International Journal of Epidemiology*, 39 (2010), pp. 675–694, doi:10.1093/ije/dyq079.

11. B. J. Marshall and J. R. Warren, "Unidentified curved bacilli in the stomach of patients with gastritis and peptic ulceration," *Lancet*, 1 (1984), pp. 1311–1315; B. J. Marshall et al., "A prospective double-blind trial of duodenal ulcer relapse after eradication of *Campylobacter pylori*," *Lancet*, 2 (1988), pp. 1437–1442; B. J. Marshall et al., "Urea protects *Helicobacter (Campylobacter) pylori* from the bactericidal effect of acid," *Gastroenterology*, 99 (1990), pp. 697–702.
12. See Note 1 for Chapter 11.
13. See Note 15 for Chapter 8.
14. G. Gregoratos et al., "ACC/AHA guidelines for implantation of cardiac pacemakers and antiarrhythmia devices: executive summary," *Circulation*, 97 (1998), pp. 1325–1335.
15. Results of SurveyUSA News polls 16192 and 16194.
16. K. Foden-Vencil, "Portland, Ore., Becomes Latest Fluoride Battleground," NPR, September 12, 2012.
17. See Note 21 for Chapter 7.
18. See Note 4 for Chapter 13.
19. C. Kopp et al., "Modulation of rhythmic brain activity by diazepam: GABAa receptor subtype and state specificity," *PNAS*, 101 (2004), pp. 3674–3679.
20. J. B. Moseley et al., "A controlled trial of arthroscopic surgery for osteoarthritis of the knee," *New England Journal of Medicine*, 347 (2002), pp. 81–88.

Chapter 16

1. D. M. Fergusson and L. J. Horwood, "Cannabis use and traffic accidents in a birth cohort of young adults," *Accident Analysis and Prevention*, 33 (2001), pp. 703–711.
2. Based on the news item "Bee off with you," *Economist*, November 2, 2002, p. 78.
3. K. F. Barnhart et al., "A peptidomimetic targeting white fat causes weight loss and improved insulin resistance in obese monkeys," *Science Translational Medicine*, 3 (2011), 108ra112, doi:10.1126/scitranslmed.3002621.
4. See Note 17 for Chapter 8.
5. See Note 32 for Chapter 1.
6. M. Park et al., "Recycling endosomes supply AMPA receptors for LTP," *Science*, 305 (2004), pp. 1972–1975.
7. H.-R. Guo, "Arsenic level in drinking water and mortality of lung cancer (Taiwan)," *Cancer Causes and Control*, 15 (2004), pp. 171–177.
8. K. E. Hobbs et al., "Levels and patterns of persistent organochlorines in minke whale (*Balaenoptera acutorostrata*) stocks from the North Atlantic and European Arctic," *Environmental Pollution*, 121 (2003), pp. 239–252.
9. M. Hack et al., "Outcomes in young adulthood for very-low-birth-weight infants," *New England Journal of Medicine*, 346 (2002), pp. 149–157.
10. See Note 20 for Chapter 3.
11. F. J. Sulloway, "Birth order and intelligence," *Science*, 316 (2007), pp. 1711–1712. The Norwegian study is Petter Kristensen and Tor Bjerkedal, "Explaining the relation between birth order and intelligence," *Science*, 316 (2007), p. 1717.

12. J. F. Swain et al., "Comparison of the effects of oat bran and low-fiber wheat on serum lipoprotein levels and blood pressure," *New England Journal of Medicine*, 322 (1990), pp. 147–152.

13. Population base from the 2005 General Household Survey, at www.statistics.gov.uk. Smoking data from Action on Smoking and Health, *Smoking and Health Inequality*, at www.ash.org.uk.

14. Statistics Canada, Census at School, 2010–2011. The data can be found at www.censusatschool.ca/data-results/2010-2011.

15. See Note 13 for Chapter 10.

Chapter 17

1. Note 1 for Chapter 13 explains the reason for this condition in the case of inference about a population mean.

2. See Note 5 for Chapter 14.

3. The degrees of freedom for the one-sample t statistic come from the sample standard deviation s in the denominator of t. We saw in Chapter 2 (page 50) that s has $n - 1$ degrees of freedom.

4. M. Mogri et al., "Testosterone concentrations in young pubertal and post-pubertal obese males," *Clinical Endocrinology*, 78 (2013), pp. 593–599, doi:10.1111/cen.12018.

5. See Note 10 for Chapter 1.

6. Z. W. Liller, "Spatial variation in brook trout (*Salvelinus fontinalis*) population dynamics and juvenile recruitment potential in an Appalachian watershed," MS thesis, West Virginia University, 2006.

7. L. B. Gordon et al., "Clinical trial of a farnesyltransferase inhibitor in children with Hutchinson-Gilford progeria syndrome," *PNAS*, 109 (2012), pp. 16666–16671, doi:10.1073/pnas.1202529109.

8. What matters is reducing the original data to a one-sample problem. Taking the difference between the two conditions is the most common approach. One could also, for instance, calculate the percent change for each pair of data points.

9. E. A. Dennis et al., "Water consumption increases weight loss during a hypocaloric diet intervention in middle-aged and older adults," *Obesity*, 18 (2010), pp. 300–307, doi:10.1038/oby.2009.235.

10. J. P. Rauschecker, B. Tian, and M. Hauser, "Processing of complex sounds in the macaque nonprimary auditory cortex," *Science*, 268 (1995), pp. 111–114.

11. For a qualitative discussion explaining why skewness is the most serious violation of the Normal shape condition, see D. D. Boos and J. M. Hughes-Oliver, "How large does n have to be for the Z and t intervals?," *American Statistician*, 54 (2000), pp. 121–128. Our recommendations are based on extensive computer work. See, for example, H. O. Posten, "The robustness of the one-sample t-test over the Pearson system," *Journal of Statistical Computation and Simulation*, 9 (1979), pp. 133–149; and E. S. Pearson and N. W. Please, "Relation between the shape of population distribution and the robustness of four simple test statistics," *Biometrika*, 62 (1975), pp. 223–241.

12. R. A. Berner and G. P. Landis, "Gas bubbles in fossil amber as possible indicators of the major gas composition of ancient air," *Science*, 239 (1988), pp. 1406–1409.
13. See Note 5 for Chapter 6.
14. R. Sakakibara et al., "Influence of body position on defecation in humans," *Lower Urinary Tract Symptoms*, 2 (2010), pp. 16–21, doi:10.1111/j.1757-5672.2009.00057.x.
15. See Note 8 for Chapter 1.
16. Data from the "wine" database in the archive of machine-learning databases at the University of California, Irvine, `ftp.ics.uci.edu/pub/machine-learning-databases`.
17. We thank Rudi Berkelhamer of the University of California at Irvine for the data. The data are part of a larger set collected for an undergraduate lab exercise in scientific methods.
18. See Note 7 for Chapter 6.
19. From Table S2 in the online supplement to A. Dell'Anno and R. Danovaro, "Extracellular DNA plays a key role in deep-sea ecosystem functioning," *Science*, 309 (2005), p. 2179.
20. E. G. van den Heuvel et al., "Oligofructose stimulates calcium absorption in adolescents," *American Journal of Clinical Nutrition*, 69 (1999), pp. 544–548.
21. Data for Cohort 2 in R. A. Morgan et al., "Cancer regression in patients after transfer of genetically engineered lymphocytes," *Science*, 314 (2006), pp. 126–129. The doubling-time data are given in the paper, and the immune response data appear in the supplementary online material.
22. Based on R. de la Fuente-Fernandez et al., "Expectation and dopamine release: mechanism of the placebo effect in Parkinson's disease," *Science*, 293 (2001), pp. 1164–1166.
23. T. W. McDowell, "An evaluation of vibration and other effects on the accuracy of grip and push force recall," PhD thesis, West Virginia University, 2006.
24. H. B. Meyers, "Investigations of the life history of the velvetleaf seed beetle, *Althaeus folkertsi* Kingsolver," MS thesis, Purdue University, 1996. The 95% t interval is 1227.9 to 2507.6. A 95% bootstrap BCa interval is 1444 to 2718, confirming that t inference is inaccurate for these data.
25. Data courtesy of Brad Hughes, Department of Ecology and Evolutionary Biology, University of California at Irvine.
26. See Note 9 for Chapter 2.
27. A. P. Melis, B. Hare, and M. Tomasello, "Chimpanzees recruit the best collaborators," *Science*, 311 (2006), pp. 1297–1300. A Normal quantile plot does not show major lack of Normality, and a saddlepoint approximation that allows for skew gives $P = 0.0039$. So the t test is reasonably accurate despite the skew and small sample size.
28. L. Yuh, "A biopharmaceutical example for undergraduate students," manuscript.

Chapter 18

1. See Note 14 for Chapter 2.
2. See Note 2 for Chapter 11.

3. E. Sanford et al., "Local selection and latitudinal variation in a marine predator-prey interaction," *Science*, 300 (2003), pp. 1135–1137.
4. See Note 7 for Chapter 5.
5. E. Courchesne et al., "Unusual brain growth patterns in early life in patients with autistic disorder: an MRI study," *Neurology*, 57 (2001), pp. 245–254.
6. See Note 25 for Chapter 1.
7. See the extensive simulation studies in H. O. Posten, "The robustness of the two-sample t-test over the Pearson system," *Journal of Statistical Computation and Simulation*, 6 (1978), pp. 295–311; and in H. O. Posten, H. Yeh, and D. B. Owen, "Robustness of the two-sample t-test under violations of the homogeneity assumption," *Communications in Statistics*, 11 (1982), pp. 109–126.
8. A. Y. Stark et al., "The effect of surface water and wetting on gecko adhesion," *Journal of Experimental Biology*, 215 (2012), pp. 3080–3086, doi:10.1242/jeb.070912.
9. See, for example, G. D. Ruxton, "The unequal variance t-test is an underused alternative to Student's t-test and the Mann-Whitney U test," *Behavioral Ecology*, 17 (2006), pp. 688–690.
10. The problem of comparing spreads is difficult even with advanced methods. Common distribution-free procedures do not offer a satisfactory alternative to the F test, because they are sensitive to unequal shapes when comparing two distributions. A recent survey of possible approaches is D. D. Boos and C. Brownie, "Comparing variances and other measures of dispersion," *Statistical Science*, 19 (2005), pp. 571–578. See also Lewis H. Shoemaker, "Fixing the F test for equal variances," *American Statistician*, 57 (2003), pp. 105–114, for adjustments to F that improve its robustness. The adjustments involve data-dependent degrees of freedom, similar in spirit to the two-sample t procedures described in this chapter.
11. V. L. Haidinger, "Feather growth, molt and sexual selection in the longtailed finch, *Poephila acuticauda*," MS thesis, University of California at Irvine, 1996.
12. E. R. Wald et al., "Addressing child nutrition and physical activity in the primary care setting," Agency for Healthcare Research and Quality, 2010 conference. Presentation downloaded from `www.ahrq.gov`.
13. Data provided by Darlene Gordon, Purdue University.
14. L. Z. Garamszegi et al., "Immune challenge mediates vocal communication in a passerine bird: an experiment," *Behavioral Ecology*, 15 (2004), pp. 148–157.
15. Data provided by Warren Page, New York City Technical College, from a study done by John Hudesman.
16. From the Avandia website, `www.avandia.com`, prescribing information.
17. D. Ziebolz et al., "Long-term effects of tongue piercing: a case control study," *Clinical Oral Investigations*, 16 (2012), pp. 231–237, doi:10.1007/s00784-011-0510-6.
18. R. E. Oldham-Cooper et al., "Playing a computer game during lunch affects fullness, memory for lunch, and later snack intake," *American Journal of Clinical Nutrition*, 93 (2011), pp. 308–313.
19. See Note 6 for Chapter 8.
20. Data provided by Marigene Arnold, Kalamazoo College.
21. *Black Newsprint Inks*, National Toxicology Program Toxicity Report Series No. 17, `ntp.niehs.nih.gov`.

22. C. Vallbona et al., "Response of pain to static magnetic fields in postpolio patients: a double-blind pilot study," *Archives of Physical Medicine and Rehabilitation*, 78 (1997), pp. 1200–1203.

23. From a graph in F. Grieco, A. J. van Noordwijk, and M. E. Visser, "Evidence for the effect of learning on timing of reproduction in blue tits," *Science*, 296 (2002), pp. 136–138.

24. See Note 6 for Chapter 3.

Chapter 19

1. See Note 13 for Chapter 9.

2. The Diabetes Control and Complications Trial Research Group, "The effect of intensive treatment of diabetes on the development and progression of long-term complications in insulin-dependent diabetes mellitus," *New England Journal of Medicine*, 329 (1993), pp. 977–986.

3. See Note 2 for Chapter 13.

4. Obtained from the Florida Department of Health's website `www.doh.state.fl.us`, Epidemiology, Antibiotic resistance.

5. See Note 10 for Chapter 7.

6. This rule of thumb is based on a study of computational results in the papers cited in Note 8 and discussion with Alan Agresti. We strongly recommend using the plus four interval. See Note 7.

7. J. S. M. Peiris et al., "Prospective study of the clinical progression and viral load of SARS associated coronavirus pneumonia in a community outbreak," available on the website of the World Health Organization, `www.who.int`.

8. This interval is proposed by A. Agresti and B. A. Coull, "Approximate is better than 'exact' for interval estimation of binomial proportions," *American Statistician*, 52 (1998), pp. 119–126. Note in particular that the plus four interval is often more accurate than the Clopper-Pearson "exact interval" based on the binomial distribution of the sample count and implemented by, for example, Minitab.

   There are several even more accurate but considerably more complex intervals for p that might be used in professional practice. See L. D. Brown, T. Cai, and A. DasGupta, "Interval estimation for a binomial proportion," *Statistical Science*, 16 (2001), pp. 101–133. A detailed theoretical study that uncovers the reason the large-sample interval is inaccurate is L. D. Brown, T. Cai, and A. DasGupta, "Confidence intervals for a binomial proportion and asymptotic expansions," *Annals of Statistics*, 30 (2002), pp. 160–201.

9. D. Fergusson et al., "Turning a blind eye: the success of blinding reported in a random sample of randomised, placebo-controlled trials," *British Medical Journal*, 328 (2004), pp. 432–436.

10. From A. Agresti and B. Caffo, "Simple and effective confidence intervals for proportions and differences of proportions result from adding two successes and two failures," *American Statistician*, 45 (2000), pp. 280–288. When can the plus four interval be safely used? The answer depends on just how much accuracy you insist on. Brown and coauthors (see Note 8) recommend $n \geq 40$. Agresti and Coull demonstrate that performance is almost always satisfactory in their eyes when $n \geq 5$. Our rule of thumb $n \geq 10$ allows for confidence levels C other than 95% and fits our

philosophy of not insisting on more exact results than practice requires. The big point is that plus four is very much more accurate than the standard interval for most values of p and all but very large n.

11. See Note 3 for Chapter 18.
12. In fact, P-values for two-sided tests are more accurate than those for one-sided tests. Our rule of thumb is a compromise to avoid the confusion of too many rules.
13. G. Ribak et al., "Adaptive aerial righting during the escape dropping of wingless pea aphids," *Current Biology*, 23 (2013), pp. R102–R103, doi:10.1016/j.cub.2012.12.010.
14. J. Fagan et al., "Performance assessment under field conditions of a rapid immunological test for transgenic soybeans," *International Journal of Food Science and Technology*, 36 (2001), pp. 357–367.
15. Pew Internet Project, "Finding answers online in sickness and in health," May 2, 2006, `www.pewinternet.org`.
16. See Note 11 for Chapter 5.
17. C. M. Willis et al., "Olfactory detection of human bladder cancer by dogs: proof of principle study," *British Medical Journal*, 329 (2004), pp. 712–717.
18. "Lead in DC drinking water," available from the EPA website, `www.epa.gov/dclead`.
19. A. Todorov et al., "Inferences of competence from faces predict election outcomes," *Science*, 308 (2005), pp. 1623–1626.
20. See Note 23 for Chapter 9.
21. See Note 7 for Chapter 1.
22. L. G. M. Bode, "Preventing surgical-site unfections in nasal carriers of *Staphylococcus aureus*," *New England Journal of Medicine*, 362 (2010), pp. 9–17.
23. See Note 3 for Chapter 9.

Chapter 20

1. See Note 2 for Chapter 19.
2. See Note 9 for Chapter 5.
3. The plus four method is due to Alan Agresti and Brian Caffo. See 8 and 10 for Chapter 19.
4. F. Lloret et al., "Fire and resprouting in Mediterranean ecosystems: insights from an external biogeographical region, the Mexican shrubland," *American Journal of Botany*, 88 (1999), pp. 1655–1661.
5. NCCAM website (`nccam.nih.gov/health/echinacea/`) and R. B. Turner et al., "An evaluation of *Echinacea angustifolia* in experimental rhinovirus infections," *New England Journal of Medicine*, 353 (2005), pp. 341–348.
6. From the Gardasil website (`www.gardasil.com`), Health Care Professionals, Efficacy Data.
7. "HIV vaccine regimen demonstrates modest preventive effect in Thailand clinical study," NIH News, September 24, 2009, `www.niaid.nih.gov`.
8. The proper count requirements for the two-sample proportions test of hypothesis should theoretically be about expected counts, as for the one-sample test (Chapter 19) and for the chi-square tests (Chapters 21 and 22). However, the

expected counts requirements become somewhat cumbersome for the two-sample procedure and we provide a simpler alternative focused on actual counts. The reason for this decision is to avoid drowning students in noncritical computational details. Here are the requirements based on expected counts:

If we have 2 samples of sizes n_1 and n_2 such that $N = n_1 + n_2$ and if we call S the total number of successes and F the total number of failures in both samples combined, then each of the following four expected counts should be greater than 5: n_1S/N, n_2S/N, n_1F/N, and n_2F/N. This implies that the total number of successes and the total number of failures should each be equal to or greater than $[5(n_1 + n_2)/(\text{larger of } n_1 \text{ or } n_2)]$. The sample counts requirements described in the text tend to be more conservative than the requirements using expected counts, especially when only one of the two samples is very small.

9. R. F. Adair and L. R. Holmgren, "Do drug samples influence resident prescribing behavior? A randomized trial," *American Journal of Medicine*, 118 (2005), pp. 881–884.

10. Steering Committee of the Physicians' Health Study Research Group, "Final report on the aspirin component of the ongoing Physicians' Health Study," *New England Journal of Medicine*, 321 (1989), pp. 129–135.

11. E. de Jonge et al., "Effects of selective decontamination of digestive tract on mortality and acquisition of resistant bacteria in intensive care: a randomised controlled trial," *Lancet*, 362 (2003), pp. 1011–1016.

12. See Note 23 for Chapter 9.

13. J. E. Tcheng, "High-dose eptifibatide (Integrilin) in elective coronary stenting: results of the ESPRIT trial; late-breaking trial," presented at the American College of Cardiology 49th Annual Scientific Session, March 12–15, 2000, Anaheim, CA.

14. F. Gaudet et al., "Induction of tumors in mice by genomic hypomethylation," *Science*, 300 (2003), pp. 489–492.

15. From an Associated Press dispatch appearing on December 30, 2002. The study report appeared in the *Journal of Adolescent Health*.

16. D. G. Altman, S. N. Goodman, and S. Schroter, "How statistical expertise is used in medical research," *Journal of the American Medical Association*, 287 (2002), pp. 2817–2820.

17. See Note 22 for Chapter 19.

18. K. deWeber et al., "Knuckle cracking and hand osteoarthritis," *Journal of the American Board of Family Medicine*, 24 (2011), pp. 169–174, doi:10.3122/jabfm.2011.02.100156.

19. See Note 3 for Chapter 5.

20. R. L. Prentice et al., "Low-fat dietary pattern and risk of invasive breast cancer," *Journal of the American Medical Association*, 295 (2006), pp. 629–642.

21. See Note 22 for Chapter 7.

22. B. Guyuron et al., "A placebo-controlled surgical trial of the treatment of migraine headaches," *Plastic and Reconstructive Surgery*, 124 (2009), pp. 461–468.

23. `www.pfizerpro.com`, Chantix, Efficacy of Chantix.

24. D. R. Focht III, C. Spicer, and M. P. Fairchok, "The efficacy of duct tape vs. cryotherapy in the treatment of *Verruca vulgaris* (the common wart)," *Archives of Pediatrics and Adolescent Medicine*, 156 (2002), pp. 971–974.

25. S. Sulheim et al., "Helmet use and risk of head injuries in alpine skiers and snowboarders," *Journal of the American Medical Association*, 295 (2006), pp. 919–924.

26. Toxicology and Carcinogenesis Studies of 60-Hz Magnetic Fields in F344/N Rats and B6C3F1 Mice (Whole-body Exposure Studies), TR-488, `ntp.niehs.nih.gov`.

27. G. Bellipanni et al., "Effects of melatonin in perimenopausal and menopausal women," *New York Academy of Sciences*, 1057 (2005), pp. 393–402.

28. "ACE inhibitors as a new frontier in cardiovascular prevention," report from the Heart Outcomes Prevention Evaluation trial, Population Health Research Institute, `www.ccc.mcmaster.ca`.

Chapter 21

1. P. A. Kulakow, H. Hauptli, and S. K. Jain, "Genetics of grain amaranths," *Journal of Heredity*, 76 (1985), pp. 27–30.

2. Based on a news item in *Science*, 305 (2004), p. 1560. The study, by Daniel Klem, appeared in the *Wilson Journal*.

3. H. Wei et al., "A dominant major locus in chromosome 9 of rice (*Oryza sativa* L.) confers tolerance to 48°C high temperature at seedling stage," *Journal of Heredity*, 104 (2013), pp. 287–294, doi:10.1093/jhered/ess103.

4. There are many computer studies of the accuracy of critical values for X^2. Our guideline goes back to Cochran (1954). Later work has shown that it is often conservative in the sense that if the expected cell counts are all similar and the degrees of freedom exceed 1, the chi-square approximation works well for an average expected count as small as 1 or 2. Our guideline protects against dissimilar expected counts. It has the added advantage that it is safe in the 2×2 case, where the chi-square approximation is least good. So our condition is helpful for beginners—there is no single condition that is not conservative and that applies to 2×2 and larger tables with similar and dissimilar expected cell counts. There are exact procedures that (with software) should be used for tables that do not satisfy our condition. See A. Agresti, "A survey of exact inference for contingency tables," *Statistical Science*, 7 (1992), pp. 131–177.

5. P. R. Arelli, L. D. Young, and V. C. Concibido, "Inheritance of resistance in soybean PI 567516C to LY1 nematode population infecting cv. Hartwig," *Euphytica*, 165 (2009), pp. 1–4.

6. G. Noé et al., "Contraceptive efficacy of emergency contraception with levonorgestrel given before or after ovulation," *Contraception*, 81 (2010), pp. 414–420, doi:10.1016/j.contraception.2009.12.015.

7. See Note 20 for Chapter 4.

8. See Note 12 for Chapter 10.

9. R. D. Jackson et al., "Calcium plus vitamin D supplementation and the risk of fractures," *New England Journal of Medicine*, 354 (2006), pp. 669–683.

10. J. L. Hoogland, "Multiple mating by female prairie dogs," *Animal Behavior*, 55 (1998), pp. 351–359.

11. E. Miura et al., "The balance between protein synthesis and degradation in chloroplasts determines leaf variegation in *Arabidopsis* yellow variegated mutants," *Plant Cell*, 19 (2007), pp. 1313–1328.

12. See Note 2 for Chapter 11.

13. W. Landauer and L. C. Dunnthe, "Frizzle characters of fowls: its expression and inheritance," *Journal of Heredity*, 21 (1930), pp. 291–305.
14. V. Buonaccorsi and A. Skibiel, "A striking demonstration of the Poisson distribution," *Teaching Statistics*, 27 (2005), pp. 8–10.

Chapter 22

1. See Note 36 for Chapter 7.
2. See Note 13 for Chapter 19.
3. See Note 10 for Chapter 21.
4. See Note 4 for Chapter 21.
5. See Note 20 for Chapter 4.
6. See Note 6 for Chapter 9.
7. Joseph P. Broderick et al., "Endovascular therapy after intravenous t-PA versus t-PA alone for stroke," *New England Journal of Medicine*, 368 (2013), pp. 893–903, doi:10.1056/NEJMoa1214300.
8. A. Sood et al., "S-Adenosyl-L-Methionine (SAMe) for smoking abstinence: a randomized clinical trial," *Journal of Alternative and Complementary Medicine*, 18 (2012), pp. 854–859, doi:10.1089/acm.2011.0462.
9. Modified from F. Barringer, "Measuring sexuality through polls can be shaky," *New York Times*, April 25, 1993.
10. See Note 16 for Chapter 6.
11. B. C. Coleman, "Study: heart attack risk cut 74% by stress management," Associated Press dispatch appearing in the Lafayette, Indiana, *Journal and Courier*, October 20, 1997.
12. See Note 2 for Chapter 5.
13. See Note 8 for Chapter 5.
14. See Note 11 for Chapter 10.
15. See Note 17 for Chapter 18.
16. `www.epiduo.com`, Full Prescribing Information.
17. See Note 10 for Chapter 5.
18. See Note 16 for Chapter 5.
19. See Note 7 for Chapter 5, and `nccam.nih.gov/health/echinacea/`.
20. See Note 12 for Chapter 10.
21. See Note 25 for Chapter 20.
22. See Note 5 for Chapter 5.

Chapter 23

1. See Note 14 for Chapter 3.
2. K. Lebret et al., "Grazing resistance allows bloom formation and may explain invasion success of *Gonyostomum semen*," *Limnology and Oceanography*, 57(2012), pp. 727–734, doi:10.4319/lo.2012.57.3.0727.
3. See Note 15 for Chapter 4.

4. From a graph in B. J. Peterson et al., "Increasing river discharge to the Arctic Ocean," *Science*, 298 (2002), pp. 2171–2173.
5. B. A. O'Brien et al., "The effect of print size on reading speed in dyslexia," *Journal of Research in Reading*, 28 (2005), pp. 332–349.
6. See Note 23 for Chapter 17.
7. From a graph in A. L. Perry et al., "Climate change and distribution shifts in marine fishes," *Science*, 308 (2005), pp. 1912–1915. The explanatory variable is the five-year running mean of winter (December to March) sea-bottom temperature.
8. C. J. Stevens et al., "Impact of nitrogen deposition on the species richness of grasslands," *Science*, 303 (2004), pp. 1876–1879.
9. See Note 26 for Chapter 4.
10. See Note 11 for Chapter 3.
11. A. P. Moller, "Rapid change in nest size of a bird related to change in a secondary sexual character," *Behavioal Ecology*, 17 (2006), pp. 108–116.
12. See Note 20 for Chapter 4.
13. See Note 5 for Chapter 2.
14. See Note 7 for Chapter 4.
15. See Note 4 for Chapter 17.
16. See Note 7 for Chapter 6.
17. See Note 3 for Chapter 3.
18. From a plot in D. de Quervain et al., "The neural basis of altruistic punishment," *Science*, 305 (2004), pp. 1254–1258. The description of the study is simplified shamelessly, though it still sounds a bit complicated.

Chapter 24

1. P. M. Johnson and P. J. Kenny, "Dopamine D2 receptors in addiction-like reward dysfunction and compulsive eating in obese rats," *Nature Neuroscience*, 13 (2010), pp. 635–641, doi:10.1038/nn.2519.
2. Based on the online supplement to P. J. Shaw et al., "Correlates of sleep and waking in *Drosophila melanogaster*," *Science*, 287 (2000), pp. 1834–1837.
3. G. T. Smith et al., "Seasonal changes in testosterone, neural attributes of song control nuclei, and song structure in wild songbirds," *Journal of Neuroscience*, 17 (1997), pp. 6001–6010.
4. See Note 15 for Chapter 2.
5. Modified from M. C. Wilson and R. E. Shade, "Relative attractiveness of various luminescent colors to the cereal leaf beetle and the meadow spittlebug," *Journal of Economic Entomology*, 60 (1967), pp. 578–580.
6. N. A. Sherif et al., "Detection of cotinine in neonate meconium as a marker for nicotine exposure in utero," *Eastern Mediterranean Health Journal*, 10 (2004), pp. 96–105.
7. C. A. Slentz et al., "The effects of aerobic versus resistance training on visceral and liver fat stores, liver enzymes and insulin resistance by HOMA in overweight adults from STRRIDE AT/RT: a randomized trial," *American Journal of Physiology: Endocrinology and Metabolism*, 301 (2011), E1033–1039, doi:10.1152/ajpendo.00291.2011.

8. M. R. McClung et al., "Denosumab in postmenopausal women with low bone mineral density," *New England Journal of Medicine*, 354 (2006), pp. 821–831.
9. C. D. Johnston et al., "Bonding to molars: the effect of etch time (an in-vitro study)," *European Journal of Orthodontics*, 20 (1998), pp. 195–199.
10. A. Adan and J. M. Serra-Grabulosa, "Effects of caffeine and glucose, alone and combined, on cognitive performance," *Human Psychopharmacology: Clinical and Experimental*, 25 (2010), pp. 310–317, doi:10.1002/hup.1115.
11. A. C. St-Pierre et al., "Insulin resistance syndrome, body mass index, and the risk of ischemic heart disease," *Canadian Medical Association Journal*, 172 (2005), pp. 1301–1305.
12. R. J. Safran et al., "Dynamic paternity allocation as a function of male plumage color in barn swallows," *Science*, 209 (2005), pp. 2210–2212.
13. Data from the online supplement to A. Kessler and I. T. Baldwin, "Defensive function of herbivore-induced plant volatile emissions in nature," *Science*, 291 (2001), pp. 2141–2144.
14. The data and the full story can be found in the Data and Story Library at `lib.stat.cmu.edu`. The original study is by F. Loven, "A study of interlist equivalency of the CID W-22 word list presented in quiet and in noise," MS thesis, University of Iowa, 1981.
15. See Note 8 for Chapter 8.
16. See Note 25 for Chapter 17.
17. See Note 10 for Chapter 1.
18. J. T. Ngai and D. S. Srivastava, "Predators accelerate nutrient cycling in a bromeliad ecosystem," *Science*, 314 (2006), p. 963. We thank Jacqueline Ngai for providing the data.
19. B. Stricanne, "Etudes d'integration multisensorielles dans la voie visuelle occipito-parietale du primate," PhD thesis, Université Paris VI, 1996.
20. See Note 13 for Chapter 2.
21. A. Alecu et al., "Effect of the homeopathic remedies *Arnica montana* and *Staphisagria* on the time of healing of surgical wounds," *Cultura Homeopatica*, 20 (2007), pp. 19–21.
22. We thank Rudi Berkelhamer of the University of California at Irvine for the data. The data are part of a larger set collected for an undergraduate lab exercise in scientific methods.
23. See Note 6 for Chapter 3.

Chapter 25

1. J. McDermott and M. D. Hauser, "Nonhuman primates prefer slow tempos but dislike music overall," *Cognition*, 104 (2007), pp. 654–668. Failure to take account of repeated measures on the same subjects is one of the most common errors observed in statistical analysis.
2. K. C. Armel and V. S. Ramachandran, "Projecting sensations to external objects: evidence from skin conductance response," *Proceedings of the Royal Society of London, Series B*, 270 (2003), pp. 1499–1506.

3. M. R. Dohm, J. P. Hayes, and T. Garland, Jr., "Quantitative genetics of sprint running speed and swimming endurance in laboratory house mice (*Mus domesticus*)," *Evolution*, 50 (1996), pp. 1688–1701.
4. See Note 11 for Chapter 6.
5. M. K. Pawlik, "The effect of ginkgo biloba on the post-lunch dip and chemosensory function," MS thesis, Purdue University, 2002.
6. Data provided by Matthew Moore.
7. K. S. Oberhauser, "Fecundity, lifespan and egg mass in butterflies: effects of male-derived nutrients and female size," *Functional Ecology*, 11 (1997), pp. 166–175.
8. See Note 9 for Chapter 16. The exercises are simplified, in that the measures reported in this paper have been statistically adjusted for "sociodemographic status."
9. See Note 2 for Chapter 24.
10. J. E. Keeley, C. J. Fotheringham, and M. Morais, "Reexamining fire suppression impacts on brushland fire regimes," *Science*, 284 (1999), pp. 1829–1831.
11. M. Xu and R. G. Palmer, "Genetic analysis of 4 new mutants at the unstable k2 Mdh1-n y20 chromosomal region in soybean," *Journal of Heredity*, 97 (2006), pp. 423–427.
12. J. L. Warner-Schmidt et al., "Antidepressant effects of selective serotonin reuptake inhibitors (SSRIs) are attenuated by antiinflammatory drugs in mice and humans," *PNAS*, 108 (2011), pp. 9262–9267, doi:10.1073/pnas.1104836108.
13. J. D. Miller, E. C. Scott, and S. Okamoto, "Public acceptance of evolution," *Science*, 313 (2006), pp. 765–766. The information in the exercise appears in the supplementary online material.
14. See Note 18 for Chapter 24.
15. The data are part of a larger data set in the *Journal of Statistics Education* archive, accessible online. The original source is P. Brofeldt, "Bidrag till kaennedom on fiskbestondet i vaara sjoear. Laengelmaevesi," in T. H. Jaervi, *Finlands fiskeriet*, Vol. 4, *Meddelanden utgivna av fiskerifoereningen i Finland*, Helsinki, 1917. The data were contributed to the archive (with information in English) by Juha Puranen of the University of Helsinki.
16. M. Banda, "Population biology of the catfish *Bagrus meridionalis* from the southern part of Lake Malawi," *Lake Malawi Fisheries Management Symposium Proceedings*, 2001, pp. 200–214.
17. See Note 2 for Chapter 2.
18. See Note 25 for Chapter 1.
19. See Note 17 for Chapter 6.
20. Data provided by Corinne Lim, Purdue University, from a student project supervised by Professor Joseph Vanable.
21. See Note 5 for Chapter 6.
22. G. S. Hotamisligil et al., "Uncoupling of obesity from insulin resistance through a targeted mutation in *aP2*, the adipocyte fatty acid binding protein," *Science*, 274 (1996), pp. 1377–1379.
23. See Note 15 for Chapter 5.
24. See Note 22 for Chapter 24.

TABLES

TABLE A Random digits

Line								
101	19223	95034	05756	28713	96409	12531	42544	82853
102	73676	47150	99400	01927	27754	42648	82425	36290
103	45467	71709	77558	00095	32863	29485	82226	90056
104	52711	38889	93074	60227	40011	85848	48767	52573
105	95592	94007	69971	91481	60779	53791	17297	59335
106	68417	35013	15529	72765	85089	57067	50211	47487
107	82739	57890	20807	47511	81676	55300	94383	14893
108	60940	72024	17868	24943	61790	90656	87964	18883
109	36009	19365	15412	39638	85453	46816	83485	41979
110	38448	48789	18338	24697	39364	42006	76688	08708
111	81486	69487	60513	09297	00412	71238	27649	39950
112	59636	88804	04634	71197	19352	73089	84898	45785
113	62568	70206	40325	03699	71080	22553	11486	11776
114	45149	32992	75730	66280	03819	56202	02938	70915
115	61041	77684	94322	24709	73698	14526	31893	32592
116	14459	26056	31424	80371	65103	62253	50490	61181
117	38167	98532	62183	70632	23417	26185	41448	75532
118	73190	32533	04470	29669	84407	90785	65956	86382
119	95857	07118	87664	92099	58806	66979	98624	84826
120	35476	55972	39421	65850	04266	35435	43742	11937
121	71487	09984	29077	14863	61683	47052	62224	51025
122	13873	81598	95052	90908	73592	75186	87136	95761
123	54580	81507	27102	56027	55892	33063	41842	81868
124	71035	09001	43367	49497	72719	96758	27611	91596
125	96746	12149	37823	71868	18442	35119	62103	39244
126	96927	19931	36809	74192	77567	88741	48409	41903
127	43909	99477	25330	64359	40085	16925	85117	36071
128	15689	14227	06565	14374	13352	49367	81982	87209
129	36759	58984	68288	22913	18638	54303	00795	08727
130	69051	64817	87174	09517	84534	06489	87201	97245
131	05007	16632	81194	14873	04197	85576	45195	96565
132	68732	55259	84292	08796	43165	93739	31685	97150
133	45740	41807	65561	33302	07051	93623	18132	09547
134	27816	78416	18329	21337	35213	37741	04312	68508
135	66925	55658	39100	78458	11206	19876	87151	31260
136	08421	44753	77377	28744	75592	08563	79140	92454
137	53645	66812	61421	47836	12609	15373	98481	14592
138	66831	68908	40772	21558	47781	33586	79177	06928
139	55588	99404	70708	41098	43563	56934	48394	51719
140	12975	13258	13048	45144	72321	81940	00360	02428
141	96767	35964	23822	96012	94591	65194	50842	53372
142	72829	50232	97892	63408	77919	44575	24870	04178
143	88565	42628	17797	49376	61762	16953	88604	12724
144	62964	88145	83083	69453	46109	59505	69680	00900
145	19687	12633	57857	95806	09931	02150	43163	58636
146	37609	59057	66967	83401	60705	02384	90597	93600
147	54973	86278	88737	74351	47500	84552	19909	67181
148	00694	05977	19664	65441	20903	62371	22725	53340
149	71546	05233	53946	68743	72460	27601	45403	88692
150	07511	88915	41267	16853	84569	79367	32337	03316

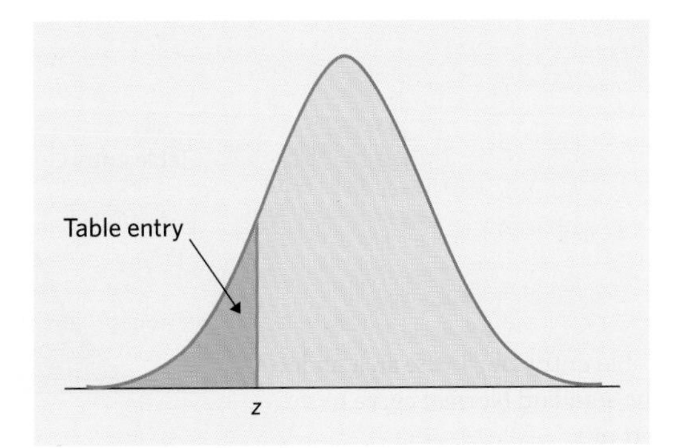

Table entry for z is the area under the standard Normal curve to the left of z.

TABLE B Standard Normal probabilities

z	.00	.01	.02	.03	.04	.05	.06	.07	.08	.09
−3.4	.0003	.0003	.0003	.0003	.0003	.0003	.0003	.0003	.0003	.0002
−3.3	.0005	.0005	.0005	.0004	.0004	.0004	.0004	.0004	.0004	.0003
−3.2	.0007	.0007	.0006	.0006	.0006	.0006	.0006	.0005	.0005	.0005
−3.1	.0010	.0009	.0009	.0009	.0008	.0008	.0008	.0008	.0007	.0007
−3.0	.0013	.0013	.0013	.0012	.0012	.0011	.0011	.0011	.0010	.0010
−2.9	.0019	.0018	.0018	.0017	.0016	.0016	.0015	.0015	.0014	.0014
−2.8	.0026	.0025	.0024	.0023	.0023	.0022	.0021	.0021	.0020	.0019
−2.7	.0035	.0034	.0033	.0032	.0031	.0030	.0029	.0028	.0027	.0026
−2.6	.0047	.0045	.0044	.0043	.0041	.0040	.0039	.0038	.0037	.0036
−2.5	.0062	.0060	.0059	.0057	.0055	.0054	.0052	.0051	.0049	.0048
−2.4	.0082	.0080	.0078	.0075	.0073	.0071	.0069	.0068	.0066	.0064
−2.3	.0107	.0104	.0102	.0099	.0096	.0094	.0091	.0089	.0087	.0084
−2.2	.0139	.0136	.0132	.0129	.0125	.0122	.0119	.0116	.0113	.0110
−2.1	.0179	.0174	.0170	.0166	.0162	.0158	.0154	.0150	.0146	.0143
−2.0	.0228	.0222	.0217	.0212	.0207	.0202	.0197	.0192	.0188	.0183
−1.9	.0287	.0281	.0274	.0268	.0262	.0256	.0250	.0244	.0239	.0233
−1.8	.0359	.0351	.0344	.0336	.0329	.0322	.0314	.0307	.0301	.0294
−1.7	.0446	.0436	.0427	.0418	.0409	.0401	.0392	.0384	.0375	.0367
−1.6	.0548	.0537	.0526	.0516	.0505	.0495	.0485	.0475	.0465	.0455
−1.5	.0668	.0655	.0643	.0630	.0618	.0606	.0594	.0582	.0571	.0559
−1.4	.0808	.0793	.0778	.0764	.0749	.0735	.0721	.0708	.0694	.0681
−1.3	.0968	.0951	.0934	.0918	.0901	.0885	.0869	.0853	.0838	.0823
−1.2	.1151	.1131	.1112	.1093	.1075	.1056	.1038	.1020	.1003	.0985
−1.1	.1357	.1335	.1314	.1292	.1271	.1251	.1230	.1210	.1190	.1170
−1.0	.1587	.1562	.1539	.1515	.1492	.1469	.1446	.1423	.1401	.1379
−0.9	.1841	.1814	.1788	.1762	.1736	.1711	.1685	.1660	.1635	.1611
−0.8	.2119	.2090	.2061	.2033	.2005	.1977	.1949	.1922	.1894	.1867
−0.7	.2420	.2389	.2358	.2327	.2296	.2266	.2236	.2206	.2177	.2148
−0.6	.2743	.2709	.2676	.2643	.2611	.2578	.2546	.2514	.2483	.2451
−0.5	.3085	.3050	.3015	.2981	.2946	.2912	.2877	.2843	.2810	.2776
−0.4	.3446	.3409	.3372	.3336	.3300	.3264	.3228	.3192	.3156	.3121
−0.3	.3821	.3783	.3745	.3707	.3669	.3632	.3594	.3557	.3520	.3483
−0.2	.4207	.4168	.4129	.4090	.4052	.4013	.3974	.3936	.3897	.3859
−0.1	.4602	.4562	.4522	.4483	.4443	.4404	.4364	.4325	.4286	.4247
−0.0	.5000	.4960	.4920	.4880	.4840	.4801	.4761	.4721	.4681	.4641

(*continued*)

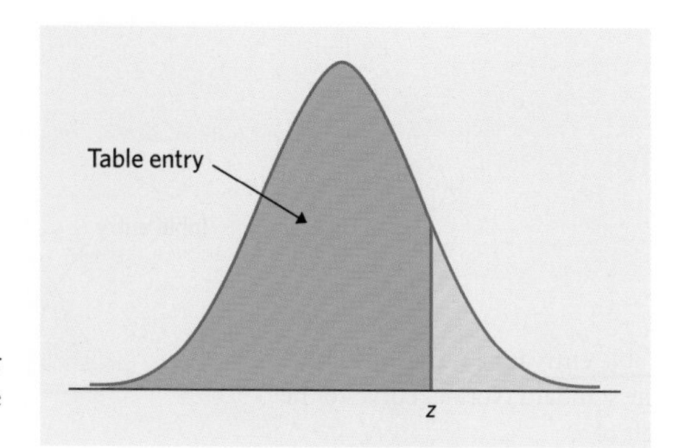

Table entry for z is the area under the standard Normal curve to the left of z.

TABLE B Standard Normal probabilities (*continued*)

z	.00	.01	.02	.03	.04	.05	.06	.07	.08	.09
0.0	.5000	.5040	.5080	.5120	.5160	.5199	.5239	.5279	.5319	.5359
0.1	.5398	.5438	.5478	.5517	.5557	.5596	.5636	.5675	.5714	.5753
0.2	.5793	.5832	.5871	.5910	.5948	.5987	.6026	.6064	.6103	.6141
0.3	.6179	.6217	.6255	.6293	.6331	.6368	.6406	.6443	.6480	.6517
0.4	.6554	.6591	.6628	.6664	.6700	.6736	.6772	.6808	.6844	.6879
0.5	.6915	.6950	.6985	.7019	.7054	.7088	.7123	.7157	.7190	.7224
0.6	.7257	.7291	.7324	.7357	.7389	.7422	.7454	.7486	.7517	.7549
0.7	.7580	.7611	.7642	.7673	.7704	.7734	.7764	.7794	.7823	.7852
0.8	.7881	.7910	.7939	.7967	.7995	.8023	.8051	.8078	.8106	.8133
0.9	.8159	.8186	.8212	.8238	.8264	.8289	.8315	.8340	.8365	.8389
1.0	.8413	.8438	.8461	.8485	.8508	.8531	.8554	.8577	.8599	.8621
1.1	.8643	.8665	.8686	.8708	.8729	.8749	.8770	.8790	.8810	.8830
1.2	.8849	.8869	.8888	.8907	.8925	.8944	.8962	.8980	.8997	.9015
1.3	.9032	.9049	.9066	.9082	.9099	.9115	.9131	.9147	.9162	.9177
1.4	.9192	.9207	.9222	.9236	.9251	.9265	.9279	.9292	.9306	.9319
1.5	.9332	.9345	.9357	.9370	.9382	.9394	.9406	.9418	.9429	.9441
1.6	.9452	.9463	.9474	.9484	.9495	.9505	.9515	.9525	.9535	.9545
1.7	.9554	.9564	.9573	.9582	.9591	.9599	.9608	.9616	.9625	.9633
1.8	.9641	.9649	.9656	.9664	.9671	.9678	.9686	.9693	.9699	.9706
1.9	.9713	.9719	.9726	.9732	.9738	.9744	.9750	.9756	.9761	.9767
2.0	.9772	.9778	.9783	.9788	.9793	.9798	.9803	.9808	.9812	.9817
2.1	.9821	.9826	.9830	.9834	.9838	.9842	.9846	.9850	.9854	.9857
2.2	.9861	.9864	.9868	.9871	.9875	.9878	.9881	.9884	.9887	.9890
2.3	.9893	.9896	.9898	.9901	.9904	.9906	.9909	.9911	.9913	.9916
2.4	.9918	.9920	.9922	.9925	.9927	.9929	.9931	.9932	.9934	.9936
2.5	.9938	.9940	.9941	.9943	.9945	.9946	.9948	.9949	.9951	.9952
2.6	.9953	.9955	.9956	.9957	.9959	.9960	.9961	.9962	.9963	.9964
2.7	.9965	.9966	.9967	.9968	.9969	.9970	.9971	.9972	.9973	.9974
2.8	.9974	.9975	.9976	.9977	.9977	.9978	.9979	.9979	.9980	.9981
2.9	.9981	.9982	.9982	.9983	.9984	.9984	.9985	.9985	.9986	.9986
3.0	.9987	.9987	.9987	.9988	.9988	.9989	.9989	.9989	.9990	.9990
3.1	.9990	.9991	.9991	.9991	.9992	.9992	.9992	.9992	.9993	.9993
3.2	.9993	.9993	.9994	.9994	.9994	.9994	.9994	.9995	.9995	.9995
3.3	.9995	.9995	.9995	.9996	.9996	.9996	.9996	.9996	.9996	.9997
3.4	.9997	.9997	.9997	.9997	.9997	.9997	.9997	.9997	.9997	.9998

Table entry for C is the critical value t^* required for confidence level C. To approximate one- and two-sided P-values, compare the value of the t statistic with the critical values of t^* that match the P-values given at the bottom of the table.

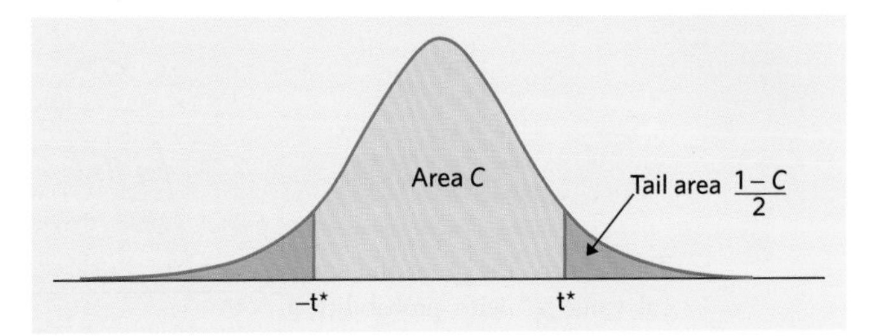

TABLE C *t* distribution critical values

Degrees of freedom	Confidence level C 50%	60%	70%	80%	90%	95%	96%	98%	99%	99.5%	99.8%	99.9%
1	1.000	1.376	1.963	3.078	6.314	12.710	15.890	31.820	63.660	127.300	318.300	636.600
2	0.816	1.061	1.386	1.886	2.920	4.303	4.849	6.965	9.925	14.090	22.330	31.600
3	0.765	0.978	1.250	1.638	2.353	3.182	3.482	4.541	5.841	7.453	10.210	12.920
4	0.741	0.941	1.190	1.533	2.132	2.776	2.999	3.747	4.604	5.598	7.173	8.610
5	0.727	0.920	1.156	1.476	2.015	2.571	2.757	3.365	4.032	4.773	5.893	6.869
6	0.718	0.906	1.134	1.440	1.943	2.447	2.612	3.143	3.707	4.317	5.208	5.959
7	0.711	0.896	1.119	1.415	1.895	2.365	2.517	2.998	3.499	4.029	4.785	5.408
8	0.706	0.889	1.108	1.397	1.860	2.306	2.449	2.896	3.355	3.833	4.501	5.041
9	0.703	0.883	1.100	1.383	1.833	2.262	2.398	2.821	3.250	3.690	4.297	4.781
10	0.700	0.879	1.093	1.372	1.812	2.228	2.359	2.764	3.169	3.581	4.144	4.587
11	0.697	0.876	1.088	1.363	1.796	2.201	2.328	2.718	3.106	3.497	4.025	4.437
12	0.695	0.873	1.083	1.356	1.782	2.179	2.303	2.681	3.055	3.428	3.930	4.318
13	0.694	0.870	1.079	1.350	1.771	2.160	2.282	2.650	3.012	3.372	3.852	4.221
14	0.692	0.868	1.076	1.345	1.761	2.145	2.264	2.624	2.977	3.326	3.787	4.140
15	0.691	0.866	1.074	1.341	1.753	2.131	2.249	2.602	2.947	3.286	3.733	4.073
16	0.690	0.865	1.071	1.337	1.746	2.120	2.235	2.583	2.921	3.252	3.686	4.015
17	0.689	0.863	1.069	1.333	1.740	2.110	2.224	2.567	2.898	3.222	3.646	3.965
18	0.688	0.862	1.067	1.330	1.734	2.101	2.214	2.552	2.878	3.197	3.611	3.922
19	0.688	0.861	1.066	1.328	1.729	2.093	2.205	2.539	2.861	3.174	3.579	3.883
20	0.687	0.860	1.064	1.325	1.725	2.086	2.197	2.528	2.845	3.153	3.552	3.850
21	0.686	0.859	1.063	1.323	1.721	2.080	2.189	2.518	2.831	3.135	3.527	3.819
22	0.686	0.858	1.061	1.321	1.717	2.074	2.183	2.508	2.819	3.119	3.505	3.792
23	0.685	0.858	1.060	1.319	1.714	2.069	2.177	2.500	2.807	3.104	3.485	3.768
24	0.685	0.857	1.059	1.318	1.711	2.064	2.172	2.492	2.797	3.091	3.467	3.745
25	0.684	0.856	1.058	1.316	1.708	2.060	2.167	2.485	2.787	3.078	3.450	3.725
26	0.684	0.856	1.058	1.315	1.706	2.056	2.162	2.479	2.779	3.067	3.435	3.707
27	0.684	0.855	1.057	1.314	1.703	2.052	2.158	2.473	2.771	3.057	3.421	3.690
28	0.683	0.855	1.056	1.313	1.701	2.048	2.154	2.467	2.763	3.047	3.408	3.674
29	0.683	0.854	1.055	1.311	1.699	2.045	2.150	2.462	2.756	3.038	3.396	3.659
30	0.683	0.854	1.055	1.310	1.697	2.042	2.147	2.457	2.750	3.030	3.385	3.646
40	0.681	0.851	1.050	1.303	1.684	2.021	2.123	2.423	2.704	2.971	3.307	3.551
50	0.679	0.849	1.047	1.299	1.676	2.009	2.109	2.403	2.678	2.937	3.261	3.496
60	0.679	0.848	1.045	1.296	1.671	2.000	2.099	2.390	2.660	2.915	3.232	3.460
80	0.678	0.846	1.043	1.292	1.664	1.990	2.088	2.374	2.639	2.887	3.195	3.416
100	0.677	0.845	1.042	1.290	1.660	1.984	2.081	2.364	2.626	2.871	3.174	3.390
1000	0.675	0.842	1.037	1.282	1.646	1.962	2.056	2.330	2.581	2.813	3.098	3.300
z^*	0.674	0.841	1.036	1.282	1.645	1.960	2.054	2.326	2.576	2.807	3.091	3.291
One-sided P	.25	.20	.15	.10	.05	.025	.02	.01	.005	.0025	.001	.0005
Two-sided P	.50	.40	.30	.20	.10	.05	.04	.02	.01	.005	.002	.001

Table entry for p is the critical value χ^* with probability p lying to its right.

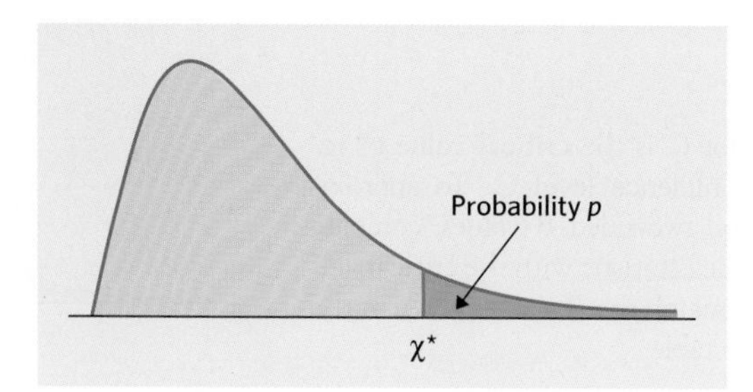

TABLE D Chi-square distribution critical values

	p											
df	.25	.20	.15	.10	.05	.025	.02	.01	.005	.0025	.001	.0005
1	1.32	1.64	2.07	2.71	3.84	5.02	5.41	6.63	7.88	9.14	10.83	12.12
2	2.77	3.22	3.79	4.61	5.99	7.38	7.82	9.21	10.60	11.98	13.82	15.20
3	4.11	4.64	5.32	6.25	7.81	9.35	9.84	11.34	12.84	14.32	16.27	17.73
4	5.39	5.99	6.74	7.78	9.49	11.14	11.67	13.28	14.86	16.42	18.47	20.00
5	6.63	7.29	8.12	9.24	11.07	12.83	13.39	15.09	16.75	18.39	20.51	22.11
6	7.84	8.56	9.45	10.64	12.59	14.45	15.03	16.81	18.55	20.25	22.46	24.10
7	9.04	9.80	10.75	12.02	14.07	16.01	16.62	18.48	20.28	22.04	24.32	26.02
8	10.22	11.03	12.03	13.36	15.51	17.53	18.17	20.09	21.95	23.77	26.12	27.87
9	11.39	12.24	13.29	14.68	16.92	19.02	19.68	21.67	23.59	25.46	27.88	29.67
10	12.55	13.44	14.53	15.99	18.31	20.48	21.16	23.21	25.19	27.11	29.59	31.42
11	13.70	14.63	15.77	17.28	19.68	21.92	22.62	24.72	26.76	28.73	31.26	33.14
12	14.85	15.81	16.99	18.55	21.03	23.34	24.05	26.22	28.30	30.32	32.91	34.82
13	15.98	16.98	18.20	19.81	22.36	24.74	25.47	27.69	29.82	31.88	34.53	36.48
14	17.12	18.15	19.41	21.06	23.68	26.12	26.87	29.14	31.32	33.43	36.12	38.11
15	18.25	19.31	20.60	22.31	25.00	27.49	28.26	30.58	32.80	34.95	37.70	39.72
16	19.37	20.47	21.79	23.54	26.30	28.85	29.63	32.00	34.27	36.46	39.25	41.31
17	20.49	21.61	22.98	24.77	27.59	30.19	31.00	33.41	35.72	37.95	40.79	42.88
18	21.60	22.76	24.16	25.99	28.87	31.53	32.35	34.81	37.16	39.42	42.31	44.43
19	22.72	23.90	25.33	27.20	30.14	32.85	33.69	36.19	38.58	40.88	43.82	45.97
20	23.83	25.04	26.50	28.41	31.41	34.17	35.02	37.57	40.00	42.34	45.31	47.50
21	24.93	26.17	27.66	29.62	32.67	35.48	36.34	38.93	41.40	43.78	46.80	49.01
22	26.04	27.30	28.82	30.81	33.92	36.78	37.66	40.29	42.80	45.20	48.27	50.51
23	27.14	28.43	29.98	32.01	35.17	38.08	38.97	41.64	44.18	46.62	49.73	52.00
24	28.24	29.55	31.13	33.20	36.42	39.36	40.27	42.98	45.56	48.03	51.18	53.48
25	29.34	30.68	32.28	34.38	37.65	40.65	41.57	44.31	46.93	49.44	52.62	54.95
26	30.43	31.79	33.43	35.56	38.89	41.92	42.86	45.64	48.29	50.83	54.05	56.41
27	31.53	32.91	34.57	36.74	40.11	43.19	44.14	46.96	49.64	52.22	55.48	57.86
28	32.62	34.03	35.71	37.92	41.34	44.46	45.42	48.28	50.99	53.59	56.89	59.30
29	33.71	35.14	36.85	39.09	42.56	45.72	46.69	49.59	52.34	54.97	58.30	60.73
30	34.80	36.25	37.99	40.26	43.77	46.98	47.96	50.89	53.67	56.33	59.70	62.16
40	45.62	47.27	49.24	51.81	55.76	59.34	60.44	63.69	66.77	69.70	73.40	76.09
50	56.33	58.16	60.35	63.17	67.50	71.42	72.61	76.15	79.49	82.66	86.66	89.56
60	66.98	68.97	71.34	74.40	79.08	83.30	84.58	88.38	91.95	95.34	99.61	102.70
80	88.13	90.41	93.11	96.58	101.90	106.60	108.10	112.30	116.30	120.10	124.80	128.30
100	109.10	111.70	114.70	118.50	124.30	129.60	131.10	135.80	140.20	144.30	149.40	153.20

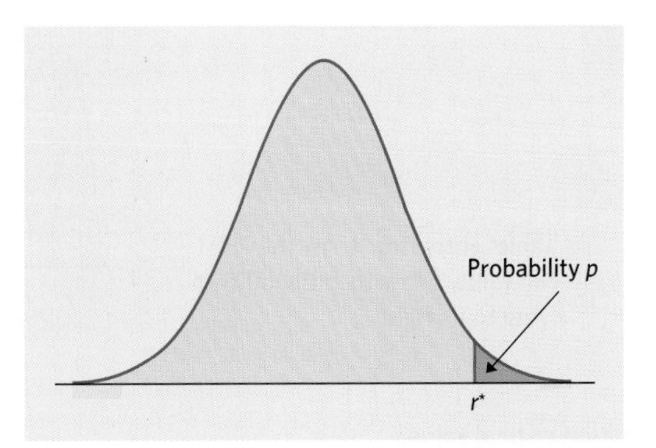

Table entry for p is the critical value r^* of the correlation coefficient r with probability p lying to its right.

TABLE E Critical values of the correlation r

	Upper tail probability p									
n	.20	.10	.05	.025	.02	.01	.005	.0025	.001	.0005
3	0.8090	0.9511	0.9877	0.9969	0.9980	0.9995	0.9999	1.0000	1.0000	1.0000
4	0.6000	0.8000	0.9000	0.9500	0.9600	0.9800	0.9900	0.9950	0.9980	0.9990
5	0.4919	0.6870	0.8054	0.8783	0.8953	0.9343	0.9587	0.9740	0.9859	0.9911
6	0.4257	0.6084	0.7293	0.8114	0.8319	0.8822	0.9172	0.9417	0.9633	0.9741
7	0.3803	0.5509	0.6694	0.7545	0.7766	0.8329	0.8745	0.9056	0.9350	0.9509
8	0.3468	0.5067	0.6215	0.7067	0.7295	0.7887	0.8343	0.8697	0.9049	0.9249
9	0.3208	0.4716	0.5822	0.6664	0.6892	0.7498	0.7977	0.8359	0.8751	0.8983
10	0.2998	0.4428	0.5494	0.6319	0.6546	0.7155	0.7646	0.8046	0.8467	0.8721
11	0.2825	0.4187	0.5214	0.6021	0.6244	0.6851	0.7348	0.7759	0.8199	0.8470
12	0.2678	0.3981	0.4973	0.5760	0.5980	0.6581	0.7079	0.7496	0.7950	0.8233
13	0.2552	0.3802	0.4762	0.5529	0.5745	0.6339	0.6835	0.7255	0.7717	0.8010
14	0.2443	0.3646	0.4575	0.5324	0.5536	0.6120	0.6614	0.7034	0.7501	0.7800
15	0.2346	0.3507	0.4409	0.5140	0.5347	0.5923	0.6411	0.6831	0.7301	0.7604
16	0.2260	0.3383	0.4259	0.4973	0.5177	0.5742	0.6226	0.6643	0.7114	0.7419
17	0.2183	0.3271	0.4124	0.4821	0.5021	0.5577	0.6055	0.6470	0.6940	0.7247
18	0.2113	0.3170	0.4000	0.4683	0.4878	0.5425	0.5897	0.6308	0.6777	0.7084
19	0.2049	0.3077	0.3887	0.4555	0.4747	0.5285	0.5751	0.6158	0.6624	0.6932
20	0.1991	0.2992	0.3783	0.4438	0.4626	0.5155	0.5614	0.6018	0.6481	0.6788
21	0.1938	0.2914	0.3687	0.4329	0.4513	0.5034	0.5487	0.5886	0.6346	0.6652
22	0.1888	0.2841	0.3598	0.4227	0.4409	0.4921	0.5368	0.5763	0.6219	0.6524
23	0.1843	0.2774	0.3515	0.4132	0.4311	0.4815	0.5256	0.5647	0.6099	0.6402
24	0.1800	0.2711	0.3438	0.4044	0.4219	0.4716	0.5151	0.5537	0.5986	0.6287
25	0.1760	0.2653	0.3365	0.3961	0.4133	0.4622	0.5052	0.5434	0.5879	0.6178
26	0.1723	0.2598	0.3297	0.3882	0.4052	0.4534	0.4958	0.5336	0.5776	0.6074
27	0.1688	0.2546	0.3233	0.3809	0.3976	0.4451	0.4869	0.5243	0.5679	0.5974
28	0.1655	0.2497	0.3172	0.3739	0.3904	0.4372	0.4785	0.5154	0.5587	0.5880
29	0.1624	0.2451	0.3115	0.3673	0.3835	0.4297	0.4705	0.5070	0.5499	0.5790
30	0.1594	0.2407	0.3061	0.3610	0.3770	0.4226	0.4629	0.4990	0.5415	0.5703
40	0.1368	0.2070	0.2638	0.3120	0.3261	0.3665	0.4026	0.4353	0.4741	0.5007
50	0.1217	0.1843	0.2353	0.2787	0.2915	0.3281	0.3610	0.3909	0.4267	0.4514
60	0.1106	0.1678	0.2144	0.2542	0.2659	0.2997	0.3301	0.3578	0.3912	0.4143
80	0.0954	0.1448	0.1852	0.2199	0.2301	0.2597	0.2864	0.3109	0.3405	0.3611
100	0.0851	0.1292	0.1654	0.1966	0.2058	0.2324	0.2565	0.2786	0.3054	0.3242
1000	0.0266	0.0406	0.0520	0.0620	0.0650	0.0736	0.0814	0.0887	0.0976	0.1039

Table entry for p is the critical value F^* with probability p lying to its right.

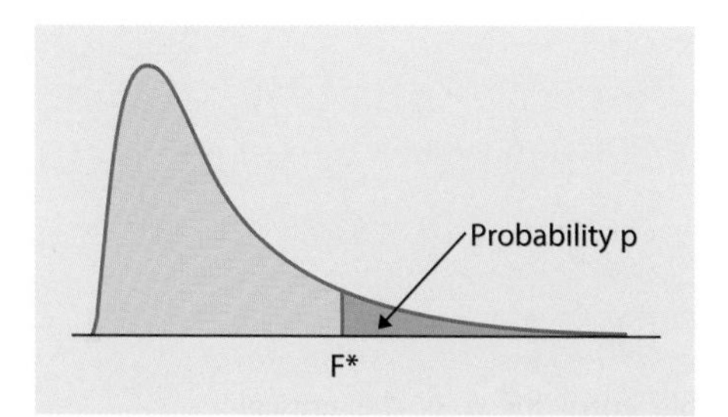

TABLE F *F* distribution critical values

Degrees of freedom in the denominator	p	Degrees of freedom in the numerator: 1	2	3	4	5	6	7	8
1	.100	39.86	49.50	53.59	55.83	57.24	58.20	58.91	59.44
	.050	161.45	199.50	215.71	224.58	230.16	233.99	236.77	238.88
	.025	647.79	799.50	864.16	899.58	921.85	937.11	948.22	956.66
	.010	4052.20	4999.50	5403.40	5624.60	5763.60	5859	5928.40	5981.10
	.001	405284.00	500000.00	540379.00	562500.00	576405.00	585937.00	592873.00	598144.00
2	.100	8.53	9.00	9.16	9.24	9.29	9.33	.35	9.37
	.050	18.51	19.00	19.16	19.25	19.30	19.33	19.35	19.37
	.025	38.51	39.00	39.17	39.25	39.30	39.33	39.36	39.37
	.010	98.50	99.00	99.17	99.25	99.30	99.33	99.36	99.37
	.001	998.50	999.00	999.17	999.25	999.30	999.33	999.36	999.37
3	.100	5.54	5.46	5.39	5.34	5.31	5.28	5.27	5.25
	.050	10.13	9.55	9.28	9.12	9.01	8.94	8.89	8.85
	.025	17.44	16.04	15.44	15.10	14.88	14.73	14.62	14.54
	.010	34.12	30.82	29.46	28.71	28.24	27.91	27.67	27.49
	.001	167.03	148.50	141.11	137.10	134.58	132.85	131.58	130.62
4	.100	4.54	4.32	4.19	4.11	4.05	4.01	3.98	3.95
	.050	7.71	6.94	6.59	6.39	6.26	6.16	6.09	6.04
	.025	12.22	10.65	9.98	9.60	9.36	9.20	9.07	8.98
	.010	21.20	18.00	16.69	15.98	15.52	15.21	14.98	14.80
	.001	74.14	61.25	56.18	53.44	51.71	50.53	49.66	49.00
5	.100	4.06	3.78	3.62	3.52	3.45	3.40	3.37	3.34
	.050	6.61	5.79	5.41	5.19	5.05	4.95	4.88	4.82
	.025	10.01	8.43	7.76	7.39	7.15	6.98	6.85	6.76
	.010	16.26	13.27	12.06	11.39	10.97	10.67	10.46	10.29
	.001	47.18	37.12	33.20	31.09	29.75	28.83	28.16	27.65
6	.100	3.78	3.46	3.29	3.18	3.11	3.05	3.01	2.98
	.050	5.99	5.14	4.76	4.53	4.39	4.28	4.21	4.15
	.025	8.81	7.26	6.60	6.23	5.99	5.82	5.70	5.60
	.010	13.75	10.92	9.78	9.15	8.75	8.47	8.26	8.10
	.001	35.51	27.00	23.70	21.92	20.80	20.03	19.46	19.03
7	.100	3.59	3.26	3.07	2.96	2.88	2.83	2.78	2.75
	.050	5.59	4.74	4.35	4.12	3.97	3.87	3.79	3.73
	.025	8.07	6.54	5.89	5.52	5.29	5.12	4.99	4.90
	.010	12.25	9.55	8.45	7.85	7.46	7.19	6.99	6.84
	.001	29.25	21.69	18.77	17.20	16.21	15.52	15.02	14.63
8	.100	3.46	3.11	2.92	2.81	2.73	2.67	2.62	2.59
	.050	5.32	4.46	4.07	3.84	3.69	3.58	3.50	3.44
	.025	7.57	6.06	5.42	5.05	4.82	4.65	4.53	4.43
	.010	11.26	8.65	7.59	7.01	6.63	6.37	6.18	6.03
	.001	25.41	18.49	15.83	14.39	13.48	12.86	12.40	12.05

Table entry for p is the critical value F^* with probability p lying to its right.

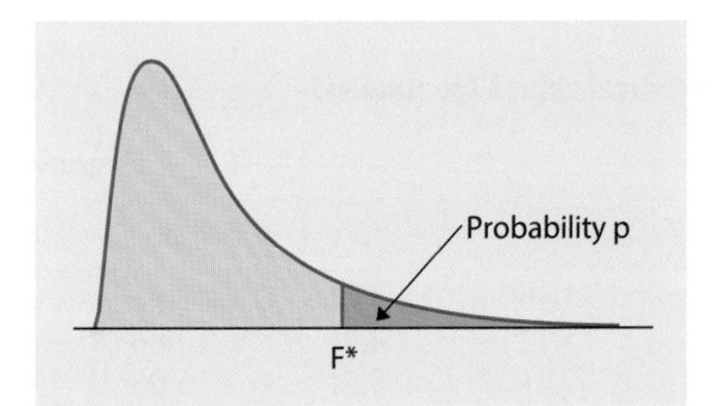

TABLE F ***F* distribution critical values (*continued*)**

Degrees of freedom in the denominator	p	9	10	15	20	30	60	120	1000
1	.100	59.86	60.19	61.22	61.74	62.26	62.79	63.06	63.30
	.050	240.54	241.88	245.95	248.01	250.10	252.20	253.25	254.19
	.025	963.28	968.63	984.87	993.10	1001.40	1009.80	10140	1017.70
	.010	6022.50	6055.80	6157.30	6208.70	6260.60	63130	6339.40	6362.70
	.001	602284.00	605621.00	615764.00	620908.00	626099.00	631337.00	633972.00	636301.00
2	.100	9.38	9.39	9.42	9.44	9.46	9.47	9.48	9.49
	.050	19.38	19.40	19.43	19.45	19.46	19.48	19.49	19.49
	.025	39.39	39.40	39.43	39.45	39.46	39.48	39.49	39.50
	.010	99.39	99.40	99.43	99.45	99.47	99.48	99.49	99.50
	.001	999.39	999.40	999.43	999.45	999.47	999.48	999.49	999.50
3	.100	5.24	5.23	5.20	5.18	5.17	5.15	5.14	5.13
	.050	8.81	8.79	8.70	8.66	8.62	8.57	8.55	8.53
	.025	14.47	14.42	14.25	14.17	14.08	13.99	13.95	13.91
	.010	27.35	27.23	26.87	26.69	26.50	26.32	26.22	26.14
	.001	129.86	129.25	127.37	126.42	125.45	124.47	123.97	123.53
4	.100	3.94	3.92	3.87	3.84	3.82	3.79	3.78	3.76
	.050	6.00	5.96	5.86	5.80	5.75	5.69	5.66	5.63
	.025	8.90	8.84	8.66	8.56	8.46	8.36	8.31	8.26
	.010	14.66	14.55	14.20	14.02	13.84	13.65	13.56	13.47
	.001	48.47	48.05	46.76	46.10	45.43	44.75	44.40	44.09
5	.100	3.32	3.30	3.24	3.21	3.17	3.14	3.12	3.11
	.050	4.77	4.74	4.62	4.56	4.50	4.43	4.40	4.37
	.025	6.68	6.62	6.43	6.33	6.23	6.12	6.07	6.02
	.010	10.16	10.05	9.72	9.55	9.38	9.20	9.11	9.03
	.001	27.24	26.92	25.91	25.39	24.87	24.33	24.06	23.82
6	.100	2.96	2.94	2.87	2.84	2.80	2.76	2.74	2.72
	.050	4.10	4.06	3.94	3.87	3.81	3.74	3.70	3.67
	.025	5.52	5.46	5.27	5.17	5.07	4.96	4.90	4.86
	.010	7.98	7.87	7.56	7.40	7.23	7.06	6.97	6.89
	.001	18.69	18.41	17.56	17.12	16.67	16.21	15.98	15.77
7	.100	2.72	2.70	2.63	2.59	2.56	2.51	2.49	2.47
	.050	3.68	3.64	3.51	3.44	3.38	3.30	3.27	3.23
	.025	4.82	4.76	4.57	4.47	4.36	4.25	4.20	4.15
	.010	6.72	6.62	6.31	6.16	5.99	5.82	5.74	5.66
	.001	14.33	14.08	13.32	12.93	12.53	12.12	11.91	11.72
8	.100	2.56	2.54	2.46	2.42	2.38	2.34	2.32	2.30
	.050	3.39	3.35	3.22	3.15	3.08	3.01	2.97	2.93
	.025	4.36	4.30	4.10	4.00	3.89	3.78	3.73	3.68
	.010	5.91	5.81	5.52	5.36	5.20	5.03	4.95	4.87
	.001	11.77	11.54	10.84	10.48	10.11	9.73	9.53	9.36

(Column headings: Degrees of freedom in the numerator)

(*continued*)

TABLE F *F* distribution critical values (*continued*)

Degrees of freedom in the denominator	p	Degrees of freedom in the numerator: 1	2	3	4	5	6	7	8
9	.100	3.36	3.01	2.81	2.69	2.61	2.55	2.51	2.47
	.050	5.12	4.26	3.86	3.63	3.48	3.37	3.29	3.23
	.025	7.21	5.71	5.08	4.72	4.48	4.32	4.20	4.10
	.010	10.56	8.02	6.99	6.42	6.06	5.80	5.61	5.47
	.001	22.86	16.39	13.90	12.56	11.71	11.13	10.70	10.37
10	.100	3.29	2.92	2.73	2.61	2.52	2.46	2.41	2.38
	.050	4.96	4.10	3.71	3.48	3.33	3.22	3.14	3.07
	.025	6.94	5.46	4.83	4.47	4.24	4.07	3.95	3.85
	.010	10.04	7.56	6.55	5.99	5.64	5.39	5.20	5.06
	.001	21.04	14.91	12.55	11.28	10.48	9.93	9.52	9.20
12	.100	3.18	2.81	2.61	2.48	2.39	2.33	2.28	2.24
	.050	4.75	3.89	3.49	3.26	3.11	3.00	2.91	2.85
	.025	6.55	5.10	4.47	4.12	3.89	3.73	3.61	3.51
	.010	9.33	6.93	5.95	5.41	5.06	4.82	4.64	4.50
	.001	18.64	12.97	10.80	9.63	8.89	8.38	8.00	7.71
15	.100	3.07	2.70	2.49	2.36	2.27	2.21	2.16	2.12
	.050	4.54	3.68	3.29	3.06	2.90	2.79	2.71	2.64
	.025	6.20	4.77	4.15	3.80	3.58	3.41	3.29	3.20
	.010	8.68	6.36	5.42	4.89	4.56	4.32	4.14	4.00
	.001	16.59	11.34	9.34	8.25	7.57	7.09	6.74	6.47
20	.100	2.97	2.59	2.38	2.25	2.16	2.09	2.04	2.00
	.050	4.35	3.49	3.10	2.87	2.71	2.60	2.51	2.45
	.025	5.87	4.46	3.86	3.51	3.29	3.13	3.01	2.91
	.010	8.10	5.85	4.94	4.43	4.10	3.87	3.70	3.56
	.001	14.82	9.95	8.10	7.10	6.46	6.02	5.69	5.44
25	.100	2.92	2.53	2.32	2.18	2.09	2.02	1.97	1.93
	.050	4.24	3.39	2.99	2.76	2.60	2.49	2.40	2.34
	.025	5.69	4.29	3.69	3.35	3.13	2.97	2.85	2.75
	.010	7.77	5.57	4.68	4.18	3.85	3.63	3.46	3.32
	.001	13.88	9.22	7.45	6.49	5.89	5.46	5.15	4.91
50	.100	2.81	2.41	2.20	2.06	1.97	1.90	1.84	1.80
	.050	4.03	3.18	2.79	2.56	2.40	2.29	2.20	2.13
	.025	5.34	3.97	3.39	3.05	2.83	2.67	2.55	2.46
	.010	7.17	5.06	4.20	3.72	3.41	3.19	3.02	2.89
	.001	12.22	7.96	6.34	5.46	4.90	4.51	4.22	4.00
100	.100	2.76	2.36	2.14	2.00	1.91	1.83	1.78	1.73
	.050	3.94	3.09	2.70	2.46	2.31	2.19	2.10	2.03
	.025	5.18	3.83	3.25	2.92	2.70	2.54	2.42	2.32
	.010	6.90	4.82	3.98	3.51	3.21	2.99	2.82	2.69
	.001	11.50	7.41	5.86	5.02	4.48	4.11	3.83	3.61
200	.100	2.73	2.33	2.11	1.97	1.88	1.80	1.75	1.70
	.050	3.89	3.04	2.65	2.42	2.26	2.14	2.06	1.98
	.025	5.10	3.76	3.18	2.85	2.63	2.47	2.35	2.26
	.010	6.76	4.71	3.88	3.41	3.11	2.89	2.73	2.60
	.001	11.15	7.15	5.63	4.81	4.29	3.92	3.65	3.43
1000	.100	2.71	2.31	2.09	1.95	1.85	1.78	1.72	1.68
	.050	3.85	3.00	2.61	2.38	2.22	2.11	2.02	1.95
	.025	5.04	3.70	3.13	2.80	2.58	2.42	2.30	2.20
	.010	6.66	4.63	3.80	3.34	3.04	2.82	2.66	2.53
	.001	10.89	6.96	5.46	4.65	4.14	3.78	3.51	3.30

TABLE F ***F* distribution critical values (*continued*)**

Degrees of freedom in the denominator	p	Degrees of freedom in the numerator: 9	10	15	20	30	60	120	1000
9	.100	2.44	2.42	2.34	2.30	2.25	2.21	2.18	2.16
	.050	3.18	3.14	3.01	2.94	2.86	2.79	2.75	2.71
	.025	4.03	3.96	3.77	3.67	3.56	3.45	3.39	3.34
	.010	5.35	5.26	4.96	4.81	4.65	4.48	4.40	4.32
	.001	10.11	9.89	9.24	8.90	8.55	8.19	8.00	7.84
10	.100	2.35	2.32	2.24	2.20	2.16	2.11	2.08	2.06
	.050	3.02	2.98	2.85	2.77	2.70	2.62	2.58	2.54
	.025	3.78	3.72	3.52	3.42	3.31	3.20	3.14	3.09
	.010	4.94	4.85	4.56	4.41	4.25	4.08	4.00	3.92
	.001	8.96	8.75	8.13	7.80	7.47	7.12	6.94	6.78
12	.100	2.21	2.19	2.10	2.06	2.01	1.96	1.93	1.91
	.050	2.80	2.75	2.62	2.54	2.47	2.38	2.34	2.30
	.025	3.44	3.37	3.18	3.07	2.96	2.85	2.79	2.73
	.010	4.39	4.30	4.01	3.86	3.70	3.54	3.45	3.37
	.001	7.48	7.29	6.71	6.40	6.09	5.76	5.59	5.44
15	.100	2.09	2.06	1.97	1.92	1.87	1.82	1.79	1.76
	.050	2.59	2.54	2.40	2.33	2.25	2.16	2.11	2.07
	.025	3.12	3.06	2.86	2.76	2.64	2.52	2.46	2.40
	.010	3.89	3.80	3.52	3.37	3.21	3.05	2.96	2.88
	.001	6.26	6.08	5.54	5.25	4.95	4.64	4.47	4.33
20	.100	1.96	1.94	1.84	1.79	1.74	1.68	1.64	1.61
	.050	2.39	2.35	2.20	2.12	2.04	1.95	1.90	1.85
	.025	2.84	2.77	2.57	2.46	2.35	2.22	2.16	2.09
	.010	3.46	3.37	3.09	2.94	2.78	2.61	2.52	2.43
	.001	5.24	5.08	4.56	4.29	4.00	3.70	3.54	3.40
25	.100	1.89	1.87	1.77	1.72	1.66	1.59	1.56	1.52
	.050	2.28	2.24	2.09	2.01	1.92	1.82	1.77	1.72
	.025	2.68	2.61	2.41	2.30	2.18	2.05	1.98	1.91
	.010	3.22	3.13	2.85	2.70	2.54	2.36	2.27	2.18
	.001	4.71	4.56	4.06	3.79	3.52	3.22	3.06	2.91
50	.100	1.76	1.73	1.63	1.57	1.50	1.42	1.38	1.33
	.050	2.07	2.03	1.87	1.78	1.69	1.58	1.51	1.45
	.025	2.38	2.32	2.11	1.99	1.87	1.72	1.64	1.56
	.010	2.78	2.70	2.42	2.27	2.10	1.91	1.80	1.70
	.001	3.82	3.67	3.20	2.95	2.68	2.38	2.21	2.05
100	.100	1.69	1.66	1.56	1.49	1.42	1.34	1.28	1.22
	.050	1.97	1.93	1.77	1.68	1.57	1.45	1.38	1.30
	.025	2.24	2.18	1.97	1.85	1.71	1.56	1.46	1.36
	.010	2.59	2.50	2.22	2.07	1.89	1.69	1.57	1.45
	.001	3.44	3.30	2.84	2.59	2.32	2.01	1.83	1.64
200	.100	1.66	1.63	1.52	1.46	1.38	1.29	1.23	1.16
	.050	1.93	1.88	1.72	1.62	1.52	1.39	1.30	1.21
	.025	2.18	2.11	1.90	1.78	1.64	1.47	1.37	1.25
	.010	2.50	2.41	2.13	1.97	1.79	1.58	1.45	1.30
	.001	3.26	3.12	2.67	2.42	2.15	1.83	1.64	1.43
1000	.100	1.64	1.61	1.49	1.43	1.35	1.25	1.18	1.08
	.050	1.89	1.84	1.68	1.58	1.47	1.33	1.24	1.11
	.025	2.13	2.06	1.85	1.72	1.58	1.41	1.29	1.13
	.010	2.43	2.34	2.06	1.90	1.72	1.50	1.35	1.16
	.001	3.13	2.99	2.54	2.30	2.02	1.69	1.49	1.22

ANSWERS TO SELECTED EXERCISES

Chapter 1

1.1 Categorical: gender, weight group, smoker. Quantitative: age, blood pressure, level of calcium.

1.3 **(b)** Yes; the three preferences make up one whole.

1.7 Somewhat symmetric and unimodal.

1.9 **(a)** Yes; there is an outlier. The group did not do so well: Only 4/18 met the goal. **(b)** Midpoint: 147.5. Spread: 78 to 359, although only one is greater than 271. **(c)** The dotplot and the stemplot are very similar, but the dotplot shows the exact location of each data point and is therefore a little bit more detailed.

1.11 The percent of births using routine episiotomy has been declining steadily over time, indicating a gradual change of medical practice away from routine episiotomy.

1.13 (c) **1.14** (c) **1.15** (a) **1.16** (b) **1.17** (c)

1.18 (a) **1.19** (b) **1.20** (b) **1.21** (b) **1.22** (a)

1.23 **(a)** Quantitative. **(b)** Quantitative. **(c)** Categorical. **(d)** Quantitative. **(e)** Categorical.

1.25 **(a)** Lakes. **(b)** 5 variables: 4 quantitative and 1 categorical (age of data).

1.27 **(a)** 130. **(b)** 20.7%. **(d)** Yes; together, these are all the categories of manatee deaths.

1.29 **(a)** All age groups have a fairly large rate of obesity. **(b)** No; each percent represents a different age group separately.

1.31 **(a)** Bar graph: categorical data. **(b)** Histogram: quantitative data. **(c)** Histogram: quantitative data. **(d)** Bar graph: categorical data.

1.33 **(a)** The two separate peaks could reflect a bimodal distribution. **(b)** The second histogram appears unimodal, but with a wide peak.

1.35 **(a)** Larger countries would almost certainly have higher emissions, even if they produce less per person. **(b)** Strong right-skew, center about 3, spread 0.1 to 19.6 metric tons per person. There are three outliers: the United States, Australia, and Canada.

1.37 **(a)** The midpoints are about 6 with buffer and about 2 with nanoparticles. Tumor increase tends to be smaller with nanoparticle treatment.

1.39 **(a)** Increasing trend starting in the mid-1970s, but no cyclical fluctuations. **(b)** No; a histogram would mask the existence of a clear trend.

1.41 **(a)** Prices tend to be highest in late summer/early fall. Prices are lowest in the winter. **(b)** Yes; prices appear to rise over time.

Chapter 2

2.1 **(a)** The distribution is irregular but somewhat symmetric except for a high outlier. The spread is from 164.0 to 740.2. **(b)** $\overline{x} = 319.25$; 12 observations have values less than the mean.

2.3 Median = 290.70. The median is less than the mean, which suggests a right-skew or high outlier(s).

2.5 Min. = 164.0, Q_1 = 260.9, M = 290.7, Q_3 = 354.95, max. = 740.2. This reflects what is seen in the plot. Note that the maximum is much farther away from the median than the minimum.

2.7 740.2 is the only outlier.

2.9 **(a)** 5.4. **(b)** 0.642.

2.11 **(a)** Standard deviation and mean should be sufficient, because the histogram is symmetric with a single peak. **(b)** Sufficient: Histogram is roughly symmetric. **(c)** May not be sufficient: Histogram is skewed.

2.13	(c)	**2.14**	(b)	**2.15**	(c)	**2.16**	(c)	**2.17**	(b)
2.18	(c)	**2.19**	(a)	**2.20**	(b)	**2.21**	(b)	**2.22**	(a)

2.23 Because the mean is larger than the median, the data might be skewed to the right, have high outliers, or both.

2.25 The data are bimodal. There is no unique center and, therefore, no simple numerical summary.

2.27 M = 14, Q_1 = 13, and Q_3 = 15.

2.29 **(a)** Symmetric distributions with no outliers. **(b)** On average, women have lower metabolic rates than men.

2.31 **(a)** Buffer: $\bar{x}$ = 6.08, s = 1.98. Nanoparticles: $\bar{x}$ = 2.020, s = 1.02. **(b)** The nanoparticle treatment was much more effective at limiting tumor growth.

2.33 **(a)** Any set of four identical numbers. **(b)** 0, 0, 10, 10 is the only possible answer.

2.35 A simple example could be 0, 0, 0, 1 (Q_3 = 0, $\bar{x}$ = 0.25).

2.37 The five-number summary is a better option here. Compressed: $\bar{x}$ = 2.908, min. = 2.68, Q_1 = 2.795, M = 2.88, Q_3 = 2.99, max. = 3.18.
Intermediate: $\bar{x}$ = 3.336, min. = 2.92, Q_1 = 3.13, M = 3.31, Q_3 = 3.45, max. = 4.26.
Loose: $\bar{x}$ = 4.232, min. = 3.94, Q_1 = 4.015, M = 4.175, Q_3 = 4.32, max. = 4.91. Soil penetration is greatest for loose soil and least for compressed soil.

2.39 Here, $\bar{x}$ and s are reasonable summaries. Group 1: $\bar{x}$ = 23.75, s = 5.07. Group 2: $\bar{x}$ = 14.08, s = 4.98. Group 3: $\bar{x}$ = 15.78, s = 5.76. Logging appears to reduce the number of trees per plot.

2.41 Aggressions received are both higher and more variable for females of lower ranks. The 4th dominance group has an extreme outlier.

2.43 **(a)** Min. = 16.8, Q_1 = 22.6, M = 23.7, Q_3 = 24.8, max. = 31.5.
(b) IQR = 2.2, 1.5 × IQR = 3.3; observations less than 19.3 or greater than 28.1 are suspected outliers. Only DC and Utah are flagged. **(c)** $\bar{x}$ = 23.743, very close to M = 23.7 (distribution is symmetric).

Chapter 3

3.1 **(a)** Explanatory: Number of calories. Response: Percent body fat. **(b)** Explore the relationship. **(c)** Explanatory: Inches of rain. Response: Yield of corn. **(d)** Explore the relationship.

3.3 Weight and different physiology/genetic makeup are two other possible factors.

3.5 Linear, negative, and reasonably strong. The number of D notes per call appears to represent predator wingspan.

3.7 **(a)** Mass is explanatory. Subject number is an index variable and thus contains no data. **(b)** Moderately strong, positive linear relationship. **(c)** The same relationship seems to hold. Men tend to have higher values and are overall more variable than women.

3.9 No; units do not affect correlation.

3.11 The relationship is clearly not linear; therefore, r is meaningless.

3.13 (a) **3.14** (a) **3.15** (c) **3.16** (b) **3.17** (c)

3.18 (a) **3.19** (b) **3.20** (a) **3.21** (b) **3.22** (c)

3.23 **(a)** The negative linear relationship suggests that it is. **(b)** $r = -0.928$. This is very strong.

3.25 **(a)** The lack of association suggests that it isn't. **(b)** $r = 0.115$. This is very weak. **(c)** *D. magna* may be an effective method but not *D. pulex*.

3.27 **(a)** r will be closer to 1. **(b)** r will decrease and could even become negative if the point on the far end is dragged down far enough.

3.29 **(a)** Linear, positive, and very strong ($r = 0.9766$). **(b)** The 3D reconstruction is a better model (r stronger).

3.31 **(a)** The order was changed in the groups. **(b)** It would look like the mirror image. There is no logical order to colors. **(c)** Numerical values wrongly suggest that the variables are quantitative when they are not.

3.33 **(a)** The negative linear association indicates habituation. **(b)** The patterns are different for the mutant and the wild type, suggesting that PVD sensory neurons are involved in nociception and habituation.

3.35 Strong, positive linear association ($r = 0.9143$). This supports the idea that beak size may play a role in cooling.

3.37 Moderately strong, positive linear association ($r = 0.8042$). This supports the idea that animal populations decline when the fish supply is low.

3.39 **(a)** $r = 1$ for a line. **(c)** Leave some space above your vertical stack. **(d)** The curve must be higher at the right than at the left.

Chapter 4

4.1 **(a)** The slope indicates that for each drink consumed, BAC increases by 0.023. The intercept is zero because if someone doesn't have any drinks, their BAC should be 0. **(b)** BAC $= 0.023 \times 4 = 0.092$.

4.3 **(a)** $\hat{y} = 2.537$. **(b)** Predicted mean seed count $= 10^{2.537} = 344.3$.

4.5 **(a)** $\bar{x} = 30.261$, $s_x = 0.4391$, $\bar{y} = 2.511$, $s_y = 0.1502$, $r = -0.8986$, $b = rs_y/s_x = -0.3074$ and $a = \bar{y} - b\bar{x} = 11.813$. **(b)** $\hat{y} = 11.811 - 0.3073x$.

4.7 Because the absolute value of the correlation is closer to 1, the data lie closer to the regression line and the regression line is a better predictor.

4.9 **(a)** Point A is a horizontal outlier, whereas Point B is a vertical outlier. **(b)** Point A has more influence on the regression line. The farther an outlier is from the center, the more leverage it has on the slope of the line.

4.11 More populated cities have a higher demand for pharmacists but also have a higher number of deaths. Everything scales with population size.

4.13 **(a)** No; there is no biological reason why education would cause breast cancer. **(b)** Women with higher education tend to have babies later in life.

4.14 (b) **4.15** (c) **4.16** (c) **4.17** (a) **4.18** (c)

4.19 (a) **4.20** (a) **4.21** (a) **4.22** (a) **4.23** (b)

4.25 **(a)** Slope = 101.98, positive relationship. **(b)** Prediction 49.8, very reliable with $r^2 = 0.95$. **(c)** Nonsense prediction 131.4; extrapolation (prediction outside range).

4.27 **(a)** $y = 2.475224x + 0.419572$. **(b)** 11.55808 cm^3. **(c)** $r^2 = 0.9537$, much better than the 2D model.

4.29 **(a)** No; curved pattern. **(b)** Yes; the pattern is now linear. **(c)** 89.

4.31 **(a)** Slope = 0.547. Intercept = 2.791. **(b)** Prediction = 30.2. A high r^2 (here, 0.9216) implies a reliable prediction.

4.33 **(a)** $y = 0.06078x - 0.1261$. Prediction = −0.00454. **(b)** 77.1%.

4.35 **(a)** They all give similar results; $y = 3.0 + 0.5x$; $r = 0.82$; very close slopes, intercepts, and correlations; $\hat{y} = 8$ for $x = 10$. **(c)** Only the first relationship is linear and appropriate for regression. The second relationship is strongly curved. The third has a strong outlier. The last has only two values for x.

4.37 **(a)** Ignoring the outlier, the plot shows a moderately strong, negative linear relationship. **(b)** For all 12 points: $r = -0.3387$. Without the outlier: $r = -0.7866$. The outlier weakens the linear pattern.

4.39 For all 12 points: $\hat{y} = -492.6 - 33.79x$. Without the outlier: $\hat{y} = -1371.6 - 75.52x$. The line is pulled toward the outlier to reduce the very large vertical deviation of this point from the line.

4.41 **(a)** There is a reasonably strong, positive linear relation between temperature and latitude. **(b)** $y = 0.818x + 52.45$, $r = 0.6033$. **(c)** $y = 0.9843x + 51.26$, $r = 0.7343$. **(d)** The outlier weakens the correlation and makes the slope less steep.

4.43 Answers will vary. For instance, mothers who smoke during pregnancy may also take drugs or drink alcohol during pregnancy.

4.45 No; it is more probable that those in critical condition are sent to large hospitals.

4.47 Strong, positive linear relationship. $\hat{y} = 10.02x + 11.846$, $r = 0.949$. The data support the hypothesis.

4.49 It makes sense to use a straight line for prediction only on the scatterplot that shows a clear linear relationship without extreme outliers.

Chapter 5

5.1 **(a)** 40 people. **(b)** 8 in the intramuscular group. **(c)** Immune response: 16 (40%). No immune response: 24 (60%).

5.3 There are an infinite number of possibilities, but $a = 50, b = 0, c = 10, d = 40$, and $a = 40, b = 10, c = 20$, and $d = 30$ are two solutions.

5.5 **(a)** Conditional distribution of alarm call: given predator nearby = 26.6%, given predator far = 41.4%. **(b)** When a predator is close, prairie dogs raise the alarm 26.6% percent of the time. When a predator is far away (making it safer for the individual to give a warning call), the prairie dogs raise the alarm 41.4% of the time.

5.7 **(a)** Open surgery: 78.0%. PCNL: 82.6%. **(b)** Small stones: 88.2%. Large stones: 72.0%. **(c)** Open surgery success for small stones: 93.1%. PCNL success for small stones: 86.7%. Open surgery success for large stones: 73.0%. PCNL success for large stones: 68.8%. This is Simpson's paradox: PCNL was used more often to treat small stones, which appear to be easier to treat (better chance of successful outcome).

5.9 (a) **5.10** (b) **5.11** (c) **5.12** (a) **5.13** (c)

5.14 (b) **5.15** (a) **5.16** (c) **5.17** (b) **5.18** (b)

5.19 Marginal for gender: males = 47.4%, females = 52.6%. Marginal for order: 1st = 28.4%, 2nd = 26.3%, 3rd = 25.3%, 4th = 20%.

5.21 Among males, 1st = 37.8%, 2nd = 35.6%, 3rd = 15.6%, 4th = 11.1%. Yes; they should total 100% aside from rounding error.

5.23 **(a)** Overall, 60.4% success ("Yes") and 39.6% failure ("No"). **(b)** Conditional distributions: percent successful outcome was 80.8% in the Botox group but only 40.7% in the placebo group. Botox was much more effective.

5.25 **(a)** Percent in 9- to 11-month age range who ate fried potatoes: 9.0%. Percent in 15- to 18-month range: 20.1%. Percent in 19- to 24-month age range: 25.9%. **(b)** Association does not imply causation.

5.27 Conditional distribution of birth outcomes, given vitamins and minerals: 1.3% with congenital malformation, 0.0% with neural-tube defect, and 98.7% with neither. Conditional distribution of birth outcomes, given minerals only: 2.3% with congenital malformation, 0.3% with neural-tube defect, and 97.4% with neither. Outcomes are more favorable with vitamins and minerals.

5.29 **(a)** 99.2% for Type 1 and 100% for Type 2 among younger patients; 54.4% for Type 1 and 58.8% for Type 2 among older patients. Overall, younger patients have a much higher survival rate. Within each age group, patients with Type 2 diabetes have a slightly higher survival rate. **(b)** 70.7% for Type 1; 59.9% for Type 2. It seems that patients with Type 1 diabetes have a higher survival rate overall. **(c)** The lurking variable here is age. Type 2 diabetes is more frequent in older patients. This is Simpson's paradox.

5.31 **(b)** Percents hatching in each cell group: cold, 59.3%; neutral, 67.9%; hot, 72.1%. The cold temperature made hatching less likely.

5.33 Conditional distribution for dead trees: 55.9% western hemlock, 38.4% Douglas fir, and only 5.7% western red cedar. Conditional distribution for live trees: 43.8% western hemlock, 33.8% Douglas fir, and 22.3% western red cedar. Conditional distribution for sapling trees: 63.1% western red cedar, 36.1% western hemlock, and only 0.8% Douglas fir. The western red cedar is in the process of displacing the Douglas fir.

5.35 Conditional distribution of no smoking given treatment: 44.0% (Chantix), 29.5% (Bupropion), 17.7% (placebo). Chantix has the best success rate after 9 weeks; the placebo group has the lowest success rate.

Chapter 6

6.1 For example: categorical = percent students with undergraduate research experience; quantitative = average number of months spent doing undergraduate research and number of professors with Nobel Prizes.

6.3 For doubling time, $IQR = 0.6$, no outliers. For immune cell increases, $IQR = 202$; the value 700 is a suspected outlier.

6.5 A scatterplot shows a moderately strong, positive association ($r = 0.7565$).

6.7 **(a)** The stemplot is roughly symmetric. **(b)** $\bar{x} = 0.2426$ and $s = 0.0476$ for all plants; without the extremes, $\bar{x} = 0.2433$ and $s = 0.0396$.

6.9 57.5%.

6.11 The distribution is clearly bimodal, so a single summary (mean and standard deviation or the five-number summary) would be misleading.

6.13 Full data set: median = 25, $\bar{x} = 25.42$, $s = 7.47$. Without the outlier (HAV = 50): median = 25, $\bar{x} = 24.76$, $s = 6.34$. Outliers have little effect (in this case, none) on medians, a stronger effect on means, and an even stronger effect on standard deviations.

6.15 **(a)** $y = 0.339x + 19.723$. **(b)** 28.198. **(c)** Not very well. Looking at the graph, the points do not seem to lie too close to the line. In numerical terms, $r^2 = 0.0913$.

6.17 **(a)** 31 neurons. **(b)** $r = 0.6386$; the relationship is moderately strong.

6.19 Yes; there is a moderate positive linear relationship between number of perch in a pen and proportion killed ($r^2 = 0.465$).

6.21 Min. = 1, $Q_1 = 11$, M = 16, $Q_3 = 19.5$, max. = 31, $\bar{x} = 15.34$, $s = 6.15$. The median date is May 5.

6.23 The overall data are quite variable but the median is declining over time, so ice is melting earlier. Five-number summaries in chronological order: Min. = 7, $Q_1 = 12$, M = 19, $Q_3 = 22$, max. = 26. Min. = 1, $Q_1 = 11.5$, M = 16, $Q_3 = 21.5$, max. = 31. Min. = 8, $Q_1 = 11$, M = 16.5, $Q_3 = 19$, max. = 23. Min. = 1, $Q_1 = 7.5$, M = 10.5, $Q_3 = 15.5$, max. = 31.

6.25 **(a)** Mean lean body mass: 8.7 kg for lean monkeys and 10.5 kg for obese monkeys. **(b)** For lean monkeys: $\hat{y} = 0.541 + 0.0826x$. For obese monkeys: $\hat{y} = 0.371 + 0.0852x$. The slope is about the same but the intercept appears to be different, so this implies that energy increases at about the same rate for both groups of monkeys but that obese monkeys expend less energy per unit of body mass.

6.27 **(a)** $y = -1.099x + 166.48$. The more lamb's-quarter in a cornfield, the less corn it will produce. **(b)** 159.9.

6.29 **(a)** $y = 93.9 + 0.778x$. The third point is A; the first point is B. **(b)** The correlation drops only slightly (from 0.6386 to 0.6101) when A is removed; it drops more drastically (to 0.4793) without B. **(c)** Without A: $y = 98.4 + 0.679x$. Without B: $y = 101 + 0.693x$.

6.31 The conditional distributions of "No recurrence" within each treatment group are 84.0% (cranberry), 61.2% (lactobacillus), 64.0% (neither). The cranberry juice preventive treatment has the highest percent of women remaining free of recurring UTIs over the six-month period.

Chapter 7

7.1 Observational. The study did not attempt to change the participants in any way.

7.3 **(a)** Observational. **(b)** Explanatory: human activity. Lurking: natural variability in the environment (for example, previous ice ages).

7.5 Actual California population.

7.7 **(a)** Label each one from 0001 to 1410. **(b)** 769, 1315, 94.

7.9 A multistage design divides the population into groups, takes a random sample of these groups, and then takes a random sample of individuals within the groups sampled. An SRS just samples from the entire population, without dividing it into subgroups. Both techniques use randomness for selecting individuals.

7.11 **(a)** The population is all physicians practicing in the United States. The sample size is 2379 (physicians for whom we do have data). Conclusions apply to all U.S. physicians. **(b)** The rate of nonresponse is 97.6%. There is a very real potential for bias. **(c)** The Physicians Foundation hoped to survey 100,000 physicians but obtained data from only 2379 physicians. The report should be called "a survey of 2379 physicians."

7.13 This is a cohort study. The population is full-term babies born in a similar time and place. The explanatory variable is antibiotic exposure. The response variable is body mass recorded over time.

7.14 (c) **7.15** (c) **7.16** (b) **7.17** (a) **7.18** (a)

7.19 (b) **7.20** (b) **7.21** (b) **7.22** (a) **7.23** (c)

7.25 **(a)** Adults living in the state of Washington. **(b)** An SRS probability sampling design. 500.

7.27 **(a)** To avoid response bias due to the disturbance of a new presence. **(b)** If animals are scared or offended by a new, unusual presence, then we cannot observe their natural behavior.

7.29 In a stratified random sample, the entire population is broken up into subgroups; then an SRS is taken from each of the individual subgroups, and the SRSs are finally recombined to create the sample. This ensures that all four forest types are represented in the sample in proportion to their representation in the population.

7.31 In the systematic sample, each student has chance 1/45 of being selected (as in an SRS), but only some combinations of students are possible (unlike in an SRS).

7.33 **(a)** The question is clear, but the wording seems to suggest that cell phone usage caused the brain cancer. **(b)** The question is slanted toward an affirmative answer. **(c)** This question is so unclear and charged with negative words that it is likely to prevent many "yes" answers.

7.35 Answers will vary.

7.37 **(a)** A case-control study. **(b)** Answers will vary. Novices, amateurs, and athletes could wear helmets in different proportions.

7.39 **(a)** A cohort study. One homogeneous group was followed for many years to track the effect of caffeine on depression. **(b)** This is an observational study, so a causal link cannot be established. **(c)** Answers will vary. Nurses with health problems could avoid caffeine, and health problems could also lead to depression and antidepressant use.

7.41 **(a)** A cohort study. A homogeneous group was followed for many years in order to track the effects of alcohol on the risk of prostate cancer. **(b)** Prostate cancer is rare enough that there might have been too few cases in the study for statistical evaluation.

Chapter 8

8.1 Subjects: the healthy people aged 18 to 40. Factor: the pill given to the subject. Treatments: ginkgo or placebo. Response variable: the number of *e*'s missed by each subject.

8.3 No and no. This is an observational study and there could be confounding variables such as weight gain. A cause-and-effect conclusion cannot be reached.

8.5 **(a)** Assign 6 plots to each treatment. **(b)** If using Table A, label 01 to 36 and take two digits at a time.

8.7 **(a)** Block design. **(b)** First, divide the crickets into two groups: 7–12 days and 15+ days. Then, within these groups, randomly sample 10 to receive parasites, and compare results.

8.9 **(a)** Randomly assign 15 students to Group 1 (easy mazes) and the other 15 to Group 2 (hard mazes). Compare each group's time estimates. **(b)** Each student does the activity twice, once with the easy mazes and once with the hard mazes, in a randomly determined order. Compare each student's easy and hard time estimates.

8.11 This was not blind, because both the subject and the experimenter knew who was receiving treatment and who was not. Subjects learning the meditation method could experience a placebo effect. More importantly, the experimenter could be biased in his assessment of reported anxiety.

8.13 Answers will vary.

8.14 (b) **8.15** (b) **8.16** (c) **8.17** (a) **8.18** (a)

8.19 (c) **8.20** (b) **8.21** (a) **8.22** (b) **8.23** (a)

8.25 **(a)** Explanatory: type of operation. Response: survival time. **(b)** Not an experiment: existing records are simply examined. **(c)** The type of operation is most likely decided based on the size of the tumor. Larger tumors would be much more dangerous to remove, regardless of type of operation.

8.27 **(a)** Completely randomized design: 12 birds randomly selected for the placebo group, the remaining 15 making up the immune challenge group (or the other way around). **(b)** Using Table A to select the 12 birds in the placebo group: 05, 19, 04, 25, 20, 16, 18, 07, 13, 02, 23, 27.

8.29 The first study is an observational, case-control study; the second is an experiment: the exercise plan is assigned randomly.

8.31 **(a)** Randomly assign the subjects to four groups: Group 1, antidepressants and stress management; Group 2, stress management only; Group 3, antidepressants only; and Group 4, control. **(b)** Using Table A, Group 1 = Chai, Hammond, Herrera, Xiang, Irwin, Hurwitz, Reed, Broden, Lucero, and Nho.

8.33 **(a)** Each subject was scanned twice (ON and OFF conditions). **(b)** Different subjects might have different brain activities. The matched pairs design allows comparison of the ON and OFF conditions for each subject. **(c)** The difference could not easily be explained by chance variations alone.

8.35 **(a)** A block design.

8.37 **(a)** Randomly sample half of the subjects and give them the treatment; everybody else gets the placebo pill. Then compare the results. **(b)** Using Table A: 170, 005, 227, 118, 007.

8.39 **(a)** This is a nonrandomized, noncomparative study. **(b)** No, because subjects were not randomly selected (they were hand-picked by Lipov) and the lack of comparison means that results could be simply a placebo effect.

8.41 **(a)** In an observational study, we observe subjects who have chosen to take supplements and compare them with others who do not take supplements. In an experiment, we assign some subjects to take supplements and others to take a placebo. **(b)** Treatments are assigned at random, and a control group is used as a basis for comparison to observe the effects of the treatment. **(c)** Subjects who choose to take supplements are more likely to make healthy lifestyle choices. When random assignment is used, some of those subjects will take the supplement and some will take the placebo.

8.43 "No significant difference" does not mean the groups are identical but that observed differences are no bigger than we might expect from true random allocation alone.

8.45 **(b)** Yes; the results appear trustworthy. The study was a comparative experiment with a placebo control group, subjects were randomly assigned to treatments, double-blinding was used to prevent bias, and the variable was clearly defined.

8.47 **(a)** The procedure might be harmful in humans. **(b)** It might be ethical to test the procedure without consent if it was felt that the patient would not survive otherwise.

8.49 The Tanzanian government and the researchers both presented serious arguments. However, the study's purpose was entirely academic. Without patients' informed consent or any benefit to them, the study was not ethically justified and should indeed have been canceled.

8.51 For example, informed consent is lacking.

8.53 Detailed outcomes will vary.

Chapter 9

9.1 This means that if you repeatedly sampled 100,000 males, on average you would have 13 diagnosed with hemophilia. For one random sample of 100,000 males, you could find more or you could find fewer than 13 individuals with hemophilia.

9.3 **(a)** 0. **(b)** 1. **(c)** 0.01. **(d)** 0.6.

9.5 **(a)** $S = \{BBB, BBG, BGB, BGG, GBB, GBG, GGB, GGG\}$. **(b)** $S = \{0, 1, 2, 3\}$. Associated probabilities $= \frac{1}{8}, \frac{3}{8}, \frac{3}{8}$, and $\frac{1}{8}$, respectively.

9.7 **(a)** 0.17. **(b)** 0.43. **(c)** 0.26.

9.9 **(a)** The sum of probabilities is 1. **(b)** The student was physically active (for at least 60 minutes) fewer than 7 days in the past week. $P(X < 7) = 0.73$. **(c)** $P(X > 0) = P(X \geq 1) = 0.85$.

9.11 **(a)** 0.4. **(b)** 0.4. **(c)** 0.2. **(d)** 0.8.

9.13 **(a)** Discrete. Answers can only be whole numbers. **(b)** Continuous. Answers can be any value between 0 and 24. **(c)** Discrete. Answers can only be whole numbers.

9.15 **(a)** Answers will vary. **(b)** You would know whether you are generally a cautious or a reckless driver, having a more accurate view of yourself than just using the population mean. **(c)** People generally tend to think they are above average; they may also underestimate the dangers of driving.

9.17 **(a)** Probability = 0.268, risk = 0.268, odds = 0.366. **(b)** By age group of increasing age, risk: 0.245, 0.448, 0.274, 0.275, 0.252, and 0.196, respectively; odds: 0.325, 0.812, 0.377, 0.379, 0.337, and 0.244. Risk and odds are greatest among 20- to 24-year-olds.

9.18	(a)	**9.19**	(b)	**9.20**	(b)	**9.21**	(b)	**9.22**	(c)
9.23	(c)	**9.24**	(b)	**9.25**	(c)	**9.26**	(a)	**9.27**	(c)

9.29 **(a)** 0.2066. **(b)** 0.7934. **(c)** 0.3801 and 0.6199, respectively.

9.31 **(a)** $S = \{\text{fails, germinates}\}$. **(b)** $S = \{\text{all positive numbers}\}$. **(c)** $S = \{\text{all numbers}\}$. **(d)** $S = \{\text{all positive numbers}\}$.

9.33 **(a)** $S = \{\text{accident, homicide, suicide, other}\}$. **(b)** 0.72 and 0.28, respectively.

9.35 **(a)** 0.04. **(b)** 0.75.

9.37 **(a)** All probabilities add to 1. **(b)** 0.62. **(c)** P(Taster A ranks higher than a 3) = 0.39, P(Taster B ranks higher than a 3) = 0.39.

9.39 **(a)** 0.14. **(b)** 0.43. **(c)** 0.75.

9.41 **(a)** $X < 2$ or $X > 4$. $P(X < 2 \text{ or } X > 4) = 0.17$. **(b)** "Eye color greater than 2 and less than or equal to 4." $P(2 < X \leq 4) = 0.72$.

9.43 **(a)** Y is continuous, because it can take any value between 0 and 2. **(b)** Height = 0.5. **(c)** $P(Y \leq 1) = 0.5$.

9.45 **(a)** $P(13 \leq X < 16) = 0.77$. **(b)** $P(X > 13) = 0.88$.

9.47 **(a)** Risk = 0.0043, odds = 0.00432. **(b)** Risk = 0.0365, odds = 0.0379. **(c)** Probability = risk = 0.127, odds = 0.145. **(d)** By decade, because risk varies over the lifespan.

9.49 Actual results will vary.

Chapter 10

10.1 Theoretical probability = 0.0199. Actual probability = 0.185. A small difference becomes more substantial when compounded.

10.3 **(a)** P(next two patients are males) = 0.5625. **(b)** P (at least one is female) = 0.7627. P(at least one is male) = 0.9990.

10.5 P (dead | RC) = 0.1, P(dead | DF) = 0.5, P(dead | WH) = 0.479. The western red cedar is relatively new.

10.7 P(CDK) = 0.091. P(CDK | diabetes) = 0.141. P(CDK | no diabetes) = 0.064.

10.9 P (belong and twice a week) = P(belong)P(twice | belong) = (0.1)(0.4) = 0.04.

10.11 **(a)** P(malaria | sickle-cell) = 0.264, P(malaria | no sickle-cell) = 0.373. **(b)** No, because knowing the sickle-cell status changes the probability of malaria. It tells us that there is a biological link between the two conditions: Malaria is more common among individuals who do not have the sickle-cell trait.

10.13 **(a)** P(blue eyes | red hair) = 0.473. P(blue eyes and red hair) = 0.0118. **(b)** P(freckles | red hair and blue eyes) = 0.857. P(freckles and red hair and blue eyes) = 0.0101.

10.15 **(a)** $P(\text{red hair}) = 0.025$, $P(\text{blue eyes}) = 0.388$, $P(\text{freckles}) = 0.253$. **(b)** $P(\text{freckles and red hair}) = 0.0203$. $P(\text{freckles} \mid \text{red hair}) = 0.811$.

10.17 **(a)** 0.9292. **(b)** 0.9998. **(c)** 0.0256 and 0.9906, respectively. **(d)** PPV depends on the rate of HIV in the population screened because it influences how many true-positives and false-positives we would get. But the sensitivity and specificity are properties of the OraQuick test that depend only on chemistry and manufacturing.

10.18 (a) **10.19** (c) **10.20** (c) **10.21** (c) **10.22** (b)

10.23 (b) **10.24** (a) **10.25** (b) **10.26** (c) **10.27** (b)

10.29 0.526.

10.31 $P(X = 0) = 0.0282$. $P(X \geq 1) = 0.9718$.

10.33 Because having brown hair and having freckles are not necessarily independent events. (The earlier tree diagram shows that they are definitely not independent.)

10.35 84%.

10.37 **(a)** 0.2400. **(b)** 0.2844, 0.3248, 0.1991, and 0.1415 (top to bottom). **(c)** The probability of damage decreases noticeably when thorny cover is 1/3 or more.

10.39 26.59%.

10.41 **(a)** 0.0651. **(b)** 0.0605. **(c)** Fourth = 0.0563, fifth = 0.0524, $k\text{th} = (0.07)(0.93)^{k-1}$.

10.43 1.

10.45 $P(X = 0) = 0.9039$, $P(X = 1) = 0.0922$, and $P(X = 2) = 0.0038$; the other three probabilities are less than 0.0001.

10.47 **(a)** $P(\text{both carriers} \mid \text{both of European Jewish descent}) = 0.00137$, $P(\text{both carriers} \mid \text{mother of European Jewish descent but father not}) = 0.000148$, $P(\text{both carriers} \mid \text{neither of European Jewish descent}) = 0.000016$. **(b)** $P(\text{child TS} \mid \text{both parents of European Jewish descent}) = 0.00034$. **(c)** $P(\text{child TS} \mid \text{neither parent of European Jewish descent}) = 0.000004$.

10.49 **(b)** $P(\text{yes}) = P(\text{yes and head}) + P(\text{yes and tail}) = 0.5p + 0.5$. **(c)** $p = (0.59/0.5) - 1 = 0.18$.

Chapter 11

11.3 **(a)** 175 ml – 625 ml. **(b)** 250 ml and less.

11.5 At age 5, z-score = 0.556. At age 7, z-score = 0.476. John has been growing slightly more slowly than average, because his z-score has gone down a little bit. A third measurement one or two years later would help our conclusion.

11.7 **(a)** 0.6915 with Table B or software. **(b)** 0.3830 with Table B, 0.3829 with software.

11.9 **(a)** z is between -0.67 and -0.68 with Table B, $z = -0.674$ with software. **(b)** z is between 0.25 and 0.26 with Table B, $z = 0.253$ with software.

11.11 It is close to Normal, because the dots appear to make a straight line.

11.13 (a) **11.14** (a) **11.15** (b) **11.16** (a) **11.17** (b)

11.18 (c) **11.19** (b) **11.20** (c) **11.21** (b) **11.22** (b)

11.23 **(a)** 50%, 0.15%, 16%, respectively. **(b)** 2.5%. **(c)** 2.5%.

11.25 **(a)** 1.22%. **(b)** 98.78%. **(c)** 3.84%. **(d)** 94.94%.

11.27 **(a)** z is between 0.84 and 0.85 with Table B, $z = 0.842$ with software. **(b)** z is between 0.38 and 0.39 with Table B, $z = 0.385$ with software.

11.29 Approximately $z > 1$ for A, $0 < z < 1$ for B, $-1 < z < 0$ for C, $z < -1$ for D and F grades.

11.31 **(a)** 0.0267 (0.0268 with Table B). **(b)** 0.00001 (less than 0.0003 with Table B).

11.33 **(a)** 0.62%. **(b)** 30.85%.

11.35 **(a)** 95.6% and 10.8%, respectively. **(b)** About 22.3 and 24.5 cm, respectively.

11.37 **(a)** 59.87%. **(b)** 54.7%. **(c)** > 279 days.

11.39 **(a)** 0.1310 (0.1314 with Table B). **(b)** 0.3327 (0.3336 with Table B).

11.41 **(a)** 97.42 to 99.78. **(b)** 96.83 to 99.57.

11.43 About 0.675, both above and below.

11.45 Atlantic acorn sizes are clearly not Normal (right-skew).

11.47 The data appear roughly Normal.

11.49 The plot for never-logged areas is nearly linear, indicating a roughly Normal distribution. The other two plots each show a low value, suggesting perhaps a slight skew to the left or a low outlier.

Chapter 12

12.1 Yes. Reach a live person or not. $n = 15$, $p = 0.2$.

12.3 Yes. $n = 100$, $p = 0.44$.

12.5 **(a)** Binomial. X is resistance; $n = 10$, $p = 0.27$. **(b)** $P(X = 1) = 0.159$, $P(X = 2) = 0.265$. **(c)** $P(X \geq 1) = 1 - P(X = 0) = 0.957$.

12.7 **(a)** $\mu = 2.7$. **(b)** $\sigma = 1.404$. **(c)** This decreases the standard deviation $\sigma_{2010} = 0.807$. As p gets closer to zero, the standard deviation gets closer to zero.

12.9 **(a)** $P(X \leq 82) = 0.0129$. **(b)** $P(X \leq 82) = 0.0145$. This is fairly close. **(c)** It is likely that many of the cases of osteopenia have been missed.

12.11 **(a)** $\{0, 1, 2, 3, \ldots\}$. **(b)** Poisson. **(c)** $\mu = 0.536$, $\sigma = 0.732$.

12.13 **(a)** $P(X = 0) = 0.585$. **(b)** $P(X = 1) = 0.314$. **(c)** $P(X > 1) = 0.101$.

12.15 (b) **12.16** (b) **12.17** (c) **12.18** (a) **12.19** (c)

12.20 (b) **12.21** (b) **12.22** (c) **12.23** (c) **12.24** (b)

12.25 **(a)** Binomial is reasonable, $n = 8$. **(b)** Binomial is reasonable, $n = 100$. **(c)** Binomial is reasonable, $n = 24$.

12.27 **(a)** $n = 20$, $p = 0.25$. **(b)** 5. **(c)** $P(X = 5) = 0.202$.

12.29 **(a)** $n = 6$, $p = 0.75$. **(b)** $S = \{0, 1, 2, 3, 4, 5, 6\}$. **(c)** $P(X = 0) = 0.00024$, $P(X = 1) = 0.0044$, $P(X = 2) = 0.03296$, $P(X = 3) = 0.1318$, $P(X = 4) = 0.2966$, $P(X = 5) = 0.3560$, $P(X = 6) = 0.1780$. **(d)** $\mu = 4.5$, $\sigma = 1.06$.

12.31 $P(X \geq 2) = 0.4275$.

12.33 The probability is nearly 1.

12.35 **(a)** P(1 of 17 vaccinated children) = 0.3741. P(1 of 3 unvaccinated children) = 0.0960. P(1 in each group) = 0.0359. **(b)** Total probability is 0.1977.

12.37 **(a)** $S = \{0, 1, 2, 3, 4, 5\}$. **(b)** $P(X = 0) = 0.2373$, $P(X = 1) = 0.3955$, $P(X = 2) = 0.2637$, $P(X = 3) = 0.0879$, $P(X = 4) = 0.0146$, $P(X = 5) = 0.0010$.

12.39 $P(X \geq 2{,}173{,}000) \approx 0$. It seems unreasonable to conclude that the probability of having a boy is the same as that of having a girl.

12.41 **(a)** Poisson. $\mu = 3.6$, $\sigma = 1.897$. **(b)** 0.0273 and 0.9727, respectively.

12.43 **(a)** 0.0111. **(b)** 0.0500. **(c)** 0.9389.

12.45 $P(X \geq 5) = 0.219$, $P(X \geq 6) = 0.105$, $P(X \geq 7) = 0.045$. Reserving 6 beds for ER admissions would be enough 95.5% of the time.

12.47 **(a)** Poisson. $\mu = 15.58$, $\sigma = 3.947$. **(b)** $P(X = 0) = 1.71 \times 10^{-7}$, $P(X \leq 5) = 0.00186$, $P(X \leq 15) = 0.509$, $P(X \leq 25) = 0.990$, $P(X > 25) = 0.01$. **(c)** $P(X \geq 48) = 3.53 \times 10^{-11}$. This is too unlikely to be due to random, isolated cases. It points to an epidemic. The contaminated food must have been distributed at least in South Dakota and Wisconsin.

12.49 **(a)** 0.0651. **(b)** 0.0605. **(c)** Fourth = 0.0563, fifth = 0.0524, kth = $(0.07)(0.93)^{k-1}$.

12.51 **(a)** 0.12705. **(b)** 0.12705, 0.12705. **(c)** 0.38115.

Chapter 13

13.1 1.3 is a parameter; 1.75 is a statistic.

13.3 **(a)** 130.2. **(b)** Sampled 1, 4, 5, 9. Values = 99, 125, 170, 147; $\bar{x} = 135.25$. **(c)** $\bar{x}$ repeated 9 times: 119.5, 125.75, 141.25, 138, 118, 119, 137.75, 123, 132.5. The center of all these is about 129. **(d)** There are way too many numbers to do by hand.

13.5 **(a)** $\bar{x}$ is not systematically higher or lower than μ. **(b)** With large samples, $\bar{x}$ is more likely to be close to μ.

13.7 **(a)** $N(188, 4.1)$. 0.5346. **(b)** 0.9792.

13.9 The mean number of moths per trap in samples of 50 traps is 0.5, and the standard deviation is 0.099. The distribution of this estimator is fairly Normal.

13.11 **(a)** 6% is a parameter, 4% is a statistic. **(b)** Mean = 0.06, standard deviation = 0.0194.

13.13 0.1512.

13.15 If 1 of the 12 workers had a serious injury, it would cost more than the premiums collected. For thousands of policies, the average claim should be close to $439.

13.17 (c) **13.18** (a) **13.19** (a) **13.20** (c) **13.21** (a)

13.22 (c) **13.23** (b) **13.24** (c) **13.25** (b) **13.26** (b)

13.27 65 inches is a statistic, 64 inches is a parameter.

13.29 **(a)** 0.2812. **(b)** Standard deviation = $2.8/\sqrt{n}$. Required sample size = 32. **(c)** 0.9566.

13.31 **(a)** 1/6. **(b)** 283.

13.33 **(a)** 0.067. **(b)** 0.0013.

13.35 About 3.64.

13.37 **(a)** Mean 86 mm, standard deviation 2.138 mm. **(b)** 0.0025.

13.39 Given the very strong right-skew, $n \geq 40$ would be enough to be able to apply the central limit theorem.

13.41 For all practical purposes, it is a parameter. The census tracks everyone (except homeless individuals).

13.43 **(a)** The mean proportion of Hispanics is 0.139. The standard deviation is 0.0089. **(b)** $P(\hat{p} \leq 0.12) = 0.0167$.

13.45 **(a)** Mean = 0.05, standard deviation = $\sqrt{0.05 \times 0.95/n}$. **(b)** $P(\hat{p} \geq 0.07) = 0.0972$. **(c)** $P(\hat{p} \geq 0.08) = 0.0258$, $P(\hat{p} \geq 0.10) = 0.0006$.

Chapter 14

14.1 **(a)** 0.4. **(b)** 0.8 cm. **(d)** 95%.

14.3 **(a)** (144, 158). **(b)** Parameter: mean weekly spending on food μ for all American adults. The procedure results in a 95% chance of capturing the true value μ.

14.5 We are 95% confident that the concentration of active ingredient in this batch is between 0.8327 and 0.8481 g/l.

14.7 **(a)** N(12 g/dl, 0.2263 g/dl). **(b)** 11.3 lies far from the middle of the curve and is therefore unlikely if in fact $\mu = 12$. 11.8 is close to the middle, so it would not be too surprising.

14.9 $H_0: \mu = 12$, $H_a: \mu < 12$.

14.11 $H_0: \mu = 64.5$ inches, $H_a: \mu \neq 64.5$ inches.

14.13 $P = 0.0209$, significant at $\alpha = 5\%$.

14.15 **(a)** $P = 0.0010$. This is significant at the $\alpha = 0.05$ and the $\alpha = 0.01$ level. **(b)** $P = 0.1884$. This is not significant at either the $\alpha = 0.05$ or the $\alpha = 0.01$ level. **(c)** A low P-value means the outcome would be very unlikely if the null hypothesis was true.

14.17 $H_0: \mu = 5$, $H_a: \mu < 5$, $z = -2.7708$, $P = 0.0028$. This is strong evidence that the mean oxygen content is less than 5 mg/l.

14.19 **(a)** Yes; 10 is within the 95% confidence interval, because the P-value is not low enough to reject at the $\alpha = 0.05$ level. **(b)** No; 10 is not in the 90% confidence interval, because the P-value is low enough to reject at the $\alpha = 0.10$ level.

14.21 **(a)** No. **(b)** Yes.

14.22 (c) **14.23** (c) **14.24** (a) **14.25** (b) **14.26** (c)

14.27 (a) **14.28** (b) **14.29** (b) **14.30** (b) **14.31** (a)

14.33 The margin of error says that 95% of surveys should contain the true population proportion, not necessarily the proportion obtained by this survey.

14.35 **(a)** 2.138. **(b)** 1.96. **(c)** 86.9 to 95.3.

14.37 124.0 to 127.2 mg/dl.

14.39 μ is the mean daily sodium intake (in mg) in the American teenage population. $H_0: \mu = 2300$, $H_a: \mu > 2300$.

14.41 **(a)** She had no expectations before she looked at the data. She cannot change her hypothesis just to fit the data. **(b)** $P = 0.0930$.

14.43 It is essentially correct.

14.45 The differences in richness and total stem densities between the two areas were so small that they could easily occur by chance if population means were identical.

14.47 Something that occurs "less than once in 100 repetitions" also occurs "less than 5 times in 100 repetitions," but not vice versa.

14.49 μ is the population mean body length of deer mice in forests. $H_0 : \mu = 86$, $H_a : \mu \neq 86$, $P = 0.0171$.

14.51 It is possible to determine statistical significance using both P-values and confidence intervals. However, P-values only compare the parameter to a hypothesis, while confidence intervals actually give a range of estimates for the parameter.

14.53 **(a)** Yes (P-value very close to zero). **(b)** 10.0021 to 10.0025. The scale is slightly biased on the heavy side. **(c)** P-values tell us only that the scale is off, while confidence intervals tell us how much it is off (not too much, in fact).

Chapter 15

15.1 **(a)** No; the poll is a voluntary survey. **(b)** No; inference is still inappropriate, because the poll is a voluntary survey.

15.3 There might be a pattern to these consecutive records, and they come from only one academic medical center—they may not be representative of other procedures performed elsewhere.

15.5 Yes, because the sample is quite large ($n = 72$), although an extreme outlier would still be problematic.

15.7 **(a)** 4.9. **(b)** 1.96. **(c)** As the sample size increases, the margin of error decreases.

15.9 **(a)** $n = 5$, $P = 0.1855$; $n = 15$, $P = 0.0607$; $n = 40$, $P = 0.0057$. **(b)** As the sample size increases, the sampling distribution gets narrower.

15.11 **(a)** 4.8 ± 0.438. **(b)** 4.8 ± 0.253. **(c)** 4.8 ± 0.155.

15.13 **(a)** No; since we are running the test so many times, we would expect about 5 individuals to get $P \leq 0.01$. **(b)** Retest these individuals. Now that there is a smaller group, low P-values are less likely to occur just by chance.

15.15 385.

15.17 **(a)** 178. **(b)** 307. A higher confidence level requires a larger sample size.

15.19 There is probability 0.57 that the test will reject the null hypothesis at significance level $\alpha = 0.05$ if the alternative hypothesis is true.

15.21 Power 0.381, 0.751, and 0.984. As the sample size increases, power increases. For a test that almost always rejects H_0 when $\mu = 4.7$, we should use $n = 40$.

15.23 **(a)** H_0: patient is healthy, H_a: patient is ill. Type I error: sending a healthy patient to the doctor. Type II error: clearing a patient who is ill.

15.24 (a) **15.25** (c) **15.26** (b) **15.27** (c) **15.28** (b)

15.29 (a) **15.30** (a) **15.31** (c) **15.32** (a) **15.33** (b)

15.35 People are likely to lie about their sex lives. The margin of error does not take this into account.

15.37 **(a)** Response bias. **(b)** No; the margin of error quantifies only variations due to random sampling.

15.39 B, A, C, respectively.

15.41 **(a)** The margin of error decreases. **(b)** The P-value gets smaller. **(c)** The outlier will have a greater effect on a small sample.

15.43 **(a)** Statistical significance is not the same as practical significance. **(b)** A significance test performed with a large sample size can be statistically significant even if the effect is small (high power).

15.45 **(a)** 166. **(b)** 97.

15.47 **(a)** There are not enough data to determine whether the outcome was due to chance under the null hypothesis. **(b)** Small sample size or large variability.

15.49 **(a)** We cannot conclude, because the studies were uncontrolled. **(b)** This experiment suggests that arthroscopic surgery is no better than a placebo. **(c)** The lack of significance cannot be explained by low power.

15.51 0.86.

15.53 **(a)** 0.812. **(b)** $n = 5$, power $= 0.557$; $n = 15$, power $= 0.927$; $n = 40$, power $= 1.0$. **(c)** $\mu = 0.5$, power $= 0.475$; $\mu = 1$, power $= 0.935$; $\mu = 1.5$, power $= 0.999$. As the effect size increases, the power increases.

15.55 **(a)** Power $= 0.924$. **(b)** Power $= 0.924$. Yes. **(c)** The power would be higher.

Chapter 16

16.1 It is likely to weaken the association, because we expect that marijuana usage will be underreported to a greater degree than accidents caused.

16.3 **(a)** The control group should have 24 trees with no beehives. **(b)** Table A gives 53, 64, 56, and 68. **(c)** The response variable is elephant damage.

16.5 **(a)** All 40 subjects are assigned at random, 20 to simply drive and the other 20 to talk on the cell phone while driving. **(b)** All 40 subjects drive both with and without using the cell phone. The order of the two drive tests is assigned at random so that 20 subjects drive first with the phone and the remaining 20 drive first without the phone. **(c)** Some subjects naturally react faster than others.

16.7 Answers will vary.

16.9 **(a)** All probabilities are between 0 and 1 and they add up to 1. **(b)** A newborn is a woman's first or second child (at most her second child). 0.719. **(c)** 0.404. **(d)** The event $X \geq 4$ (or $X > 3$). 0.116.

16.11 **(a)** It is on or above the horizontal axis everywhere, and the total area beneath the curve is 1. **(b)** One-third. **(c)** One-tenth.

16.13 **(a)** 0.5899. **(b)** 384.2 mg.

16.15 **(a)** Roughly 99.7% of the individual weights x should be within 195 to 855 mg. **(b)** Roughly 99.7% of the average weights $\bar{x}$ should be within 492 to 558 mg.

16.17 **(a)** A response at least as large at that observed would be seen almost half the time (just by chance) even if there were no effect in the entire population of rats. **(b)** The differences observed were so small that they would often happen just by chance.

16.19 **(a)** $H_0 : \mu = 188$, $H_a : \mu < 188$, $z = -1.46$, $P = 0.0721$; significant only at the $\alpha = 0.10$ level. **(b)** The value 188 doesn't fall within the confidence interval; therefore, the results are not significant at the $\alpha = 0.05$ level for the one-sided alternative.

16.21 $Z = -2.92$ and $P = 0.0018$; small differences can be significant with large sample sizes.

16.23 **(a)** Yes; the sample is an SRS (although undercoverage of individuals without a telephone is an issue). **(b)** Nonresponse and undercoverage (individuals without a telephone). The "personally" phrasing is a little odd and might incite people to disagree. **(c)** The Harris Poll was conducted online rather than over the phone and is likely to reach a somewhat different population (younger and with more education, for example). Wording is more academic and comprehensive for the Harris Poll.

16.25 Placebos can provide genuine pain relief. (Believing that one will experience relief can lead to actual relief.)

16.27 **(a)** Factors: drink type (beer or soft drink) and glass type (straight or curved). This makes 4 treatments (drink-glass combinations). Response variable: time to finish drink. **(b)** Randomly allocate n participants to each of the 4 treatments, and then compare finishing times. **(c)** Yes; this is a comparative randomized experiment. The treatments are the only logical explanation besides chance variation.

16.29 $H_0: \mu = 100$ ng/g, $H_a: \mu > 100$ ng/g, $z = 14.54$, P is close to 0.

16.31 84.8 to 90.4.

16.33 **(a)** The treatment was not assigned. **(b)** Socioeconomic and hereditary factors may also affect IQ.

16.35 **(a)** There aren't enough data to show that the outcome was very unlikely by chance alone. **(b)** It is a relatively small effect size. This would explain why previous results failed to reject the null hypothesis. **(c)** The older siblings in that age range most likely have more education and therefore a slightly higher IQ score.

16.37 A Type I error would be to conclude that mean IQ is less than 100 when it really is 100 (or more). A Type II error would be to conclude that the mean IQ is 100 (or more) when it truly is less than 100.

16.39 **(a)** In each group, all probabilities are between 0 and 1 and they add up to 1. **(b)** $X < 3$ is "intense physical activity on fewer than 3 days last week." $P(X < 3 \mid \text{girl}) = 0.43$. $P(X < 3 \mid \text{boy}) = 0.31$. **(c)** This is $X \geq 1$. $P(X \geq 1 \mid \text{boy}) = 0.92$. **(d)** They are dependent because, for example, $P(X < 3 \mid \text{girl}) \neq P(X < 3 \mid \text{boy})$.

16.41 0.036.

16.43 $P(C|A) = 0.10 = 10\%$.

16.45 **(a)** Binomial, $n = 5$, $p = 0.5$. **(b)** $P(5 \text{ boys}) = 0.03125$, $P(4 \text{ boys, } 1 \text{ girl}) = 0.15625$. **(c)** $P(1 \text{ girl} \mid 4 \text{ boys}) = 0.5$. This is a conditional probability.

16.47 **(a)** Binomial, $n = 500$, $p = 0.26$. **(b)** $P(100 \text{ or fewer}) = 0.0010$, $P(150 \text{ or fewer}) = 0.9805$, using software.

Chapter 17

17.1 $s = 3.79$.

17.3 **(a)** 2.015. **(b)** 2.518.

17.5 **(a)** 2.262. **(b)** 2.861. **(c)** 1.440.

17.7 A dotplot or stemplot does not suggest any strong departures from Normality. We find $\bar{x} = 25.67$ and $s = 8.324$. With df = 17, $t^* = 2.110$, which gives the interval 25.67 ± 4.14. We are 95% confident that the mean healing rate for all newts of this species is between 21.53 and 29.81 micrometers per hour.

17.9 **(a)** 24. **(b)** t is between 1.059 and 1.318, so $0.20 < P < 0.30$. Technology gives $P = 0.2738$. **(c)** $t = 1.12$ is not significant at either level.

17.11 Data are strongly skewed to the left, but the sample size is relatively large ($n = 29$), so the test should be relatively accurate; $P = 4.2 \times 10^{-10}$, extremely significant.

17.13 This is a matched pairs setting; compute the call minus pure-tone differences. $H_0 : \mu = 0$, $H_a : \mu > 0$, $t = 4.84$, df = 36, $P < 0.0001$.

17.15 **(a)** The outlier is the only significant departure from Normality. **(b)** $t = 2.076$, df = 11, $P = 0.031$ (all newts); $t = 1.788$, df = 10, $P = 0.052$ (without the outlier).

17.17 (b) **17.18** (c) **17.19** (a) **17.20** (a) **17.21** (b)

17.22 (c) **17.23** (b) **17.24** (a) **17.25** (b) **17.26** (c)

17.27 **(a)** $\bar{x} = 15.59$ feet, $s = 2.550$ feet. **(b)** 14.81 to 16.37 feet. We reject the claim that $\mu = 20$ feet. **(c)** What population are we examining? Full-grown sharks? Male sharks? Can this be considered an SRS?

17.29 10.04 ± 0.49.

17.31 The data are slightly skewed to the right, but the sample is large enough for the 95% t confidence interval.

17.33 **(a)** 0.2450 to 0.3113 g/m^2. **(b)** Yes, because the sample size is large enough.

17.35 **(a)** A stemplot shows no major causes for concern. **(b)** 0.9211 to 1.4244 days.

17.37 **(a)** A subject's responses to the two treatments would not be independent. **(b)** Yes ($t = -4.41$, $P = 0.0069$).

17.39 **(a)** Only if the students were randomly chosen from the population of all male college students. No major departure from Normality. **(b)** 425.4 ± 32.4.

17.41 $H_0 : \mu = 1$, $H_a : \mu > 1$, $t = 8.402$, $P = 0.0002$. However, the data may be skewed to the left, in which case the sample size would not be large enough for inference.

17.43 This is a matched pairs setting; compute the differences for collaboration required minus not required. $H_0 : \mu = 0$, $H_a : \mu > 0$, $t = 7.37$, df = 7, $P < 0.0005$.

17.45 μ is the mean difference in healing rates, experimental minus control. $H_0 : \mu = 0$, $H_a : \mu < 0$. The stemplot is roughly Normal. $t = -2.02$, df = 13, $P = 0.032$.

Chapter 18

18.1 Matched pairs.

18.3 Single sample.

18.5 **(a)** Oregon: $n = 6$, $\bar{x} = 26.9$, $s = 3.82$. California: $n = 7$, $\bar{x} = 11.9$, $s = 7.09$. **(b)** df = 9.45.

18.7 $H_0 : \mu_1 = \mu_2$, $H_a : \mu_1 \neq \mu_2$, $t = 4.55$, df = 25.3, $P < 0.001$, significant.

18.9 **(a)** For the HSAM group, $\bar{x} = 6.00$, $s = 1.63$, SE = $0.52 = 1.63/\sqrt{10}$. For the control group, $\bar{x} = 3.61$, $s = 1.97$, SE = $0.47 = 1.97/\sqrt{18}$. $t = (6.00 - 3.61)/(\sqrt{0.52^2 + 0.47^2}) = 3.44$. **(b)** There is significant evidence

($t = 3.44$, $P = 0.001$, df $= 21$) that individuals with HSAM have higher visual memory test scores than typical individuals, on average.

18.11 **(a)** Breast-feeding women and non-breast-feeding–nonpregnant women. **(b)** $H_0: \mu_C = \mu_{BF}$, $H_a: \mu_C > \mu_{BF}$, $t = 8.50$, $P < 0.001$.

18.13 (a) **18.14** (c) **18.15** (b) **18.16** (b) **18.17** (b)

18.18 (c) **18.19** (a) **18.20** (c) **18.21** (b) **18.22** (c)

18.23 **(a)** The female data are a bit skewed, but the combined sample size is large enough. **(b)** 10.4 to 22.9 mm (df $= 32$).

18.25 **(b)** $H_0: \mu_G = \mu_B$, $H_a: \mu_G \neq \mu_B$, $t = 1.64$, $P = 0.1057$.

18.27 -1.12 to 11.35 (df $= 56.9$).

18.29 **(a)** Plotting the data reveals no major causes for concern. **(b)** $t = 2.57$, df $= 24$, $P = 0.008$, one-sided.

18.31 $H_0: \mu_1 = \mu_2$, $H_a: \mu_1 < \mu_2$, $t = -10.29$, $P(\text{df} = 36.89) = 1.07 \times 10^{-12}$, $P(\text{df} = 31) < 0.0005$.

18.33 **(a)** $H_0: \mu_{TP} = \mu_{no}$, $H_a: \mu_{TP} > \mu_{no}$, $t = 5.09$, df $= 57$, $P = 0.000002 < 0.01$, significant. There is very strong evidence that young adult males with tongue piercing have more enamel cracks, on average, than similar individuals with no tongue piercing. **(b)** No; this is an observational study, so causation cannot be established.

18.35 **(a)** $H_0: \mu_a = \mu_w$, $H_a: \mu_a < \mu_w$, $t = -2.31$, $P(\text{df} = 106.75) = 0.0115$. On average, people receiving acupuncture had fewer migraines than those on the wait list. **(b)** 1.2 ± 1.03. On the low end the reduction is almost nothing, but on the high end it is about 50%.

18.37 If those who had fewer migraines dropped out the most, then those remaining would have a higher migraine rate. Since the wait list had the highest dropout rates, this would imply that the wait list group would have a disproportional increase in migraines. If those who believed least in the acupuncture dropped out, the remaining subjects might feel that only acupuncture can help them.

18.39 **(a)** Two-sample t test, $P = 0.992$. **(b)** Compute the weight difference for each group, then find a two-sample confidence interval: 11.42 ± 6.88 with df rounded to 17. The control group gained significantly more weight than the group receiving ink applications. The ink appears to be toxic.

18.41 **(a)** $H_0: \mu_m = \mu_p$, $H_a: \mu_m \neq \mu_p$, $t = 0.4594$, $P(\text{df} = 39.6) = 0.6484$. **(b)** Were the data skewed or bimodal? Were there any outliers?

18.43 $t = -3.74$; $0.01 < P < 0.02$.

18.45 **(a)** 95% confidence interval: 38.88 ± 4.95. **(b)** 62.54 ± 5.97. **(c)** 95% confidence interval for inactive − active: 23.66 ± 7.56 (df $= 45.11$).

Chapter 19

19.1 **(a)** The population was cultures from individuals diagnosed with strep. p = proportion resistant to antibiotic in the population. **(b)** $\hat{p} = 0.568$.

19.3 **(a)** $\mu = 0.184$, $\sigma = 0.0039$. **(b)** Yes; the sample is an SRS, and it is large enough to ensure that $\hat{p}$ is approximately Normally distributed.

19.5 0.568 ± 0.023.

19.7 The lowest count, 11, is too small for the large-sample method.

19.9 **(a)** The number of "successes" is only 9. **(b)** 0.1078 ± 0.0505, or 0.0573 to 0.1584.

19.11 1537.

19.13 $H_0 : p = 0.5$, $H_a : p > 0.5$, $z = 0.21$, $P = 0.4174$, not significant.

19.15 **(a)** $np_0 = 2 < 10$. **(b)** $n(1 - p_0) = 8 < 10$.

19.16 (b)	**19.17** (c)	**19.18** (c)	**19.19** (b)	**19.20** (c)
19.21 (a)	**19.22** (b)	**19.23** (a)	**19.24** (a)	**19.25** (a)

19.27 **(a)** The number of successes is less than 15. However, $n > 10$, so the plus four confidence interval is appropriate. **(b)** 0.125 ± 0.0636.

19.29 683.

19.31 Sample and confidence interval sizes are large enough. 0.0922 ± 0.0220.

19.33 **(a)** $H_0 : p = 1/7$, $H_a : p > 1/7$. **(b)** $z = 5.556$, $P < 0.0001$. The dogs do significantly better than expected by chance alone.

19.35 0.2752 ± 0.0086.

19.37 0.169 to 0.171.

19.39 The conditions for inference are met. We test $H_0 : p = 0.5$, $H_a : p > 0.5$ and find $z = 4.74$, $P = 0.000001$, highly significant. The plus four 95% confidence interval is 0.7328 to 0.9490. There is very strong evidence that young children prefer graham cracker snacks packaged with popular cartoon images ($P < 0.001$). With 95% confidence we estimate that between 73.3% and 94.9% of young children show such a preference.

19.41 0.176 to 0.194.

Chapter 20

20.1 0.1941 ± 0.0663.

20.3 **(a)** Not enough of the subjects taking echinacea failed to develop a cold. **(b)** 0.0524 ± 0.1079.

20.5 $H_0 : p_{\text{accept}} = p_{\text{decline}}$, $H_a : p_{\text{accept}} < p_{\text{decline}}$, $z = -2.878$, $P = 0.0020$. Yes; there is significant evidence.

20.7 **(a)** $\hat{p}_{asp} = 0.000906$, $\hat{p}_{plac} = 0.002356$. **(b)** $RR = 0.3845$. Individuals in the aspirin group were 38% less likely to suffer a fatal heart attack compared with individuals in the placebo group. **(c)** $OR = 0.3840$. In this case, OR and RR are very close.

20.9 **(a)** $ARR = 0.00145$, $RRR = 0.6154$. ARR is the difference in risk between treatment and control; RRR is the reduction in risk in the treatment group relative to the risk of the control group. **(b)** 690. This means that 690 physicians would need to take aspirin daily for 5 years to save 1 of them from a fatal heart attack compared with taking a placebo daily for 5 years.

20.11 (a)	**20.12** (c)	**20.13** (a)	**20.14** (c)	**20.15** (c)
20.16 (b)	**20.17** (b)	**20.18** (b)	**20.19** (b)	**20.20** (c)

20.21 **(a)** There were no "successes" in the control group. **(b)** Treatment group: $n = 35$ with 24 mice developing tumors. Control group: $n = 20$ with 1 mouse developing a tumor. 0.636 ± 0.238. **(c)** The difference is significant at the 1% level because the 99% confidence interval does not include zero.

20.23 **(a)** $H_0: p_{AZT} = p_{plac}$, $H_a: p_{AZT} < p_{plac}$. There are enough patients with and without AIDS to use a z procedure for this test. **(b)** $P = 0.0017$. **(c)** This means that neither the doctors nor the patients knew who had received the drug and who had received the placebo.

20.25 $H_0: p_1 = p_2$, $H_a: p_1 \neq p_2$, $z = 3.39$, $P = 0.0007$.

20.27 0.0631 to 0.2179 (plus four: 0.0614 to 0.2158).

20.29 The two samples are not independent, because the same 40 children were used for both conditions.

20.31 $H_0: p_n = p_f$, $H_a: p_n < p_f$, $z = -2.398$, $P = 0.0082$.

20.33 $z = 1.59$, $P = 0.1118$, not significant.

20.35 **(a)** $z = 2.462$, $P = 0.0069$, significant. **(b)** 0.331 to 0.663.

20.37 **(a)** Both parties will know whether liquid nitrogen or duct tape was applied. **(b)** A plus four 95% confidence interval is the only valid test: 0.2288 ± 0.233. Because the interval does not contain the value zero, we can conclude that the two treatments yield significantly different results ($P < 0.05$), with duct tape being more efficient.

20.39 $H_0: p_m = p_f$, $H_a: p_m \neq p_f$, $z = 0.5258$, $P = 0.599$, no significant difference.

20.41 **(a)** $z = -2.79$, $P = 0.0052$, significant. **(b)** $z = 2.16$, $P = 0.0308$, significant. **(c)** For women in their 40s, the proportion with increased LH levels is significantly greater in the placebo group. For women in their 50s and 60s, the proportion with increased LH levels is significantly greater in the melatonin group. The effect is reversed in the two age groups. **(d)** $z = -0.68$, $P = 0.4988$, not significant. Pooling the two groups cancels out the two separate effects.

20.43 $ARR = 0.1746$, $RRR = 0.7255$, $NNT = 6$.

20.45 **(a)** $OR = 0.4249$, $RR = 0.4474$. **(b)** $ARR = 0.0483$, $RRR = 0.5526$. **(c)** $NNT = 21$.

Chapter 21

21.1 $H_0: p_0 = p_{20} = p_{40} = 1/3$, H_a: H_0 is not true.

21.3 Expected $= 53/3 = 17.67$, $X^2 = 16.11$.

21.5 Not all window angles are equally likely to produce accidental bird strikes.

21.7 Data fall into a multinomial setting; expected counts are all larger than 5.

21.9 Yes; data fall into a multinomial setting; expected counts are all 6.

21.11 df $= 2$, closest critical value is 15.20, $P < 0.0005$.

21.13 **(a)** $z = -0.847$, $X^2 = 0.717 = z^2$. **(b)** $P = 0.3972$ in both cases (two-sided z test).

21.14 (c) **21.15** (c) **21.16** (a) **21.17** (c) **21.18** (c)

21.19 (c) **21.20** (b) **21.21** (a) **21.22** (b) **21.23** (c)

21.25 **(a)** Yes; counts are high enough. **(b)** $H_0: p_{YR} = 9/16$, $p_{YW} = 3/16$, $p_{GR} = 3/16$, $p_{GW} = 1/16$, H_a: H_0 is not true. **(c)** 312.75, 104.25, 104.25, and 34.75. **(d)** $P = 0.9254$. The data are consistent with Mendel's second law.

21.27 **(a)** Pregnancy: 0 observed, 13.2 expected. No pregnancy: 87 observed, 73.8 expected. Cohort can be considered a random sample. Expected counts are large enough. $X^2 = 15.56$, df = 1, $P = 0.00008$, highly significant. **(b)** Pregnancy: 6 observed, 7.1 expected. No pregnancy: 29 observed, 27.9 expected. Expected counts are large enough. $X^2 = 0.21$, df = 1, $P = 0.6438$, not significant. **(c)** There is very strong evidence that Plan B prevents pregnancy when taken before ovulation ($X^2 = 15.56$, $P = 0.00008$) but not when taken after ovulation ($X^2 = 0.21$, $P = 0.6438$).

21.29 $H_0: p_H = p_T = p_U = p_L = 1/4$, H_a: H_0 is not true. $X^2 = 107.83$, $P < 0.0005$. Highly significant evidence that melanoma sites are not all equally likely in women.

21.31 H_0: probabilities match U.S. percents, H_a: H_0 is not true. $X^2 = 1709.8$, df = 4, $P < 0.0005$. Highly significant evidence that the ethnic breakdown in the population targeted by this study does not match the U.S. ethnic breakdown.

21.33 H_0: values of p_i are given by the model of random breeding; H_a: H_0 is not true. $X^2 = 9.377$, df = 7, $P = 0.2267$. The data are consistent with the hypothesis that prairie dogs breed randomly.

21.35 $H_0: p_w = 12/16$, $p_v = 3/16$, $p_t = 1/16$, H_a: H_0 is not true. $X^2 = 0.0980$, df = 2, $P = 0.952$, not significant. The data are consistent with a dominant epistatic model.

21.37 $H_0: p_F = 1/4$, $p_{SF} = 1/2$, $p_S = 1/4$, H_a: H_0 is not true. $X^2 = 0.7204$, df = 2, $P = 0.6975$. The data are consistent with a codominance model of inheritance.

Chapter 22

22.1 **(a)** By groups of increasing age: 31.07%, 16.50%, 9.32%, 17.48%, 15.92%, 9.71%. **(b)** By groups of increasing age: 34.99%, 7.40%, 8.96%, 16.93%, 17.21%, 14.51%. **(c)** The 20 to 24 age group makes up a substantially greater percent of the positive test results than of the negative results. The inverse is true for the 50 to 59 age group. There might be an association between age group and HPV status.

22.3 **(a)** 3 cm (0.486, 0.808), 5 cm (0.587, 0.884), 10 cm (0.734, 0.972), 20 cm (0.816, 1.007). **(b)** Individual confidence intervals ensure only that one parameter falls within the confidence interval with the chosen confidence level.

22.5 **(a)** 174.8 expected positives, 477.2 expected negatives. Sums up to 652. **(b)** This age group is slightly less likely to have a positive test result than expected under H_0.

22.7 **(b)** $X^2 = 40.554$, $P < 0.0005$ (rounded to 0.000 by Minitab). **(c)** The 20 to 25 age group is the largest contributor: they are much more likely to be HPV positive.

22.9 All expected cell counts are well over 5.

22.11 A chi-square test is not appropriate, because there are cells with expected counts less than 1.

22.13 The study used a single SRS, and the researchers hope that it represents a larger population.

22.15 **(a)** $(6-1)(2-1)=5$. **(b)** $X^2 > 22.11$, so $P < 0.0005$.

22.17 **(a)** The conditions are met for both. For chi-square, all cells have expected counts above 5. For the z test, the number of failures and successes is more than 5 in all groups (we do not have expected counts in the two-sample z test). **(b)** $H_0: p_{t\text{-PA}} = p_{\text{combined}}$, $H_a: p_{t\text{-PA}} \neq p_{\text{combined}}$, $z = -0.506$, $P = 0.6131$. **(c)** The observed counts are 86, 136, 177, 257. $X^2 = 0.256 = z^2$, $P = 0.6131$. **(d)** The data fail to show a significant difference between the two treatments ($P = 0.6131$). Endovascular therapy does not appear to increase the effectiveness of t-PA when treating patients with an acute ischemic stroke.

22.18 (b) **22.19** (a) **22.20** (a) **22.21** (c) **22.22** (a)

22.23 (b) **22.24** (a) **22.25** (a) **22.26** (c) **22.27** (b)

22.29 **(b)** Yes. H_0: no relationship between answer type and interview type, df = 2. **(c)** $X^2 = 10.619$, $P = 0.0049$. **(d)** There is significant evidence that interviewing method influences respondent answers.

22.31 **(a)** Stress = 0.909, exercise = 0.794, regular treatment = 0.700. **(b)** Expected heart attacks: stress = 6.78, exercise = 6.99, regular treatment = 8.22. All cells have expected counts above 5. **(c)** $X^2 = 4.84$ and $P = 0.0889$. This is not significant at the 0.05 level.

22.33 $X^2 = 11.975$, df = 3, $P = 0.007$. There is a significant relationship between birth order and sex. Males are more represented among eggs that hatch first and second.

22.35 H_0: no relationship between tongue piercing and the presence of enamel cracks. $X^2 = 7.39$, df = 1, $P = 0.0066$, significant. There is strong evidence that, in young adult males, enamel cracks are more common when the tongue is pierced (82.6%) than when it isn't (56.5%). However, causation cannot be established because this is an observational study.

22.37 The data suggest that desipramine is more successful. H_0: no relationship between relapse and treatment type, $X^2 = 10.5$, df = 2, $P = 0.0052$. The evidence is very significant.

22.39 The chi-square test found no significant relationship between treatment type and death following a stroke during the two-year treatment period ($P = 0.701$). It appears that none of the active treatments would do better than a placebo in preventing stroke-related death.

22.41 Neither test was statistically significant ($P = 0.577$ and $P = 0.3099$, respectively). The echinacea extracts used as treatment or preventively do not appear to do better than a placebo at preventing the common cold or reducing the number of cold symptoms.

22.43 H_0: no association between gender and tumor rate, $X^2 = 0.2765$, df = 1, $P = 0.5990$. Not significant.

22.45 **(a)** Small stones: open surgery = 93.1%, PCNL = 86.7%. Large stones: open surgery = 73.0%, PCNL = 68.8%. Open surgery appears to perform better for both small and large stones. **(b)** Small stones: $X^2 = 2.626$, $z = 1.621$, $P = 0.1051$. Large stones: $X^2 = 0.5507$, $z = 0.7421$, $P = 0.4580$. The differences in outcomes are not significant for the two procedures, whether the stones are small or large. **(c)** This is Simpson's paradox: PCNL was used more often for small stones, which have a higher rate of cure regardless of treatment.

Chapter 23

23.1 **(b)** The slope β is the average additional algal net growth for each unit increase in the number of grazers. We estimate $b = -0.5286$ and $a = -1.30$. **(c)** Residuals: $-0.071, -0.143, 0.686, -0.486, -0.157, 0.171$. The estimated standard deviation is $s = 0.443$.

23.3 **(a)** Positive linear trend. $r^2 = 21.95\%$. **(b)** $\hat{y} = -3385.4 + 2.6445x$, $s = 109.4\ \text{km}^3$ of water.

23.5 $b = 2.6445$ and $SE_b = 0.5836$. Therefore, $t = b/SE_b = 4.53$, $df = n - 2 = 73$, $P = 0.00002$ (two-sided), significant.

23.7 **(a)** $t = -4.99$, df = 4, $0.0025 < P < 0.005$ (Table C). **(b)** $r = 0.9282$, $n = 6$, $0.0025 < P < 0.005$ (Table E).

23.9 With df = 4, the 95% confidence interval is -0.8227 to -0.2344 net growth rate per grazer.

23.11 With df = 73, the 90% confidence interval is 1.672 to 3.617 km^3/year. Yes, because this interval is entirely positive and does not contain 0.

23.13 **(b)** (48.406, 51.093).

23.15 **(a)** Residuals are more or less randomly spread, but there are few observations for the lower temperatures. **(b)** The histogram shows Normality except for one low outlier.

23.16 (c) **23.17** (c) **23.18** (a) **23.19** (c) **23.20** (c)

23.21 (c) **23.22** (b) **23.23** (c) **23.24** (a)

23.25 **(a)** $b = -0.408$. For every additional kilogram of nitrogen per hectare, the species richness measure decreases by 0.408, on average. **(b)** $r^2 = 0.55$. The linear relationship accounts for 55% of the observed variation in species richness. **(c)** $H_0: \beta = 0$ versus $H_a: \beta \neq 0$ (or $\beta < 0$). The small P says that there is very strong evidence that the population slope β is negative, that is, that species richness does decrease as the amount of nitrogen deposited increases.

23.27 **(b)** $r = 0.6769$. **(c)** $t = 2.60$, $P = 0.0316$, significant.

23.29 Excel's 95% confidence interval for β is 0.4739 to 7.8856; this agrees (except for roundoff error) with $b \pm t^* SE_b = 4.1797 \pm (2.306)(1.607)$.

23.31 **(a)** 12.997 to 14.165 cm^3. **(b)** 11.088 to 16.074 cm^3. **(c)** The prediction interval. Probably not.

23.33 **(a)** 12.520 to 13.071 cm^3. **(b)** 11.439 to 14.153 cm^3. **(c)** The prediction interval. The 3D system is more accurate than the 2D version but maybe still not accurate enough for practical use.

23.35 The scatterplot shows a positive association. The regression equation is $\hat{y} = 0.1523 + 8.1676x$, and $r^2 = 0.606$. The slope is significantly different from 0 ($t = 13.25$, df = 114, $P < 0.001$).

23.37 **(a)** There is considerably greater scatter for larger values of the explanatory variable. **(b)** The distribution of residuals looks reasonably Normal.

23.39 **(a)** Negative linear relationship. The regression equation is $\hat{y} = 0.6471 - 0.0104x$. **(b)** The observations are independent. The standard deviation appears roughly constant, although there are fewer observations for large BMIs. The distribution of residuals is irregular but not markedly non-Normal. **(c)** -0.0155 to -0.0053.

23.41 0.0259 to 0.6434 nmol/l.

23.43 **(b)** $r = 0.8382$, $P = 0.0006$, significant. **(c)** Amount taken is the response variable, but it takes only a few values. For inference purposes, the response variable should vary continuously.

23.45 $P = 0.332$. No significant evidence that the intercept is not zero.

Chapter 24

24.1 **(a)** Randomly allocate 36 flies to each of four groups, which receive varying dosages of caffeine; observe length of rest periods. **(b)** H_0: all groups have the same mean rest period, H_a: at least one group has a different mean rest period. The significant P-value leads us to conclude that caffeine reduces the length of the rest period.

24.3 **(a)** The stemplots show no extreme outliers or strong skewness (given the small sample sizes). **(b)** The means suggest that logging reduces the number of trees per plot and that recovery is slow. **(c)** $F = 11.43$, $P = 0.000205$, $H_0: \mu_1 = \mu_2 = \mu_3$, H_a: not all the means are the same.

24.5 **(a)** When the three means are similar, F is very small and P is near 1. **(b)** Moving any mean increases F and decreases P.

24.7 **(a)** The ratio of standard deviations is about 1.16. **(b)** The ratio of standard deviations is about 1.20.

24.9 **(b)** These standard deviations do not follow our rule of thumb: The smallest standard deviation is 24.23; the largest is 143.67. The distributions appear skewed to the right. The value 700 is an extreme outlier. Thus, ANOVA results would not be reliable. **(c)** Even after removing 700, the standard deviation is still more than twice the standard deviation of the nonsmokers.

24.11 **(a)** 3 for treatment, 140 for error. **(b)** 2.70. **(c)** 5.86.

24.13 **(a)** Yes; the ratio of standard deviations is 1.8. **(b)** The data set is large (110 observations). **(c)** $\bar{x} = 20.25$, MSG $= 4504.4$. MSE $= 1765.0$. **(d)** $F = 2.552$. With df $= 2$ and 107 (use 2 and 100 in Table F), $0.05 < P < 0.1$ ($P = 0.0827$ with technology). There is some moderate evidence that the three exercise programs do not all yield the same mean subcutaneous fat reduction ($P = 0.0827$); however, this is not statistically significant at a typical significance level of 0.05.

24.15 (c) **24.16** (b) **24.17** (c) **24.18** (b) **24.19** (a)

24.20 (a) **24.21** (a) **24.22** (c) **24.23** (c) **24.24** (a)

24.25 Populations: morning people, evening people, neither. Response variable: difference in memorization scores. $k = 3$, $n_1 = 16$, $n_2 = 30$, $n_3 = 54$, $N = 100$; df $= 2$ and 97.

24.27 Populations: normal-weight men, overweight men, obese men. Response variable: triglyceride level. $k = 3$, $n_1 = 719$, $n_2 = 885$, $n_3 = 220$, $N = 1824$; df $= 2$ and 1821.

24.29 **(a)** Yes; the mean control emission rate is half of the smallest of the others. **(b)** H_0: all groups have the same mean emission rate, H_a: at least one group has a different mean emission rate. **(c)** Are the data Normally distributed? Were these random samples? **(d)** $s = \text{SEM} \times \sqrt{8}$. The rule-of-thumb ratio is 1.4755.

24.31 **(a)** ANOVA is risky. The standard deviation ratio is nearly 3 and the samples are small. **(b)** We test whether the mean biomasses from the three conditions are the same versus at least one population mean is different. **(c)** $F = 27.52$, df $= 2$ and 15, $P < 0.0005$ (significant), but to use with caution as noted in (a).

24.33 H_0: relative fitness is not influenced by previous pH exposure. Full data set: standard deviations = 0.058, 0.023, and 0.115; $F = 48.571$, df $= 2$ and 15, $P = 2.8 \times 10^{-7}$. Without the low outlier (0.56): standard deviations = 0.058, 0.023, and 0.055; $F = 93.329$, df $= 2$ and 14, $P = 8.1 \times 10^{-9}$, highly significant.

24.35 The differences among the groups were significant ($F = 3.44$, df $= 3$ and 27, $P = 0.031$). Nitrogen had a positive effect, the phosphorus and control groups were similar, and the plants that got both nutrients fell between the others.

24.37 Means = 2.908, 3.336, and 4.232; standard deviations = 0.1390, 0.3193, and 0.2713. ANOVA: df $= 2$ and 57, $F = 140.53$, $P < 0.00001$. If we remove the outliers: means = 2.908, 3.287, and 4.157; standard deviations = 0.1390, 0.2401, and 0.1545. ANOVA: df $= 2$ and 54, $F = 229.24$, $P < 0.00001$.

24.39 Means = 21.5, 15.5, 15.7, 15.5, and 15.6; standard deviations = 0.74, 0.52, 0.49, 0.52, and 0.51. ANOVA conditions are satisfied. (*Note:* Time to heal is continuous, but the researchers rounded the values to the nearest day. We treat these values as continuous). ANOVA: $F = 332.50$, df $= 4$ and 70, $P < 0.0005$, significant. The placebo group performed worse than all the other treatment groups.

24.41 $\bar{x} = 21.585$, MSG = 745.5, MSE = 460.2, df $= 3$ and 28, $F = 1.62$, not significant.

Chapter 25

25.1 **(a)** Two-sample z for proportions. **(b)** Two-sample t for means.

25.3 **(a)** t for means. **(b)** Matched pairs.

25.5 **(a)** A chi-square test. **(b)** ANOVA. **(c)** ANOVA.

25.7 **(a)** Matched pairs. **(b)** We need to know the standard deviation of the differences, not the two individual standard deviations.

25.9 $H_0: \mu_f = \mu_m$, $H_a: \mu_f > \mu_m$, $t = 2.21$, $0.01 < P < 0.02$ (df $= 100$) or $P = 0.0143$ (df $= 186.02$).

25.11 **(a)** The residuals add up to 0 (as expected). **(b)** There is no obvious nonlinearity or change in spread. **(c)** Their distribution is reasonably Normal.

25.13 **(a)** A placebo allows researchers to account for any psychological benefit (or detriment) the subject might get from taking a pill. **(b)** Neither the subjects nor the researchers who worked with them knew who was getting ginkgo extract; this prevents expectations or prejudices from affecting the evaluation of the effectiveness of the treatment. **(c)** $t = 2.147$, for which $P = 0.0387$ (df $= 35.35$). Those who took gingko extract had significantly more misses per line.

25.15 $H_0: \mu_1 = \mu_2$, $H_a: \mu_1 > \mu_2$, $t = 10.4$, $P < 0.0001$ (df $= 32.39$).

25.17 $H_0: \mu_1 = \mu_2$, $H_a: \mu_1 < \mu_2$, $t = -3.50$, $P < 0.0005$ (df $= 100$ or 216.4).

25.19 $H_0: p_1 = p_2$, $H_a: p_1 \neq p_2$, $z = 6.79$, P is very small.

25.21 $H_0: p_{yellow} = 3/4$; $p_{tan} = 1/4$, $H_a: H_0$ is not true. $X^2 = 1.165$, df $= 1$, $P = 0.281$, not significant. The data are consistent with Mendel's first law.

25.23 The large-sample interval is 0.3753 to 0.4252.

25.25 **(a)** The relationship is clearly linear and very strong; r^2 is 0.95. **(b)** 4.270 ± 0.118. **(c)** Fish 143 is very wide for its length and has a larger residual than other points; it can make inference less accurate.

25.27 **(a)** The sample size is very large. **(b)** 39.3 to 40.5 cm.

25.29 $H_0: \mu_{HSAM} = \mu_{control}$, $H_a: \mu_{HSAM} > \mu_{control}$. The two samples are independent. The control group is skewed but the total sample size is reasonably large (27). The two-sample t test should be safe. We find $t = 0.18$ (df $= 23.98$), $P = 0.429$, not significant. The study failed to find evidence that individuals with HSAM have higher scores, on average, in abstract visual reproduction tasks than typical individuals ($t = 0.18$, $P = 0.429$).

25.31 **(a)** The diabetic potentials appear to be large. **(b)** $H_0: \mu_D = \mu_N$, $H_a: \mu_D \neq \mu_N$, $t = 3.077$, $P = 0.0039$ (df $= 36.60$). **(c)** $t = 3.841$, $P = 0.0005$ (df $= 37.15$).

25.33 $H_0: p_{\text{acet}} = p_{\text{ibu}}$, $H_a: p_{\text{acet}} \neq p_{\text{ibu}}$, $z = 3.802$, $P = 0.00014$.

25.35 $H_0: \mu_G = \mu_I$, $H_a: \mu_G \neq \mu_I$, $t = 5.59$, and either df $= 9.89$, $P = 0.00012$, or df $= 9$, $P = 0.00017$, highly significant.

25.37 H_0: no relationship between temperature and hatching outcome, $X^2 = 1.703$, df $= 2$, $P = 0.427$, not significant.

25.39 $H_0: \mu_C = \mu_N = \mu_H$, $F = 0.080$, df $= 2$ and 126, $P = 0.923$, not significant.

25.41 The data are appropriate for ANOVA. ANOVA: $F = 47.19$, df $= 4$ and 85, $P < 0.0005$, significant. Mean ventilation rate is not the same at all temperatures. A graph of the data suggests that ventilation rate is particularly high at the highest temperature (25°C).

SOME DATA SETS RECURRING ACROSS CHAPTERS

Animal behavior and physiology

Aphid righting behavior: pages 475, 476, 535, 537, 540, 545, 546, 549, 551

Chickadee alarm calls: pages 69, 71, 79, 96, 103, 594

Commercial and transgenic chickens: pages 276, 445, 530

Grading oysters: pages 84 (x2), 113 (x2), 561, 565, 566, 572, 576, 585, 590 (x3)

Manatee deaths: pages 32, 67, 105, 149, 231

Monitoring rabies in Florida: pages 216, 310

Neural mechanism of nociception: page 85; Exercises 28-9, 28-10, 28-12

Paw preference in tree shrews: pages 7, 28, 145, 159, 367

Rats on a cafeteria-style diet: pages 597, 600, 610, 613; Examples 26-1, 26-2, 26-3; Exercise 26-6

Sex and longevity in male *Drosophila*: pages 72, 283, 287, 462, 630; Examples 28-1, 28-2, 28-3, 28-6, 28-10, 28-11

Spider silk: pages 40, 43, 47, 48, 63, 362, 412, 417, 643

Ecology and plant physiology

Cicadas as fertilizer?: pages 144, 361; Exercise 27-43

Comparing tropical flowers: pages 144, 645

Ecological approach to algal bloom control: pages 82 (x2), 567, 572, 574, 576, 580

Mycorrhizal colonies and plant nutrition: page 205; Example 26-12; Exercise 26-39; Exercise 28-37

Weeds among the corn: pages 149 (x2), 433; Examples 27-1, 27-2, 27-3, 27-4, 27-10, 27-11; Exercise 27-8

Health

Antibiotic resistance (gonorrhea): pages 294, 296, 297, 332

Antibiotic resistance (strep): pages 465, 468, 471

Blood types: pages 212, 230 (x2), 231, 259 (x2), 261, 290, 292, 309 (x2), 310, 327

Children with progeria: pages 418, 423; Exercise 27-41

Echinacea for the common cold?: pages 133, 387, 448, 490, 558

HPV infections in women: pages 228, 463, 468, 535, 540, 544, 546, 549, 551

Men's physiology, women's physiology: page 105; Exercise 28-11

Nanomedicine: pages 35, 61; Exercise 27-46

Smoking cessation: pages 137, 508

Testosterone and obesity in adolescent males: pages 415, 458, 593, 594

The overweight problem: pages 33 (x2), 298, 530

Who gets the flu?: pages 163, 207, 315, 325

Human behavior

Children's food choices: pages 13 (x2), 481, 507

Daily activity and obesity: pages 62, 440, 444; Exercises 27-1, 27-3, 27-7, 27-9

Does fidgeting keep you slim?: pages 70, 90, 93, 588

Highly Superior Autobiographical Memory, HSAM: pages 33, 34, 57, 58, 451 (x2), 644

Shoe support and foot health: pages 121, 122, 126, 128, 554

The polarizing taste of cilantro: pages 234, 258, 481 (x2), 505

Studying drinking behavior: pages 194, 401

Smoking on campus: pages 150, 645

Tongue piercing and dental health: pages 459, 556

Large Data Sets

Elderly health: pages 64, 88, 119, 436, 462, 630

Estimating body fat in men: pages 37, 64, 88, 120, 288, 435, 482, 595

Lung capacity of children: pages 37, 64, 88, 119, 288, 435, 462, 595, 630

INDEX

SYMBOLS

μ	population mean
$\overline{x}$	sample mean
σ^2	population variance
s^2	sample variance
σ	population standard deviation
s	sample standard deviation
Q_1, Q_3	first and third quartiles in a sample
p	population proportion
$\hat{p}$	sample proportion
$\mu_y = \alpha + \beta x$	α slope and β intercept of a population regression line; μ_y mean response for a given value of x
$y = a + bx$	a slope and b intercept of a sample regression line
r	correlation coefficient for a sample linear relationship
r^2	square of the correlation coefficient
$P(A)$	probability of event A
$P(A\|B)$	probability of event A, given that event B has occurred or is true
$N(\mu, \sigma)$	Normal distribution with mean μ and standard deviation σ
z-score	standardized score under a Normal distribution
SEM	standard error of the mean
SE	standard error of an estimate
df	degrees of freedom
m	margin of error in a confidence interval
H_0	null hypothesis
H_a	alternative hypothesis
α	chosen significance level and probability of a Type I error
β	probability of a Type II error
z	z statistic
t	t statistic
X^2	chi-square statistic
F	F statistic